S0-CDW-962

Polynuclear Aromatic Hydrocarbons:

A Decade of Progress

MARCUS COOKE
Chemistry and Biomedical Sciences
Department
BATTELLE
Columbus, Ohio

ANTHONY J DENNIS
Biotechnology
BATTELLE
Columbus, Ohio

Tenth International Symposium

Sponsored by:

Electric Power Research Institute
U.S. Environmental Protection Agency
American Petroleum Institute

Library of Congress Cataloging-in-Publication Data

Polynuclear aromatic hydrocarbons.

Includes bibliographies and index.

I. Cooke, Marcus, 1943- .
II. Dennis, Anthony J., 1948- .
III. United States. Environmental Protection Agency.
IV. Electric Power Research Institute.
V. Battelle Memorial Institute. Columbus Laboratories.
VI. International Symposium on Polynuclear Aromatic Hydrocarbons (10th: 1985: Columbus, Ohio)

615.9'511 79-642622

ISBN 0-935470-34-4

505 King Avenue
Columbus, Ohio 43201
614-424-6393

PREFACE

This monograph comprises sixty-three papers presented at the tenth annual, International Symposium on Polynuclear Aromatic Hydrocarbons. The meeting was held October 21 through October 23, 1985, in Columbus, Ohio, at the Battelle Memorial Institute. This book contains a series of original research papers addressing this important class of compounds, the proceedings of the "PAH Symposium". The tenth annual PAH Symposium and this book are dedicated to Dr. Louis F. Fieser, his memory and his pathfinding investigations of PAH reactivity.

The program for the polynuclear aromatic hydrocarbon (PAH) symposium includes contributions from physical as well as biological scientists. Synergism generated by the interaction of multidisciplinary researchers continues to be the strength of this conference.

Featured papers in this book include the Louis F. Fieser Memorial Address given by Dr. James W. Flesher (University of Kentucky) who not only reviews Dr. Fieser's most important papers, but includes a detailed discussion of predictive structure-activity models for PAH. A useful reference work included in this proceedings is a nomenclature guide for naming metabolic PAH products. The guide was prepared by scientists at the Nomenclature Division of the American Chemical Society. A review of risk assessment models for PAH and the underlying data to support these models is included in this volume. A paper of general interest describes the efforts of a group of interested scientists to compile a relational data base that includes important reference information about a large number of PAH compounds.

Several papers on metabolic processes involving PAH are included in this book. Metabolism products of benzo[a]pyrene (BaP) were studied after intratracheal instillation of [^{3}H]-BaP. Free radicals as intermediates in the metabolism of PAH are investigated through oxy-radicals of BaP and its derivatives. Activation and metabolic pathways of two nitro-substituted PAH that are important air pollutants, 3-nitro-fluoranthene, and 2-nitroanthracene, were reported.

Metabolism of characteristic thiaarenes (sulfur-containing PAH), quinolines, and benzoquinolines was presented at the meeting and are included in this monograph.

A number of PAH sources were discussed at the PAH symposium, including several combustion sources. Wood stoves and their impact on indoor air pollution, characterization of naphthenic distillate oil, bitumen and bitumen fumes, and flue gas from coal-fired power plants are subjects of investigation in the meeting papers. The extent of chemical conversion of PAH to nitro-substituted PAH during sampling diesel vehicle emissions was studied using ammonia injection techniques.

Mutagenicity as an indicator of physiological significant chemical species was extensively studied for use with PAH compounds. Among the compound types studied by mutagenic assay are 1,8-dinitropyrene, chloropyrenes, methylated pyrenes, and chloromethyl-BaP. Complex mixtures were included in mutagenicity testing, examples include photocopier toner, photocopies, unleaded gasoline, and fractionated petroleum. Ultraviolet light acting on 2-aminofluorene, a primary aromatic amine, utilized mutagenic screening as a tool to assess potential photoactivation into a potent promutagen.

Measurement of nitrofluoranthenes and nitropyrenes in ambient air, and atmospheric reactivity of fluoranthene and pyrene in the presence of N_2O_5 deal with new classes of environmentally significant PAH compounds. The effect of daytime temperature on decay of PAH when associated with airborne soot is presented in this proceedings.

Carcinogenic effects such as dermal carcinogenicity of petroleum products, ferric oxide cocarcinogenicity, and transformation of diploid human fibroblasts by 1-nitropyrene and 1-nitrosopyrene are important works documented in this volume.

PAH in the aquatic environment continues, too, as a major research thrust for the PAH meeting. Mutagenicity of contaminated marine sediments, and the existence of seasonal PAH concentrations in the amphipod *Pontoporeia hoyi*, are significant findings addressed at the 1985 PAH meeting.

A number of new analytical techniques to detect PAH were discussed at the meeting and are published in this proceedings document. On-the-fly fluorescence lifetime measurement, and fiber optic immunofluorescence spectroscopy are methods proposed to characterize PAH species. Microwave desorption of PAH found on particulate matter is used to enhance recoveries.

Synthesis is a critical component of progress in PAH research. Synthetic routes for cyclopenta epoxides of PAH with peripherally fused cyclopenta rings, PAH with "Bay Region" sulfur substitution, chloropyrenes, model synthesis of fluorene and nitro-fluorene PAH, and potentially tumorigenic cyclopenta-fused PAH are included in the papers from this meeting.

Models for the chemical and biochemical reactions of PAH continue to improve as fundamental knowledge of structure-activity relationships grows. Bioalkylation is used as a predictor of PAH reactivity, while the mutagenicity and tumorigenicity of methylene-bridged, "Bay Region" PAH were discussed at the meeting.

The search for detoxification routes for PAH continues to interest many researchers. Inhibition of PAH-caused chromosomal breakage, inhibition of DNA polymerase alpha by PAH, and detoxification mechanisms for BaP using the CHO/HG PRT assays are included in this book. Inhibitory characteristics of a number of chemical agents are presented, including plant phenols, BaP antibodies, Chinese tea, 2-chlorophenothiazine, and resorufin.

This book marks ten years for the annual PAH symposium. Since its inception this meeting has sought to foster progress in understanding these complex molecules through the mixing of scientific disciplines. Over the years we have observed researchers intensely discussing their findings, and many times collaboration was fostered by the exposure to new ideas provided by the meeting format.

CONTRIBUTORS

E. A. ADAMS
Naylor Dana Institute for Disease Prevention
American Health Foundation
Valhalla, NY 10595
USA

K. R. AMBROSE
Advanced Monitoring Development Group
Health and Safety Research Division
Oak Ridge National Laboratory
P.O. Box X
Oak Ridge, TN 37831
USA

P. A. ANDREWS
Department of Chemistry
McMaster University
Health Sciences Centre
1200 Main Street, W.
Hamilton, Ontario L8M 3Z5
CANADA

B. ANTHELME
Groupe de Microfluorimétrie Quantitative et Pharmacocinétique Cellulaire
Laboratoire de Chimie Physique
Université de Perpignan
66025 Perpignan Cedex
FRANCE

P. ASOKAN
Department of Dermatology
University Hospitals of Cleveland
Case Western Reserve University
Veterans Administration Medical Center
Cleveland, OH 44106
USA

T. B. ATHERHOLT
Department of Microbiology
Institute for Medical Research
Camden, NJ 08103
USA

R. ATKINSON
Statewide Air Pollution Research Center
University of California
Riverside, CA 92521
USA

W. M. BAIRD
Department of Medicinal Chemistry and Pharmacognosy
School of Pharmacy and Pharmacal Sciences
Purdue University
West Lafayette, IN 47907
USA

E. BALFANZ
Gesellschaft für Arbeitsplatz- und Umweltanalytik GbR
Nottulner Landweg 102
D-4400 Münster-Roxel
FEDERAL REPUBLIC OF GERMANY

J. C. BALL
Ford Motor Company
Research Staff
Dearborn, MI 48121
USA

L. M. BALL
Department of Environmental Sciences and Engineering
School of Public Health
University of North Carolina at Chapel Hill
Chapel Hill, NC 27514
USA

F. BANNOURA
Department of Chemistry
Ohio University
Athens, OH 45701
USA

Z. P. BAO
Institute of Environmental Science
Beijing Normal University
Beijing
CHINA

A. W. BARTCZAK
Department of Environmental
Sciences and Engineering
Lineberger Cancer Research Center
University of North Carolina
at Chapel Hill
Chapel Hill, NC 27514
USA

G. BECHER
National Institute of
Public Health
0462 OSLO 4
NORWAY

F. A. BELAND
National Center for
Toxicological Research
Jefferson, AR 72079
USA

D. BELL
Department of Environmental
Sciences and Engineering
School of Public Health
University of North Carolina
Chapel Hill, NC 27514
USA

C. BENESTAD
Center for Industrial
Research
P.O. Box 350
0314 OSLO 3
NORWAY

D. R. BEVAN
Virginia Polytechnic Institute
and State University
Blacksburg, VA 24061
USA

D. R. BICKERS
Department of Dermatology
University Hospitals of
Cleveland
Case Western Reserve University
Cleveland, OH 44106
USA

D. P. BIK
Department of Dermatology
University Hospitals of
Cleveland
Case Western Reserve University
Cleveland, OH 44106
USA

G. R. BLACKBURN
Mobil Environmental and
Health Science Laboratory
P.O. Box 1029
Princeton, NJ 08540
USA

J. W. BLAKE
Department of Pharmacology
Albert B. Chandler Medical Center
University of Kentucky
Lexington, KY 40536
USA

B. M. BOULOS
School of Public Health
University of Illinois
at Chicago
Chicago, IL 60608
USA

D. W. BRYANT
Department of Biochemistry
McMaster University
Health Sciences Centre
1200 Main Street, W.
Hamilton, Ontario L8M 3Z5
CANADA

J. BURG
Department of Health and
Human Services
U.S. Public Health Service
Centers for Disease Control
National Institute for
Occupational Safety and Health
Division of Biomedical and
Behavioral Science
4676 Columbia Parkway
Cincinnati, OH 45226
USA

D. BUSBEE
Departments of Anatomy and
Physiology and Pharmacology
College of Veterinary Medicine
Texas A&M University
College Station, TX 77843
USA

X. -H. CHEN
Department of Chemistry
Ohio University
Athens, OH 45701
USA

Y. -K. CHENG
Institute of Industrial Health
Anshan Iron and Steel Company
THE PEOPLE'S REPUBLIC OF CHINA

S. E. CHIUN
Department of Chemistry
University of Arkansas
at Little Rock
Little Rock, AR 72204
USA

A. L. COLMSJÖ
University of Stockholm
Department of Analytical Chemistry
S-106 91
Stockholm, SWEDEN

F. DARACK
Department of Chemical
Engineering and Chemistry
New Jersey Institute of
Technology
Newark, NJ 07102
USA

B. S. DAS
Applied Chemistry Division
Ontario Research Foundation
Mississauga, Ontario L5K 1B3
CANADA

M. DAS
Department of Dermatology
University Hospitals of Cleveland
Case Western Reserve University
Cleveland, OH 44106
USA

G. H. DAUB
Departments of Chemistry
and Biochemistry
University of New Mexico
Albuquerque, NM 87131
USA
(Deceased)

L. M. DAVISON
Department of Biochemistry
McMaster University
Health Sciences Centre
1200 Main Street, W.
Hamilton, Ontario L8M 3Z5
CANADA

M. C. DeFLORIA
Naylor Dana Institute for
Disease Prevention
American Health Foundation
Valhalla, NY 10595
USA

R. A. DEITCH
Mobil Environmental and
Health Science Laboratory
Box 1029
Princeton, NJ 08540
USA

D. J. DESILETS
Department of Chemistry
Purdue University
West Lafayette, IN 47907
USA

G. DETTBARN
Biochemical Institute for
Environmental Carcinogens
2070 Ahrensburg
Sieker Landstraße 19
FEDERAL REPUBLIC OF GERMANY

P. DI RADDO
Ben May Laboratory for
Cancer Research
University of Chicago
Chicago, IL 60637
USA

J. F. DOOLEY
Mobil Environmental and
Health Science Laboratory
P.O. Box 1029
Princeton, NJ 08540
USA

B. J. EADIE
Great Lakes Environmental
Research Laboratory
NOAA
2300 Washtenaw Avenue
Ann Arbor, MI 48104
USA

N. T. EDWARDS
Environmental Sciences Division
Oak Ridge National Laboratory
Oak Ridge, TN 37831
USA

W. C. EISENBERG
IIT Research Institute
Chemistry and Chemical Engineering
Research Department
Chicago, IL 60616
USA

W. R. FAUST
Great Lakes Environmental
Research Laboratory
NOAA
2300 Washtenaw Avenue
Ann Arbor, MI 48104
USA

E. K. FIFER
National Center for
Toxicological Research
Jefferson, AR 72079
USA

J. W. FLESHER
Department of Pharmacology
Graduate Center for Toxicology
Albert B. Chandler Medical Center
University of Kentucky
Lexington, KY 40536
USA

B. F. FOWLER
Seakem Oceanography Ltd.
2045 Mills Road
Sidney, British Columbia V8L 3S1
CANADA

S. FOXALL-VANAKEN
Ford Motor Company
Research Staff
Dearborn, MI 48121
USA

P. P. FU
National Center for
Toxicological Research
Jefferson, AR 72079
USA

J. FULCHER
Department of Environmental
Sciences and Engineering
School of Public Health
University of North Carolina
Chapel Hill, NC 27514
USA

W. FUNCKE
Gesellschaft für Arbeitsplatz-
und Umweltanalytik GbR
Nottulner Landweg 102
D-4400 Münster-Roxel
FEDERAL REPUBLIC OF GERMANY

N. G. GEDDIE
Naylor Dana Institute for
Disease Prevention
American Health Foundation
Valhalla, NY 10595
USA

A. GOLD
Department of Environmental
Sciences and Engineering
Lineberger Cancer Research Center
University of North Carolina
at Chapel Hill
Chapel Hill, NC 27514
USA

J. M. GOLDRING
Department of Environmental
Sciences and Engineering
School of Public Health
University of North Carolina
at Chapel Hill
Chapel Hill, NC 27514
USA

A. GREENBERG
Department of Chemical
Engineering and Chemistry
New Jersey Institute of
Technology
Newark, NJ 07102
USA

A. GRIEFE
National Institute for Occupational
Safety and Health
Cincinnati, OH 45226
USA

G. D. GRIFFIN
Advanced Monitoring Development Group
Health and Safety Research Division
Oak Ridge National Laboratory
P.O. Box X
Oak Ridge, TN 37831
USA

G. GRIMMER
Biochemical Institute for
Environmental Carcinogens
2070 Ahrensburg
FEDERAL REPUBLIC OF GERMANY

Z. GUO
Department of Environmental
Sciences and Engineering
School of Public Health
University of North Carolina
Chapel Hill, NC 27514
USA

C. GUTHENBERG
Department of Biochemistry
Arrhenius Laboratory
University of Stockholm
S-106 91 Stockholm
SWEDEN

W. R. HAHN
Guelph-Waterloo Centre for
Graduate Work in Chemistry
University of Waterloo
Department of Chemistry
Waterloo, Ontario N2L 3G1
CANADA

W. P. HARGER
Statewide Air Pollution
Research Center
University of California
Riverside, CA 92521
USA

R. HARKOV
Office of Science
and Research
New Jersey Department of
Environmental Protection
Trenton, NJ 08625
USA

C. C. HARRIS
Laboratory of Human
Carcinogenesis
National Cancer Institute
Bethesda, MD 20205
USA

C. JOE
Departments of Anatomy and
Physiology and Pharmacology
College of Veterinary Medicine
Texas A&M University
College Station, TX 77843
USA

S. W. JOHNSON
Mobil Environmental and
Health Science Laboratory
P.O. Box 1029
Princeton, NJ 08540
USA

R. KAMENS
Department of Environmental
Sciences and Engineering
School of Public Health
University of North Carolina
Chapel Hill, NC 27514
USA

M. KATZ
Department of Chemistry
York University
4700 Keele Street
North York (Toronto), Ontario M3J 1P3
CANADA

H. KAZARIANS-MOGHADDAM
Department of Chemistry
York University
Downsview, Ontario M3J 1P3
CANADA

A. KEMENA
Institute of Hygiene and
Occupational Medicine
University Medical Center
Essen University
Hufelandstr.55
D-4300 Essen
FEDERAL REPUBLIC OF GERMANY

P. T. KISSINGER
Department of Chemistry
Purdue University
West Lafayette, IN 47907
USA

C. V. KNIGHT
The University of Tennessee
School of Engineering
Chattanooga, TN 37402
USA

M. J. KOHAN
Genetic Bioassay Branch
Health Effects Research Laboratory
U.S. Environmental Protection Agency
Research Triangle Park, NC 27711
USA

J. KÖNIG
Gesellschaft für Arbeitsplatz-
und Umweltanalytik GbR
Nottulner Landweg 102
D-4400 Münster-Roxel
FEDERAL REPUBLIC OF GERMANY

D. J. KOWBEL
Environmental Health Directorate HPB
Health and Welfare Canada
Ottawa K1A 0L2
CANADA

S. LAHMY
Groupe de Microfluorimétrie
Quantitative et Pharmacocinétique
Cellulaire
Laboratoire de Chimie Physique
Université de Perpignan
66025 Perpignan Cedex
FRANCE

I. B. LAMBERT
Department of Biochemistry
McMaster University
Health Sciences Centre
1200 Main Street, W.
Hamilton, Ontario L8M 3Z5
CANADA

P. F. LANDRUM
Great Lakes Environmental
Research Laboratory
NOAA
2300 Washtenaw Avenue
Ann Arbor, MI 48104
USA

D. A. LANE
Environment Canada
Atmospheric Environment Service
4905 Dufferin Street
Downsview, Ontario M3H 5T4
CANADA

D. LAUTIER
Groupe de Microfluorimétrie
Quantitative et Pharmacocinétique
Cellulaire
Laboratoire de Chimie Physique
Université de Perpignan
66025 Perpignan Cedex
FRANCE

E. J. LaVOIE
Naylor Dana Institute for
Disease Prevention
American Health Foundation
Valhalla, NY 10595
USA

P. R. LEBRETON
Department of Chemistry
University of Illinois
at Chicago
Chicago, IL 60680
USA

E. LEE-RUFF
Department of Chemistry
York University
Downsview, Ontario M3J 1P3
CANADA

A. A. LEON
Departments of Chemistry
and Biochemistry
University of New Mexico
School of Medicine
Albuquerque, NM 87131
USA

G. J. LEPINSKE
Calumet Industries, Inc.
14000 Mackinaw Avenue
Chicago, IL 60633
USA

J. LEWTAS
Genetic Bioassay Branch
Health Effects Research Laboratory
U.S. Environmental Protection Agency
Research Triangle Park, NC 27711
USA

K. L. LOENING
Nomenclature Division
Chemical Abstracts Service
P. O. Box 3012
Columbus, OH 43210
USA

J. B. LOUIS
Office of Science and Research
New Jersey Department of
Environmental Protection
Trenton, NJ 08625
USA

F. E. LYTLE
Department of Chemistry
Purdue University
West Lafayette, IN 47907
USA

C. R. MACKERER
Mobil Environmental and
Health Science Laboratory
P.O. Box 1029
Princeton, NJ 08540
USA

V. M. MAHER
Carcinogenesis Laboratory
Michigan State University
East Lansing, MI 48824
USA

M. MALAIYANDI
Environmental Health Directorate HPB
Health and Welfare Canada
Ottawa K1A 0L2
CANADA

B. MANNERVIK
Department of Biochemistry
Arrhenius Laboratory
University of Stockholm
S-106 91 Stockholm
SWEDEN

D. R. McCALLA
Department of Biochemistry
McMaster University
Health Sciences Centre
1200 Main Street, W.
Hamilton, Ontario L8M 3Z5
CANADA

B. E. McCARRY
Department of Chemistry
McMaster University
Health Sciences Centre
1200 Main Street, W.
Hamilton, Ontario L8M 3Z5
CANADA

J. J. McCORMICK
Carcinogenesis Laboratory
Michigan State University
East Lansing, MI 48824
USA

G. D. MCCOY
Department of Environmental
Health Sciences
School of Medicine
Case Western Reserve University
Cleveland, OH 44106
USA

D.M.A. McCURVIN
Environment Canada
Atmospheric Environment Service
4905 Dufferin Street
Downsview, Ontario M3H 5T4
CANADA

M. McMANIS
Great Lakes College Association
Science Semester Student
The University of Tennessee
Knoxville, TN 37916
USA

K. T. MENZIES
Arthur D. Little, Inc.
Acorn Park
Cambridge, MA 02140
USA

J. E. MERRITT
Nomenclature Division
Chemical Abstracts Service
P. O. Box 3012
Columbus, OH 43210
USA

D. W. MILLER
National Center for
Toxicological Research
Jefferson, AR 72079
USA

C. E. MOSS
Department of Health
and Human Services
U.S. Public Health Service
Centers for Disease Control
National Institute for
Occupational Safety and Health
Division of Biomedical and
Behavioral Science
Experimental Toxicology Branch
4676 Columbia Parkway
Cincinnati, OH 45226
USA

H. MUKHTAR
Department of Dermatology
University Hospitals
of Cleveland
Case Western Reserve University
and the VA Medical Center
Cleveland, OH 44106
USA

C. M. MURCHISON
Great Lakes College Association
Science Semester Student
The University of Tennessee
Knoxville, TN 37916
USA

R. W. MURRAY
Department of Chemistry
University of Missouri-St. Louis
St. Louis, MO 63121
USA

S. R. MYERS
Department of Pharmacology
Albert B. Chandler Medical Center
University of Kentucky
Lexington, KY 40536
USA

D. NATSIASHVILI
Department of Chemical
Engineering and Chemistry
New Jersey Institute of
Technology
Newark, NJ 07102
USA

K. -W. NAUJACK
Biochemical Institute for
Environmental Carcinogens
2070 Ahrensburg
Sieker Landstraße 19
FEDERAL REPUBLIC OF GERMANY

E. R. NESTMANN
Environmental Health Directorate HPB
Health and Welfare Canada
Ottawa K1A 0L2
CANADA

M. J. NEWMAN
Department of Pathology
Uniformed Services University
of the Health Sciences
Bethesda, MD 20814
USA

J. W. NICKOLS
Genetics Group
Life Sciences Division
Los Alamos National Laboratory
Los Alamos, NM 87545
USA

R. W. NIEMEIER
Department of Health
and Human Services
U.S. Public Health Service
Centers for Disease Control
National Institute for
Occupational Safety and Health
Division of Biomedical and
Behavioral Science
Experimental Toxicology Branch
4676 Columbia Parkway
Cincinnati, OH 45226
USA

U. NILSSON
Department of Analytical Chemistry
Arrhenius Laboratory
University of Stockholm
S-106 91 Stockholm
SWEDEN

M. NISHIOKA
Battelle Columbus Laboratories
Columbus, OH 43201
USA

J. NORMAN
Veterinary Toxicology and
Entomology Research Laboratory
USDA
College Station, TX 77841
USA

K. H. NORPOTH
Institute of Hygiene and
Occupational Medicine
University Medical Center
Essen University
Hufelandstr.55
D-4300 Essen
FEDERAL REPUBLIC OF GERMANY

R. B. ODENSE
Seakem Oceanography Ltd.
46 Fielding Avenue
Dartmouth, Nova Scotia B3B 1E4
CANADA

R. T. OKINAKA
Genetics Group
Life Sciences Division
Los Alamos National Laboratory
Los Alamos, NM 87545
USA

J. C. ORR
Department of Biochemistry
McMaster University
Health Sciences Centre
1200 Main Street, W.
Hamilton, Ontario L8M 3Z5
CANADA

C. E. ÖSTMAN
University of Stockholm
Department of Analytical Chemistry
S-106 91
Stockholm, SWEDEN

J. D. PATTON
Carcinogenesis Laboratory
Michigan State University
East Lansing, MI 48824
USA

J. N. PITTS, JR.
Statewide Air Pollution
Research Center
University of California
Riverside, CA 92521
USA

J. D. POPHAM
Seakem Oceanography Ltd.
2045 Mills Road
Sidney, British Columbia V8L 3S1
CANADA

A. S. PRAKASH
Department of Chemistry
University of Illinois
at Chicago
Chicago, IL 60680
USA

D. PRUESS-SCHWARTZ
Department of Medicinal Chemistry
and Pharmacognosy
School of Pharmacy and Pharmacal Science
Purdue University
West Lafayette, IN 47907
USA

A. S. RAJ
Department of Biology
York University
4700 Keele Street
North York (Toronto), Ontario M3J 1P3
CANADA

T. RAMDAHL
Statewide Air Pollution
Research Center
University of California
Riverside, CA 92521
USA

A. RANNUG
Department of Genetic Toxicology
Wallenberg Laboratory
University of Stockholm
S-106 91 Stockholm
SWEDEN

U. RANNUG
Department of Genetic Toxicology
Wallenberg Laboratory
University of Stockholm
S-106 91 Stockholm
SWEDEN

L. RECIO
University of Kentucky Graduate
Center for Toxicology
Lexington, KY 40536
USA

J. E. RICE
Naylor Dana Institute for
Disease Prevention
American Health Foundation
Valhalla, NY 10595
USA

I.G.C. ROBERTSON
Department of Toxicology
Karolinska Institute
S-104 01 Stockholm
SWEDEN

H. S. ROSENKRANZ
Department of Environmental
Health Sciences
School of Medicine
Case Western Reserve University
Cleveland, OH 44106
USA

T. A. ROY
Mobil Environmental and
Health Science Laboratory
Box 1029
Princeton, NJ 08540
USA

J.-M. SALMON
Groupe de Microfluorimétrie
Quantitative et Pharmacocinétique
Cellulaire
Laboratoire de Chimie Physique
Université de Perpignan
66025 Perpignan Cedex
FRANCE

R. SANGAIAH
Department of Environmental
Sciences and Engineering
Lineberger Cancer Research Center
University of North Carolina
at Chapel Hill
Chapel Hill, NC 27514
USA

A. SCHMOLDT
Institute for Forensic Medicine
University Hamburg
2000 Hamburg 54
FEDERAL REPUBLIC OF GERMANY

R. SCHOENY
University of Cincinnati Medical Center
Department of Environmental Health
Cincinnati, OH 45267-0056
USA

C. A. SCHREINER
Mobil Environmental and
Health Science Laboratory
P.O. Box 1029
Princeton, NJ 08540
USA

M. J. SEPANIAK
Department of Chemistry
University of Tennessee
Knoxville, TN 37916
USA

K. SHANKARAN
Guelph-Waterloo Centre for
Graduate Work in Chemistry
University of Waterloo
Department of Chemistry
Waterloo, Ontario N2L 3G1
CANADA

A. SHIGEMATSU
Naylor Dana Institute for
Disease Prevention
American Health Foundation
Valhalla, NY 10595
USA

M. P. SIBI
Guelph-Waterloo Centre for
Graduate Work in Chemistry
University of Waterloo
Department of Chemistry
Waterloo, Ontario N2L 3G1
CANADA

M. J. SKINNER
Mobil Environmental and
Health Science Laboratory
P.O. Box 1029
Princeton, NJ 08540
USA

V. SNIECKUS
Guelph-Waterloo Centre for
Graduate Work in Chemistry
University of Waterloo
Department of Chemistry
Waterloo, Ontario N2L 3G1
CANADA

G. F. STRNISTE
Genetics Group
Life Sciences Division
Los Alamos National Laboratory
Los Alamos, NM 87545
USA

P.G.R. ST. WECKER
Department of Chemistry
The University of Tennessee
Knoxville, TN 37916
USA

P. D. SULLIVAN
Department of Chemistry
Ohio University
Athens, OH 45701
USA

J. A. SWEETMAN
Statewide Air Pollution Research Center
University of California
Riverside, CA 92521
USA

V. SYLVIA
Departments of Anatomy and Physiology and Pharmacology
College of Veterinary Medicine
Texas A&M University
College Station, TX 77843
USA

K. TAYLOR
IIT Research Institute
Chemistry and Chemical Engineering Research Department
Chicago, IL 60616
USA

P. S. THAYER
Arthur D. Little, Inc.
Acorn Park
Cambridge, MA 02140
USA

R. N. THOMASON
High School Teacher Summer Research Program Participant
The University of Tennessee
Knoxville, TN 37916
USA

G. E. TRIVERS
Laboratory of Human Carcinogenesis
National Cancer Institute
Bethesda, MD 20205
USA

B. J. TROMBERG
Advanced Monitoring Development Group
Health and Safety Research Division
Oak Ridge National Laboratory
Oak Ridge, TN 37831
USA

A. F. TUCCI
ERCO
A Division of ENSECO, Inc.
Cambridge, MA 02138
USA

K. VAHAKANGAS
Laboratory of Human Carcinogenesis
National Cancer Institute
Bethesda, MD 20205
USA

D. L. VANDER JAGT
Departments of Chemistry and Biochemistry
University of New Mexico
Albuquerque, NM 87131
USA

P. VIALLET
Groupe de Microfluorimétrie Quantitative et Pharmacocinétique Cellulaire
Laboratoire de Chimie Physique
Université de Perpignan
66025 Perpignan Cedex
FRANCE

T. VO-DINH
Advanced Monitoring Development Group
Health and Safety Research Division
Oak Ridge National Laboratory
P.O. Box X
Oak Ridge, TN 37831
USA

A. VON SMOLINSKI
College of Pharmacy
Health Sciences Center
University of Illinois
at Chicago
Chicago, IL 60608
USA

P. VON THUNA
Arthur D. Little, Inc.
Acorn Park
Cambridge, MA 02140
USA

L. S. VON TUNGELN
National Center for
Toxicological Research
Jefferson, AR 72079
USA

F. -L. WANG
Institute of Labor Protection
Ministry of Labor and Personnel
Box 4711
Beijing
CHINA

G. -H. WANG
Institute of Labor Protection
Ministry of Labor and Personnel
Box 4711
Beijing
CHINA

Y. WANG
Department of Chemical
Engineering and Chemistry
New Jersey Institute of Technology
Newark, NJ 07102
USA

Z. -Y. WANG
Institute of Labor Protection
Ministry of Labor and Personnel
Box 4711
Beijing
CHINA

D. WARSHAWSKY
University of Cincinnati Medical Center
Department of Environmental Health
Cincinnati, OH 45267-0056
USA

E. H. WEYAND
Virginia Polytechnic Institute
and State University
Blacksburg, VA 24061
USA

T. W. WHALEY
Toxicology Group
Life Sciences Division
Los Alamos National Laboratory
Los Alamos, NM 87545
USA

K. WILLIAMS
Genetic Bioassay Branch
Health Effects Research Laboratory
U.S. Environmental Protection Agency
Research Triangle Park, NC 27711
USA

A. M. WINER
Statewide Air Pollution
Research Center
University of California
Riverside, CA 92521
USA

Y. WU
Beijing Normal University
CHINA

D.T.C. YANG
Department of Chemistry
University of Arkansas at Little Rock
Little Rock, AR 72204
USA

P. ZACMANIDIS
Ford Motor Company
Research Staff
Dearborn, MI 48121
USA

Y. U. ZEBÜHR
University of Stockholm
Department of Analytical Chemistry
S-106 91, Stockholm
SWEDEN

Z. -C. ZHOU
Beijing Medical University
CHINA

B. ZIELINSKA
Statewide Air Pollution
Research Center
University of California
Riverside, CA 92521
USA

ACKNOWLEDGEMENTS

Participating in the Polynuclear Aromatic Hydrocarbons (PAH) Symposium during the past ten years has been a pleasure for everyone involved in the meeting. Four groups deserve special recognition for their support and individual efforts to make the International Symposium on PAH a success. The sponsoring organizations, the technical coordination group, the conference planning staff, and the manuscript preparation team deserve special acknowledgements for making this series function so well.

Sponsoring Organizations -- The PAH Symposium has been hosted by the Battelle Memorial Institute in Columbus, Ohio every year since the inaugural conference in 1977. During the ten years of this meeting the U.S. Environmental Protection Agency (U.S. EPA), the Electric Power Research Institute (EPRI), and the American Petroleum Institute have been continuing and sustaining sponsors. The International Joint Commission between the United States and Canada also provided financial support for the meeting.

Technical Coordination Group -- Several scientists gave freely of their time and hard work to make the meeting a success. The concept of the PAH meeting started with Dr. Peter Jones (Battelle) along with Dr. Larry D. Johnson (EPA), and Dr. Jacques Guertin (EPRI). Drs. Johnson and Guertin continued to encourage, direct, and participate in the PAH meeting during the past ten years. The positive effect of their guidance, foresight, and encouragement can not be overstated. Dr. Alf Bjorseth (1978-1979), and Dr. Gerald L. Fisher (1980-1981) are senior editors who gave generously of their time to co-direct the PAH Symposium.

Special thanks go to Dr. Kenneth T. Knapp, Dr. Robert G. Lewis, Dr. Stephen Nesnow, Dr. William E. Wilson, Dr. Roy L. Bennett, all from the U.S. EPA, and Dr. Charles P. Holdsworth from the American Petroleum Institute.

Conference Coordination Staff -- The Battelle Conference Coordination Staff is especially commended for their planning, coordination, and administrative participation in the meeting. Susan R. Armstrong and her professional staff won the admiration of everyone who attended the meeting. Denise C. Sheppard, and Ruth Anne Gibson directed the meeting coordination staffs that included Joan K. Purvis, Betty Persons, David K. Daulton, and Carolyn S. Johnson.

Manuscript Preparation -- Throughout the PAH series, one person labored behind the scenes, managing myriad details, and collating hundreds of manuscripts into the PAH monographs that scientists in many countries now use. Karen S. Rush edited, documented, and organized over 7,000 pages of technical material in the combined PAH proceedings. Few people can claim this accomplishment. Everyone who uses the PAH books can thank Karen for these high quality, valuable reference works.

DEDICATION

The Tenth International Symposium on Polynuclear Aromatic Hydrocarbons is dedicated to Professor Louis Frederick Fieser and his coworkers in recognition of their many fundamental contributions to our understanding of the relation between the chemistry and carcinogenic activity of this important class of compounds.

In 1935, Professor Fieser and his students at Harvard began their now classic studies on the relationship of carcinogenic activity to chemical structure. He had begun his investigations on polynuclear aromatic hydrocarbons (PAH) in the 1920s by synthesizing dibenz[a,h,]anthracene and related compounds with Emma Dietz at Bryn Mawr College in Pennsylvania. That synthesis was published in Berichte in 1929 shortly after Clar's paper appeared describing the same ring structures. In 1930, Kennaway and his associates announced the discovery that dibenz[a,h,]-anthracene produced cancer in experimental animals. This was an exciting discovery that encouraged Fieser and his colleagues at Harvard, in particular E. B. Hershberg, Melvin Newman, and Arnold Seligman, to continue this line of investigation.

The goal of Fieser's work as to understand the arrangement of atoms which caused an aromatic molecule to become carcinogenic. He concluded that a methyl or methylene group, attached in the meso-anthracenic position, greatly increases the carcinogenicity of unsubstituted PAH. Further, he demonstrated that N-methyl-formanilide formylates carcinogenic, unsubstituted PAH by attack mainly at the reactive meso-anthracenic positions. These formyl groups increase carcinogenicity of the parent ring system.

Fieser's research team searched for reactive centers in PAH ring systems using lead tetraacetate, which attacks both in the nucleus and on side chains, causing oxidation reactions that were later used as models for biological oxidation.

The keynote address at the Tenth International Symposium on Polynuclear Aromatic Hydrocarbons was delivered jointly by Dr. James W. Flesher from the Albert B. Chandler Medical Center at the University of Kentucky, and Dr. Melvin S. Newman from the Ohio State University. Dr. Flesher discussed the background of Fieser's research, and discussed new developments in the study of biochemical reactions of PAH that are modeled from Fieser's work with N-methylformanilide and lead tetraacetate reactions. Dr. Flesher's talk is included in this book, the keynote paper that appears as the first published work in this monograph. Dr. Newman recounted his experiences working with Louis Fieser and told of the probing wisdom and careful experimental technique he learned from Dr. Fieser, tools that served Dr. Newman throughout his own distinguished career.

This tribute to Dr. Louis Fieser for his contributions to the field of knowledge on PAH gives only a partial insight into his extensive career in research and teaching. Louis Fieser with his wife and coauthor Mary produced some of the classic books used by synthetic organic chemists. Among their numerous books are "Reagents for Organic Synthesis", a series of several volumes that summarize the development of new chemical reagents, the classic text "Organic Chemistry", and "Steroids". "Organic Chemistry" was published in ten languages, and four editions of "Steroids" have been published. Dr. Louis Fieser received the Nichols Medal, the Norris Award, and the American Chemical Society Award in Chemical Education.

Louis Fieser led a generation of organic chemists, biochemists, and pharmacologists into the exciting study of structure-activity relationships of carcinogenic PAH. Louis Fieser was an inspiring research leader and teacher. As we pause to remember him and dedicate this volume to his memory, we are awed by the magnitude of his contributions, and the length of Dr. Louis Fieser's influence on the study of chemical carcinogens.

CONTENTS

LOUIS F. FIESER MEMORIAL ADDRESS

GENERAL PAPERS

LOUIS F. FIESER MEMORIAL ADDRESS

LOUIS F. FIESER MEMORIAL ADDRESS

"FOUNDATION OF PAH RESEARCH-CONTRIBUTIONS OF LOUIS F. FIESER"

JAMES W. FLESHER
Department of Pharmacology, Albert B. Chandler Medical Center, University of Kentucky, Lexington, Kentucky 40536 USA.

INTRODUCTION

The Louis F. Fieser Memorial Address was presented October 21, 1985 on the Tenth Anniversary of the International Symposium on Polynuclear Aromatic Hydrocarbons, to recognize and honor Professor Louis Frederick Fieser and his coworkers for their many fundamental contributions to our understanding of the chemistry and carcinogenic activity of these remarkable substances. This occasion marks the 50th Anniversary since Professor Fieser and his group, which included his wife Mary and his most distinguished post-doctoral pupil, Professor Melvin S. Newman, who has been an active investigator in this field for half a century, began their classic studies of the relationship between structure, chemical reactivity and carcinogenic activity of polynuclear aromatic hydrocarbons. I assume that everyone has read the classic papers of Fieser and his collaborators from the period 1935 through 1941. In case you have not, I urge you to do so not only because of the valuable information and insights they contain but also for their clear language and logical thought. Although no summary could possibly do justice to these excellent papers, let me try to pull together some of the principal results of these brilliant investigations which have contributed so much in laying the foundation and providing the conceptual framework to the eventual solution of the problem of PAH carcinogenesis. I should also like to say, at the outset, that I wish to acknowledge our indebtedness to Professor Fieser and his group for giving us, through his published work, inspiration and guidance in our efforts to contribute to an understanding of carcinogenesis by polycyclic aromatic hydrocarbons.

In 1930, dibenz(a,h)anthracene (Fig. 1), 2, was discovered to be carcinogenic by the London workers of Kennaway and associates at the Royal Cancer Hospital, in connection with their research to identify the carcinogen present in coal tar (56). The compound had been prepared by Clar (8) in Germany in 1929 and also by Fieser and Dietz (17) at Bryn Mawr College in Pennsylvania. The London workers isolated from coal tar in 1933 the potent carcinogen benzo(a)pyrene, 3, and established the identity of the compound by comparison with an authentic compound prepared by synthesis (12). They also demonstrated that 3-methylcholanthrene, 4, the only hydrocarbon of demonstrated carcinogenic activity which has been obtained by chemical methods (70) from substances normally present in the body, surpasses even this substance in the rapidity of tumor induction (2,13).

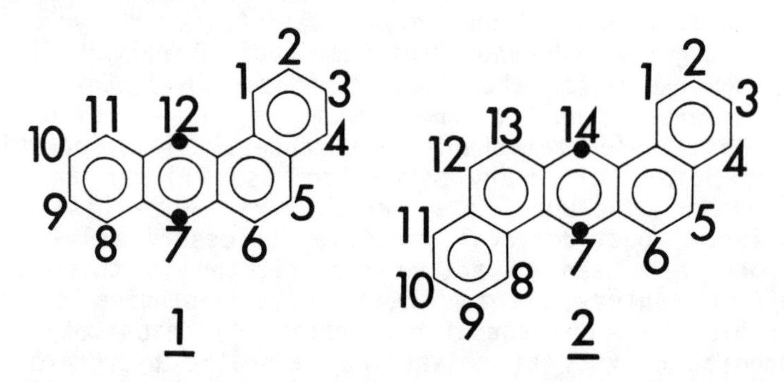

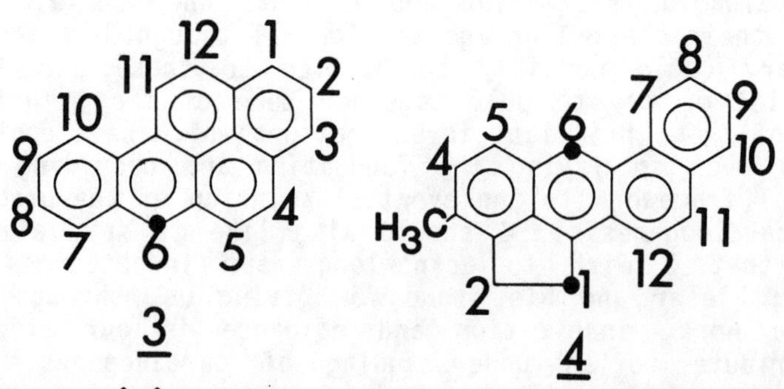

FIGURE 1. Structures and numbering system of some polynuclear aromatic hydrocarbons. Reactive centers are indicated by a heavy dot. Abbreviations: Benz(a)anthracene, BA; Dibenz(a,h)anthracene, DBA; Benzo(a)pyrene, BP; and 3-Methylcholanthrene, 3-MC.

One objective of the work undertaken by Fieser at Harvard in 1935 on the basis of these important advances was an attempt to define the features of structure responsible for the pronounced carcinogenic activity of these remarkable substances. Using the technic of injecting the crystalline material subcutaneously in 5-10 mg dosage into pure-strain mice, Shear (28,60) observed significant differences in the rapidity of action of the hydrocarbons, although all eventually produced tumors in a large proportion of the animals tested. The average time of the appearance of tumors was as follows: 3-methylcholanthrene, 2.5 months; benzo(a)pyrene, 3.5 months; dibenz(a,h)anthracene, 7 months. Fieser sought to determine the reason for the high potency of 3-methylcholanthrene. A point of interest is that these hydrocarbons may be considered to be derivatives of either anthracene or phenanthrene (Fig. 2).

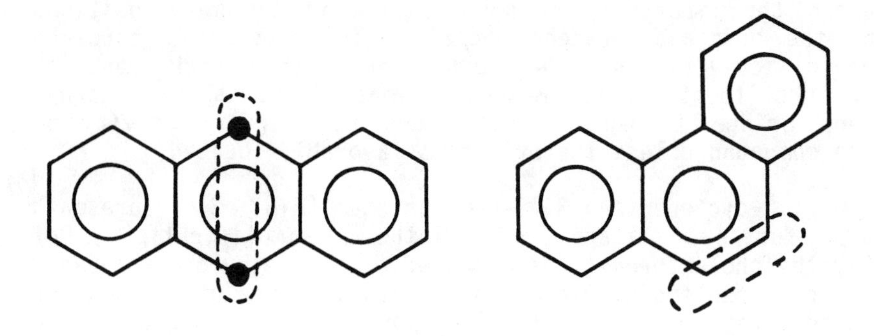

FIGURE 2. Structures of anthracene and phenanthrene showing the L-region of anthracene and the K-region of phenanthrene. Reactive centers are indicated by a heavy dot.

RESULTS AND DISCUSSION

As shown in Table 1, substitution of a methyl group in the 1-position, 5, of 3-methylcholanthrene or of an elaborate t-butyl substituent in the 3 position, 8, essentially abolishes activity, but some activity remains with the 2-methyl compound, 6, which had an average time of appearance of tumors of about 7.5 months. The fusing of an additional ring to the cholanthrene side chain, 7, practically destroys the activity of the compound. 3-Ethylcholanthrene, 9, however, does possess appreciable activity.

In order to understand more fully the features of structure that are necessary for carcinogenic activity, simplifications of the 3-methylcholanthrene molecule 4, were investigated. After removal of the methyl group of 3-methylcholanthrene to form cholanthrene or 7,8-dimethyleneBA, 17, there is little loss of activity, indicating that the methyl group in the 3 position of cholanthrene does not contribute much to carcinogenic activity. The 6,7-dimethylene, derivative 16, and 11,12-dimethylene, derivative, 18, isomers are also active. The 6,7-dimethyleneBA, derivative 16, appears to be quite toxic but to exhibit striking activity at low dosage and the 11,12 dimethylene compound, 18, is comparable with dibenz(a,h)anthracene, 2, in activity. The results demonstrate again the importance of substitution in the meso-positions of the benz(a)anthracene nucleus. The absence of carcinogenic activity in 4,5-dimethyleneBA, 15, which contains no substituents in the meso-positions, shows that a dimethylene bridge in the molecule does not confer activity on the compound unless the bridge is favorably located.

1,2-cyclopenteno-5,10-aceanthrene (Fig. 3) represents a further stage in the simplification of 3-methylcholanthrene. As compared with 3-methylcholanthrene, 1,2-cyclopenteno-5,10-aceanthrene has in place of the angular benzo ring fused to the 1,2 positions of anthracene, a five-membered cyclopenteno ring. Replacement of the angular benzo ring by a cyclopenteno ring results in a very great diminution of activity but not in a complete loss (35). It seems evident therefore that the angular benzo ring is not altogether indispensible for carcinogenic activity.

Of the various hydrocarbons known to possess cancer producing properties, the most potent are methylcholanthrene, 4, and cholanthrene, 17. In the regularity and rapidity with which they produce tumors in mice, these substances far outshadow all of the many other derivatives of benz(a)anthracene, 1, thus far investigated (14). It is a matter of considerable interest to attempt to define the features of structure responsible for their striking activity. Clearly the methyl group present in 3-methylcholanthrene in the 9-position of the benz(a)anthracene nucleus is

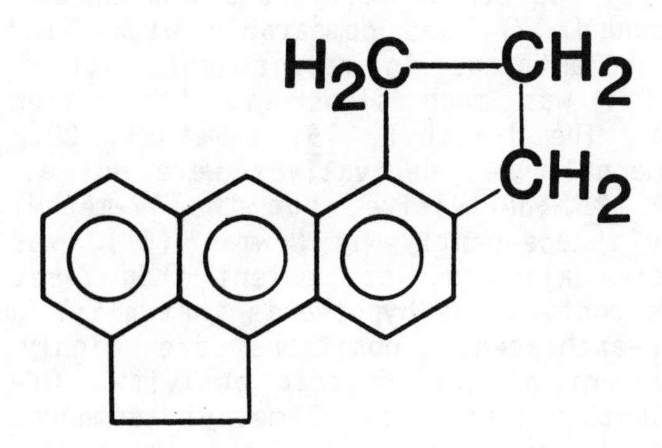

FIGURE 3. Structure of 1,2-cyclopenteno-5,10-aceanthrene.

relatively unimportant, for cholanthrene is as active, or very nearly as active as 3-methylcholanthrene. Of other carcinogenic derivatives of the weakly active benz(a)-anthracene, the 8-methyl compound, 22, is weakly active, the 8,9-dimethyl (2,13) and the 11,12-dimethylene, 18, derivatives are somewhat more active, but none of these hydrocarbons are comparable with cholanthrene, 17, in potency. It should be possible to determine whether the special activity of cholanthrene is associated with the presence of the five-membered ring including carbon atoms 7 and 8, or merely with the presence of alkyl substituents at these positions, by investigating the biological actions of 7,8-dimethylbenz(a)anthracene, 13. With this end in view Fieser and Newman (27) undertook the synthesis of the latter hydrocarbon and of the likewise interesting 7-methylbenz(a)anthracene, 21.

Of the various dimethyl BAs investigated for carcinogenic activity, 7,12-DimethylBA, 14, is a very potent carcinogen, but the 1,7-dimethylBA, 10, is far less active and the 6,7-dimethylBA, 12, and 6,12-dimethylBA, 11, are less active than the 7,8-dimethyl compound, 13, that was prepared for comparison with cholanthrene, 17. The high activity of the 7,8-dimethyl compound, 13, indicates that the intact ring of cholanthrene is by no means essential to activity. Of the two methyl or methylene groups, clearly the one attached to the meso-anthracenic 7-position is the more important. Interestingly, benzo(a)pyrene, 3, may be regarded as a benz(a)anthracene derivative with a carbon substituent attached to the meso-position 12. The results of carcinogenicity testing showed that the 7-methyl compound, 21, was comparable with 3-methylcholanthrene and cholanthrene in carcinogenic activity whereas 8-methylBA, 22, was much slower in its action than the cholanthrenes. The 1-methyl, 19, 6-methyl, 20, 10-methyl, 23, and 11-methyl, 24, derivatives were not as active as the 7-methyl, 21, derivative, but the 12-methyl compound, 25, prepared independently by Newman (57), was found to be quite active although less potent than 7-methylBA, 21. These results confirm the hypothesis that alkyl substitution in the meso-anthracenic positions are highly favorable to the attainment of carcinogenic activity. Of the monomethylbenz(a)anthracenes, the 7-methyl compound, 21, is the most potent, and lengthening the chain to ethyl, 26, moderates but does not abolish activity whereas increasing the side chain attached to the 7-position with propyl, 27, isopropyl, 28, allyl, 29, butyl, 30, and amyl, 31, groups respectively, causes a complete loss of activity.

Although BA is itself but very weakly active under most conditions of testing, the introduction in the meso-anthracenic 7-position of formyl, 34, hydroxymethyl, 35, or acetoxymethyl, 36, confers moderate activity on the new compounds. By contrast, a hydroxyethyl group, 37, confers little or no activity on BA while 7-methylBA, 21, is highly carcinogenic and 7-ethylBA, 26, somewhat less so, whereas the introduction of a methoxy group in the K-region moderates but does not abolish activity of the new compounds, 38 and 39, respectively. While a hydroxy, 40, or a methoxy group, 41, in the 5-position of the K-region of BA has perhaps about the same activity as the parent hydrocarbon, hydroxy, 42, nitro, 44, and cyano, 46, groups, respectively in the 7-position of BA are distinctly unfavorable to carcinogenic activity. On the other hand, the 7-methoxy, 43, and 7-amino, 45, derivatives of BA do possess weak activity.

The 3-methyl derivative, of dibenz(a,h)anthracene, 76, prepared by Fieser and Dietz in 1929 (17) was weakly active, whereas the 7-methyl compound, 77, was somewhat more active and the 7,8-methylene derivative, 78, which has a methylene group attached to both the angular ring and the meso-anthracenic position was carcinogenic, but the latent period of about 11 months is considerably longer than other closely related compounds.

Among the numerous derivatives of benz(a)anthracene, 1, which have been tested for carcinogenic activity, those exhibiting the highest potency are cholanthrene, 17, 3-methylcholanthrene, 4, and 7,12-dimethylbenz(a)anthracene, 14. The 7-methyl derivative falls somewhat below these hydrocarbons in general ability to evoke tumors rapidly in various tissues and among the other isomers 12-methylbenz-(a)anthracene comes next in the order of potency. Whereas the 1-acetoxy, 47, 1-hydroxy 48, and 1-keto, 49, derivatives of 3-MC are highly active compounds, 8-hydroxy-3-MC, 50, and 8-methoxy-3-MC, 51, are devoid of activity. The 9-chloro, 52, and 9-cyano, 53, derivatives, respectively, and the 11-chloro, 54, and 11-cyano, 55, derivatives, respectively, are also devoid of activity.

A knowledge of the center or centers of chemical reactivity of an active carcinogen may be of value in understanding the biological actions of the compound. The 7-methyl BA, a potent carcinogen, was converted into the meso-dihydro derivative, 56, by the alcoholysis of the disodium compound (30). The position of the added hydrogens were definitely established by the observation that the hydrocarbon yields benzanthraquinone on oxidation.

The product with the hydrogen atoms added to the meso positions is devoid of carcinogenic activity, whereas some activity remains when the angular ring is saturated in the 1,2,3,4 positions, 57. Activity is also absent when the side ring of the anthracene nucleus is hydrogenated, 58, or when both the meso-positions and the angular ring are saturated with hydrogen, 59. Activity is reduced when hydrogens are added to the meso positions of 3-MC, 61. Activity is abolished by the addition of hydrogens to the K-region of 3-MC, 63, or to the meso positions and the angular ring, 62. When the angular ring of 6,7-dimethyleneBA, 60, is saturated, activity is reduced but not abolished as found for 7-methylBA when the angular ring positions are saturated with hydrogen.

The powerful carcinogen benzo(a)pyrene, 3, contains five aromatic rings, four of which are arranged in such a way so as to form benz(a)anthracene, 1. It is possible that the hydrocarbon is properly classified with the members of the BA series. For the attainment of the strongest activity, a methyl group must be introduced in the meso-anthracenic position, 65, while a methyl group attached to the positions on either side of the meso-anthracenic position, 64, and 66, respectively, confer only moderate activity on the hydrocarbon. Hydrogenation of the 7,8,9,10 positions, of BP, 69, greatly reduces activity of this compound. Similarly, 7-methyl-9,10-dihydroBP, 68, is weakly active. Whereas 6-hydroxyBP, 74, and 6-acetoxyBP, 75, are inactive, the 6-formyl derivative, 73, is a very potent carcinogen.

Although, Fieser remarked that there is no *a priori* reason for supposing that the carcinogenic action of a hydrocarbon in inducing malignant growth is not of an entirely physical character, he clearly considered it more likely that the action of the hydrocarbon is dependent primarily upon the occurrence of a definite chemical reaction (35). If such a reaction occurs, it is important to discover the nature of the reaction undergone by the hydrocarbon, particularly the first step. The work of Boyland and Levi (3) on the metabolism of anthracene provided a significant clue to the nature of a metabolic change. These investigators found that anthracene undergoes a metabolic reaction to form 1,2-dihydroxy-1,2-dihydroanthracene and commented on this attack on a side ring rather than at the meso positions. Dobriner *et al*. found a phenolic metabolite of dibenz(a,h)anthracene to be definitely less carcinogenic than the original hydrocarbon (15,16). On this basis, they suggested that resistance to carcinogenesis by hydrocarbons may be dependent upon the ability of the organism to convert the carcinogen into a phenolic derivative. Cason and Fieser combined the observations into a general pathway of detoxification the initial step of which was called "perhydroxylation." They considered that the reaction was an addition of the elements of hydrogen peroxide which inactivated the hydrocarbon in much the same way as addition of hydrogen abolished carcinogenic activity (7). Furthermore, the idea that a phenolic metabolite represents an avenue for the elimination of carcinogens in a detoxified form has highly important implications (Fig. 4).

Although a pathway of detoxification leading to phenolic metabolites seemed to have been established, the reactions leading to carcinogenesis were not known. However, chemical substitution reactions, had received little attention, and although Fieser investigated several such reactions, two in particular, seemed especially significant.

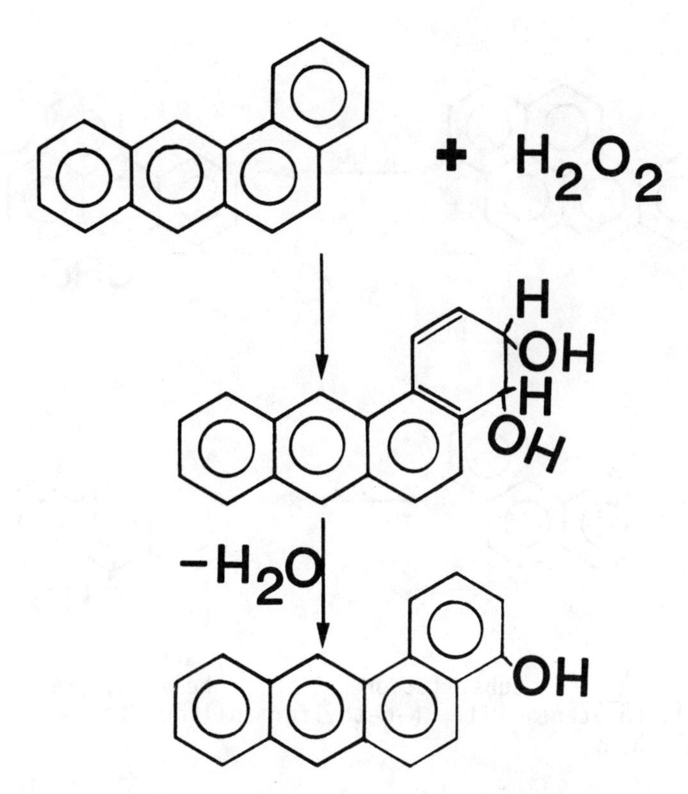

FIGURE 4. "Perhydroxylation" of benz(a)anthracene by the addition of the elements of hydrogen peroxide followed by elimination of water to form a phenol.

The first is the reaction with N-methylformanilide which converts a compound possessing a sufficiently reactive center into the formyl derivative (Fig. 5). The reaction is a means of probing for reactive centers, and it is a convenient method of preparing formyl derivatives and related compounds. Following the observations of Vollmann et al. (68), that pyrene-1-aldehyde can be obtained in excellent yield by the methylformanilide method, the reaction was applied in Fieser's laboratory to other hydrocarbons. BP was found to react as readily as pyrene or anthracene to afford the aldehyde in 90% yield (34). Benz(a)anthracene was converted into the 7-aldehyde in 64% yield (37) whereas DBA failed to react. In exploring other cases, they found no trace of aldehyde could be obtained from phenanthrene or chrysene and the solid hydrocarbons were recovered in

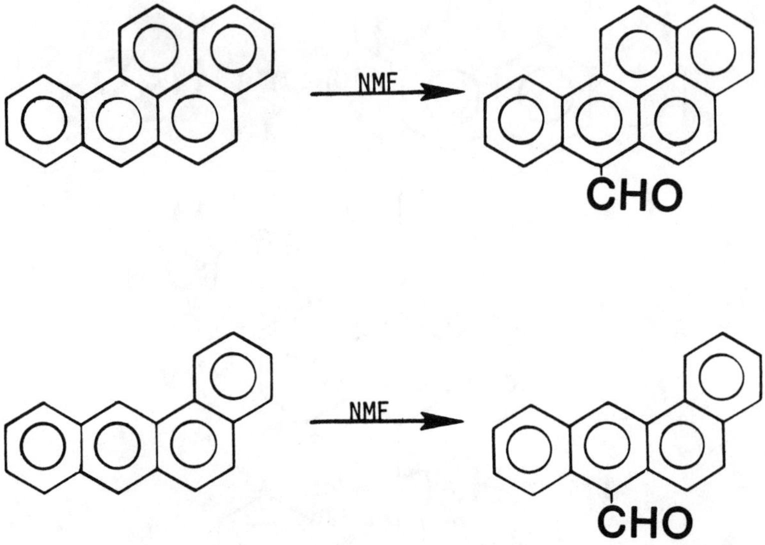

FIGURE 5. Substitution of benzo(a)pyrene or benz(a)anthracene with N-methylformanilide to yield the meso-aldehyde.

good yield.

However, it is in studies of hydroxylation reactions with lead tetraacetate that Fieser found special significance (Fig. 6). With lead tetracetate the hydroxy group is introduced in an acetylated condition. In a reaction with BA, the chief point of attack in the oxidation is the less hindered meso-position 7 (40). On the other hand when 7-methylBA was investigated, the reaction occurred mainly on the methyl group (40). The lead tetraacetate reaction offers promise of providing information of value concerning the reactivity of various types of polynuclear aromatic hydrocarbons with or without side chains.

This specific hydroxylation reaction provides a means of distinguishing between hydrocarbons of varying degrees of reactivity and of locating a reactive center in the molecule whether this is in an aromatic nucleus or a side chain. The reaction of 3-methylcholanthrene with lead tetraacetate reveals the site in the molecule most

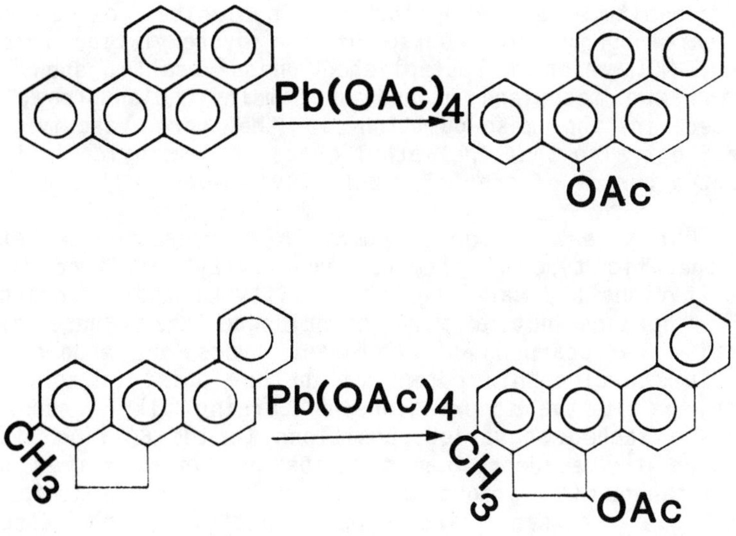

FIGURE 6. Acetoxylation of benzo(a)pyrene or 3-methylcholanthrene with lead tetraacetate to yield 6-acetoxybenzo(a)pyrene and 1-acetoxy-3-methylcholanthrene. The reaction is a means of probing for reactive centers.

susceptible to hydroxylation. The special significance of these observations is that there are certain points of correspondence between the chemical reactivity of hydrocarbons in their substitution reactions and their carcinogenic activity. These unusual substances (3-MC and BP) both belong to a small group of hydrocarbons that show a high susceptibility to hydroxylation with lead tetraacetate even though substitution occurs in one case in the nucleus and the other in the side chain. The reaction with lead tetraacetate reveals that the methyl or alkyl group is a reactive center, and it now appears possible that the function of the methyl group in conferring carcinogenic activity on a comparatively inert parent hydrocarbons is not to accentuate the reactivity at some other position, but to provide a suitable reactive center for further metabolism and activation of the hydrocarbon. Fieser concluded that the reaction with lead tetraacetate constitutes about as close an approximation to a direct hydroxylation of the hydrocarbon as can be conducted in a practical manner in the laboratory. Since hydroxylation in the body is conceivable the reaction under discussion may constitute a model for the biological oxidation of the carcinogen. The course of the hydroxylation reaction with 3-MC, which is established unequivocally, shows that of the two methylene groups of 3-methylcholanthrene, that located in the meso-position is the more reactive. It is evident also that the methyl group of 3-methylcholanthrene is not a center of special reactivity.

There are a number of points of correspondence between the specific type of chemical reactivity displayed in the substitutions and carcinogenic activity of the hydrocarbons. It is perhaps not a mere coincidence that these highly reactive hydrocarbons of different types are endowed with superior carcinogenic potency. Another point of correspondence, is in the observations concerning alkyl groups and chains attached at various positions to the BA ring system. Although the evidence needs to be put on a firmer basis and extended in various directions, the relationships now observable between carcinogenic activity and chemical reactivity of a specific nature suggest a causative associated between this chemical property and hydrocarbon carcinogenesis. A high degree of susceptibility to substitutions of a type exemplified by reaction with lead tetraacetate may be one factor of importance in determining carcinogenic activity of a hydrocarbon. A given compound may have adequate chemical reactivity of the type required but lack certain other necessary attributes (35).

If a carcinogen administered to a test animal undergoes some substitution reaction as the first step in a chain of events leading eventually to carcinogenesis, the reaction might be one of hydroxylation, or of the substitution of some related group. The introduction of an active functional group, in the first step of the process, might provide a mechanism for the conjugation of the carcinogen with substances present in the organism. The postulated reaction may not correspond with the reaction suggested from the hydroxylation reaction as constituting the primary step, but may represent a step further along in the chain (35). However, the earlier suggestion (35) that the first step in the metabolic activation of the carcinogen consists in the introduction of a simple group such as hydroxyl, sulfhydryl or amino was largely discounted by Wood and Fieser (71) since such simple derivatives as were investigated did not show appreciable carcinogenic activity. Furthermore, these investigators thought it unlikely that a benzo(a)pyrene derivative of the aromatic type ArX could function in the same manner as the structurally different type $ArCH_2X$ derived from substituted hydrocarbons such as 7-methylbenz(a)anthracene and 3-methylcholanthrene.

NEW DEVELOPMENTS

Since 1965, there has been a slow but steady accumulation of fresh evidence which is quite consistent with Fieser's theory (35) and in the past few years new discoveries have been reported which further extend the theory. The principal developments are that the biochemical equivalents of the N-methylformanilide alkylation and lead tetraacetate hydroxylation reactions have been discovered to take place readily when carcinogenic hydrocarbons are incubated with tissue cytosol preparations, or are administered to whole animals (47-53,69). The situation becomes understandable in the light of the unified hypothesis (47,50) which predicts that the chemical or biochemical introduction of an alkyl group, most favorably in the reactive meso-anthracenic position(s) is a structural requirement for carcinogenic activity, at least for compounds that are strong carcinogens. The unified hypothesis is consistent with Fieser's hypothesis (35) that the function of the alkyl group is not to enhance the reactivity of some other position but to serve as a reactive center for further metabolism and activation. This new hypothesis lends itself to certain predictions, offering a means of testing its validity, in addition to the obvious course of searching for the postulated substitution products _in vitro_ and _in vivo_. Thus the hypothesis predicts that compounds of the aromatic type ArX are preprocarcinogens when X = H is located at a reactive meso-anthracenic center and will be inactive or at best weakly active as carcinogens when X serves as a blocking group to prevent the biochemical introduction of an alkyl group into the molecule at reactive centers. Compounds of the aromatic type $ArCH_2X$ are procarcinogens when X is H and the alkyl group is attached to a meso-anthracenic position. When X is OH the compound is a proximate carcinogen. The formyl group, which also confers activity on the unsubstituted hydrocarbon, is readily reduced to the hydroxyalkyl group (67). When X is Cl, Br, I, or the acetate, phosphate, and sulfate esters of the proximate carcinogen, the compound functions as an ultimate carcinogen. However, when the hydroxyalkyl group of the proximate carcinogen is "blocked" by the formation of stable ethers _e.g._, X = OCH_3, OC_2H_5, carcinogenic activity is greatly reduced or abolished (67). If these new predictions are confirmed, as seems highly likely, a completely consistent theory may be possible which is capable of predicting the carcinogenic activity of most, if not all, substituted and unsubstituted polynuclear aromatic hydrocarbons.

Whatever the outcome of these future developments, all workers in the field of polynuclear aromatic hydrocarbons are indebted to Professor Louis Frederick Fieser for the fundamental contributions of this brilliant investigator, who, perhaps more than anyone else, laid the foundation for the eventual solution to the problem of the manner by which certain hydrocarbons in contact with tissue are able to induce malignant growth.

SUMMARY OF THE FUNDAMENTAL CONTRIBUTIONS OF FIESER AND HIS COWORKERS

1) Developed new methods for the synthesis and purification of numerous polycyclic aromatic hydrocarbons and their derivatives.

2) Recognized the importance of an alkyl substituent in the meso-anthracenic position(s) to the attainment of carcinogenic activity.

3) Postulated that the function of the alkyl group is not to enhance the reactivity at some other position but to serve as a reactive center for further metabolism of the molecule.

4) Identified reactive centers by studies of chemical substitution reactions.

5) Postulated that hydroxylation with lead tetraacetate may be a model for the biological oxidation of the hydrocarbon.

6) Formulated a highly original theory of the nature and significance of competing reaction pathways involved in the metabolism of aromatic hydrocarbons. The one in which specific substitution reactions take place at reactive centers are the reactions leading to carcinogenesis and the other in which generalized addition reactions take place at reactive regions are the reactions leading to detoxification.

TABLE 1

Column I - Compound Number; Column II - Substituent or Compound, Column III - Synthesis, Column IV - Tests for carcinogenic activity and Column V - Relative carcinogenic activity.[b,c]

I	II	III	IV	V
1	benz(a)anthracene (BA)	17	64,56	+
2	dibenz(a,h)anthracene (DBA)	17	56	++
3	benzo(a)pyrene (BP)	18	12	+++
4	3-methylcholanthrene (3-MC)	23	61	+++
	Derivatives of Cholanthrene			
5	1,3-dimethyl	5	66	+
6	2,3-dimethyl	25	61	++
7	1,2-benzdehydro	26	61	-
8	3-t-butyl	43	66	+
9	3-ethyl	6	62	++
	Dimethyl and Dimethylene(ace) Benz(a)anthracenes			
10	1,7-dimethyl	42	62	-
11	6,12-dimethyl	41	65	++
12	6,7-dimethyl	41	64	++
13	7,8-dimethyl	27	62	+++
14	7,12-dimethyl	58	62	+++
15	4,5-dimethylene	54	9	-
16	6,7-dimethylene	32	62	++
17	7,8-dimethylene	26	60	+++
18	11,12-dimethylene	26	61	+++
	Monomethyl Benz(a)anthracenes			
19	1-methyl	42	62	±
20	6-methyl	41	62	+
21	7-methyl	27,33,37	62	+++
22	8-methyl	27	62	++
23	10-methyl	10	62	+
24	11-methyl	44	62	-
25	12-methyl	58	62	++

	Monoalkyl Benz(a)anthracenes			
26	7-ethyl	33	64	+
27	7-n-propyl	33	64	-
28	7-isopropyl	33	64	-
29	7-allyl	33	64	-
30	7-n-butyl	33	64	-
31	7-n-amyl	33	64	-
	Other Derivatives of Benz(a)anthracene			
32	7-iodomethyl-12-methyl	46,59	46	+++
33	7-methoxymethyl-12-methyl	59	46	-
34	7-formyl	37	64	+++
35	7-hydroxymethyl	72	64	++
36	7-acetoxymethyl	73	64	+++
37	7-α-hydroxyethyl	37	64	-
38	5-methoxy-7-methyl	33	64	+
39	5-methoxy-7-ethyl	33	64	-
40	5-hydroxy	31	64	+
41	5-methoxy	31	64	+
42	7-hydroxy	73	64	-
43	7-methoxy	73	64	+
44	7-nitro	40	64	-
45	7-amino	40	64	+
46	7-cyano	37	64	-
	Derivatives of 3-Methylcholanthrene			
47	1-acetoxy	34	ND[a]	+++
48	1-hydroxy	34	66	++
49	1-keto	34	66	++
50	8-hydroxy	36	66	-
51	8-methoxy	36	66	-
52	9-chloro	72	66	-
53	9-cyano	72	66	-
54	11-chloro	36	66	-
55	11-cyano	36	66	-
	Hydrogenated Derivatives of Benz(a)anthracene			
56	7,12-dihydro-7-methyl	30	64	-
57	1,2,3,4,-tetrahydro-7-methyl	30	64	+
58	8,9,10,11-tetrahydro-7-methyl	30	64	-

59	1,2,3,4,-7,12-hexahydro-7-methyl	30	64	-
60	1,2,3,4-tetrahydro-6,7-dimethylene	32	62	+
61	7,12-dihydro-7,8-dimethylene-9-methyl	38	4	++
62	1,2,3,4,7,12-hexahydro-7,8-dimethylene-9-methyl	38	61	-
63	5,6-dihydro-7,8,-dimethylene-9-methyl	38	66	-
	Derivatives of Benzo(a)pyrene			
64	5-methyl	39	ND	
65	6-methyl	39	64	+++
66	7-methyl	18,22	63	++
67	9-methyl	39	ND	
68	7-methyl-9,10-dihydro	18	63	+
69	7,8,9,10-tetrahydro	18	63	+
70	7-hydroxy	21,31	63	-
71	7-methoxy	21	63	-
72	7-acetoxy	21	63	-
73	6-formyl	39	64	+++
74	6-hydroxy	39	64	-
75	6-acetoxy	34	67	-
	Derivatives of Dibenz(a,h)anthracene			
76	3-methyl	17	11	+
77	7-methyl	45	64	+++
78	7,8-methylene	20	9	++

a The carcinogenic activity of 1-acetoxy-3-MC (Compound #47) has apparently not been reported. A theory, developed in the author's laboratory, predicts the compound to be strongly carcinogenic and to be a model for the ultimate carcinogen of 3-MC.

b Strong (+++), moderate (++), weak (+), inactive (-), ND - not determined.

c Carcinogenic activities were determined from the percentage of treated animals that developed sarcomas at the site of injection, i.e., 5-33%, weak; 34-66%, moderate; and 67-100%, strong.

ACKNOWLEDGEMENT

Acknowledgement is made to the University of Kentucky, The American Cancer Society, The Department of Agriculture, and The National Cancer Institute, NIH, for generous support of investigations conducted in the author's laboratory over the past 25 years.

REFERENCES

1. Bachman, W.E.: The reaction of alkali metals with polycyclic hydrocarbons: 1,2-benzanthracene, 1,2,5,6-dibenzanthracene, and methylcholanthrene. J. Org. Chem., 1:347-353, 1936.
2. Barry, G., Cook, J.W., Haslewood, G.A.D., Hewett, C.S., Hieger, I., and Kennaway, E.L.: Production of Cancer by hydrocarbons. III. Proc. Roy. Soc. Ser. B., 117:318-351, 1935.
3. Boyland, E. and Levi, A.A.: CCCXVII. Metabolism of polycyclic compounds I. Production of dihydroxydihydroanthracene from anthracene. Biochem. J., 29:2679-2693, 1935.
4. Bradbury, J.T., Bachmann, W.E., and Lewisohn, M.G.: The production of cancer by some new chemical compounds. Factors affecting latent period of tumor production. Cancer Research, 1:685-694, 1941.
5. Bruce, W.F. and Fieser, L.F.: Carcinogenic hydrocarbons I. 15,20-dimethylcholanthrene. J. Amer. Chem. Soc., 59:479-480, 1937.
6. Bruce, W.F. and Kahn, S.J.: Carcinogenic hydrocarbons II. Ethylcholanthrene. J. Amer. Chem. Soc. 60:1017-1019, 1938.
7. Cason, J. and Fieser, L.F.: Synthesis of 4',8'-dihydroxy-1,2,5,6-dibenzanthracene and its relation to products of metabolism of the hydrocarbon. J. Amer. Chem. Soc., 62:2681-2687, 1940.
8. Clar, E.: Zur kenntnis mehrkernigen aromatisher kohlenwasserstoffe und Ihrer Abkömmlinge, IV Mitteil. Naphthophenanthrene und ihrer chinone. Berichte, 62:1574-1582, 1929.
9. Cook, J.W.: The production of cancer by pure hydrocarbons. Proc. Roy. Soc., Ser. B., 111:485-496, 1932.
10. Cook, J.W.: Polycyclic aromatic hydrocarbons IX. Synthesis of methyl and isopropyl homologs of 1,2-benzanthracene. J. Chem. Soc., pp. 456-472, 1932.

11. Cook, J.W., Hieger, I., Kennaway, E.L. and Mayneord, W.V.: Production of cancer by pure hydrocarbons, I. Proc. Roy. Soc. Ser. B, 111:455-485, 1933.
12. Cook, J.W., Hewett, L.L. and Hieger, I.: Isolation of a cancer-producing hydrocarbon from coal tar, I. Concentration of the active substance. J. Chem. Soc., pp. 395-396, 1933.
Cook, J.W. and Hewett, C.L.: Synthesis of 1,2 and 4,5-benzopyrene. J. Chem. Soc., pp. 398-405, 1933.
13. Cook, J.W. and Haslewood, G.A.D.: Synthesis of 5,6-dimethyl-1,2-benzanthraquinone, a degradation product of desoxycholic acid. J. Chem. Soc., pp. 428-433, 1934.
14. Cook, J.W.: Chemische Beiträge zum Krebs-problem. Berichte 69A, 38, 1936.
15. Dobriner, K., Rhoads, C.P. and Lavin, G.I.: Conversion of 1,2,5,6-dibenzanthracene by rabbits, rats, and mice. Significance in carcinogenesis of this conversion. Proc. Soc. Exp. Biol. Med., 41:67-69, 1939.
16. Dunlap, C.E. and Warren, S.: Chemical configuration and carcinogenesis. Cancer Research 1:730-731, 1941.
17. Fieser, L.F. and Dietz, E.M.: Beitrag zur kenntnis der synthese von mehrkernigen anthracenen. Berichte 62:1827-1833, 1929.
18. Fieser, L.F. and Fieser, M.: 1,2-benzpyrene. J. Amer. Chem. Soc., 57:782-783, 1935.
19. Fieser, L.F. and Hershberg, E.B.: 1',9-methylene-1,2,5,6-dibenzanthracene. J. Amer. Chem. Soc., 57:1681-1683, 1935.
20. Fieser, L.F. and Hershberg, E.B.: A new phenanthrene synthesis. J. Amer. Chem. Soc., 57:1508-1509, 1935.
21. Fieser, L.F., Hershberg, E.B. and Newman, M.S.: 4'-Hydroxy-1,2-benzpyrene. J. Amer. Chem. Soc., 57:1509-1510, 1935.
22. Fieser, L.F. and Newman, M.S.: The choleic acids of certain carcinogenic hydrocarbons. J. Amer. Chem. Soc., 57:1602-1604, 1935.
23. Fieser, L.F. and Seligman, A.M.: The synthesis of methylcholanthrene. J. Amer. Chem. Soc., 57:942-956, 1935.
24. Fieser, L.F. and Seligman, A.M.: Methylcholanthrene from cholic acid. J. Amer. Chem. Soc., 57:961, 1935.
25. Fieser, L.F. and Seligman, A.M.: 16,20-Dimethylcholanthrene. J. Amer. Chem. Soc., 57:1377-1378, 1935.

26. Fieser, L.F. and Seligman, A.M.: Cholanthrene and related hydrocarbons. J. Amer. Chem. Soc., 57:2174-2176, 1935.
27. Fieser, L.F. and Newman, M.S.: The synthesis of 1,2-benzanthracene derivatives related to cholanthrene. J. Amer. Chem. Soc., 58:2376-2382, 1936.
28. Fieser, L.F., Fieser, M., Hershberg, E.B., Seligman, A.M. and Shear, M.J.: Carcinogenic activity of the cholanthrenes and of other 1,2-benzanthracene derivatives. Am. J. Cancer, 29:260-268, 1937.
29. Fieser, L.F. and Hershberg, E.B.: Aceanthrene derivatives related to cholanthrene. J. Amer. Chem. Soc., 59:394-398, 1937.
30. Fieser, L.F. and Hershberg, E.B.: Reduction and Hydrogenation of compounds of the 1,2-benzanthracene series. J. Amer. Chem. Soc., 59:2502-2509, 1937.
31. Fieser, L.F., Hershberg, E.B., Long, L. and Newman, M.S.: Hydroxy derivatives of 3,4-benzpyrene and 1,2-benzanthracene. J. Amer. Chem. Soc., 49:475-478, 1937.
32. Fieser, L.F. and Seligman, A.M.: 4,10-ace-1,2-benzanthracene. J. Amer. Chem. Soc., 59:883-887, 1937.
33. Fieser, L.F. and Hershberg, E.B.: 10-Substituted 1,2-benzanthracene derivatives. J. Amer. Chem. Soc., 59:1028-1036, 1937.
34. Fieser, L.F. and Hershberg, E.B.: The oxidation of methylcholanthrene and 3,4-benzpyrene with lead tetraacetate; Further derivatives of 3,4-benzpyrene. J. Amer. Chem. Soc., 60:2542-2548, 1937.
35. Fieser, L.F.: Carcinogenic activity, structure, and chemical reactivity of polynuclear aromatic hydrocarbons. Am. J. Cancer, 34:37-124, 1938.
36. Fieser, L.F. and Desreux, V.: ; The synthesis of 2 and 6-substituted derivatives of 20-methylcholanthrene. J. Amer. Chem. Soc., 60:2255-2262, 1938.
37. Fieser, L.F. and Hartwell, J.L.: Mesoaldehydes of anthracene and 1,2-benzanthracene. J. Amer. Chem. Soc., 60:2555-2559, 1938.
38. Fieser, L.F. and Hershberg, E.B.: Reduction and hydrogenation of methylcholanthrene. J. Amer. Chem. Soc., 60:940-946, 1938.
39. Fieser, L.F. and Hershberg, E.B.: A new synthesis of 3,4-benzpyrene derivatives. J. Amer. Chem. Soc., 60:1658-1665, 1938.
40. Fieser, L.F. and Hershberg, E.B.: Substitution reactions and meso derivatives of 1,2-benzanthracene. J. Amer. Chem. Soc., 60:1893-1896, 1938.

41. Fieser, L.F. and Jones, N.R.: Synthesis of 4,9- and 4,10-dimethyl-1,2-benzanthrances. J. Amer. Chem. Soc., 60:1940-1945, 1938.
42. Fieser, L.F. and Seligman, A.M.: 1'-Methyl and 1',10-dimethyl-1,2-benzanthracenes. J. Amer. Chem. Soc., 60:170-176, 1938.
43. Fieser, L.F. and Snow, D.K.: 20-t-butylcholanthrene. J. Amer. Chem. Soc., 60:176-177, 1938.
44. Fieser, L.F. and Johnson, W.S.: 8-Methyl-1,2-benzanthracene. J. Amer. Chem. Soc., 61:168-171, 1939.
45. Fieser, L.F., Kilmer, G.W.: 9-Methyl-1,2,5,6-dibenzanthracene. J. Amer. Chem. Soc., 61:862-865, 1939.
46. Flesher, J.W. and Sydnor, K.L.: Carcinogenicity of derivatives of 7,12-dimethylbenz(a)anthracene. Cancer Res., 31:1951-1954, 1971.
47. Flesher, J.W. and Sydnor, K.L.: Possible role of 6-hydroxymethylbenzo(a)pyrene as a proximate carcinogen of benzo(a)pyrene and 6-methylbenzo(a)pyrene. Intern. J. Cancer, 11:433-437, 1973.
48. Flesher, J.W., Stansbury, K.H. and Sydnor, K.L.: S-adenosyl-L-methionine is a carbon donor in the conversion of benzo(a)pyrene to 6-hydroxymethylbenzo(a)pyrene. Cancer Letters, 16:91-94, 1982.
49. Flesher, J.W., Kadry, A.M., Stansbury, K.H., Gairola, C. and Sydnor, K.L.: Metabolic activation of carcinogenic hydrocarbons in the meso-position (L-region). In: The 7th International PAH Symposium: Polycyclic Aromatic Hydrocarbons, W.M. Cooke and A.J. Dennis (eds.), Battelle Press, pp. 505-515, 1983.
50. Flesher, J.W., Myers, S.R. and Blake, J.W.: Biosynthesis of the potent carcinogen 7,12-dimethylbenz(a)anthracene. Cancer Letters, 24:335-343, 1984.
51. Flesher, J.W. and Myers, S.R.: Oxidative metabolism of 7-methylbenz(a)anthracene, 12-methylbenz(a)anthracene, and 7,12-dimethylbenz(a)anthracene by rat liver cytosol. Cancer Letters, 26:83-88, 1985.
52. Flesher, J.W., Myers, S.R. and Lovelace, G.: Biotransformation of benzo(a)pyrene to 6-methylbenzo(a)pyrene and related metabolites in rat liver and lung. In: The 8th International PAH Symposium on Polynuclear Aromatic Hydrocarbons, M. Cooke and A.J. Dennis (eds.), Battelle Press, pp. 423-436, 1984.

53. Flesher, J.W., Myers, S.R. and Blake, J.W.: Bioalkylation of polynuclear aromatic hydrocarbons: A predictor of carcinogenic activity. In: The 9th International Symposium on Polynuclear Aromatic Hydrocarbons, Battelle Press, W.M. Cooke and A.J. Dennis (eds.), (in press).
54. Geyer, B.P. and Zuffanti, S.: A pyrolytic synthesis of 2,3-(Naphtho-2',3')acenaphthene. J. Amer. Chem. Soc., 57:1787-1788, 1935.
55. Grimmer, G., Jacob, J., Schmoldt, A., Raab, G., Mohr, U. and Emura, M.: Metabolism of benz(a)anthracene in hamster lung cells in culture in comparison to rat liver microsomes. In: The 8th International Symposium on Polynuclear Aromatic Hydrocarbons: Mechanisms, Methods, and Metabolism. M. Cooke and A. J. Dennis (eds.), Battelle Press, pp. 521-532, 1984.
56. Kennaway, E.L. and Hieger, I.: Carcinogenic substances and their fluorescence spectra. Brit. Med. J., 1044-1046, 1930.
57. Newman, M.S.: The synthesis of 1,2-benzanthracene derivatives related to 3,4-benzpyrene. J. Amer. Chem. Soc., 59:1003-1006, 1937.
58. Newman, M.S.: The synthesis of 9,10-dimethyl-1,2-benzanthracene. J. Amer. Chem. Soc., 60:1141-1142, 1938.
59. Sandin, R.B. and Fieser, L.F.: Synthesis of 9,10-dimethyl-1,2-benzanthracene and of a thiophene isolog. J. Amer. Chem. Soc., 62:3098-3105, 1940.
60. Shear, M.J.: Studies in carcinogenesis I. The production of tumors in mice with hydrocarbons. Am. J. Cancer, 26:322-332, 1936.
61. Shear, M.J.: Studies in carcinogenesis III. Isomers of cholanthrene and methylcholanthrene. Am. J. Cancer, 28:334-344, 1936.
62. Shear, M.J.: Studies in carcinogenesis V. Methyl derivatives of 1,2-benzanthracene. Am. J. Cancer, 33:499-537, 1938.
63. Shear, M.J.: Studies in carcinogenesis VII. Compounds related to 3:4-benzpyrene. Am. J. Cancer, 36:211-228, 1939.
64. Shear, M.J. and Leiter, J.: Studies in carcinogenesis XIV. 3-Substituted and 10-substituted derivatives of 1,2-benzanthracene. J. Natl. Cancer Inst., 1:303-336, 1940.
65. Shear, M.J.: Studies in carcinogenesis XVI: Production of subcutaneous tumors in mice by miscellaneous polycyclic compound. J. Natl. Cancer Inst., 2:241-258, 1941.

66. Shear, M.J. and Leiter, J.: Studies in carcinogenesis XV. Compounds related to 20-methylcholanthrene. J. Natl. Cancer Inst., 2:99-113, 1941.
67. Sydnor, K.L., Bergo, C.H., and Flesher, J.W.: Effect of various substituents in the 6-position on the relative carcinogenic activity of a series of benzo(a)pyrene derivatives. Chem.-Biol. Interacts., 29:159-167, 1980.
68. Vollman, H., Becker, H., Corell, M., Streek, H. and Langbein, G.: Pyrene and its derivatives. Annalen der chemie, 531:1-159, 1937.
69. Watabe, T., Ishizuka, T., Fujida, T., Hiratzuka, A. and Ogura, K.: Sulfate esters of hydroxymethyl-methylbenz(a)anthracene as active metabolites of 7,12-dimethylbenz(a)anthracene. Japanese J. Cancer Res., 76(8):684-698, 1985.
70. Wieland, H. and Dane, E.: The constitution of the bile acids L11. The place of attachment of the side chain. Ztschr. für physiol. Chem., 219:240-244, 1933.
71. Wood, J.L. and Fieser, L.F.: Sulfhydryl and cysteine derivatives of 1,2-benzanthracene, 10-methyl-1,2-benzanthracene, and 3,4-benzpyrene. J. Amer. Chem. Soc., 62:2676-2691, 1940.
Fieser, L.F.: Hydrocarbon Carcinogenesis, In: A.A.A.S. Research Conference on Cancer, F.R. Moulton (ed)., pp. 108-116, 1945.
72. Fieser, L.F. and Riegel, B.: The synthesis of 3-substituted derivatives of methylcholanthrene. J. Amer. Chem. Soc., 59:2561-2565, 1937.
73. Kamp, E.: Uber die Einwirkung von Formaldehyd Anthracens. Inaugural Dissertation, Univ. of Frankfort a. M., 109 pp (1936).

56. Newman, M.S. and [illegible], [illegible] Studies in Carcinogenesis XV. Compounds related to 20-methylcholanthrene. J. Natl. Cancer Inst. 2:189-193, 1941.

57. [illegible], K.L., [illegible], and [illegible], J.H. Effect of various substituents in the 4-position on the relative carcinogenic activity of a series of benz(a)pyrene derivatives. Chem.-Biol. Interacts. 29:139-146, 1980.

58. [illegible], H., [illegible], [illegible], [illegible], H. and [illegible] [illegible] Chem. 1956 [illegible].

60. [illegible] [illegible], [illegible] and [illegible] [illegible] of [illegible] [illegible] [illegible] J. [illegible] 1961.

61. [illegible] [illegible] and [illegible] [illegible] [illegible] of the [illegible] [illegible] [illegible] 218 [illegible] 1954.

62. [illegible] and [illegible] [illegible] and [illegible] derivatives [illegible] 10-methyl-1,2-[illegible] and 3,4-benzopyrene. J. Amer. Chem. Soc. 62:2637-2691, 1940.

[illegible] L.F. [illegible] hydrocarbon carcinogenesis. In: A.A.A.S. Research Conference on Cancer, [illegible] 1945.

62. [illegible] and [illegible] The synthesis of [illegible] derivatives of [illegible] methylcholanthrene. J. Amer. Chem. Soc. 59:[illegible] 1937.

63. [illegible] J. [illegible] [illegible] [illegible] [illegible] Frankfort a. M. 10 [illegible] 1934.

GENERAL PAPERS

THE FATE OF 1,8-DINITROPYRENE IN Salmonella typhimurium:

METABOLISM AND ADDUCT FORMATION

P.A. ANDREWS[1], J.C. ORR[2], I.B. LAMBERT[2], D.W. BRYANT[2], L.M. DAVISON[2], B.E. McCARRY[1], and D.R. McCALLA[2]
(1)Department of Chemistry, McMaster University; (2)Department of Biochemistry, McMaster University, Health Sciences Centre, 1200 Main St.W., Hamilton, Ontario CANADA L8N 3Z5

INTRODUCTION

Nitroarenes have recently become a focus of attention in environmental research due to their high mutagenic activity and their apparent facile formation in combustion processes. They have been identified in fly ash, diesel engine emissions, soot from wood burning stoves, air particulates from urban environments, and originally (but no longer) in xerographic toners (1). One of the most prominent nitroarenes in terms of its mutagenic activity in short term tests with Salmonella typhimurium is 1,8-dinitropyrene. While bacteria which are partially deficient in nitroreductase (TA98NR) (2) are resistant to 1-nitropyrene, niridazole, and some other nitro-compounds, they remain exquisitely sensitive to the mutagenic effects of both 1,6- and 1,8-dinitropyrene (DNP). Conversely, a strain (TA98/1,8DNP) selected for resistance to 1,8-dinitropyrene retains its sensitivity to 1-nitropyrene. The differences in mutagenic response manifest by these otherwise isogenic strains led McCoy et al. (3) to postulate that more than one nitroreductase might be involved in activating nitrated polycyclic compounds. Further investigations revealed a reduced mutational response by (TA98/1,8DNP) to 2-acetylaminofluorene (2-AAF) or N-hydroxy-2AAF. N-acetoxy-AAF on the other hand showed limited but equal mutagenicity in TA98, TA98NR, and TA98/1,8DNP (4). It was proposed that TA98/1,8DNP is deficient in a specific esterifying enzyme and that esterification of the arylhydroxylamines derived from nitro- and aminoarenes results in the formation of the potent electrophiles which are responsible for the genotoxicity of these compounds.

We have examined the possibility that more than one nitroreductase exists S. typhimurium, and whether any of these is specifically responsible for metabolism of dinitropyrenes. Although TA98NR clearly lacked a major nitroreductase activity (5) , there is no difference between a strain sensitive (TA98) or resistant (TA98/1,8DNP) to dinitropyrenes so far as detectable levels of this enzyme are concerned. Reverse phase HPLC detected a number of DNP metabolites when sensitive strains were used as a source of crude extract. One of these was absent when resistant strains were incubated with [^{3}H]-1,8-dinitropyrene. We have identified this metabolite as N-acetyl-8-aminopyrene. Recently, Orr (6) independently reported results which confirm observations by Saito et al. (7). TA98/1,8DNP is deficient in an acetyl-CoA acetyltransferase activity distinct from the nitroreductase function, and it is the deficiency of this enzyme which is responsible for the resistance of TA98/1,8DNP to 1,8-dinitropyrene.

We have identified an adduct which can be detected in the DNA of sensitive bacteria treated with 1,8-dinitropyrene (8). The adduct identified in hydrolysates of calf thymus DNA treated with N-hydroxy-1-amino-8-nitropyrene at pH 5.0 was the same C8 product found in analogous preparations of bacterial DNA after in vivo treatment of S.typhimurium with [^{3}H]-1,8-dinitropyrene. Beland and his group (9) using essentially similar methods have identified the same covalently bound adduct in their chemically synthesized preparations. This adduct (1-N-(2'-deoxyguanosin-8-yl)-amino-8-nitropyrene) is not acetylated so the deficiency responsible for resistance to DNP does not produce a stable product that appears in DNA.

In this report we wish to present progress in three areas of dinitropyrene metabolism. (1) Mutagenesis studies have been and continue to be quite valuable in this research, providing indications of the critical intermediate steps in DNP metabolism. An analysis of the relative mutagenic potential of the isolated metabolic and putative chemical intermediates of 1,8-dinitropyrene plainly shows that the first reductive step is crucial to formation of the ultimate mutagen. (2) Continuing examination of the enzymes involved DNP metabolism through the use of genetic mutants also shows that some kind of complex of nitroreductase and acetyltransferase may possibly be involved in the in vivo activation process. Furthermore, the enzyme directly implicated in the generation of the potent electrophile is quite distinct from the nitroreductase previously identified in S.typhimurium. (3) The preparation of synthetic adducts of 1,6- and 1,3-dinitropyrene through the treatment of DNA in vitro with arylhydroxylamine derivatives provides a direct method of measuring and identifying in vivo adducts.

MATERIALS AND METHODS

Bacterial Strains and Assays

Sallmonella typhimurium strain TA98 was obtained from Dr.B.N. Ames, University of California, Berkley, CA. Strains TA98NR, TA98NR/1,8DNP, and TA98/1,8DNP were supplied by Dr. H.S. Rosenkranz, Case Western Reserve University, Cleveland, OH, and have been previously described (2,3,10). The method used for the Salmonella mutagenicity test (Ames assays) was as described by Maron and Ames (10).

Chemicals

1,8-Dinitropyrene was isolated by LC Services (Boston, MA) from a mixture of dinitropyrene isomers prepared by the nitration of pyrene with HNO_3 in acetic anhydride. Preparation of [^{3}H]-1,8-dinitropyrene (6 Ci/mmole) has been described elsewhere (5). Reduction of the 1,8-DNP with $H_2/PtO_2/CH_3OH$ gave 1,8-diaminopyrene (1,8-DAP) directly. Synthesis and chemical characterization of 1-amino-8-nitropyrene (ANP), 1-nitroso-8-nitropyrene, 1-nitrosopyrene, 1-N-acetylamino-8-aminopyrene (N-acetyl-DAP) and 1,8-N-N'-diacetyldiaminopyrene has been described in other publications (6 and 7). Further details of the synthesis and characterization of 1,6- or 1-nitroso-3-nitropyrene and their hydroxylamines will be published elsewhere. Alternating copolymer duplexes were supplied by Pharmacia Molecular Biologicals; enzymes and calf thymus DNA from Sigma; and ^{32}P-ATP (3000Ci/mmol) was supplied by NEN/Dupont

Preparation and assay of crude extracts of S. typhimurium:

The preparation of crude extracts from TA98NR and assay of nitroreductase has been described elsewhere (5). Although nitroreductase activity as measured by the reduction of nitrofurazone (5-nitro-2-furaldehyde semicarbazone) was stable for months at -80°C, the additional nitroreductase activity described in this communication was much less stable, and extracts were routinely prepared and used within a week. A detailed description of the assays used to identify and isolate metabolites has been published (6).

HPLC analysis

Metabolites of 1,8-DNP were analysed using a reverse-phase ODS column with an elution program of acetonitrile and 0.02M phosphate (pH 7.6) (5). Metabolites were monitored by absorbance at 254nm and by fluorescence using a Varian Fluorichrom detector equipped with 355nm excitation and 430 emission filters. DNA adducts were analysed on a Spectra-Physics SP-8000.

Results and Discussion

(i) MUTAGENESIS

Metabolites of [^{3}H]-1,8-DNP accumulate in cultures, or crude extracts of strains of S.typhimurium which are sensitive to dinitropyrenes (5; 6). Figure 1 indicates the relative positions of synthesized DNP standards eluted from reverse phase HPLC. Comparison of the enzymatically generated intermediates of metabolism with these standards by HPLC elution and mass spectrometry confirmed their in vivo and in vitro presence.

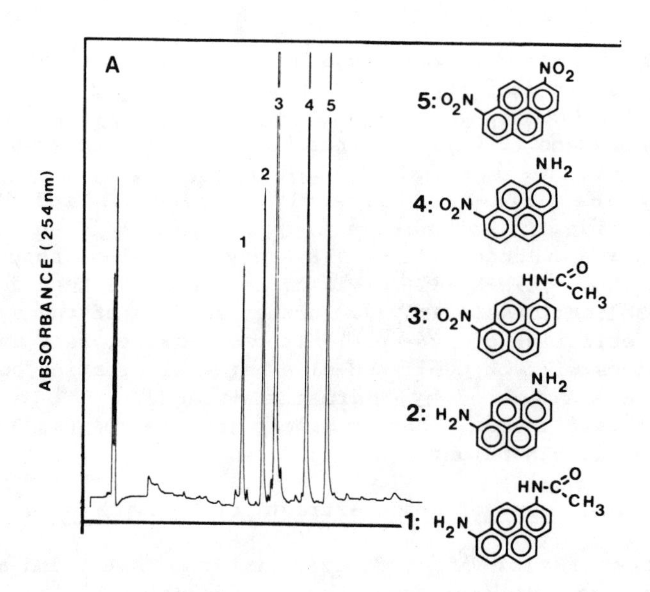

FIGURE 1. HPLC profile of synthetic standards of 1,8-DNP metabolites.

Ames assays were performed using each of the metabolites found or suspected to occur in incubations of bacterial crude extracts and DNP. The addition of S9 to the assays invariably eliminated almost all mutagenic activity. Table 1 shows the results of these experiments.

1-N-acetylamino-8-aminopyrene or 1,8-N-N'-diacetyldiaminopyrene like ANP and DAP, are only weakly mutagenic and thus cannot be considered candidates for the ultimate mutagen responsible for the potent genotoxicity of 1,8-DNP. Both compounds appear to be stable end products formed by complete reduction and subsequent acetylation of the 1,8-DNP. Nevertheless, the fact that an acetylated compound is formed in the strains sensitive to dinitropyrenes but not in resistant ones provides an important clue to the metabolic activation of this mutagen. The implication of our results is that acetylation is involved in the formation of the ultimate genotoxic product. This is consistent with the demonstration by McCoy, et al. (4,10) that DNP sensitive strains are able to acetylate 2-aminofluorene to 2-acetylaminofluorene and that this activity is lacking in the resistant mutants.

We anticipate that the actual mutagen will prove to be a highly reactive (unstable) compound. Since both ANP and DAP are weak mutagens, the ultimate genotoxic compound is probably at an oxidation level between dinitropyrene and the diamine. Aminonitropyrene, which fits this criterion, has a considerably reduced mutagenic activity, though it is more mutagenic than the fully reduced diamine in all the tester strains (Table 1).

By analogy with aromatic amines, the obvious candidate is 1-N-acetoxy-8-nitropyrene. Both reduction of the nitro-group and subsequent acetylation would be required to generate this compound. There is currently no direct evidence that 1-N-acetoxy-8-nitropyrene is formed by S.typhimurium or by cell free extracts of the bacteria. The extreme electrophilicity associated with N-acetoxy derivatives which makes it an attractive candidate as an ultimate mutagen also means that such a compound would likely undergo rapid decomposition.

Some nitro-PAH compounds are metabolized by a series of sequential two electron reduction steps which progress from the nitro via the nitroso to the arylhydroxylamine (1). Such is the case with 1-nitropyrene (1-NP) which readily forms adducts with DNA when it is reduced by xanthine oxidase under anaerobic conditions (12). Bacteria deficient in nitroreductase respond weakly to 1-NP, but it is possible to obviate the requirement for full nitroreductase activity by supplying the bacteria directly with nitrosopyrene. The lack of the major reductase in TA98NR has little effect on either the potency of 1,8-DNP, or on the formation of N-acetyl-DAP. This implies that the minor nitroreductase component remaining in TA98NR (5) which is markedly less effective in activating 1-NP, could provide sufficient reduction potential for the metabolic activation of 1,8-DNP. On the other hand, the nitroreductase detected through the reduction of nitrofurazone may have little to do with the pathway involved in DNP activation. Another possibility is that the association of the enzyme with an acetyltransferase is an obligatory condition for the generation of the electrophile which binds to DNA.

Table 1 shows the relative mutagenic potencies of the two electron reduction products of 1-NP and 1,8-DNP which are 1-nitrosopyrene and 1-nitroso-8-nitropyrene respectively. 1-Nitrosopyrene, which is much more mutagenic than 1-NP itself (11 and Table 1), does not react directly with DNA but via the hydroxylaminopyrene (generated by reduction of the nitroso with such substances as ascorbate) which reacts to produce a stable adduct (14). 1-Nitrosopyrene is somewhat less genotoxic in TA98NR or TA98/1,8DNP than in TA98. Analogous results for TA98 have been reported by El-Bayoumy and Hecht (13) and I. Salmeen (personal communication).

TABLE 1

S. typhimurium Response to Various Nitroarenes

Chemical	Revertants per Nanomole (S.D.) without S9 activation			
Strain:	TA98	TA98NR	TA98/1,8DNP	TA98NR/1,8DNP
1-nitropyrene	157(6)	16(1.0)	129(8.0)	3(.05)
1-nitrosopyrene	6827(314)	6152(161)	2898(228)	1794(70)
1-aminopyrene	47	87	nd*	
*	*	*	*	*
1,8-dinitropyrene	171,210(5542)	105,619(6944)	2434(123)	1251(114)
1-nitroso-8-nitropyrene	5919(563)	3131(287)	119(6)	60(5)
1-amino-8-nitropyrene	122(5)	102(5)	1.9(0.1)	1.73(0.11)
1-N-acetylamino-8-nitropyrene	0.04(0.01)	nd	nd	nd
1,8-diaminopyrene	0.01(0)	nd	nd	nd
1-N-acetylamino-8-aminopyrene	0.16(0.01)	nd	nd	nd
1,8-N,N'-diacetyldiaminopyrene	0.01(0)	nd	nd	nd

* The amount of mutagen required was beyond the limit of solubility of the compound.

In contrast, 1-nitroso-8-nitropyrene is a powerful mutagen in those strains which are sensitive to 1,8-DNP but shows a greatly decreased genotoxicity in TA98/1,8DNP and TA98NR/1,8-DNP. Distinct from observations with 1-nitrosopyrene, the mutagenicity of the 1,nitroso-8-nitropyrene is an order of magnitude less than that of the parent 1,8-dinitropyrene. The explanation of this behavior is unclear, but two possibilities arise: the 1-nitroso-8-nitropyrene may be reduced much faster than 1,8-DNP, rapidly saturating the acetylase and pushing more of the material toward reduction to the diamine. Alternatively, the uptake of 1-nitroso-8-nitropyrene into bacteria may be slower than that of 1,8-DNP. What is clear from our results is that addition of a second nitro group to the pyrene ring system leads to a compound whose mutagenic potency is roughly equal to 1-NP in the absence of acetylation (TA98/1,8DNP), but is two hundred fold greater in strains capable of performing acetylation.

ENZYMES and METABOLISM

An HPLC analysis of metabolites of 1,8-DNP or 1,8-DAP treated with a partially purified extract of S.typhimurium gave the results shown in Table 3. The concentration of each product was determined from the areas under each peak in the chromatogram (corrected for variation in absorbance by using the extinction coeficient at 254nm) and then calculated as a mole percent of the starting material. Substrate concentrations were approximately equimolar for the experiments.

TABLE 2

Metabolism by DEAE Extracts in the Presence of Acetyl-CoA

Subtrate:	1,8-dinitropyrene		1,8-diaminopyrene	
Metabolite:	TA98NR	TA98/1,8DNP	TA98NR	TA98/1,8DNP
DNP	4.5*	19.9	-	-
ANP	31.5	42.2	-	-
DAP	.08	15.3	22.6	96.4
Monoacetyl-DAP	63.0	22.6	71.0	3.6

* Mole percent total substrate

All the results shown are for assays carried out in the presence of acetyl-CoA. Although some acetylation occurred in the absence of an acetyl donor, the addition of acetyl-CoA to the incubation mix significantly increased the level of acetylated metabolites observed. The TA98/1,8-DNP strain appeared to have much less acetyltransferase activity than did TA98NR when either substrate was used. The facility with which TA98NR extracts convert diaminopyrene into its acetylated derivative is unequal for the experiments shown. These experiments show a rapid utilization of substrate (as measured by acetylated products) consistantly occurs when dinitropyrene rather than diaminopyrene is used in the assay. Further examination of the rates of metabolism of the two substrates indicates that the nitroreductase and transacetylase activities may possibly be subunits of a single enzyme resulting in a failure of the reduced diamine to accumulate, and the direct conversion of dinitropyrene to its acetylated diamine. Further evidence that the enzyme activities form a unique complex is indicated by the specific inhibition of in vitro DNP metabolism by N-ethylmaleimide. Treatment with this thiol inhibitor does not appear to alter the rate either of nitroreduction of ANP to DAP nor the mono- and diacetylation of DAP in crude preparations (manuscript in preparation).

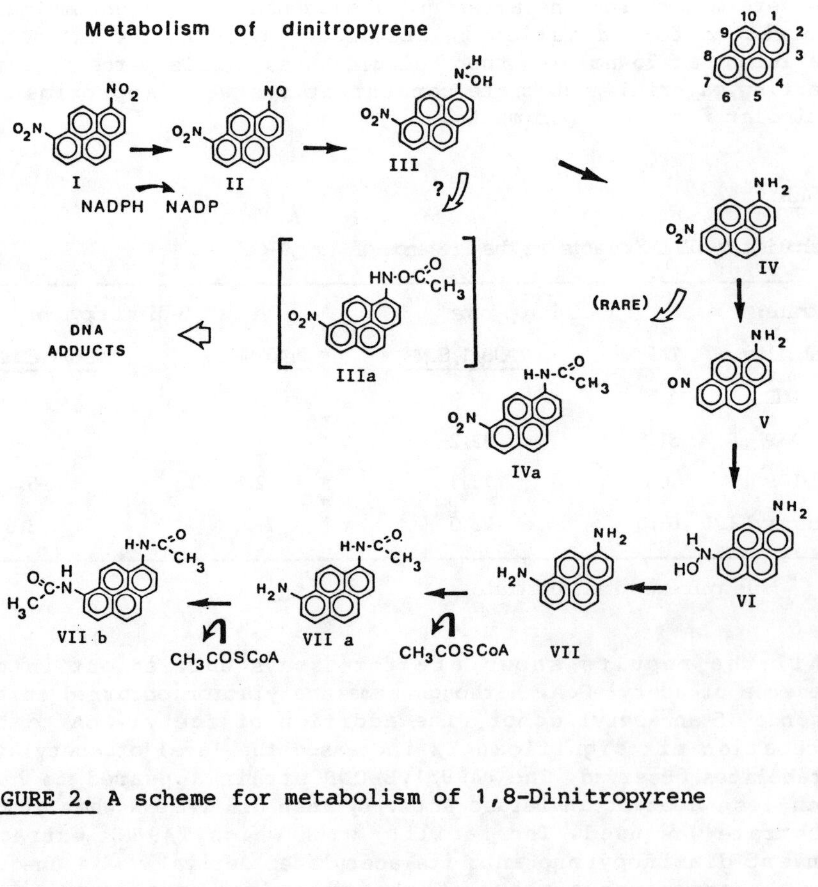

Metabolism of dinitropyrene

I → II → III → IV → V → VI → VII → VII a → VII b

NADPH → NADP

III ? → IIIa → DNA ADDUCTS

IV (RARE) → IVa

$CH_3COSCoA$

$CH_3COSCoA$

FIGURE 2. A scheme for metabolism of 1,8-Dinitropyrene

Figure 2 is a presentation our current speculation regarding metabolism of 1,8-dinitropyrene in S.typhimurium . In the presence of functional nitroreductase and transacetylase activities, DNP (I) is reduced to 1-amino-8-nitropyrene (IV) in a series of steps which require NADPH as a cofactor. The first of these steps in the generation of the nitroso-nitropyrene (II) which with the loss of two electrons and the addition of two H^+ generates the N-hydroxylamino-8-nitropyrene (III). A further loss of two electrons, a molecule of water and the addition of two H^+ yeilds ANP. We believe that it is between steps III and IV that the activity of the transacetylase is essential to the formation of the potent electrophile that is the ultimate mutagen (IIIa). From (IV), the nitrosoamino- (V), the 1-amino-8-N-hydroxylamino- (VI) and the diamine (VII) are produced through nitroreduction by means analogous to steps I-IV. Finally, the free diamine (in a reaction quite distinct from what occurs with the dinitro-compound) can be either mono- or diacetylated in the presence of an acetyl donor such as acetylphosphate or acetyl-CoA by the transacetylase enzyme. With the exception of the N-acetoxy-8-nitropyrene (IIIa) all of the compounds in figure 2 have been synthesized in our laboratory. This has simplified the identification of metabolites formed in vivo or in our in vitro systems. It has also allowed the testing of the intermediate compounds for potential mutagenicity. As seen above, none of the compounds for which we present data is as mutagenic as the parent DNP. Because no intermediate we have been able to make is potent enough to generate DNA adducts in vivo in sufficient quantities for chemical analysis, we have turned to the preparation of adducts by purely chemical means as a method of producing standards by which in vivo products may be measured.

Preparation of Synthetic Adducts of DNA with 1,8-, 1,6- and 1,3-DNP.

Howard et al. (12) have shown that chemical reduction of the 1-nitrosopyrene with ascorbate generates a reactive electrophile which forms covalent adducts with DNA. This adduct has been identified as 1-N-(2-deoxyguanosin-8-yl)-aminopyrene, which corresponds to the product formed by bacteria in vivo. In similar experiments Andrews et al. (8) prepared an adduct in calf thymus DNA by generating the N-hydroxylamino-8-nitropyrene (1,8-HANP) from the N-nitroso-8-nitropyrene (1,8-NONP) through ascorbate reduction. Characterization of this adduct from DNA hydrolysates was accomplished by FAB mass spectrometry, NMR, ultraviolet/visible and fluorescence spectroscopy. The major product detected was identified as 1-N-(2'-deoxyguanosin-8-yl)-amino-8-nitropyrene (Figure 3). Further experiments with in vivo exposures of S. typhimurium to [^{3}H]-1,8-DNP resulted in an adduct which displayed behavior similar to the synthetic adduct on HPLC.

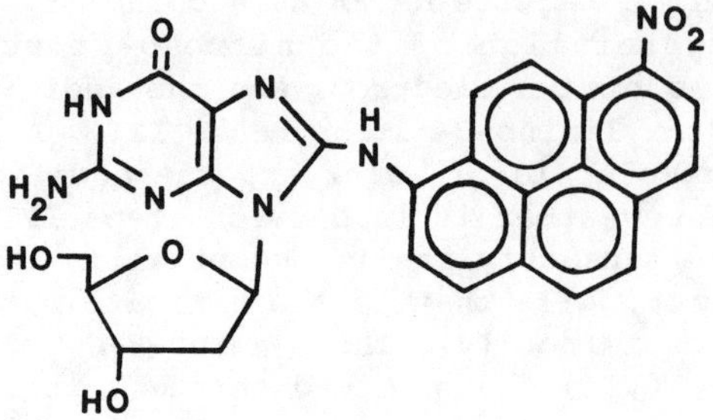

FIGURE 3. 1-N-(2'-deoxyguanosin-8-yl)-amino-8-nitropyrene identified in DNA treated with 1,8-HANP.

We have recently utilized a highly sensitive enzymatic ^{32}P-postlabelling technique (15,16) to study adduct formation in calf thymus (CT-) DNA treated with N-hydroxylamino-derivatives of 1,8-, 1,6-, and 1,3-dinitronitropyrene (1,8-HANP; 1,6-HANP; and 1,3-HANP respectively).

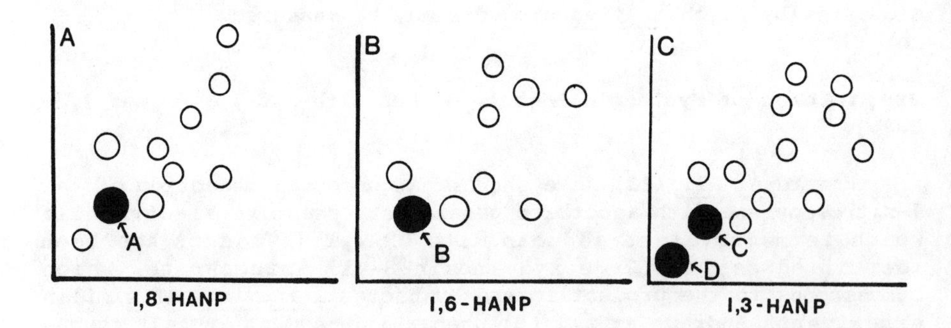

FIGURE 4. Maps of ^{32}P-labelled digests of calf thymus DNA treated in vitro with 1,8-HANP (Panel A); 1,6-HANP (Panel B); or 1,3-HANP (Panel C). Resolution of adducts from unaltered nucleotides is performed by four directional polyethyleneimine-cellulose thin-layer chromatography.

Figure 4 gives a diagramatic depiction of the unique fingerprinting patterns elicited in DNA hydrolysates by exposure to each compound. 1,8-HANP gave rise to ten distinct spots after PEI cellulose thin layer chromatography (TLC) in four dimensions. The major product (A) contained 70% of the recovered label. We suspect this is the same C8 adduct described by Andrews, et al. (8) which is found both in vivo and in vitro (also see Figure 3). 1,6-HANP treated CT-DNA produced a similar but not identical result, giving rise to nine spots, again with one major product (B) accounting for 73% of the label. Reaction with 1,3-HANP, on the other hand, gave a distinctly different pattern of labelling. Of the thirteen recognizable products, two (C and D) contained 38% and 34% respectively. The combination of relatively high adduct concentration generated by the synthetic method for production of the hydroxylamines (8) and the extreme sensitivity afforded by the ^{32}P-postlabelling technique have combined to reveal putative adducts not previously observed. These minor products may indeed be new adduct types produced in vitro, or alternatively represent artifacts of the assay procedure. The latter include oligonucleotides from incomplete hydrolysis, products of 3'-dephosphorylation, direct chemical modification by the buffer systems used or other post exposure processes. Similar artifacts have been reported by others (17,18). Identification of the minor spots are the subject of continuing investigation in this laboratory.

In an analogous series of reactions 1,8-HANP was used to generate adducts in poly(dG-dC).poly(dG-dC) and poly(dA-dT).poly(dA-dT) alternating copolymer duplexes (Figure 5). When compared with CT-DNA (panel A), the adduct formation with duplexes gave interesting results.

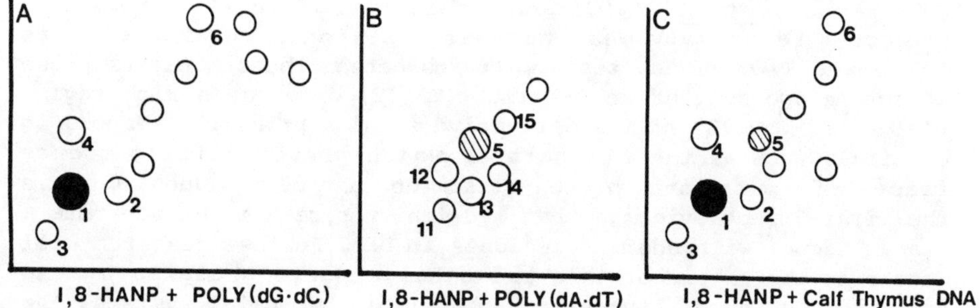

FIGURE 5. Adducts produced in alternating copolymer duplexes treated with 1,8-HANP.

Table 3 quantifies the results of relative adduct labelling in the different preparations. With the exception of one lightly labelled spot, labelled hydrolysates of CT-DNA contained adducts which co-chromatographed with similar hydrolysates of poly(dG-dC). By contrast, only the most heavily labelled spot occurring in hydrolysates of poly(dA-dT) co-chromatographed with an identifiable CT-DNA product. This A/T unique adduct comprised only 1.5% of the total CT-DNA adducts. These results suggest that 1,8-HANP preferentially reacts with dG and possibly dC in DNA. By comparison reactions with dA or dT appear to be much more infrequent.

TABLE 3

Relative Adduct Labelling* by 1,8-HANP: (Adducts per 10^6 nucleotides)

Spot #	poly(dG-dC)	poly(dA-dT)	CT-DNA
1	504	0	807
2	137	0	80
3	48	0	82
4	12	0	80
5	0	172	17
6	202	0	41
remainder	418	4	67
11	0	51	0
12	0	133	0
13	0	82	0
14	0	43	0
15	0	4	0

$$* \text{ relative adduct labelling} = \frac{\text{cpm in adducts}}{(\text{cpm in normal nucleotides})} \times \text{dilution factor}$$

Massaro et al. (19) noted that nitroarenes were able to produce a relatively weak mutagenic response in Ames strains TA102 and TA96. These tester strains detect those mutagens whose affect is upon either an A:T rich (TA102) or on an A rich region (TA96)(20). TA98 and its derivative strains primarily respond to chemicals such as the nitroarenes which preferentially induce frameshift mutations through attack on guanine residues. Ours is the first direct evidence that 1,8-dinitropyrene might generate a stable adduct with adenine residues in DNA. Further research will determine the nature of this and other adducts produced by the nitroarenes. Though powerful mutagens in bacterial systems, the potency of nitroarenes as possible carcinogens is still being assessed. In addition to in vitro experiments we expect to quantitate these or any additional adducts which occur in bacteria or mammalian systems in vivo.

BIBLIOGRAPHY

(1) Rosenkranz, H.S. and R. Mermelstein (1983): Mutagenicity and genotoxicity of nitroarenes. All nitro-containing compounds were not created equal, Mutat. Res., 114 : 217-267.

(2) Speck, W.T.,J.L. Blumer, E.J. Rozenkranz and H.S. Rosenkranz (1981): Effect of genotype on mutagenicity of niridazole in nitroreductase deficient bacteria, Cancer Res., 41 : 2305-2307.

(3) McCoy, E.C., H.S. Rosenkranz, and R. Mermelstein (1981): Evidence for the existance of a family of bacterial nitroreductases capable of activating nitrated polycyclics to mutagens, Environ. Mutagen., 3: 421-427.

(4) McCoy, E.C., G.D. McCoy and H.S. Rosenkranz (1982): Esterification of arylhydroxylamines: Evidence for a specific gene product in mutagenesis, Biochem. Biophys. Res. Commun.,108: 1362-1367.

(5) Bryant D.W., D.R. McCalla, P. Lultschick, M.A. Quilliam and B.E. McCarry (1984): Metabolism of 1,8-dinitropyrene by Salmonella typhimurium, Chem. Biol. Interactions, 49: 351-368.

(6) Orr, J.C., D.W. Bryant, D.R. McCalla and M.A. Quilliam (1985): Dinitropyrene-resistant Salmonella typhimurium are deficient in an acetyl CoA acetyltransferase. Chemico-Biol. Interact. (in press).

(7) Saito, K., A. Shinohara, T, Kamataki and R. Kato (1985): Metabolic activation of mutagenic N-hydroxylamines by O-acetyltransferase in Salmonella typhimurium TA98. Arch. Bioch. Biophys.,239: 286-295.

(8) Andrews,P.J., M.A. Quilliam, B.E. McCarry, D.W. Bryant and D.R. McCalla (1985): Identification of the DNA adduct formed by metabolism of 1,8-dinitropyrene in Salmonella typhimurium. Carcinogenesis (in press).

(9) Fifer, E.K., R.H. Heflich, Z. Djuric', P.C. Howard and F.C. Beland (1985): Synthesis and mutagenicity of 1-nitro-6-nitrosopyrene and 1-nitro-8-nitrosopyrene, intermediates in the metabolic activation of 1,6- and 1,8-Dinitropyrene. Carcinogenesis (in press).

(10) Maron, M.M. and B.N. Ames (1983): Revised methods for the Salmonella mutagenicity test, Mutat. Res., 113: 173-215.

(11) McCoy, E.C., M. Anders and H.S. Rosenkranz (1983): The basis of the insensitivity of Salmonella typhimurium strain TA98/1,8DNP to the mutagenic action of nitroarenes, Mutat. Res.,121: 17-23.

(12) Howard, P.C., R.H. Heflich, F.E. Evans, and F.A. Beland (1983): Formation of DNA adducts in vivo and in Salmonella typhimurium upon metabolic reduction of the environmental mutagen 1-nitropyrene. Cancer Res.,43: 2052-2058.

(13) El-Bayoumy, K., and S.S. Hecht (1983): Identification and mutagenicity of metabolites of 1-nitropyrene formed by rat liver, Cancer Res.,43: 3132-3137.

(14) Heflich, R.H., P.C. Howard, F.A. Beland, G.L. White, and D.T. Beranek (1983): 14th annual meeting Environmental Mutagen Society (abstract), 157

(15) Gupta, R.C., M.V. Reddy and K. Randerath (1982): ^{32}P-postlabelling analysis of non-radioactive aromatic carcinogen-DNA adducts, Carcinogenesis, 3: 1081-1092.

(16) Reddy, M.V., R.C. Gupta, E. Randerath and K. Randerath (1984): ^{32}P-postlabelling test for covalent DNA binding of chemicals in vivo: application to a variety of aromatic carcinogens and methylating agents, Carcinogenesis, 5: 231-243.

(17) Randerath, K., E. Randerath, H.P. Agrawal, and M.V. Reddy (1984): Biochemical (postlabelling) methods for analysis of carcinogen-DNA adducts, Monitoring Human Exposure to Carcinogenic and mutagenic Agents (IARC Scientific Publications No. 59), A. Berlin, M. Draper, K. Hemminiki and V. Vainio, eds.,Lyon International Agency for Research on Cancer, 217-231.

(18) Gupta, R.C. (1984): Nonrandom binding of the carcinogen N-hydroxy-2-acetylaminofluorene to repetitive sequences of rat liver DNA in vivo, Proc. Natl. Acad. Sci.,USA,81: 6943-6947.

(19) Massaro, M., M. McCartney, E.J. Rozenkranz, M. Anders, E.C. McCoy, R. Mermelstein and H.S. Rosenkranz (1983): Evidence that nitroarene metabolites form mutagenic adducts with DNA-adenine as well as with DNA-guanine, Mutat. Res., 122: 243-249.

(20) Levin, D.E., M. Hollstein, M.F. Christman, E.A. Schwiers and B.N. Ames (1982): A new Salmonella tester strain (TA102) with A:T base pairs at the site of mutation detects oxidative mutagens, Proc. Natl. Acad. Sci.,(USA), 79 : 7445-7449.

Acknowledgement: This work was supported by a "strategic" grant from the National Science and Engineering Research Council of Canada.

MUTAGENICITY OF CHLOROMETHYLBENZO[a]PYRENES IN THE AMES ASSAY AND IN CHINESE HAMSTER V79 CELLS

JAMES C. BALL[1*], ALBERTO A. LEON[2], SUSAN FOXALL-VANAKEN[1], PANOS ZACMANIDIS[1], GUIDO H. DAUB[2,3], AND DAVID L. VANDER JAGT[2].
(1)Ford Motor Company, Research Staff, Dearborn, MI 48121; (2)Departments of Chemistry and Biochemistry, Univ. of New Mexico, Albuquerque, NM 87131; (3) Deceased April 14, 1985

INTRODUCTION

The biological effects and chemical analysis of complex environmental pollutants, such as cigarette smoke (1), forest fire smoke (2), spark engine (3) and diesel engine (4) exhaust particles are important because of the unknown human health effects of such pollutants. However, the analysis of these mixtures is difficult because of the thousands of compounds known to be present in these samples. One approach to the analysis of these complex mixtures is to identify classes of compounds which are responsible for genetic damage (e.g. mutations or cancer). Methylated polynuclear aromatic hydrocarbons (Me-PAH) represents a class of compounds found in all of the above combustion products. In the case of diesel engine exhaust particles approximately 2% of the extractable mass is methylated or dimethylated PAH and their derivatives (4). Some of these Me-PAH are animal carcinogens (5) and mutagenic in bacteria (6) and mammalian cells (7).

The metabolism of Me-PAH to a reactive intermediate(s) which can react with DNA is thought to proceed by oxidation of the aromatic ring to an arene oxide by analogy with the well known metabolism of BaP. However, the oxidation of the methyl group to a hydroxymethyl substituent, and the further metabolism to a reactive ester (e.g. sulfate or phosphate ester) represents a mechanism for the activation of Me-PAH distinctly different from that proposed for BaP or other PAH. In support of this mechanism there are several activated esters of hydroxymethyl-PAH (HyMe-PAH) which have been found to be genotoxic. For example, the sulfate esters of 1-hydroxymethylpyrene and 7-hydroxymethyl-12-methylbenz[a]anthracene are mutagenic in bacteria cells (8), and 7-bromomethylbenz[a]anthracene is both carcinogenic (9,10,) and mutagenic (11,12). In order to test the hypothesis that esters of hydroxymethylbenzo[a]pyrenes (HyMe-BaP) are mutagenic intermediates in the metabolism of Me-BaP we have synthesized several chloromethylbenzo[a]pyrenes (ClMe-BaP, Figure 1) as model compounds for the reactive metabolites. Herein we report the mutagenicity and solvolytic reactivity of ClMe-BaP substituted in the 1,4,5,6, 10,11, and 12 positions.

MATERIALS AND METHODS

Chemicals

The ClMe-BaP used in this study were synthesized as described previously (13-16).

Solvolysis Studies

The rate constants for the solvolysis of ClMe-BaP used in this study were determined in 50% acetone/water at 25 C. The reactions were followed by measuring the change in absorbance between the ClMe-BaP and the corresponding HyMe-BaP at a particular wavelength. The wavelengths used to follow the solvolysis of ClMe-BaP to HyMe-BaP are summarized in Table 1. The rate constant for the solvolysis of 6-ClMe- BaP was measured in 80% acetone/water (v/v) because of the

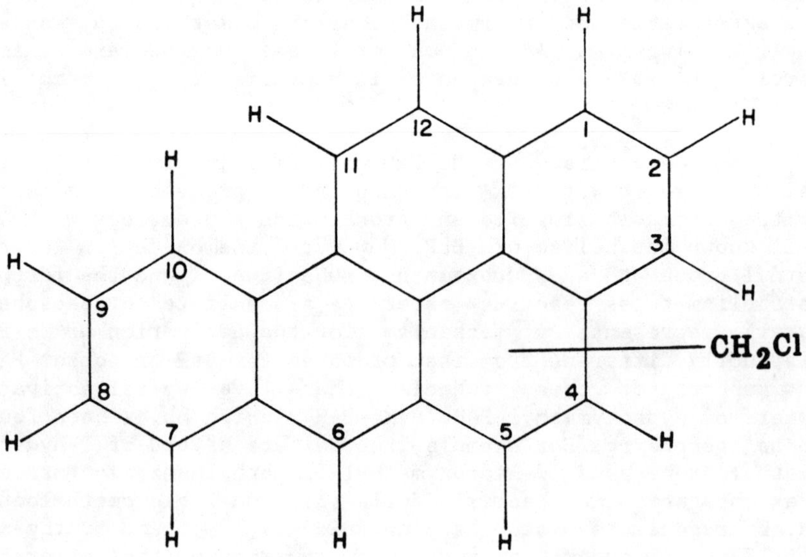

FIGURE 1. Structures of the chloromethylbenzo[a]pyrenes studied in this report. The chloromethyl group was located in the following positions: 1,4,5,6,10,11, and 12.

TABLE 1

WAVELENGTH USED TO MONITOR THE SOLVOLYSIS OF ClMe-BaP TO HyMe-BaP

Position	WAVELENGTH (nm)
1	392
4	380
5	389
6	409
10	390
11	390
12	387

reactivity of this compound(17). The rate constant for the solvolysis of this compound in 50% acetone/water was calculated from the rate constant determined in 80% acetone/-water and the rate constant of 1-ClMe-BaP in 50% and 80% acetone/water. All solvolysis reactions were followed for at least two half lives, and the correlation coefficients for the first order plots were >.95. The rate constants were determined in triplicate and were reproducible to ± 10%.

Ames Assay

The plate incorporation assay was used as described by Maron and Ames (18). A compound was considered to be mutagenic if the number of revertants per plate exceeded the 99.9% confidence limit and a concentration-dependent increase in the mutagenicity was observed. The 99.9% confidence limit was 98 revertants above the spontaneous revertant levels for TA100 (spontaneous revertants = 193 ± 29, N = 49) and 46 revertants above the spontaneous revertant level for TA98 (spontaneous revertants = 41 ± 10, N = 61). The 99.9% confidence limit was calculated by multiplying the standard deviation by the appropriate t value from the Student's t distribution (one-sided) requiring 99.9 % confidence(19). The concentration-dependent mutagenicity of methyl methane sulfonate (TA100, 141 ± 41 revertants/umole, N = 23) and 2-nitrofluorene (TA98, 28.9 ± 5.3 revertants/nmole, N = 34) were determined in each experiment as a positive control. Toxicity in this assay was determined by examination of photomicrographs of the background lawn.

V79 Cell Mutagenicity Assay

V79 cells were grown at 37 C in 5% CO2 in air using Dulbecco's Modified Eagles medium supplemented with 5% fetal calf serum and 50 ug/ml gentamycin. Exponentially growing cells were plated out 24 h prior to treatment with a mutagen. Cells were treated with varying concentrations of the ClMe-BaP for 5 h in serum free media at 37 C. The cells were replated at 100 and 300 cells per T25 flask to determine the surviving fraction. Aprox. $1X10^6$ surviving cells (estimated from previous cytotoxicity data) were plated out in 4 T75 flasks and allowed to continue phenotypic expression with subculturing as necessary to maintain exponential growth. After a period of 7 days, $1x10^6$ cells were selected for 6-thioguanine resistance in growth medium supplemented with 30 uM 6-thioguanine (20). Thioguanine resistant cells were stained with methylene blue 7 days later. Cloning efficiencies (>85%) were determined at the time of selection for 6-thioguanine resistance cells. Ethyl methanesulfonate (EMS) was used as a positive control in each experiment.

RESULTS

Solvolysis of ClMe-BaP

The rate constants for the solvolysis of ClMe-BaP are shown in Table 2. There was a 460,000 fold range in reactivity with 6-ClMe-BaP being the most reactive and 11-ClMe-BaP being the least reactive. The variation in reactivity among these compounds can be subdivided into two groups. The first group is characterized by half-lives on the order of seconds, and this would include 1- and 6-ClMe-BaP. The second group is characterized by half-lives on the order of days, and this includes 4-,5-,10-,11-, and 12-ClMe-BaP.

Ames Assay Mutagenicity

The concentration-dependent mutagenicity curves for the ClMe-BaP tested are shown in Figures 2 and 3. The specific mutagenicity (revertants/nm) for each compound was calculated from the linear portion of the curves and is summarized in Table 3. Some of the concentration-dependent mutagenicity curves were non-linear at higher concentrations of mutagen, and these plates were examined for toxicity. Toxicity was measured by counting the background lawn from photomicrographs and comparing these values to the solvent control. In each case where a non-linear curve was observed toxicity was seen on those plates responsible for the deviation from

TABLE 2

RATE CONSTANTS FOR THE SOLVOLYSIS OF ClMe-BaP IN 50% ACETONE/WATER, 25 C

Position	k (sec^{-1})	t 1/2	Relative k
1	1.9×10^{-2}	36 sec	9,000
4	6.6×10^{-6}	1.0 day	3.1
5	2.8×10^{-6}	2.5 day	1.3
6	*9.7×10^{-1}	0.7 sec	460,000
10	2.7×10^{-6}	2.6 day	1.3
11	2.1×10^{-6}	3.8 day	1.0
12	4.2×10^{-6}	1.6 day	2.0

* Calculated from the rate constant determined in 80% acetone/water.

linearity. The relative mutagenicity of these compounds in strain TA98 was 11>12>4>5,10>1>6. The order is somewhat different in TA100 with the major difference being a switch in the relative mutagenicity of 11- and 12-ClMe-BaP (12>11). In general, these compounds were more mutagenic in strain TA98 than in strain TA100 with the exceptions of 6- and 10-ClMe-BaP. We also tested the corresponding HyMe-BaP for mutagenicity and these data are shown in Tables 4 and 5.None of the HyMe-BaP were mutagenic. 6-HyMe-BaP was somewhat toxic at the higher concentrations tested. Consequently, this compound was reassayed at lower concentrations and found to be non mutagenic.

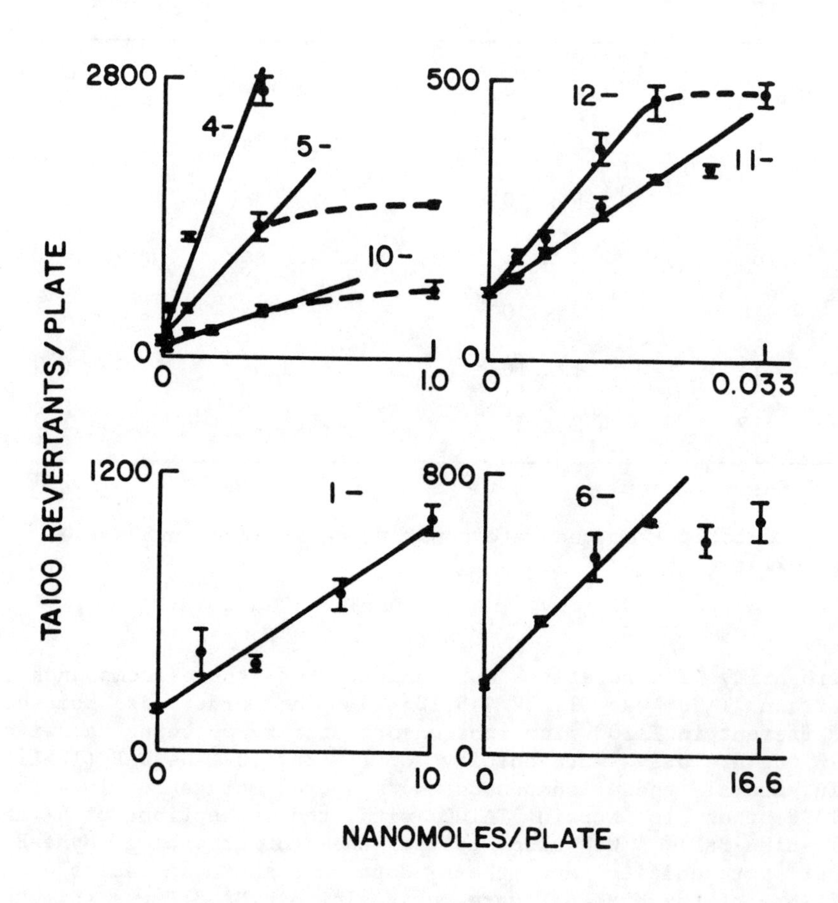

FIGURE 2. Mutagenicity of chloromethylbenzo[a]pyrenes in the Ames assay using strain TA100 without metabolic activation.

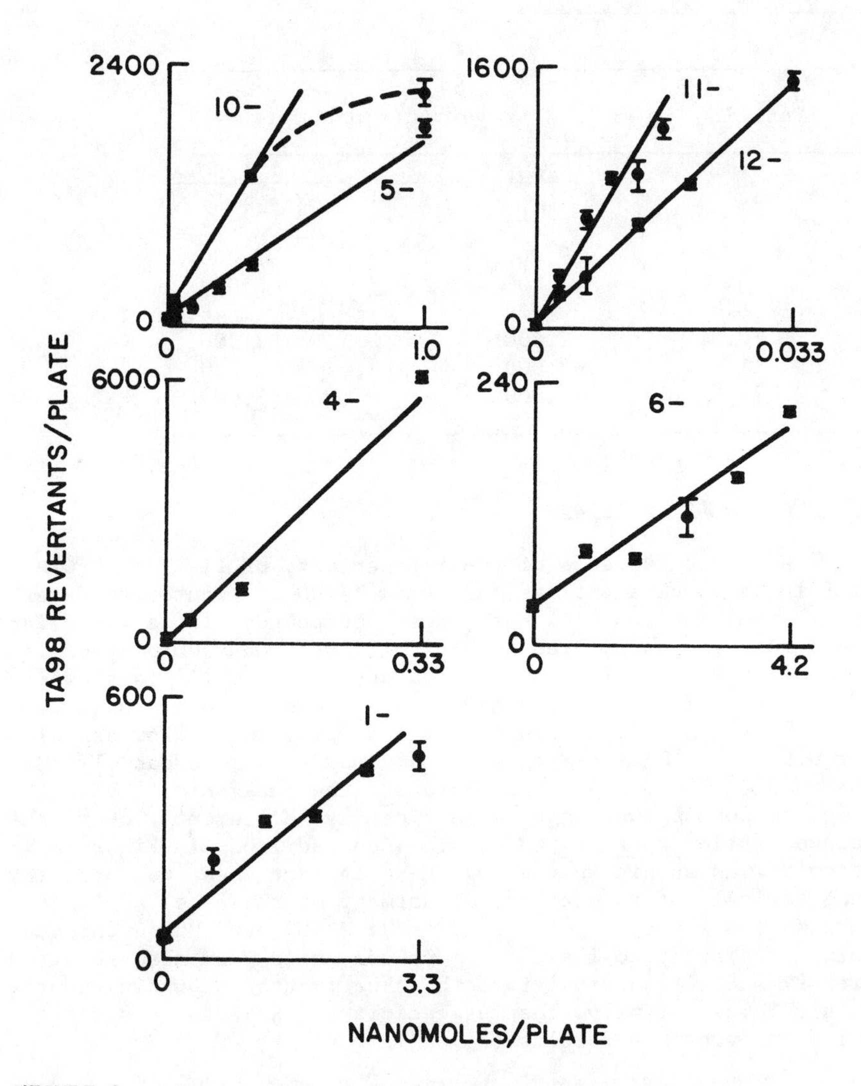

FIGURE 3. Mutagenicity of chloromethylbenzo[a]pyrenes in the Ames assay using strain TA98 without metabolic activation.

TABLE 3

SUMMARY OF THE MUTAGENICITY OF ClMe-BaP

IN STRAINS TA98 AND TA100

Position	Revertants/nanomole	
	TA98	TA100
1	120	75
4	19,000	7,200
5	1,800	1,000
6	39	48
10	2,000	3,300
11	69,000	12,000
12	43,000	17,000

V79 Cell Mutation Assay

As a consequence of the mutagenicity of 11- and 12-ClMe-BaP in Salmonella strains TA98 and TA100 we wanted to determine the mutagenicity of these compounds in a mammalian cell mutation assay. The cytotoxicity and mutagenicity of 11- and 12-ClMe-BaP in Chinese hamster V79 cells is shown in Figures 4 and 5. 12-ClMe-BaP was somewhat more toxic than 11-ClMe-BaP at equal concentrations of mutagen. For example, at 0.5 um 12-ClMe- BaP showed a 5% survival whereas 11-ClMe-BaP had a 20-25% survival. However, the mutagenicity of these two compounds was not significantly different over the concentration range studied. The curved line in Figure 4 is provided as an aid in viewing the data and does not have any theoretical interpretation. Treatment of these cells with 8.1 mM EMS ,as a positive control, induced 999 and 952 6-thioguanine resistant cells/10^6 clonable cells, and also resulted in 49% and 56% survival for the experiments shown in Figures 4 and 5 respectively. The mutagenicity of EMS was consistent with our previous results (21).

DISCUSSION

These results show a wide range of bacterial mutagenicity for these compounds depending on the position of the chloromethyl functional group. While some of the ClMe-BaP

TABLE 4

MUTAGENICITY OF HYDROXYMETHYLBENZO(a)PYRENES IN STRAIN TA100 REVERTANTS/PLATE[1]

	Nanomoles hydroxymethylbenzo(a)pyrene/plate											
Isomer	0.004	0.007	0.014	0.021	0.028	0.035	0.106	0.212	1.06	3.5	10.6	17.7
1										126	130	
4							112		122			
5								97	118			
6							155[3]	188[3]	178[3]	230[3]	T	T
10							100		105			
11	115	114	105	108	110							
12	111	124	114	81		100						

1 Spontaneous revertants - 114 except where noted

2 T - toxicity observed as a reduction in the background lawn

3 Spontaneous revertants - 190

TABLE 5

MUTAGENICITY OF HYDROXYMETHYLBENZO(a)PYRENES IN STRAIN TA98: REVERTANTS/PLATE[1]

	Nanomoles Hydroxymethylbenzo(a)pyrene/Plate													
Isomer	0.004	0.007	0.011	0.014	0.018	0.021	0.035	0.106	0.212	1.06	1.77	2.12	3.5	4
1												40	31	
4								38		34				
5									46	41				
6											37			
10								37		42				
11	42	50	47	44	40									
12	38	36		32		44	36							

[1] Spontaneous revertants = 36

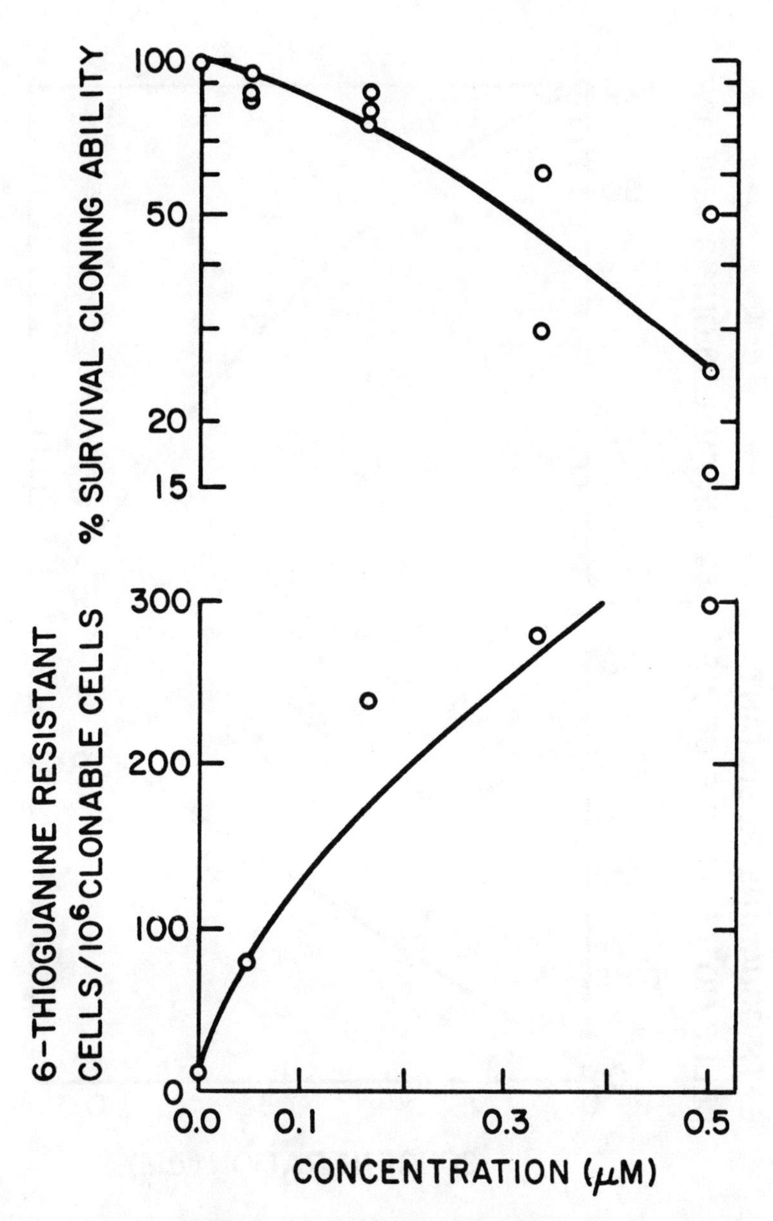

FIGURE 4. Mutagenicity and cytotoxicity of 11-chloromethylbenzo[a]pyrene in Chinese hamster V79 cells.

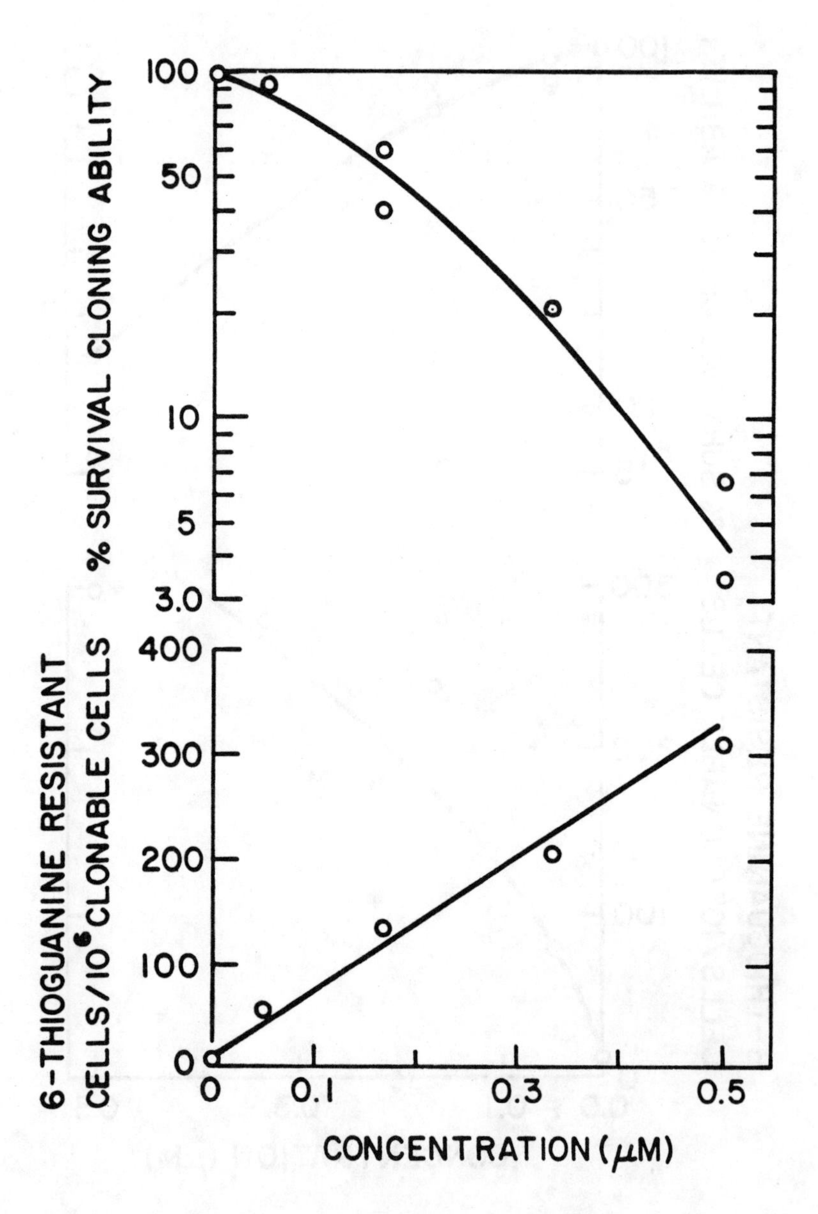

FIGURE 5. Mutagenicity and cytotoxicity of 12-chloromethylbenzo[a]pyrene in Chinese hamster V79 cells.

were potent mutagens in the Ames assay others were weak. In order to help explain the variablity between isomers we measured the solvolytic reactivity of these compounds in 50% aqueous acetone at 25 C. It is difficult to understand the relationship between the solvolytic reactivity and the mutagenicity of these ClMe-BaP in the Ames Assay. The two most reactive compounds (1- and 6-ClMe-BaP) were the least mutagenic, while the least reactive compound was the most mutagenic (11-ClMe-BaP in strain TA98). The lack of mutagenicity of 1- and 6-ClMe-BaP can be explained by their reactivity. For example, 6-ClMe-BaP has a half-life of 0.7 sec in 50% aqueous acetone at 25 C. Presumably, a large fraction of this compound would be hydrolyzed prior to coming into close proximity with DNA. In fact, it is surprising that this compound is mutagenic at all in the Ames Assay. The rate of solvolysis of 6-ClMe-BaP would be much faster in 100% water and 45 C, the conditions of the liquid top agar in the plate incorporation assay. Therefore , it is likely that most of the 6-ClMe-BaP would be hydrolyzed to 6-HyMe-BaP, and this latter compound maybe responsible for the mutagenicity of the former compound. However, none of the HyMe-BaP tested were mutagenic suggesting that enough 6-ClMe-BaP survived long enough to induce mutations in the bacteria.

The mutagenicity of the other ClMe-BaP is more difficult to explain since they all have similar half-lives (ranges from 1-4 days). It is interesting to compare the mutagenicity of ClMe-BaP and para-substituted benzyl derivatives (chlorides and tosylates) in the Ames assay (22). Depending on the para substituent, benzyl derivatives can be reactive (p-methylbenzyl tosylate) or relatively inert (p-bromobenzyl chloride) (23,24). With the exception of the nitrobenzyl derivatives and p-acetoxybenzyl chloride, which were mutagenic due to endogenous bacterial nitroreductases and hydrolysis of the acetoxy group respectively, none of the other benzyl derivatives were mutagenic. Therefore, there are two classes of compounds with the same reactive functional group and similar reactivities, but markedly different mutagenicities in the Ames assay. The only apparent difference between the benzyl chlorides and the ClMe-BaP is the size of the ring system. One explanation for the mutagenicity of these larger ring systems is that the ClMe-BaP molecules can non-covalently bind to DNA (possibly via intercalation) and then react with a DNA base. This might also account for the differences in mutagenicity between the ClMe-BaP isomers. A preliminary binding to DNA may allow one isomer to be in a sterically favored position to covalently bind with a DNA base. Presumably, the mono ring systems would bind much less efficiently to DNA and would not be in a favorable position to react with a DNA nucleophile.

It is possible that the ClMe-BaP were mutagenic due to the formation of HyMe-BaP followed by endogenous bacterial esterification , and subsequent reaction of these esters with DNA. None of the HyMe-BaP ,however, are mutagenic in the absence of exogenous metabolic activation and this suggests that ClMe-BaP are compounds which can react directly with DNA. This is a reasonable suggestion since these compounds would be expected to react with nucleophiles. In addition, 6-ClMe-BaP has been shown to react with guanosine and adenosine (25). If one excludes those compounds which require metabolic activation (*i.e.* exogenous S9 or endogenous bacterial activation) then 11- and 12-ClMe-BaP are the most potent direct acting mutagens yet reported in strains TA98 and TA100 respectively.

As a result of the mutagenic potency of 11- and 12-ClMe-BaP in bacterial cells we wanted to see if these compounds showed any unusual mutagenicity in mammalian cells. The data in Figures 4 and 5 show that these compounds are cytotoxic and mutagenic in Chinese hamster V79 cells. Mutagenic potency in V79 cells has been defined as the concentration of mutagen required to increase the number of mutant colonies 10 fold over the spontaneous mutant levels. The background mutant fractions for the mutation experiments shown in Figures 4 and 5 were 10.5 and 7.2 6-thioguanine resistant cells/10^6 clonable cells. The mutagenic potency using the concentration of mutagen at 105 and 72 mutants is 0.08 and 0.11 uM for 11- and 12-ClMe-BaP respectively. The concentration of 11- and 12-ClMe-BaP required to cause a 10 fold increase in the mutant frequency is among the lowest reported by the Gene-Tox program for direct acting mutagens in V79 cells using 6-thioguanine as the genetic marker (26).

In summary, ClMe-BaP are potent direct acting mutagens in bacterial and mammalian cells. The mutagenicity of these compounds cannot be explained in terms of their solvolytic reactivity, and other factors, possibly a preliminary binding step prior to reaction with DNA, must account for the mutagenicity of these compounds. These results suggest that activated esters of HyMe-BaP formed during metabolism of methylated PAH may be important genotoxic metabolites of methylated PAH.

REFERENCES

1) Severson, R.F., Snook, M.E., Akin, F.J., and Chortyk, O.T. (1978): Correlation of biological activity with polynuclear aromatic hydrocarbon content of tobacco smoke fractions, Carcinogenesis, Vol. 3: Polynuclear Aromatic Hydrocarbons, edited by P.W. Jones and R.I. Freudenthal, Raven Press, New York, pp 115-130.
2) McMahon, C.K., and Tsoukalas, S.N. (1978): Polynuclear aromatic hydrocarbons in forest fire smoke, Carcinogenesis, Vol. 3: Polynuclear Aromatic Hydrocarbons, edited by P.W. Jones and R.I. Freudenthal, Raven Press, New York, pp 61-73.
3) Gross, G.P. (1973): Fourth annual report on gasoline composition and vehicle exhaust gas polynuclear aromatic content, Coordinating Research Council of the Air Pollution Research Advisory Committee, New York, NY, pp 1-134.
4) Schuetzle, D. (1983): Sampling of vehicle emissions for chemical analysis and biological testing, Environ. Health Perspectives, 47: 65-80.
5) Iyer, R.P., Lyga, J.W., Secrist, J.A., Daub, G.H., and Slaga, T.J. (1980): Comparative tumor initiating activity of methylated benzo(a)pyrene derivatives in mouse skin, Cancer Res., 40: 1073-1076.
6) Barfknecht, T.R., Hites, R.A., Cavaliers, E.L., and Thilly, W.G. (1982): Human cell mutagenicity of polycyclic aromatic hydrocarbon components of diesel emissions. In: Toxicological Effects Of Emissions From Diesel Engines, edited by J. Lewtas, pp. 277-294, Elsevier Biomedical, New York.
7) Santella, R., Kinoshita, T., and Jeffery, A.M. (1982): Mutagenicity of some methylated benzo(a)pyrene derivatives, Mutation Res., 104: 209-213.
8) Watabe, T., Ishizuka, T., Isobe, M., and Ozawa, N. (1982): A 7-hydroxymethyl sulfate ester as an active metabolite of 7,12-dimethylbenz[a]anthracene, Science, 215: 403-405.
9) Roe, F.J.C, Dipple, A., and Mitchley, B.C.V. (1972): Carcinogenic activity of some benz(a)anthracene derivatives in newborn mice, Br. J. Cancer, 26: 461-465.
10) Dipple, A., and Slade, T.A. (1970): Structure and activity in chemical carcinogenesis. Reactivity and carcinogenicity of 7-bromomethylbenz(a)anthracene and 7-bromomethyl-12-methylbenz(a)anthracene, Eur. J. Cancer, 6: 417-423.
11) Duncan, M.E. and Brookes P. (1973): Induction of azaguanine resistant mutants in cultured Chinese hamster

cells by reactive derivatives of carcinogenic hydrocarbons, Mutation Res., 21: 107-118.

12) Maher,V.M., McCormick,J.J., Grover, P.L., and P. Sims (1977): Effect of DNA repair on the cytotoxicity and mutagenicity of polycyclic hydrocarbon derivatives in normal and Xeroderma pigmentosum human fibroblasts, Mutation Res., 43: 117-138.
13) Leon, A.A., Daub, G.H. and Vander Jagt, D.L. : Synthesis of 4-,5-,11-, and 12-chloromethylbenzo[a]pyrene, J. Org. Chem., in press.
14) Royer, R.E., Daub, G.H. and Vander Jagt, D.L. (1976): Synthesis of carbon-13 labelled 6-substituted benzo-[a]pyrenes, J. Labelled Compd. Radiopharm., 12: 377.
15) Deck, L.M. and Daub, G.H. (1983): Synthesis of 10-(chloromethyl)- benzo[a]pyrene, J. Org. Chem., 48: 3577.
16) Moy, G.G., Daub, G.H. and Vander Jagt, D.L.: Synthesis of 1-chloromethylbenzo[a]pyrene and selectivity of solvolytic trapping of its arylmethyl cation by nucleophiles, in preparation.
17) Royer, R.E., Daub, G.H., and Vander Jagt, D.L. (1979): Solvolytic reactivity of 6-chloromethylbenzo[a]pyrene and selectivity of trapping of the arylmethyl cation by added nucleophiles, J. Org. Chem., 44: 3196.
18) Maron, D.M. and Ames,B.N. (1983): Revised methods for the Salmonella mutagenicity test, Mutation Res., 113: 173-215.
19) Skopek,T.R., Liber,H.L., Kaden, D.A. and W.G. Thilly (1978): Relative sensitivities of forward and reverse mutation assays in Salmonella typhimurium, Proc. Natl. Acad. Sci. (USA), 75: 4465-4469.
20) Ball, J.C., McCormick, J.J., and Maher, V.M. (1983): Biological effects of incorporation of O-6-methyldeoxyguanosine in Chinese hamster V79 cells, Mutation Res., 110: 423-433.
21) Ball, J.C., Zacmanidis, P., and Salmeen, I.T. (1985):The reduction of 1-nitropyrene to 1-aminopyrene does not correlate with the mutagenicity of 1-nitropyrene in V79 Chinese hamster cells. In: Polynuclear Aromatic Hydrocarbons: Mechanisms, Methods, And Metabolism, edited by Cooke, M. and Dennis, A.J., pp.113-120, Battelle Press, Columbus, Ohio
22) Ball, J.C., Foxall-VanAken, S., and Trescott, T.E. (1984): Mutagenicity studies of p-substituted benzyl derivatives in the Ames salmonella plate incorporation assay, Mutation Res., 138: 145-151.
23) Kochi, J.K., and Hammond, G.S. (1953): Benzyl Tosylates. II. The application of the Hammett equation to the

rates of their solvolysis, J. Am. Chem. Soc., 75: 3445-3451.
24) Bennett, G.M., and Jones, B. (1935): 422. Velocities of reaction of benzyl chlorides in two reactions of opposed polar types, J. Chem.Soc., 1815-1819.
25) Royer,R.E., Lyle, T.A., Moy, G.G., Daub, G.H., and Vander Jagt, D.L. (1979): Reactivity-selectivity properties of reactions of carcinogenic electrophiles with biomolecules. Kinetics and products of the reaction of benzo[a]pyrenyl-6-methyl cation with nucleosides and deoxynucleosides, J. Org. Chem., 44: 3202-3207.
26) Bradley, M.O., Bhuyan, B., Francis, M.C., Lagenbach, R., Peterson,A., and Huberman, E. (1981) Mutagenesis by chemical agents in V79 Chinese hamster cells: A review and analysis of the literature. A report of the Gene-Tox program, Mutation Res., 87: 81-142).

METABOLISM AND ACTIVATION PATHWAYS OF THE ENVIRONMENTAL MUTAGEN 3-NITROFLUORANTHENE.

L.M. BALL, M.J. KOHAN[1], K. WILLIAMS[1], M.G. NISHIOKA[2], A. GOLD and J. LEWTAS[1]
Department of Environmental Sciences and Engineering, University of North Carolina, Chapel Hill, NC 27514, U.S.A., (1) Genetic Bioassay Branch, Health Effects Research Laboratory, U.S. Environmental Protection Agency, Research Triangle Park, NC 27711, U.S.A., and (2) Battelle Columbus Laboratories, Columbus, OH 43201, U.S.A.

INTRODUCTION

The presence of nitrated polycyclic aromatic hydrocarbons (NO_2PAH) in photocopier toners (1), in diesel and woodsmoke combustion emissions (2,3,4), and in ambient air (4,5), has generated concern about possible adverse health effects arising from the widespread environmental occurrence of these compounds many of which are potent mutagens in the Ames Salmonella typhimurium plate incorporation assay (6). 1-Nitropyrene (1-NP), abundant in extracts of diesel particulate matter, has been extensively characterised as a representative NO_2PAH, with respect to mutagenicity in Salmonella (6,7) and mammalian (8) assay systems, metabolism both *in vivo* (9,10) and *in vitro* (10, 11,12,13,14), formation of adducts with DNA (13,14), and animal carcinogenicity (15,16). 3-Nitrofluoranthene (3-NFA) has been reported as a potent bacterial mutagen (17) and possibly an animal carcinogen (15), but little was known of its mechanisms of activation and to what extent nitroreductase enzymes (as is the case for other NO_2PAH [7]) were involved in generation of genotoxic intermediates. On account of the particular abundance of fluoranthene and its derivatives in ambient air particulate matter (5), we have undertaken to investigate further the mutagenicity, metabolism and activation pathways of 3-NFA.

MATERIALS AND METHODS

Chemicals

3-Nitro[3-^{14}C]fluoranthene (^{14}C-3-NFA) and unlabelled 3-NFA were synthesised by Midwest Research Institute, Kansas City, MO. 3-Aminofluoranthene (3-AFA) was purchased from Aldrich Chemical Co., Milwaukee, WI, and recrystallised from aq. EtOH until yellow flakes (m. pt. 107-108°C; lit. 113-115°C [18]) were obtained which were demonstrated by HPLC analysis to be over 99.5% pure (judged by monitoring UV absorbance at 254 nm). 3-Acetamidofluoranthene (3-AAFA) was prepared by acetylation of 3-AFA and recrystallised from aq. EtOH to yield orange-yellow needles, m. p. 243.5-245 °C (lit. 244-245 C [18]); by HPLC analysis (254 nm) over 99.5 % of the product chromatographed as a single peak. MS and proton NMR spectra were consistent with the proposed structures. Other chemicals and solvents were bought from Sigma Chemical Co., St. Louis, MO, Burdick and Jackson Inc., Muskegon, MI and Fisher Scientific, Raleigh, NC.

Mutagenicity Assays

The Salmonella typhimurium plate incorporation assay was carried out as described by Ames *et al.* (19), with the following modifications: minimal histidine was added to the base agar rather than to the soft agar overlay, and the number of histidine-prototrophic revertant colonies was counted (Artek Colony Counter, Farmingdale, NY) after 72 rather than 48 h. Salmonella strain TA98 was obtained from Dr. B. Ames, University of California, Berkeley, CA, USA, and strains TA98NR and TA98/1,8-DNP_6 from Dr. H.S. Rosenkranz, Case Western Reserve University, Cleveland OH, USA. The S9 fraction (9000xg supernatant) used to provide metabolic activation was prepared from the livers of Aroclor 1254-treated male CD-1 rats (Charles River, Wilmington, MA, USA), stored at -70°C until used, and supplemented with NADPH-generating co-factor mix as described (19). The amount of S9 protein added (1.2 to 1.5 mg per plate) was determined by optimisation with benzo(a)pyrene and 2-aminoanthracene for each batch of S9 prepared. Each compound was assayed in triplicate, on three separate occasions, at doses ranging from 0.01 to 10 μg per plate, both with and without exogenous metabolic activation (S9). Compounds were freshly dissolved in DMSO before each assay. Specific mutagenicity was calculated from the linear portion of the dose-response curve.

The forward mutation assay to 8-azaguanine resistance was performed scaled down ten-fold from the method of Skopek et al. (20), as described by Goto *et al.*(21), in a bacterial strain cloned from a culture of Salmonella typhimurium TM677 supplied by Dr. W.G. Thilly, Massachusetts Institute of Technology, Cambridge, MA, and selected for maximum activity towards benzo-(a)pyrene. 3-NFA dissolved in DMSO was incubated with bacteria (and S9 mix) for 1 h, then triplicate plates (for each dose) were poured for scoring of resistant mutants and of survivors.

Incubation Of 3-Nitro[3-^{14}C]fluoranthene With S9

^{14}C-NFA (or unlabelled 3-NFA; final concentration 50 μM) dissolved in DMSO was added to Aroclor-treated rat liver S9 fraction (1 mg/ml in 0.1M HEPES buffer, pH 7.4; final vol. 80 ml) prepared as described (19). The reaction was started by addition of NADPH (1mM) and terminated (after 1 h at 37 °C) by addition of 1 vol. of ice-cold ethyl acetate/acetone (2:1,v/v). The reaction products were extracted with a further 2 x 1 vol. of ethyl acetate/acetone. The pooled organic extracts were concentrated (to approx. 3 ml) by rotary evaporation, then taken to dryness under a stream of N_2 gas. The residue was dissolved in MeOH (1 ml) for further analysis. Recovery of ^{14}C was monitored throughout by liquid scintillation counting.

Analysis and Quantitation of 3-Nitrofluoranthene Metabolites

The profile of ^{14}C organic-extractable metabolites was determined by HPLC with a gradient from 70 to 100 % MeOH in H_2O (rising at 2 %/min) and liquid scintillation counting of 30-sec fractions of the eluate. The H_2O-soluble metabolites were similarly analysed over a gradient from 0 to 100% MeOH in H_2O (2% /min); fractions were collected and counted every min. The remainder of the ^{14}C-extractable fraction and of parallel incubations conducted with unlabelled 3-NFA were chromatographed over a gradient from 50 % MeOH in H_2O to 100 % MeOH; fractions were collected according to the UV (254 nm) absorbance of the eluate.

Instrumentation

Proton NMR spectra were recorded in acetone-d_6 on a Brucker WM-250 instrument. Mass spectra were obtained in the direct probe insertion EI mode (70 eV) on a VG7070H (Vacuum Generators, Altrincham, Cheshire) mass spectrometer and by HRGC/NCI-MS on a Finnegan gas chromatograph/EI-CI mass spectrometer (Finnegan Instrument Co., Sunnyvale, CA). Melting points were determined on a Hoover capillary melting point apparatus (A.H.Thomas Co., Philadelphia, PA) and are uncorrected throughout. UV-vis spectra were recorded on a Cary 2200 spectrophotometer (Varian Techtron Pty Ltd, Mulgrave, Australia), and fluorescence spectra on an in-line Perkin-Elmer MPF-2A fluorometer (Perkin-Elmer, Norwalk, CT). Analytical HPLC was performed on a Varian 5050 (Varian Instrument Co., Palo Alto, CA) chromatogragraph equipped with a Zorbax ODS (4.6 x 250 mm) column (Dupont Instrument Co., Wilmington, DE) whose eluate was monitored for UV absorbance (UV-100, Varian Instrument Co.) and fluorescence (Ex. 360 nm, Em. 430 nm; MPF-2A, Perkin-Elmer, Norwalk, CT). Metabolites were isolated and UV spectra determined on-line with a Dupont 850 HPLC fitted with a Zorbax 9.4 x 250 mm ODS column and a Perkin-Elmer LC-85B spectrophotometric detector.

RESULTS

Mutagenicity of 3-Nitrofluoranthene and its Reduced Derivatives

The mutagenicity of 3-NFA (Fig. 1) towards Salmonella typhimurium strain TA98 (Fig. 2) is predominantly direct-acting (4 200 His^+ rev/nmole) and is decreased by the presence of S9 to 250 rev/nmole. In strain TA98NR, deficient in the "classical" nitroreductase enzyme (7) and in the variant TA98/1,8-DNP_6 apparently deficient in a transacetylase enzyme (7), activity is lower than in the fully competent TA98 (800 and 475 rev/nmol respectively), and is not restored by addition of S9 (125 rev/nmole in TA98NR; no activity above background in TA98/1,8-DNP_6)

3-AFA, the final product of 6-electron reduction of 3-NFA, is less active in all three variants of TA98 than is 3-NFA (highest values 100 rev/nmole in TA98 and TA98NR in the presence of S9; see Fig. 3). The mutagenicity of 3-AAFA is even lower, 60 rev/nmole or less, and is entirely S9-dependent in all three variants (Fig. 4).

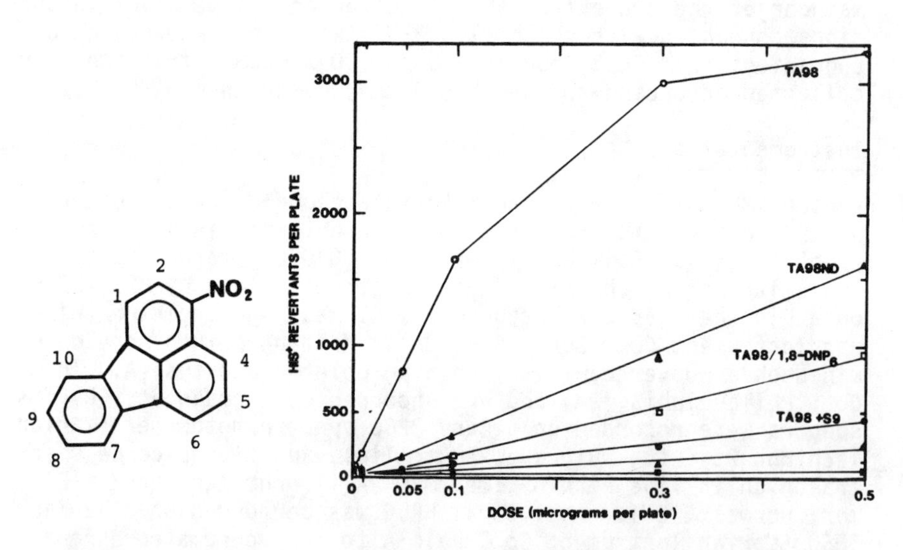

FIGURE 1.
3-Nitrofluoranthene.

FIGURE 2. Mutagenicity of 3-Nitrofluoranthene towards Salmonella typhimurium strains TA98, TA98NR and TA98/1,8-DNP_6 with (closed symbols) and without (open symbols) metabolic activation (S9). Assays were conducted as described under Materials and Methods.

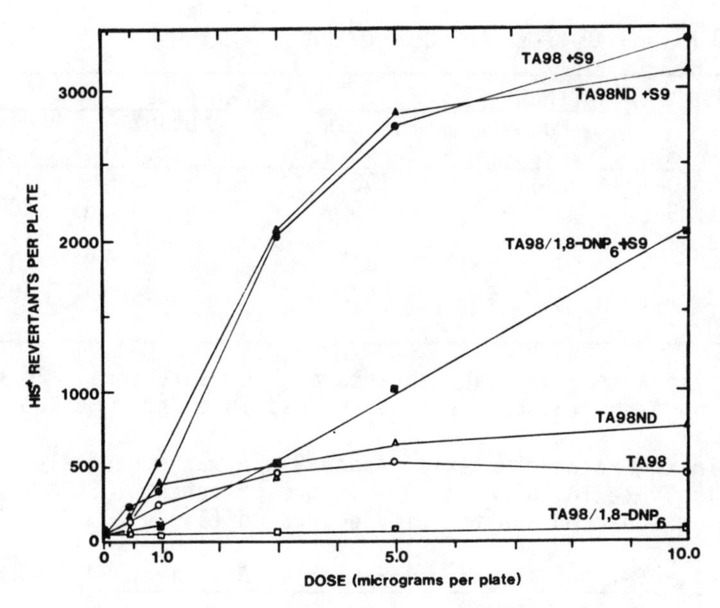

FIGURE 3. Mutagenicity of 3-Aminofluoranthene in the Ames plate incorporation assay (see legend to Figure 2).

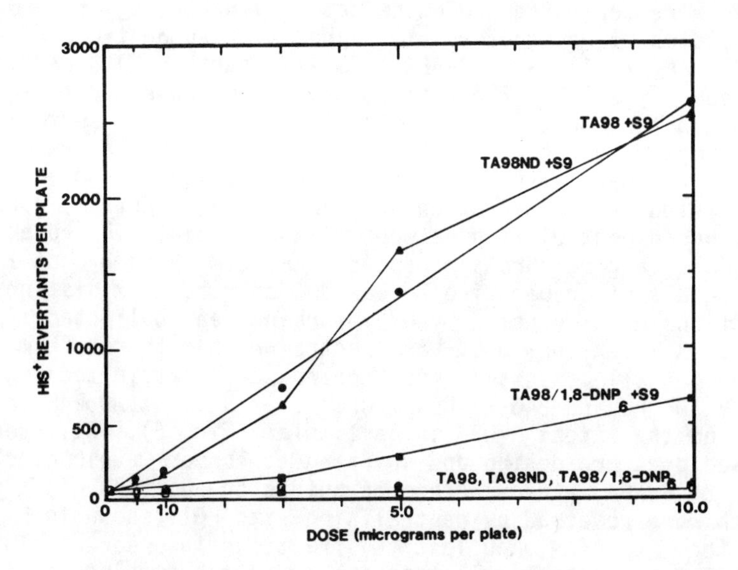

FIGURE 4. Mutagenicity of 3-Acetamidofluoranthene towards Salmonella typhimurium in the Ames plate incorporation assay (see legend to Figure 2).

TABLE 1

MUTAGENICITY OF 3-NITROFLUORANTHENE TOWARD TM677

Dose (μg per ml)	Without S9 Mutant Fraction[a]	Without S9 Percent Survival	With S9 Mutant Fraction	With S9 Percent Survival
0 (DMSO alone)	1	100	3	100
1.25	1	55	6	200
2.5	1	37	41	115
5	2	43	88	87
10	4	68	94	103
20	1	37	4	47

a 8-Azaguanine-resistant mutants per 10^5 survivors. Forward Mutation Assay was carried out as described under Methods

In the Forward Mutation Assay 3-NFA was moderately active, and highly toxic, without S9. In the presence of S9 activity was much enhanced and toxicity decreased (Table 1).

Metabolism of 3-NFA by Aroclor-treated Rat Liver S9

After 1 h of incubation at 37°C, over 40 % of the substrate 3-NFA had undergone metabolic transformation (20 nmole/mg protein in 1 h). The ethyl acetate/acetone-extractable metabolites were separated and quantitated by HPLC (Fig. 5). Minor quantities of ^{14}C co-chromatographed with authentic 3-AAFA, and hardly any (< 1%) with 3-AFA. Small amounts of ^{14}C eluting between the solvent front and 3-AAFA were converted to material with a retention time around 17 min (in this HPLC system) by treatment with 1 M HCl for 30 min at 50°C. The major portion of the radioactivity (other than parent compound, which was identified by its retention time and UV-vis spectrum) consisted of a broad peak eluting between 17 and 18 min. This peak was collected and re-chromatographed at 85 % MeOH in H_2O at 2 ml/min on a semi-preparative Zorbax ODS column. Six distinct fractions were detected by UV absorbance and collected separately. All six showed UV-vis spectra which differed from each other but all possessed broad absorbance peaks in the regions 260 - 290 nm and 340 - 400 nm characteristic of NO_2PAH in general and the parent 3-NFA in particular (Fig. 6). All spectra showed peak broadening and shifting to longer wavelengths relative to 3-NFA, and bathochromic shifts on addition of NaOH which were reversed by neutralising with HCl (shown in Fig. 7 for the first (AA) and last (E) fractions in order of elution). HRGC/MS analysis of this fraction indicated a molecular ion of m/e 263, corresponding to the expected molecular weight of 3-NFA-OH (Fig. 8). The loss of 16 mass units (M^+ - 16, to

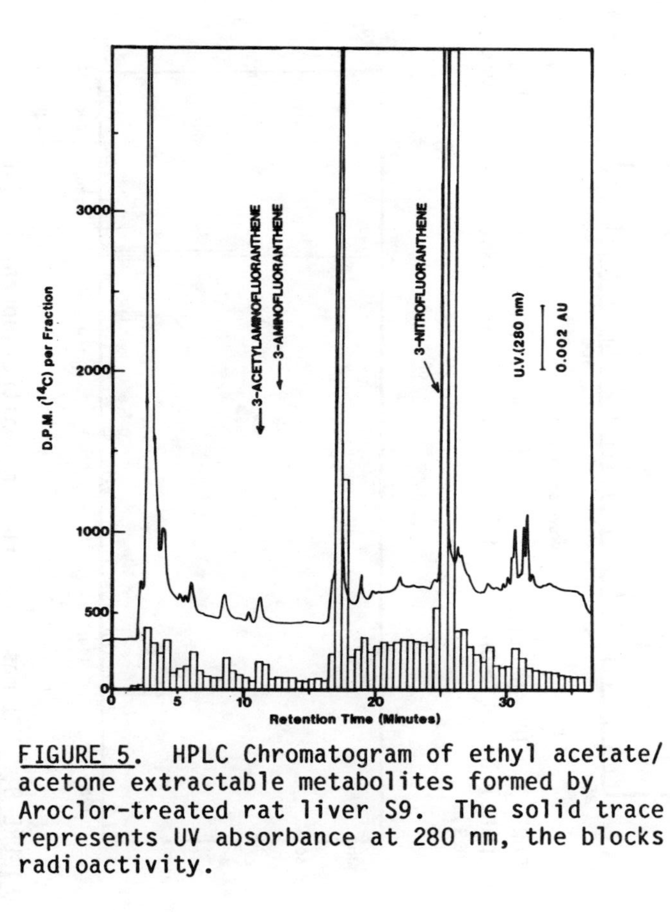

FIGURE 5. HPLC Chromatogram of ethyl acetate/acetone extractable metabolites formed by Aroclor-treated rat liver S9. The solid trace represents UV absorbance at 280 nm, the blocks radioactivity.

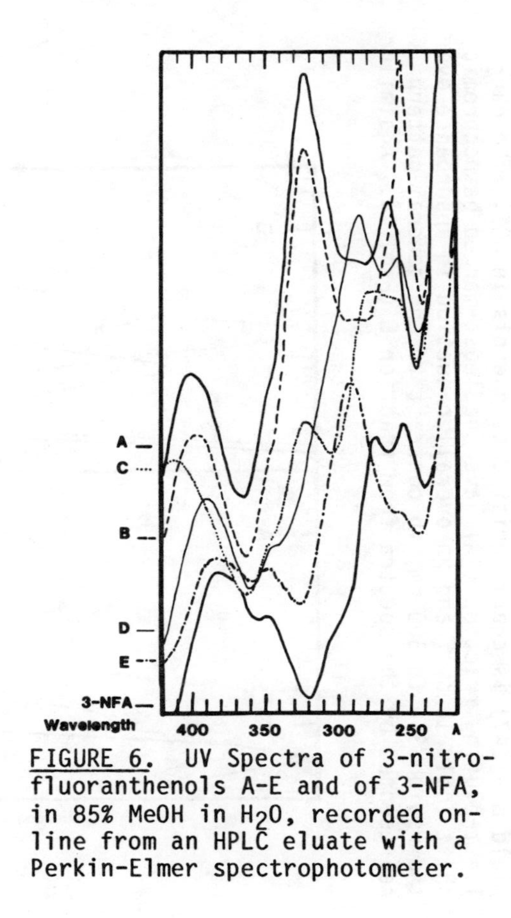

FIGURE 6. UV Spectra of 3-nitrofluoranthenols A-E and of 3-NFA, in 85% MeOH in H_2O, recorded on-line from an HPLC eluate with a Perkin-Elmer spectrophotometer.

yield m/e 247) is characteristic of phenols in NCI. The earliest-eluting fraction (AA) gave the least-marked bathochromic shift (only 10 nm), and an overall reduction in absorbance over the region 350 to 500 nm, in contrast to the general pattern represented by the spectra from fraction E (Fig. 7). A similar

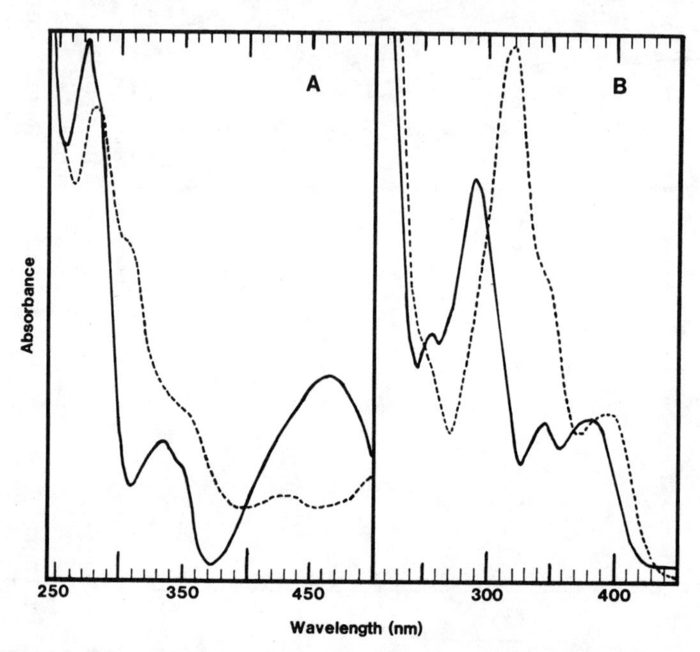

FIGURE 7. UV spectra of 3-nitrofluoranthenols AA (panel A) and E (panel B) in MeOH (solid lines) and in 0.1M NaOH/MeOH (broken lines).

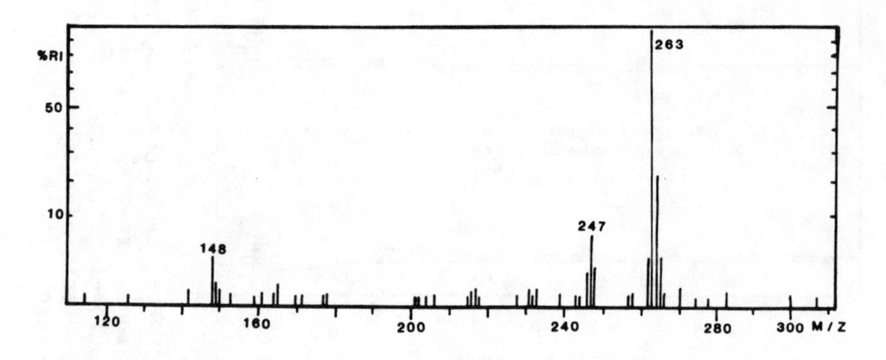

FIGURE 8. NCI mass spectrum of 3-nitrofluoranthenol E

divergence from the general pattern of the nitropyrenols is exhibited by 1-nitropyren-10-ol (not shown), where the hydroxyl group is peri to the nitro group and the hydroxyl proton can interact with a nitro oxygen to form a stabilised pseudo-six-membered ring. By analogy, it can be inferred that the nitro and hydroxyl groups of AA would be peri to each other, i.e. occupy positions 3 and 4 respectively.

The last phenolic fraction to elute (E) was also the most abundant; its NMR spectrum (Fig. 9) shows clearly the down-field signals due to the protons ortho and peri to the nitro substituent (8.60 ppm, d, H2, and 8.45 ppm, d, H4 respectively) identified by their coupling to H1 (8.16 ppm, d, J = 7.5 Hz) and H5 (7.81 ppm, t, J = 8.2 Hz, also coupled to H6, 7.98 ppm, d, J = 6.9 Hz). The signal at 7.34 ppm is due to an impurity in the solvent acetone-d_6. The presence of an upfield singlet (7.56 ppm, showing meta coupling 2.5 Hz) and a coupled pair of doublets (7.84, J = 8.1, and furthest upfield, 6.96, J = 8.1 Hz, further split (2.5 Hz) by meta coupling, indicates that hydroxylation had occurred on the distal benzo ring, at one of positions 8 or 9. Consideration of the strongly directing effects of the nitro group on subsequent electrophilic substitutions (22) suggests that the hydroxyl group would be expected to be found on position 9. Calculation of the effects of the conflicting up-field (due to a hydroxyl group: - 0.4 ppm for

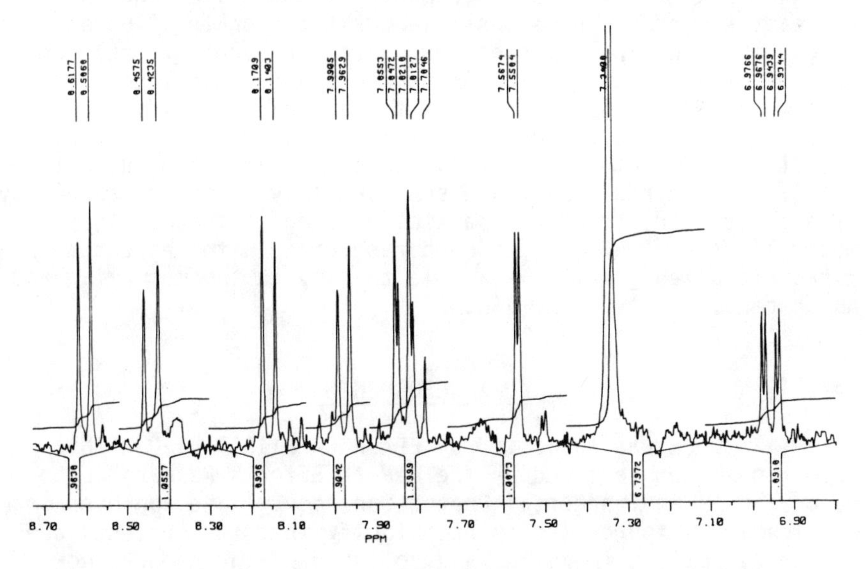

FIGURE 9. 250 MHz Proton NMR spectrum of 3-nitrofluoranthenol E in acetone-d_6. Signals are in ppm downfield from Me_4Si.

ortho, meta or para protons [23]) and down-field (from the long-range deshielding effect of the nitro group) shifting influences on the signals due to the protons at positions 7 and 10 (8.06 and 8.12 ppm respectively in 3-NFA; cf 8.01 ppm in the parent fluoranthene) yields chemical shift values more consistent with the occurrence of hydroxylation on position 8, though since the difference in shifts of H7 and H10 (in 3-NFA) is less than 0.1 ppm, and the variation in substituent effect shifts is of the order of 0.1 ppm, this argument is not conclusive in favour of 3-NFA-8-OH.

DISCUSSION

Whereas conversion through 1-aminopyrene to 1-acetamidopyrene vastly increased the S9-dependent mutagenicity of 1-NP (12), the present study shows that in the case of 3-NFA metabolism to 3-AAFA would result in detoxication rather than increase the genotoxicity of this compound. Only a small proportion of the mutagenicity of 3-NFA is mediated by the classical nitroreductase enzyme, as shown by its low activity in strain TA98/1,8-DNP_6 which is proficient in this enzyme. In the Forward Mutation Assay 3-NFA is activated most extensively in the presence of S9, suggesting that S9 may itself generate mutagenic species. 3-NFA appears to be reduced less readily by S9 than is 1-NP (12). S9 is capable of hydroxylating 3-NFA on at least six of the nine possible positions on the fluoranthene rings. Epoxidation is suggested by the presence of small amounts of metabolite which chromatographs as a phenol on treatment with acid, and is therefore tentatively identified as a dihydrodiol. Bioassay-directed fractionation studies of ambient air particulate matter (24) indicate that nitrofluoranthenol-like compounds contribute substantially to the mutagenicity of organic extracts of air particles, suggesting that these metabolites of 3-NFA as yet uncharacterised with respect to mutagenic potency may have a role to play in the mammalian and human genotoxicity of NO_2PAH.

ACKNOWLEDGMENTS

This work was supported by EPA # CR-811817-01-0. This report has been reviewed by the Health Effects Research Laboratory, U.S. Environmental Protection Agency, and approved for publication. Approval does not signify that the contents necessarily reflect the views and policies of the Agency, nor does mention of trade names or commercial products constitute endorsement or recommendation for use.

REFERENCES

1. Löfroth, G., Hefner, E., Alfheim, I., and Møller, M. (1980): Mutagenic activity of photocopies, Science, 209: 1037-1039.
2. Nishioka, M.G., Peterson, B.A., and Lewtas, J. (1982): Comparison of nitroaromatic content and direct-acting mutagenicity of diesel emissions. In: W.M. Cooke, A.J. Dennis and G.L. Fisher (eds.), Polynuclear Aromatic Hydrocarbons: Physical and Biological Chemistry, Battelle Columbus Press, Columbus, OH, USA, pp. 603-613.
3. Schuetzle, D., Riley, T.L., Prater, T.J., Harvey, T.M. and Hunt, D.F. (1982): Analysis of nitrated polycyclic hydrocarbons in diesel particulates, Analyt. Chem., 54:265-271.
4. Gibson, T.L. (1982): Nitroderivatives of polynuclear aromatic hydrocarbons in airborne and source particulate matter. Atmos. Environ. 16:2037-2040.
5. Wise, S.A., Chesler, S.N., Hilpert, L.R., May, W.E., Rebbert, R.E., Vogt, C.R., Nishioka, M.G., Austin, A., and Lewtas, J. (1985): Quantification of polycyclic aromatic hydrocarbons and nitro-substituted polycyclic aromatic hydrocarbons and mutagenicity testing for the characterization of ambient air particulate matter. Environ. Int., 11:147-160.
6. Mermelstein, R., Kiriazides, D.K., Butler, M., McCoy, E. and Rosenkranz, H.S. (1981): The extraordinary mutagenicity of nitropyrenes in bacteria, Mutat. Res., 89: 187-196.
7. Rosenkranz, E.J., McCoy, E.C., Mermelstein, R., and Rosenkranz, H.S. (1982), Evidence for the existence of distinct nitroreductases in Salmonella typhimurium: Roles in mutagenesis, Carcinogenesis, 3:121-123.
8. Li, A.P. and Dutcher, J.S. (1983): Mutagenicity of mono-, di- and tri-nitropyrenes in Chinese hamster ovary cells, Mutat. Res. Lett., 119:387-392.
9. Ball, L.M., Kohan, M.J., Inmon, J., Claxton, L.D., and Lewtas, J. (1984): Metabolism of 1-nitro[^{14}C]pyrene in vivo in the rat and mutagenicity of urinary metabolites. Carcinogenesis, 5:1557-1564.
10. Howard, P.C., Flammang, T.J., and Beland, F.A. (1985): Comparison of the in vitro and in vivo hepatic metabolism of the carcinogen 1-nitropyrene. Carcinogenesis, 6:243-249.
11. El-Bayoumy, K. and Hecht, S.S. (1983): Identification and mutagenicity of metabolites of 1-nitropyrene formed by rat liver. Cancer Res., 43, 3132-3137.
12. Ball, L.M., Kohan, M.J., Claxton, L.D., and Lewtas, J. (1984): Mutagenicity of derivatives and metabolites of

1-nitropyrene: Activation by rat liver S9 and bacterial enzymes, Mutat. Res., 138:113-125.

13. Messier, F., Lu, C., Andrews, P., McCarry, B.E., Quilliam, M.A., and McCalla, D.R. (1981): Metabolism of 1-nitropyrene and formation of DNA adducts in Salmonella typhimurium. Carcinogenesis, 2:1007-1011.
14. Howard, P.C., Heflich, R.H., Evans, F.E., and Beland, F.A. (1983): Formation of DNA adducts in vitro and in Salmonella typhimurium upon metabolic reduction of the environmental mutagen 1-nitropyrene, Cancer Res., 43:2052-2058.
15. Ohgaki, H., Matsukara, N., Morino, K., Kawachi, T., Sugimura, T., Morita, K., Tokiwa, H., and Hirota, T. (1982): Carcinogenicity in rats of the mutagenic compounds 1-nitropyrene and 3-nitrofluoranthene, Cancer Lett., 15:1-7.
16. Hirose, M., Lee, M.S., Wang, C.Y., and King, C.M. (1984): Induction of rat mammary glands tumors by 1-nitropyrene, a recently-recognized environmental mutagen, Cancer Res., 44:1158-1162.
17. Vance, W.A., and Levin, D.E. (1984): Structural features of nitroaromatics which determine mutagenic activity in Salmonella typhimurium. Environ. Mutagen., 6:797-811.
18. Campbell, N., Leadill, W.K., and Wilshire, J.F.K. (1951): Friedel-Crafts reaction with fluoranthene. J. Chem. Soc., 1404-1406.
19. Ames, B.W., McCann, J.M. and Yamasaki, E. (1975): Methods for detecting carcinogens and mutagens with the Salmonella/mammalian microsome mutagenicity test. Mutat. Res., 31:347-364.
20. Skopek, T.R., Liber, H.L., Krolewski, J.J., and Thilly, W.G. (1978): Quantitative forward mutation assay in Salmonella typhimurium using 8-azaquanine resistance as a genetic marker. Proc. Natl. Acad. Sci. (U.S.A.), 75:410-414.
21. Goto, S., Williams, K., Claxton, L. and Lewtas, J. (1986): Further development and application of a micro-forward mutation assay in Salmonella typhimurium TM677, Manuscript in preparation.
22. Campbell, N. and Keir, N.H. (1955): The orientation of disubstituted fluoranthene derivatives. J. Chem. Soc., 1233-1237.
23. Dyke, S.F., Floyd, A.J., Sainsbury, M. and Theobald, R.S. (1971): Organic Spectroscopy. Penguin Books Ltd, Harmondsworth, Middlesex, U.K., p. 151.
24. Nishioka, M.G., Howard, C.C., and Lewtas, J. (1986): Identification and Quantification of OH-NO_2 PAHs and NO_2-PAHs in an Ambient Air Particulate Extract by NCI HRGC/MS. Manuscript in preparation.

SYNTHESIS AND BIOLOGICAL ACTIVITY OF CYCLOPENTA EPOXIDES OF PAH CONTAINING PERIPHERALLY FUSED CYCLOPENTA RINGS

ANDI WEISS BARTCZAK, RAMIAH SANGAIAH, LOUISE M. BALL AND AVRAM GOLD
Department of Environmental Sciences and Engineering and Lineberger Cancer Research Center, University of North Carolina at Chapel Hill, Chapel Hill, NC 27514

INTRODUCTION

Polycyclic aromatic hydrocarbons (PAH) containing peripherally fused cyclopenta rings have been identified in soots from many different combustion sources (1,2). The biologically active cyclopenta-PAH studied in detail to date appear to be activated in large part by the cytochrome P450 mixed function oxidase system via epoxidation of the cyclopenta ring, even when a bay region is present in the molecule (3,4,5).

Metabolism and mutagenicity studies have been reported on the series of four cyclopenta-fused isomers derived from benzanthracene. On the basis of metabolite profiles and S9 concentration dependence of bacterial mutagenicity, it has been suggested that Aroclor 1254-induced rat liver S9 activates benz(e)aceanthrylene, benz(j)aceanthrylene and benz(l)aceanthrylene predominantly via epoxidation of the cyclopenta ring, while other routes must be involved in the case of benz(k)acephenanthrylene.

To test these conclusions, the cyclopenta epoxides of all four isomers of the cyclopenta benzanthracene derivatives have been synthesized and their biological activity evaluated in the Ames assay.

MATERIALS AND METHODS

Instrumentation

HPLC was performed with a DuPont Model 850 gradient pump and controller (DuPont Instrument Co., Wilmington, DE), Perkin Elmer LC Autocontrol and LC-85B spectrophotometer detector and Spectra-Physics Model SP-4270 integrator.

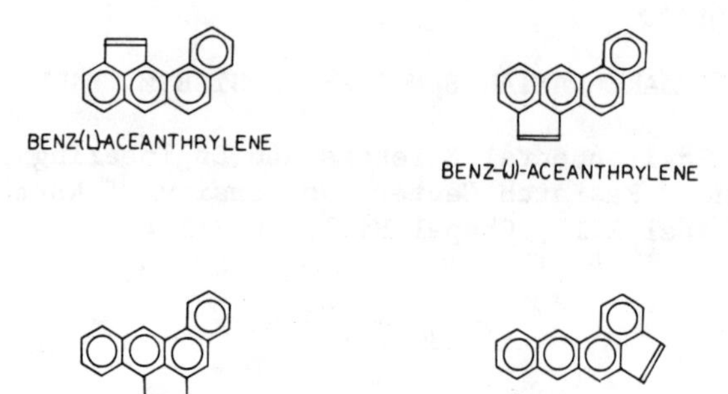

FIGURE 1. Structures of the four peripherally-fused cyclopenta isomers of benzanthracene.

UV-visible spectra were recorded on a Cary Model 15 spectrophotometer. Mass spectra were obtained on a VG Micromass 7070F mass spectrometer operated in the electron impact mode at 70 eV. ^{1}H NMR spectra were determined in acetone-d_6 on a Bruker WM250 spectrometer. Chemical shifts are given in ppm relative to TMS.

Synthesis

All four isomers of benzanthracene containing a peripherally fused cyclopenta ring were synthesized according to the procedures of Sangaiah, et al. (6). The epoxides were obtained from the corresponding bromohydrins (Figure 2) by procedures similar to those published for cyclopenta(cd)-pyrene (7). The cyclopenta ring bromohydrins were generated by treating 10 mg of the cyclopentaPAH isomer, dissolved in 1.4 mL DMSO containing 0.1 mL H_2O, with 10 mg N-bromosuccinimide. After stirring for 10 min. in an ice-water bath, the mixture was filtered through a sintered glass medium frit (ASTM 10-15) and kept on ice. The resulting bromohydrin was purified by HPLC in two portions on a Zorbax C-8 9.4x250 mm column equilibrated at 70:30::MeOH:H_2O and eluted with a concave gradient to 100% methanol (exponent +2) over 10 min. The colorless bromohydrin, monitored by absorbance at 254 nm, eluted between 3 and 6 min., at a flow rate of 5-6 mL/min. The fractions from both runs were pooled and concentrated under vacuum at 45° until solid precipitated. The suspension was filtered through a 10 mL sintered glass fine frit (ASTM 4-5.5), the collected solid dried under vacuum and stored at

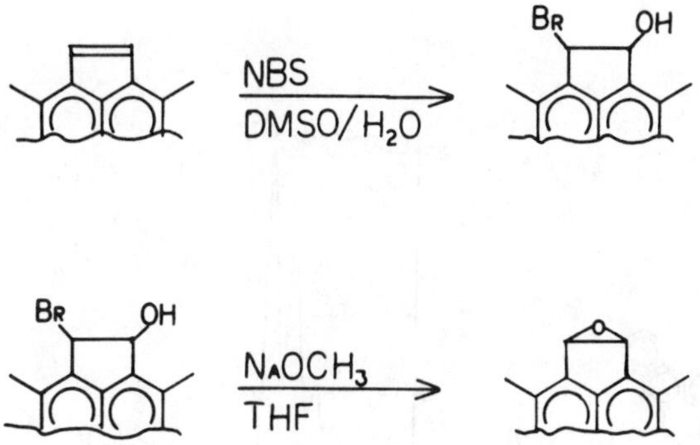

Br OH

NBS

$DMSO/H_2O$

Br OH

$NaOCH_3$

THF

FIGURE 2. Reaction scheme for generation of cyclopenta epoxide via the bromohydrin.

-20°C. The collected solids from 3 or 4 preparations were dissolved directly from the frit with dried, distilled THF. Sodium methoxide (10 mg/10 mg cyclopentaPAH reacted) was added to the THF solution with stirring. After reacting at room temperature for 10 min., the reaction mixture was quickly percolated through a short column (1.5x2 cm) of grade IV neutral alumina (Woelm Eschwege, West Germany). Removal of the THF under a stream of nitrogen yielded the pale yellow solid epoxide.

Bacterial Mutagenicity Assays

Mutagenicity of the epoxides was assayed by the histidine plate incorporation method as described by Ames *et al*. (8), using strain TA98, at the Health Effects Research Laboratory of the Environmental Protection Agency, Research Triangle Park, North Carolina. The nutrient broths and TA98 were prepared as described in the EPA standard protocol (9). Testing was done without addition of exogenous activation.

RESULTS

The mass spectra of the four epoxides were consistent with the structural assignments. An accurate mass measurement of the molecular ion confirmed the elemental composition of $C_{20}H_{12}O$ in each case and the low resolution electron impact spectrum was characterized by a base peak at M/z 268 ($M^{\cdot +}$) with a major fragment at M/z 239 (M^+), consistent with an arene oxide.

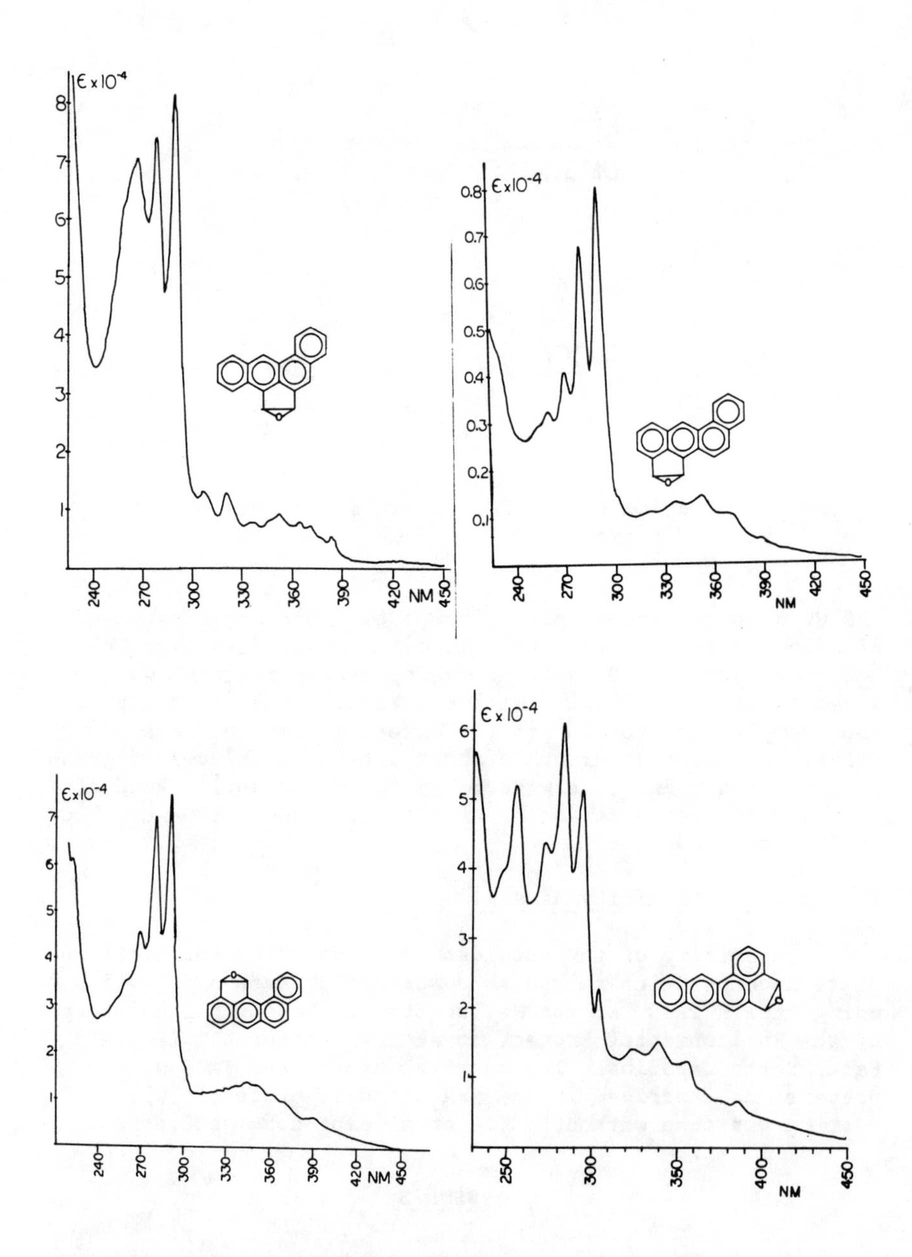

FIGURE 3. UV-visible Spectra (MeOH).
a) benz(e)aceanthrylene-5,6-oxide, b) benz(j)aceanthrylene-1,2-oxide, c) benz(l)aceanthrylene-1,2-oxide, d) benz(k)-acephenanthrylene-4,5-oxide

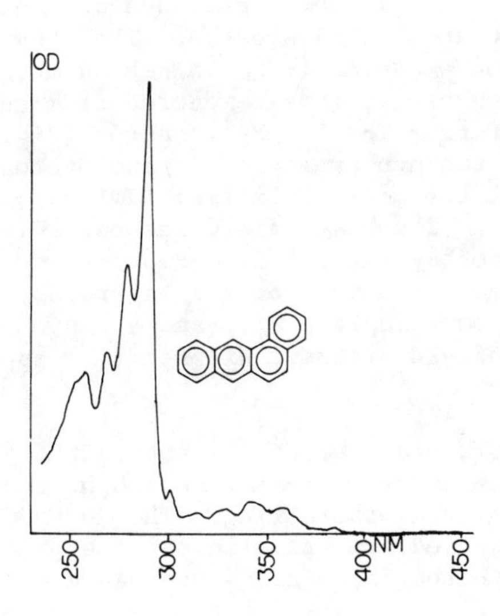

FIGURE 4. UV-vis spectrum (THF) of benzanthracene.

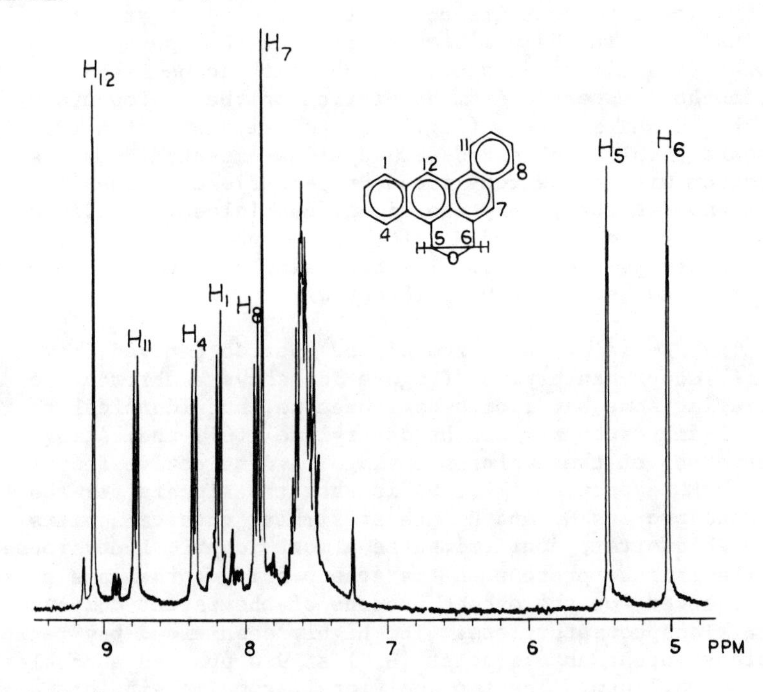

FIGURE 5. ^{1}H NMR spectrum (250 MH_z, acetone-d_6) of benz(e)aceanthrylene-5,6-oxide.

Structures of the isomeric epoxides were established by their UV-visible and ^{1}H NMR spectra. The UV-visible spectrum of the epoxidized e-isomer (Fig. 3a) shows the benzanthracene-like chromophore (Fig. 4) expected from saturation of the cyclopenta ring. The ^{1}H NMR spectrum (Fig. 5) confirms the presence of two non-aromatic oxirane protons at 5.1 ppm and 5.6 ppm with the small splitting (J<1 Hz) expected for eclipsed protons. The lower field resonance is assigned to H_5 in the "psuedo bay region" formed by the cyclopenta ring. Highly deshielded resonances of the bay region protons are present at 9.2 ppm (singlet, H_{12}) and 8.9 ppm (doublet, H_{11}). An additional upfield aromatic singlet (H_7) appears at 8.0 ppm.

The UV-visible spectrum of the epoxidized j isomer (Fig. 3b) shows the benzanthracene-like chromophore expected from saturation of the cyclopenta ring. The ^{1}H NMR spectrum (Fig. 6) shows the expected non-aromatic oxirane protons at 5.1 ppm and 5.5 ppm, with the lower field resonance assigned to H_2 in the "pseudo bay". The highly deshielded bay region protons appear as a singlet (H_6) at 9.5 ppm and doublet (H_7) at 9.0 ppm. Consistent with the benz(j)aceanthrylene structure, the ^{1}H NMR spectrum contains no additional singlet aromatic resonances. The UV-visible spectrum of the epoxidized l-isomer (Fig. 3c) also shows the benzanthracene-like chromophore expected from saturation of the cyclopenta ring. In the ^{1}H NMR spectrum (Fig. 7), one oxirane proton (H_1) appears highly deshielded (5.9 ppm) as expected from its location within the four-sided "psuedo fjord". Consistent with expectations, only one highly deshielded aromatic signal appears, the doublet of H_{12} (9.3 ppm), which is located within the psuedo fjord. The only singlet aromatic resonance is from the meso proton H_6 (8.4 ppm).

The UV-visible spectrum of the epoxide derived from benz(k)acephenanthrylene (Figure 3d) shows a chromophore differing somewhat from benzanthracene, but identical to that of 4,5-dihydrobenz(k)acephenanthrylene (6), indicating saturation of the cyclopenta ring. A distinctive feature of the ^{1}H NMR spectrum (Fig. 8) is that the signals for the oxirane protons H_4 and H_5 are at similar chemical shifts as a true AB quartet, confirming the almost identical environment of the oxirane protons on the etheno bridge: the same pattern is observed for the etheno protons of the parent compound benz(k)acephenanthrylene. The highly deshielded bay region protons appear as a singlet (H_{12}) at 9.4 ppm and a doublet (H_1) at 8.7 ppm. The two additional aromatic singlets at 8.6 and 8.1 ppm from the meso proton (H_7) and K-region (H_6) respectively, unequivocally establish the assigned structure, since no other isomer can have three aromatic singlets.

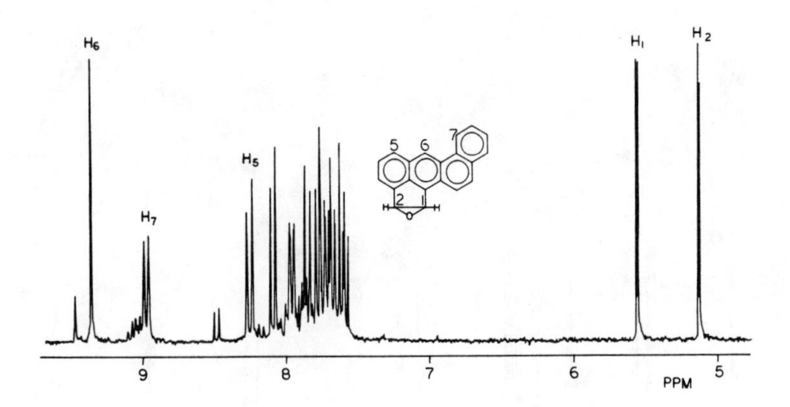

FIGURE 6. ^{1}H NMR spectrum (250 MH_z, acetone-d_6) of benz(j)aceanthrylene-1,2-oxide.

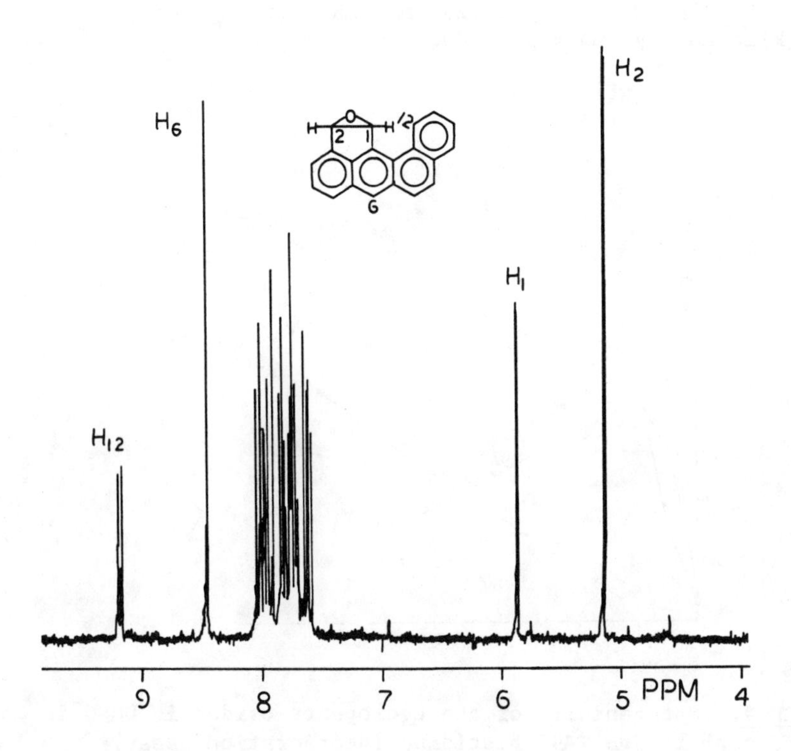

FIGURE 7. ^{1}H NMR spectrum (250 MH_z, acetone-d_6) of benz(l)aceanthrylene-1,2-oxide.

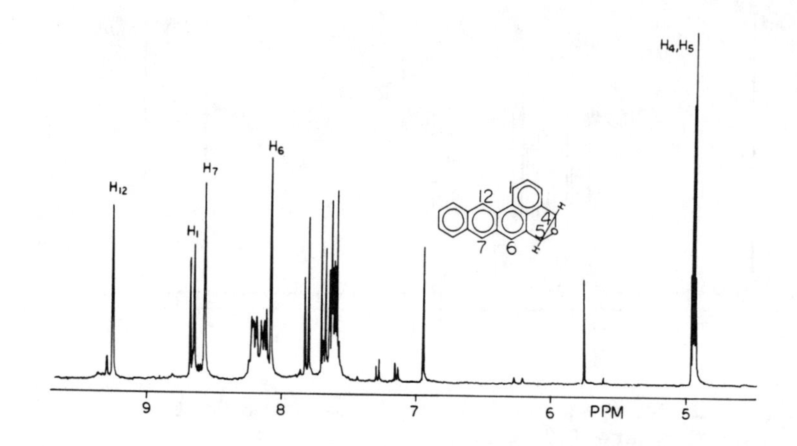

FIGURE 8. ^{1}H NMR spectrum (250 MH_z, acetone-d_6) of benz(k)aceanthrylene-4,5-oxide.

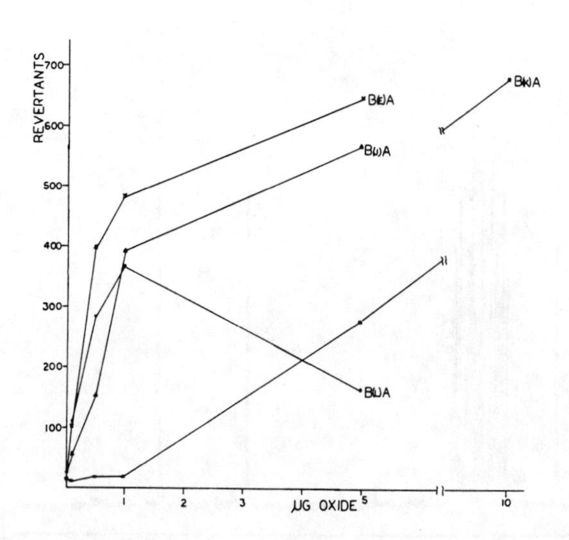

FIGURE 9. Mutagenicity of the cyclopenta oxides in DMSO in the **S. typhimurium** TA98 histidine incorporation assay.

Ames Assay

The cyclopenta oxides of benzanthracene appear to be potent direct acting mutagens (Fig. 9). The e-, j- and l-oxides are 10 to 20 times more active than their parent hydrocarbons on a per-mole basis (unpublished data), while the k-oxide is less active relative to its parent hydrocarbon and much less active than the other three cyclopenta oxides.

DISCUSSION

The cyclopenta isomers of benzanthracene are of interest because of their possible occurrence in the environment and also because initial metabolism and mutagenicity studies on these PAH suggest that activation to potent mutagenicity can proceed by pathways different from bay region diolepoxides. Syntheses of the cyclopenta oxides of the four cyclopenta-fused isomers of benzanthracene via the bromohydrins can provide pure epoxides suitable for biological testing as possible ultimate mutagenic metabolites. Preliminary Ames assays have confirmed their direct-acting mutagenicity and enabled us to establish that the cyclopenta oxides contribute to the overall mutagenicity of each parent PAH.

The results of the Ames assay on the series of four epoxides suggest that the cyclopenta oxides of the e-, j- and l-isomers of benzaceanthrylene can account for much of the mutagenic activity of the parent PAH observed when Aroclor 1254 induced rat liver S9 is employed as an exogenous activation system. This conclusion is based on: the high level of direct acting mutagenicity of the epoxides, which exceeds the activity of the parent PAH by 10 to 20-fold; the occurrence of the corresponding cyclopenta dihydrodiols as major isolated metabolites, implying that the oxides are the initially formed species; and the previously reported low concentration optima of the S9 concentration-dependence curves of mutagenicity. This type of concentration-dependence has been interpreted as indicating activation via a one-step epoxidation (3) as opposed to the multiple reaction sequence involved in formation of a diolepoxide.

The cyclopenta epoxides of the e-, j- and l-isomers are capable of opening to yield highly resonance-stabililzed carbonium ions (Table 1).

TABLE 1

STABILIZATION ENERGIES OF BENZYLIC CARBONIUM IONS DERIVED FROM RING-OPENED EPOXIDES

Epoxide	Carbonium ion	$\Delta E_{deloc}/\beta$
Cyclopenta(cd)pyrene	C_3	0.794
Aceanthrylene	C_1	0.931
Acephenanthrylene	C_5	0.664
B(j)A-1,2-oxide	C_1	0.879
B(e)A-5,6-oxide	C_5	0.879
B(l)A-1,2-oxide	C_1	0.833
B(k)A-4,5-oxide	C_5	0.722
BP-7,8-diol-9,10-oxide	C_{10}	0.794

The PMO delocalization energies ($\Delta E_{deloc}/\beta$) (10), which are hypothesized to correlate with S_N1 reactivity towards the DNA bases, are larger than those of carbonium ions derived from the ultimate mutagenic/carcinogenic epoxides/diolepoxides of other biologically active PAH. This observation lends additional support to the involvement of the cyclopenta epoxides in the mutagenic activity of the parent PAH.

The cyclopenta epoxide of the fourth isomer, benz(k)acephenanthrylene, is much less active than those of the benzaceanthrylenes and also less active relative to the parent PAH. Although the cyclopenta dihydrodiol of benz(k)acephenanthrylene is the major metabolite (hence epoxidation of the five-membered ring is the major metabolic pathway), $\Delta E_{deloc}/\beta$ of the most stable carbonium ion derived from this epoxide is smaller than $\Delta E_{deloc}/\beta$ values commonly associated with high levels of mutagenicity. The lower mutagenic activity of the k- oxide can therefore be anticipated from its $\Delta E_{deloc}/\beta$. Also consistent with the predicted lower activity for this cyclopenta epoxide is the lack of a low-concentration optimum in the S9 concentration-dependence curve of mutagenicity, suggesting that metabolites from additional pathways of activation are important contributors to mutagenicity.

The correlation observed in this study between the value of $\Delta E_{deloc}/\beta$ and the mutagenic activity of the corresponding epoxides in the Ames assay strongly supports the use of $\Delta E_{deloc}/\beta$ as a predictor of mutagenic activity. The results of this study also support the suggestion that the cyclopenta oxides of the benzaceanthrylene isomers are major contributors to mutagenicity of the parent compounds in the Ames assay, while other metabolic pathways must be important in determining the mutagenicity of benz(k)acephenanthrylene.

ACKNOWLEDGEMENT

This work was supported in part by EPA Grant CR811817 and American Cancer Society Grant No. BC388.

REFERENCES

(1) Grimmer, G., Naujack, K.-W., Dettbarn, G., Brune, H., Deutsch-Wenzel, R. and Misfeld, J., "Characterization of polycyclic aromatic hydrocarbons as essential carcinogenic constituents of coal combustion and automobile exhaust using mouse-skin-painting as a carcinogen-specific detector, **Toxicol. Environ. Chem**, **6**: 97-107 (1983).

(2) Bjorseth, S., "Analysis of polycyclic aromatic hydrocarbons in environmental samples by glass capillary chromatography." In: **Carcinogenesis, A Comprehensive Survey,** edited by P.W. Jones and R.I. Freudenthal, Vol. 3, pp. 75-83, Raven Press, New York (1978).

(3) Eisenstadt, E. and Gold, A., "Cyclopenta(cd)pyrene: A highly mutagenic polycyclic aromatic hydrocarbon", **Proc. Natl. Acad. Sci., USA 75**, 1667-1669 (1978).

(4) Gold, A., Nesnow, S.. Moore, M., Garland, H., Curtis, G., Howard, B., Graham, D. and Eisenstadt, E., "Mutagenesis and morphological transformation of mammalian cells by a non-bay-region polycyclic cyclopenta(cd)pyrene and its 3,4-oxide", **Cancer Res.**, **40**, 4482-4484 (1980).

(5) Nesnow, S., Leavitt, S., Easterling, R., Watts, R., Toney, S.H., Claxton, L., Sangaiah, R., Toney, G.E., Wiley, J., Fraher, P. and Gold, A., "Mutagenicity of cyclopenta-fused isomers of benz(a)anthracene in bacterial and rodent cells and identification of the major rat liver microsomal metabolites", **Cancer Res., 44**, 4993-5004 (1984).

(6) Sangaiah, R., Gold, A. and Toney, G.E., "Synthesis of a series of noval polycyclic aromatic systems: isomers of benz(a)anthracene containing a cyclopenta-fused ring", **J. Org. Chem., 48**, 1632-1638 (1983).

(7) Gold, A., Brewster, J. and Eisenstadt, E., "Synthesis of cyclopenta(cd)pyrene-3,4-epoxide, the ultimate mutagenic metabolite of the environmental carcinogen, cyclopenta(cd)pyrene, **J. Chem. Soc., Chem. Commun.**, 903-904 (1979).

(8) Ames, B.N., McCann, J., Yamasaki, E., "Methods for detecting carcinogens and mutagens with the Salmonella/mammalian-microsome mutagenicity test, **Mut. Res., 31**: 347-364 (1975).

(9) Claxton, C.D., Kohan, M., Austin, A.C., Evans, C., "The Genetic Bioassay Branch Protocol for Bacterial Mutagenesis Including Safety and Quality Assurance Procedures. HERL-0323, U.S. EPA, Research Triangle Park, NC, 147 pp.

(10) Jerina, D.M., Lehr, R.E., Yagi, H., Hernandez, O., Dansette, P.M., Wislocki, P.G., Wood, A.W., Chang, R.L., Levin, W. and Conney, A.H., "Mutagenicity of benzo(a)pyrene derivatives and the description of a quantum mechanical model which predicts the use of carbonium ion formation from diol-epoxides", In:"In Vitro Metabolic Activation in Mutagenesis Testing", F.J. DeSerres, J.R. Fouts, J.R. Bend and R.M. Philpot (eds.), Elsevier North Holland Biomedical Press (Amsterdam), pp. 159-177 (1976).

ESTIMATION OF THE DERMAL CARCINOGENIC POTENCY OF PETROLEUM FRACTIONS USING A MODIFIED AMES ASSAY

GARY R. BLACKBURN, ROBIN A. DEITCH, TIMOTHY A. ROY, SUSAN W. JOHNSON, CEINWEN A. SCHREINER, CARL R. MACKERER Mobil Environmental and Health Science Laboratory, P.O. Box 1029, Princeton, New Jersey 08540.

INTRODUCTION

Both experimental (1) and epidemiological (2,3) studies have demonstrated that certain mineral oils can induce skin cancers in animals and humans (review, 4). At least part of the carcinogenic activity of these oils is associated with their content of polynuclear aromatics (PNA) (4). Removal of PNA during the refining process, usually by extraction with furfural (solvent refining), eliminates carcinogenic activity (5).

However, for oils that do not undergo solvent refining, carcinogenic potential can be determined in mouse skin-painting bioassays. These tests usually provide a good indication of tumorigenic potential, but are occasionally complicated by severe skin irritation or systemic toxicity. Furthermore, they are expensive and time-consuming, often requiring two years to obtain a definitive answer.

For these reasons, there has been an attempt during the past ten years to apply short-term, *in vitro* mutagenicity assays to the screening of petroleum-derived materials. In general, the strong correlation between mutagenicity and carcinogenicity that exists for most pure chemical compounds (6-9), has not held when the assays are applied to the extremely complex, water-insoluble mixtures that comprise mineral oils (10,11). For example, in a study commissioned by the American Petroleum Institute (API) (11), thirteen oils previously evaluated in skin-painting bioassays were tested in the standard bacterial mutagenesis assay of Ames et al. (12); virtually no correlation was found between their activities in the two test systems.

In previous publications (13,14), we reported that modifications in sample preparation and delivery, metabolic activation, and data analysis lead to a marked improvement in the predictability of the Ames assay for a variety of mineral oils. These modifications include (a) extraction of the test oil with DMSO to yield a water-miscible, PNA-enriched fraction; (b) an eight-fold increase in the concentration of liver metabolizing mixture (S9) to 400 µl/plate, together with substitution of Syrian golden hamster for the standard rat S9; and (c) a parallel increase in NADP cofactor concentration to 8 mM. These experimental modifications were combined with a change in the method of data analysis, i.e. the use of nonlinear regression (15) for determination of the slopes of the dose response curves for mutagenesis. The values for these slopes were used as a measure of the relative potency of the mutagenic response.

In the present study, the modified Ames assay was applied to a subset of the oil samples that were tested in the mouse skin-painting and standard Ames assay. The results of this testing are compared with those obtained in the earlier studies (11), and the predictability of the test for these representative petroleum fractions is evaluated and discussed.

MATERIALS AND METHODS

Oil Samples

The oil samples tested in this study were laboratory preparations obtained from the American

Petroleum Institute, Washington, D.C. Detailed methods for their preparation, and an extensive discussion of their physical and chemical properties were previously published (16). Table 1 describes the oils and their subfractions with reference to the parent crudes, boiling ranges, and their correspondence to typical fractions obtained by actual refinery distillation of crude oil.

TABLE 1

PROPERTIES OF TESTED PETROLEUM FRACTIONS

OIL FRACTION	DISTILLATION RANGE °F	EQUIVALENT REFINERY FRACTION	CAS NUMBER
C-3[a]	350-550	STRAIGHT RUN KEROSENE	8008-20-6
C-4	550-700	STRAIGHT RUN GAS OIL	64741-43-1
C-5	700-1070	HEAVY VACUUM GAS OIL	64741-53-3
C-6	>1070	VACUUM RESIDUUM	64741-55-6
D-2[a]	120-350	LIGHT STRAIGHT RUN NAPHTHA	64741-46-4
D-3	350-550	STRAIGHT RUN KEROSENE	8008-20-6
D-4	550-700	STRAIGHT RUN GAS OIL	64741-43-1
D-5	700-1070	HEAVY VACUUM GAS OIL	64741-53-3
D-6	>1070	VACUUM RESIDUUM	64741-55-6

a "C" fractions are derived from a domestic, high-sulfur, naphthenic crude; "D" fractions from a foreign, low-sulfur, paraffinic crude.

Dermal Carcinogenicity Assays

Dermal carcinogenicity assays were performed under the sponsorship of the American Petroleum Institute at the Department of Environmental Health, Kettering Laboratory, University of Cincinnati Medical Center, Cincinnati, Ohio. Groups of 50 male C3H/HeJ mice were dosed twice weekly by application of 50 mg of oil to the shaved interscapular region of the back. For four of the fractions, high viscosity (C6 and D6) or excessive toxicity (C4 and D4) necessitated a 50% dilution of the oil into toluene prior to application. Details of the skin-painting assays have been published elsewhere (11).

Mutagenicity Assays

Standard Ames assays were performed at three contract laboratories, referred to as X, Y, and Z in Table 2. All samples were tested with and without Aroclor 1254-induced rat liver S9 in both pre-incubation and plate incorporation assays. Laboratories X and Z used all five strains of *Salmonella*, while Laboratory Y omitted strain TA1538. The criterion for significant mutagenicity was a dose response with at least a three-fold increase in revertant colonies at the highest considered dose. Other aspects of the tests were consistent with the recommendations of Ames et al. (12), and are reported in detail in a previous publication (14).

Correlation of Mutagenic and Carcinogenic Potency

Two determinants were used to quantify relative carcinogenic potency: the inverse of latent period and the Carcinogenicity Index (CI). Both were dose-adjusted for samples C4 and D4 (delivered at half dose) by multiplying by 2.

The CI is defined as follows:

$$CI = \frac{1}{\text{Latent Period}} \times \frac{\text{\# Mice With Tumors}}{\text{FEN}} \times 1000$$

FEN, the Final Effective Number, which corrects for mortality that is unrelated to tumorigenesis, is the number of mice surviving at the time of appearance of the median tumor (or 60 weeks, whichever is earlier) plus any animals that died with tumor(s) before that time. Multiplication by 1000 converts the index to a convenient scale. The degree of correlation was determined by linear regression analysis.

RESULTS

The 9 oils listed in Table 1 are a subset of the 13 used in the original API-sponsored mutagenicity and carcinogenicity studies (11). Although the oils are laboratory distillation cuts, boiling ranges were selected so that the samples would be representative of typical refinery fractions. Thus the group includes a naphtha (D2, gasoline range), two kerosenes (C3 and D3, kerosene, jet fuel range), two straight run gas oils (C4 and D4, diesel-light fuel oil range), two heavy vacuum gas oils (C5 and D5, heavy fuel oil range), and two vacuum residua (C6 and D6, asphalt range). As such, they represent the entire spectrum of distillation fractions derived from two distinctly different crude oils (C and D), and are well suited to the determination of inherent biological activity of minimally refined petroleum products.

Tables 2 and 3 present a comparison of the results of the original skin-painting and Ames assays of the nine oils used in this study. In these early

TABLE 2

RESULTS OF STANDARD AMES TESTING

OIL FRACTION	TEST RESULT USING STANDARD PROCEDURE — CONTRACT LABORATORY —			
	X	Y	Z-1[a]	Z-2
C-3	NEG	NEG	NEG	NEG
C-4	NEG			
C-5	4.0[b]	NEG	NEG	
C-6	NEG			
D-2	NEG			
D-3	NEG			
D-4	NEG	NEG	NEG	
D-5	NEG	3.1[b]	NEG	
D-6	NEG			

a Laboratory Z tested materials in two separate studies.

b Numerical values indicate fold increases in mean revertants relative to solvent controls. Increases were not dose-responsive or reproducible.

studies, all of the fractions, except the vacuum residua, C6 and D6, were significantly tumorigenic (induced tumors in more than one mouse). However, only the two highly tumorigenic heavy vacuum gas oils, C5 and D5, produced a greater than threefold increase in revertants in any of the standard Ames tests. Since these increases were neither dose-responsive nor reproducible, C5 and D5 were

TABLE 3

SUMMARY OF SKIN-PAINTING DATA

OIL FRACTION	PERCENT MICE WITH TUMORS	MEAN LATENT PERIOD (WKS)	CARCINO-GENICITY INDEX[a]
C-3	30	70	4.0
C-4[b]	34	85	8.0
C-5	31	50	16
C-6[b]	0	—	0
D-2	25	85	3.0
D-3	15	62	2.4
D-4[b]	24	92	5.2
D-5	91	33	28
D-6[b]	2	70	0[c]

a $CI = \frac{1}{\text{Latent Period}} \times \frac{\text{No. Mice with Tumors}}{\text{FEN}} \times 1000$

FEN - Number of mice surviving at the time of appearance of the median tumor (or 60 wks, whichever is earlier) plus any mice that died with tumor(s) before that time. Values for C4 and D4 are dose-adjusted (see note b).

b C-4, D-4, C-6, and D-6 were delivered at half the dose of the other oils, i.e. 25 mg/application.

c Oils inducing tumors in fewer than two mice out of 50 are considered inactive.

judged to be nonmutagenic (11).

These results are contrasted with those obtained in the present study using the modified Ames assay

(13,14). Figure 1 shows the dose response curves for the 9 oils, and Table 4 the Mutagenicity Indices calculated from nonlinear regression analysis of those curves. For comparison purposes, Table 4 also includes dose-adjusted determinants of tumorigenic potency derived from the data in Table 3.

TABLE 4

INDICES OF MUTAGENIC AND CARCINOGENIC POTENCY

OIL FRACTION	MI[a]	RECIPROCAL OF LATENT PERIOD X 100[b]	CI[a,b]
C-3	1.1	1.43	4.0
C-4	3.4	2.35	8.0
C-5	6.2	2.00	16
C-6	0.0	—	0.0
D-2	1.3	1.18	3.0
D-3	1.9	1.61	2.4
D-4	5.2	2.17	5.2
D-5	7.5	3.03	28
D-6	0.0	—[c]	0.0[c]

a MI - Mutagenicity Index, slope of dose-response curve for mutagenicity. CI - Carcinogenicity Index (see Table 3, note a).

b Indices for fractions C4 and D4 were dose-adjusted by multiplication by 2.

c D-6 is categorized as non-tumorigenic since one animal with tumor out of 50 on study is not considered significant.

Mutagenicity indices range from 0 (C6 and D6) to

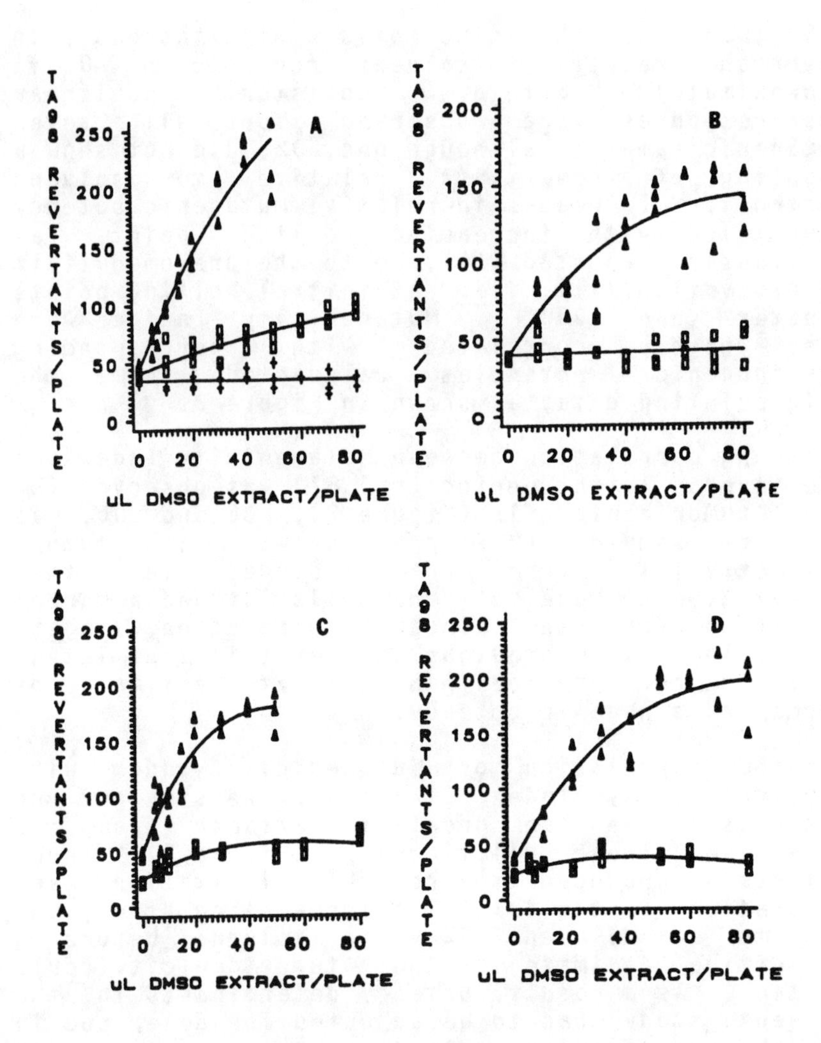

FIGURE 1. Dose-response curves in the modified Ames assay. Panel A - (△) C-5, slope 6.2; (□) C-3, slope 1.1; (+) D6, slope 0.0. Panel B - (△) C-4, slope 3.4; (□) C-6, slope 0.0. Panel C - (△) D-5, slope 7.5; (□) D-3, slope 1.9. Panel D - (△) D4, slope 5.2; (□) D-2, slope 1.3. Curves were fit and slopes determined using nonlinear regression analysis according to the method of Myers et al. (16).

7.5 (D5) for the nine oils, with increases in revertants relative to solvent controls of 0 to approximately 5-fold. Consistent, nonlinear dose-responses were observed for all seven mutagenic samples, although one, D2, did not show a doubling of revertants relative to solvent controls. Increases in relative mutagenic potency correlated with increasing boiling point, as previously reported (14), up to the presumed limit of bioavailability, i.e. at initial boiling points greater than 1000°F. Mutagenicity Indices were also highly correlated with corresponding carcinogenic potencies calculated from the skin-painting data, as shown in Figure 2.

A strong correlation between Mutagenicity Index and the inverse latent period (r=0.87) was observed for the 7 tumorigenic oils (Figure 2). C6 and D6, as inactive samples (0 and 1 animal with tumor, respectively), were not included in this correlation. None of the oils showed a marked deviation from the linear correlation, again indicating, as in previous studies (14) that latent period alone provides an accurate estimate of dermal carcinogenic potency.

For the correlation of Mutagenicity Index with Carcinogenicity Index, C4 and D4, the straight run gas oils, showed the greatest variance from the otherwise linear relationship (Figure 2). These materials produced severe skin irritation and ulceration in the initial skin-painting study, and had to be diluted 50% into toluene before a definitive estimate of their tumorigenicity could be made. As a result, potency determinants in the present study had to be adjusted for dose; and in the absence of sufficient data relating dose to latent period and tumor yield, this was done by simply multiplying the determinants by two. Thus, part of the deviation from correlation probably relates to the complicating factor of toxicity in the skin-painting study, and part to the dose-adjustment in the potency parameters. Even with these difficulties, the correlation coefficient for the 9 samples is 0.89.

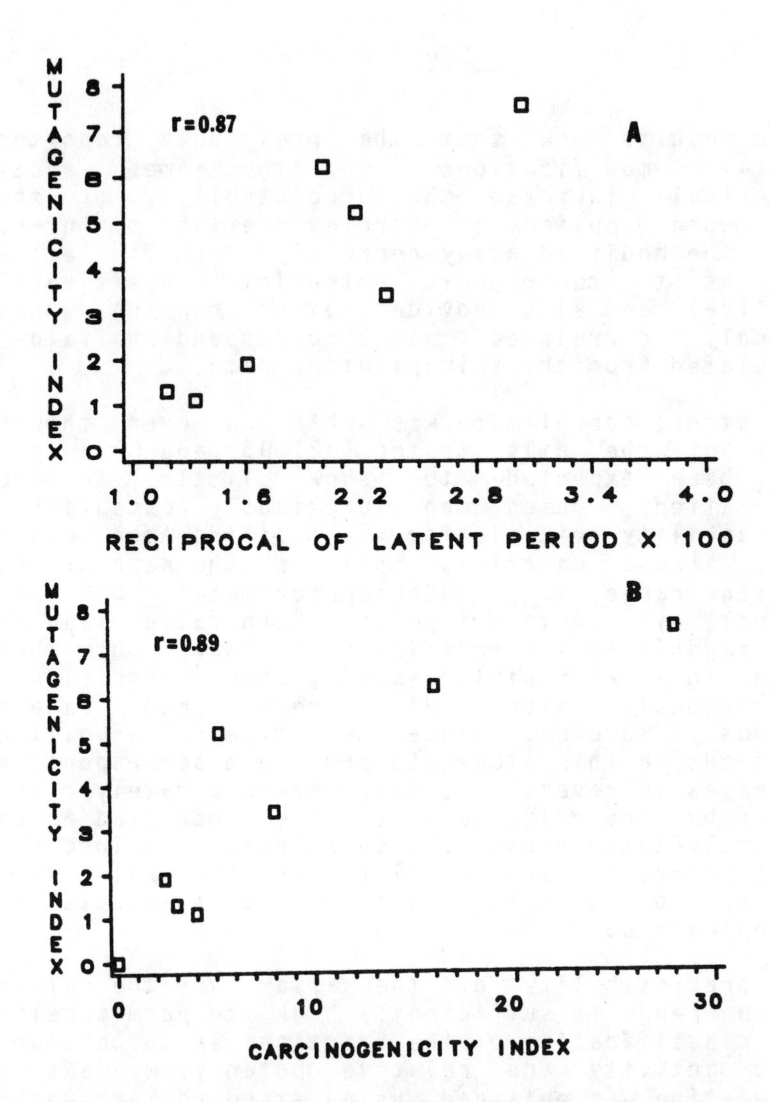

FIGURE 2. Correlation between mutagenic and carcinogenic potencies. Panel A, the correlation of Mutagenicity Index (MI) with the reciprocal of the latent period; Panel B, the correlation of MI with Carcinogenicity Index. For definitions of parameters, see notes to Tables 3 and 4. Correlation coefficients (r-values) were determined using linear regression analysis.

DISCUSSION

These results show that the previously reported (13,14) modifications to the Ames assay dramatically increase the predictability of the test when applied to petroleum-derived mixtures. Thus, the modified assay correctly classified all 9 oils as to tumorigenic potential (7 positive, 2 negative), and also provided potency rankings that strongly correlated with corresponding values calculated from the skin-painting data.

This strong correlation was obtained even though three of the oils tested (D2, D3, and C3) might have been expected to show little or no correlation, based on previously established applicability criteria for the modified Ames assay (14). These materials boil in the naphtha and kerosene range, i.e. under approximately 500° F. Several oils of this type have been categorized as nonmutagenic in the modified Ames Assay, but when tested in a skin-painting study, showed significant tumorigenesis, albeit with very long latent periods. However, since the three low-boiling fractions in this study did produce dose-responsive increases in revertants, they were accurately classified by the criteria of the modified assay as likely tumorigens. The correlation obtained for these materials may be related to the relatively wide span of distillation temperatures selected to define each cut.

The predictability of the assay over the entire boiling range is sufficiently high to permit reliable classification of the fractions as to carcinogenic activity and relative potency, whereas no correlation was obtained using standard test methods (11). Moreover, the mutagenicity rankings follow the expected trend toward increasing PNA content with increasing boiling range (4), and do not appear to be complicated by undue toxicity, a problem evident in the skin-painting studies of several of these materials.

One other finding in this study demonstrates the strength of the correlation between mutagenicity and carcinogenicity obtained using the modified

Ames assay: the absence of mutagenicity for the two vacuum residua, C6 and D6. DMSO extracts of these fractions contain greater than 2% 3-7 ring PNA, a level that would imply a weak to moderate carcinogenic response in lower-boiling fractions. Yet neither was tumorigenic in the skin painting study. Apparently, PNA with boiling points in this range are not bioavailable, or biologically active, in either the *in vitro* or *in vivo* test system, possibly because of their large molecular size or high level of alkylation.

In summary, the modified Ames assay provides an effective means for assessing the carcinogenic activity of distillation fractions derived from petroleum. The assay is most reliable for fractions boiling above 500° F, but as shown by the results of this study, can be useful for lower boiling materials, provided some mutagenic activity is observed. Because of the trend to increasing mutagenicity with increasing boiling range (up to about 1000° F) and hence increasing PNA content, it seems likely that the predictability of the assay is based on its sensitivity to PNA-mediated mutagenicity. This hypothesis is currently being tested by detailed chemical analysis of the DMSO extracts of the petroleum fractions used in this study.

ACKNOWLEDGMENTS

The authors express their appreciation to the American Petroleum Institute for its cooperation in providing the petroleum distillation fractions used in this study. Mutagenicity assays were performed with the expert technical assistance of Nancy P. Hoe, Patricia L. Burnett, and Claire L. Marshall.

REFERENCES

1. Twort, C.C. and Twort, J.M. (1931): The carcinogenic potency of mineral oils. *J. Indust.*

Hyg. Toxicol., 13:204-226.
2. Cruickshank, C.N.D. and Squire, J.R. (1950): Skin cancer in the engineering industry from the use of mineral oil, Br. J. Indust. Med., 7: 1-11.
3. Cruickshank, C.N.D. and Gourevitch, A. (1952): Skin cancer of the hand and forearm, Br. J. Indust. Med., 9:74-79.
4. Bingham, E., Trosset, R.P., and Warshawsky, D. (1979): Carcinogenic potential of petroleum hydrocarbons. J. Environ. Path. Toxicol., 3: 483-563.
5. Bingham, E. and Horton, A.W. (1969): Environmental Carcinogenesis: Experimental observations related to occupational cancer. In: Advances in Biology of the Skin, Vol. VII, pp. 183-193, Pergamon Press, Elmsford, N.Y.
6. McCann, J., Choi, E., Yamasaki, E., and Ames, B.N. (1975): Detection of mutagens in the Salmonella microsome test: Assay of 300 chemicals. Proc. Natl. Acad. Sci., (USA), 72:5135-5139.
7. Coombs, M.M., Dixon, C., and Kissonerghis, A.-M. (1976): Evaluation of the mutagenicity of compounds of known carcinogenicity, belonging to the benz[a]anthracene, chrysene, and cyclopenta[a]phenanthrene series, using Ames's test. Cancer Res., 36:4525-4529.
8. Purchase, I.F.H., Longstaff, E. Ashby, J. Styles, J.A., Anderson, D., Lefevre, P.A., and Westwood, F.R. (1978): An evaluation of six short-term tests for detecting organic chemical carcinogens. Br. J. Cancer, 37:873-944.
9. Rinkus, S.J. and Legator, M.S. (1979): Chemical characterization of 465 known or suspected carcinogens and their correlation with mutagenic activity in the Salmonella typhimurium system. Cancer Res., 39:3289-3318.
10. Conaway, C.C., Schreiner, C.A., and Cragg, S.T. (1982): Mutagenicity evaluation of petroleum hydrocarbons. In: The Toxicology of Petroleum Hydrocarbons, edited by H.N. MacFarland, C.E. Holdsworth, J.A. MacGregor, R.W. Call, and M.L. Kane, pp. 128-138, American Petroleum Institute, Washington, D.C.
11. MacGregor, J.A., Conaway, C.C., and Cragg, S.T. (1982): Predictivity of the Salmonella/microsome assay for carcinogenic and noncarcinogenic

complex petroleum hydrocarbon mixtures. Ibid., pp. 149-161.
12. Ames, B.N., McCann, J., and Yamasaki, E. (1975): Methods for detecting carcinogens and mutagens with the Salmonella/microsome mutagenicity test. Mutat. Res., 31:347-363.
13. Blackburn, G.R., Deitch, R.A., Schreiner, C.A., Mehlman, M.A., and Mackerer, C.R. (1984): Estimation of the dermal carcinogenic activity of petroleum fractions using a modified Ames assay. Cell Biol. Toxicol., 1:67-80.
14. Blackburn, G.R., Deitch, R.A., Schreiner, C.A., and Mackerer, C.R. (1986): Predicting carcinogenicity of petroleum distillation fractions using a modified Salmonella mutagenicity assay. Cell Biol. Toxicol., 2:63-84.
15. Myers, L.E., Sexton, N.H., Southerland, L.I., and Wolff, T.J. (1981): Regression analysis of Ames test data. Env. Mut., 3:575-586.
16. King, R.W. (1982): Skin carcinogenic potential of petroleum hydrocarbons. 1. Separation and characterization of fractions for bioassay. In: The Toxicology of Petroleum Hydrocarbons, edited by H.N. MacFarland, C.E. Holdsworth, J.A. MacGregor, R.W. Call, and M.L. Kane, pp. 171-184, American Petroleum Institute, Washington, D.C.

RISK ASSESSMENT OF POTENTIATING FACTORS IN POLYNUCLEAR AROMATIC HYDROCARBONS (PAH) TOXICITY

BADI M. BOULOS*[1], ALFRED VON SMOLINSKI[2]
(1) School of Public Health, University of Illinois at Chicago, Chicago, Illinois 60608, USA; (2) College of Pharmacy, Health Sciences Center, University of Illinois at Chicago, Chicago, Illinois 60608.

INTRODUCTION

Polynuclear aromatic hydrocarbons occur in several environmental situations in air, such as tobacco smoke, automobile and industrial exhaust and other combustion procedures in leaf burning. PAH from incomplete combustion occur in several foods such as charcoal broiled and smoked goods. The levels of PAH in these situations may be so small that the hazards to human health cannot be assessed. Such evaluations do not take into account the synergistic effects of other substances which may be present in the environment.

PAH comprises a group of several hundred compounds which are strongly suspected in inducing cancer in man (1-7). Some of them have been shown to cause cancer in animals not only by skin applications but also by oral administration (8-11). Synergistic action between PAH members has been demonstrated in animal experiments. Wynder and Hoffman (12) reported that there were 100 fold increases in skin cancer in mice when painted by cigarette condensate comparable to equivalent amounts of benzo(a)pyrene (B(a)P) alone. Epidemiological data on synergistic effects of other chemicals on PAH carcinogenicity have been reported. Examples of these are the increased rate of lung cancer among workers in asbestos and uranium ore industries. Workers who smoke in these industries have nearly ten times higher incidence of lung cancer than exposed workers who do not smoke (13-20). It is important to note that prediction of human cancer risks cannot be made from knowledge of levels of PAH exposure in the environment alone. It should include synergistic effects, as well as metabolic pathways, species differences and genetic factors. Berenblum and Schoental (21) and Harper (22-24) have demonstrated that some of the hydroxyderivatives of B(a)P have no carcinogenic effect if painted on mice skin; however, B(a)P and 6-hydroxymethyl B(a)P exhibited the ability to induce local sarcomas in rats. Dibenz(a,h)anthracene [DB(a,h)A] appeared to be equally potent as B(a)P on the skin of mice (24). In a dose-response study of subcutaneous carcinogenicity with DB(a,h)A, B(a)P, and methylcholanthrene (MC); DB(a,h)A was shown to be effective at a lower dose than B(a)P or MC. However, the latent period for tumor production was longer (25). DB(a,h)A induced local sarcomas and increased the incidence of lung adenomas following a single subcutaneous injection in newborn mice at dose levels which were ineffective with MC (26-27).

Factors essential in toxicological evaluation are the stressor, the receptors (host) and the stressor-host interactions. Deposition of particles, absorption and reserve functional capacity of the organ system, should be explored in assessing the risks pertaining to the host. Stressor-receptor interactions are the determining means in the ultimate effects of these compounds. Metabolic pathways, genetic deficiencies are other important aspects in assessing human risk to PAH exposure. Public health awareness by educational programs as well as stricter legislative action to limit levels of exposure are needed criteria to reduce health hazards of PAH compounds.

MATERIAL AND METHODS

In assessing risks of human exposure to PAH, evaluation of potentiating factors should deal with the different stressors, i.e., groups of compounds as present in the environment, host with all genetic deficiencies that affect metabolic pathways and host-stressor interactions.

Source of Stressors

Several sources have been identified to emit PAH in the atmosphere. Most of the estimates for PAH were done by determining the levels of one of its members, B(a)P (28-31). Gasoline-powered automobiles, industrial plants such as coke ovens, are sources which emit PAH. Cigarette smoke, fresh vegetables, vegetable oils, and smoked meat are other sources for PAH (31). Nanogram levels of PAH compounds were found in cosmetics and medicinal preparations such as soft paraffins (32). Environmental monitoring for PAH uses extraction techniques from the matrix, followed by individual identification of the components by gas chromatography and mass spectrometry (GC/MS). Harvath (33) described techniques using high performance liquid chromatography as a more sensitive approach for detecting several PAH members than the use of GC/MS alone. Other investigators have used another substance such as lead, and showed correlation between Pb levels and environmental PAH levels (34). There was a linear correlation between such levels on microgram per cubic meter of air basis in New Jersey.

Receptor (Host) Monitoring

Assessing exposure to PAH can be monitored by analysis of urine samples, for both parent compounds and metabolites, fat and liver samples, or hair follicles (35-38). PAH are oxidized to epoxides by mixed-function oxidases. P450 and P448 are among the most active group of these enzymes (39-43). "Toxification" and "detoxification" processes can be detected by aryl epoxide hydrolases, and monooxygenases (44-49).

Stressor-Host Interactions

The usual pathway or "detoxification" indicates that the parent compound may alter cell repair mechanisms and the end product excreted (50). Yet, the host may transform a non-harmful compound into a harmful metabolite which may cause damage to the cells and tissues (51). Understanding the metabolic pathways for each PAH is essential in assessing ultimate hazards.

RESULTS

Stressors

Considerable amounts of PAH are present in the environment. In the work place, PAH compounds are adsorbed on particulate matter, especially in coke production, gas work, iron and steel manufactures, and aluminum electrolytic smelting operations (52). B(a)P and perylene (Per) are formed together with several other PAH during incomplete combustion of carbon-containing fuels (53). These PAH are easily nitrated under various conditions producing, for example, 6-nitrobenzo(a)pyrene, 3-nitroperylene, and 1- and 3-nitrobenzo(a)pyrene (54). Frying ground beef at 250°C results in the formation of 2-amino-3,8-dimethylimidazo[4, 5-f]quinoxaline (MeI Qx) and 2-amino-3,-methylimidazo[4,5-f]-quinoxaline (IQ) (54). N-phenyl-2-naphthylamine (PBNA) has been found to be produced in rubber manufactures, pitch and coal tar (55).

Carcinogenic Classification of Stressors

Several investigators attempted to classify members of PAH as highly strong carcinogens, (+++), strong carcinogens (++) and weak carcinogens (+) (9, 11). Highly strong carcinogens include: B(a)P, dibenzo(a,h)anthracene, dibenzo-(a,h)pyrene, benzo(c)phenanthrene, dibenzo(a,i)pyrene, 3-methylcholanthrene (3MC), and 7,12-dimethylbenz(a)-anthracene (DMBA). Strong carcinogens include: benzo(b)-fluoranthene and benzo(j)fluoranthene, while weak carcinogens include: benzo(e,l)pyrene, indeno(1,2,3-d)pyrene, dibenzo-(a,j)anthracene and 2-methylfluorene. Such classifications are based on epidemiological data. However, more recent developments using _in vitro_ assays such as genotoxic effects, sister chromatid exchange and hypoxanthine - guanine phosphoribosyltransferase locus mutation tests have added more sensitive and alternative criteria for classification of compounds (56). As an example, 5 naphthofurans were studied (56) and revealed that all biological activities, _i.e._, genotoxic and mutagenic effects, are directly linked to the presence of a NO_2 group in position 2. Also, activities are enhanced by methoxy group in position 7 or 8. However, such methoxy group decreases or suppresses cancer skin tests. Venegas _et al._ (57) rated the compounds as follows with descending effects: 2-nitro-7-methoxynaphtho[2,1-b]furan is highest in its mutagenic effect followed by 2-nitro-8-methoxynaphthol[2,1-b]furan, 2-nitronaphthol[2,1-b]furan, 2-nitro-7-bromonaphthol[2,1-b]furan, and 7-methoxynaphthol-[2,1-b]furan being the least. Wood _et al._ (58) showed that pseudodiequatorial conformations of diolepoxide isomers of benzo(c)phenanthrene may explain their high mutagenicity in both bacterial and mammalian cells. They found that for optically active isomers a range from two to four times the mutagenic activity. Briant and Jones (59) found that the more the benz(a)anthracene nucleus is planar the more carcinogenicity it causes.

The presence of compounds along with PAHs could act as inducer and increase mutagenicity of the parent compound. As an example, it was found that benz(a)anthracene in the presence of Clophen A50 showed 70% of mutagenic activity compared to B(a)P (60). The 8,9-dihydroxy-10,11-epoxide of benzo(a)anthracene showed nearly the same mutagenic activity in a direct Ames test as the K-region epoxide (61).

Host (Receptor) Factors-
Stressor Interactions

The carcinogenic activity of B(a)P is expressed only after metabolic formation of electrophilic intermediates (62). Recently, it is reported that a specific metabolite such as 9-OH-BP-4, 5-oxide is the reactive DNA-binding intermediate (63). Some of the metabolic processes for DMBA were found to depend on the presence of sufficient amounts of β-glucuronidase. If this enzyme, β-glucuronidase, is inhibited, the tumorigenic activity of DMBA in rats decreased more than 70% (64). Pyruvate is very important to the basic metabolism of normal tracheal epithelial cells, yet carcinogen-altered cells change the cell metabolism and can do without pyruvate during growth (65). On the other hand, glutathione plays an important role in detoxification of some PAHs members (66). Some transformed (oncogenic) cells have been shown to use protease and polyphosphate instead of glutamine in the synthesis of DNA and RNA (67). The absence of glucose-6-phosphate dehydrogenase affected the conversion of B(a)P to water soluble metabolites, and hence decreased the toxicity and tumorigenicity among human skin fibroblast cultures (68). Other intermediate metabolites have been reported for several members of the PAH group. Trans-1, 2-dihydro-1,2-dihydroxy-6- nitrochrysene was found to be the major proximate mutagen for 6-nitrochrysene in S. typhimurium TA 100 (69). Dinitropyrenes are known to cause mutagenic effects in in vitro cell cultures and have been shown recently to induce sarcomas in rats (70). These compounds have the same carcinogenic activity as B(a)P and 3MC. Some association of promoters was demonstrated to be site (organ) specific. Hexachlorocyclohexane (HCH) was found to be specific for liver tumor promotion by DMBA, however, it failed to show such effects on DMBA-induced skin tumors in mice (71). Long term administration of adrenal steroid, dehydroepiandrosterone inhibits the development of spontaneous mammary tumors in mice. Also this steroid inhibits DMBA-induced skin papillomas in mice (72). In hamsters, Beems and Beck (73) found that a high fat diet, especially in unsaturated forms, enhanced B(a)P-induced respiratory tract carcinogenesis. The Ca^{2+}-calmodulin system was found responsible for tumor promotion of DMBA-induced skin tumors in mice (74). Chromosomal aberrations of bone marrow cells has been used as an in vivo method to detect mutagenic and carcinogenic activity. Increased levels of hepatic glutathione transferase before DMBA-induction, significantly diminished chromosomal aberrations in bone marrow of rats (75).

Not only differences in enzyme levels can modify induction of carcinogenic and mutagenic activity, but also, age, sex, species, and pregnancy can modify such actions. Male mice showed 2-3 fold more lung tumors after fluoranthene and B(a)P injections than females (76). Cultures of mammary cells from virgin rats produced more metabolites when DMBA was added than cells from pregnant rats (77). This has been explained assuming that the metabolic processes of the same tissues undergo modification by age and endocrine effects. Spontaneously hypersensitive rats (SHR) showed more chromosomal aberrations to DMBA than Wiston Kyoto (WKY) rats indicating strain specificity (78). Comparison of metabolic activation of DMBA by two different cell lines; one was human hepatoma, the other was hamster embryo cells; showed that different pathways were the predominate observation in enzyme activity (79). Species differences may explain the higher susceptibility to carcinogenic action in one species than the other. This has been explained by the fact that marked reduction in DNA-binding metabolites in one cell species will reduce cancer incidence. Hamster embryo cell cultures produce more soluble metabolites of DMBA and hence more mutagenic effects than human hepatoma cell culture (79). Stressor-host interactions can be manifested along with tumorigenic activities. An example of this was found during treatment of pregnant mice by B(a)P. The progeny showed suppression of defense mechanisms up to 18-24 months of age beside increased incidence of tumors (80). The T-cell compartment exhibited immunodeficiency and altered component development. This may aid in growth of nascent neoplasms later in life. Another type of reaction to PAH, beside carcinogenicity and mutagenicity, is histologic abnormalities in skeletal tissues of rainbow trout alevins. It was reported that exposure of rainbow trout alevins to different concentrations of B(a)P showed nuclear pycnosis and karyorrhexis in skeletal muscles near the skull and vertebral column (81). A third type of reaction is the genotoxic effects of some PAH on peripheral human lymphocytes. 2,4,7-Trinitro-9-fluorenone was found to cause sister chromatid exchange and chromosomal aberrations on human peripheral lymphocytes cultures (82).

Dealing with the ultimate carcinogenic effects of PAH, other environmental contaminants may potentiate this reaction. Asbestos (crocidolite type) with B(a)P increased the extent of squamous metaplasia in hamster tracheal epithelial cultures (83). This supports the epidemiological observation indicating that bronchogenic carcinoma was significantly higher in asbestos workers who smoke than those who did not. Compounds which compete for the same enzymological pathway could increase or decrease the amounts of ultimate carcinogenic metabolites. Lawrence, et al. (84) reported that patients with psoriasis using anthralin showed decreased irritation from the drug after application of a coal tar solution to non-affected skin areas. A possible explanation is that, both coal tar and anthralin use aryl hydrocarbon hydroxylase (AHH) in their metabolic pathways. As coal tar was applied, AHH was induced and increased the removal of anthralin, and hence its irritancy.

Chemical modification of ultimate carcinogens may shed some light on the study of carcinogenesis as an aberrant gene expression. Some investigators have shown that inactivation of the thymidine kinase gene after B(a)P-diol-epoxide modification can stop its carcinogenic binding to DNA (85). Therefore, gene transfer of modified eukaryotic cells challenges the repair enzymes. Fluoranthene and pyrene are proved to be potent carcinogens when applied together with B(a)P. Both compounds are found to potentiate the tumorigenicity and production of DNA-adducts caused by B(a)P (86). An ideal model to characterize PAH has been suggested by White et al. (87). The investigators reported that immunosuppression activity of mouse spleen cells can be caused by B(a)P and not by B(e)P. They indicated that metabolic activation is not needed to cause such effects. Cells after acquiring carcinogenic transformation show changes in their enzymological behavior. Carcinogen-altered tracheal epithelial cells do not achieve a growth advantage early in progression but show higher hexose uptake and aerobic glycolysis than normal cells (88,89). A reverse correlation between tumors and adenylate cyclase activity in activation of B(a)P is needed, however, for the carcinogenic action. Co-oxidation with lipoxygenases has been suggested in B(a)P activation, and a possible role in the arachidonic acid cascade system, and in the NADPH-dependent cytochrome P-450 system. Nemato and Takayama (90) found that linoleic acid was more effective than arachidonic acid in activating B(a)P with microsomes and cytosols. Prostaglandin H synthase was found to be needed in the co-oxidation of arachidonic acid in explants of human and hamster tracheobronchial tissues (91). Gonadectomy in male rats affected the metabolism and activation of some PAH metabolites, and hence reduced hepatocarcinogenesis when done before introduction of N-fluorenylacetamide (92). Epidemiologic studies, sometimes, reveal the validity of *in vitro* colony assays (93). This may help in extrapolation of *in vitro* studies to risk assessments. Another area of development in oncology research is the application of *in vivo* techniques to shorten time of tumor induction. Using air pouch techniques in rats, induction time for mammary induced tumors with DMBA, was reduced from 180 days to 80 days (94).

The use of more sensitive techniques to monitor exposure to PAH compounds is desirable. A dose-response effect on urinary metabolites may help identify variables in low-level exposure. One of these products has been recently investigated (95). At low dose levels of B(a)P by percutaneous treatment, excretion of 3-OH-B(a)P occurred after oral administration in rats. This may be useful in predicting human cutaneous exposures to PAH compounds. In interpreting data for PAH exposure one has to take in consideration transient DNA lesions which may undergo repair and hence exhibit no tumorigenic effect. This depends on the period of exposure. The shorter the exposure time, the more transient the effect will be, and rapid repair will occur (96). Some deletions may occur in DNA which may include target and non-target nucleotide. These effects have been studied by the use of short DNA oligonucleotides (97). Some bioantioxidants have been shown to inhibit B(a)P induced mutagenesis by binding with free radicals and by linkage to cytochrome P-450, thus preventing further metabolic activation of B(a)P (98). On the other hand, dietary administration of butylated hydroxyanisole caused changes in glutathione S-transferase activity but no protection against PAH (99). This may suggest the synergistic action between arsenic and smoking observed in smelter workers. More recently, some investigators have found that the transformed cells, *e.g.*, tumor tissue, produce a special amino acid which increases capillary formation to supply nutrients and blood to early cancer cells (100-102).

DISCUSSION

Assessment of hazards to humans from PAH exposure is seen to be a complex mixture of currently available scientific data, assumptions and judgments based on prevailing studies. The process involves the identification of stressors, exposed host and stressor-host interactions.

Dealing with PAH, several members of the group have been chemically identified. The problem exists in determining the exact level of each compound either in the air, water, or food. Some investigators determined one member of the group, such as B(a)P and related toxicological and carcinogenic effects of other members as a ratio of the B(a)P. In dealing with these data, one has to consider the whole group, as was presented; for some may potentiate the effect of others in the biological system. Not only members of the group show these potentiating effects, but also unrelated compounds such as asbestos (18,83) and arsenic (100), could potentiate the effects of PAH. Emerging areas, such as more sensitive techniques to identify low levels of PAH, will have an impact on regulatory actions. Results of total risk assessment should be expressed in a way that a clear distinction is drawn between a chemical for which qualitative evidence is overwhelming, and a chemical for which the qualitative evidence is only marginal or unknown. It may be desirable to consider all members of the group using individual quantitative determinations (103).

Considering the host, toxicological evaluation for PAH compounds has dealt with in vitro bioassays, whole animal studies, and epidemiological data. In vitro studies include: gene mutation, chromosome effects, DNA damage, and transformation. These may be referred to as short-term studies. The consensus of available information suggests that short-term tests, if properly used, may provide strong indications of potential carcinogenicity. However, due to limited knowledge of the mechanisms of cancer induction, there are classes of compounds that will not be detected by available short-term tests. Recent discoveries of "negative" and "positive" markers may help in identifying deficient steps by this technique. Animal studies, or long-term carcinogen bioassay, deal with dose, frequency, diet, sex, and genetic-host deficiencies, as well as experimental design. The usual use of simple fractions of maximum tolerated doses (MTD) should be based on the results of adequate subchronic studies, pharmacokinetic data, and on anticipated human exposure (104). The interpretation of data as to whether a compound is carcinogenic or not, has been based on either of the following definitions. A substance can be considered a carcinogen if it significantly increases the incidence of cancer in animals or humans, or decreases significantly the time it takes a naturally-occurring (spontaneous) tumor to develop (105). Several statistical methods used in analysis of long-term tests, have been used, yet the questions of false negatives and positives, and the use of control, suggest that evaluation of long-term tests should not be a routine statistical exercise. This should be a multidisciplinary process with toxicologist, pathologist, and statistician interacting. Epidemiological data comprise one of the major strategies in creating the scientific base necessary for the regulatory decision-making. Case-control and cohort studies are useful in testing etiological hypotheses, and to identify and quantify carcinogenic risks to man. Intervention studies (or experimental studies) are especially useful in confirming causal relationships suggested by case-control or cohort studies. The drawback in epidemiological studies is that they deal with the morbidity and mortality after occurrence. These may be useful to establish preventative measures.

Stressor-host interactions are the end-result of the toxicological process. Most PAH members are carcinogenic, yet some may need bioactivation by several metabolic pathways. High-risk populations include changes which may stimulate the conversion of compounds to ultimate carcinogens, and hence DNA adduction and cell transformation. Several models have been used to assess host-stressor and stressor-host interactions at low-dose levels. Low-dose extrapolation techniques based on a tolerance-distribution model have been described. The Mantel-Bryan (106), one-hit (107), and the multistage model of Armitage and Doll (108), have been described. These have their limitations and once were popular, now their use for quantitative risk assessment has markedly declined. To decrease risk to the population, public health awareness by educational programs, as well as stricter legislative action to limit levels of exposure to PAH, are needed.

REFERENCES

1. Pott, F. and Oberdorster, G. (1983): Intake and distribution of PAH. In: Environmental Carcinogens: Polycyclic Aromatic Hydrocarbons, edited by Gernot Grimmer, pp. 130-136, CRC Press, Inc., Boca Raton, Florida.

2. Andrews, A.W., Thibault, L.H. and Lijinsky, W. (1978): The relationship between carcinogenicity and mutagenicity of some polynuclear hydrocarbons. Mutat. Res. 51: 311-315.
3. Wisloki, P.G., Juliana, M.M., MacDonald, J.S., Chou, M.W., Yang, S.K. and Lu, A.Y.H. (1981): Tumorigenicity of 7,12-dimenthylbenz(a)anthracene, its hydroxy methylated derivatives and selected dihydrodiols in newborn mouse. Carcinogenesis (London) 2 (6): 511-514.
4. Raikow, R.B., Okunewick, J.P., Buffo, M.J., Jones, D.L., Brozovich, B.J., Seeman, P.R. and Koval, T.M. (1981): Potentiating effect of benzo(a)pyrene and caffeine on Friend viral leukemogenesis. Carcinogeneses (London) 2 (1): 1-6.
5. Campbell, J., Grumplin, G.C., Garner, J.V., Garner, R.C., Martin, C.N. and Rutter, A. (1981): Nitrated polycyclic aromatic hydrocarbons: potent bacterial mutagens and stimulators of DNA repair synthesis in cultured cells. Carcinogenesis (London) 2 (6): 559-564.
6. Cavalier, E., Rogan, E., Toth, B. and Munhall, A. (1981): Carcinogenicity of the environmental pollutants cyclopenteno[cd]pyrene and cyclopentano[cd]-pyrene in mouse skin. Carcinogenesis (London) 2 (4): 277-281.
7. Doll, R. (1977): Environmental factors. In: Origin of Human Cancer, edited by H.H. Hiatt and J.D. Watson, pp. 1-9, Cold Spring Harbor Laboratory, London.
8. Pierce, W.E.H. (1961): Tumor promotion by lime oil in the mouse forestomach. Nature (London) 89: 497-501.
9. Dunn, B.P. (1982): Polycyclic aromatic hydrocarbons. In Carcinogens and Mutagens in The Environment, volume I, Food Products, edited by Hans, F. Stich, pp. 176-181, CRC Press Inc., Boca Raton, Florida.
10. Lo, M.T. and Sandi, E. (1978): Polycyclic aromatic hydrocarbons (polynuclear) in foods. Residue Rev. 69: 35-39.
11. Dunn, B.P. and Fee, J. (1979): Polycyclic aromatic hydrocarbon carcinogens in commercial seafood. J. Fish. Res. Board Can. 36: 1469-1474.
12. Wynder, E.L. and Hoffmann, D. (1967): Tobacco and Tobacco Smoke Studies in Experimental Carcinogenesis, pp. 1-30, Academic Press, London, New York.
13. International Agency for Research on Cancer (1972): IARC on the Evaluation of Carcinogenic Risk of Chemicals to Man, volume 1, pp. 9-36, Lyon.
14. International Agency for Research on Cancer (1973): IARC on the Evaluation of Carcinogenic Risk of Chemicals to Man, volume 3, pp. 7-45, Lyon.

15. Loyd, J.W. (1971): Long-term mortality study of steel-workers. V. Respiratory cancer in coke plant workers. J. Occup. Med. 13: 53-62.
16. Kawai, M., Amamoto, H. and Harada, K. (1967): Epidemiologic study of occupational lung cancer. Arch. Environ. Health, 14:859-862.
17. Kotin, P. and Falk, H.L. (1959): The role and action of environmental agents in the pathogenesis of lung cancer. I. Air pollutants: Cancer 12: 47-57.
18. Selikoff, I.J., Seidman, H. and Hammond, E.C. (1980): Mortality effects of cigarette smoking among amosite asbestos factory workers. J. Natl. Cancer Inst. 66: 507-513.
19. Woodworth, C., Mossman, B.T. and Craighead, J.E. (1983): Interaction of asbestos with metaplastic squamous epithelium developiing in organ cultures of hamster trachea. Environ. Health Perspect. 51: 27-33.
20. Wang, H., You, X. Qu, Y., Wang, W., Wang, D., Long, Y. and Ni, J. (1984): Investigation of cancer epidemiology and study of carcinogenic agents in the Shanghai rubber industry. Cancer Research 44: 3101-3105.
21. Berenblum, I. and Schoental, R. (1943): Carcinogenic constituents of shale oil. Brit. J. Expt. Path. 24: 232-237.
22. Harper, K.H. (1957): The carcinogenicity of benzpyrene metabolites. A.R. Brit. Emp. Cancer Campgn. 35: 151-157.
23. Harper, K.H. (1958): The carcinogenicity of benyo-pyrene metabolites. A.R. Brit. Emp. Campgn. 36: 180-185.
24. Flesher, J.W. and Syndor, K.L. (1973): Possible role of 6-hydroxymethylbenzo(a)pyrene as a proximate carcinogen of benzo(a)pyrene and 6-methylbenzo-(a)pyrene. Int. J. Cancer 11: 433-439.
25. Boyland, E. and Sims, P. (1965): The metabolism of benzo(a)anthracene and dibenz(a,h)anthracene and their 5,6-epoxy-5,6-dihydroderivatives by rat liver homogenates. Biochem. J. 97: 7-11.
26. Grover, P.L., Sims, P., Huberman, E., Marquardt, H., Kuroki, T. and Heidelberger, C. (1971): In vitro transformation of rodent cells by K- region derivatives of polycyclic hydrocarbons. Proc. Nat. Acad. Sci. (Washington) 689: 1096-1098.
27. Sims, P. (1970): Quanlitative and quantitative studies on the metabolism of a series of aromatic hydrocarbons by rat-liver preparations. Biochem. Pharmacol. 19: 795-799.

28. Scheutzle, D. and Perez, J.M. (1983): Factors influencing the emissions of nitrated-polynuclear aromatic hydrocarbons (nitro-PAH) from diesel engines. Journal of The Air Pollution Control Association 33 (8): 751-755.
29. Hamburg, F.C. (1969): Economically feasible alternatives to open burning in railroad freight car dismantling. Journal of The Air Pollution Control Association 19: 477-480.
30. Hangebrauck, R.P., VonLehmden, D.J. and Meeker, J.E. (1964): Emission of polynuclear hydrocarbons and other pollutants from heat generation and incineration. Journal of The Air Pollution Control Association 14: 267-275.
31. Walker, E.A. (1977): Some facts and legislation concerning polycyclic aromatic hydrocarbons in smoked foods. Pure Applied Chemistry 49: 1673-1675.
32. Conning, D.M. and Lasndown, A.B.G. (1983): Toxic Hazards in Food, pp. 127-161, Raven Press, New York.
33. Harvath, P.V. (1983): Quantitative analysis of multiple PAH's in the coal conversion atmosphere. Am. Ind. Hyg. Assoc. J. 44 (10): 739-745.
34. Greenberg, A., Bozzell, J.W., Cannova, F., Forstner, E., Giorgio, P., Stout, D. and Yokoyama, R. (1981): Correlations between lead and coronene concentrations at urban, suburban and industrial sites in New Jersey. Environmental Science and Technology 15 (5): 566-571.
35. Becher, G. and Bjorseth A. (1983): Determination of exposure to polyclic aromatic hydrocarbons by analysis of human urine. Cancer Letters 17: 301-311.
36. Obana, H., Hori, S., Kashimoto, T. and Kunita, N. (1981): Polycyclic aromatic hydrocarbons in human fat and liver. Bull. Environ. Contam. Toxicol. 27: 23-27.
37. Hutcheon, D.E., Kantrowitz, J., Van Gelder, R.N. and Flynn, E. (1983): Factors affecting plasma benzo(a)pyrene levels in environmental studies. Environmental Research 32: 104-110.
38. Hukkelhoven, M.W.A.C., Vromans, L.W.M., Vermorken, A.J.M., Van Diepen, C.B. and Bloemendal, H. (1982): Determination fo phenolic benzo(a)pyrene metabolites formed by human hair follicle. Analytical Biochemistry. 125: 370-375.
39. Coon, M.J., Haugen, D.A., Guengerich, F.P., Vermillion, J.L. and Dean, W.L. (1976): Liver microsomal membranes. In: The Structural Basis of Membrane Function, edited by Y. Hatef et al., pp. 409-420, Academic Press, New York.

40. Gram, T.E. (1973): Comparative aspects of mixed function oxidation by lung and liver of rabbits. Drug Metabolism Rev. 2: 1-8.
41. Jacob, J., Schmoldt, A. and Grimmer, G. (1981): Time course of oxidative benz(a)anthracene metabolism by liver microsomes of normal and PCB-treated rats. Carcinogenesis 2:395-399.
42. Lu, A.Y.H. and Levin, W. (1972): Partial purification of cytochrome P450 and P448 from rat liver microsomes. Biochem. Biophys. Res. Commun. 46: 1334-1339.
43. Lu, A.Y.H., Kuntzman, R., West, S., Jacobson, M. and Conney, A.H. (1972): Reconstituted liver microsomal enzyme system that hydroxylates drugs, other foreign compounds and endogenous substrates. J. Biol. Chem. 247: 1727-1735.
44. Bentley, P. and Oesch, F. (1975): Purification of rat liver epoxide hydratase to apparent homogeneity. FEBS Lett. 59: 291-295.
45. Knowles, R.G. and Burchell, B. (1977): A simple method for purificatioin of epoxide hydratase from rat liver. Biochem. J. 163: 381-385.
46. Lu, A.Y.H., Levin, W., Thomas, P.E., Jerina, D.M. and Conney, A.H. (1978): Enzymological properties of purified liver microsomal cytochrome P450 system and epoxide hydrase. In: Polynuclear Aromatic Hydrocarbons, edited by P.W. Jones and R.I. Freudenthal, pp. 243-248, Raven Press, New York.
47. Jacob, J., and Grimmer, G. (1983): Metabolism of polycyclic aromatic hydrocarbons. In: Environmental Carcinogens: Polycyclic Aromatic Hydrocarbons, edited by Gernor Grimmer, pp. 137-152, CRC Press Inc., Boca Raton, Florida.
48. Oesch, F. and Daly, J. (1972): Conversion of naphthalene to trans-naphthalene - dihydrodiol. Evidence for the presence of a coupled aryl monoxygenase-epoxide hydrase system in hepatic microsomes. Biochem. Biophys. Res. Commun. 46: 1713-1718.
49. Heimann, R. and Rice, R.H. (1983): Polycyclic aromatic hydrocarbon toxicity and induction of metabolism in cultivated esophageal and epidermal keratinocytes. Cancer Research 43: 4856-4862.
50. Boulos, B.M. (1985): Risk assessment to exposure to polynuclear aromatic hydrocarbons: an overview. In: Polynuclear Aromatic Hydrocarbons: Mechanisms, Methods and Metabolism, edited by Marcus Cooke and Anthony Dennis, pp. 215-225, Battelle Press, Columbus, Ohio.

51. Boulos, B.M. and VonSmolinski, A.W. (1984): New perspectives in evaluating health hazards to polynuclear aromatic hydrocarbons (PAH) exposure. Proceedings of the 1984 International Chemical Congress of Pacific Basin Societies 1: 01G08.
52. Becher, G., Haugen, A. and Bjoresth, Alf. (1984): Multimethod determination of occupational exposure to polycyclic aromatic hydrocarbons in an aluminum plant. Carcinogenesis (London) 5(5): 647-651.
53. Pitts, J.N. Jr., Van Cauwenberghe, K.A., Grosjean, D., Schmid, J.P., Fitz, D.R., Belser, W.L. Jr., Knudson, G.B. and Hynds, P.M. (1978): Atmospheric reactions of polycyclic aromatic hydrocarbons: Facile formation of mutagenic nitroderivatives. Science (Wash.) 202: 515-519.
54. Lofroth, G., Toftgard, R., Nilsson, L., Agurell, E. and Gustafsson, J.A. (1984): Short-term bioassays of nitroderivativesof benz(a)pyrene and perylene. Carcinogenesis (London) 5(7): 925-930.
55. Felton, J.S., Knize, M.C., Wood, C., Wuebbles, B.J., Healy, S.K., Stuermer, D.H., Bjeldanes, I.F., Kimble, B.J. and Hatch, F.T. (1984): Isolation and characterization of new mutagens from fried ground beef. Carcinogenesis (London) 5(1): 95-102.
56. Wang, H.W., You, X-j, Qu, Y-h, Wang, W-fu, Wang, D., Long, Y-m and Ni, J-a. (1984): Investigation of Cancer Epidemiology and Study of Carcinogenic Agents in the Shanghai Rubber Industry. Cancer Research 44: 3101-3105.
57. Venegas, W., Sala, M., Buisson, J., Royer, R. and Chouroulikov, I. (1984): Relationship between the chemical structure and the mutagenic and carcinogenic potentials of five naphtho-furans. Cancer Research 44: 1969-1975.
58. Wood, A.W., Chang, R.L., Levin, W., Thakker, D.B., Yagi, H. and Sayer, J.M. (1984): Mutagenicity of the enantiomers of the diastereomeric bay-region benzo(c)phenanthrene3,4-diol-1,2-epoxides in bacterial and mammalian cells. Cancer Research 44: 2320-2324.
59. Briant, C.E. and Jones, D.W. (1984): The molecular structure of 11-methylbenz(a)anthracene: an almost planar substituted benz(a)anthracene. Carcinogenesis (London) 5(3): 363-365.
60. Norpoth, K., Kemena, A., Jacob, J. and Schumann, C. (1984): The influence of 18 environmentally relevant polycyclic aromatic hydrocarbons and Clophen A50, as liver monoxygenase inducers, on the mutagenic activity

of benz(a)anthracene in the Ames test. Carcinogenesis (London) 5(6): 747-752.

61. Malaveille, C., Bartsch, H., Grover, P.L. and Sims, P. (1975): Mutagenicity of Benz(a)anthracene and Benzo(a)-pyrene. Biochem. Biophys. Res. Commun. 66: 693-700.
62. Gelboin, H.V. (1980): Benzo(a)pyrene metabolism, activation and carcinogenesis: role and regulation of mixed function oxidases and related enzymes. Physiol. Rev. 60: 1107-1166.
63. Robertson, J.A., Nordenskjold, M. and Jernstrom, B. 1984): The identification of bases in DNA involved in covalent binding of the reactive metabolite from 9-hydroxybenzo(a)pyrene. Carcinogenesis (London) 5(6): 821-826.
64. Walaszek, Z., Hanausek-Walaszek, M. and webb, T.E. (1984): Inhibition of 7,12-dimethylbenzanthracene-induced rat mammary tumorigenesis by 2,5-di-O-actyl-D-glucaro-1,4:6,3-dilactone, an in-vivo ß-glucaronidase inhibitor. Carconogensis (London) 5(6): 767-772.
65. Marchok, A.C., Huang, S.F. and Martin, D.H. (1984): Selection of carcinogen-altered rat tracheal epithelial cells preexposed to 7,12-dimethylbenz(a)anthracene by their loss of a need for pyruvate to survive in culture. Carcinogenesis (London) 5(6): 789-796.
66. Moller, M.E., Glowinski, I.B. and Thorgeirsson, S.S. (1984): The genotoxicity of aromatic amines in primary hepatocytes isolated from C57BL/6 and DBA/2 mice. Carcinogenesis (London) 5(6): 797-804.
67. Deffie, A.M. and LeJohn, H.B. (1984): Involvement of protease in ℓ-glutamine control of nucleic acid and polyphosphate metabolism in cells transformed by 9,10-dimethyl-1,2-benzanthracene, SV40 and H-ras oncogene. Biochem. Biophys. Res. Commun. 124(1): 6-13.
68. Feo, F., Pirisi, L., Pascale, R., Daino, L. Frassetto, S., Carcea, R. and Gaspa, L. (1984): Modulatory effect of glucose-6-phosphate dehydrogenese deficiency on benzo(a)pyrene toxicity and transforming activity for in-vitro-cultured human skin fibroblasts. Cancer Research 44: 3419-3425.
69. EL. Bayoumy, K. and Hecht, S.S. (1984): Identification of trans-1,2-dihydro-1,2-dihydroxy-6-nitrochrysene as a major mutagenic metabolite of 6-nitrochrysene. Cancer Research 44: 3408-3413.
70. Ohgaki, H. Negishi, C., Wakabayashi, K., Kusama, K., Sato, S. and Sugimura, T. (1984): Induction of sarcomas in rats by subcutaneous injection of dinitropyrenes. Carcinogenesis (London) 5(5): 583-585.

71. Munir, K.M., Rao, K.V.K., and Bhide, S.V. (1984): Effect of hexchlorocyclohexane on diethylnitrosamine-induced hepatocarcinogenesis in rat and its failure to promote skin tumors on dimethylbenz(a)anthracene initiation in mouse. Carcinogenesis (London): 5(4): 479-481.
72. Pashko, L.L., Rovito, R.J., Williams, J.R., Sobel, E.L., and Schwartz, A.G. (1984): Dehydroepiandrosterone (DHEA) and 3-methylandrost-5-en-17-one: inhibitors of 7,12-dimethylbenz(a)anthracene(DMBA)-initiated and 12-0-tetradecanoyl-phorbol-13-acetate-(TPA)-promoted skin papilloma formation in mice. Carcinogenesis (London) 5(4): 463-466.
73. Beems, R.B. and van Beek, L. (1984): Modifying effect of dietary fat on benzo(a)pyrene-induced respiratory tract tumors in hamsters. Carcinogenesis (London) 5(3): 413-417.
74. Nishino, H., Iwashima, A., Na Kadate, T., Kato, R., Fujiki, H. and Sugimura, T. (1984): Potent antitumor promoting activity of N-(6-aminohexyl)-5-chloro-1-naphalenesulfonamide, a calmodulin antagonist, in mouse skin tumor formation induced by 7,12-dimethylbenzo(a)-anthracene plus teleocidin. Carcinogenesis (London) 5(2): 283-285.
75. Ito, Y., Maeda, S. Souno, K., Ueda, N. and Suglyama, T. (1984): Induction of hepatic glutathione transferase and suppression of 7,12-dimethylbenz(a)anthracene-induced chromosome aberrations in rat bone marrow cells be Sudan III and related azo dyes. J. Natl. Cancer Instit. 75(1): 177-183.
76. Moore, C.J. and Gould, M.N. (1984): Differences in mediated mutagenesis and polycyclic aromatic hydrocarbon metabolism in mammary cells from pregnant and virgin rats. Carcinogenesis (London) 5(1): 103-108.
77. Busby, W.F. Jr., Goldman, M.E., Newberne, P.M. and Wogan, G.N. (1984): Tumorigenicity of fluoranthene in a newborn mouse lung adenoma bioassay. Carcinogenesis (London) 5(10): 1311-1316.
78. Ueda, N. and Kondo, M. (1984): Chromosome aberrations induced by 7,12-dimethylbenz(a)anthracene in bone marrow cells of spontaneously hypersensitive rats (SHR) and control Wistar Kyoto (WKY) rats: time course and site specificity. J. Natl. Cancer Instit. 73(2): 525-528.
79. DiGiovanni, J., Singer, J.M. and Diamond, L. (1984): Comparison of the metabolic activation of 7,12-dimethylbenz(a)anthracene by a human hepatoma cell line (HEPG2)

and low passage hamster embryo cells. Cancer Research, 44:2878-2884.
80. Urso, P. and Gengozian, N. (1984): Subnormal expression of cell-mediated and humoral immune responses in progeny disposed toward a high incidence of tumors after in-utero exposure to benzo(a)pyrene. J. of Toxicology and Environmental Health, 14:569-584.
81. Hose, J.E., Hannah, J.B., Puffer, H.W. and Landolt, M.L. (1984): Histologic and skeletal abnormalities in benzo(a)pyrene-treated rainbow trout alevins. Arch. Environ. Contam. Toxicol. 13: 675-684.
82. Tucker, J.D. and Ong, Tong-man (1984): Induction of sister-chromatid exchange and chromosome aberrations in human peripheral lymphocytes by 2,4,7-trinitro-9-fluorenone. Mutation Research 138: 181-184.
83. Mossman, B.T., Eastman, A. and Bresnick, E. (1984): Asbestos and benzo(a)pyrene act synergistically to induce squamous metablasia and incorporation of [^{3}H] thymidine in hamster tracheal epithelium. Carcinogenesis (London) 5(11): 1401-1404.
84. Lawrence, C.M., Finnen, M.J. and Shuster, S. (1984): Effect of coal tar cutaneous aryl hydrocarbon hydroxylase induction and anthralin irritancy. British J. of Dermatology 110: 671-675.
85. Schaefer-Ridder, M., Moeroey, T. and Englhardt, U. (1984): Inactivation of the thymidine kinase gene after in vitro modification with benzo(a)pyrene-diol-epoxide and transfer to LTK cells as a eukaryotic test for carcinogens. Cancer Research 44: 5861-5866.
86. Rice, J.E., Hosted, T.J. Jr. and Lavoie, E.J. (1984): Fluoranthene and pyrene enhance benzo(a)pyrene-DNA adduct formation in vivo in mouse skin. Cancer Letters 24: 327-333.
87. White, K.L. and Holsapple, M.P. (1984): Direct suppression of in-vitro antibody production by mouse spleen cells by the carcinogen benzo(a)pyrene but not by the noncarcinogen benzo(e)pyrene. Cancer Research 44: 3388-3393.
88. Wasilneko, W.J. and Marchock, A.C. (1984): Hexose uptake in 7,12-dimethylbenz(a)-anthracene-preexposed rat tracheal epithelial cells during the progression of neoplasia. Cancer Research 44: 3081-3089.
89. Goldman, P.R. and Vogel, W.H. (1984): Striatal dopamine-stimulated adenylate cyclose activity reflects susceptibility of rats to 7,12-dimethylbenz(a)anthracene-induced mammary tumor development. Carcinogenesis (London) 5(7): 971-973.

90. Nemoto, N. and Takayama, S. (1984): Arachidonic acid-dependent activation of benzo(a)pyrene to bind to proteins with cytosolic and microsomal fractions from rat liver and lung. Carcinogensis (London) 5(7): 961-964.
91. Reed, A.G., Grafstrom, R.C., Krauss, R.S., Autrup, H. and Eling, T.E. (1984): Prostaglandin H synthase-dependent co-oxygenation of (±)-7,8-dihydroxy-7,8-dihydrobenzo(a)pyrene in hamster trachea and human bronchus explants. Carcinogenesis (London) 5(7): 955-960.
92. Katayama, S. Ohmori, T., Maeura, Y., Croci, T. and Williams, G.M. (1984): Early Stages of N-2-fluorenylacetamide-induced hepatocarcinogenesis in male and female rats and effect of gonadectomy on liver neoplastic conversion and neoplastic development. J. Natl. Cancer Instit. 75(1): 141-149.
93. Levin, M., Pandya, K.J., Khandekar, J.D., Horton, J., Glick, J.H., Bennett, J.M., Muggia, F.M. and Falkson, G. (1984): Phase II study of mitoxantrone in advanced breast cancer: An estern cooperative oncology group pilot study. Cancer Treatment Reports 68(12): 1511-1519.
94. Arun, B., Udayachander, M. and Meenakshi, A. (1984): 7,12-Dimethylbenzanthracene induced mammary tumors in Wistar rats by "air-pouch" technique - a new approach. Cancer Letters 24: 187-194.
95. Jongeneelen, F.J., Leijdekkers, C.M. and Henderson, P. TH. (1984): Urinary excretion of 3-hydroxy-benzo(a)-pyrene after percutaneous penetration and oral absorption of benzo(a)pyrene in rats. Cancer Letters 25: 195-201.
96. Blakey, D.H. and Douglas, G.R. (1984): Transient DNA lesions induced by benzo(a)pyrene in chinese hamster ovary cells. Mutation Research 140: 141-145.
97. Wei, S.C., Desai, S.M., Harvey, R.G. and Weiss, S.B. (1984): Use of short DNA oligonucleotides for determination of DNA sequence modificatioins induced by benzo(a)pyrene diol epoxide. Proc. Natl. Acad. Sci. (U.S.A.) 81: 5936-5940.
98. Paschin, Y.V. and Bahitova, L.M. (1984): Inhibition of the mutagenicity of benzo(a)pyrene in the V79/HGPRT system by bioantioxidants. Mutation Research 137: 57-59.
99. Dock, L., Martinez, M. and Jernstrom, B. (1984): Induction of hepatic glutathione s-transferase activity by butylated hydroxyanisole and conjugation of benzo(a)-

(a)pyrene diol-epoxide. Carcinogenesis (London) 5(6): 841-844.
100. Pershagen, G., Nordberg, G. and Bjorklund, N. (1984): Carcinomas of the respiratory tract in hamsters given arsenic trioxide and/or benzo(a)pyrene by the pulmonary route. Environmental Research 34: 227-241.
101. Boulos, B.M., MacDougall, M., Shoeman, D.W. and Azarnoff, D.L. (1972): Evidence that inhibition of hepatic drug oxidation by tumors is mediated by a circulating humor. Proc. Soc. Exp. Biol. Med. 139: 1353-1357.
102. Bloom, G.S., Luca, F.C. and Vallee, R.B. (1985): Identification of high molecular weight microtubule - associated proteins in anterior pituitary tissue and cells using Taxol-dependent purification combined with microtubule-associated protein specific antibodies. Biochemistry 24: 4185-4191.
103. Squire, R.A. (1981): Ranking animal carcinogens. A proposed regulatory approach. Science 214: 877-880.
104. Dayton, P.G. and Sanders, J.E. (1983): Dose-dependent pharmacokinetics: Emphasis on Phase I metabolism. Drug Metabol. Rev. 14: 347-408.
105. Food safety council (1978): Chronic toxicity testing. Proposed system for food safety assessment. Food Cosmet. Toxicol. 16(suppl. 2): 97-108.
106. Mantel, N., Bohidar, N.R., Brown, C.C., Ciminera, J.L. and Tukey, J.W. (1975): An improved "Mantel-Bryan" procedure for "safety testing" of carcinogens. Cancer Research 35: 865-872.
107. Krewski, D., Van Ryzin, J. (1981): Dose response models for quantal response toxicity data. In: Statistics and Related Topics edited by Caorgo, D., Dawson, R. and Saleh, J.N.K. pp. 201-231, Amsterdam, North Holland.
108. Armitage, P. and Doll, R. (1981): Stochastic models for carcinogenesis. In: Proceedings of the Fourth Berkely Symposium on Mathematical Statistics and Probability, edited by Lecam, L. and Neyman, J. pp. 419-438, University of California Press, Berkley, California.

INHIBITION OF DNA POLYMERASE ALPHA BY PAH

DAVID BUSBEE*, CHEOL JOE, JAMES NORMAN[1], VICTOR SYLVIA
Departments of Anatomy and of Physiology and Pharmacology, College of Veterinary Medicine, Texas A&M University, College Station, TX 77843 and (1) Veterinary Toxicology and Entomology Research Laboratory, USDA, College Station, TX 77841.

INTRODUCTION

Chemical modification of DNA by the formation of DNA-mutagen adducts has been widely quantitated and accepted to be a primary event leading to initiation of a mutation which may ultimately result in cell transformation. DNA-adduct formation has also been generally accepted to inhibit DNA synthesis, constituting a, if not the, major mechanism of cytotoxicity common to the reactive hydrocarbon carcinogens (1,2). While there is little doubt that decreased DNA synthesis results from inhibition of DNA polymerase activity at the site of DNA adducts produced by compounds such as N-acetylaminofluorene (3,4), there is a growing body of information suggesting that direct alteration of polymerases by reactive chemicals may also result in significant inhibition of DNA synthesis. Cleaver reported that methyl methanesulfonate (MMS) inhibition of UV-light-initiated DNA excision repair and MMS-modification of DNA are separate phenomena, and that MMS inhibition of UV-initiated excision repair apparently occurs subsequent to the repair-associated endonuclease nicking step (5-7). DNA nucleotidyltransferase, EC 2.7.7.7 (DNA pol-α), is a major enzyme of DNA strand elongation, has been proposed to function in excision repair subsequent to oligonucleotide hydrolysis by endo- and exonucleases (8,9,10), and is a logical candidate for modification resulting in inhibition of DNA synthesis. Aphidicolin, a tetracyclic diterpene tetraol mycotoxin, inhibits DNA pol-α -associated excision repair subsequent to the endonuclease nicking step by binding to the enzyme and competitively inhibiting incorporation of dCTP into the growing DNA strand (9-14). Busbee et al. (15) reported that 7,8-dihydroxy-9,10-epoxy-7,8,9,10-tetrahydrobenzo(a)pyrene (BPDE) inhibition of DNA excision repair appears similar to that of aphidicolin in that BPDE binds to DNA pol-α resulting in the competitive inhibition of dGTP utilization and of DNA synthesis by the enzyme.

DNA pol-α has been extensively investigated and shown to be a major eukaryotic DNA elongation enzyme apparently functioning in both scheduled and unscheduled DNA synthesis (8-20). In this paper we present the results of an investigation of DNA pol-α inhibition by a variety of DNA-interactive and non-interactive chemicals known to be immunosuppressive and to inhibit DNA excision repair.

MATERIALS AND METHODS

Materials

Unlabeled 2'-deoxynucleoside 5'-triphosphates, corticosteroids, and proteinase K were purchased from Sigma. Tritiated deoxynucleotides were obtained from ICN Chemical and Radioisotope Division and are as follows: ^{3}H-dTTP (20 Ci/mmole); ^{3}H-dCTP (30 Ci/mmole); ^{3}H-dATP (9 Ci/mmole); and ^{3}H-dGTP (5 Ci/mmole). Benzo(a)pyrene, benz(a)anthracene, and methyl methanesulfonate were obtained from Aldrich Chemical Co. The metabolites of benzo(a)pyrene and benz(a)anthracene, 7,8-dihydroxy-9,10-epoxy-7,8,9,10-tetrahydrobenzo-(a)pyrene (BPDE), 7,8-dihydroxy-7,8dihydrobenzo(a)pyrene (BPDO), 8,9-dihydroxy-10,11-epoxy-8,9,10,11-tetra- hydrobenz(a)anthracene (BADE), and 8,9-dihydroxy-8,9-dihydrobenz(a)anthracene (BADO) were obtained from the Division of Cancer Cause and Prevention, National Cancer Institute, National Institutes of Health, Bethesda, MD. BPDE was obtained either unlabeled or tritiated (403 mCi/mmole). Labeled MMS (^{3}H-MMS, 58 Ci/mmole) was obtained from Amersham International. Aphidicolin was a gift from Imperial Chemical Industries, England. Phytohemagglutinin, HA16, was obtained from Wellcome Reagents Ltd., Beckenham, England.

DNA Polymerase Isolation

Deoxyribonucleic acid nucleotidyltransferase, EC 2.7.7.7, DNA pol-α was purified using a modification of the method of Mechali et al. (21). To each sample of approximately 10^{8} packed lymphocytes or mouse sarcoma cells, 1 ml of buffer A containing 1 mM dithiothreitol, 0.25 M sucrose, 50 mM Tris-HCl (pH 8.0), 25 mM KCl, and 7.5 mM $MgCl_2$ was added. Cells were disrupted by 35 strokes of a Teflon pestle in a glass homogenizer. The homogenate and homogenizer wash were pooled and centrifuged at 35,000 rpm in a Beckman 56 rotor for 1 hr. Finely ground aluminum ammonium sulfate (0.4 g/ml) was added to the supernatant, and the mixture was stirred for 20 min and centrifuged at 24,000 x g for 25 min. The precipitate was resuspended in 100 ml of buffer A with 20% glycerol and 0.22 g ammonium sulfate/ml, and centrifuged at 24,000 x g for 25 min. The precipitate was washed in buffer A containing, in succession, 0.19, 0.17, 0.14, 0.12, 0.10, and 0.08 g ammonium sulfate/ml. The supernatant collected from each wash was reprecipitated with ammonium sulfate (70% saturation) for 1 hr at 4° C and centrifuged at 30,000 x g for 20 min. The precipitates were independently dissolved in buffer A and dialyzed against buffer A in 30% glycerol. Each dialyzed fraction was assayed for enzyme activity, the active fractions were pooled, concentrated if necessary through an XM-50 ultrafiltration membrane (Amicon Corp), and applied to a DEAE-cellulose column equilibrated with buffer A. The column was eluted with 0.02 to 0.04 M K_2HPO_4/KH_2PO_4 (pH 7.5) with 20 %

glycerol over 24 h, with collection of 200 x 5 min fractions. DNA pol-α eluted just before the phenol red tracking dye. Active fractions from the first column were applied to a phosphocellulose column (1.5 x 18 cm) equilibrated in Buffer B (50 mM Tris-HCl (pH 7.5), 5 mM KCl, 5 mM DTT, 0.1 mM EDTA, 20% glycerol). The column was eluted with a salt gradient of 0.15 to 0.4 M NaCl in buffer B over an 8 hr period. Active fractions from the phosphocellulose column were concentrated, applied to a DNA-cellulose affinity column, 1.5 x 3 cm, and eluted with 0.1 to 0.4 M KCl in buffer B over an 8 hr period with collection of 5 ml fractions. Active fractions from the DNA-cellulose column were concentrated by membrane filtration as given above and subjected to polyacrylamide gel electrophoresis for analysis of homogeneity (21).

Cell Culture and Lymphocyte Preparation

CFW mouse sarcoma cells were maintained in monolayer culture in Dulbecco's modified Eagles Minimal Essential Medium supplemented with 10% fetal calf serum. Logarithmically growing cells were synchronized with 1 mM hydroxyurea (HU) for 16 hr, and harvested in S phase 3 hr after removal of HU. Peripheral lymphocytes (PBL) from venous blood were prepared using the Ficoll-Hypaque centrifugation technique of Rankin et al (22). Cells were counted in a Neubauer hemocytometer, and dispensed into 30 mm culture dishes at 1-2 x 10^6 cells/ml. Lymphocytes to be mitogen-stimulated received phytohemagglutinin, 50 ng/ml (PHA) at this time. Cells were diluted to a concentration of 1-2 x 10^6 cells/ml in PBS for treatment with test compounds dissolved in dimethylsulfoxide (0.5% maximum final DMSO conc.).

^{3}H-Thymidine Incorporation

^{3}H-Thymidine (methyl-^{3}H, 1.22 x 10^{14} dpm/mmole, 0.0182 mM; ICN Chemical and Radioisotope Division), was added at 1 uCi/ml to PHA-treated cells. The cell suspension was divided into 1 ml aliquots and incubated at 37° C under 5% CO_2. Each sample group was examined in triplicate. Hydroxyurea, 3 mM, was added to preclude scheduled DNA synthesis. After completion of treatment periods, aliquots were removed from each sample and cells were examined for Trypan Blue exclusion and adenylate charge (23) as measures of viability. Cell preparations with Trypan Blue exclusion above 95%, and with an adenylate charge between 0.5 and 0.8, were considered to show no cytotoxicity. Cell samples were centrifuged at 700 x g for 3 min, the supernatants were aspirated, pellets were resuspended in PBS and stored at -70° C in 1 ml aliquots. Within 48 h of freezing, cell samples were thawed, incubated with protease (0.1 mg/ml of Sigma Type XIV), and radiometrically assessed (24,25).

Assay For DNA Polymerase Activity

Each 0.2 ml of reaction buffer for DNA pol-α assays contained 50 mM Tris-HCl (pH 8.0), 7.5 mM $MgCl_2$, 0.5 mM dithiothreitol, 70 ug of bovine serum albumin (A grade Sigma), 20 ul of enzyme preparation, 100 ug DNA template/primer, and all four deoxynucleoside triphosphates (dNTP) as specified, of which only one was labeled. Different enzyme preparations varied in specific activity determined as a function of ^{3}H-dTMP incorporation into 0.1 mg DNA template/20 ul of enzyme preparation (diluted to contain 0.01 mg of total enzyme protein/20 ul). The pol- activity determined for a large preparation of enzyme on which the MMS study was completed was 8.91 x 10^3 dpm ^{3}H-dTMP incorporated/0.01 mg of enzyme. The reaction mixture was incubated with or without inhibitory compounds (added in 2 ul of DMSO), at 37° for 1 hr. The reaction was stopped by adding 5 ml of cold 10% TCA, and ^{3}H-2'-deoxyribonucleoside 5'-monophosphate incorporation into acid insoluble material was radiometrically determined (24).

RESULTS

Human lymphocytes (PBL) treated with BPDE were assessed for ^{3}H-Tdr incorporation to determine DNA synthesis as a function of (1) blastogenesis in mitogen-stimulated cells, and (2) UDS in non-mitogen-activated PBL. BPDE inhibition of DNA pol-α activity in non-mitogen-activated PBL closely correlated with decreased blastogenesis in mitogen-activated cells (Fig. 1), suggesting that direct enzyme inhibition by BPDE may be associated with decreased blastogenesis in mitogen-activated PBL. To examine the relative differences in incorporation of ^{3}H-Tdr into DNA containing different concentrations of BPDE adducts, DNA from PBL was isolated, treated with BPDE at varying concentrations, enzymatically activated to produce template/primer, and the concentration of BPDE adducts in each template/primer sample was quantitated (Fig. 2A). DNA template containing unlabeled BPDE was used with purified pol-α to assess the capacity of pol-α to utilize adducted DNA as substrate. DNA pol-α incorporation of dTMP into substrate DNA did not differ between templates containing 0, 0.95, and 1.84 pmoles of BPDE/ug of substrate DNA (Fig. 2B). These data could be interpreted to suggest that BPDE adducts did not inhibit *in vitro* pol-α activity, or that there were not enough adducts in the template DNA to significantly inhibit pol-α activity, or that pol-α, encountering an adduct, dissociated from the template and bound at another available template primer site. To address this point, an examination of the kinetics of pol-α inhibition by BPDE was completed varying the DNA template concentration. These data (Fig. 3) show Lineweaver-Burk plots consistent with uncompetitive inhibition in which the inhibitor does not combine with free enzyme or affect its normal

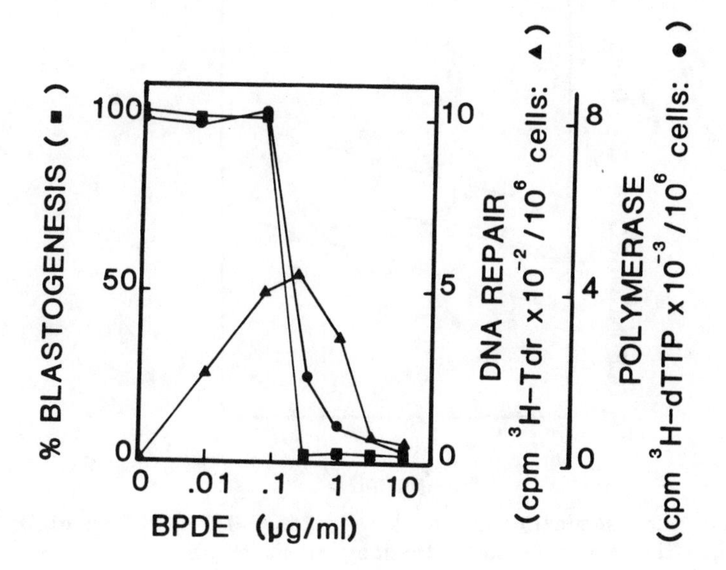

FIGURE 1. An examination of the BPDE inhibition of blastogenesis in PHA-treated lymphocytes corresponding to inhibition of unscheduled DNA synthesis and DNA polymerase in non-mitogen-stimulated cells.

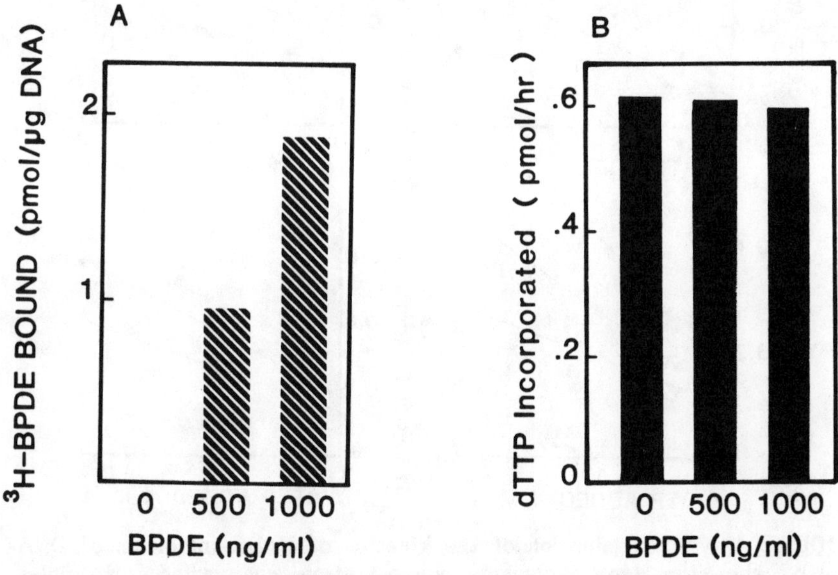

FIGURE 2. (A) Quantitation of BPDE adducts in DNA treated with 0, 500, and 1,000 ng of BPDE/ml. (B) Incorporation of ^{3}H-dTTP into DNA template containing three BPDE adduct levels (as per in A).

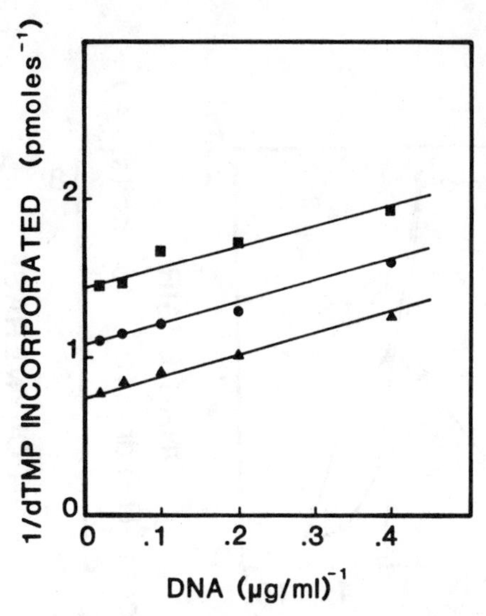

FIGURE 3. an examination of the kinetics of BPDE inhibition of DNA pol-α when the substrate concentration was limiting.

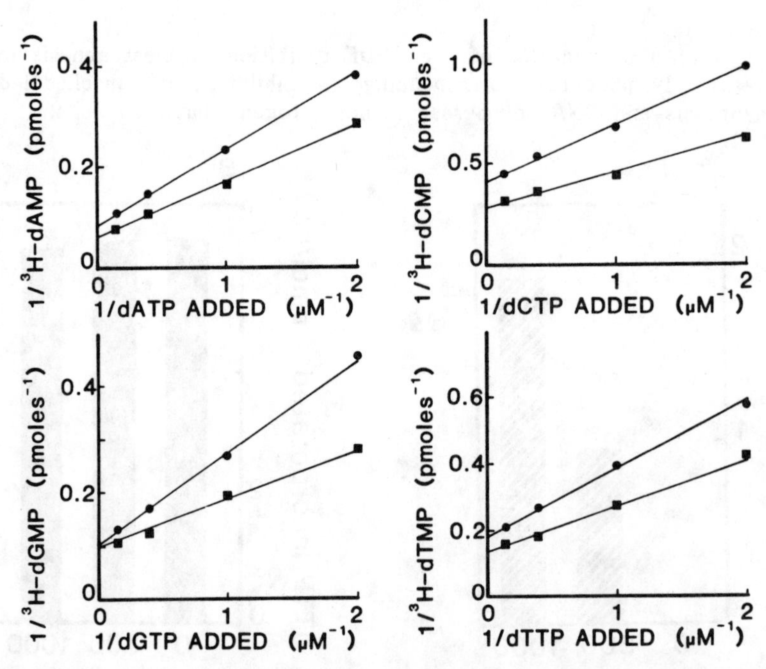

FIGURE 4. An examination of the kinetics of BPDE inhibition of DNA pol-α when the deoxynucleoside concentration was varied. DNA pol-α activity is expressed as a function of ^{3}H-dNMP incorporated into DNA template (^{3}H-dNMP/0.01 mg enzyme/hr).

reaction with the substrate. Since DNA pol-α follows an ordered two substrate reaction sequence (26) in which template serves as the initial substrate, these data suggested that inhibition of the enzyme was not substrate-related. Kinetic studies in which the second substrate was varied indicated that dATP, dTTP, and dCTP incorporation were each noncompetitively inhibited, but that dGTP incorporation was competitively inhibited (Fig. 4), suggesting that BPDE inhibition of DNA pol-α activity was, under these circumstances, due to direct interaction of the inhibitor with the enzyme. When pol-α was treated with BPDE prior to association of the enzyme with DNA template, BPDE bound to the enzyme but was not inhibitory. These data are consistent with the suggestion that BPDE does not bind a critical vnzyme site until after the enzyme binds an initial DNA template, after which the inhibitor may bind to a site inhibiting enzymatic activity.

A number of PAH structurally related to BaP are known to bind DNA and to inhibit DNA synthesis. We examined the effects of a reactive benz(a)anthracene (BA) metabolite on pol-α to determine whether a PAH structurally similar to BaP exerts similar inhibitory effects on DNA synthesis and on pol-α activity. When PBL were treated with BADE, but not with BADO or BA, the PAH inhibited both ^{3}H-Tdr incorporation as a function of blastogenesis and pol-α activity in non-mitogen-stimulated cells (Fig. 5). An examination of the kinetics of pol-α inhibition by BADE showed noncompetitive inhibition when

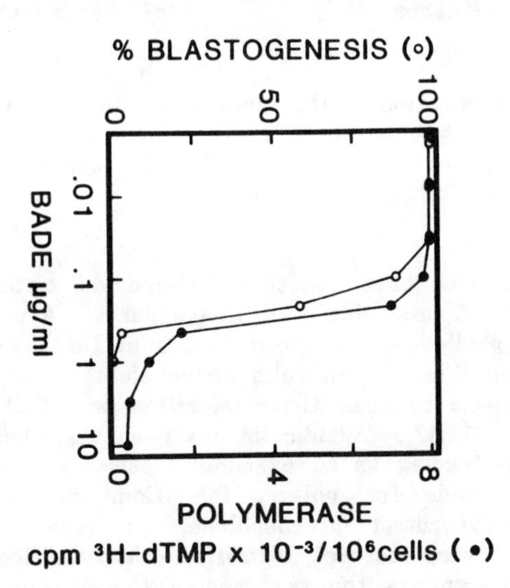

FIGURE 5. An examination of BPDE inhibition of blastogenesis and of DNA pol-α activity in PHA-stimulated in human lymphocytes. Lymphocytes were PHA-treated in RPMI-1640.

dATP, dTTP, and dCTP were varied, and competitive inhibition when dGTP was varied (Fig. 6). These data suggested that BADE inhibits DNA synthesis in the same manner as does BPDE, by the competitive inhibition of dGMP incorporation into DNA template.

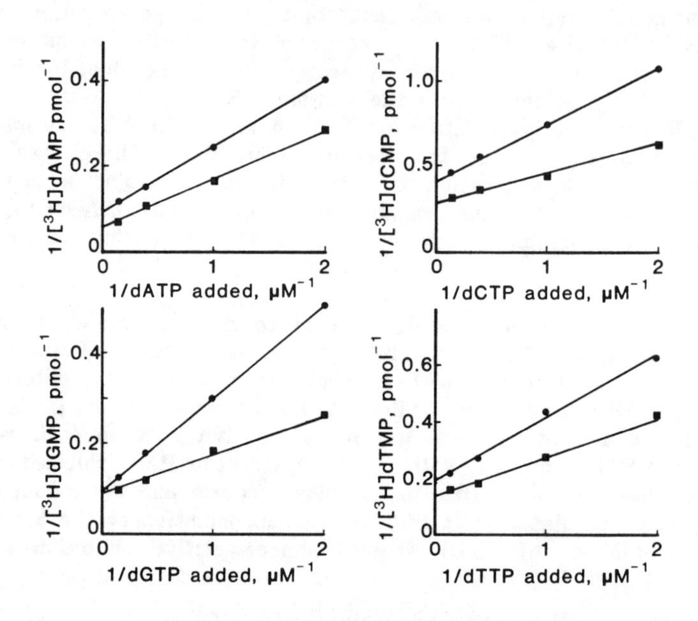

FIGURE 6. An examination of the kinetics of BADE inhibition of DNA pol-α when the deoxynucleoside concentration was varied. DNA pol-α activity is expressed as a function of ^{3}H-dNMP incorporated into DNA template (^{3}H-dNMP/0.01 mg enzyme/hr).

BaP and BA both have reactive diol-epoxide metabolites which covalent bind to DNA producing PAH-DNA adducts. A number of compounds, such as aphidicolin, are known to inhibit DNA synthesis but do not covalently bind DNA. Aphidicolin noncovalently binds pol-α , decreasing DNA synthesis by competitive inhibition of dCMP incorporation into DNA template (11,12). Aphidicolin has been described as "steroid-like" in structure, leading us to examine a series of the immunosuppressive corticosteroids for pol- α inhibition. An examination of progesterone, its precursor pregnenolone, and five corticosteroids synthesized from progesterone identified deoxycorticosterone and 11-dehydrocorticosterone as the two most DNA-synthesis inhibitory of the corticosteroids (Table 1). An analysis of pol-α inhibition by

TABLE 1

STEROID INHIBITION OF dTMP INCORPORATION INTO DNA TEMPLATE-PRIMER IN VITRO

Adrenal Steroids	Steroid Concentration ($M \times 10^{-6}$)	Percent of Control Activity
deoxycorticosterone	10	80%
	30	64%
	100	62%
11-dehydrocorticosterone	10	81%
	30	67%
pregnenolone	10	84%
	30	68%
cortisone	10	92%
	30	80%
corticosterone	10	95%
	30	86%
progesterone	10	97%
	30	92%
cortisol	10	100+%
	30	100+%

deoxycorticosterone produced Lineweaver-Burk plots showing dATP, dGTP, and dTTP incorporation into template to be noncompetitively inhibited by the steroid, while dCTP incorporation was competitively inhibited (Fig. 7). When pol-α was treated with deoxycorticosterone and the steroid was removed prior to addition of DNA substrate, no inhibition of dNTP incorporation was observed.

These data established the probability that different classes of pol- α -inhibitory compounds, some of which do not form DNA adducts, may be similar in their modes of inhibition of the enzyme. An examination of yet another DNA synthesis-inhibitory chemical, MMS, produced somewhat different data than that seen for BaP, BA, or the corticosteroids. PBL treated with MMS showed a reduction in mitogen-stimulated ^{3}H-Tdr incorporation coincident with inhibition of pol-α activity (Fig. 8). DNA pol-α treated with MMS prior to addition of template was significantly methylated, and showed decreased enzyme activity. This differs from BPDE, BADE, and steroid inhibition of enzyme activity, which were dependent on enzyme binding to DNA template prior to interaction of inhibitor with the enzyme. When MMS-treated pol-α was reacted with control DNA template, an MMS concentration-dependent inhibition of ^{3}H-dTMP incorporation into template DNA was observed (Table 2:A). When control pol-α was

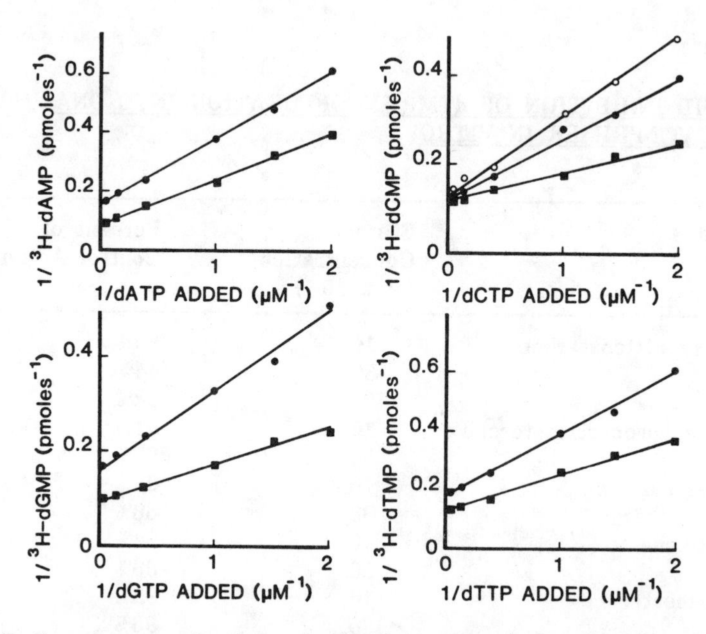

FIGURE 7. An examination of the kinetics of deoxycorticosterone inhibition of DNA pol-α when the deoxynucleoside concentration was varied. DNA pol-α activity is expressed as a function of ^{3}H-dNMP incorporated into DNA template (^{3}H-dNMP/0.01 mg enzyme/hr).

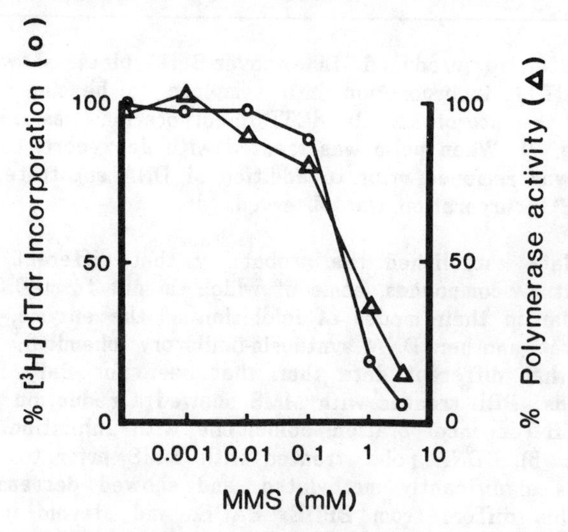

FIGURE 8. An examination of MMS inhibition of PHA-stimulated blastogenesis and DNA pol-α activity in human lymphocytes. Blastogenesis was measured as ^{3}H-Tdr incorporation. DNA pol-α was measured as ^{3}H-dTMP incorporated/0.01 mg enzyme/hr.

TABLE 2

INTERACTION OF MMS-MODIFIED OR UNMODIFIED DNA WITH MMS-MODIFIED OR UNMODIFIED DNA POLYMERASE ALPHA

Enzyme + DNA Template			^{3}H-dTMP incorporated / mg of enzyme protein
Constant		Variable	(percent of control)
A	control DNA	control enzyme	100.0 % +/- 0.4
	control DNA	MMS (1 mM) enzyme *	41.5 % +/- 5.4
	control DNA	MMS (10 mM) enzyme *	3.7 % +/- 0.9
B	control enzyme	control DNA	100.0 % +/- 0.4
	control enzyme	MMS (1 mM) DNA	103.1 % +/- 6.9
	control enzyme	MMS (10 mM) DNA	99.6 % +/- 4.7
C	MMS (1 mM) enzyme*	control DNA	41.5 % +/- 5.4
	MMS (1 mM) enzyme*	MMS (1 mM) DNA	17.3 % +/- 3.2
D	MMS (1 mM) enzyme**	control DNA	100.0 % +/- 5.9
	MMS (1 mM) enzyme **	MMS (1 mM) DNA	47.1 % +/- 7.2
	MMS (10 mM) enzyme**	control DNA	1.9 % +/- 0.4
	MMS (10 mM) enzyme**	MMS (1 mM) DNA	0.3 % +/- 0.1

N = the average of 5 values from two separate experiments, N=10.
* DNA pol-α isolated and then treated with MMS.
** DNA pol-α isolated from cells treated with MMS.

reacted with control DNA or with modified DNA isolated from MMS-treated mouse sarcoma cells, there was no difference between ^{3}H-dTMP incorporation into DNA without MMS adducts and DNA modified with 1 mM or 10 mM MMS (Table 2:B). This appears to suggest that MMS adducts do not inhibit pol-α activity; however, the data in Table 2:C show that treatment of both enzyme and template with MMS produces a synergistic inhibition of pol-α activity below that seen for MMS-treated enzyme alone. DNA isolated from mouse sarcoma cells treated with 1 mM MMS contained about 0.45 adducts/kb, while cells treated with 10 mM MMS yielded DNA with about 4.8 adducts/kb. MMS-modified DNA was used as template without enzymatic hydrolysis to activate it. Modified pol-α was also isolated from 1 mM or 10 mM MMS-treated mouse sarcoma cells. When MMS-DNA was used as the template for MMS-pol-α , inhibition of ^{3}H-dTMP incorporation was significantly greater than was seen for

^{3}MMS-pol-α reacted with control DNA template. Lastly, the inhibition of ^{3}H-dNMP incorporation catalyzed by MMS-treated pol-α was examined. Lineweaver-Burk plots of the data are shown in Fig. 9, and indicate that MMS noncompetitively inhibits incorporation of all four deoxynucleotides by pol-α. This pattern of inhibition suggests the possibility that MMS randomly methylates the enzyme, initiating steric changes that block or inactivate nucleotide acceptor sites.

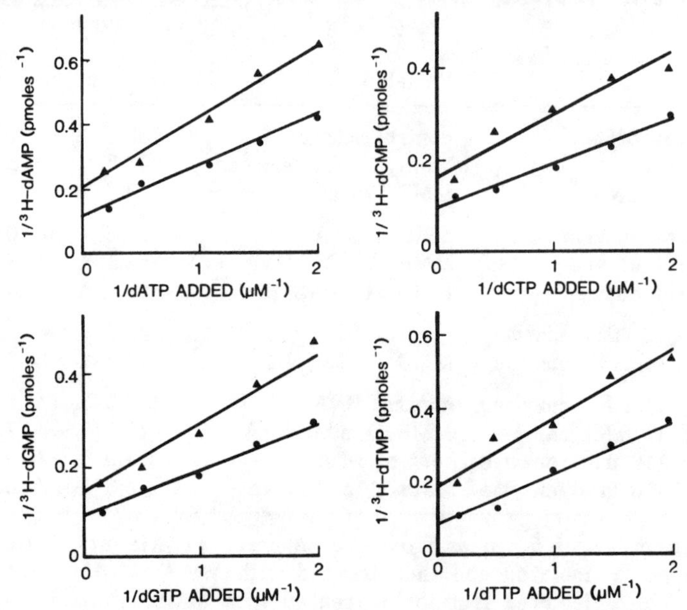

FIGURE 9. An examination of MMS inhibition of DNA pol-α activity when the deoxynucleotide concentration was varied. Enzyme activity was measured as ^{3}H-dNMP incorporation into template DNA.

CONCLUSIONS

Reactive metabolites of chemical carcinogens covalently bind DNA, RNA, and proteins, and inhibit both DNA and protein synthesis. While carcinogen binding to DNA has been extensively quantitated and is known to inhibit DNA synthesis (1-4), the consequences of RNA and protein adduction by carcinogens are less well understood. Aphidicolin, which does not bind DNA, binds pol-α, competitively inhibiting incorporation of dCMP into DNA template (9-14). The anthracyclines are also reported (30) to interact with pol-α, resulting in inhibition of DNA synthesis. In this study we have shown that BPDE and BADE bind DNA pol-α and inhibit DNA synthesis, while the parent compounds and the partially activated diol forms of the compounds are not inhibitory. Our data show that immunosuppressive corticosteroids also reversibly bind pol-α, competitively inhibiting pol-α activity in much the same manner as does the steroid-like diterpenoid, aphidicolin.

Structural configuration of the inhibitory compounds appears important to the mode of pol-α inhibition. Hydroxylation of the C11 position of deoxycorticosterone produces cortisone, which is less DNA synthesis-inhibitory than is deoxycorticosterone. Hydroxylation of cortisone at the C17 position produces cortisol, which does not inhibit pol-α at the concentrations we used. The PAH also appeared to have a structural basis for enzyme inhibition. BaP, BA, BPDO, and BADO did not inhibit pol-α , while BPDE and BADE were quite inhibitory. This generalization can be carried a step farther to show that steroids and a steroid-like diterpenoid competitively inhibited dCMP incorporation, while the PAH competitively inhibited dGMP incorporation. Both the PAH and the corticosteroids required pol-α to complex with DNA template before pol-α activity was affected, suggesting the possibility that the compounds interact with a site(s) on the enzyme not available unless the enzyme has formed a template complex. This was not necessary for MMS interaction with and inhibition of DNA pol-α . This might be expected of a methylating agent capable of modification of protein sites which might not be available for interaction with larger molecules such as the PAH or steroids.

These data show that a variety of compounds are capable of binding with and inhibiting the activity of DNA pol-α . The data do not suggest that pol-α is not inhibited by formation of DNA adducts. Rather, the MMS data support the findings documented for a number of DNA-adducting agents that show adduct inhibition of DNA synthesis. The data do, however, support the premise that DNA synthesis may be inhibited by interaction of chemicals with pol-α , and suggest that a great deal of the inhibition of DNA synthesis presumed to be due to DNA adduct formation may well be due to alteration of DNA pol-α .

ACKNOWLEDGEMENTS

Supported in part by NIH Grant HL31973, Council For Tobacco Research grant 1448, TAMU ORR and BSRG grants, by the Texas Agriculture Experiment Station, and by the United States Department of Agriculture.

REFERENCES

1. Crathorn, A.R. and Roberts, J.J. (1966): Mechanism of the cytotoxic action of alkylating agents in mammalian cells and evidence for removal of alkylated groups from deoxyribonucleic acid. Nature, 211, 150-152.

2. Painter, R.B. (1977): Inhibition of HeLa cell replicons by methyl methanesulfonate. Mutation Res., 42, 299-304.
3. Scudiero, D. and Strauss, B. (1974): Accumulation of single strand regions in DNA and the block to replication in a human cell line alkylated with methyl methanesulfonate. J. Mol. Biol., 83, 17-34.
4. Moore, P.D., Bose, K.K., Rabkin, S.D. and Strauss, B.S. (1981): Sites of termination if in vitro DNA synthesis on ultraviolet- and N-acetylaminofluorene-treated ØX174 templates by prokaryotic and eukaryotic DNA polymerases. Proc. Natl. Acad. Sci., 78(1), 110-114.
5. Cleaver, J.E. (1979): Inactivation of ultraviolet repair in normal and xeroderma pigmentosum cells by methyl methanesulfonate. Cancer Res., 42, 860-863.
6. Park, S.D., Choi, K.H., Hong, S.W. and Cleaver, J.E. (1981): Inhibition of excision repair of ultraviolet damage in human cells by exposure to methyl methanesulfonate. Mutation Res., 82, 365-371.
7. Cleaver, J.E. (1982): Inactivation of ultraviolet repair in normal and xeroderma pigmentosum cells by methyl methanesulfonate. Cancer Res., 42, 860-863.
8. Mosbaugh, D.W. and Linn, S. (1984): Gap-filling DNA synthesis by HeLa DNA polymerase-a in an in vitro base excision DNA repair system. J. Biol. Chem., 259, 10247-10251.
9. Waters, R. (1981): Aphidicolin: an inhibitor of DNA repair in human fibroblasts. Carcinogenesis, 2, 795-797.
10. Cleaver, J.E. (1983): Structure of repaired sites in human DNA synthesized in the presence of inhibitors of DNA polymerases alpha and beta in human fibroblasts. Biochim. Biophys. Acta., 739, 301-311.
11. Oguro, M., Suzuki-Hori, C., Nagano, H., Mano, Y. and Ikegami, S. (1979): The mode of inhibitory action by aphidicolin on eukaryotic DNA polymerase-a. Eur. J. Biochem., 97, 603-607.
12. Oguro, M., Shioda, M., Nagano, H. and Mano Y. (1980): The mode of action of aphidicolin on DNA synthesis in isolated nuclei. Biochem. Biophys. Res. Commun., 92, 13-19.
13. Longiaru, M., Ikeda, J., Jarkovsky, Z., Horwitz, S., and Horwitz, M. (1979): The effect of aphidicolin on adenovirus DNA synthesis. Nucleic Acids Res., 6, 3369-3386.
14. Hanaoka, F., Kato, H., Ikegami, S., Ohashi, M. and Yamada, M. (1979): Aphidicolin does inhibit repair replication in HeLa cells. Biochem. Biophys. Res. Commun., 87, 575-580.
15. Busbee, D. Joe, C., Norman, J. and Rankin, P. (1984): Inhibition of DNA synthesis by an electrophilic metabolite of benzo(a)pyrene. Proc. Natl. Acad. Sci., 81, 5300-5304.
16. Loeb, L. and Agarwal, S. (1971): DNA polymerase: correlation with DNA replication during transformation of human lymphocytes. Exptl. Cell Res., 66, 299-304.
17. Hanawalt, P., Cooper, P., Ganesan, A. and Smith, C. (1979): DNA repair in bacteria and mammalian cells. Ann. Rev. Biochem., 48, 783-836.

18. Ciarrochi, G., Jose, J. and Linn, S. (1979): Further characterization of a cell-free system for measuring replicative and repair synthesis with cultured human fibroblasts and evidence for the involvement of DNA polymerase-a in DNA repair. Nucleic Acid Res., 7(5), 1205-1219.
19. Miller, M. and Chinault, D. (1982): The roles of DNA polymerases alpha, beta, and gamma in DNA repair synthesis induced by different DNA damaging agents. J. Biol. Chem., 257, 10204-10209.
20. Detera, S., Becerra, S., Swack, J. and Wilson, S. (1981): Studies on the mechanism of DNA polymerase alpha: nascent chain elongation, steady state kinetics, and the initiation phase of DNA synthesis. J. Biol. Chem., 256, 6933-6943.
21. Mechali, M. Abadiedebat, J. and de Recondo, A. (1980): Eukaryotic DNA polymerase-a. J. Biol. Chem., 255(5), 2114-2122.
22. Rankin, P., Jacobson M., Mitchell, V. and Busbee, D. L. (1980): Reduction of nicotinamide adenine dinucleotide levels by ultimate carcinogens in human lymphocytes. Cancer Res., 40, 1803-1807.
23. Chapman, A., Fall, L. and Atkinson, D. (1971): A measure of the total adenylate energy charge in logarythmically growing cells. J. Bacteriol., 108, 1072-1086.
24. Joe, C., Rankin, P. and Busbee, D. (1984): Human lymphocytes treated with r-7,t-8-dihydroxy-t-9,10-epoxy-7,8,9,10-tetrahydrobenzo(a)pyrene require low density lipoproteins for the initiation of excision repair. Mutation Res., 131, 37-43.
25. Joe, C., Norman, J. Irvin, T.R. and Busbee, D. (1985): DNA polymerase activity in a repair-deficient human cell line. Biochem. Biophys. Res. Commun., 128, 754-759.
26. Fisher, P. and Korn, D. (1981): Ordered sequential mechanism of substrate recognition and binding by KB cell DNA polymerase-a. Biochem., 20, 4560-4569.
27. Sturrock, J. and Crathorn, A. (1971): Quantitative aspects of the repair of alkylated DNA in cultured mammalian cells. Chem. Biol. Interact., 3, 29-47.
28. Peterson, A., Peterson, H. and Heidelberger, C. (1979): Oncogenesis, mutagenesis, DNA damage, and cytotoxicity in cultured mammalian cells treated with alkylating agents. Cancer Res., 39, 131-138.
29. Singer, B. (1975): The chemical effects of nucleic acid alkylation and their relation to mutagenesis and carcinogenesis. Prog. Nucl. Acid Res. Mol. Biol., 15, 219-284.
30. Tanaka, M. and Yoshida, S. (1980): Mechanism of the inhibition of calf thymus DNA polymerases alpha and beta by daunomycin and adriamycin. J. Biochem., 87, 911-918.

A METHOD FOR THE SYNTHESIS OF BAY-REGION SULFUR SUBSTITUTED POLYAROMATIC HYDROCARBONS.

ANDERS L. COLMSJÖ, YNGVE U. ZEBÜHR and CONNY E. ÖSTMAN.
University of Stockholm
Department of Analytical Chemistry
S-106 91, Stockholm, Sweden

ABSTRACT

A method is described where bay region sulfur-substituted polyaromatic hydrocarbons are formed as secondary compounds to the parent compounds when sulfur is used as a dehydrogenating agent at elevated temperatures. The conditions for the reaction to take place are primarily that the carbon atoms adjacent to the bay region are saturated and that the temperature applied is high during the period of dehydrogenation.

INTRODUCTION

Sulfur heterocyclic polyaromatic compounds (S-PAC) have during the last years gathered increasing interest (1-4,15,17-20,22,23). Larger molecules with three or more fused rings have been detected in different types of samples and some of the compounds have been shown to exhibit strong mutagenic effects (4,5). Even though the requirement for standard compounds has been accentuated, few methods for preparing sulfur heterocyclics with large fused ring systems have been reported (6,7, 8). This paper presents a general method for the introduction of a sulfur atom at a bay-position of a polyaromatic hydrocarbon.

Sulfur has for some time been known as a strong dehydrogenating agent (9,10). The use of sulfur has in many cases been avoided because of its ability to participate in the reaction to form undesired compounds. Starting from this fact, we decided to investigate if advantage could be taken of this property to form S-PAC. Indeed, if the two outermost carbon atoms in a bay-region of a polyaromatic hydrocarbon (PAH) are saturated, the dehydrogenation with sulfur will also give the sulfur substituted PAH as an impurity in variable yields, depending on a number of parameters.

FORMATION OF THE COMPOUNDS

Dodecahydrotriphenylene 1 was chosed in order to examinate the formation of triphenyleno(1,12-bcd)thiophene or 1,12-epithiotriphenylene 3. The reaction was carried out in sealed glass ampoules. Sub-mg amounts of sublimed sulfur and 1 were mixed and the ampoule was heated under controlled times and temperatures. Maximum yields were at first determined as a function of the relative quantity of sulfur added. This factor is governed by the fact that on the one hand, the more sulfur present at the time of reaction the larger the probability of achieving a sulfur substituted compound (which is true to a certain concentration of sulfur) and on the other hand, the more sulfur present at the time of reaction the more destructive the effect on the hydrocarbon (strong decomposition). This factor was approximately at a maximum when the quantities by weight of sulfur and hydrocarbon were equal. Secondly, the reaction was carried out in vacuum, air and nitrogen atmospheres. These different conditions had no significan influence on the yields and thus air was used for the continued study. Thirdly,

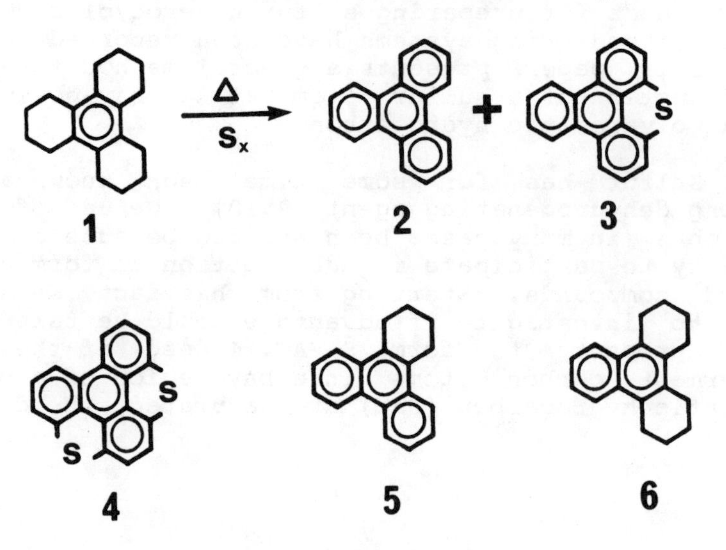

catalysts as $AlCl_3$, $SnCl_4$ and $FeCl_3$ were used, which could either be shown to have any significant influence on the yields.

Temperature and time were the most important factors for the formation of 3. The time of reaction is again a function of the quantity of reactants (or the heat propagation in the ampoule reactant system). The reaction was completed within a minute after the temperature reached equilibrium if

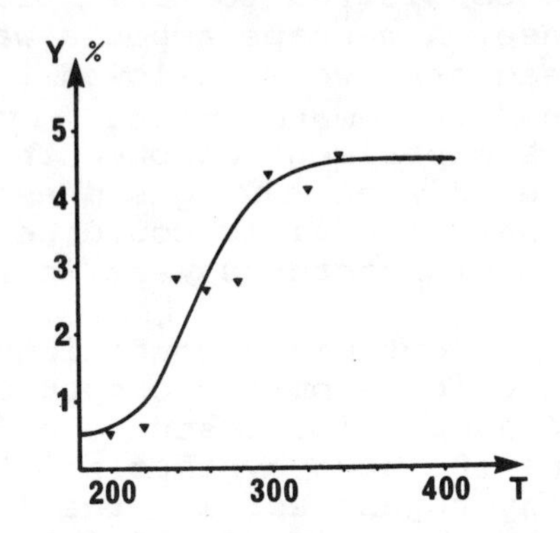

FIGURE 1. The yield of 1,12-epithiotriphenylene on heating dodecahydrotriphenylene and sulfur in sealed glass ampoules.

the ampoule was sealed at NTP. The temperature is of predominant importance. The yields of 3 calculated as a percentage of the triphenylene formed is plotted in figure 1 for temperatures up to 400°C. A maximum yield can be observed when the temperature exceeds 320°C. This characteristic temperature does not apply to the reacton of sulfur with other compounds. For example, when hydrogenated chrysene 8 is treated with sulfur at 320°C, no sulfur substitution takes place.

Fast, uncontrolled heating in a flame, where the products formed are allowed to sublime to the cooler areas of a glass tube, has been shown to triple the yields of 3. Under these vigorous conditions, the disulfur-substituted triphenylene, 1,12-4,5-diepithiatriphenylene 4 was also formed at a maximum yield of 2 percent relative to the hydrocarbon. The tri-sulfur substituted triphenylene was not formed in detectable quantities in any of these experiments. Under the same conditions, two partially hydrogenated compounds with approximately the same concentrations as 3 were formed, namely 1,2,3,4-tetrahydrotriphenylene 5 and 1,2,3,4,5,6,-7,8-octahydrotriphenylene 6.

Based on the fact that volatile heating increased the yield of bay-region sulfur substitution and that increased amounts of reagents decreased this yield, the reaction giving the sulfur substituted compounds were considered to take place primarily in the gas phase. A w-shape ampoule was constructed where the reagents were allowed to be heated in separate "knees" and were mixed by gas phase transfer. This experiment showed that the highest yield was achieved when the gas-phase reaction was applied and that the yield could be held constant when the amount of reactants were increased.

Under these optimized conditions, a number of sulfur heterocyclic aromatic compounds were synthesized. Thus 4,5-epithiochrysene 9 and 4,5-10,11-diepithiochrysene 10 were synthesized from chrysene, which was dehydrogenated in the 4, 5, 10 and 11 positions with the aid of lithium and ethylenediamine.

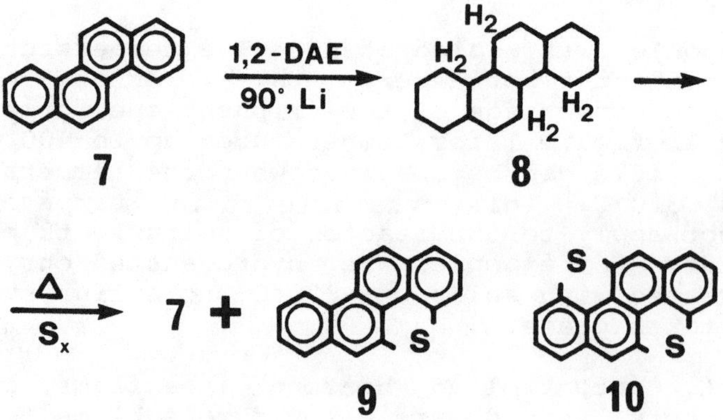

1,12-epithioperylene 13 was formed by the treatment of a mixture of hexa- and octahydroperylene (11) with sulfur. Disulfur-substituted perylene was not formed in detectable quantities.

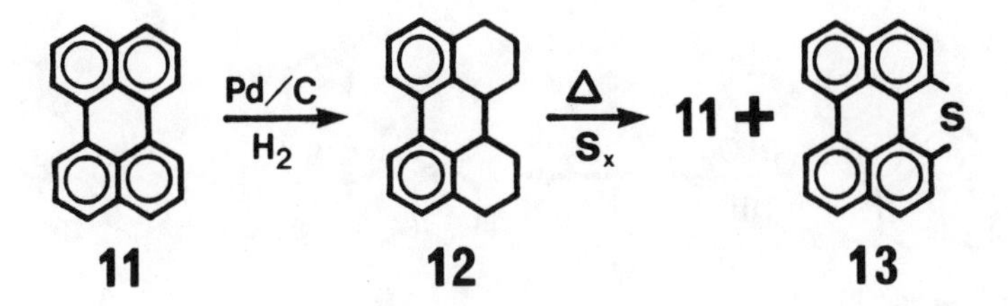

1,12-epithiobenzo(e)pyrene <u>16</u> was formed when 1,2,3,6,7,8,9,10,11,12-decahydrobenzo(e)pyrene <u>14</u> (13) was treated with sulfur. This compound is also present as an impurity in commercial benzo(ghi)perylene (18).

It was found necessary to synthesise 10,11-epithiobenzo(a)pyrene <u>19</u> according to the general method described from benzo(a)pyrene hydrogenated at the 10 and 11 positions. The synthesis of benzo(a)pyrene hydrogenated at the 11-position is decribed indirectly by Jacobs and Scott (14) as a synthesis of perhydrobenzo(a)pyrene. The mass spectrum of this compound showed that hydrogenation had not taken place at four positions of benzo(a)pyrene. Treatment with sulfur at elevated temperatures gave only benzo(a)pyreneas the major component, indicating that the carbon atom at the 11-position was unsaturated. Nevertheless, a small fraction from the product of synthesis, containing <u>19</u>, could be isolated by semipreparative HPLC. This implies that either a small part of the described perhydrobenzo(a)pyrene <u>18</u>, less than 0.1%, is hydrogenated at the 11-position or that the sulfur substitution can take place to a small extent even though the compound is unsaturated at the 11-position. The latter possibility has been demonstrated by us for similar compounds, <u>e.g.</u> benzo(ghi)perylene. The overall yield of 19 was only 0.01%. The fluorescence spectrum of <u>19</u> is shown in figure 5.

PtO_2, 100°, H_2
100 atm

17 18 19

EXPERIMENTAL SECTION

A Spectra Physics HPLC chromatograph model 3500, equipped with a semi preparative ODS-column and a Schoeffel variable wavelength UV-detector monitoring the effluent at 289 nm was used. The ambient temperature fluorescence spectra were recorded

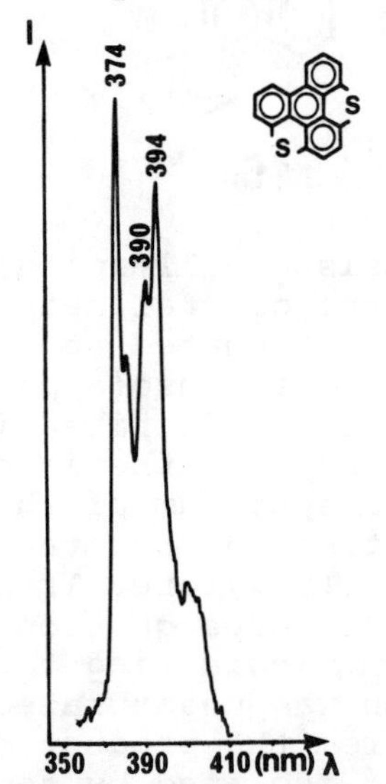

FIGURE 2. Fluorescence spectrum of 1,12-4,5-diepithiotriphenylene in n-hexane.

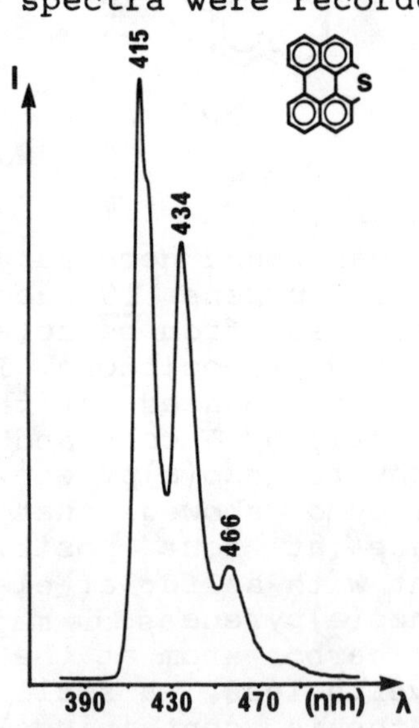

FIGURE 3. Fluorescence spectrum of 1,12-epithioperylene in n-hexane

by a Perkin-Elmer MPF-2A fluorescence spectrophotometer and low temperature fluorescence spectra were recorded by equipment especially designed for the registration of Shpol'skii fluorescence (15).

Subsequent analysis with gas chromatography was carried out using a Varian 3700 gas chromatograph equipped with a flame ionization detector (FID). Mass spectrometric analysis was performed on a Jeol JMS-D300 mass spectrometer interfaced to a Finnegan INCOS computer. A Carlo Erba Fractovap 2150 chromatograph was connected to the mass spectrometer.

IDENTIFICATION OF THE COMPOUNDS

The sulfur heterocyclic compounds were separated from their parent compounds by HPLC and identified primarily by their mass spectra. Mass spectra from sulfur heterocyclic aromatic compounds are characteristic due to the loss of a 45 amu fragment (3-11%, EI-70 eV) and the enhanced M+2 peak (19). The compounds are assumed to be the only possible isomers, based on the assumption that no rearrangements have taken place if the parent compound is formed simultaneously. Furthermore, S-PAC exhibit characteristic retention times

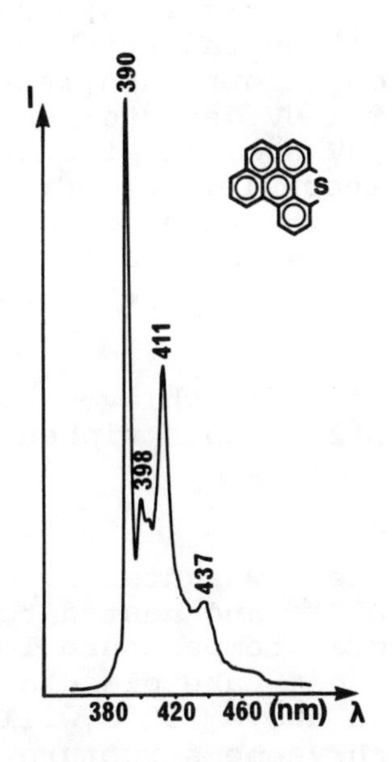

FIGURE 4. Fluorescence spectrum of 1,12-epithiobenzo(e)pyrene in n-hexane.

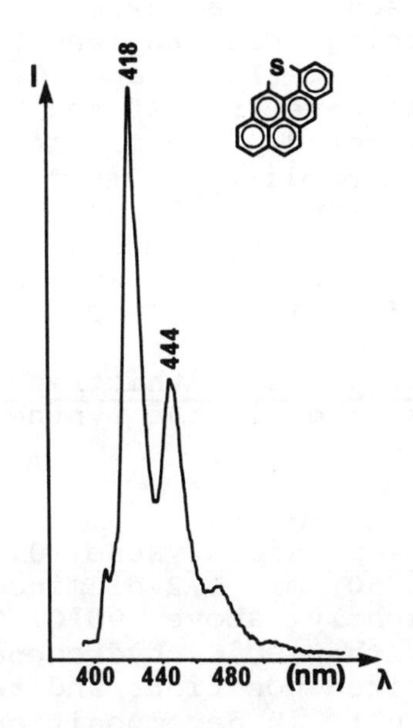

FIGURE 5. Fluorescence spectrum of 10,11-epithiobenzo(a)pyrene in n-hexane.

by gas chromatography on a unpolar stationary phase and reversed phase liquid chromatography (20). Most of the synthesised compounds also exhibit well resolved Shpol'skii spectra (20,22) which can be used for positive identification. The structure of epithiobenzo(e)pyrene 16 has been verified by NMR-spectroscopy (21).

1,12-epithiotriphenylene: Dodecahydrotriphenylene 1 was synthesized from cyclohexanone and methanolic sulfuric acid according to Mannich (16). 50 mg of 1 and 50 mg of sublimed sulfur were introduced in separate "knees" in a w-shaped pyrex tube which was sealed. The ampoule was heated above 400°C for five minutes, opened and extracted three times with boiling acetone. The extract containing the crude product and sulfur was centrifuged and the acetone phase transferred to a vessel and evaporated. The crude product was dissolved in methanol and one droplet of mercury was added in order to remove elemental sulfur. The precipitate was removed and the solution was injected on a semi-preparative reversed phase HPLC-column. The main fractions, containing triphenylene 2 (80%), 1,12-epithiatriphenylene 3 (18%) and 1,12-4,5-diepithiotriphenylene (2%) were registered by means of a UV detector and collected. The fluorescence and UV spectra of 3 has been published elsewhere (4). Incomplete dehydrogenation of 1 due to short heating time or cold parts in the ampoule where compounds can sublime, also yields the two compounds 5 and 6 which will interfere in the HPLC separation.

1,12-4,5-diepithiotriphenylene, 4 was achieved as described in the synthesis of 1,12-epithiotriphenylene.

4,5-epithiochrysene: 0.5 g chrysene 7 was dissolved in 50 ml 1,2-diaminoethane (DAE) and heated to slightly above 90°C. On heating at temperature lower than 90°C, hydrogenation does not take place at desired positions and temperatures above 100°C will result in decomposition of the chrysene structure. 0.5 g lithium was added in small portions under vigorous stiring. Cold water was added, the hydrogenated chrysene 8 was extracted with cyclohexane and the extract was washed with water. The solvent was evaporated and the hydrogenated chrysene was treated with sulfur as in the synthesis of 3. The yield was 6% relative to the chrysene formed. The overall yield of chrysene, when dodecahydrochrysene is treated with sulfur, has been reported to be 25% (17). The fluorescence and UV-spectra have been published elsewhere (4).

4,5-10,11-diepithiochrysene was formed at the same time as 9. The yield was 1.8% relative to the formed chrysene. Three main peaks were registered in the HPLC-chromatogram derived from 7, 9 and 10. The compound 10 is non-fluorescent.

1,12-epithioperylene: 50 mg of perylene was hydrogenated to 12 according to Zinke *et al.* (11) and treated with sulfur and heated as in the synthesis of 3. The HPLC-separated crude product contained 26% of 13 relative to the perylene formed. The overall yield was 18%. The fluorescence spectrum of 13 is shown in figure 3.

1,12-epithiobenzo(e)pyrene: 31 mg of the hydrobenzo(e)pyrene 14 was synthesized according to Cook and Hewett (14) and treated with sulfur as in the synthesis of 3. The HPLC-chromatogram showed two main peaks deriving from 15 and 16 at a ratio of 2.9 (mean value of seven dehydrogenations), *i.e.* compound 16 constituted 26% of the heavier PAC:s formed. The fluorescence spectrum of 16 is shown in figure 4.

ACKNOWLEDGEMENT

We wish to thank Beryl Holm for reviewing the manuscript.

REFERENCES

1. Lee, M.L. and Hites, R.A. (1976): Characterization of sulfur-containing polycyclic aromatic compounds in carbon blacks. *Anal. Chem.* 48, 1840.

2. Lee, M.L., Willey, C., Castle, R.N. and White, C.M. (1980): Separation and identification of sulfur heterocycles in coal-derived products, In: Polynuclear Aromatic Hydrocarbons, Edited by A. Björseth and A. Dennis, p. 59, Battelle Press, Columbus, Ohio.
3. Colmsjö, A.L. and Ostman, C.E. (1982): Shpol'-skii spectra of polycyclic aromatic compounds in samples from carbon black and soil. In: Polynuclear Aromatic Hydrocarbons, Edited by M. Cooke and A. Dennis, p. 201, Battelle Press, Columbus, Ohio.
4. Karcher, W., Nelen, A., Depaus, R., van Eijk, J., Glaude, P. and Jacobs, J. (1981): New results in the detection, identification and mutagenic testing of heterocyclic polycyclic aromatic hydrocarbons. In: Polynuclear Aromatic Hydrocarbons, Edited by M. Cooke and A. Dennis, p. 317, Battelle Press, Columbus, Ohio.
5. Tilak, B.D. (1960): Carcinogenesis by thiophene isosters of polycyclic hydrocarbons, Tetrahedron, 9:76,
6. Rao, D.S. and Tilak, B.D. (1959): Thiophenes and thiapyrans: Part XVIII. Benzo(2,1-b,3,4-b')dithiophene and benzodithionaphthenes. J. Sci. Ind. Res., 18B, Feb:77.
7. Pandya, L.J. and Tilak, B.D. (1959): Thiophenes and thiapyrans: Part XXI. A new synthesis of aryl w-dimethylethyl sulphides and synthesis of thieno(2,3-b,5,4-b')dithiophene and thieno(3,4-b)-pyrene, J. Sci. Ind. Res., 18B, Sept:371.
8. Badger, G.M., Christie, B.J., Pryke, J.M., and Sasse, W.H.F. (1957): Synthetic applications of activated metal catalysts. Part V. The Desulfurisation of Flavophene and Tetraphenylthiophene, J. Chem. Soc., 4417.
9. Dzienwonski, K. (1903): Ueber Decacyclen (Trinaphthylbenzol), Einen neuer hochmolekularen aromatischen kohlenwasserstoff, und über Dinaphthylenthiophen, einer rothen Thiokörper. Ber. Dtsch. Chem. Ges., 36:962.
10. Plattner, P.A. (1963): Dehydrierungen mit schwefel, selen und platin metallen. In: Neuere Methoden der Praparativen Organischen Chemie, W. Foerst, Verlag Chemie, Weinheim / Bergstr. Bd. I, p. 39.
11. Zinke, A. and Schniederschitsch, N. (1929): Untersuchungen über perylen und seine derivate, Mh. Chem., 51:280.

12. Zinke, A. and Benndorf, O. (1931): Untersuchungen uber perylen und seine derivate, Mh. Chem., 59:241.
13. Cook, J.W. and Hewett, C.L. (1933): Synthesis of 1:2- and 4:5-benzpyrenes, J. Chem. Soc., 401.
14. Jacobs, T.L. and Scott, W.R. (1953): Acetylenic ethers, VI, phenylalkoxyacetylenes, J. Amer. Chem. Soc., 75:5497.
15. Colmsjö, A. and Stenberg, U. (1979): Identification of polynuclear aromatic hydrocarbons by Shpol'skii low temperature fluorescence, Anal. Chem., 51:145.
16. Mannich, C. (1970): Ueber triphenylen, Ber. Dtsch. Chem. Ges., 40:159.
17. Zander, M. and Grundmann, C. Aromaten. aus cyclischen verbindungen. In: Houben-Weyl, Methoden der Organischen Chemie, Bd 5, Teil 2b, p. 119.
18. Karcher, W., Depaus, R., van Eijk, J. and Jacobs, J. (1979): Separation and identification of sulfur containing polycyclic aromatic hydrocarbons (thiophene derivates), from some PAH, In: Polycyclic Aromatic Hydrocarbons, Edited by P. Jones and P. Leber, p. 341, Ann Arbor Science Publ., Inc., Ann Arbor, Michigan.
19. Gallegos, E.J. (1975): CHS sulfur compound analysis by gas chromatography-mass spectroscopy. Anal. Chem., 47:1150.
20. Colmsjö, A.L., Zebühr, Y.U., and Östman, C.E. (1982): Shpol'skii effect in the analysis of sulfur-containing heterocyclic aromatic compounds, Anal. Chem., 54:1673.
21. Personal communication with W. Karcher, commission of the European Communities, Joint Res. Center, Petten Establishment, The Netherlands.
22. Colmsjö, A.L., Zebühr, Y.U. and Östman, C.E. (1984): The utility of Shpol'skii fluorescence in the analysis of polynuclear aromatic compounds containing condensed thiophene rings, Chemica Scripta, 24:95.
23. Kong, R.C., Lee, M.L., Tominga, Y., Pratap, R., Iwao, M., Castle, R.N. and Wise, S.A. (1982): Capillary column gas chromatographic resolution of isomeric polycyclic aromatic sulfur heterocycles in a coal liquid, J. Chromatogr. Sci. 20:502.

THE SYNTHESIS AND MUTAGENICITY OF CHLORODERIVATIVES OF PYRENE

ANDERS COLMSJÖ, ULRIKA NILSSON, AGNETA RANNUG[1] and ULF RANNUG[1]
Department of Analytical Chemistry, Arrhenius Laboratory and (1) Department of Genetic Toxicology, Wallenberg Laboratory, University of Stockholm, S-106 91 Stockholm, Sweden

INTRODUCTION

Several polynuclear aromatic hydrocarbons (PAH) are known carcinogens and mutagens. Experimental data on pyrene, however, indicate that this compound is a non-carcinogen as well as a non-mutagen. Certain derivatives of pyrene, on the other hand, have shown to be active in different short-term mutagenicity tests, the most obvious example being the high mutagenic potency of 1-nitro- and 1,8-dinitropyrene in Ames Salmonella assay.

In our first chlorination experiments pyrene was treated with chlorine in carbon tetrachloride in the presence of $AlCl_3$ as a catalyst. The reaction took place at room temperature and was terminated after one minute. The crude product of synthesis was tested for mutagenicity on *Salmonella typhimurium* strains TA98 and TA100 in the absence and presence of a metabolizing system (liver S9 fractions from Aroclor 1254 pretreated male Sprague-Dawley rats). A mutagenic effect was seen both in the absence and presence of S9 (1). Thus, from a compound with no or very low mutagenic effect two types of mutagens were formed upon chlorination.

The reaction products of pyrene and chlorine were described in 1883 by Goldschmiedt and Wegscheider (2). The products of synthesis are numerous and difficult to separate. In 1937, Vollman *et al*.(3) identified the main products that are formed when pyrene and chlorine react. As the reaction continues pyrene is transformed to 1-chloropyrene, which reacts further to give 1,6-dichloropyrene, 1,8-dichloropyrene and 1,3-dichloropyrene in minor quantities. The three latter compounds react with chlorine, giving 1,3,6-tri- chloropyrene from which 1,3,6,8-tetrachloropyrene is formed on further chlorination. Apart from these compounds, a number of chloro-substituted and chloro-added products are formed as minor constituents.

Since the crude product showed a relatively strong direct mutagenic effect, 3.5 revertants/ng (1) we intended to isolate and characterize at least some of the mutagenic compounds formed during the synthesis. The first part of this work has been published elsewhere (1,3).

MATERIAL AND METHODS

Chemicals

Pyrene, 99% pure, was purchased from EGA Chemicals

Synthesis

1-chloropyrene was synthesized according to Vollman *et al*.(3). To a solution of 1g pyrene in CCl_4 (10 ml), at room temperature, 0.67g SO_2Cl_2 in 2 ml of CCl_4 was added. The temperature of the solution was increased to 75°C and kept constant until no more HCl was formed. The crude product was recrystallized in glacial acetic acid and further purified by semi-preparative reversed phase HPLC.

4-chloropyrene was synthesized in the following way. Chlorine (Cl_2, 11 ml at NTP) and catalytic amounts of $AlCl_3$ was added to a solution of 1,2,3,6,7,8-hexahydropyrene (0.1g) in chloroform (10 ml). After a reaction time of one hour, at room temperature, the solution was washed with water and the chloroform subsequently evaporated. The crude product was dissolved in nitrobenzene (25 ml) and I_2 (0.38g) was added to the solution which was kept at 90°C for two hours. After removal of nitrobenzene by steam distillation, 4-chloropyrene was separated from the crude product by semi-preparative reversed phase liquid chromatography.

The preparation of 4,5-dichloro-4,5-dihydropyrene and 1,4,5-trichloro-4,5-dihydropyrene was carried out in the following way. Chloro-addition at the 4,5-positions was conducted using the parent compound, chlorine and ultraviolet light. After addition of Cl_2 (22 ml at NTP) to a solution of pyrene (0.2g) in chloroform (10 ml) the solution was irradiated for one hour using a 300W Hg-arc lamp. The solvent was then evaporated and the crude product dissolved in cyclohexane. Separation of unreacted pyrene and the fairly polar chloro-added products was performed on a short, open silica gel column. The unreacted material was eluted with cyclohexane. The chloro-added compounds was then eluted with acetone.

Mutagenicity tests

Salmonella typhimurium strains TA98 and TA100 were kindly provided by B. N. Ames (University of California, Berkeley, CA). The plate incorporation assay according to Maron and Ames (4) was followed with minor modifications (1). All compounds were dissolved in acetone. Liver S9-preparations from Aroclor 1254 pretreated male Sprague-Dawley rats were used as metabolizing systems (50 ul S9 fraction/plate)

Instrumentation

A Liqrosorb reversed phase ODS column was used with methanol as mobile phase for separation of the products of synthesis. Identification of the compounds was carried out by means of retention indices, UV-spectroscopy (PYE-UNICAM SP8-150), mass spectrometry (JEOL D-300), proton NMR (JEOL 400 MHz-instrument) and cryo fluorescence.

RESULTS

Although the two isomers, 1-chloro- and 4-chloropyrene are difficult to separate, having the same retention time in the GC- and HPLC-systems, they give different Shpol´skii spectra as reported earlier (5). The structures of these two isomers were confirmed with 400 Hz proton NMR spectroscopy. Shpol´skii spectroscopy of the same samples revealed a difference in quantum efficiency, which could be used to quantify the two isomers in a mixture by means of Shpol´skii spectroscopy. The two compounds were tested for mutagenicity in Ames´ assay on strain TA100 in the presence of S9. The results are shown in Fig. 1 A and B, respectively.

1-Chloropyrene gave 8 revertants per microgram, which is equivalent to approx. 2 rev/nmol. 4-Chloropyrene showed a five times higher mutagenicity, i. e. 41 revertants/µg or approx. 10 rev/nmol. None of the compounds gave a significant mutagenic effect on strain TA100 in the absence of the metabolizing system in accordance with our earlier data on 1-chloropyrene (1). Although the third possible monochloropyrene, 2-chloropyrene, has not been synthesized or tested for mutagenicity, the results with 1-chloropyrene and 4-chloropyrene confirmed that chloro-substituted pyrenes could not be responsible for the direct mutagenic effects seen with the crude product after chlorination of pyrene (1,6).

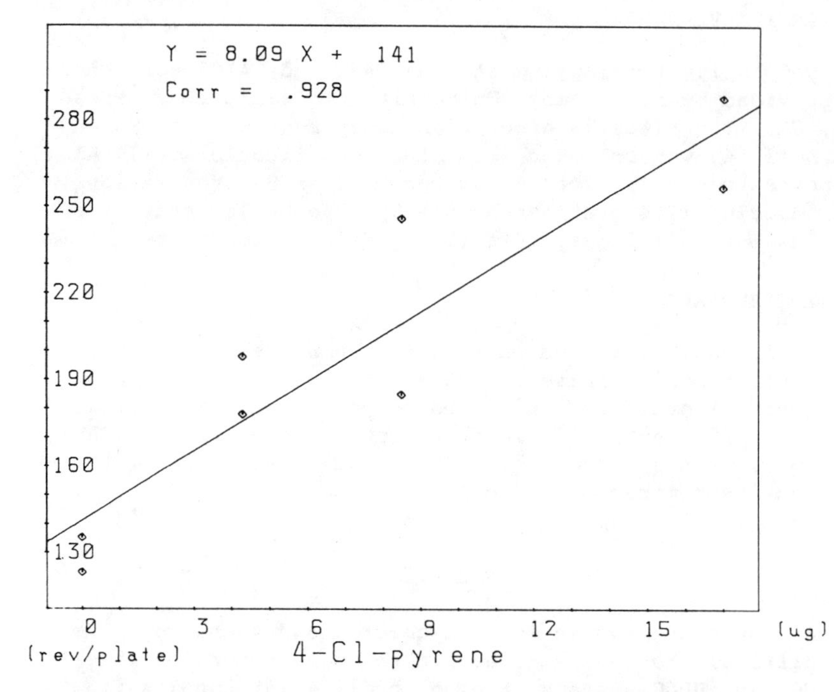

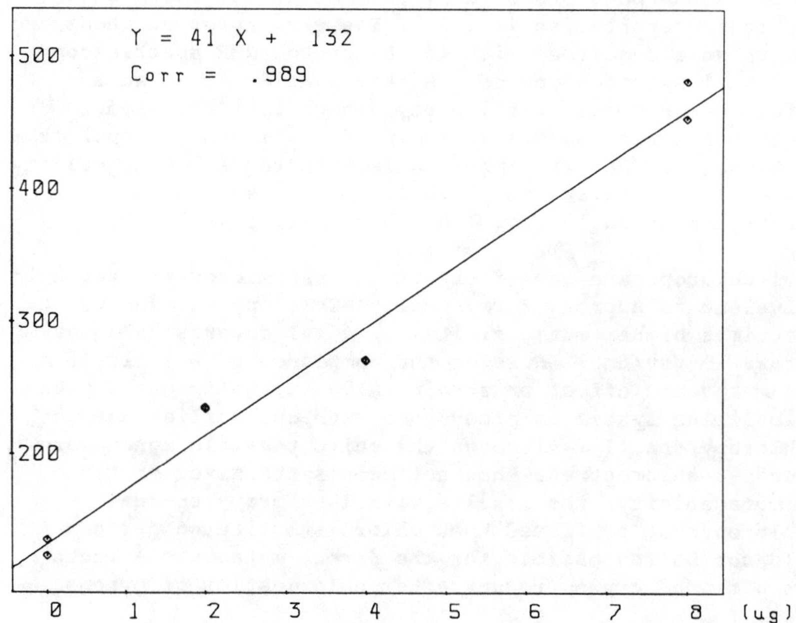

FIGURE 1. The mutagenic effect of 1-chloropyrene (A) and 4-chloropyrene (B) on TA100 in the presence of S9.

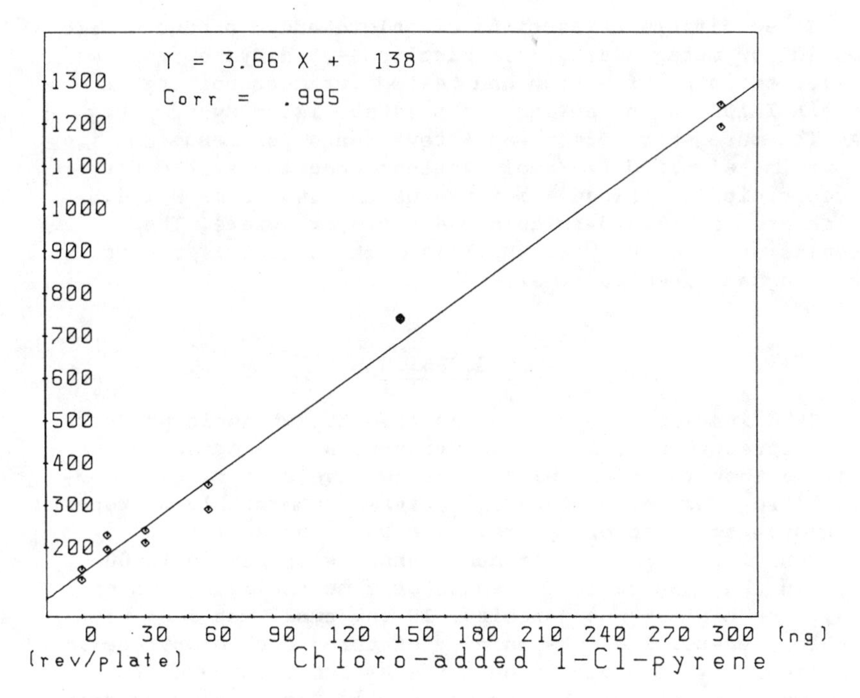

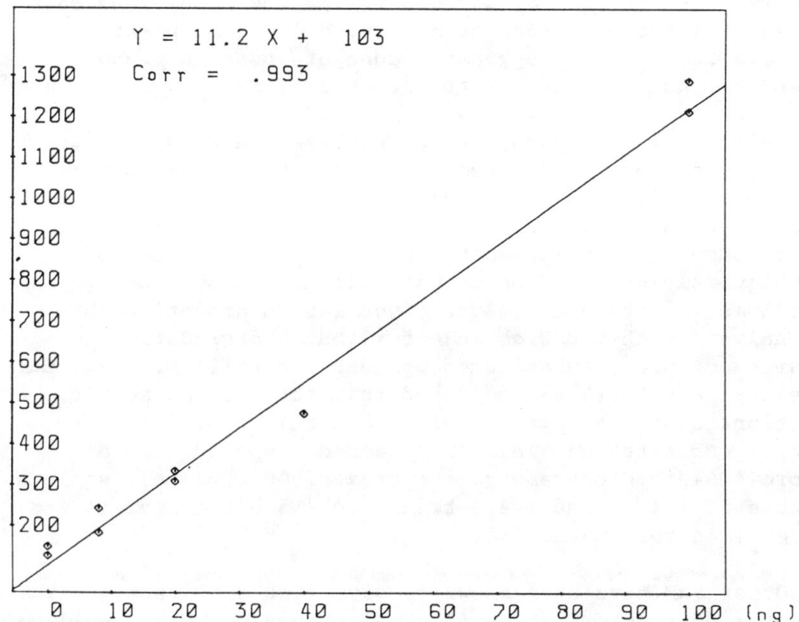

FIGURE 2. **The mutagenic effect of 4,5-dichloro-4,5-dihydropyrene (A) and a mixture of 1,4,5- and 3,4,5-trichloro-4,5-dihydropyrene (B) on strain TA100 in the absence of S9.**

SYNTHESIS AND MUTAGENICITY OF CHLOROPYRENES

Three different fractions of chloro-added products were tested for mutagenicity. 4,5-Dichloro-4,5-dihydropyrene was collected in one fraction and tested for mutagenicity on strain TA100 in the absence of a metabolizing system, Fig 2A. The mutagenic effect was 4 revertants per nanogram, i.e. approximately 1000 rev/nmol. Another fraction tested for mutagenicity contained a mixture of the two isomers 1,4,5-trichloro- and 3,4,5-trichloro-4,5-dihydropyrene. The results are shown in Fig. 2B. This mixture gave 11 rev/ng, which equals 3400 rev/nmol.

DISCUSSION

Chlorination of pyrene gives rise to mutagenic products of different types. The major products are monochloropyrenes. Further reactions with chlorine give rise to dichloro- trichloro- and tetrachloroderivatives. Two monochloropyrenes, 1-chloro- and 4-chloropyrene, have been tested for mutagenicity in Ames´ Salmonella assay with strain TA100 both in the absence and presence of a metabolizing system (S9). The highest mutagenicity, 10 rev/nmol, was seen with 4-chloropyrene. The mutagenicity of three dichloropyrenes, 1,3-dichloro-, 1,6-dichloro- and 1,8-dichloropyrene, have been reported earlier by our group (1). The dichloropyrenes showed at least one order of magnitude lower mutagenicity than did the monochloropyrenes. None of these compounds showed any mutagenicity in the absence of S9.

Chlorination of pyrene also results in a complex mixture of compounds more polar than pyrene and the chlorosubstituted pyrene derivatives. These polar compounds exhibit a rapid degradation in polar solvents or at elevated temperatures. For this reason these compounds cannot be unambiguously detected using gas chromatography. The on-column injection technique gives less degradation than GC-analysis with the Grob injector, but a degradation to pyrene and chloro-substituted pyrenes can still be seen. Our previous results (1,6) indicated that the most mutagenic fractions contained pyrenes with both chloro-addition and chloro- substitution. The chloro added compound, 4,5-dichloro-4,5-dihydropyrene gave approx. 1000 rev/nmol, while a mixture of 1,4,5- and 3,4,5-trichloro-4,5-dihydropyrene was three times more mutagenic.

Other PAH have been shown to give rise to direct acting mutagens upon chlorination (1,6). The complexity, as well as the mutagenicity, varies from PAH to PAH. Although it is likely that halogenated PAHs, especially the more stable chloro-substituted derivatives, can be found in the environment it is impossible, at least until further data are available, to know if this implies a risk from a toxicological point of view.

REFERENCES

1. Colmsjö, A., Rannug, A. and Rannug, U. (1984): Some chloro derivatives of polynyclear aromatic hydrocarbons are potent mutagens in Salmonella typhimurium. Mutation Res., 135: 21-29.
2. Goldschmiedt, G. and Wegscheider, R. (1883): Uber Derivate des Pyrens. Monatsh. Chem., 4: 666-690.
3. Vollman, H., Becker, H., Corell, M. and Streeck, H. (1937): Beiträge zur Kenntnis des Pyrens und seiner Derivate; Halogenpyrene und substituierte Pyrenchinone. Ann. Chem., 531, 1-159.
4. Maron, D. M. and Ames, B. N. (1983): Revised methods for Salmonella mutagenicity test. Mutation Res., 113: 173-215
5. Colmsjö, A. L., Östman, C. E. and Zebühr, Y. U. (1985): Identification of chlorinated polyaromatic compounds by Shpol´skii spectroscopy. In: Polynuclear Aromatic Hydrocarbons: Mechanisms, Methods and Metabolism, edited by M. Cooke, and A. J. Dennis, pp 273-280, Battelle Press, Columbus.
6. Rannug, U. Nilsson, U. and Colmsjö, A. (1986): The mutagenicity and chemical structure of certain chloroderivatives of polynuclear aromatic hydrocarbons. In: Genetic Toxicology of Environmental Chemicals, Part B: Genetic effect and Applied Mutagenesis, edited by C. Ramel, B. Lambert and J. Magnusson, pp 157-164, Alan Liss, New York.

INHIBITION OF BENZO(A)PYRENE-DNA ADDUCT FORMATION BY NATURALLY OCCURRING PLANT PHENOLS IN EPIDERMIS OF SENCAR MICE.

MUKUL DAS*, PARTHASARATHY ASOKAN, DANIEL P. BIK, DAVID R. BICKERS AND HASAN MUKHTAR
Department of Dermatology, University Hospitals of Cleveland, Case Western Reserve University, and Veterans Administration Medical Center, Cleveland, Ohio 44106, USA.

INTRODUCTION

Benzo(a)pyrene (BP), a ubiquitous environmental pollutant, is relatively inert and to elicit its carcinogenicity it must first undergo extensive metabolism by microsomal enzymes which convert them to their biologically active ultimate carcinogenic forms (1,2). Studies on the mutagenicity, tumorigenicity and metabolism of BP have indicated that bay region diol-epoxides (BPDE) are the ultimate carcinogenic species and that enzyme mediated binding of BPDE to DNA is essential for its carcinogenicity (1,2). The metabolism of BP has been shown to be stereoselective, producing a mixture of 4 enantiomers of BPDE, with the formation of the (+)-anti enantiomer predominating over the others (1,2). Stereoselective binding to DNA was observed in intact cultured cells, with the (+)-anti enantiomer binding at a level 6 times higher than that observed for the (-)-anti enantiomer (3). The (+)-anti isomer has been shown to react almost exclusively with the N^2 position of dGua in cultured cells (3). A positive correlation between the levels of binding of polycyclic aromatic hydrocarbons (PAH) such as BP to DNA of mouse skin and its carcinogenic potency is established (4).

Knowledge that metabolic activation and subsequent binding of the reactive metabolite to DNA is a critical determinant of tumor induction by BP, has prompted a search for non-toxic inhibitors of these pathways. For example, in recent years there has been growing interest in identifying naturally occurring dietary compounds that can protect against chemical carcinogenesis. Wood _et al._ (5) have shown that ellagic acid, a degradation product of certain tannins, possesses exceptionally high ability to inhibit the mutagenicity of the bay-region diol-epoxide of BP. Subsequent studies from our laboratory have shown that ellagic acid is a potent inhibitor of epidermal microsomal aryl hydrocarbon hydroxylase (AHH) activity and of enzyme-mediated BP-binding to both calf thymus DNA in vitro and to epidermal DNA in vivo (6). Dixit et al. (7) have shown that ellagic acid inhibits the metabolic activation and the binding of BP and BP-7,8-diol to DNA in cultured mouse lung. It has also been shown that topical application of ellagic acid to skin of Balb/C mice affords protection against 3-methylcholanthrene (MCA) induced skin carcinogenesis (8). Lesca (9) has shown that ellagic acid inhibits BP-induced lung tumor formation and 7,12-dimethylbenz(a)anthracene (DMBA) induced skin carcinogenesis. Recently, ellagic acid has also been found to protect against BPDE induced skin carcinogenesis in newborn mice (10). Since the binding of BPDE to target tissue DNA is critical for BP carcinogenicity we have analyzed a series of plant phenols including tannic acid, quercetin, myricetin and anthraflavic acid on in vitro and in vivo BP-DNA adduct formation in skin of SENCAR mice.

MATERIALS AND METHODS

Chemicals

[G-^{3}H]BP (specific activity, 25 Ci/mmol), was purchased from Amersham Searle, Chicago, IL. Radiolabelled BP, was purified by a silica gel (Partisil 10-μ, Waters) column with hexane as the eluting solvent and subsequently by reverse phase HPLC using a DuPont Zorbax ODS column (76.2 mm x 25 cm) eluted with methanol:water (19:1, v/v). The purity of these compounds was greater than 99% as judged by HPLC. Phenol (> 99% pure), tannic acid, myricetin and anthraflavic acid were obtained from Aldrich Chemical Co., Milwaukee, WI. Quercetin, MCA, NADPH, protease (Type XI), m-cresol, 8-hydroxyquinoline and ribonuclease A (Type III-A) were purchased form Sigma Chemical Co., St. Louis, MO. Sephadex LH-20 was a product from Pharmacia Fine Chemicals, Piscataway, NJ. All solvents and other chemicals were obtained in the purest form commercially available.

Treatment of Animals

Six week-old female SENCAR mice (25 ± 5 g body weight), obtained from NCI-Frederick Cancer Research Facility, Frederick, MD were used. The mice were shaved with electric clippers and Nair depilatory was applied 1 day before the beginning of the experiment. For the in vitro DNA-binding studies, one group of animals received a single topical application of MCA (50 mg/kg) in 0.1 ml of acetone while the other group received an equal volume of the vehicle and served as control. For studies on the in vivo effect of plant phenols on PAHs binding to DNA, animals were topically treated with different doses (1.5-12.0 μmol/animal) either in 0.1 ml of acetone and/or DMSO. Animals treated with an identical volume of vehicle served as controls. One hour later all the animals were treated with 5 nmol of [^{3}H]-BP.

Preparation of Epidermal Homogenates

Animals treated with or without plant phenols followed by [^{3}H]BP were killed by cervical dislocation 24 hours after PAH application. Skin was removed and washed in ice cold 0.1 M phosphate buffer, pH 7.4. The epidermis was separated from dermis by heat treatment of the skin at 52°C for 30 sec (11) as described earlier (12). All subsequent operations were carried out at 0-4°C. The epidermis was minced with scissors in 0.1 M phosphate buffer pH 7.4 containing 10 mM EDTA as described earlier (13). The minced tissue was homogenized with a Polytron Tissue Homogenizer (Brinkmann Instruments, Westbury, NY) and used for the isolation of DNA. For *in vitro* studies, epidermal microsomes were prepared from control and MCA pretreated animals as described previously (13).

DNA Extraction from Epidermis

The DNA from minced epidermal homogenates was extracted, purified and estimated essentially as described earlier (6,8). Covalent binding is expressed as the amount of [^{3}H]BP bound to tissue DNA.

HPLC Analysis of BPDE-I-dGua Adducts

Purified DNA samples from epidermis were hydrolyzed sequentially using DNase I, snake venom phosphodiesterase and alkaline phosphatase and the DNA hydrolyzates were subsequently analyzed by HPLC. Prior to injection on to HPLC all DNA samples were processed through a small column of Sephadex LH-20 (prepared in Pasteur pipettes) and washed with 20 ml of water. The less polar BP-DNA adducts were then eluted with 30 ml of methanol. The methanol-soluble material from this cleanup step represent 80-85 of th total DNA-bound radioactivity, dried under nitrogen and dissolved in 0.1 ml of methanol.

A Waters Associates model 204 liquid chromatograph, fitted with a Waters Associates μBondapak C18 column (3.9 mm x 15 cm) was used for the analysis of the radiolabelled BP-DNA adducts. Identification of BPDE-I-dGua adduct was based on the reference standard, kindly supplied by Dr. R.M. Santella (Columbia University, New York). The column was eluted at ambient temperature with a 40-min linear gradient of 50-75% methanol in water at a flow rate of 0.5 ml/min. The eluates were monitored at 254 nm, fractions of approximately 0.2 ml were collected dropwise and the radioactivity of each fraction was determined on a Packard TriCarb 460 CD liquid scintillation spectrometer.

In Vitro BP-DNA-Binding Studies

The incubation system employed was similar to that described by Hesse *et al.* (14). After incubation, the reaction mixtures were centrifuged at 105,000 xg to isolate the microsomes, and digested with SDS to extract any traces of protein. The extraction procedure was similar to that described by Lesca *et al.* (15). DNA estimation and binding to BP was then carried out as described earlier (6,8).

RESULTS

Effect of Plant Phenols on Epidermal Microsomal Enzyme-Mediated Binding of [^{3}H]-BP to Calf Thymus DNA

Epidermal microsomes prepared from animals pretreated with MCA enhanced the binding of BP to calf thymus DNA by 75-91% as compared to microsomes prepared from control animals. The *in vitro* addition of tannic acid, quercetin, myricetin and anthraflavic acid, at a concentration of 25 μM to epidermal microsomes from either control or MCA-treated animals resulted in 63-64%, 38-43%, 36-37% and 27-33% inhibition of BP-binding to calf thymus DNA, respectively (Table 1).

In Vivo Effect of Plant Phenols on Covalent Binding of [^{3}H]-BP to Epidermal DNA

The effect of a single topical application of varying doses of tannic acid, quercetin, myricetin and anthraflavic acid on the binding of BP to epidermal DNA is shown in Table 2. Each of the four plant phenols tested substantially inhibited the binding of BP to epidermal DNA. The observed inhibitory effects were dose-dependent. Topical application of 12 μmol of tannic acid, quercetin, myricetin and anthraflavic acid to mice resulted in 73, 64, 63 and 48% inhibition of BP-DNA adduct formation in the epidermis (Table 2). The effect of three repeated topical applications of the plant phenols on *in vivo* BP binding to epidermal DNA is shown in Table 3. Animals received 12 μmol of plant phenols every 6 hours for 3 doses and then were utilized

TABLE 1

IN VITRO EFFECT OF PLANT PHENOLS ON EPIDERMAL MICROSOMAL ENZYME MEDIATED COVALENT BINDING OF [^{3}H]-BP TO CALF THYMUS DNA.

	Covalent binding (pmol/mg DNA)[a]			
Plant Phenols	Control microsomes	% Inhibition	MCA induced microsomes	% Inhibition
Acetone	62.6 ± 4.9[b]	--	119.6 ± 10.6	--
DMSO	67.0 ± 4.3	--	114.0 ± 9.9	--
Tannic acid	23.5 ± 2.1	63	43.5 ± 4.1	64
Quercetin	35.7 ± 2.7	43	74.5 ± 5.8	38
Myricetin	40.2 ± 4.2	36	74.9 ± 6.9	37
Anthraflavic acid	44.7 ± 3.3	33	82.6 ± 6.8	27

[a]The incubation mixture contained 12 mg calf thymus DNA, 10 mg epidermal microsomal protein, 8.8 mg NADPH, 5 nmol of [^{3}H]-BP and 25 μM plant phenols in a final volume of 10 ml of 50 mM Tris buffer containing 25 mM KCl, 5 mM $MgCl_2$, pH 7.5. After 1 hr of incubation the DNA was isolated and purified as described in "Materials and Methods".

[b]Data represent mean ± SEM of the three individual values.

TABLE 2

DOSE DEPENDENT EFFECT OF PLANT PHENOLS ON COVALENT BINDING OF BP TO EPIDERMAL DNA IN SENCAR MICE.

Concentration of Plant phenols (μmol)	DNA-Binding (pmol/mg DNA)			
	Tannic acid	Quercetin	Myricetin	Anthraflavic acid
0	13.5 ± 1.1	13.5 ± 1.1	13.5 ± 1.1	13.5 ± 1.1
1.5	9.8 ± 0.7	10.3 ± 0.6	11.5 ± 0.5	12.0 ± 0.7
3.0	8.5 ± 0.7	8.8 ± 0.3	9.4 ± 0.6	10.3 ± 0.4
6.0	5.8 ± 0.3	6.5 ± 0.4	7.1 ± 0.4	8.9 ± 0.5
9.0	4.5 ± 0.1	5.5 ± 0.3	5.7 ± 0.2	7.7 ± 0.3
12.0	3.6 ± 0.2	4.8 ± 0.1	5.0 ± 0.3	7.0 ± 0.2

Data represent mean ± S.E.M. of 3 values.
Animals were topically applied with a single application of plant phenols as described in "Methods".

TABLE 3

EFFECT OF REPEATED TOPICAL APPLICATION OF PLANT PHENOLS ON COVALENT BINDING OF BP TO EPIDERMAL DNA IN SENCAR MICE.

Treatment	DNA-Binding (pmol/mg DNA)	% Inhibition
Acetone	11.0 ± 1.2	-
DMSO	11.6 ± 1.1	-
Tannic acid	4.8 ± 0.3	56
Quercetin	2.5 ± 0.3	77
Myricetin	4.0 ± 0.5	63
Anthraflavic acid	6.4 ± 0.5	45

Data represent mean ± S.E.M. of 3 values.
Animals were topically applied with three repeated applications of plant phenols as described in "Methods".

as described above. Repeated topical applications of 12 µmol of tannic acid, quercetin, myricetin and anthraflavic acid resulted in 56, 77, 64 and 45% inhibition in BP-DNA adduct formation in epidermis (Table 3). The inhibitory effects of repeated applications were not significantly greater than that of single application of the plant phenols (12 µmol).

In Vivo Effect of Plant Phenols on BPDE-I-dGua Adduct Formation in Epidermis

The effect of a single topical application of plant phenols on BPDE-I-dGua adduct formation in epidermis of SENCAR mice is shown in Table 4. In this study the epidermal DNA was isolated from animals treated with different plant phenols or the vehicles. Equal amounts of DNA from each sample were digested and BPDE-I-dGua adduct formation was analyzed on HPLC. The HPLC profile of BP-

DNA adducts showed a major peak typical of BPDE-I-dGua adduct and other unidentified peak which eluted after BPDE-I-dGua adduct. In this study only the BPDE-I-dGua adduct was quantitated and the results are shown in Table 4. A single topical application of 12 umol of tannic acid, quercetin, myricetin and anthraflavic acid resulted in 79, 62, 86 and 80% inhibition of BPDE-I-dGua adduct formation in epidermis. The inhibitory effects of each plant phenol on BPDE-I-dGua adduct formation was greater than that observed for total BP-DNA adducts formation shown in Table 2.

TABLE 4

EFFECT OF SINGLE TOPICAL APPLICATION OF PLANT PHENOLS ON BPDE-I-dGua ADDUCT FORMATION IN EPIDERMIS OF SENCAR MICE.

Treatment	BPDE-I-dGua-adduct (pmol/mg DNA)	% Inhibition
Control	6.9 ± 0.4	-
Tannic acid	1.5 ± 0.2	78
Quercetin	2.5 ± 0.2	68
Myricetin	1.0 ± 0.1	86
Anthraflavic acid	1.4 ± 0.1	80

Data represent mean ± S.E.M. of 3 values.
Animals were treated with a single topical application of plant phenols as described in "Methods".

DISCUSSION

Metabolic conversion of inert precursor chemicals into reactive species that bind to DNA is now believed to be a prerequisite for chemical carcinogenesis. It has been suggested that the persistence of specific carcinogen-DNA adducts in target tissue correlates with susceptibility to neoplasia. For example, the persistence of BPDE-I-dGua adduct and the conformational change in the DNA molecule appears to correlate with the susceptibility of mouse skin to tumors induced by topically applied BP (4). Furthermore, Slaga and Bracken (16) have reported that antioxidants including butylated hydroxyanisole inhibits tumor initiation by DMBA in a two stage skin tumorigenesis which correlates with the inhibition of DMBA-DNA adduct formation in skin.

Skin is perhaps the most susceptible organ for BP tumorigenicity and in the present study was designed to assess the capacity of topically applied plant phenols such as tannic acid, quercetin, myricetin and anthraflavic acid on the covalent binding of BP, to the DNA of SENCAR mice epidermis. Tannic acid had the maximal inhibitory effects of the plant phenols tested on the covalent binding of BP to epidermal DNA. Our unpublished results indicate that tannic acid, quercetin, myricetin and anthraflavic acid can inhibit epidermal AHH activity in vitro and in vivo. Tannic acid was found to be the most effective inhibitor of AHH activity with a I_{50} of 4.4×10^{-5}M. The decrease in covalent binding of BP to epidermal DNA by plant phenols may be due to a decrease in the production of activated metabolites of BP. Tannic acid, quercetin, myricetin and anthraflavic acid are highly active inhibitors of the mutagenicity of many PAHs, while several other flavanoids are known to be less effective in this regard (17,18). It has been postulated that flavanoids that lack a free phenolic group are inactive as inhibitors. An examination of the structure-activity relationship of the hydroxylated anthraquinones, cinnamic acid derivatives and flavanoids reveals that phenolic hydroxyl groups on the phenyl substituent are important for the antimutagenic activity (17,18). The inhibitory effects of tannic acid, quercetin, myricetin and anthraflavic acid on the binding of BPDE-I to DNA may be due to the interaction of BPDE to plant phenol (17,18). In this regard antimutagenic activity of these compounds against PAHs has been shown to be due to the disappearance of bay-region diol epoxides by interacting with plant phenols (17,18). The half life for disappearance of BPDE in aqueous solution has been shown to be 43 min in the absence of antagonist and 9, 4.5 and 1.2 min in the presence of 10 μM quercetin, myricetin and tannic acid, respectively (17,18). Previous studies have shown that ellagic acid, another plant polyphenol, is a strong antagonist of bay-region diol epoxides of PAHs (5). Ellagic acid is two times more antimutagenic than myricetin but less than tannic acid (17,18).

In the present study we have examined the effects of plant phenols when the binding of topically applied carcinogen is optimal. Tannic acid, quercetin, myricetin and anthraflavic acid not only inhibited BP-DNA adduct formation but also decreased the binding of

BP-7,8-diol and DMBA to epidermal DNA (unpublished results). We have observed 11-13 pmol BP-DNA adducts in epidermis of SENCAR mice 24 hours after the topical application of the hydrocarbon. These adduct levels are consistence with recent studies of Digiovanni et al.(19) who showed that when mice are topically treated with [^{3}H]-BP, the covalent binding of the hydrocarbon to epidermal DNA reached a peak 24 hour after treatment. In addition these investigators have also shown that between 24-48 hours after application of the hydrocarbon there occurred a rapid drop in the level of bound BP to a DNA. The drop was 50% in 48 hours and thereafter, the level of bound BP disappeared with much slower rate (19). More recent studies have indicated that all the DNA-adducts disappeared from epidermal DNA with a biphasic decay curve (19). It was studied that DNA adducts from (+)anti BPDE with d-adenosine disappeared more rapidly than the major adduct [(+)anti BPDE-dGua]. Kekefuda and Yamamoto (20) provided evidence that binding of BPDE to adenine in DNA caused local denaturation, where as binding to guanine did not produce such denaturation, even though there was an 10 fold greater binding to guanine residues. This is because that the BPDE-I-dGua adduct lies in the minor groove of DNA helix causing little, if any, distortion and thus may be more efficiently recognizable to repair processes (21). However, work by Abbott et al. (22,23) using the skin carcinogen 15,16-dihydro-11-methylcylopenta(a)phenanthrene-17-one suggested that skin cells had little capacity to repair DNA-adducts derieved from this carcinogen. Several studies have provided evidence for the repair of adducts and also persistence of adducts for as much as one year in whole skin preparations following application of PAH at high doses (24). The decrease in epidermal BPDE-I-dG adduct formation by tannic acid, quercetin, myricetin and anthraflavic acid may be due to the increase in repair processes and needs further elucidation.

In conclusion our studies suggest that topically applied tannic acid, quercetin, myricetin and anthraflavic acid are potent inhibitors of BPDE-I-dGua adduct formation in skin and that some of these plant phenols could prove useful in modifying the risk of carcinogenicity of BP in skin.

ACKNOWLEDGEMENTS

Supported in part by NIH Grants ES-1900, CA-38028, AM-34368 and research funds from the Veterans Administration. Mukul Das is a recipient of Burroughs Wellcome Fund fellowship award from Dermatology Foundation. Thanks are due to James D. Steele for technical assistance and to Sandra Evans for typing the manuscript.

REFERENCES

1. Conney, A.H. (1982): Induction of microsomal enzymes by foreign chemicals and carcinogenesis by polycyclic aromatic hydrocarbons: G.H.A. Clowes Memorial Lecture. Cancer Res., 42, 4875-4917.
2. Gelboin, H.V. (1980): Benzo(a)pyrene metabolism, activation, and carcinogenesis: role and regulation of mixture function oxidases and related enzymes. Physiol. Rev., 60, 1107-1166.
3. Brookes, P. and Osborne, M.R. (1983): Mutation in mammalian cells by stereoisomers of anti-benzo(a)pyrene-diol epoxide in relation to the extent and nature of the DNA reaction products. Carcinogenesis, 3, 1223-1226.
4. Nakayama, J., Yuspa, S.H. and Poirier, M.C. (1984): Benzo(a)pyrene-DNA adduct formation and removal in mouse epidermis in vivo and in vitro: Relationship of DNA binding to initiation of skin carcinogenesis. Cancer Res., 44, 4087-4095.
5. Wood, A.W., Haung, M.T., Chang, R.L., Newmark, H.L., Lehr, R.E., Yagi, H., Sayer, J.M., Jerina, D.W. and Conney, A.H. (1982): Inhibition of mutagenicity of bay-region diol-epoxides of polycyclic aromatic hydrocarbons by naturally occurring plant phenols: Exceptional activity of ellagic acid. Proc. Natl. Acad. Sci. USA, 79, 5513-5517.
6. Mukhtar, H., Das, M., DelTito, B.J. and Bickers, D.R. (1984): Epidermal benzo(a)pyrene metabolism and DNA-binding in Balb/C mice: inhibition by ellagic acid. Xenobiotica, 14, 527-531.
7. Dixit, R., Teel, R.W., Daniel, F.B. and Stoner, G.D. (1985): Inhibition of benzo(a)pyrene and benzo(a)pyrene-trans-7,8-diol metabolism and DNA binding in mouse lungs explants by ellagic acid. Cancer Res., 45, 2951-2956.
8. Mukhtar, H., Das, M., DelTito, B.J. and Bickers, D.R. (1984): Protection against 3-methylcholanthrene-induced skin tumorigenesis in BALB/C mice by ellagic acid. Biochem. Biophys. Res. Commun., 119, 751-757.
9. Lesca P. (1983): Protective effects of ellagic acid and other plant phenols on benzo(a)pyrene-induced neoplasia in mice. Carcinogenesis, 4, 1651-1653.
10. Chang, R.L., Huang, M.T., Wood, A.W., Wong, C.Q., Newmark, H.L., Yagi, H., Sayer, J.M., Jerina, D.M. and Conney, A.H. (1985): Effect of ellagic acid and hydroxylated flavonoids on the tumorigenicity of benzo(a)pyrene and ($\pm$)7β,8α-dihydroxy-9α,10α-epoxy-7,8,9,10-tetrahydrobenzo(a)pyrene on mouse skin and in the newborn mouse. Carcinogenesis, 6, 1127-1133.

11. Slaga, T.J., Das, S.B., Rice, J.M. and Thompson, S. (1974): Fractionation of mouse epidermal chromatin components. J. Invest. Dermatol., 63, 343-349.
12. Das, M., Bickers, D.R. and Mukhtar, H. (1985): Inhibition of mouse epidermal polycylic aromatic hydrocarbon metabolism and carcinogenicity by clotrimazole. In: M. Cooke, A.J. Dennis and G.L. Fisher (eds.), Polynuclear Aromatic Hydrocarbon, Battalle Press, Columbus, in press.
13. Das, M., Bickers, D.R. and Mukhtar, H. (1985): Effect of ellagic acid on hepatic and pulmonary xenobiotic metabolism in mice: studies on the mechanism of its anticarcinogenic action. Carcinogenesis, in press.
14. Hesse, S., Jernstrom, B., Martinez, M., Moldeus, P., Christodulides, L. and Ketterer, B. (1982): Inactivation of DNA-binding metabolites of benzo(a)pyrene and benzo(a)pyrene-7,8-dihydrodiol by glutathione and glutathione-S-transferases. Carcinogenesis, 3, 757-760.
15. Lesca, P., Lecointe, P., Paoletti, C. and Mansuy, D. (1979): Ellipticine as potent inhibitors of microsome-dependent chemical mutagenesis. Chem. Biol. Interact., 25, 279-286.
16. Slaga, T.J. and Bracken, W.M. (1977): The effect of antioxidants on skin tumor initiation and ayl hydrocarbon hydroxylase. Cancer Res., 37, 1631-1635.
17. Huang, M.T., Wood, A.W., Newmark, H.L., Sayer, J.M., Yagi, H., Jerina, D.M. and Conney, A.H. (1983): Inhibition of the mutagenicity of bay-region diol-epoxides of polycyclic aromatic hydrocarbons by phenolic plant flavonoids. Carcinogenesis, 4, 1631-1637.
18. Huang, M.T., Chang, R.L., Wood, A.W., Newmark, H.L., Sayer, J.M., Yagi, H., Jerina, D.M. and Conney, A.H. (1985): Inhibition of mutagenicity of bay-region diol epoxides of polyclic aromatic hydrocarbons by tannic acid, hydroxylated anthraquinones and hydroxylated cinnamic acid derivatives. Carcinogenesis,
19. DiGiovanni, J., Decina, P.C., Prichett, W.P., Fisher, E.P. and Aalfs, K.K. (1985): Formation and disappearance of benzo(a)pyrene DNA-adducts in mouse epidermis. Carcinogenesis, 6, 741-747.
20. Kakefuda, T. and Yamamoto, H.A. (1978): Modification of DNA by the benzo(a)pyrene metabolite diol epoxide r-7,t-8-dihydroxy-t-9,10-oxy-7,8,9,10-tetrahydrobenzo(a) pyrene. Proc. Natl. Acad. Sci. USA, 75, 415-419.

21. Grunberger, D. and Weinstein, I.B. (1979): Conformational changes in nucleic acids modified by chemical carcinogens. In: P.L. Grover (ed.), Chemical Carcinogens and DNA, Vol. 2, CRC Press Inc, Boca Raton, Florida, pp 153-201.
22. Abbott, P. and Crew, F. (1981): Repair of DNA adducts of the carcinogen 15,16-dihydro-11-methylcylclopenta(a)phenanthren-17-one in mouse tissues and its relation to tumor production. Cancer Res., 41, 4115-4120.
23. Abbott, P. (1983): Strain specific tumorigenesis in mouse skin induced by the carcinogen, 15,16-dihydro-11-methylcyclopenta(a)phenanthren-17-one, and its relation to DNA adduct formation and persistence. Cancer Res., 43, 2261-2266.
24. Ashurst, S.W. and Cohen, G.M. (1982): The formation of benzo(a)pyrene-deoxyribonculeoside adducts in vivo and in vitro. Carcinogenesis, 3, 267-273.

IDENTIFICATION OF POLYCYCLIC AROMATIC HYDROCARBONS USING LIQUID CHROMATOGRAPHY WITH ON-THE-FLY FLUORESCENCE LIFETIME DETERMINATION

DAVID J. DESILETS, PETER T. KISSINGER, FRED E. LYTLE*
Department of Chemistry, Purdue University, West Lafayette, Indiana 47907.

INTRODUCTION

Trace analysis of polycyclic aromatic hydrocarbons (PAH) is important because many of these compounds are potent carcinogens. Liquid chromatography (LC) is often used as the instrumental method in these analyses because of its high sample throughput, ability to handle complex mixtures, and relative ease of use. However, in most cases, compound identification is based solely on LC retention time since more selective detectors such as linear photodiode arrays lack the sensitivity needed for trace analysis.

This paper will describe a new technique for identifying low levels of PAH. The method utilizes on-the-fly measurement of fluorescence lifetime, that is, without resorting to stopped flow, as the additional selective factor required for a positive identification. We will show that chromatographic retention time combined with fluorescence lifetime estimates can yield sufficient information for an unambiguous identification of PAH in most cases.

MATERIALS AND METHODS

Standard Compounds

PAH standards were obtained from a variety of commercial sources. Their purity was checked using fluorescence and UV absorption spectroscopy, and by liquid chromatography. The standard mixture used contained the following compounds at approximately 100 ng/mL each in acetonitrile: fluoranthene (Fln), pyrene (Pyr), benz[a]anthracene (Baa), benzo[b]fluoranthene (BbFln), benzo[k]fluoranthene (BkFln), benzo[a]pyrene (BaP), and benzo[ghi]perylene (BghiP).

Solvents

Acetonitrile (J. T. Baker) was organic residue analysis grade, and was fractionally distilled in glass prior to use.

Deionized water was distilled in glass as well. For LC mobile phase, each solvent was filtered through a 0.2-μm nylon membrane filter before mixing. Dichloromethane was of the type used for pesticide residue analysis (Fisher Scientific).

Sample Preparation

Fly ash from a biomass combustion furnace was used as a source of anthropogenic PAH. Samples were collected on 20 x 25-cm glass-fiber absolute filters. These filters were extracted for six hours with 200 mL of dichloromethane in a Soxhlet extractor (approximately 30 cycles). Each dichloromethane extract was subjected to preparative column chromatography according to a previously published procedure in order to isolate the PAH from other combustion products (1).

Liquid Chromatography

Chromatography was performed on a Bioanalytical Systems Model LC-154 liquid chromatograph. The mobile phase consisted of an 80/20 (v/v) mixture of acetonitrile and water. The flow rate was 1.1 mL/min, the temperature was ambient, and a 20-μL injection loop was used.

Fluorescence Detection System

The 337.1-nm line of a pulsed nitrogen laser (Princeton Applied Research Model 2100) was used as the excitation source. Pulse widths were approximately 1.5 ns FWHM. The laser was operated at a repetition rate of 10 Hz. A portion of the beam was split off with a quartz plate and used to trigger a sampling oscilloscope via a photodiode (Texas Instruments TIED 56). The main part of the beam was focused into a vertical flow cell from the underside, thus eliminating much of the scatter, and providing a vertical image matching the emission monochromator entrance slit. Emission centered at 420 nm was collected and focused into a J-Y H-20 monochromator, which passed the desired wavelength range to a fast-wired RCA 931A photomultiplier tube (2). An 8-nm bandpass was used. For the chromatograms shown here, a neutral-density filter was placed between the flow cell and the collection optics so that a 30-fold attenuation in the signal was obtained. This was necessary to keep the signal on scale.

A two-channel sampling oscilloscope (Tektronix 5103N mainframe and a 5S14N dual-channel sampling head with two 350-ps apertures) was used to measure the photomultiplier anode current (Figure 1). This instrument could monitor two channels of input simultaneously upon each trigger from the laser. The oscilloscope was not scanned, but rather, both apertures monitored the same fixed point at some arbitrary time after triggering. Before input into the oscilloscope, the signal from the photomultiplier was split evenly with a power divider (General Radio). Half of the signal was made to pass

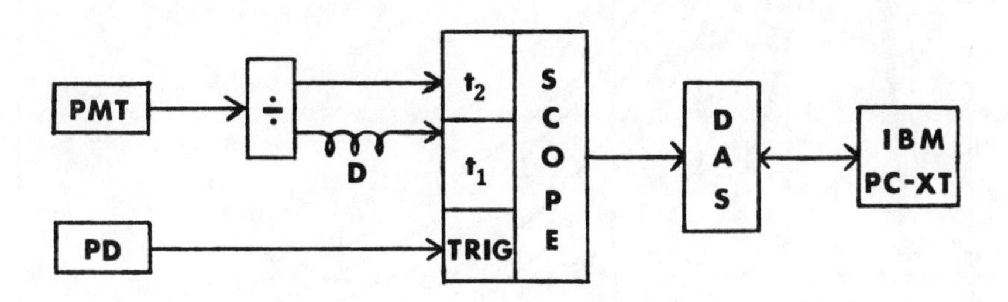

FIGURE 1. Apparatus for 2-point decay measurements. PMT, photomultiplier; PD, fast photodiode; ÷, power divider; D, 10-ns coaxial delay cable; t_2 and t_1, channel 2 and channel 1 inputs to the sampling head; SCOPE, 2-channel sampling oscilloscope; DAS, computer interface.

through a length of coaxial cable such that a 10-ns delay was added to the signal in this channel (t_1) before it was seen by the oscilloscope. The other half of the signal was sent directly to the oscilloscope with no delay (t_2). Because there is no delay on the channel 2 input, the fluorescence seen at this channel will have decayed away somewhat compared to that at channel 1 which arrives later. The result was that the signals being sampled by the two channels of the oscilloscope were out of phase by 10 ns, so two points on each fluorescence decay were being measured for every laser pulse (Figure 2) (3).

The oscilloscope outputs from channels 1 and 2 were digitized by means of a Keithley DAS Series 500 interface, and the digitized data were then sent to an IBM PC-XT microcomputer for storage and display.

Determination of Fluorescence Lifetimes

Notice from Figure 2 that for all fluorescence decays expected, the signal at t_2 is lower than that at t_1. Therefore, a ratio of the signals, t_2/t_1, will approach zero for short lifetimes, and will asymptotically approach unity

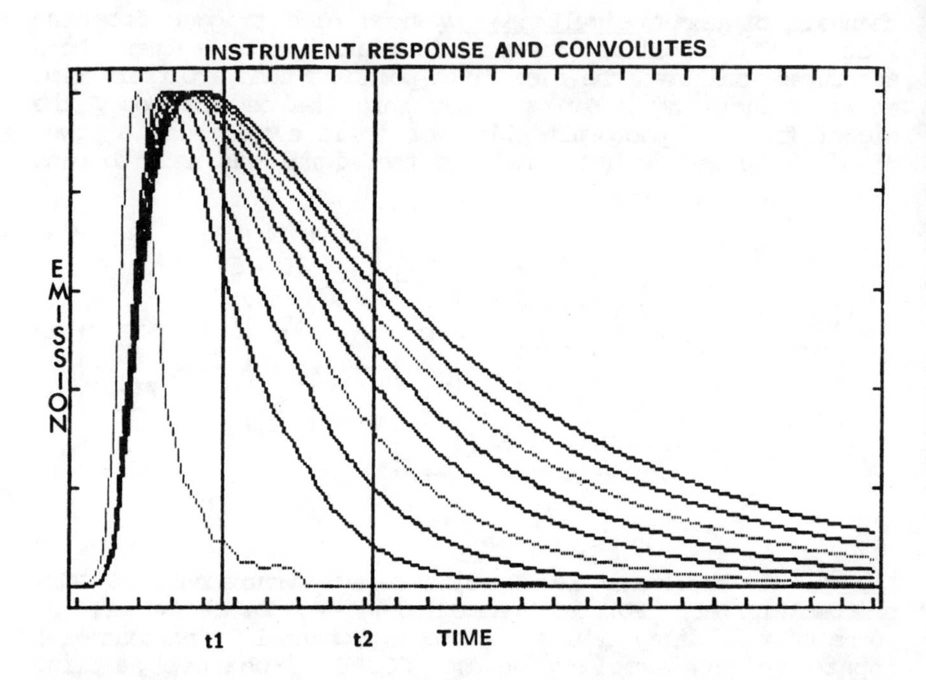

FIGURE 2. Positioning of the sampling apertures. The decays shown here were computer generated from the excitation function shown at the extreme left. These decays represent the expected spread in lifetimes (approximately 4 to 30 ns) for the PAH examined. The apertures were placed so as to yield the maximum spread in the ratios without having t1 before the peak of the longest decay, and without placing t2 beyond measurable emission from the shortest decay.

as the fluorescence lifetime increases toward infinity. In principle, a continuous (smooth) calibration curve could be constructed of the ratio, t_2/t_1, vs. fluorescence lifetime. Such a curve is shown in Figure 3. Lifetimes for unknown compounds can be obtained from measured ratios using this kind of calibration curve.

RESULTS

A portion of a typical two-channel chromatogram of PAH standards is shown in Figure 4. This expanded view also shows the ratio of the two chromatograms. It is important to note that the "ratiogram" is flat across each peak regardless of the

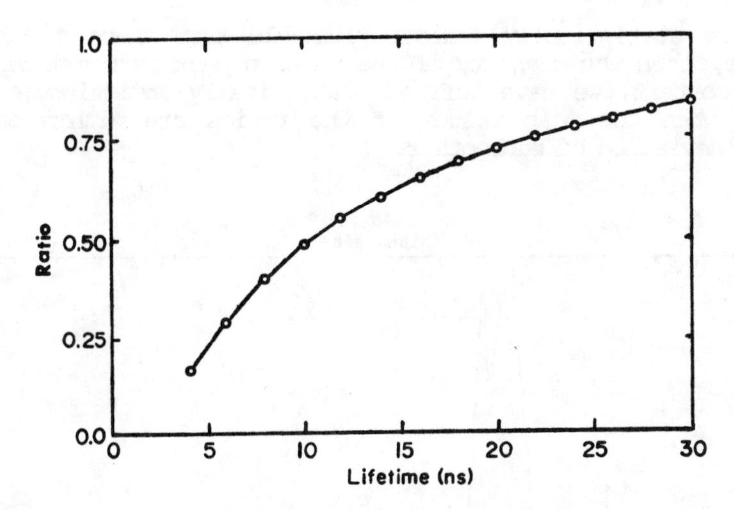

FIGURE 3. Calibration curve of ratios expected for a given fluorescence lifetime. This curve was constructed by calculating the ratio of t_2/t_1 for computer-generated decays such as those shown in Figure 2 (4 to 30-ns decays, at 2-ns intervals). The ratio was then plotted vs. the "lifetime" of the decay (3).

continuous changes in concentration and the somewhat severe noise encountered in the raw data. Most of this noise is due to fluctuations in laser intensity (3). Since the two channels of data are obtained simultaneously, both chromatograms will exhibit the same response to fluctuations in source intensity. Should the laser misfire, both chromatograms will exhibit a sharp drop in signal intensity. If an exceptionally strong pulse is encountered, both channels will experience a marked increase in signal strength. These phenomena are especially evident in the middle peak of Figure 4. When a ratio of the two chromatograms is obtained, these peaks and valleys correlate, and much of the noise is eliminated. The resulting ratio is related to the fluorescence lifetime as mentioned above.

Figure 5 illustrates one channel each for two chromatograms obtained sequentially. The first consists of PAH standards, and the second is an unknown sample consisting of an extract of fly ash. On the basis of retention time alone, tentative identifications can be made for most of the major peaks in the unknown mixture. Confirmation of peak identity is obtained by comparing the "ratiograms" from these two samples. For a given retention time, if the ratios for the fly ash sample are statistically indistinguishable from those of the standards, then the identity of the unknown peak is confirmed. In this context, we have defined statistically indistinguishable as when the mean values of the ratios are within one standard deviation of each other.

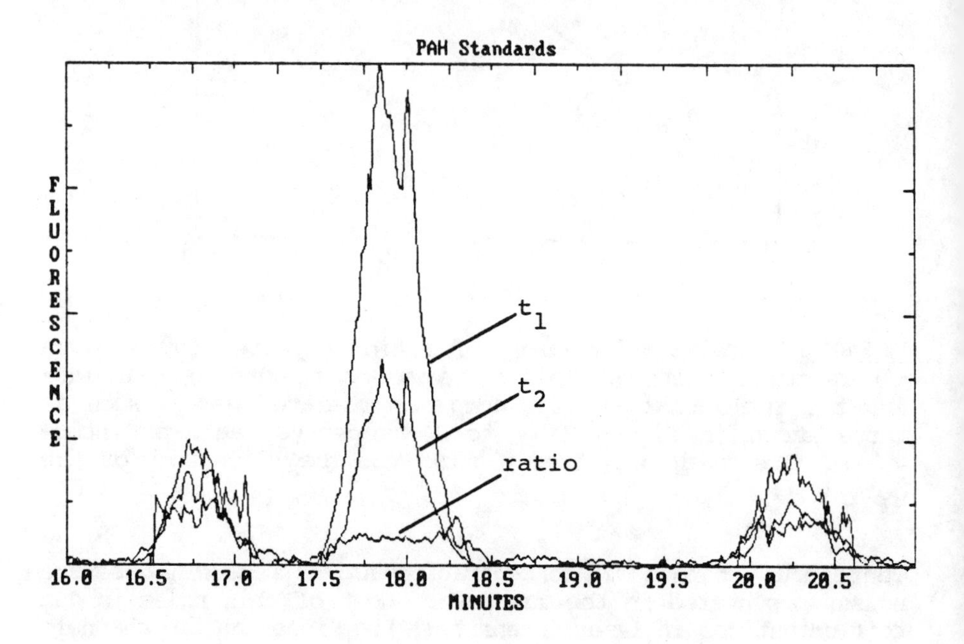

FIGURE 4. Portion of a two-channel chromatogram and "ratiogram" for some PAH standards (BbFln, BkFln, and Bap).

Table 1 contains a list of positive identifications which were made on the basis of ratio comparison. In practice, actual lifetimes are not determined since the ratios contain all the information necessary for the comparison. benzo[a]-fluorene (BaFl), chrysene (Chry), benzo[j]fluoranthene (BjFln), benzo[e]pyrene (BeP), perylene (Pery), and indeno[1,2,3-cd]-pyrene (IcdP) were not present in the standard mixture shown in Figure 5. Their confirmation in the fly ash sample was based on comparison to retention times and lifetimes of the pure compounds.

TABLE 1

RATIOS, LIFETIMES, AND PEAK IDENTITIES FOR THE CHROMATOGRAMS IN FIGURE 5

Peak	Standards R $\pm\sigma$	Fly Ash R $\pm\sigma$	τ (ns)	Compound
A,1	0.893±0.111	0.838±0.083	32	Fln
B,2	0.784±0.089	0.804±0.090	25	Pyr
3	-	0.343±0.067	6	BaFl*
C,4	0.797±0.134	0.586±0.069	26,12	Baa,Chry*
5	-	0.634±0.134	13	BjFln*
D,6	0.810±0.141	0.749±0.101	27,21	BbFln,Bep*
7	-	0.465±0.039	9	Pery*
E,8	0.369±0.063	0.378±0.045	6.5	BkFln
F,9	0.564±0.098	0.543±0.112	11	Bap
G,10	0.504±0.128	0.486±0.068	10	BghiP
11	-	0.402±0.081	7.5	IcdP*

* Compounds present in the fly ash sample, but not present in the standard mixture. Their identities were determined by comparison to retention times and lifetimes of the pure compounds. All other compounds listed were present in both the fly ash sample and the standard set.

DISCUSSION

We have determined elsewhere that the 2-point decay ratios measured in the manner described above yield accurate estimates of fluorescence lifetime (3). Furthermore, the lifetime estimate is concentration independent since, by the very nature of the ratio, effects of concentration on emission intensity tend to cancel out as a result of the division. The technique is thus ideally suited for peak identification in liquid chromatography because the changing elution profile has no effect on the measured lifetime.

A close examination of Table 1 reveals that for at least two of the peaks in the unknown chromatogram (4 and 6), an identification based on retention time alone would have been only partially correct. For cases where correct identification of isomeric PAH species is essential, such as in carcinogenic risk assessment, the additional effort and expense incurred by the ratio method are justified.

Since the sampling apertures are only 350 ps wide, a significant amount of the emission is rejected. As with any method providing enhanced selectivity, tradeoffs in sensitivity are to be expected. While the system described here uses an emission monochromator with an 8-nm bandpass, a simple filter fluorimeter could be used, yielding an increased sensitivity with no loss in selectivity. However, the monochromator allows spectral selectivity in the unlikely event that two overlapping peaks would have the same fluorescence lifetime to within the precision of the ratio method. It should also be noted that the sensitivity could be increased by removing the neutral-density filter from the collection optics. Ultimately, the detection limit of the technique would be limited by background fluorescence from the mobile phase and from the flow cell itself. We estimate a practical concentration limit for usable ratios at about 10 ng/mL (200 pg injected) depending on the compound.

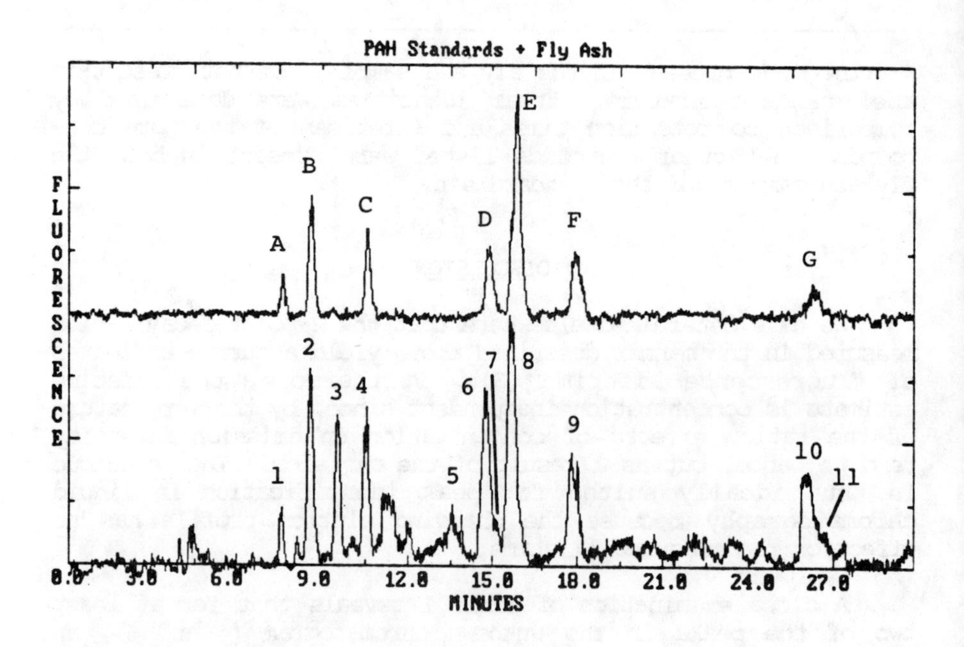

FIGURE 5. Chromatograms of PAH standards (top) and a fly ash extract (bottom). For peak identification, see Table 1.

For peaks that are not well resolved, the ratio method begins to break down. No arbitrary degree of resolution can be cited as the limit beyond which a two-point decay cannot be determined. This is because the relative peak heights and the difference in lifetimes also play a role in the precision of the measurement. Still, poorly resolved peaks can be identified. Figure 6 shows a small shoulder nearly buried under the right side of the BbFln/Bep peak of the fly ash sample. The shoulder is due to the presence of a small amount of perylene. The ratiogram clearly shows an abrupt drop as the shoulder is encountered. This behavior is consistent for perylene, which has a short fluorescence lifetime under these conditions.

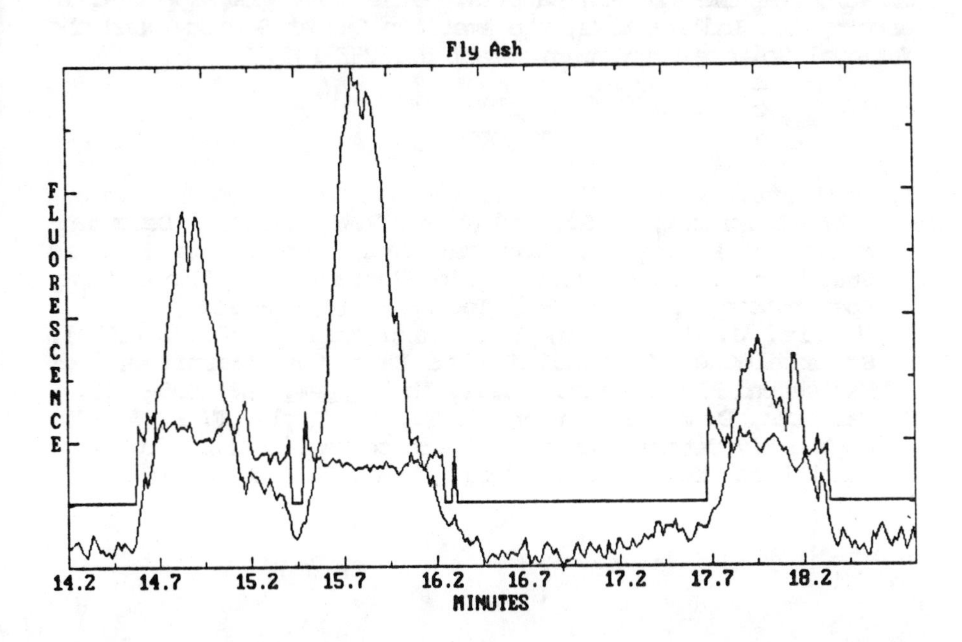

FIGURE 6. Expanded view of one channel of the fly ash chromatogram shown in Figure 5, and the corresponding ratiogram. The shoulder on the leftmost peak causes an abrupt change in the ratio, enabling identification of this poorly resolved minor component of the mixture.

Although further development is still required, it is safe to say that the two-point decay method has the potential for becoming a useful tool for analysis of PAH at trace levels.

Studies concerning the optimization of instrumental variables such as the aperture delays and their spacing are needed in order to derive the maximum benefit from this technique. Similarly, an examination of the precision, accuracy, and reproducibility of the lifetime estimates must be made in order to determine with what degree of confidence peak assignments can be made. Studies along these lines are presently being conducted.

ACKNOWLEDGMENTS

We thank R. B. Jacko, M. A. Horne, and M. S. Ludwiczak for donation of the fly ash samples. This work was supported in part by the Indiana Elks, the American Cancer Society, and the National Science Foundation, grant No. CHE-8320158.

REFERENCES

1. Desilets, D. J., Kissinger, P. T., Lytle, F. E., Horne, M.A., Ludwiczak, M. S., and Jacko, R. B. (1984): Determination of Polycyclic Aromatic Hydrocarbons in Biomass Gasifier Effluents with Liquid Chromatography/Diode Array Spectroscopy, Environ. Sci. Technol., 18:386-391.
2. Harris, J. M., Lytle, F.E., and McCain, T. C. (1976): Squirrel-Cage Photomultiplier Base for Measurement of Nanosecond Fluorescence Decays, Anal. Chem., 48:2095-2098.
3. Desilets, D. J., Kissinger, P. T., and Lytle, F. E. (1986): On-the-Fly Determination of Fluorescence Lifetimes from Two-Point Decay Measurements, Anal. Chem., 58 (accepted).

EVALUATION OF THE GENOTOXICITY OF API REFERENCE UNLEADED GASOLINE

JOHN F. DOOLEY, MICHAEL J. SKINNER, TIMOTHY A. ROY, GARY R. BLACKBURN, CEINWEN A. SCHREINER, CARL R. MACKERER. Mobil Environmental and Health Science Laboratory, P.O. Box 1029, Princeton, New Jersey 08540

INTRODUCTION

A series of studies sponsored by the American Petroleum Institute (API) demonstrated that a reference unleaded gasoline (API PS-6) was not mutagenic in the Ames *Salmonella* assay or the L5178Y mouse lymphoma assay and, when administered orally, did not induce dominant lethal mutations in mice or chromosome aberrations in rat bone marrow cells (1,2). However, Farrow *et al.* (3) recently reported that unleaded gasoline obtained from Phillips Petroleum Company was mutagenic in the L5178Y mouse lymphoma assay. This finding prompted a re-examination of the mutagenic potential of unleaded gasoline. Mutagens often initiate a carcinogenic response in animals and API PS-6 gasoline has previously been reported to produce renal tumors in male rats and liver tumors in female mice (4).

In this study, we evaluated the ability of API PS-6 unleaded gasoline to induce mutations in L5178Y mouse lymphoma cells and chromosome aberrations in bone marrow cells of orally dosed rats. In addition, we tested two fractions of API PS-6, a DMSO extract and a six-hour evaporative residue, in a modified Ames assay (6,7) and in the L5178Y mouse lymphoma assay. These fractions were tested to determine the presence of small quantities of mutagenic components, whose activity was being masked by the toxicity of the mixture. The DMSO extraction was used to isolate both polar compounds and aromatics and the six-hour evaporation to concentrate the higher boiling hydrocarbons, such as polynuclear aromatics (PNAs). The DMSO extract and the evaporative residue of API PS-6 were also analyzed by gas chromatography/mass spectrometry

(GC/MS). The results of the mutagenicity testing are discussed in relation to the chemical analyses of these two fractions.

MATERIALS AND METHODS

Sample Preparation

DMSO Extract/Mutagenicity Testing. The DMSO extract of API PS-6 unleaded gasoline used in the mutagenicity assays was prepared by successively extracting 2.0 ml gasoline five times with 2.0 ml DMSO and then pooling the extracts (10.4 ml total). Approximately 20% of the unleaded gasoline (*i.e.*, 0.4 ml) was extracted into the DMSO.

DMSO Extract/Chemical Analyses. The DMSO extract of API PS-6 unleaded gasoline used in the PNA analyses was prepared differently from the extract used for mutagenicity testing. The changes in extraction procedures were implemented to maximize PNA recovery and facilitate GC/MS analysis. Duplicate 5 ml aliquots of API PS-6 were pipetted into 7.5 ml of cyclohexane in 60 ml separatory funnels. The samples were fortified with toluene-d_8, naphthalene-d_8, and anthracene-d_{10} at approximately 400 µg each. The samples were extracted with 2 x 12.5 ml DMSO and the PNA-enriched extracts were collected in 125 ml separatory funnels. The DMSO fraction was diluted 1:2 with distilled water (50 ml) and was back extracted with pentane (3 x 45 ml). The pentane extracts were then transferred to 45 ml graduated centrifuge tubes and were evaporated under nitrogen to approximately 8 ml. The samples were back extracted with 1 ml distilled water to remove any residual DMSO. The sample extracts were then passed through a prewashed sodium sulfate funnel with a glass wool plug to remove the residual water. The centrifuge tube and the funnel were rinsed with pentane and the rinsings along with the extract were collected in 15 ml graduated centrifuge tubes. The sample extracts were then evaporated to a 5 ml final volume under nitrogen. A recovery standard containing the same concentration of the three internal standards was also prepared.

Evaporative Residue. Residue samples were prepared by adding 4 ml of API PS-6 unleaded gasoline (2.85 - 2.99 gm) to preweighed scintillation vials and evaporating under a gentle stream of nitrogen for 6 hours at ambient temperature. The residues (106.3 - 116.5 mg), which ranged from 3.54% to 4.08% (and averaged 3.8%) of the weight of the whole unleaded gasoline samples prior to evaporation, were resuspended in ethanol for the mutagenicity testing and chemical analyses.

Assay Procedures

Metaphase Analysis. Male Sprague Dawley rats (175 - 225 gm; 5/group) were dosed orally with API PS-6 unleaded gasoline (neat at 500, 750 and 1000 mg/kg/day) or with isotonic saline (10 ml/kg/day) for five consecutive days. Colchicine (4.0 mg/kg) was administered intraperitoneally four hours after the last treatment and the animals were sacrificed two hours later. The positive control animals were dosed orally with cyclophosphamide (45.0 mg/kg) twenty-four hours before sacrifice. Bone marrow was harvested and processed as described by Brusick (5). The data were analyzed using a one-tailed Mann-Whitney test.

Ames Assay. We have developed modifications to the Ames assay that markedly improve the predictability of this assay for petroleum-derived materials (6,7). These modifications include (a) extraction of the test material with DMSO to yield a water-miscible, PNA-enriched fraction; (b) an eightfold increase in the concentration of Aroclor 1254-induced liver S-9 to 400 µl/plate; (c) substitution of Syrian golden hamster liver S-9 for the standard rat liver S-9; and (d) a twofold increase in NADP cofactor concentration to 8 mM. The DMSO extract of API PS-6 was tested over a range of concentrations from 5 to 200 µl/plate with an eightfold increase in Aroclor 1254-induced rat liver S-9. The evaporative residue was tested over a range of concentrations from 50 to 10,000 µg/plate with an eightfold increase in Aroclor 1254-induced hamster liver S-9 and a twofold increase in NADP. The assays were conducted as pre-incubation assays as previously described (6 and 7, respectively).

Mouse Lymphoma Mutagenesis Assay. The mouse lymphoma mutagenesis assays of API PS-6 gasoline, the DMSO extract of PS-6 and the evaporative residue were conducted according to the procedures of Clive et al. (8,9). The concentrations tested were based on the results of preliminary cytotoxicity assays and are listed in Tables 2, 3 and 4, respectively. Each assay was performed (in duplicate) with and without an Aroclor 1254-induced rat liver S-9 activation system. Ethyl methanesulfonate (EMS) was used as the positive control for the nonactivated assays and 7,12-dimethylbenz(a)anthracene (DMBA) was used for the S-9 activated assays. Ethanol was used as the solvent and as the negative control (1%) for the mutagenicity assays of API PS-6 and the evaporative residue of PS-6. DMSO was used as the solvent and the negative control (1%) for the mutagenicity assays of the DMSO extract of PS-6.

Chemical Analyses. Qualitative and quantitative analyses of the DMSO extract and the evaporative residue of API PS-6 unleaded gasoline were performed on a Hewlett-Packard 5985B gas chromatograph/mass spectrometer/data system.

RESULTS

Metaphase Analysis

Slight oral irritation (blood around the mouth and nose) was noted in several of the rats dosed with API PS-6. No other adverse pharmacologic effects were observed in any of the treated animals. No chromosome aberrations were observed in 300 bone marrow cells from negative control animals, in 200 cells from animals treated with the lowest dose of PS-6 (500 mg/kg/day) or in 250 cells from animals treated with the intermediate dose of PS-6 (750 mg/kg/day). Only two aberrations, a chromatid break and a minute chromosome, were observed in 245 cells from animals treated with the highest dose of PS-6 (1000 mg/kg/day). These aberrations did not represent a statistically significant increase over control values. In contrast, a highly significant increase ($p<0.001$) in aberrations (14 chromatid breaks, 44 exchanges, 3 fragmented chromosomes and 30 pulverized chromosomes) was observed in 54 of 128 cells from positive control animals. These data indicate that API PS-6 unleaded gasoline does not induce chromosome damage in the bone marrow of rats following oral administration.

Ames Assay

The DMSO extract of API PS-6 unleaded gasoline [DE(PS-6)] was tested at concentrations of 5, 10, 20, 30, 50, 75, 100 and 200 µl/plate in our modified assay (strain TA98, 400 µl rat liver S-9/plate). The background bacterial lawn was reduced in a concentration-dependent manner at concentrations above 20 µl/plate and microcolonies were observed at the highest concentration of extract. The mean number of revertant colonies in extract-treated cultures (27 to 40 revertants/plate) was not significantly different from control values (30, 30 and 35 revertants/plate at 0, 100 and 200 µl DMSO/plate), except at the highest concentration where a slight reduction was observed (12 revertants/plate). In contrast, a significant increase in revertant recovery was observed in the positive control cultures (295 revertants/plate at 5.0 µg B(a)P/plate and 426 revertants/plate at 2.0 µg 2-AA/plate). These data indicate that while the DMSO extract of API PS-6 gasoline is toxic, it is not mutagenic in our modified assay.

The results of testing the evaporative residue of API PS-6 unleaded gasoline [PS-6(RES)] in the modified Ames assay (strain TA 98, 400 µl hamster liver S-9/plate and 8 mM NADP) are presented in Table 1. A slight increase (1.4-fold) in revertant recovery was observed for two non-toxic concentrations of residue in the first assay. Similar increases (1.7- to 1.9-fold) were observed over a range of concentrations in a repeat assay but the number of revertants did not reach a doubling of control values. These data indicate that the evaporative residue, like the DMSO extract, is not mutagenic in this modified assay.

TABLE 1

MUTAGENICITY OF THE EVAPORATIVE RESIDUE OF API PS-6 UNLEADED GASOLINE IN THE MODIFIED AMES ASSAY[a]

Treatment	Mean TA98 Revertants/Plate (n=3) Trial 1	Trial 2
μg PS-6(RES)/plate		
10,000	27 (mc)[b]	-
5,000	51	-
4,000	-	31 (er)
3,000	-	38 (sr)
2,500	63	46
2,000	-	39
1,500	-	59
1,250	60	-
1,000	-	54
750	-	54
500	54	52
250	46	-
125	46	-
50	41	-
2.0 μg 2-AA/plate	655	1115
5.0 μg B(a)P/plate	357	525
100 μl EtOH/plate	45	-
50 μl DMSO/plate	-	30

[a]8x Aroclor-induced hamster liver S-9/2x NADP.
[b]mc - microcolonies; er and sr - extremely and slightly reduced bacterial lawn.

Mouse Lymphoma Mutagenesis Assay

The results of the mouse lymphoma mutagenesis assay of API PS-6 unleaded gasoline are listed in Table 2. PS-6 did not produce a significant increase in mutant frequency (twofold or greater) at concentrations with total growths of 10% or greater in the presence or absence of metabolic activation. However, at concentrations yielding total growths of less than 10%, PS-6 did produce

TABLE 2

MUTAGENICITY OF API PS-6 UNLEADED GASOLINE IN THE MOUSE LYMPHOMA ASSAY

Treatment	S-9	Mutant Frequency ($\times 10^{-4}$)	% Total Growth[a]
PS-6 (µl/ml)			
0.070	-	0.61	5.1
0.065	-	0.32	21.2
0.060	-	0.57	2.7
0.055	-	0.26	60.0
0.050	-	0.31	83.3
0.045	-	0.23	106.3
EMS (µl/ml)			
1.0	-	15.33	6.3
0.5	-	4.76	61.0
EtOH (µl/ml)			
10.0	-	0.19	100.0
10.0	-	0.24	100.0
PS-6 (µl/ml)			
0.175	+	0.44	5.3
0.163	+	0.20	40.7
0.150	+	0.48	4.9
0.138	+	0.30	101.4
0.125	+	0.28	126.2
DMBA (µg/ml)			
5.0	+	5.31	1.1
2.5	+	1.33	107.2
EtOH (µl/ml)			
10.0	+	0.28	100.0
10.0	+	0.20	100.0

[a]calculated as described by Clive _et al._ (8).

twofold or greater increases in mutant frequency with and without metabolic activation. The increases in mutant frequency reflected increases in the total number of mutants and were not caused by reduction in cloning efficiency. Thus, PS-6 unleaded gasoline is not mutagenic to mouse lymphoma cells by standard evaluation criteria (8,9); however, the increased number of mutant colonies suggests that PS-6 may contain a weakly mutagenic component or small quantities of mutagenic components with mutagenicity masked by the toxicity of the total mixture.

The results of testing the DMSO extract and the 6-hour evaporative residue of API PS-6 unleaded gasoline in the mouse lymphoma mutagenesis assay are presented in Tables 3 and 4, respectively. The extract produced a significant, near dose-dependent increase in mutant frequency over a narrow range of concentrations in the absence of metabolic activation. The increases in mutant frequency ranged from approximately twofold to fourfold and were paralleled by increases in the total number of mutant colonies. In contrast, the extract did not produce a significant increase in mutant frequency in the presence of metabolic activation, even at fairly toxic concentrations (2.3 - 2.5 μl/ml). The residue, on the other hand, produced a significant, dose-dependent increase in mutant frequency in the presence of metabolic activation, but not in the absence of metabolic activation. In the nonactivated assay, the residue did produce an approximate threefold increase in mutant frequency at the highest concentration tested (90.0 μg/ml). However, this increase was not considered indicative of a mutagenic response because the total growth of this culture was less than 10% and because the increase in mutant frequency was due to a reduction in cloning efficiency and not to an increase in the total number of mutant colonies.

In summary, the DMSO extract of PS-6 is mutagenic without metabolic activation, whereas the evaporative residue is mutagenic with metabolic activation. These results demonstrate the presence of at least two different mutagenic components, one that is direct-acting and soluble in DMSO and one that requires metabolic activation and has a relatively high boiling point. Neither of these mutagens was present in sufficient quantity in the whole gasoline to be observed without the enrichments provided by DMSO extraction or evaporation. The DMSO extract contained, by weight, approximately 20% of the gasoline and the residue approximately 4%.

TABLE 3

MUTAGENICITY OF THE DMSO EXTRACT OF API PS-6 UNLEADED GASOLINE IN THE MOUSE LYMPHOMA ASSAY

Treatment	S-9	Mutant Frequency ($\times 10^{-4}$)	% Total Growth[a]
µl DE(PS-6)/ml			
1.75	-	1.12	6.8
1.70	-	1.40*	11.4
1.60	-	1.34*	17.1
1.50	-	0.82*	28.2
1.40	-	0.61	47.4
1.30	-	0.81	69.7
1.20	-	0.76	91.9
EMS (µl/ml)			
1.0	-	7.80	22.1
0.5	-	5.37	46.5
DMSO (µl/ml)			
10.0	-	0.36	100.0
µl DE(PS-6)/ml			
2.5	+	0.33	7.0
2.4	+	0.36	16.5
2.3	+	0.35	17.2
2.2	+	0.24	75.5
2.1	+	0.18	60.2
2.0	+	0.26	79.5
DMBA (µg/ml)			
5.0	+	4.71	9.6
2.5	+	1.35	93.7
DMSO (µl/ml)			
10.0	+	0.24	100.0
10.0	+	0.23	100.0

[a]calculated as described by Clive _et al_. (8).
*twofold or greater increase with acceptable total growth ($\geq$ 10%).

TABLE 4

MUTAGENICITY OF THE EVAPORATIVE RESIDUE OF API PS-6 UNLEADED GASOLINE IN THE MOUSE LYMPHOMA ASSAY

Treatment	S-9	Mutant Frequency ($x 10^{-4}$)	% Total Growth[a]
μg PS-6(RES)/ml			
90.0	-	0.63	3.2
80.0	-	0.43	15.7
70.0	-	0.27	42.6
60.0	-	0.28	80.5
50.0	-	0.19	85.5
EMS (μl/ml)			
1.0	-	15.05	7.0
0.5	-	5.34	57.0
EtOH (μl/ml)			
10.0	-	0.27	100.0
10.0	-	0.22	100.0
μg PS-6(RES)/ml			
200.0	+	1.14	1.0
175.0	+	0.71*	13.1
150.0	+	0.61*	28.1
125.0	+	0.41	31.5
100.0	+	0.36	46.3
50.0	+	0.45	79.1
DMBA (μg/ml)			
5.0	+	8.44	1.0
2.5	+	1.33	77.0
EtOH (μl/ml)			
10.0	+	0.26	100.0
10.0	+	0.24	100.0

[a]calculated as described by Clive _et al_. (8).
*twofold or greater increase with acceptable total growth ($\geq$ 10%).

Chemical Analyses

Qualitative and quantitative analyses of the DMSO extract and the evaporative residue of API PS-6 unleaded gasoline revealed different component profiles (Table 5). Forty-six peaks were integrated for the DMSO extract, whereas sixty-four were integrated for the evaporative residue (with 25 of the 64 accounting for 75% of the residue). The DMSO extract contained a high concentration of alkylated benzenes (73%) and relatively low concentrations of non-aromatics, alkylated naphthalenes and indans. PNAs with three or more rings were present at very low concentrations in the extract, but they could not be identified by the procedures used. In contrast, the evaporative residue contained a high concentration of alkylated naphthalenes (41%) and low concentrations of non-aromatics, alkylated benzenes and indans. Several PNAs, containing three or more rings, were detected at low levels in the evaporative residue. These included anthracene, phenanthrene, methyl fluorenes, dimethyl fluorenes, methyl anthracenes, C2-alkyl phenanthrenes, pyrene and fluoranthene. The components identified in the DMSO extract and the evaporative residue suggest that the mutagens detected in the mouse lymphoma assay are in the heavy ends of unleaded gasoline, most of which boil at temperatures above 212° F (100° C).

DISCUSSION

The results of the mutagenicity testing of API PS-6 reference unleaded gasoline are summarized in Table 6. These results demonstrate that whole API PS-6 unleaded gasoline is not mutagenic at the thymidine kinase locus of mouse lymphoma cells or clastogenic in the bone marrow cells of rats following oral administration, thus confirming the results of previous API sponsored studies (1,2). However, PS-6 gasoline does contain at least two mutagenic components; one of these is a direct-acting mutagen while the other requires metabolic activation to be mutagenic. It is likely that both of these are aromatic in nature. The direct-acting mutagen was extracted with DMSO, which quantitatively removes aromatics from oil samples. The mutagen requiring activation, which was found

TABLE 5

CLASSIFICATION OF COMPONENTS IN THE PNA-ENRICHED DMSO EXTRACT OF API PS-6 AND THE EVAPORATIVE RESIDUE OF API PS-6

Compound Class	Percentage in Extract		Percentage in Residue
Non-aromatics (alkanes and cycloalkenes)	12.0	(8200)[a]	8.0
Alkyl and alkenyl Benzenes	73.0	(52000)	9.0
Alkyl Indans	5.9	(4200)	10.0
Tetrahydro-naphthalenes	0.2	(150)	2.0
Alkyl Naphthalenes	9.3	(6600)	41.0
Unidentified (minor peaks)	-		4.0
Unidentified (major peaks)	-		26.0

[a] numbers in parentheses are concentrations (ppm) of each compound class in API PS-6 unleaded gasoline. These concentrations were based on internal standardization and were used to determine the percentage of each compound class in the DMSO extract.

in the evaporative residue, is likely to be a PNA present at low concentration in the whole gasoline.

The DMSO extract and the evaporative residue of API PS-6 are not mutagenic in our modified Ames assay. In previous studies, we have demonstrated an excellent correlation (r=0.92) between mutagenicity of petroleum distillation fractions in this modified assay and carcinogenicity of these

TABLE 6

SUMMARY OF MUTAGENICITY TESTING OF API PS-6 REFERENCE UNLEADED GASOLINE

Assay	Sample	In Vitro -S9	In Vitro +S9	In Vivo
metaphase	whole gasoline	nt[a]	nt	-
Ames	whole gasoline[b]	-	-	nt
	DMSO extract	-	-	nt
	residue	-	-	nt
Mouse Lymphoma	whole gasoline	-	-	nt
	DMSO extract	-	+	nt
	residue	+	-	nt

[a] NT - not tested; (+) - positive; (-) - negative.
[b] API Medical Research Publication Number 28-31344.

fractions in mouse dermal bioassays (6,7). In addition, we have shown a good correlation (r=0.86) between mutagenicity of these fractions in this modified assay and their total PNA concentration, for PNAs with three to seven rings (see T. A. Roy et al, this volume). It is likely that the levels of mutagens present in the two enriched preparations were too low to be detected in our modified assay.

PNAs are known to require metabolic activation for their genotoxic expression. Although both classes of mutagenic components were probably extracted by DMSO, only the direct-acting mutagen was present in sufficient concentration to be observed in the mouse lymphoma assay of the extract. The direct-acting mutagen was then converted in the presence of the S-9 activation mixture to a non-mutagenic species by biological transformation or physical sequestration. PNAs were sufficiently concentrated in the evaporative residue to produce mutagenesis. However, only two of the PNAs identified in the evaporative residue, anthracene and pyrene, have been evaluated for mutagenicity in the mouse lymphoma assay. Contradictory results were reported for both of these compounds. Anthracene was reported to be non-mutagenic with metabolic activation in one study (10) and mutagenic in another (11). Likewise, pyrene was evaluated as being non-mutagenic in several studies (10,12) and mutagenic in another (13). By our criteria, and those recommended by Clive _et al._ (8,9), anthracene would have been evaluated as mutagenic and pyrene as non-mutagenic in these published studies. Additional studies would be necessary to identify the components responsible for both the direct-acting and S-9 dependent mutagenicity of the DMSO extract and the evaporative residue in the mouse lymphoma assay. Recent research at CIIT indicated that unleaded gasoline did not induce unscheduled DNA synthesis in hepatocytes of rats treated with a single oral dose (14) and was not mutagenic to human cells in culture (15). Gibson and Bus (16) have demonstrated the involvement of a sex specific protein, α-2u globulin, in gasoline-induced nephrotoxicity in male rats, suggesting that factors other than genotoxicity can be responsible for the carcinogenic effect seen in rats and mice.

Based on the polar or aromatic nature of the mutagenic components in the DMSO extract and the evaporative residue, it is probable that these components are present in the heavy ends of unleaded gasoline and have boiling points in excess of 212° F. Because of the higher boiling points, it is likely that there would be negligible human exposure to the mutagenic components in occupational or consumer settings.

ACKNOWLEDGEMENT

The authors would like to acknowledge the excellent technical assistance of the following individuals; R. A. Deitch, A. C. Garnier, J. G. Harper, S. E. Irwin and S. W. Johnson.

REFERENCES

1. Conaway, C.C., Schreiner, C.A. and Cragg, S.T. (1982): Mutagenicity evaluation of petroleum hydrocarbons. In: Toxicology of Petroleum Hydrocarbons, pp. 128-138, API Press, USA.
2. API Medical Research Publication Number 28-31344.
3. Farrow, M.G., McCarroll, N., Cortina, T., Draus, M., Munson, A., Steinberg, M., Kirwin, C., and Thomas, W. (1983): In vitro mutagenicity and genotoxicity of fuels and paraffinic hydrocarbons in the Ames, sister chromatid exchange and mouse lymphoma assays, Toxicologist, 3(1):36a.
4. MacFarland, H.N., Ulrich, C.E., Holdsworth, C.E., Kitchen, D.N., Halliwell, W.H., and Blum, S.C. (1984): A chronic inhalation study with unleaded gasoline vapor, J. American College of Toxicology, 3:231-248.
5. Brusick, D. (1980): Protocol 13: Bone Marrow Cytogenetic Analysis in Rats. In: Principles of Genetic Toxicology, pp. 235-239, Plenum Press, New York.
6. Blackburn, G.R., Deitch, R.A., Schreiner, C.A., and Mackerer, C.R. (1986): Predicting carcinogenicity of petroleum distillation fractions using a modified Salmonella mutagenicity assay, Cell Biology and Toxicology, 2(1):63-84.
7. Blackburn, G.R., Deitch, R.A., Schreiner, C.A., Mehlman, M.A., and Mackerer, C.R. (1984): Estimation of the dermal carcinogenic activity of petroleum fractions using a modified Ames assay, Cell Biology and Toxicology, 1:67-80.
8. Clive, D., and Spector, J.F.S. (1975): Laboratory procedure for assessing specific locus mutations at the TK locus in cultured L5178Y mouse lymphoma cells, Mutation Research, 31(1):17-29.
9. Clive, D., Johnson, K.O., Spector, J.F.S., Batson, A.G., and Brown, M.M.M. (1979):

Validation and characterization of the L5178Y/TK+/- mouse lymphoma mutagen assay system, Mutation Research, 59:61-108.

10. Amacher, D.E., Paillet, S.C., Turner, G.N., Ray, V.N., and Salsburg, D.S. (1980): Point mutations at the thymidine kinase locus in L5178Y mouse lymphoma cells, Mutation Research, 72:447-474.
11. Short-term in vitro mammalian cell assay results. NTP Technical Bulletin 9:7, 1983.
12. Amacher, D.E., and Turner, G.N. (1982): Mutagenic evaluation of carcinogens and non-carcinogens in the L5178Y/TK assay utilzing postmitochondrial fractions (S9) from normal rat liver, Mutation Research, 97:49-65.
13. Jotz, M.M., and Mitchell, A.D. (1980): Effects of 20 coded chemicals on the forward mutation frequency at the thymidine kinase locus in L5178Y mouse lymphoma cells. In: Evaluation of Short-Term Tests for Carcinogens: Report of the International Collaborative Program, Progress in Mutation Research, 1:580-593.
14. Loury, D.J., Smith-Oliver, T., and Butterworth, D.E. (1985): Measurements of DNA repair and cell replication in hepatocytes from rats exposed to 2,2,4-trimethylpenatne or unleaded gasoline, Environmental Mutagenesis, 7(S3):70.
15. Richardson, K.A., Wilmer, J.L., Smith-Simpson, D., and Skopek, T.R. (1986): Assessment of the genotoxic potential of unleaded gasoline and 2,2,4-trimethylpentane in human lymphoblasts in vitro, Toxicology and Applied Pharmacology, 82: 316-322.
16. Gibson, J.E., and Bus, J.S. (1986): Current perspectives on gasoline (light hydrocarbon) induced male rat nephropathy. Annals N.Y. Academy Sciences, in press.

EXISTENCE OF A SEASONAL CYCLE OF PAH CONCENTRATION IN THE AMPHIPOD PONTOPOREIA HOYI.

BRIAN J. EADIE*, PETER F. LANDRUM, WARREN R. FAUST
Great Lakes Environmental Research Laboratory, NOAA,
2300 Washtenaw Avenue, Ann Arbor, Michigan 48104, USA.

INTRODUCTION

Polycyclic aromatic hydrocarbons (PAH) have low water solubilities and sorb to the settling particulate matter in proportion to their solubilities and their concentration. The settling organic material and sediment detritus in turn act as food and PAH sources for benthic invertebrates (1,2). High levels of PAH have been measured in various benthic organisms of the Great Lakes (3-5). Concentrations in the PPM range have been measured in the dominant benthic invertebrate, the amphipod *Pontoporeia hoyi* (6).

The abundant biomass of benthic organisms is the base of an important food web in the Great Lakes. Gardner, *et al.* (7) estimate that approximately 13000 $cal.m^{-2}$ (2.1 g ash free dry $weight.m^{-2}$) of *P. hoyi* (greater than 70% of benthos caloric content) is consumed annually in southern Lake Michigan. This level of consumption coupled with the elevated concentrations of PAH makes *P. hoyi* an important vector in PAH cycling.

In contrast to fish, *P. hoyi* do not appear to biotransform PAH (1,8). This fact, coupled with their ease of collection and high PAH concentrations, has led to considering *P.hoyi* as sentinal organisms for monitoring PAH in the Great Lakes.

Considering the importance of *P. hoyi* both as a food source for fish and as a monitoring tool, we decided to examine their seasonal PAH levels. Laboratory measurements of uptake and depuration rates of PAH, coupled with models (1,8), suggested that there should be a strong seasonal signal, with maximum concentrations predicted for spring. Our objectives in this study were to verify these predicted seasonal changes in *P. hoyi* PAH concentration through field measurements and to examine the potential role of food quality and lipid storage in this seasonal cycle.

MATERIALS AND METHODS

Pontoporeia hoyi were collected from Lake Michigan approximately three miles south of Grand Haven, MI at a depth of 29m by PONAR grab. Animals were removed gently from the sediment and transported to our laboratory in lake water (transport time 4-5 h). Benthic organisms were hand-picked from the residue, depurated in lake water overnight, separated into triplicates, blotted dry on paper towels, weighed and stored frozen in aluminum foil. After thawing, the organisms were dried to constant weight in a desicator. After weighing, the samples were Soxhlet extracted for 18 hours in methanol:hexane (1:1).

Water samples (18 L) were collected in a precleaned 30 L teflon lined Niskin bottle and drained into a clean glass carboy. They were transported to the laboratory, in black plastic covered carboys and filtered through a Gelman glass fiber filter within 30 hours of collection. The water was then liquid-liquid extracted with (3x300 ml) hexane.

Sample extracts were partitioned by chromatography on silica gel (9,10). PAH compounds in the cleaned extracts were measured on a gas chromatograph (GC; 30 m SE54 WCOT fused silica capillary column; 120°-260°C @ 4°/min) equipped with a photoionization detector. Extract yields were calculated by adding a known spike of ^{14}carbon labeled anthracene and tritium labeled BaP to each matrix prior to extraction and counting a known fraction of the final extract. Triplicate extractions and analyses yielded coefficients of variation for phenanthrene (PH; 1-36%), fluoranthene (FL; 16-62%), pyrene (PY; 1-49%) and benzo[a]pyrene (BaP; 8-77%). All lab work was performed under gold fluorescent light to minimize photodegradation.

Lipids were $CHCl_3$:MEOH (2:1;V:V) extracted from individual P. hoyi by a microtechnique (7,11) and quantified gravimetrically. A large number (ca. 10) of individual lipid extractions were analyzed for each time point.

RESULTS

Water

There were 11 sample collections between April 9 and December 10, 1984. The concentration ranges of phenanthrene (43-320 ng L^{-1}), fluoranthene (4.3-28), pyrene (17-36), and benzo[a]pyrene (1-21) in filtered water are shown in Figure 1.

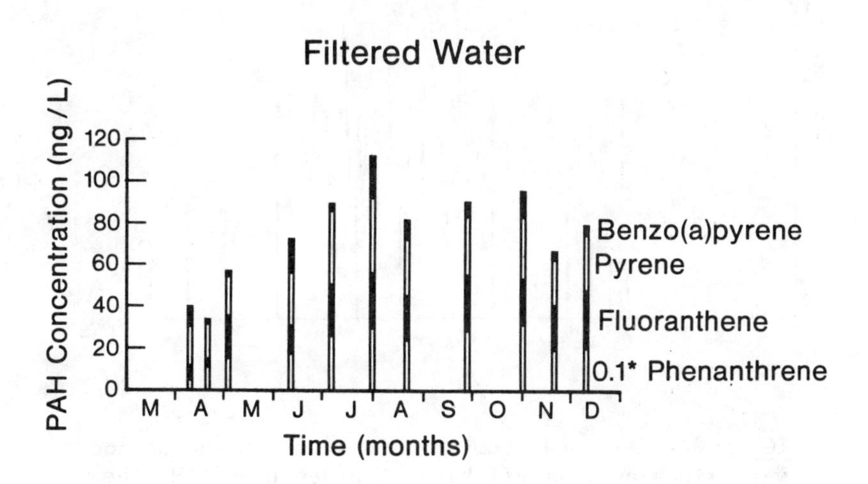

FIGURE 1. Seasonal concentrations (ng/ml) of PAH in filtered lake water from mid-depth at the 29M station. PH values were multiplied by 0.1 to fit plot.

Concentrations were generally 2-3 times lower in the spring than between June and October. The change in concentration of "dissolved" PH corresponds to the measured change in solubility in distilled water (12) over the temperature range at this station. Daily mean temperatures from a nearby water intake at a depth of 16 M (Figure 2) illustrates both the seasonal cycle and the local frequency of upwelling of cold water in this area. The P. hoyi at 29 M depth would experience a somewhat more buffered thermal environment. Peak temperatures occur at the same time as peak concentrations in dissolved PH. Except for BaP, which varied irregularly, the seasonal concentrations of the other PAH in water were similar in behavior to PH.

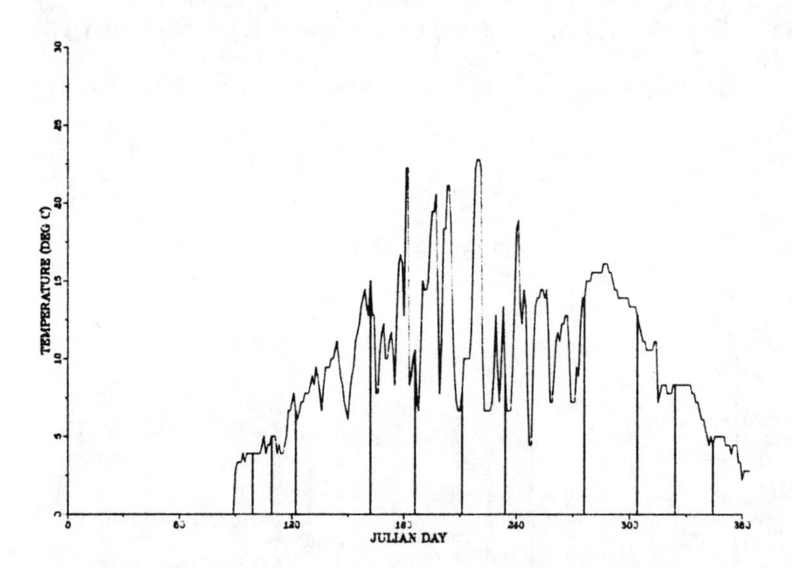

FIGURE 2. Daily mean temperature at the Grand Rapids water intake, 2 Km offshore at a depth of 16M. The vertical lines represent sample collection dates.

Pontoporeia hoyi

The PAH concentrations in P. hoyi show a more complicated seasonal trend (Figure 3). Concentrations of all four PAH are low in the early spring and rise to maximum values in summer. The ranges in concentration are approximately a factor of 2 for FL, PY and BaP but reach a factor of 7 (225-1585 ng/g) for the much more soluble PH. The concentrations of PAH in P. hoyi are generally in phase with those in filtered water, resulting in bioconcentration factors for the 4 PAH that show seasonal patterns (Figure 4) that range over a factor of approximately 5 for each compound. The patterns for PH and FL show a decline from spring through fall, while those for PY and BaP appear to have a mid-summer minimum. In the case of BaP, an anomalously high water concentration is the cause. The large range in BCF can be due

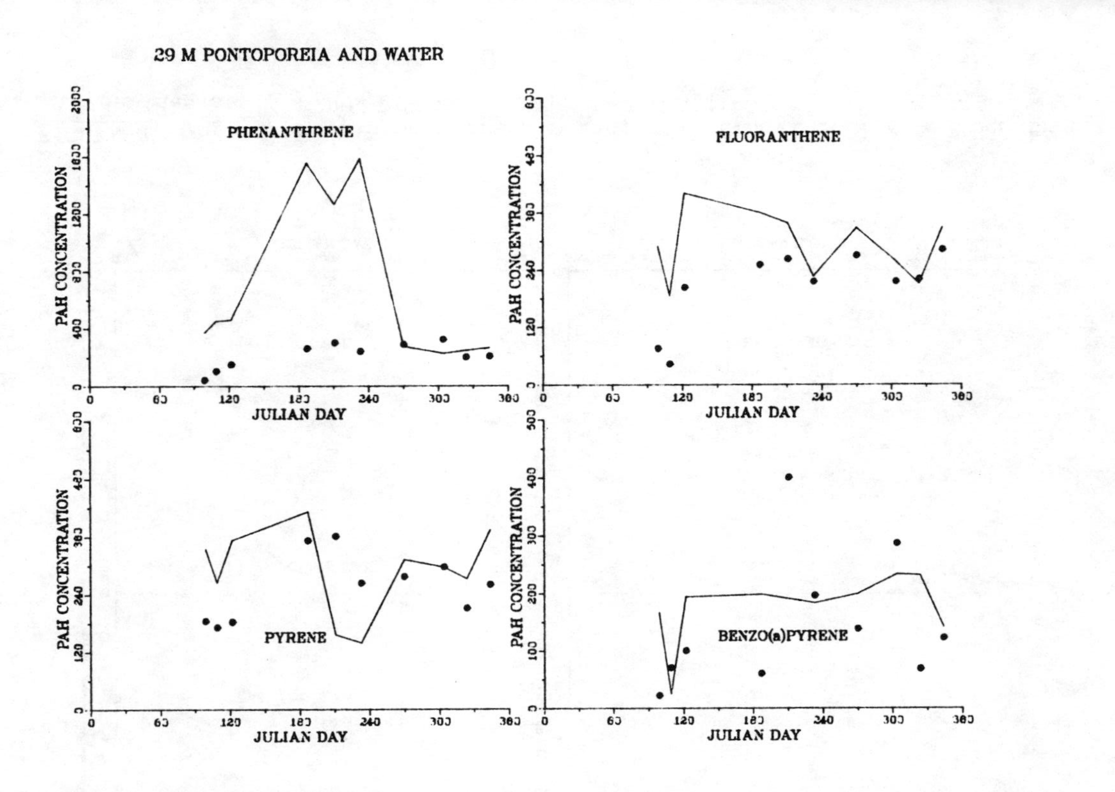

FIGURE 3. The mean PAH in P.hoyi (solid line; ng/g dry wt) throughout 1984. The solid circles represent water concentrations (ng/ml), scaled to fit the plot. Data were multiplied by: PH(1), FL(10), PY(10), and BaP(20).

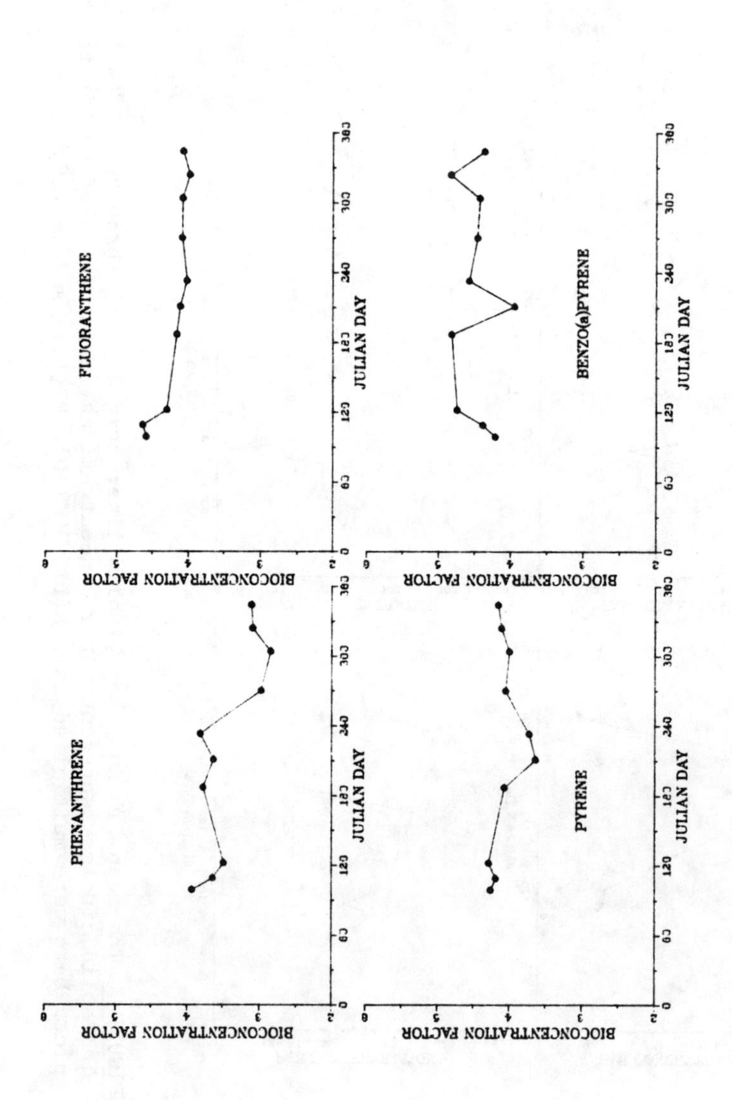

FIGURE 4. Log of the bioconcentration factor in P.hoyi in 1984. Biocontration is defined as [g PAH/g P.hoyi] / [g PAH/ ml].

to the propagation of errors from dividing two numbers with large variances, or to the fact that the concentration of trace contaminents in organisms is a complex response to potentially rapid changes in water concentration. If the latter is correct, then measured ratios do not represent steady state.

These organisms live at the sediment-water interface and feed both on interfacial sediments and on the high concentrations of suspended matter (TSM) in the few meter thick benthic nepheloid layer. This layer is created and maintained through local resuspension of surfacial sediments, downslope transport of fine grained, organic rich particulate matter (13,14) and input from local productivity. Bulk and the fine grain fraction of local sediments and the total suspended matter (TSM) were collected and analyzed for their PAH composition (Table 1).

TABLE 1

MEAN PAH (NG/G) IN 29m SAMPLES

Matrix	Phen	Fl	Py	BaP
P.hoyi	669+/-564.	292+/-70.	293+/-88.	176+/-60.
Water[1]	0.208+/-.086	0.020+/-.008	0.026+/-.007	0.009+/.006
Sediments[2]	30.+/-26.	27.+/-23.	24.+/-20.	19.+/-18.
Sed(<53um)	295.	520.	434.	222.
TSM[3]	4023.	1193.	2865	797.

1. Water filter through Gelman H/A glass fiber filter.
2. Bulk surfacial (0-1 cm) sediments.
3. Collected July 6, 1984.

The fine grain (<53 um) sediments, approximately 10% by weight in this area (15), have a significantly higher concentration of PAH than the bulk sediments. The TSM, which ranges in concentration from 1-6 mg L^{-1} at this site, has a very high PAH concentration and if P. hoyi use this as a food source to any significant extent, it should be an important source of contaminants.

Considering that these organisms are most abundant in the nearshore, it follows that the quantity and quality of their food will vary seasonally. High terrestrial runoff due to spring snow melt and rain combined with the spring peak of phytoplankton will result in the highest food values in spring. This seasonal abundance of organic rich food results in a clear seasonal response in the $CHCl_3$:MEOH extractable lipid content of the P. hoyi (Figure 5).

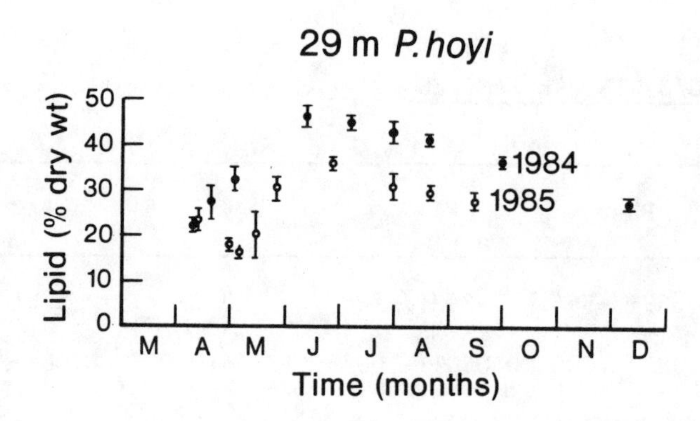

FIGURE 5. Lipids in the 29M P.hoyi. Solid circles represent 1984 and the open circles 1985 data. Bars are standard errors.

In both 1984 and 1985, organism lipids were lowest in spring, and rapidly increased in May/June to peak levels followed by a slow decline. There is a significant difference between the two years, with the 1985 animals approximately 10% lower in lipid content. This appears to correspond to a reduced spring phytoplankton bloom in 1985 (16), probably due to cool spring weather, resulting in a lower food flux to the benthos. This seasonal signal is similar to, perhaps slightly earlier than, that of the PAH concentrations measured in the organisms (Figure 3). In many organisms, hydrophobic contaminant concentrations tend to correlate with organism lipid content. This does not appear to be true for P.hoyi. The PAH concentrations in these organisms does not correlate significantly (P>0.2) with lipid content (figure 6).

DISCUSSION

Except for a few (replicated) numbers, the small (ca 2X) seasonal change observed for FL, PY and BaP is not much larger than replicates measured from the same collection and could represent inherent heterogeneity in the population. The low PY values for the July 6th and 30th samples and the low BaP value for the May 2nd sample were well replicated (cv = 11, 30 and 21% respectively) and are significantly lower than PAH in these organisms collected at other times. The cause of these aberrant concentrations is not clear, and could not be related to low lipid levels (Figure 6). The two low PY concentrations occurred at high lipid levels, while the BaP anomaly was found in the sample with a relatively low lipid content.

The apparent lack of correlation between lipid content and hydrophobic PAH levels is confirmational evidence of earlier findings (1,2) that water and not food is the major source of the P. hoyi's body burden of PAH. The rich food in the spring, derived from the plankton bloom and fresh terrestrial runoff should contain relatively high concentrations of PAH (see TSM in Table 1). Although P. hoyi apparently consumes a large amount of this material, resulting in the rapid increase in lipid levels, the associated PAH apparently are not retained. Lack of precision complicates our interpertation; the concentration in the lipid portion of the organism would have to be about 5 times the concentration in the rest of the organism in order to see a clear signal. Therefore, we cannot say, from these data, whether or not the lipids act as a significant PAH storage pool.

In contrast to the other PAH, a strong seasonal signal was observed for PH and the maximum concentration corresponds to the highest lipid levels. The relationship is very non-linear, virtually a constant PH concentration until the lipid level exceeds 36% organism dry weight, then a 4-7 fold increase in PH concentration was measured. The increased PAH and the increase in lipid are coincident with both a temperature increase and fresh runoff. These latter factors result in a significant increase in the "dissolved" PH, that may be the source of the increased PH in P. hoyi.

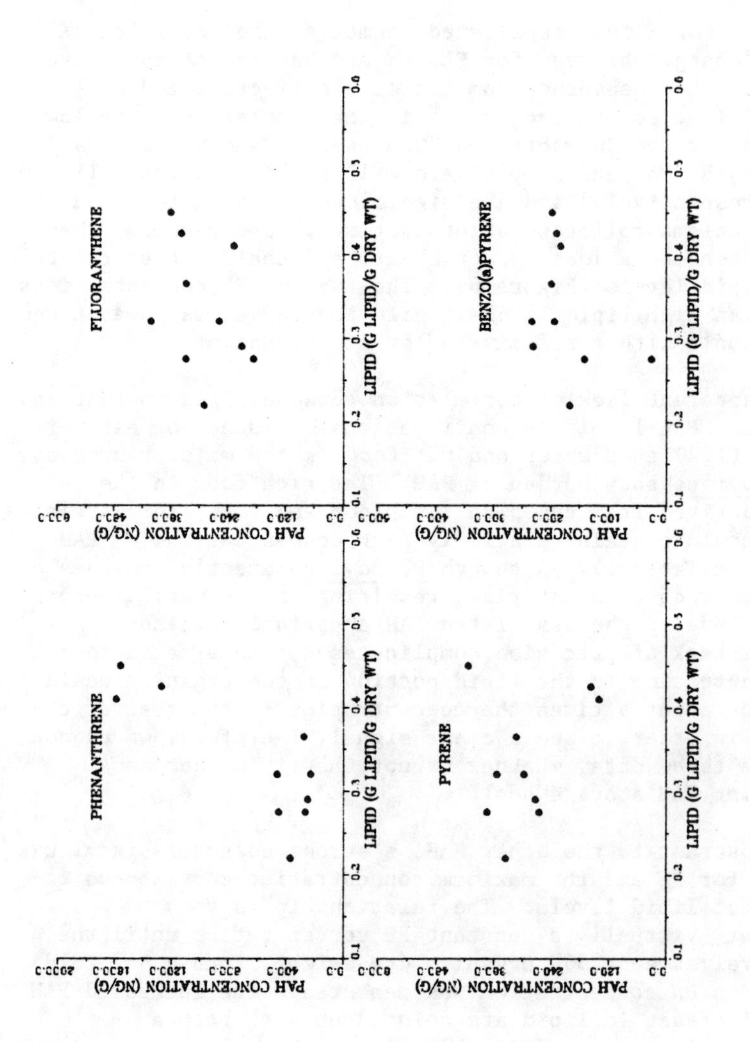

FIGURE 6. PAH concentration (ng/g dry wt) in P. hoyi vs. lipid concentration. Correlations are not significant.

Studies of the toxicokinetics of PAH in P.hoyi over the same season clearly demonstrate that the bioaccumulation process is complex (17) and lipid accumulation is only one factor. Increased lipids were found to reduce both the elimination and uptake rate constants for BaP.

A first order contaminant model, incorporating both seasonal and thermal changes in the toxicokinetic rate constants (1) was modified to include variable PAH water concentration. The model was parameterized from laboratory toxicokinetic rate data measured for PH and BaP in P. hoyi. Measured water and bulk sediment PH and BaP concentrations were used as input. The resultant output for PH (Figure 7) compares favorably to the measured values in the spring.

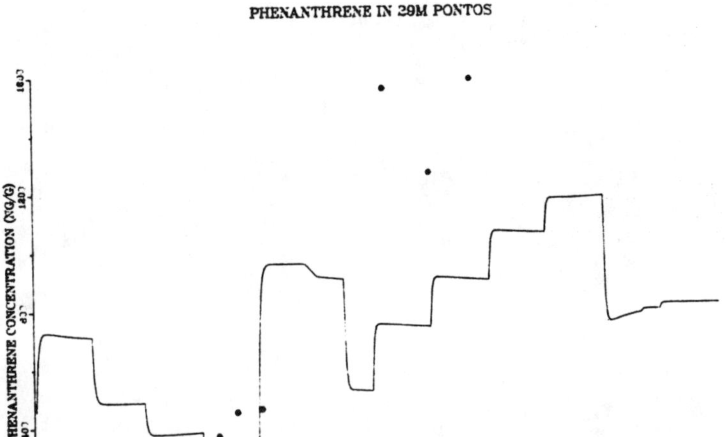

PHENANTHRENE IN 29M PONTOS

PHENANTHRENE CONCENTRATION (NG/G)

JULIAN DAY

FIGURE 7. Comparison of model (solid line) and data for the concentration of PH in P.hoyi.

The model shows a summer increase, but underpredicts the concentration, and does poorly in the fall when measured values decline rapidly to a constant value. This may be due to the fact that the model doesn't account for the P. hoyi's consumption of stored lipid in the fall as observed in field collected animals (Figure 5).

In the simulation of BaP (Figure 8), the model does show a slight minimum in spring, followed by seasonal transients not observed in the data. These transients are largely the result of the wide range in the measured values for BaP in water.The model overestimates the mean BaP concentration in P.hoyi by about a factor of 2. If the 3 highest BaP water values are removed from the data set, the annual mean concentration is reduced by 38%. Although the transformation isn't linear, this would reduce the model output to within 1 SD of the measured concentrations in P.hoyi.

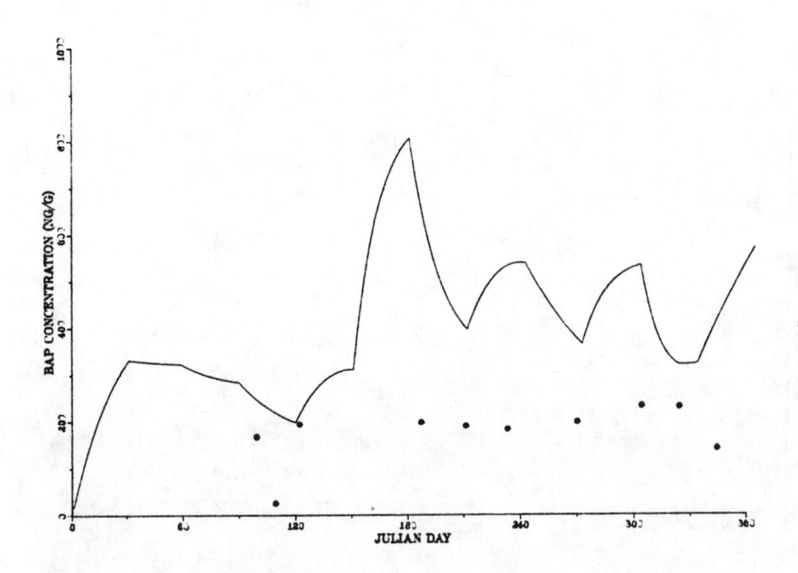

FIGURE 8. Comparison of model (solid line) and data for the concentration of BaP in P.hoyi.

The model predicts that only 6.5% and 4.9% of the P.hoyi body burden is aquired from sediment for BaP and PH respectively. Because the water term dominates, uncertainty in accurately measuring the water concentration will have a major effect on the predictions. In addition, the thermal data used may poorly represent the temperatures experienced by the organisms due to thermal buffering in the sediments. Refinement in measurements of the water concentrations as well as in model parameterization should permit improved prediction of the PAH concentrations in P.hoyi on a seasonal basis.

SUMMARY

Analysis of our data supports the following:

1. a strong seasonal cycle exists in the phenanthrene concentration of P. hoyi,

2. seasonal changes of fluoranthene, pyrene and benzo a pyrene in P.hoyi are small (if they are real), although there are occasional significant changes in concentrations,

3. the lack of a strong, positive relationship between PAH and lipid levels in P.hoyi indicates that the major uptake pathway is through water,

4. more accurate and precise PAH measurements are needed in water, and

5. existing contaminant models need modification to account for growth and consumption of internal lipids.

REFERENCES

1. Landrum, P.F., Eadie, B.J., Faust, W.F., Morehead, N.R., and McCormick, M.J. (1985): Role of sediment in the bioaccumulation of benzo(a)pyrene by the amphipod Pontoporeia hoyi. In: Polynuclear aromatic hydrocarbons: Eighth international symposium on mechanisms, methods, and metabolism, ed. W.M. Cooke and A.J.Dennis, pp. 799-812. Columbus, OH: Battelle Press.
2. Eadie, B.J., Faust, W.R., Landrum, P.F., and Morehead, N.R. (1985): Factors affecting bioconcentration of PAH by Great Lakes benthos. In: Polynuclear Aromatic Hydrocarbons: Mechanisms, Methods and Metabolism. M. Cooke and A.J. Dennis, eds. Battelle Press, Columbus, OH, pp. 363-378.

3. Eadie, B.J., Faust, W.R., Gardner, W.S., and Nalepa, T. (1982): Polycyclic aromatic hydrocarbons in sediments and associated benthos in Lake Erie. Chemosphere 11:185-191.

4. Eadie, B.J., Landrum,P.F., and Faust, W.R. (1982): Polycyclic aromatic hydrocarbons in sediments, pore water and the amphipod Pontoporeia hoyi from Lake Michigan. Chemosphere 11:847-858.

5. Eadie, B.J., Faust, W.R., Landrum, P.F., Morehead, N., Gardner, W.S., and Nalepa, T. (1982): Bioconcentrations of PAH by some benthic organisms. In Polycyclic Aromatic Hydrocarbons: Seventh international symposium on formation, metabolism and measurement, ed. W.M. Cooke and A.J. Dennis, pp. 437-449. Columbus, OH: Battelle Press.

6. Nalepa, T.F., Quigley, M.A., Childs, K.F., Gauvin, J.M., Heatlie, T.S., Parker, M.P., and Vanover, L. (1985): Macrobenthos of Southern Lake Michigan, 1980-81, NOAA Data Report ERL GLERL-28 273 pp/

7. Gardner, W.S., Nalepa, T.F., Frez, W.A., Cichocki, E.A., and Landrum, P.F. (1985): Seasonal patterns in lipid content of Lake Michigan macroinvertebrates. Can. J. Fish and Aquat. Sci. 42 November.

8. Landrum, P.F. (1982): Uptake, depuration and biotransformation of anthracene by the scud, Pontoporeia hoyi. Chemosphere 11:1049-1057.

9. Geiger, W. and Blumer, M. (1974): Polycyclic aromatic hydrocarbons in the environment: Isolation and characterization by chromatography, visible, ultraviolet and mass spectometry. Anal. Chem. 46:1663-1668.

10. Geiger, W. and Scheffner, C. (1979): Determination of polycyclic aromatic hydrocarbons in the environment by glass capillary gas chromatography. Anal. Chem. 50:243-249.

11. Gardner, W.S., Frez, W.A., Cichocki, E.A., and Parish, C.C. (1985): Micromethod for lipids in aquatic invertebrates. Limnol. and Ocn. 30:1099-1104.

12. May, W.E., Wasik, S.P., and Freeman, D.H. (1978): Determination of the Solubility Behavior of Some Polycyclic Aromatic Hydrocarbons in Water. Anal. Chem. 50:997-1000.

13. Chambers, R.L. and Eadie, B.J. (1981): Nepheloid and suspended particulate matter in southeastern Lake Michigan. Sedimentology 28:439-447.

14. Sandiland, R.G. and Mudroch, A. (1983): Nepheloid layer in Lake Ontario. J. Great Lakes Res. 9:190-200.

15. Chambers, R.L. and Eadie, B.J. (1980): Nearshore chemistry in the vicinity of the Grand River, Michigan. NOAA Tech. Memo ERL GLERL-28, NOAA/GLERL, Ann Arbor, MI.
16. Fahnensteil, G. (Personal Communication)
17. Landrum, P.F., Eadie, B.J., Faust, W.R.(1984): Role of environmental factors affecting the toxicokinetics of organic xenobiotics in P.hoyi. Presented at the 5th Annual Meeting, SETAC; Nov 4-7, Arlington,VA.

ASSIMILATION AND METABOLISM OF POLYCYCLIC AROMATIC HYDROCARBONS BY VEGETATION - AN APPROACH TO THIS CONTROVERSIAL ISSUE AND SUGGESTIONS FOR FUTURE RESEARCH

NELSON T. EDWARDS
Environmental Sciences Division, Oak Ridge National Laboratory, Oak Ridge, Tennessee 37831

INTRODUCTION

Concentrations of compounds in plant tissues (leaves, stems, roots, *etc.*) reflect a myriad of processes, including uptake rates, chemical transformations and metabolism, and translocation into and out of the tissues. The rates of polycyclic aromatic hydrocarbon (PAH) metabolism and the translocation rates of PAH transformation products are as important as uptake rates in determining the role of vegetation in bioconcentrating or degrading a particular PAH.

Relatively few experiments have been conducted that address questions relating to the uptake, metabolism, and translocation of PAHs in vegetation. Also, there are conflicting reports in the literature concerning the degree of plant uptake and translocation of PAHs. For example, Graf and Nowak (1) demonstrated growth stimulation of tobacco, rye, and radishes by a number of PAHs, including benzo[a]pyrene (BaP), and concluded that the compounds were assimilated through the roots. However, Harms (2) reported negligible translocation of BaP from wheat roots to shoots, and Gunther *et al.* (3) found no translocation of several PAHs from orange rind (point of application) to other plant parts. Ellwardt (4) concluded from field experiments with fresh compost containing a number of PAHs and with several agricultural crops that little or no uptake by plant roots occurred. Yet Durmishidze *et al.* (5), using ^{14}C-labelled PAHs, showed translocation from leaves to roots as well as from roots to upper plant organs in ryegrass (*Lolium multiflorum* Lam.), chick pea (*Cicer arietinum* L.), alfalfa (*Medicago sativa* L.), cucumber (*Cucumis sativus* L.), and vetch (*Vicia faba* L.). While some plants such as ryegrass translocated PAHs from roots to leaves more readily than from leaves to roots, the opposite was true for other plants (e.g., cucumber). Edwards *et al.* (6) demonstrated uptake of ^{14}C-anthracene (ANTH) from nutrient solution into soybean (*Glycine max* Merr.) roots and translocation to stems and leaves. ANTH uptake by soybeans from nutrient solution was proportional to ANTH concentration in the solution. Field collections of vegetation grown in PAH-contaminated environments (7,8,9,10,11,12) lend support to laboratory studies that relate PAH contamination of vegetation to concentrations in the environment.

Factors that affect PAH uptake rates by plants include the nature of the substrate, degree of PAH solubility in water, whether the PAH is in the vapor phase or adsorbed onto particles, and molecular weight. Dörr (13) found no uptake of BaP by wheat (Triticum spp.) and rye (Secale cereale L.) from nutrient solution or soil when BaP was applied in relatively insoluble form (i.e., not dissolved in oil or some other organic solvent). However, uptake and translocation did occur when BaP was dissolved in oil before applying it to the substrate. Müller (14) reported greater uptake of BaP by carrots (Daucus carota L.), radishes (Raphanus sativus L.) and spinach (Spinacia oleracea L.) growing in sand culture when the BaP was applied dissolved in benzene than when dissolved in plant oil or solubilized by a detergent, and greater uptake of BaP by carrots and radishes growing in sand than in soil and compost. Dörr (13) found greater uptake of BaP by rye from nutrient solution than from soil but reported no effect of different soil types on uptake rates.

Also, polycyclic aromatic hydrocarbons may concentrate more in some plant tissues than in others. Wagner and Siddiqi (15), Shabad and Cohan (16), and Stevcevsk and Jovanovic-Kolar (11) reported higher concentrations of BaP in the vegetative portion of wheat than in wheat seeds of plants growing in substrates containing BaP. Siegfried (17) found more BaP in carrot and lettuce (Lactuca sativa L.) leaves than in roots, while Müller (14) found that most of the BaP in carrots remained associated with the roots after 123 d of exposure. Then again, Linne and Martens (18) reported more PAH (a total of 11 PAHs including BaP) in the foliage of carrots than in the roots after 150 d of growth.

In addition, the chemical fate of these compounds within plants is not well resolved. Dörr (13) found a decline in BaP concentrations in rye plants after 30 d of growth, following a period (20 d) of increasing concentrations in the plants due to uptake. The decline in BaP concentration was attributed to degradation or chemical changes of the BaP. Durmishidze (19) and Durmishidze et al. (5), demonstrated chemical transformations of BaP and BaA (mostly to organic acids) within a number of different plant species. The amount of BaP catabolized over a 14-d period varied from 2 to 18% of the BaP assimilated and depended on plant species.

Finally, an extensive review of the literature on PAHs in terrestrial ecosystems (20) revealed that almost all the PAH research with terrestrial vegetation has been with BaP only. Accordingly, more field research is needed, along with controlled laboratory experiments with PAH representing a wide range of physico-chemical characteristics, to clarify environmental factors affecting plant uptake and metabolism of PAHs.

In this paper, the results of laboratory studies on the uptake, translocation, and metabolism of ^{14}C-anthracene (ANTH) and ^{14}C-benz[a]anthracene (BaA) in bush bean (*Phaseolus vulgaris* L.) are presented. A comparative study with BaP is in progress.

MATERIALS AND METHODS

Bush beans (cv. Blue Lake 274) were germinated and grown for two weeks, physically supported by glass beads, in a liquid growth medium with a pH of 5.5 [Hoaglands #2 nutrient solution, (21)] before being transferred to culture bottles. Nutrient solution (1.8 L) was placed in each of six 2-L, amber-colored, narrow-mouth glass bottles with a side port near the top for addition of PAH and replacement of water lost by transpiration. The side port was closed with a ground-glass stopper. A cotton plug was wrapped around the stem of a single plant and the roots submersed in the solution.

Three bottles of nutrient solution were dosed with ANTH and three with BaA dissolved in 1 mL of acetone before adding to the nutrient solution. Plants were kept in a growth chamber with controlled temperature (23°C at night, 26°C during day), irradiance, and photoperiod (15 h). The first doses, about 20 µg of ANTH or BaP, with specific activities of 2.33×10^3Bq µg^{-1} and 7.81×10^3 Bq µg^{-1} (1 Bq = 1 disintergration/second), respectively, (Amersham, Arlington Heights, Illinois) were added to the nutrient solution when flower buds were beginning to develop. Carbon-14 ANTH activity in the nutrient solution was measured 2 h after the initial dose, 1 d after the initial dose, and then once per week for the duration of the experiment. Enough ANTH was added on the second day and then at weekly intervals to raise the ^{14}C activity to 23.3×10^3 Bq L^{-1} (0.01 mg ANTH L^{-1}), assuming no transformation of the parent compound. Average total dosage (i.e., the amount added to the nutrient solution during the experiment) was 75 µg per plant, with a 4% coefficient of variation.

Carbon-14 BaA activity in the nutrient solution was measured 2 h after the initial dose and at 2, 6, 7, 10, 11, 22, and 30 d after the initial dose. By the second day the BaA concentration had dropped to nearly 5% of the initial dose. A second dose was then added to bring the concentration back to 0.01 mg BaA L^{-1}. By the end of 6 d the concentration had again dropped to about 5% of the total dosage. Quick measurements (<5-min counts of 1-mL samples) of ^{14}C activity after each dose revealed that within 2 h the concentration typically dropped to about 20% of the dose. Average total ^{14}C-BaA dosage per bottle was 41.9 µg (32.7×10^4 Bq), with a 3% coefficient of variation. Plants were harvested at maturity but while seed pods were still green. Exposure time was 30 d.

After harvest, the plants were separated into leaves, stems, roots, and pods, freeze-dried, and ground to pass through a 2-mm-mesh screen. Replicate subsamples of each fraction (~0.5 to 1 g each) were oxidized with a Packard Model 306-C Tri-Carb Oxidizer (Packard Instrument Co., Downers Grove, Illinois), and levels of radioactivity were determined by scintillation spectrometry (Packard Model 460-CD Tri-carb Liquid Scintillation spectrometer). The remaining samples were stored at -20°C for extraction later.

The procedure for extracting PAH from plant tissues was a modification of that described by Santoro et al. (22). Plant tissues (1 to 2.5 g) were placed in 30 mL of 20% KOH to which had been added 5 mL of ethanol (ETOH), and shaken gently for 2 h in an 80°C water bath. The residue was removed by centrifugation, and the KOH-ETOH extract decanted into a separatory funnel. Cyclohexane (10 mL) was added to the extract, shaken gently for 3 min, and allowed to separate. The cyclohexane was collected and the step repeated three times, saving the cyclohexane each time.

Radioactivity of a 1-mL sample of the KOH-ETOH extract was determined and the remaining volume measured. Compounds containing ^{14}C in the KOH-ETOH extracts were collectively assumed to be polar metabolites of ANTH or BaA. The cyclohexane was washed with 12 mL of water four times by shaking the water-cyclohexane mixtures together for 3 min. The water was discarded, and the cyclohexane was dried over about 100 mg of anhydrous sodium sulfate. A 1-mL sample of the cyclohexane extract was analyzed for ^{14}C. Carbon-14 compounds in the cyclohexane extract were assumed to be PAH (22) and nonpolar metabolites of PAH.

The remaining cyclohexane was evaporated to ~0.1 mL under dry, flowing nitrogen. One µg of unlabelled ANTH (about tenfold the expected ^{14}C-ANTH) was added to the cyclohexane extracts having low ^{14}C activity (i.e., extracts of stems and leaves) before the concentration procedure to reduce evaporative losses of ^{14}C-ANTH. Recovery was increased from 88% to >95% with this procedure. The extract (0.1 mL) was then cochromatographed with known amounts of ^{14}C PAH (ANTH or BaA) standards by thin-layer chromotography (TLC). The TLC eluant used was chloroform;acetic acid:methanol [90:5:5(V/V)]. The PAH spots (one spot from the sample and one spot from the standard PAH) were identified under ultraviolet light and scraped into separate scintillator vials. Bray's liquid-scintillation cocktail (Research Products International Corp., Elk Grove, Illinois) was added, and the mixture was shaken to extract the PAH present. The solids were permitted to settle and then the ^{14}C activity was determined by liquid-scintillation spectroscopy.

The remainder of each TLC column was scraped into vials and ^{14}C activity was determined as described above. Corrections for losses due to evaporation during sample concentration and TLC separations were made on each sample by comparing ^{14}C activity in the cyclohexane extract with ^{14}C activity in the TLC scrapings. All work was performed under yellow lights (General Electric Gold Fluorescent) to minimize photolytic degradation.

RESULTS AND DISCUSSION

Carbon-14 Distributions

An average of 60% of the total ANTH dose (i.e., the total added to each bottle) was associated with roots at harvest, compared with 66% for BaA (Table 1). Washing the roots three times with a 1:1 mixture of acetone and water removed only 6% of the ANTH and 5% of the BaA. Therefore, it was assumed that about 54% of the ANTH and about 60% of the BaA had been incorporated (i.e., not adsorbed) into root tissues. Based on an earlier single-dosage 16-d experiment (N. T. Edwards, unpublished data) most of the ANTH and BaA

TABLE 1

BUDGET OF ^{14}C ACTIVITY IN NUTRIENT SOLUTION AND BUSH BEAN PLANTS AT HARVEST (30 d AFTERINITIAL EXPOSURE TO ^{14}C-ANTH OR ^{14}C-BaA). VALUES ARE MEANS ± 1 STD DEV. (n = 3)

	ANTH		BaA	
	Thousands of Bq	Percentage of Dose	Thousands of Bq	Percentage of Dose
Nutr. solution[a]	29.4±17.8	16.8	28.6±3.6	8.7
Roots	105.4±38.4	60.3	215.9±29.6	66.0
Stems	5.9±1.6	3.4	1.3±0.2	0.4
Leaves	5.5±0.7	3.1	1.3±0.3	0.4
Pods	0.3±0.06	0.14	0.1±0.02	0.04
Unaccounted for	28.3±28.9	16.2	79.9±25.6	24.4

[a]Based on activity in cyclohexane extracts + aqueous extracts of the nutrient solution

in the root tissues had been absorbed by the roots within 24 h after dosage (Figs. 1 & 2). The rapid uptake of ^{14}C by the roots coincided with a rapid decrease in nutrient-solution ^{14}C. The decrease occurred within 24 h in the single-dose 16-d

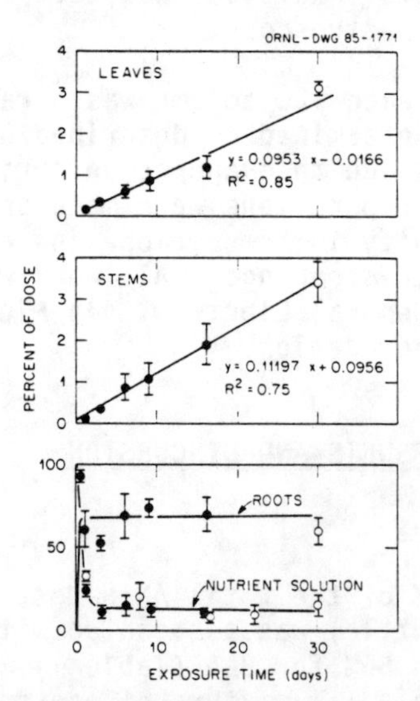

FIGURE 1. Carbon-14 in bush bean plant tissues and in nutrient solution after various lengths of exposure to ^{14}C-ANTH. Dark circles are values for plants exposed for 16 d to an acute dose of 1.43 µg ^{14}C-ANTH. Light circles are values for plants chronically exposed for 30 d to a total of 75 µg ^{14}C-ANTH (x ± SE; n = 3).

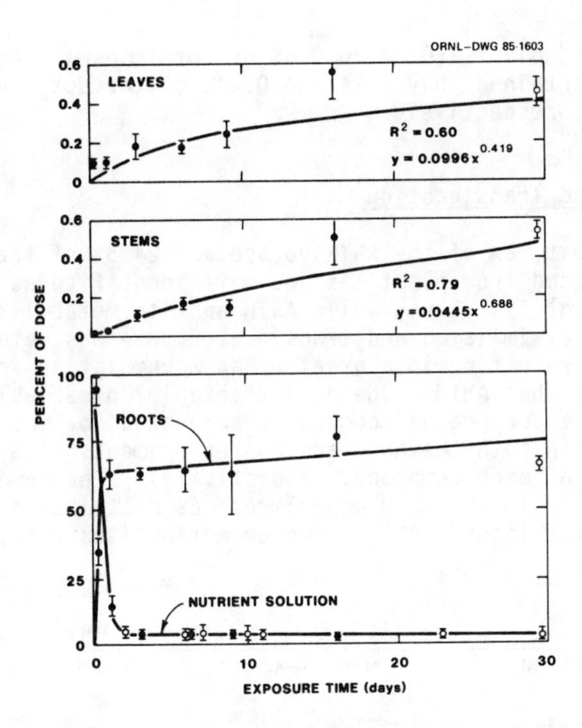

FIGURE 2. Carbon-14 in bush bean plant tissues and in nutrient solution after various lengths of exposure to ^{14}C-BaA. Dark circles are values for plants exposed for 16 d to 2.07 µg ^{14}C-BaA. Light circles are values for plants chronically exposed for 30 d to 41.9 µg ^{14}C-BaA ($\bar{x} \pm$ SE; n = 3).

experiment as well as after each dose in the multiple-dose 30-d experiments. Thus, concentrations of ^{14}C in nutrient solutions remained about the same throughout the 30-d exposure period except for very brief peaks in activity immediately after each dose.

In both ANTH- and in BaP-treated plants, ^{14}C was present in much lesser amounts in upper plant organs than in the roots (Table 1). Amounts of ^{14}C in upper plant organs increased linearly over the 30-d exposure period, in contrast to the increase for roots, which reached an equilibrium after only a few hours of exposure (Figs. 1 and 2). Greater amounts of ^{14}C were found in stems and leaves of ANTH-treated plants (0.14 to 3.4% of total ^{14}C dose) than in the same organs of BaA-treated plants (0.04 to 0.4% of total dose). Pods plus seeds contained only 0.14 and 0.04% of the dose for ANTH and BaA, respectively.

Metabolism and Translocation

At harvest, 9% of the ANTH dosage and 24.5% of the BaA dosage extracted from plant tissues were identified as parent compounds (Fig. 3). Thus, while ANTH and its metabolites were more readily assimilated and translocated, BaA was metabolized more slowly, resulting in a greater BaA accumulation in the plant tissues than ANTH. The nonextractable, presumably structurally incorporated, compounds accounted for most of the activity in both the ANTH and BaA experiments (31% and 32% of the doses of each compound, respectively). The remainder of the activity in the ANTH experiment was distributed among the polar metabolites (18%), nonpolar metabolites (9%), and

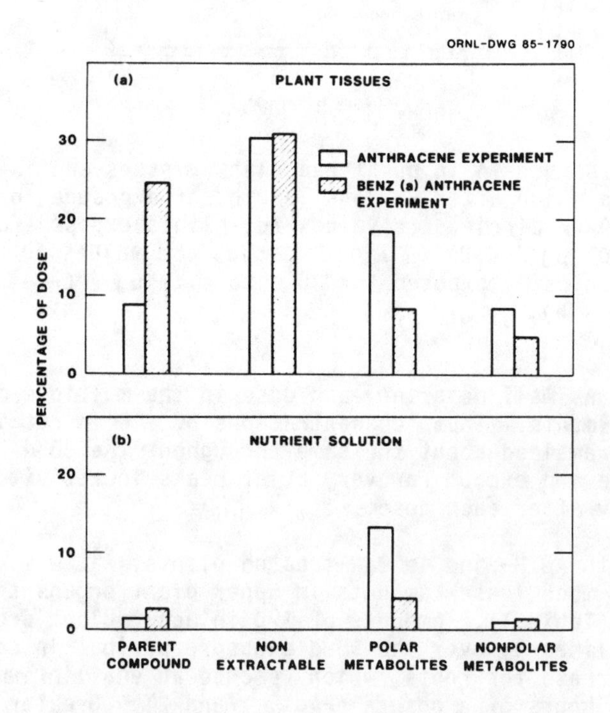

FIGURE 3. Relative amounts of ^{14}C-ANTH and its metabolites vs ^{14}C-BaA and its metabolites in nutrient solution and bush bean plants after 30 d of exposure to the PAHs.

parent compound (9%). The distribution was substantially different in the BaA experiment, where activity was found in the parent compound (25%), polar metabolities (9%), and nonpolar metabolites (5%).

Polar metabolites were the most dominant group of PAH-derived compounds in the nutrient solution at harvest (13 and 9% of the ANTH and BaA dosages, respectively). This relatively large amount of polar metabolites in nutrient solution may be due to their exudation from plant roots following transformation of the parent compounds within the plants, or to their transformation by microorganisms in the nutrient solution. While both of the above are likely, the greatest amounts are thought to be due to exudation because of our efforts to maintain near-sterile conditions in the nutrient solution. Also, if metabolism of the parent compounds had occurred at appreciable rates in the nutrient solution, the polar metabolites should have entered the plant roots more readily than the parent compounds and nonpolar metabolites, and therefore they should have been present in the nutrient solution in smaller quantities than the parent compounds or their nonpolar transformation products. But, the opposite was true. Only 1.5% and 2.7% of the ^{14}C in nutrient solution at harvest was due to ANTH and BaA, respectively, with only 1.2% and 1.5% due to nonpolar metabolites.

There was also a nonuniform distribution of the parent compounds and transformation products among the plant organs. More than 90% of the extractable ^{14}C in the leaves of both ANTH- and BaA-treated plants was due to polar metabolites (Fig. 4). An equally great proportion of the ^{14}C in stems of BaA-dosed plants and almost as great a proportion of the ^{14}C in stems of ANTH-dosed plants (78% of the extractable radioactivity) was due to polar metabolites. By contrast, roots contained smaller amounts of polar compounds and greater amounts of parent compounds and nonpolar transformation products. For example, about 75 and 28% of the extractable radioactivity in roots of BaA- and ANTH-dosed plants, respectively, was due to the parent compounds. Thus, while translocation of the parent compounds did occur, most of the parent compound remained in the roots, where it was stored or metabolized. As the more-water-soluble transformation products were formed, they were translocated out of the roots acropetally (*i.e.*, to stems and leaves) with lesser amounts exuded into the support medium.

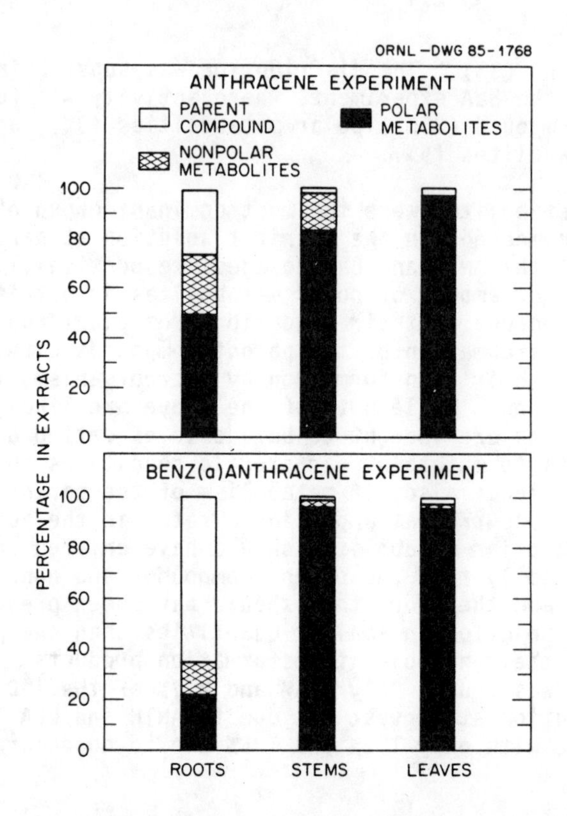

FIGURE 4. Percentages of parent compounds, polar metabolites, and nonpolar metabolites in ^{14}C compounds extracted from roots, stems, and leaves of plants exposed for 30 d to ^{14}C-ANTH or ^{14}C-BaA.

The distribution of ^{14}C associated with the PAH and their transformation products are summarized in Table 2. For comparative purposes the ANTH values were multiplied by 3.35 to account for differences in specific activities (radioactivity per µg of PAH) between the ANTH and BaA used in the experiments. Activity levels of ^{14}C in aboveground plant parts ranged from about three to six times greater for ANTH than for BaA, while radioactivity levels in the roots were about equal. By contrast, activity levels of ^{14}C due to parent compounds were more than twice as great in the roots of BaA-dosed plants compared with that for ANTH-dosed plants. The concentration of ANTH in stems was about twentyfold greater than BaA concentrations, while BaA concentrations in leaves were about sevenfold greater than

TABLE 2

DISTRIBUTION OF PAHs AND UNIDENTIFIED METABOLITES AMONG PLANT ORGANS AND NUTRIENT SOLUTION AFTER 30-d ROOT EXPOSURE TO ^{14}C-ANTH OR ^{14}C-BaA. VALUES IN PARENTHESES SHOW ^{14}C-ANTH AFTER NORMALIZATION OF ^{14}C-ANTH SPECIFIC ACTIVITY TO THAT OF BaA

	ANTH		BaA
	-----thousands of Bq kg^{-1}-----		
Carbon-14			
Roots	28583	(95753)	92883
Stems	650	(2178)	350
Leaves	400	(1340)	367
Pods + seeds	67	(224)	74
Nutr. solution	19	(64)	22
PAH			
Roots	4275	(14321)	33217
Stems	3.3	(11)	0.5
Leaves	0.2	(0.7)	4.8
Nutr. solution	1.7	(5.7)	7.2
Nonpolar metabolites[a]			
Roots	3992	(13373)	6783
Stems	23	(77)	5.2
Leaves	2	(7)	4.2
Nutr. solution	1	(3)	4.0
Polar metabolites			
Roots	7967	(26689)	11417
Stems	127	(425)	145
Leaves	158	(529)	190
Nutr. solution	16	(54)	11
Nonextractable[b]			
Roots	12350	(41373)	41467
Stems	497	(1665)	200
Leaves	240	(804)	167
Nutr. solution	--		--

[a]Bq in cyclohexane extract minus Bq in PAH.

[b]Total ^{14}C activity minus ^{14}C activity in PAHs plus polar and nonpolar metabolities.

ANTH concentrations. One possible explanation for this pattern is that ANTH moved into the stems more readily than BaA; but, after entering the leaf tissue, ANTH was very quicky transformed; while BaA, with a slower transformation rate in the leaves, accumulated more than ANTH.

Whole-plant activity levels of parent compounds and metabolites (based on weighted averages of root, stem, and leaf mass and normalized ^{14}C activities per unit weight of plant tissues) are presented in Fig. 5. For both ANTH- and BaP-dosed plants, the dominant component in plant tissues was the nonextractable compounds. Because of the relatively high nonextractable ^{14}C in stems and leaves (e.g., 1.5 to 4 times greater than ^{14}C associated with ANTH polar metabolites), it is proposed that they represent structurally incorporated compounds formed from simple (i.e., low-molecular-weight) transformation products, contrasted to conjugated forms of the parent compounds. Such large molecules would not be translocated as readily as the simpler polar metabolites. The second most dominant component in the BaA-dosed plants was the parent compound, while polar metabolites were the second most dominant components in the ANTH-dosed plants.

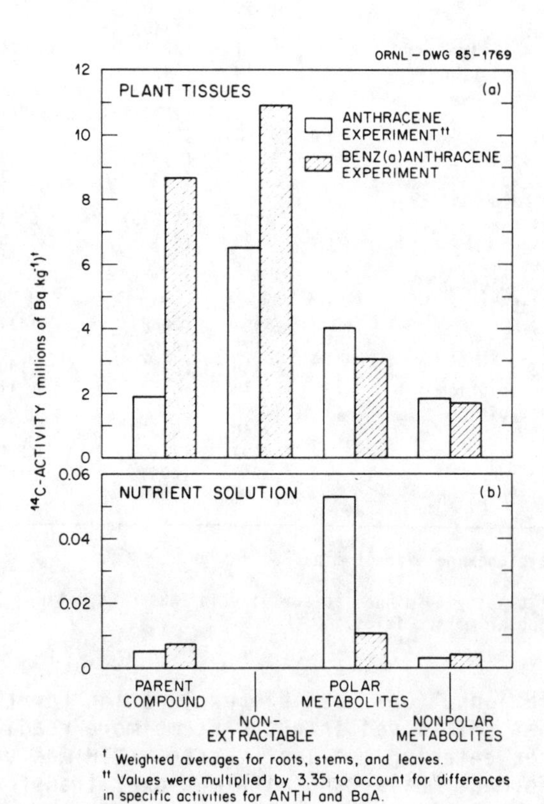

FIGURE 5. Comparative concentrations of parent compounds, polar and nonpolar metabolites, and nonextractable compounds in nutrient solution and in bush bean plants after 30 d of exposure to ^{14}C-ANTH or ^{14}C-BaA.

Using the ^{14}C counts associated with parent compounds and the respective specific activities of ANTH and BaA, the quantity of each parent compound per unit weight of plant tissue or nutrient solution at harvest was calculated. Leaves contained 0.1 µg of ANTH per kg of tissue, and 0.6 µg of BaA per kg of tissue. Stems contained 1.4 and 0.06 µg kg^{-1} ANTH and BaA, respectively. Concentrations of BaA in roots (4254 µg kg^{-1}) was 2.3 times greater than ANTH concentrations in roots (1834 µg kg^{-1}). Concentrations in nutrient solution were only 0.7 µg ANTH L^{-1} and 0.9 µg BaA L^{-1}.

Bioaccumulation

Bioconcentration Factors (BCFs) were calculated for parent compounds and for total ^{14}C activity [i.e., parent compounds plus metabolites] (Table 3). These calculations were based on ^{14}C activity levels at harvest. Calculations were performed on separate plant organs and on whole plants (using weighted averages). Values >1 are evidence for bioaccumulation. Bioaccumulation of ^{14}C (not PAH) occurred in all plant organs. BCFs for ^{14}C were consistently greater in

TABLE 3

VEGETATION:SUBSTRATE CONCENTRATION RATIOS (BIOCONCENTRATION FACTORS) AT HARVEST (30-d EXPOSURE) FOR TOTAL ^{14}C ACTIVITY (^{14}C RATIOS) AND FOR PARENT COMPOUND ONLY (PAH RATIOS)

	Vegetation:substrate concentration ratios			
	PAH ratios		^{14}C ratios	
	ANTH	BaA	ANTH + Metabolites	BaA + Metabolites
Pods + seeds	--	--	4	4
Leaves	0.1	0.7	21	17
Stems	1.9	0.1	34	16
Roots	2515.0	4613.0	1504	4222
Whole plant[a]	303	1015	200	940

[a]Weighted averages

aboveground plant parts in ANTH-dosed plants (ranging from 4 to 34) than in BaA dosed plants (ranging from 3 to 17), while the converse was true for roots (4222 for BaA versus 1504 for ANTH). Because of the very high BCF for roots, the whole-plant ^{14}C BCF was greater for BaA (940) than for ANTH (200).

Bioaccumulation of parent compounds occurred only in the roots of BaA-dosed plants (BCF = 4613) but in both stems (BCF = 1.9) and roots (BCF = 2515) of ANTH-dosed plants. Parent-compound BCF was greater in leaves of BaA-dosed plants than in ANTH-dosed plants. As in ^{14}C BCFs, whole-plant parent-compound BCFs were heavily influenced by concentrations in roots. Whole-plant ^{14}C BCFs did not differ greatly from whole-plant parent-compound BCFs. This was because of the rapid assimilation and retention of parent compounds in the roots and the rapid translocation of ^{14}C transformation products acropetally.

Unrestricted extrapolation of these BCFs to field situations should not be undertaken because of a number of complicating variables. They include differences between PAH degradation rates in soil and in nutrient solution, and the adsorptive qualities of soil, which are absent in nutrient solution. Differences between adsorptive qualities also exist among soil types because of variation in the amounts of clay, sand, silt, and organic matter present in the soil. Other variables that can affect BCF in terrestrial vegetation are the modes of contamination (i.e., whether from atmospheric deposition to vegetation and soil, leaching from organic waste-disposal sites, or floodplain sedimentation from PAH-contaminated streams and rivers). Such variables can affect the availability of PAH for assimilation by plants.

In order to explore some of these problems, the roots of bean plants were exposed to ^{14}C-ANTH which was dissolved in water or vapor-phase-coated onto stack ash and then added to an Emory silt loam soil (N. T. Edwards, unpublished data). ANTH concentrations and other variables were kept the same as in the nutrient solution experiments. Preliminary results (only ^{14}C oxidation data is presently completed) show that BCFs for ANTH and/or its transformation products are significantly ($P < .001$) greater if plants are grown in nutrient solution than if grown in soil. Whole-plant BCFs were 86 times as great in nutrient-solution experiments than in soil experiments and fifty-five times greater than in soil containing ANTH-coated stack ash. The ^{14}C BCFs in leaves and stems of plants grown in soil containing dissolved ANTH were fivefold and eightfold greater, respectively, than for plants grown in soil containing ANTH-coated stack ash. Thus the bioavailability of ANTH to the plants was influenced not only by the plant-growth medium but also by the form in which the ANTH was applied (i.e., whether dissolved in water or adsorbed onto stack ash).

SUMMARY AND CONCLUSIONS

Results of this study suggest that no simple relationship exists between the PAH accumulation rates in plant tissues and the physico-chemical characteristics of the individual compounds studied. ANTH, the smaller and more water-soluble of the two compounds was assimilated and translocated in bean plants more readily than BaA, but ANTH was much more readily transformed than BaA. Therefore, the BCF for BaA in bean plants was greater than the BCF for ANTH.

Both ANTH and BaA were translocated within bean plants. However, most (79%) of the extractable ^{14}C compounds found in stems and leaves were water-soluble metabolites produced in the roots and preferentially translocated, or they were produced as a result of very rapid transformation of the PAH after reaching the stem and leaf tissue. These data suggest that nearly half the polar metabolites were exuded into the growth medium from the roots, with very little exudation of nonpolar transformation products.

Based on ^{14}C activity only, bean plants assimilated and translocated ANTH and/or its metabolites much more readily from nutrient solution than from soil. Also, the method of ANTH application (whether applied to soil as an aqueous solution or as a vapor-phase-deposited coating on stack ash) influenced the rates of uptake and translocation. Concentrations of ^{14}C in roots of "soil-stack-ash-ANTH" treated plants were about fourfold greater than those of "soil-dissolved-ANTH" treated plants. However, translocation rates acropetally were greater in the "soil-dissolved-ANTH" treated plants.

If one assumes that bean plants are representative of other plant species in terms of PAH assimilation rates, these results indicate that PAH bioaccumulation in vegetation growing in contaminated soil can be expected in roots only. Concentrations typically reported in the literature for BaP in highly contaminated soils range from ≈100 to 1000 μg/kg.

It is reasonable to assume that in such highly contaminated soils the total PAH concentration would be 10 to 20 times the concentration of BaP alone. Edwards et al. (6) reported that, within a narrow concentration range (3 to 12 µg/kg), ANTH assimilation rates by bean plants were linearly correlated with concentrations in the growth medium. If this holds true over a wide range of concentrations and PAH, then PAH concentrations in vegetation growing in highly contaminated soil would be high relative to ambient levels even without bioaccumulation.

The results presented here and the limited and varied results reported in the literature illustrate the need for continuing research on the fate of PAH in the environment. Studies should include at least those PAH with demonstrated carcinogenic potential and which are also commonly released into the environment. Of 21 PAH frequently found in the environment (23), ten have exhibited carcinogenic potential (24). They are BaP, BaA, benzo[k]fluoranthene, benzo[b]fluoranthene, benzo[i]fluoranthene, benzo[ghi]perylene, cyclopenta[cd]pyrene, chrysene, indeno[1,2,3-cd]pyrene, and coronene. There is a need for integrated field research, designed to investigate the rates that these PAH move from soil to vegetation and animals. In addition degradation rates should be determined in soils and in vegetation. Concurrent with field studies, laboratory studies are needed to determine the metabolic pathways of each of these PAH in plants and in soil microorganisms.

ACKNOWLEDGMENTS

I thank Monty Ross-Todd for technical assistance in collecting the data. Also, thanks to Dr. Charles T. Garten, Dr. Bruce A. Tomkins, and Dr. George E. Taylor, Jr., for their constructive criticisms of the manuscript.

Research sponsored by the Office of Health and Environmental Research, U.S. Department of Energy, under Contract No. DE-AC05-84OR21400 with Martin Marietta Energy Systems, Inc. Publication No. 2599, Environmental Sciences Division, ORNL.

REFERENCES

1. Graf, W., and Nowak, W. (1966): Promotion of growth in lower and higher plants by carcinogenic polycyclic aromatics, Arch. Hyg. Bakteriol., 150:513-528. English translation, ORNL/Tr-4111, Oak Ridge National Laboratory, Oak Ridge, Tennessee.
2. Harms, H. (1975): Metabolism of benzo(a)pyrene in plant cell cultures and wheat seedlings. Landbauforsch. Voelkenrode, 25(2):83-90.
3. Gunther, F. A., Buzzetti, F. and Westlake, W. E. (1967): Residue behavior of polynuclear hydrocarbons on and in oranges, Residue Rev., 17:81-104.
4. Ellwardt, P. (1977): Variation in content of polycyclic aromatic hydrocarbons in soil and plants by using municipal waste composts in agriculture. pp. 291-298. In: Proc. Symp. on Soil Organic Matter Studies, Vol. II. International Atomic Energy Agency, Vienna.
5. Durmishidze, S. V., Devdorian, T. V., Kavtaradze, L. K. and Kuartskhava, L, Sh. (1974): Assimilation and conversion of 3,4-benzopyrene by plants under sterile conditions. Translated from Dokl. Akad. Nauk SSSR, 218(6):1368-1471. Available from Plenum Publishing Corporation, New York.
6. Edwards, N. T., Ross-Todd, R. M. and Garver, E. G. (1982): Uptake and metabolism of ^{14}C anthracene by soybean (Glycine max), Environ. Exp. Bot., 22(3):349-357.
7. Audere, A. K., Lindberg, Z. Y. Smirnov, G. A. and Shabad, L. M. (1973): Experiment in studying the influence of an airport located within the limits of a city on the level of environmental pollution by benzo(a)pyrene, Gig. Sanit., 38(9):90-92.
8. Kveseth, K., Sortland, B. and Stobet, M. B. (1981): Polycyclic aromatic hydrocarbons in leafy vegetables, a comparison of the Nordic results. Nordic PAH-project, Report No. 8, Central Inst. for Industrial Research, Oslo, Norway, 20pp.
9. Larsson, B., and Sahlberg, G. (1981): Polycyclic aromatic hydro-carbons in lettuce; influence of a highway and an aluminum smelter. In: Polynuclear Aromatic Hydrocarbons; Physical and Biological Chemistry, edited by M. Cooke, A. J. Dennis, and G. L. Fisher, pp. 417-426, Battelle Press, Columbus, Ohio.
10. Pyysalo, H. (1979): Analysis of polycyclic aromatic hydrocarbons (PAH) in Finnish leafy vegetables. Nordic PAH-project, Report no. 3, Central Inst. for Industrial Research, Oslo, Norway. 13 pp.

11. Stevcevsk, V., and Jovanovic-Kolar, J. (1974): Influence of air pollution on the content of 3,4-benzopyrene in nonrefined oil obtained from sunflower seeds grown in differently polluted areas, Arh. Hig. Rada, 23:191-196.
12. Wang, D. T., and Meresz, O. (1981): Occurrence and potential uptake of polynuclear aromatic hydrocarbons of highway traffic origin by proximally grown food crops. In: Polynuclear Aromatic Hydrocarbons; Physical and Biological Chemistry, edited by M. Cooke, A. J. Dennis, and G. L. FIsher, pp. 885-896, Battelle Press, Columbus, Ohio.
13. Dorr, R. (1970): Absorption of 3,4-benzopyrene by plant roots. Landwirtsch. Forsch., 23(4):371-379. English translation, ORNL/TM-4122, Oak Ridge National Laboratory, Oak Ridge, Tennessee.
14. Müller, V. H. (1976): Uptake of 3,4 Benzopyrene by food plants from artificially enriched substrates, Z. Pflanzenernaehr. Bodenkd., 6:685-695.
15. Wagner, K. H., and Siddiqi, I. (1970): The metabolism of 3,4-benzopyrene and benzo(e)acephenanthrylene in summer wheat, Z. Pflanzenernaehr. Bodenkd., 127:211-218.
16. Shabad, L. M., and Cohan, Y. L. (1972): The contents of benzo(a)pyrene in some crops, Arch. Geschwulstforsch., 40(3):237-243.
17. Siegfried, R. (1975): Effect of garbage compost on the 3,4-benzo-pyrene content of carrots and head lettuce, Naturwissenschaften, 62:300. English translation, ORNL/TM-4124, Oak Ridge National Laboratory, Oak Ridge, Tennessee.
18. Linne, C., and Martens, R. (1978): Examination of the risk of contamination by polycyclic aromatic hydrocarbons in the harvested crops of carrots and fungi after the application of composted municipal waste, Z. Pflanzenernaehr. Bodenkd. 141(3):265-274.
19. Durmishidze, S. V. (1977): Metabolism of some organic air pollutants in plants. Academy of Sciences of The Georgian SSR Institute of Plant Biochemistry. Mektlsniereba, Tbilisi, USSR. pp. 27-48. Available from National Agricultural Library, Beltsville, Maryland.
20. Edwards, N. T. (1983): Polycyclic aromatic hydrocarbons (PAHs) in the terrestrial environment - a review, J. Env. Qual., 12(4):427-441.
21. Bonner, J. and Galston, A. W. (1959): Principles of plant physiology. W. H. Freeman, San Francisco, 499 pp.
22. Santoro, R. M., Paglialunga, S. and Bartosek, I. (1979): Rapid determination of polycyclic aromatic hydrocarbons (PAH) in yeasts grown on n-paraffins and molasses, Toxicol. Lett., 3:85-93.

23. Grimmer, G. (1983): Choice of relevant polycyclic aromatic hydrocarbons. In: Environmental Carcinogens: Polycyclic Aromatic Hydrocarbons, edited by G. Grimmer, pp. 27-31, CRC Press Inc. Boca Raton, Florida.
24. Lee, L. L., Novotny, M. V., and Bartle, K. D., editors: Analytical Chemistry of Polycyclic Aromatic Compounds. Academic Press, New York, 462 pp.

ANALYSIS OF POLYCYCLIC AROMATIC HYDROCARBONS IN NAPHTHENIC DISTILLATE OILS BY HIGH PERFORMANCE LIQUID CHROMATOGRAPHY

WALTER C. EISENBERG*[1], KEVIN TAYLOR[1], GERALD J. LEPINSKE[2]
(1) IIT Research Institute, Chemistry and Chemical Engineering Research Department, Chicago, Illinois 60616;
(2) Calumet Industries, Inc., 14000 Mackinaw Avenue, Chicago, Illinois 60633

INTRODUCTION

The characterization of minor components of certain commercial products derived from crude oil is of increasing interest since the passage of the "right to know" laws. One of these components is the class of compounds called polycyclic aromatic hydrocarbons (PAH). Their measurement in petroleum crudes and processed oils involves a variety of analytical methods including liquid chromatography (1), gas chromatography (2), and gas chromatography-mass spectrometry (GC-MS) (3). In all cases, one or more steps are required to isolate the PAH fraction prior to the analysis. The present work concerns a high performance liquid chromatography (HPLC) method that was developed and applied to the measurement of eleven PAH in naphthenic distillate oils. The PAH identified in the samples were fluorene, phenanthrene, anthracene, fluoranthene, pyrene, benz[*a*]anthracene, chrysene, benzo[*b*]fluoranthene, benzo[*k*]fluoranthene, benzo[*a*]pyrene, and benzo[*g,h,i*]perylene. The method involves a solvent partition to isolate the PAH fraction, a semipreparative HPLC fractionation to obtain subfractions of a specific ring number, and a fluorescence HPLC analysis to measure the individual PAH in the subfractions. The concentrations of the PAH in one sample were verified in an independent analysis that was similar to the original method in all respects except that the subfraction PAH were measured by GC-MS.

MATERIALS AND METHODS

The oil samples were obtained from a commercial petroleum refining stream of Louisiana low cold test, low-sulfur naphthenic crude. The process consists of double distillation over caustic soda under vacuum. The entire lubricating oil boiling range is represented by these samples, which by convention are described by viscosity rather than boiling range. Pyrene and chrysene for the fractionation standard was obtained from Aldrich Chemical Company. The standard used for the quantitation of PAH in the HPLC fractions was SRM 1647 obtained from the National Bureau of Standards.

Solvent Partition

A processed oil sample (~1 g) was dissolved in 50 ml pentane and placed in a 500 ml separatory funnel. An equal volume of dimethylsulfoxide (DMSO) was added to the separatory funnel and the solvent mixture was shaken. After the solvent layers reseparated, the pentane layer was removed. Milli-Q water (50 ml) was added to a second separatory funnel containing the DMSO fraction. Hexane (100 ml) was added, and the solvent mixture was shaken. After the solvent layers reseparated, the hexane layer was removed. The DMSO-water layer were extracted with two additional 50 ml volumes of hexane. The combined hexane extracts were dried over anhydrous sodium sulfate and then concentrated in an Organovap at 50°C under a gentle stream of ultra high purity argon. The final extract volume was 500 µl.

Extract Fractionation

The heptane extract (~100 µl) was fractionated using semipreparative HPLC. A Waters Model 244 liquid chromatograph was used with a Bondapak NH_2 column to achieve the separation. Hexane was the eluant at a flow rate of 3 ml/min. The fractionation was monitored using a Waters Model 440 absorbance detector at $\lambda = 254$ nm. The column may be backflushed following the elution of the sample by reversing the solvent flow using a Model 721 rheodyne valve and a linear gradient to 100% methylene chloride at a rate of 10% methylene chloride per minute. The methylene chloride flow was continued for 10 minutes, and then an identical reverse gradient to 100% hexane was run. At 100% hexane, the system was allowed to equilibrate for 15 minutes. The frequency of backflushing required depends on the nature of the sample.

HPLC Analysis

The samples were analyzed on a Waters automated tri-module gradient elution liquid chromatograph consisting of a Waters Model 730 system controller, a Model 720 data module, and Model 710B WISP. It is connected to a Waters Model 440 dual-channel absorbance detector and a Schoeffel Model FS970 fluorescence detector. Fluorescence emission spectra were obtained with a Perkin-Elmer 650-40 spectrofluorometer equipped with an HPLC cell.

The three PAH fractions were analyzed using reverse phase HPLC. A Vydac ODS column was used with specific chromatographic elution and selective fluorescence detection for each of the fractions.

Fraction 1. This elution scheme is a linear gradient (flow rate, 1 ml/min). The initial solvent system is acetonitrile/water (50/50, v/v). During the gradient, the solvent composition is changed to acetonitrile/water (70/30, v/v) in 30 minutes, then to 100% acetonitrile in 10 minutes.

Fraction 2. The analysis of fraction 2 involves an isocratic elution (flow rate, 1 ml/min) with acetonitrile/water (70/30, v/v) for 15 minutes. This is followed by a linear gradient to 100% acetonitrile in 15 minutes. The elution at 100% acetonitrile is continued for 15 minutes.

Fraction 3. This elution scheme involves a linear gradient (flow rate, 1 ml/min) from acetonitrile/water (50/50, v/v) to 100% acetonitrile in 40 minutes, followed by 100% acetonitrile for 20 minutes.

Initial conditions were reestablished in each case by a reverse gradient of 5% acetonitrile per minute. The system was then allowed to reequilibrate for 15 minutes.

GC-MS Analysis

The GC-MS analyses were performed on two samples of distillate oil 3-1006-3. The samples were prepared in a manner identical to the HPLC procedure, and were analyzed using a Finnigan MAT 44S GC-MS system operated in a software-controlled selected ion monitoring (SIM) mode. Electron impact mass spectra were measured at 70 eV, and the data were acquired and processed using a Finnigan MAT SpectroSystem SS-200/M data system.

The gas chromatograph, a Varian 3700, was equipped with an unheated on-column injector and a 30 meter x 0.32 mm ID fused-silica, chemically bonded DB5 capillary column. The column was connected to the mass spectrometer via a heated (285°C), open-split interface; the helium carrier-gas flow rate was 1.1 ml/min. The chromatographic conditions were modified according to the fraction being analyzed:

Fraction 1. Samples were eluted isothermally for 5 minutes at 75°C. A linear program of 3°C/min was run to 250°C, and this temperature was held for 15 minutes.

Fraction 2. Samples were eluted isothermally for 5 minutes at 50°C. A linear program was then run at 3°C/min to 300°C. This temperature was held for 15 minutes.

Fraction 3. Samples were eluted isothermally at 200°C for 5 minutes. A linear program of 3°C/min was run to 310°C. This temperature was held for 20 minutes.

Initial conditions were reestablished in each case by a reverse temperature program of ~10°C/min. The system was then allowed to reequilibrate for 15 minutes.

RESULTS

The method that was used to measure PAHs in naphthenic distillate oils is outlined in Figure 1. Samples are prepared for analysis using a liquid-liquid partition to isolate and enrich the PAH fraction, and a fractionation on semipreparative Bondapak NH_2 to separate the PAH fraction according to ring size. The fractionation provides three subfractions: one containing up to three-ring parent PAH, one containing four-ring parent PAH, and another containing

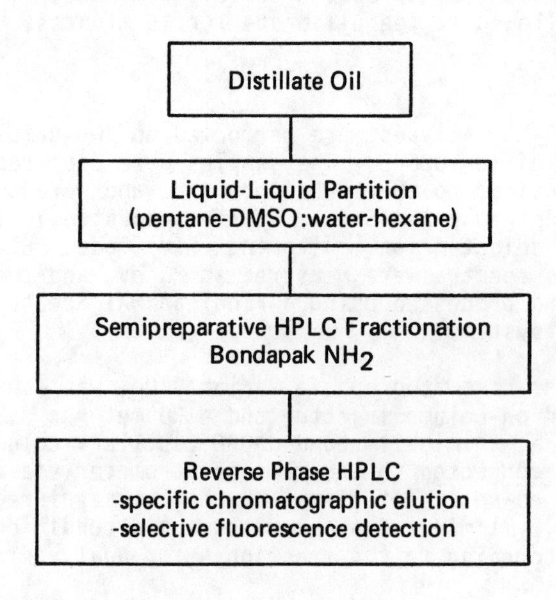

FIGURE 1. Outline of analytical method.

five-ring parent PAH and larger ring compounds. Each fraction was analyzed by reverse phase HPLC using specific chromatographic elution and selective fluorescence detection. Grimmer (2) and Eyres (4) used a similar but more complex method to measure PAH in oils. It involved two liquid-liquid partition steps, a chromatographic purification on silica gel and then Sephadex LH-20, and analysis by gas chromatography. One of us (5) used a two-step procedure involving a fractionation on Bondapak NH_2 and reverse phase fluorescence HPLC analysis to measure ten PAH in diesel emission extracts. This simpler two-step procedure was unsatisfactory for the analysis of PAH in naphthenic distillate oils. The first two fractions contained substances that interfered with the measurement of PAH by reverse phase fluorescence HPLC. May (1), however, has used this two-step procedure to measure benzo[a]pyrene in a number of lubricating oils.

The oil samples were initially extracted by the partition of the PAH between DMSO and pentane, a procedure first developed by Natusch and Tomkins (6). The oil was first dissolved in pentane, then extracted with an equal volume of DMSO. The PAH are extracted into the DMSO phase while most of the aliphatic hydrocarbons remain in pentane. An equal volume of water was added to the DMSO, which was then reextracted with hexane. This step allows the partitioning of the PAH into the hexane phase with the polar materials remaining in the DMSO/H_2O mixture.

Table 1 lists the eleven PAH eluted on Bondapak NH_2 by hexane. The times may vary depending on various factors (the individual column differences, sample type, etc.). It is therefore important to run a PAH standard periodically. The standard contained pyrene and chrysene, and was used to establish collection times.

Figure 2 is the chromatogram of the fractionation of the enriched PAH extract. The arrows indicate the points at which the fraction cuts were made. The intense peak at the beginning of the chromatogram suggests that the bulk of the PAH fraction consists of two- to four-ring PAH. It was found that the most reproducible way to fractionate the extract was to monitor the elution profile and to take fractions at the points indicated by the arrows. A procedure for backflushing the column was selected to remove polar organic materials. The frequency of backflushing depends on the nature of the sample. However, it should be used whenever a change in the retention times of the profile or the PAH

TABLE 1

ELUTION OF POLYCYCLIC AROMATIC HYDROCARBONS ON A BONDAPAK NH_2 COLUMN

Fraction	Compound	Relative Elution Times (min)
1	Fluorene	0.55
	Anthracene	0.75
	Phenanthrene	0.78
2	Fluoranthene	1.00
	Pyrene	0.98
	Benz[a]anthracene	1.41
	Chrysene	1.43
3	Benzo[b]fluoranthene	1.85
	Benzo[k]fluoranthene	1.70
	Benzo[a]pyrene	1.80
	Benzo[g,h,i]perylene	2.51

standard is observed. The NH_2 column could be regenerated or cleaned by reverse elution with tetrahydrofuran. Warner (7) also provides a procedure for the reactivation of the Bondapak NH_2 column.

Each PAH fraction was chromatographed usng specific elution conditions. The objective was to maximize the resolution of the parent PAH that would be measured in each of the fractions. Further enhancement of this resolution was obtained using selective wavelength fluorescence. Table 2 presents the fluorescence conditions and the elution times for the PAH in their respective fractions, as well as their detection limits. Figures 3 to 5 are the fluorescence emission chromatograms of a standard PAH mixture, SRM 1647 (National Bureau of Standards), and fractions 1 to 3 using the elution and detection conditions chosen for each fraction. A coeluting interference was observed for benz[a]-anthracene using the fluorescence condition λ_{ex} = 260 nm with a 370 nm cut-off filter (see Figure 5b). However, monitoring the elution of this compound at λ_{ex} = 280 nm with a 389 nm

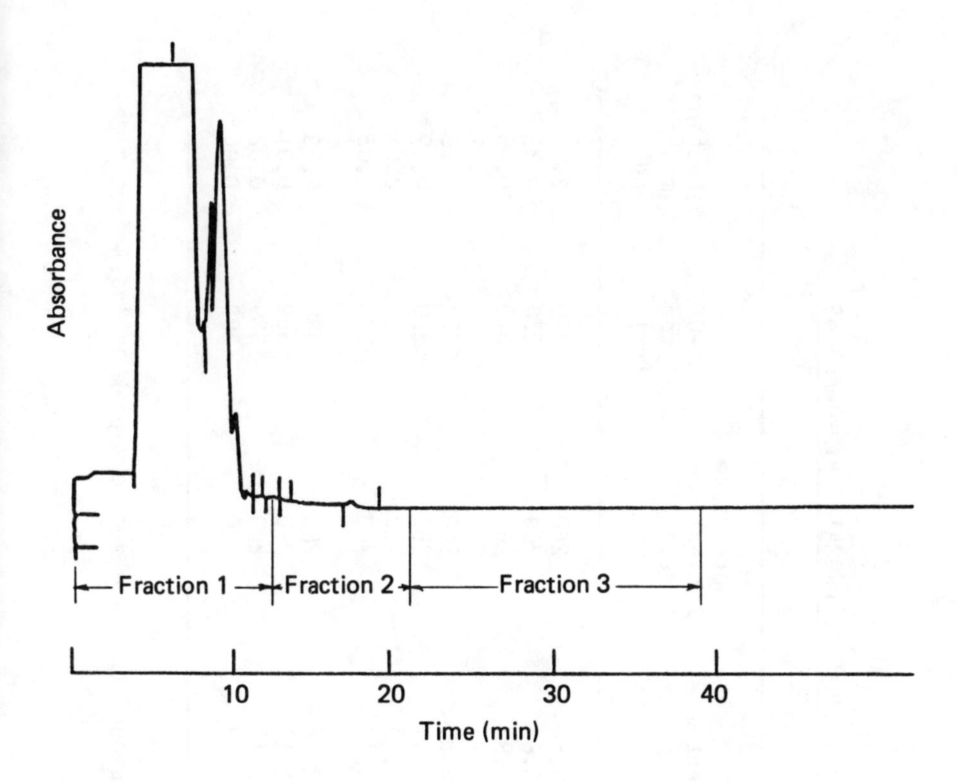

FIGURE 2. Fractionation of the distillate oil extract.

cut-off filter minimizes this interference. Figures 6a-b are the fluorescence emission chromatograms for SRM 1647 and fraction 2 under these conditions.

Table 3 shows the results of four analyses of distillate oil sample 3-1006-3 using the HPLC method, along with the mean and standard deviation. Quantitation was achieved using the external standard method. The percent relative standard deviation ranged from 10.3% for phenanthrene to 78.3% for benzo[a]pyrene. Considering the level of PAH in the sample, the precision was acceptable. The authenticity of the peaks was determined using retention times and peak enhancement. A method blank was also performed, and this analysis demonstrated that interferences were not arising from materials (glassware, solvents, *etc.*) that were used in the analysis.

TABLE 2

HIGH PERFORMANCE LIQUID CHROMATOGRAPHY OF POLYCYCLIC AROMATIC HYDROCARBONS

Elution Scheme	Compound	Retention Time (min)	Fluorescence Excitation Wavelength[a] (nm)	Cut-Off Filter (nm)	Detection Limit[a,b] (ng)
1	Fluorene	16.50	260	370	2.50
	Phenanthrene	18.85	260	370	0.30
	Anthracene	20.92	260	370	0.15
2	Fluoranthene	10.30	260	370	0.13
	Pyrene	11.97	260	370	0.19
	Benz[a]anthracene	17.40	280	389	0.04
	Chrysene	18.72	260	370	0.05
3	Benzo[b]fluoranthene	32.34	280	389	0.03
	Benzo[k]fluoranthene	34.11	280	389	0.01
	Benzo[a]pyrene	35.73	280	389	0.02
	Benzo[g,h,i]perylene	40.82	280	389	0.06

a Detector range 1.0, minimum range 0.01.

b The detection limit is the on-column amount that would give a peak at least five times the baseline noise.

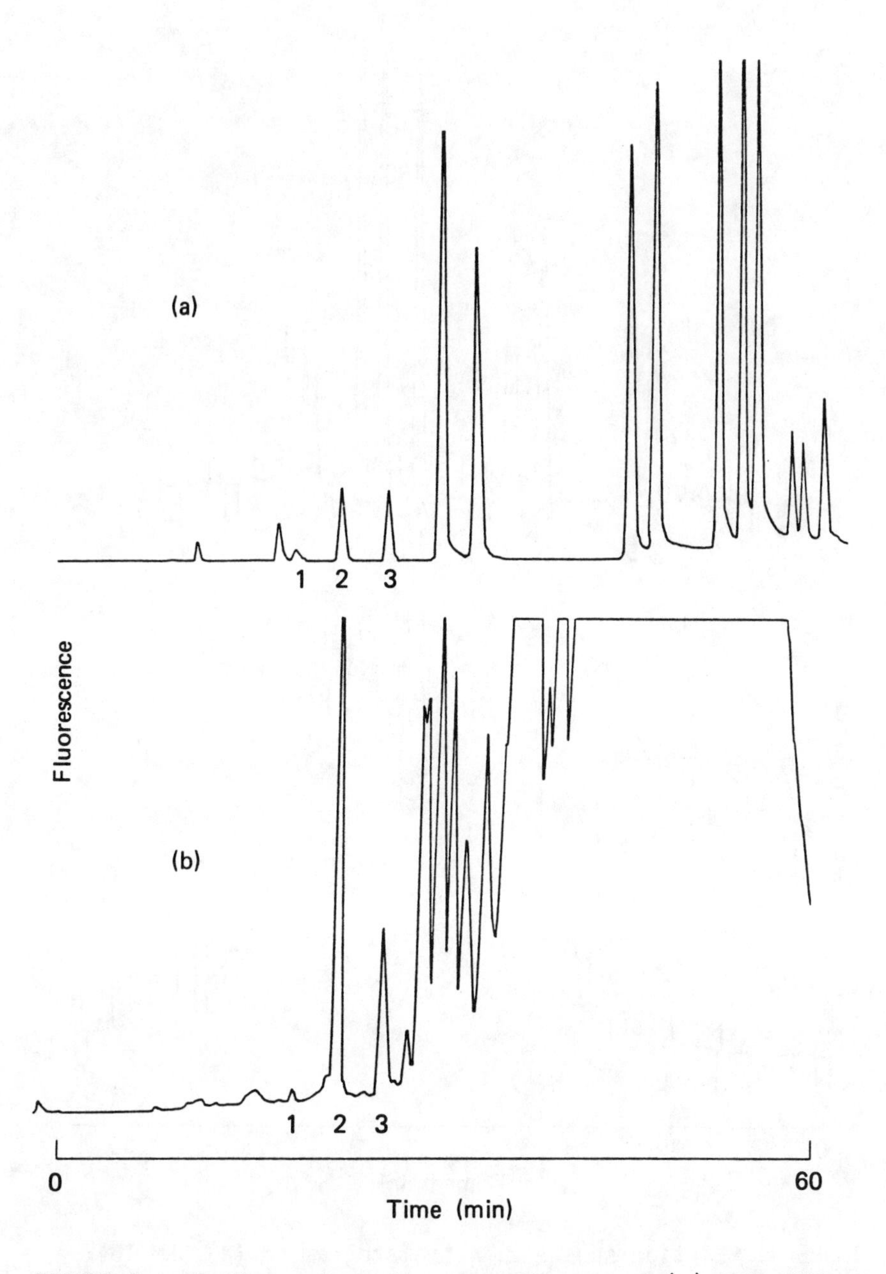

FIGURE 3. Elution scheme 1: chromatogram of (a) SRM 1647 and (b) extract fraction 1. Fluorescence: λ_{ex} = 260 nm, cut-off filter = 370 nm. (1) Fluorene, (2) phenanthrene, (3) anthracene.

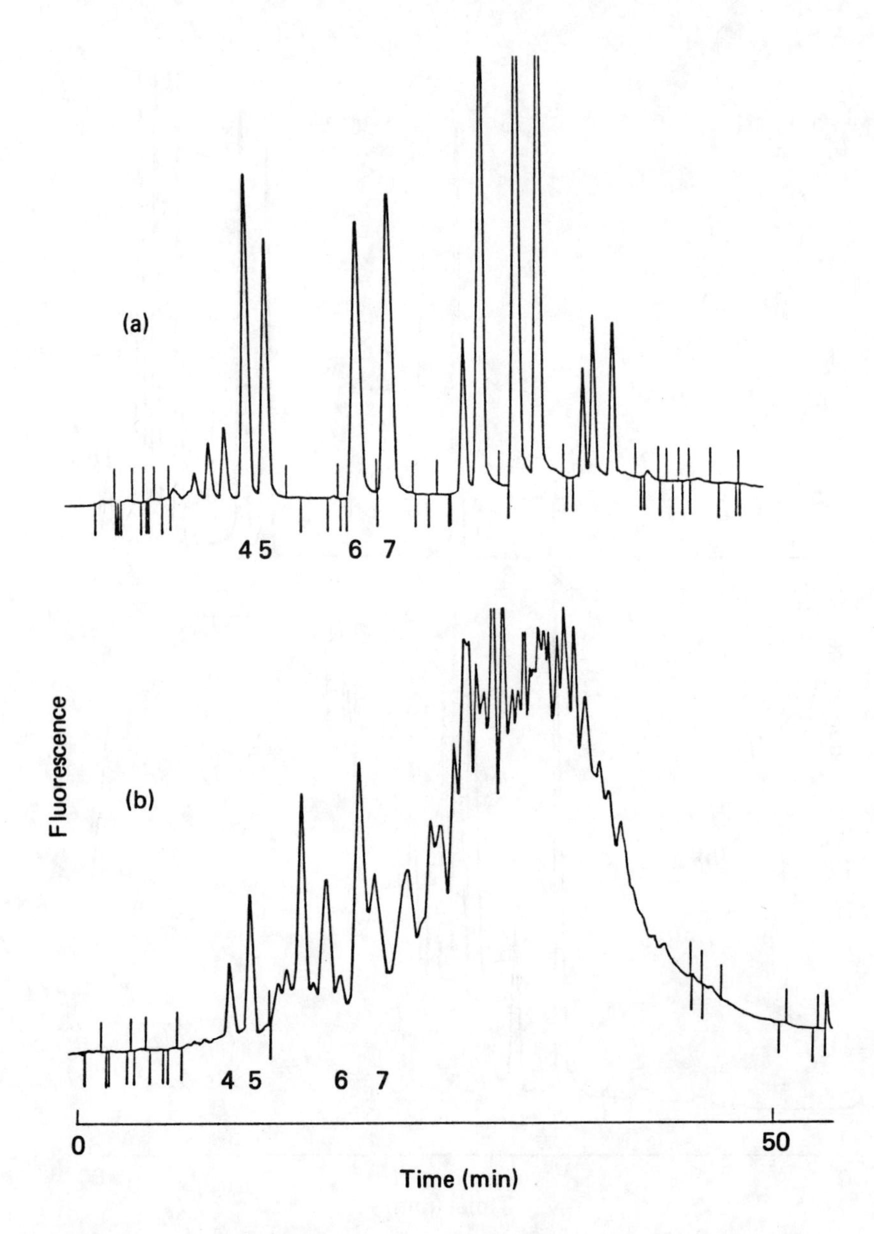

FIGURE 4. Elution scheme 2: chromatogram of (a) SRM 1647 and (b) extract fraction 2. Fluroescence: λ_{ex} = 260 nm, cut-off filter = 370 nm. (4) Fluoranthene, (5) pyrene, (6) benz[a]anthracene, (7) chrysene.

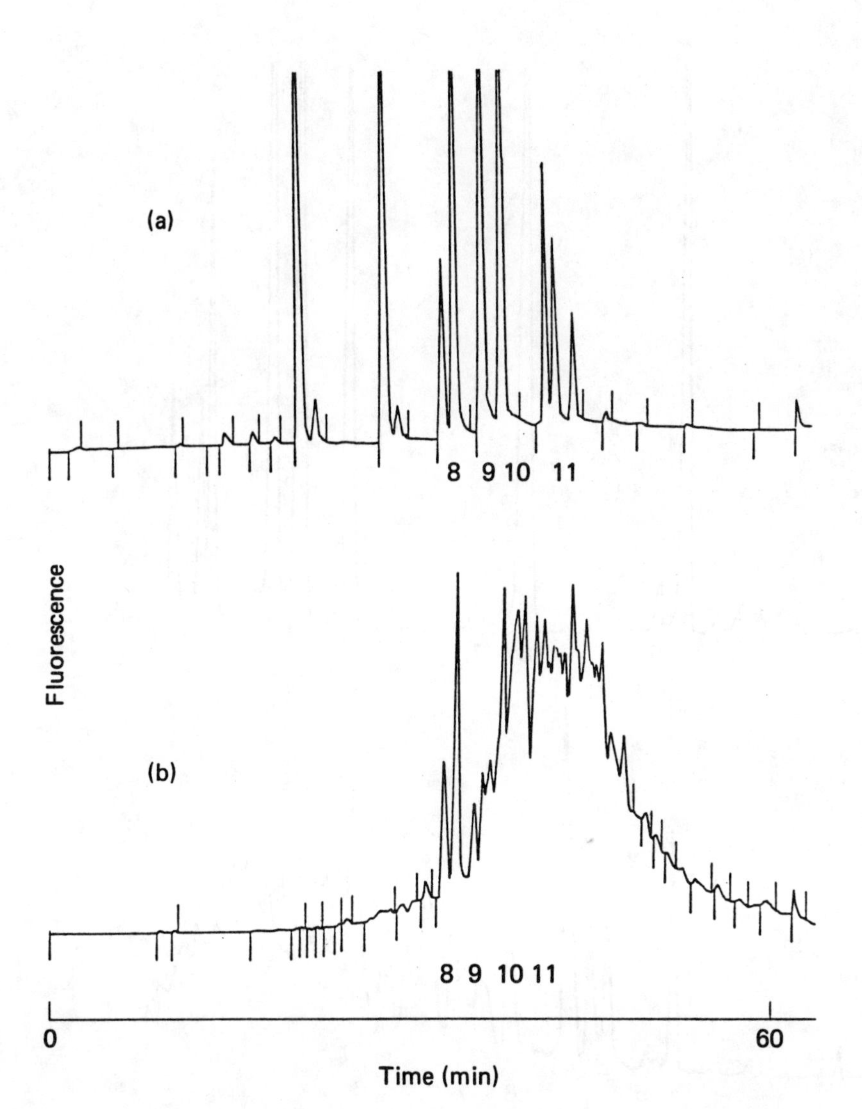

FIGURE 5. Elution scheme 3: Chromatogram of (a) SRM 1647 and (b) extract fraction 3. Fluorescence: λ_{ex} = 280 nm, cut-off filter = 389 nm. (8) Benzo[*b*]fluoranthene, (9) benzo[*k*]fluoranthene, (10) benzo[*a*]pyrene, (11) benzo[*g,h,i*]perylene.

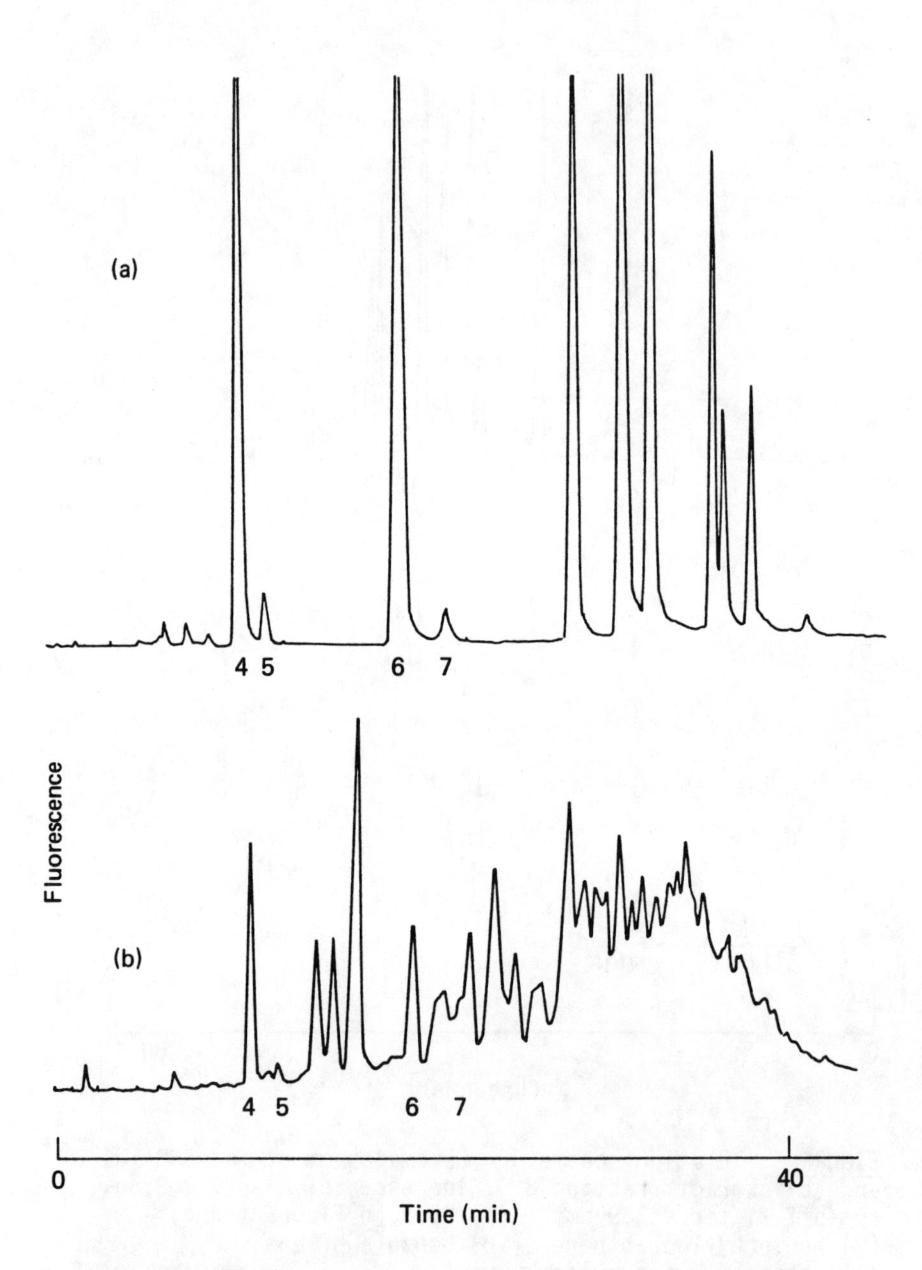

FIGURE 6. Elution scheme 2: chromatogram of (a) SRM 1647 and (b) extract fraction 2. Fluorescence: λ_{ex} = 280 nm, cut-off filter = 389 nm. (4) Fluoranthene, (5) pyrene, (6) benz[a]anthracene, (7) chrysene.

TABLE 3

CONCENTRATION OF PAH IN DISTILLATE OIL SAMPLE 3-1006-3 BY HPLC ANALYSIS

Compound	Analysis (ppm)[a]					
	1	2	3	4	Mean ± SD	Blank
Fluorene	9.00	16.4	10.4	11.1	11.7 ± 3.2	0.00
Phenanthrene	48.0	52.4	44.4	41.1	46.5 ± 4.8	0.00
Anthracene	7.26	13.3	8.23	9.32	9.53 ± 2.65	0.00
Fluoranthene	2.86	1.48	2.21	1.41	1.99 ± 0.68	0.01
Pyrene	3.15	1.451	2.40	2.84	2.46 ± 0.74	0.00
Benz[a]anthracene	0.53	0.57	1.23	0.400	0.68 ± 0.32	0.00
Chrysene	3.72	4.26	2.75	2.23	3.24 ± 0.92	0.00
Benzo[b]fluoranthene	NA[b]	0.419	0.337	0.525	0.43 ± 0.09	0.00
Benzo[k]fluoranthene	NA	0.222	0.078	0.277	0.19 ± 0.10	0.00
Benzo[a]pyrene	NA	0.423	0.173	0.083	0.23 ± 0.18	0.00
Benzo[g,h,i]perylene	NA	1.28	0.665	0.605	0.85 ± 0.37	0.00

a Not corrected for recovery.
b NA - not analyzed.

Additional support for the validity of the HPLC method comes from an analysis of distillate oil sample 3-1006-3 using GC-MS. In this analysis the sample preparation (*i.e.*, the solvent partition and the HPLC fractionation) was identical to that used in the HPLC method. Each fraction was analyzed using specific temperature programs. The objective was to maximize the resolution of the parent PAH that were measured in each fraction. The measurement of the selected PAH was further enhanced by operating the mass spectrometer in the specific ion monitoring mode. Table 4 lists the PAH, the fractions in which they are found, and the specific ion monitored for each PAH. Peak authenticity was determined by the combined use of retention times and specific ion monitoring.

TABLE 4

PAH STANDARDS USED IN SAMPLE QUANTITATIONS

Fraction	Compound[a]	Ion Monitored
1	Fluorene	166
	Phenanthrene	178
	Anthracene	178
2	Fluoranthene	202
	Pyrene	202
	Benz[*a*]anthracene	228
	Chrysene	228
3	Benzo[*b*]fluoranthene	252
	Benzo[*k*]fluoranthene	252
	Benzo[*a*]pyrene	252
	Benzo[*g,h,i*]perylene	276

a Compounds are listed in their elution order on a DB5 capillary column.

Table 5 presents the average of two analyses of distillate oil sample 3-1006-3 using the GC-MS method. Benzo[*b*]-fluoranthene and benzo[*k*]fluoranthene coeluted in the GC-MS analysis and were measured as one compound. The mean and

TABLE 5

CONCENTRATION OF PAH IN DISTILLATE OIL SAMPLE 3-1006-3

Compound	Analysis (ppm)	
	GC-MS[a]	HPLC[b]
Fluorene	8.55	11.7 ± 3.2
Phenanthrene	191	46.5 ± 4.8
Anthracene	9.43	9.53 ± 2.65
Fluoranthene	1.47	1.99 ± 0.68
Pyrene	1.73	2.46 ± 0.74
Benz[b]anthracene	1.00	0.68 ± 0.37
Chrysene	5.31	3.24 ± 0.92
Benzo[b]fluoranthene)	0.62	0.43 ± 0.09
Benzo[k]fluoranthene)		0.19 ± 0.10
Benzo[b]pyrene	0.17	0.23 ± 0.18
Benzo[g,h,i]perylene	0.11	0.85 ± 0.37

a Average of two determinations.
b Mean ± standard deviation.

standard deviation of the PAH concentration of distillate oil sample 3-1006-3 obtained using HPLC are also included in Table 5 for comparison. With the exception of phenanthrene the distillate oil PAH concentrations obtained by the two analytical methods show excellent agreement.

Fluorescence emission spectra of the PAH peaks in the standard and samples were obtained using stopflow fluorescence spectroscopy. The emission spectra of all the PAH except phenanthrene showed one or more additional emissions indicating the presence of one or more minor impurities in the PAH peak. The excellent agreement of the HPLC and GC-MS data demonstrates that by carefully choosing the fluorescence wavelengths, one can, at least in these cases, accurately measure the PAH in a sample despite the presence of a minor coeluting impurity.

A recovery study was performed by dissolving a standard containing the eleven PAH of the study in 50 ml pentane, and analyzing this solution as one would analyze an oil sample dissolved in pentane. The concentration of the PAH in the spike was approximately the same as the PAH concentration in the oil samples. The average recovery of the PAH in two experiments is given in Table 6. It ranged from 19.9% for fluorene to 80.7% for benzo[g,h,i]perylene.

TABLE 6

RECOVERY STUDY

Compound	Recovery (%)		
	Sample 1	Sample 2	Average
Fluorene	19.6	20.2	19.9
Phenanthrene	50.2	34.5	42.4
Anthracene	28.9	39.7	34.3
Fluoranthene	73.6	67.0	70.3
Pyrene	48.0	38.4	43.2
Benz[a]anthracene	38.2	53.4	45.8
Chrysene	39.1	61.2	50.2
Benzo[b]fluoranthene	61.8	60.1	61.1
Benzo[k]fluoranthene	59.6	55.0	57.3
Benzo[a]pyrene	62.3	57.9	60.1
Benzo[g,h,i]perylene	74.6	86.8	80.7

The HPLC method was used to measure the concentration of eleven PAH in six distillate oil samples representative of the lubricating oil boiling range. These results are shown in Table 7. Each oil is identified by its viscosity. The oil samples in the viscosity range of 10-20 centistokes contain the highest concentration of PAH, although the levels are still quite low.

DISCUSSION

The method as described in the paper has been validated for the measurement of ten of the eleven PAH studied in naphthenic distillate oils. The ten PAH are fluorene, anthracene, fluoranthene, pyrene, benz[a]anthracene, chrysene, benzo[b]fluoranthene, benzo[k]fluoranthene, benzo[b]-pyrene, and benzo[g,h,i]perylene. One advantage of the

TABLE 7

CONCENTRATION OF PAH IN DISTILLATE OIL SAMPLES

Compound	HPLC Analysis (ppm)						
	3-1006-1	3-1006-2	3-1006-3	3-1006-4	3-1006-5	3-1006-6	Blank
Fluorene	<0.935	30.9	56.8	5.78	<0.935	<0.935	<0.185
Phenanthrene	<0.028	28.1	103	169.6	0.745	0.038	0.006
Anthracene	<0.014	5.33	26.5	14.2	0.962	<0.014	<0.003
Fluoranthene	<0.006	0.316	2.74	0.047	0.037	0.006	<0.001
Pyrene	<0.014	1.09	9.30	2.48	0.035	<0.014	<0.002
Benz[a]anthracene	<0.002	0.203	1.23	2.11	0.216	<0.002	<0.004
Chrysene	0.004	1.01	8.31	5.06	1.31	0.018	<0.0006
Benzo[b]fluoranthene	0.029	0.762	0.466	0.862	<0.002	<0.002	<0.0003
Benzo[k]fluoranthene	0.030	0.272	0.087	0.141	<0.002	0.002	<0.0002
Benzo[a]pyrene	0.002	0.013	0.363	0.356	0.674	<0.001	<0.0002
Benzo[g,h,i]perylene	<0.009	0.004	0.706	0.605	1.08	<0.009	<0.0002
Viscosity, at 40°C (centistokes)	3.5	5	10	20	260	630	

method is that it offers the option of analyzing samples by either HPLC or GC-MS. The precision of the method is such that it will allow one to distinguish changes in PAH concentrations in oils, resulting from various refining and treatment processes, as well as to distinguish different PAH concentrations accurately in various feedstocks.

ACKNOWLEDGMENT

The authors are indebted to Dr. Sydney M. Gordon and Mr. Michael Miller of IITRI for performing the GC-MS analysis.

REFERENCES

1. Brown, J. M., Wise, S. A., and May, W. E. (1980): Determination of benzo[a]pyrene in recycled oils by a sequential HPLC method. J. Environ. Sci. Health, A15(6):613-623.
2. Grimmer, G., and Bohnke, H. (1976): Enrichment and gas chromatographic profile-analysis of polycyclic aromatic hydrocarbons in lubricating oils, Chromatographia, 9:30-40.
3. Grimmer, G., Jacob, J., Naujack, K. W., and Dettborn, G. (1981): Profile of the polycyclic aromatic hydrocarbons from used engine oil, Fresenius Z. Anal. Chem., 309:13-19.
4. Eyres, A. R. (1981): Polycyclic aromatic hydrocarbon contents of used metalworking oils, Institute of Petroleum Technical Paper IP 81-002, London, England, 13 pp.
5. Eisenberg, W. C., and Cunningham, D.L.B. (1983): Analysis of polycyclic aromatic hydrocarbons in diesel emissions using high performance liquid chromatography. In: Polycyclic Aromatic Hydrocarbons, edited by W. M. Cooke and A. Dennis, pp. 379-393, Battelle Press, Columbus, Ohio.
6. Natusch, D. F., and Tomkins, B. A. (1978): Isolation of polycyclic organic compounds by solvent extraction with dimethylsulfoxide, Anal. Chem., 50:1429-1434.
7. Warner, I., Karlesky, D., and Skelly, D. C. (1981): Reactivation of amino bonded phase liquid chromatographic column, Anal. Chem., 53:2146-2147.

IN VITRO METABOLISM OF 2-NITROANTHRACENE BY RAT LIVER MICROSOMES

E. KIM FIFER[1†], DOMINIC T.C. YANG[1,2], SHUN E. CHIUN[2], LINDA S. VON TUNGELN[1], DWIGHT W. MILLER[1], FREDERICK A. BELAND[1], PETER P. FU[1]
(1)National Center for Toxicological Research, Jefferson, Arkansas 72079; (2) Department of Chemistry, University of Arkansas at Little Rock, Little Rock, Arkansas 72204.

INTRODUCTION

Nitro polycyclic aromatic hydrocarbons (nitro PAH) are genotoxic environmental pollutants found in soil, fly ash, diesel emission and urban air [reviewed in (1,2)]. Since these compounds require metabolism in order to exert their mutagenicity and tumorigenicity (1,2), we have been determining how they are metabolized and identifying structural features which can affect their activation and detoxification. In this study, we report the rat liver microsomal metabolism of a mutagenic component of diesel exhaust, 2-nitroanthracene (3), under both aerobic and hypoxic conditions. The results are compared to those previously obtained with 9-nitroanthracene (4), a weakly mutagenic constituent of diesel emission (3), in order to determine the factors which contribute to the differences in their mutagenicities.

MATERIALS AND METHODS

Materials

2-Nitroanthracene was synthesized by oxidation of 2-aminoanthracene with *m*-chloroperoxybenzoic acid by the method of Scribner *et al.* (5). The product was purified by chromatography on alumina by eluting with benzene and then recrystallized from benzene-hexane to give yellow needles: mp, 181-182° [lit. (5) 179-181°]. [G-^{3}H]2-Nitroanthracene (120 mCi/mmol) was prepared by direct exchange in tritiated trifluoroacetic acid by Robert W. Roth of Midwest Research Institute, Kansas City, MO.

---

†Present Address: Department of Basic Pharmaceutical Sciences, West Virginia University School of Pharmacy, Morgantown, West Virginia 26506.

In Vitro Metabolism of 2-Nitroanthracene with Rat Liver Microsomes

Liver microsomes from 150 g male Sprague-Dawley rats, pretreated for three days with 3-methylcholanthrene (25 mg/kg, ip), were prepared as described by Schenkman and Cinti (6). Metabolites were obtained by incubation of 2-nitroanthracene (40 μmol in 10 ml acetone) under aerobic conditions at 37° for 1 hr in a 500-ml reaction mixture containing 25 mmol Tris-HCl buffer (pH 7.5), 1.5 mmol $MgCl_2$, 50 units glucose-6-phosphate dehydrogenase (Type XII; Sigma, St. Louis, MO), 48 mg $NADP^+$, 280 mg glucose-6-phosphate and 500 mg microsomal protein. After quenching the incubation with 300 ml acetone, 2-nitroanthracene and its metabolites were partitioned into 800 ml ethyl acetate. The organic layer was evaporated under reduced pressure. The residue was washed with acetone (2 X 5 ml) and the insoluble material, which contained microsomal protein, was removed by centrifugation. After removal of the acetone under reduced pressure, the residue was dissolved in 0.5 ml methanol for analysis by reversed-phased high performance liquid chromatography (HPLC).

Hypoxic incubations (3% oxygen as measured by mass spectrometry) were similarly conducted, except a closed system filled with argon was employed.

HPLC Separation of Metabolites

HPLC was performed with a Beckman system consisting of two 100A pumps, a 210 injector, a 420 solvent programmer and a Waters Associates (Milford, MA) 440 absorbance detector operated at 254 nm. Metabolites from the aerobic incubations were separated using a 15-min linear gradient of 55-80% methanol followed by a 5-min linear gradient of 80-100% methanol at a flow rate of 2.8 ml/min on a DuPont Zorbax ODS column (9.4 X 250 mm). These metabolites were further purified with a DuPont Zorbax SIL column (9.4 X 250 mm) operated isocratically with 35-50% THF in hexane at a flow rate of 2.8 ml/min. Metabolites from hypoxic incubations were separated by reversed-phase HPLC using a 30-min linear gradient of 60-100% methanol at a flow rate of 2.8 ml/min.

Physicochemical Properties of Metabolites

Ultraviolet-visible absorption spectra of the metabolites were measured in methanol with a Varian CARY 219 spectrophotometer. Mass spectra of the metabolites were recorded with a Finnigan model 4023 system, and proton NMR spectra were obtained with a Bruker WM 500 spectrometer. The metabolites were dissolved in acetone-d_6 or dimethylsulfoxide-d_6 with a trace of D_2O. Chemical shifts are in parts per million (δ) relative to tetramethylsilane.

RESULTS

Separation and Characterization of 2-Nitroanthracene Metabolites from Aerobic Incubation

Metabolites formed from incubation of 2-nitroanthracene with rat liver microsomes were separated by reversed-phase HPLC (Figure 1). The material contained in the chromatographic peaks eluting prior to peak 1 and in the unlabeled chromatographic peaks between peaks 1 and 2 were observed in control incubations. Peak 5 contained recovered substrate. Chromatographic peak 4 was identified as 2-nitro-9,10-anthraquinone based on comparison of its UV-visible and mass spectra and HPLC retention time with those of an authentic sample. Peak 3 contained more than one metabolite. Further separation by a normal-phase HPLC system (Figure 2A) resulted in two baseline-resolved chromatographic peaks. The components of each peak had similar mass spectra with molecular ions at m/z 257 and a characteristic fragment at m/z 239 (due to loss of a water molecule) which suggested that both metabolites were trans-dihydrodiols of 2-nitroanthracene. Their structures were determined by analysis of their high resolution 500 MHz proton NMR spectra (Table 1). The metabolite with the shorter HPLC retention time (Figure 2A) was identified as 2-nitroanthracene trans-7,8-dihydrodiol and the one with the longer retention time was 2-nitroanthracene trans-5,6-dihydrodiol. Based on their coupling constants, both trans-dihydrodiols adopted conformations with quasiequatorial hydroxyl groups (4,7).

Chromatographic peak 2 in Figure 1 also contained two metabolites which were partially separated by a normal-phase HPLC (Figure 2B). Both of these metabolites had similar mass spectra with molecular ions at m/z 273. Proton NMR spectral analysis (Table 1) indicated that the peak with the shorter retention time was 2-nitroanthracene 7-keto-5,6,7,8-tetrahydro-trans-5,6-diol while the peak with the longer retention time was 2-nitroanthracene 6-keto-5,6,7,8-tetrahydro-trans-7,8-diol. These dihydrodiol-ketone metabolites were also formed during incubation of the respective 2-nitroanthracene trans-5,6- or 7,8-dihydrodiol with microsomes.

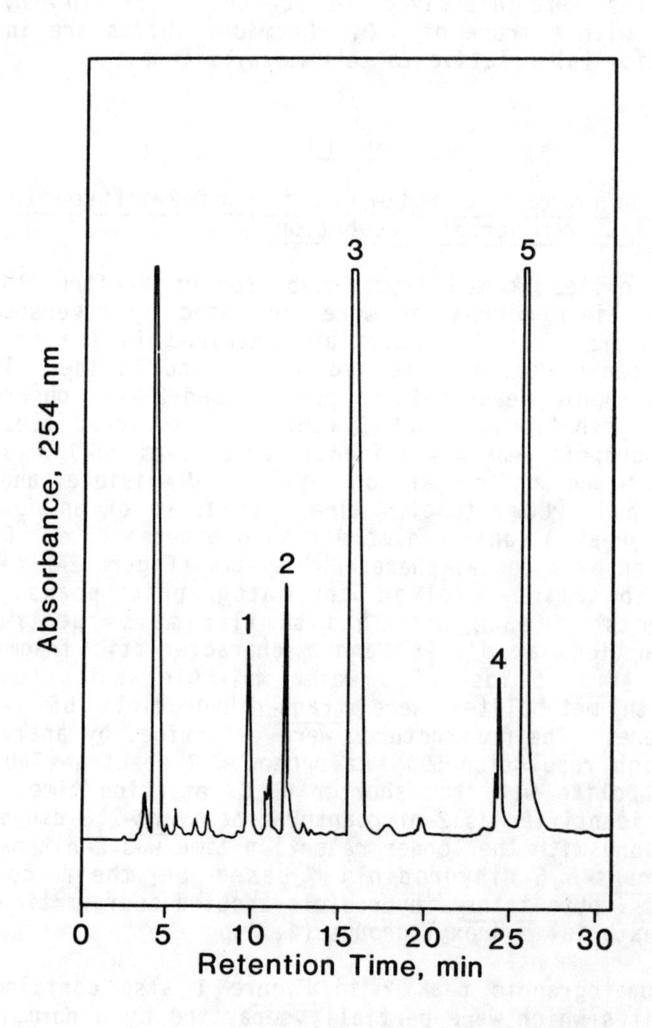

FIGURE 1. Reversed-phase HPLC profile of aerobic microsomal metabolites of 2-nitroanthracene. The identity of the metabolite in each peak is as follows: 1) 2-nitroanthracene 5,6,7,8-tetrahydrotetrol (tentative); 2) 2-nitroanthracene 7-keto-5,6,7,8-tetrahydro-trans-5,6-diol and 2-nitroanthracene 6-keto-5,6,7,8-tetrahydro-trans-7,8-diol; 3) 2-nitroanthracene trans-5,6- and 7,8-dihydrodiol; 4) 2-nitro-9,10-anthraquinone; and 5) 2-nitroanthracene.

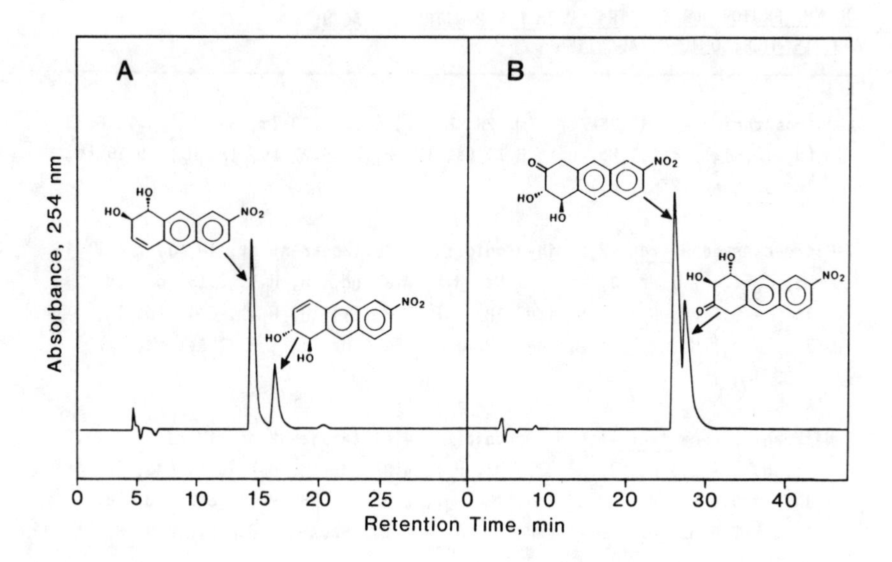

FIGURE 2. Normal-phase HPLC separation of (A) the dihydrodiol metabolites in peak 3 of Figure 1, eluted with 40% tetrahydrofuran in hexane, and (B) the dihydrodiol-ketone metabolites in peak 2 of Figure 1, eluted with 35% tetrahydrofuran in hexane.

The metabolite contained in peak 1 (Figure 1) was further purified by normal phase HPLC using 50% tetrahydrofuran in hexane and exhibited a UV-visible spectrum similar to that of 2-nitronaphthalene which suggested that it was a tetrahydrotetrol. This was supported by analysis of its NMR spectrum; however, because of its small quantity and lack of purity, the conformation of this metabolite could not be assigned. Thus, peak 1 was tentatively assigned as a 5,6,7,8-tetrahydrotetrol.

The relative quantities of the trans-dihydrodiol, trans-dihydrodiol-ketone and tetrahydrotetrol metabolites (peak 3, peak 2, and peak 1 in Figure 1) were determined by incubation of [G-^{3}H]2-nitroanthracene under identical conditions. The ratio of these three types of metabolites was 4:8:1, respectively. Since 2-nitro-9,10-anthraquinone was also formed chemically via air-oxidation of 2-nitroanthracene, the extent of its enzymatic formation was not determined.

TABLE 1

500 MHZ PROTON NMR SPECTRAL DATA FOR 2-NITROANTHRACENE AND ITS MICROSOMAL METABOLITES[†]

2-Nitroanthracene: δ7.75-7.70 (m, 2H, $H_{6,7}$), 8.18-8.23 (m, 3H, $H_{3,5,8}$), 8.33 (d, 1H, $J_{3,4}$ = 9.0 Hz, H_4), 8.75 (s, 1H, H_{10}), 8.98 (s, 1H, H_9), 9.15 (s, 1H, $J_{1,3}$ = 2.2 Hz, H_1).

2-Nitroanthracene trans-7,8-dihydrodiol: δ4.48 (apparent dt, 1H, $J_{7,8}$ = 10.4 Hz, $J_{6,7}$ = 2.0 Hz, $J_{5,7}$ = 2.0 Hz, H_7), 4.87 (d, 1H, H_8), 6.18 (dd, 1H, $J_{5,6}$ = 9.7 Hz, H_6), 6.67 (dd, 1H, H_5), 7.78 (s, 1H, H_{10}), 8.08 (d, 1H, $J_{3,4}$ = 9.1 Hz, H_4), 8.20 (dd, 1H, $J_{1,3}$ = 2.4 Hz, H_3), 8.32 (s, 1H, H_9), 8.85 (d, 1H, H_1).

2-Nitroanthracene trans-5,6-dihydrodiol: δ4.49 (apparent dt, 1H, $J_{5,6}$ = 10.2 Hz, $H_{6,7}$ = 2.3 Hz, $J_{6,8}$ = 2.3 Hz, H_6), 4.87 (d, 1H, H_5), 6.14 (dd, 1H, $J_{7,8}$ = 9.8 Hz, H_7), 7.68 (dd, 1H, H_8), 7.92 (s, 1H, H_9)*, 8.13 (d, 1H, $J_{3,4}$ = 8.9 Hz, H_4), 8.21 (dd, 1H, $J_{1,3}$ = 2.0 Hz, H_3), 8.22 (s, 1H, H_{10})*, 8.82 (d, 1H, H_1).

2-Nitroanthracene 6-keto-5,6,7,8-tetrahydro-trans-7,8-diol: δ3.81 (d, 1H, $J_{7,8}$ = 3.9 Hz, H_7), 3.94 (d, 1H, $J_{5-a,5-e}$ = 9.0 Hz, H_{5-a}), 4.25 (d, 1H, H_8), 4.80 (d, 1H, H_{5-e}), 8.17 (d, 1H, $J_{3,4}$ = 9.0 Hz, H_4), 8.26 (dd, 1H, $J_{1,3}$ = 2.5 Hz, H_3), 8.26 (s, 1H, H_{10})*, 8.43 (s, 1H, H_9)*, 8.94 (d, 1H, H_1).

2-Nitroanthracene 7-keto-5,6,7,8-tetrahydro-trans-5,6-diol: δ3.81 (d, 1H, $J_{5,6'}$ = 4.4 Hz, H_6), 3.95 (d, 1H, $J_{8-a,8-e}$ = 8.9 Hz, H_{8-a}), 4.25 (d, 1H, H_5), 4.80 (d, 1H, H_{8-e}), 8.19 (d, 1H, $J_{3,4}$ = 8.9 Hz, H_4), 8.27 (dd, 1H, $J_{1,3}$ = 2.5 Hz, H_3), 8.31 (s, 1H, H_{10})*, 8.39 (s, 1H, H_9)*, 8.91 (d, 1H H_1).

2-Aminoanthracene**: δ6.89 (d, 1H, $J_{1,3}$ = 2.2 Hz, H_1), 7.04 (dd, 1H, $J_{3,4}$ = 9.0 Hz, H_3), 7.28 (apparent td, 1H, $J_{5,6}$ = 8.2 Hz, H_6), 7.36 (apparent td, 1H, $J_{7,8}$ = 8.6 Hz, H_7), 7.81 (d, 1H, H_4), 7.86 (d, 1H, H_8), 7.90 (d, 1H, H_5), 8.02 (s, 1H, H_9), 8.27 (s, 1H, H_{10}).

[†]The samples were dissolved in acetone-d_6 (except where otherwise noted) with a trace of D_2O and are reported in ppm (δ) downfield from tetramethylsilane. Assignments were made after extensive homonuclear decoupling and nuclear Overhauser experiments.

*The assignments may be reversed.

**The solvent was dimethylsulfoxide-d_6.

Separation and Characterization of 2-Nitroanthracene Metabolites from Hypoxic Incubation

Metabolites formed from incubation of 2-nitroanthracene with rat liver microsomes under hypoxic (3% oxygen) conditions were separated by reversed-phase HPLC (Figure 3). Peak 3 contained the recovered substrate. The metabolite contained in peak 2 was identified as 2-aminoanthracene by comparison of its UV-visible, mass and NMR spectra and its HPLC retention time with those of an authentic sample. The metabolite in peak 1 has not been identified.

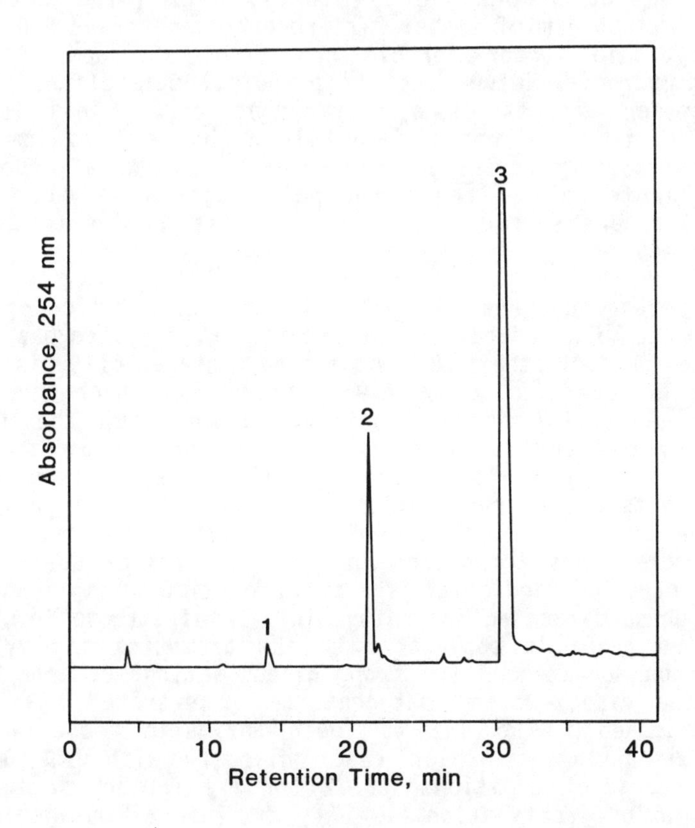

FIGURE 3. Reversed-phase HPLC profile of hypoxic microsomal metabolites of 2-nitroanthracene. The metabolite in each peak was identified as follows: 1) unknown; 2) 2-aminoanthracene; and 3) 2-nitroanthracene.

DISCUSSION

Based on the results we obtained from the aerobic and hypoxic metabolism of 2-nitroanthracene, the metabolic pathways shown in Figure 4 are proposed. For comparison, the previously reported metabolic pathways of 9-nitroanthracene (4) are illustrated in Figure 5. As indicated in these figures, two distinct differences were observed between the metabolism of these isomeric nitroanthracenes. First, 2-nitroanthracene could be reduced to 2-aminoanthracene under hypoxic (3% oxygen) incubation conditions, while nitroreduction of 9-nitroanthracene was not detected even under anaerobic conditions (0.2% oxygen) (4). Second, further metabolism of either 2-nitroanthracene *trans*-5,6- and 7,8-dihydrodiol under aerobic conditions afforded vicinal *trans*-dihydrodiol-ketones as the principal metabolites along with lesser amounts of a tetrahydrotetrol. This is in contrast to the results obtained by metabolism of 9-nitroanthracene *trans*-3,4-dihydrodiol, which afforded a tetrahydrotetrol as the principal metabolite (4). The molecular basis for these differences has yet to be determined.

In several classes of nitro PAH, the orientation of the nitro group with respect to the aromatic ring system has been used to predict their direct-acting mutagenicity in the *Salmonella* reversion assay (8,9). Nitro PAH which have two *peri* substituents adopt a conformation in which the nitro group is perpendicular or nearly perpendicular to the aromatic moiety, whereas nitro PAH which have one or no *peri* substituents adopt conformations in which the nitro group is coplanar or nearly coplanar with the aromatic system (8). Furthermore, those compounds in which the nitro substituent is oriented perpendicular to the aromatic moiety exhibit little or no direct acting mutagenicity, while those in which the nitro group is coplanar with the aromatic ring system range from nonmutagens to strong direct-acting mutagens (8). Since the direct-acting mutagenicity of nitrated PAHs has been proposed to be due to their enzymatic reduction to *N*-hydroxy arylamines which react directly with DNA under slightly acidic conditions (10-12) or are further activated by enzymatic *O*-acetylation (13-17), the lack of mutagenicity in those compounds in which the nitro group is oriented perpendicular to the aromatic ring system may be due to an inability of the active site of the nitroreductase to accommodate these compounds. 9-Nitroanthracene as well as its dihydrodiol metabolites have been previously shown to adopt this perpendicular orientation, and the lack of

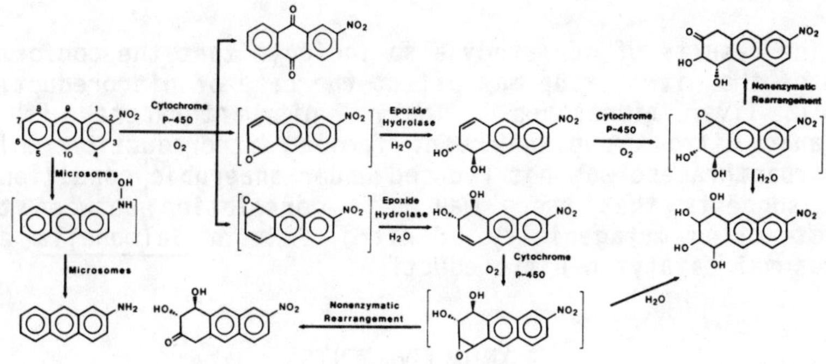

FIGURE 4. Proposed metabolic pathways of 2-nitroanthracene.

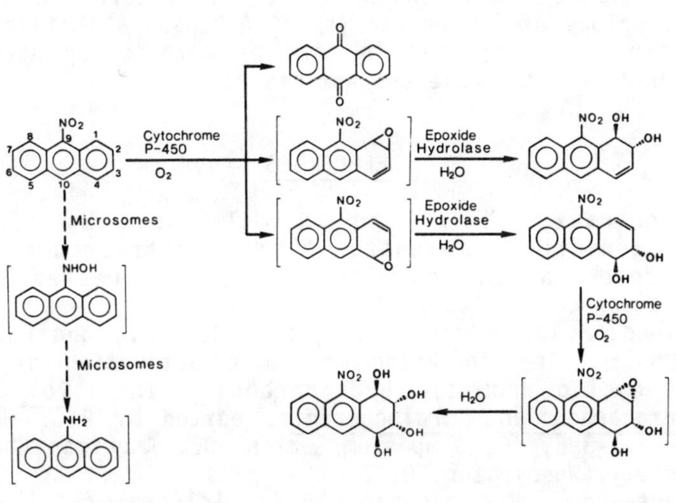

FIGURE 5. Proposed metabolic pathways of 9-nitroanthracene.

mutagenicity of each supports the above hypothesis (4,5). In contrast, the nitro groups of 2-nitroanthracene and its oxidized metabolites adopt conformations which are coplanar with the anthracene ring system, as indicated by the induced downfield shift of the two ortho protons H_1 and H_3 (see Table 1). Thus, the high level of mutagenicity of 2-nitroanthracene is in agreement with the above hypothesis. Mutagenicity studies are currently being conducted with its dihydrodiol metabolites.

The results of our study also indicate that the conformation of the nitro group may affect the ease of nitroreduction by rat liver microsomes. Thus, 2-nitroanthracene with a coplanar nitro group underwent facile nitroreduction while 9-nitroanthracene was not reduced under anaerobic conditions. This suggests that there may be a correlation between the direct-acting mutagenicity of nitro PAH in Salmonella and microsomal catalyzed nitroreduction.

ACKNOWLEDGEMENTS

We thank L. Unruh for obtaining the mass spectra and Cindy Hartwick for assistance in preparing this manuscript. D.T.C. Yang and S.E. Chiun were partially supported by the Faculty Research Fund and the Office of Research in Science and Technology of the University of Arkansas at Little Rock. E.K. Fifer was supported in part by Interagency Agreement 224-82-002 with the Veterans Administration.

REFERENCES

1. Rosenkranz, H.S. and Mermelstein, R. (1983): Mutagenicity and genotoxicity of nitroarenes: all nitro-containing chemicals were not created equal. Mutation Res., 114:217-267.
2. Beland, F.A., Heflich, R.H., Howard, P.C., and Fu, P.P. (1985): The in vitro metabolic activation of nitro polycyclic aromatic hydrocarbons. In: Polycyclic Hydrocarbons and Carcinogenesis, edited by R.G. Harvey, pp. 371-396, ACS Symposium Series 283, American Chemical Society, Washington, D.C.
3. Paputa-Peck, M.C., Marano, R.S., Schuetzle, D., Riley, T.L., Hampton, C.V., Prater, T.J., Skewes, L.M., Jensen, T.E., Ruehle, P.H., Bosch, L.C., and Duncan, W.P. (1983): Determination of nitrated polynuclear aromatic hydrocarbons in particulate extracts by capillary column gas chromatography with nitrogen selective detection. Anal. Chem., 55:1946-1954.
4. Fu, P.P., Von Tungeln, L.S., and Chou, M.W. (1985): Metabolism of 9-nitroanthracene by rat liver microsomes: identification and mutagenicity of metabolites. Carcinogenesis, 6:753-757.

5. Scribner, J.D., Fisk, S.R., and Scribner, N.K. (1979): Mechanisms of action of carcinogenic aromatic amines: an investigation using mutagenesis in bacteria. Chem.-Biol. Interact., 26:11-25.
6. Schenkman, J.B. and Cinti, D.L. (1978): Preparation of microsomes with calcium. In: Methods in Enzymology, edited by S. Fleischer and L. Packer, Vol. LII, pp. 83-89, Academic Press, New York.
7. Zacharias, D.E., Glusker, J.P., Fu, P.P., and Harvey, R.G. (1979): Molecular structures of the dihydrodiols and diol epoxides of carcinogenic polycyclic aromatic hydrocarbons. X-ray crystallographic and NMR analysis. J. Am. Chem. Soc., 101:4043-4051.
8. Fu, P.P., Chou, M.W., Miller, D.W., White, G.L., Heflich, R.H., and Beland, F.A. (1985): The orientation of the nitro substituent predicts the direct-acting bacterial mutagenicity of nitrated polycyclic aromatic hydrocarbons. Mutation Res., 143:173-181.
9. Vance, W.A. and Levin, D.E. (1984): Structural features of nitroaromatics that determine mutagenic activity in Salmonella typhimurium. Environ. Mutagenesis, 6:797-811.
10. Mermelstein, R., Kiriazides, D.K., Butler, M., McCoy, E.C., and Rosenkranz, H.S. (1981): The extraordinary mutagenicity of nitropyrenes in bacteria. Mutation Res., 89:187-196.
11. Howard, P.C., Heflich, R.H., Evans, F.E., and Beland, F.A. (1983): Formation of DNA adducts in vitro and in Salmonella typhimurium upon metabolic reduction of the environmental mutagen 1-nitropyrene. Cancer Res., 43:2052-2058.
12. Heflich, R.H., Howard, P.C., and Beland, F.A. (1985): 1-Nitrosopyrene: an intermediate in the metabolic activation of 1-nitropyrene to a mutagen in Salmonella typhimurium TA1538. Mutation Res., 149:25-32.
13. McCoy, E.C., McCoy, G.D., and Rosenkranz, H.S. (1982): Esterification of arylhydroxylamines: evidence for a specific gene product in mutagenesis. Biochem. Biophys. Res. Commun., 108:1362-1367.
14. McCoy, E.C., Anders, M., and Rosenkranz, H.S. (1983): The basis of the insensitivity of Salmonella typhimurium strain $TA98/1,8-DNP_6$ to the mutagenic action of nitroarenes. Mutation Res., 121:17-23.
15. Bryant, D.W., McCalla, D.R., Lultschik, P., Quilliam, M.A., and McCarry, B.E. (1984): Metabolism of 1,8-dinitropyrene by Salmonella typhimurium. Chem.-Biol. Interact., 49:351-368.

16. Djurić, Z., Fifer, E.K., and Beland, F.A. (1985): Acetyl coenzyme A-dependent binding of carcinogenic and mutagenic dinitropyrenes to DNA. Carcinogenesis, 6:941-944.
17. Fifer, E.K., Heflich, R.H., Djurić, Z., Howard, P.C., and Beland, F.A. (1986): Synthesis and mutagenicity of 1-nitro-6-nitrosopyrene and 1-nitro-8-nitrosopyrene, potential intermediates in the metabolic activation of 1,6- and 1,8-dinitropyrene. Carcinogenesis, 7 (in press).

BIOALKYLATION OF POLYNUCLEAR AROMATIC HYDROCARBONS *IN VIVO*: A PREDICTOR OF CARCINOGENIC ACTIVITY

JAMES W. FLESHER*[1,2], STEVEN R. MYERS[1] AND JERRY W. BLAKE[2]
(1) Department of Pharmacology, (2) Graduate Center for Toxicology, Albert B. Chandler Medical Center, University of Kentucky, Lexington, KY 40536, USA.

INTRODUCTION

In previous accounts of investigations in this laboratory of a possible relationship between biochemical reactivity and carcinogenic activity of aromatic molecules, we emphasized the importance of reactive centers in the meso-anthracenic position(s) or L-region of unsubstituted carcinogens such as benz(a)anthracene and benzo(a)pyrene. We reasoned from structure-activity relationships, known to exist in this class of compounds, that biochemical substitution reactions might play an important role in conferring carcinogenic activity on the hydrocarbon, since many alkyl substituted compounds are known that are more carcinogenic than the corresponding unsubstituted compound.

We postulate that one of the essential steps, probably the first step, in the biochemical activation of unsubstituted preprocarcinogens to an ultimate carcinogen, consists in the reaction between the preprocarcinogen and S-adenosyl-L-methionine, and that it takes place most favorably in the meso-anthracenic reactive center(s) or L-region to form the corresponding procarcinogen (9,12,14,15,16,17). Although the bioalkylation of preprocarcinogens to form procarcinogens has been demonstrated in certain hydrocarbons *in vitro* (12,14,15,16,17) it is obviously desirable to demonstrate that the reaction also occurs *in vivo*.

In the present paper, evidence is presented that the bioalkylation reaction occurs *in vivo* and that the results are useful in making predictions of carcinogenic activity. A unified hypothesis, which incorporates specific rules of molecular geometry, appears to correctly predict the carcinogenic activity of most of the polynuclear aromatic hydrocarbons that have been tested for carcinogenic activity.

Chemicals

Benzo(a)pyrene (BP) and benz(a)anthracene (BA) were purchased from Eastman Organic Chemicals, Rochester, NY. 7-Methylbenz(a)anthracene (7-methylBA) was purchased from Schuchardt, Munich, West Germany. Pyrene was purchased from Aldrich Chemical Co., Milwaukee, WI. 6-Methylbenzo(a)-pyrene, 6-formylbenzo(a)pyrene (7), 6-hydroxymethylbenzo(a)-pyrene (24), 7-hydroxymethylbenz(a)anthracene (12), and 7-hydroxymethyl-12-methylbenz(a)anthracene (8) were prepared according to previously published methods. 1-Methylpyrene was purchased from K and K Laboratories, Plainview, NY. All compounds used were found to be free of contamination by HPLC and GC/MS analysis.

Animals

Male and female Sprague-Dawley rats (∿ 200 gms) were purchased from Harlan Sprague-Dawley, Indianapolis, IN. Animals were housed in polyurethane cages and provided with food and water *ad libitum*.

METHODS

Bioalkylation of polycyclic aromatic hydrocarbons following subcutaneous administration.

The hydrocarbon (0.4 µmol in 200 µl sesame oil) was injected into the dorsal subcutis tissue and 24 hours later the animal was sacrificed by cervical fracture and the tissue in contact with the hydrocarbon was excised under uv light. The excised tissue was minced and homogenized in 70% acetone and the homogenate was centrifuged in a clinical centrifuge. The supernatant was removed and extracted twice with ethyl acetate, and the combined extracts washed with water and evaporated under reduced pressure. The residue was stored at -20°C until analysis by HPLC and GC/MS.

Bioalkylation of benzo(a)pyrene in liver following intraperitoneal injection.

Benzo(a)pyrene (5 mg in 0.5 ml DMSO) was administered intraperitoneally to rats and the animals were sacrificed and the livers removed, two or four hours later. The livers were minced and homogenized in 70% acetone and centrifuged at low speed to separate the acetone soluble and insoluble material. The supernatant solution was separated and extracted twice with ethyl acetate, washed with water and evaporated under reduced pressure. The residue was stored at -20°C until analysis by HPLC and GC/MS.

Bioalkylation of benzo(a)pyrene in liver following oral administration.

Benzo(a)pyrene (50 mg in 0.5 ml sesame oil) was administered orally through a rubber catheter. Twenty-four hours later the animals were sacrificed by cervical fracture and the livers removed, rinsed with 0.1 M phosphate buffer (pH 7.4), blotted on filter paper, minced, and homogenized in 70% acetone and extracted as above.

Analysis of products by High Pressure Liquid Chromatography

The ethyl acetate extracted products were dissolved in 100 µl of methylene chloride and aliquots (2-5 µl) were analyzed by HPLC. The injections were made into a Water's U6K injector connected to the HPLC system. The products were eluted from an ALTEX 25 cm x 4.6 mm ID C18 column packed with ultrasphere ODS-5 µm connected to a Water's M6000 solvent pump using 100% methanol as the mobile phase at a temperature of 20°C and a flow rate of 1.0 ml/min. Ultraviolet absorbance was monitored at 254 nm using an ISCO type 6 ultraviolet detector. The absorbance profiles of the various metabolites were compared to the absorbance profiles of the authentic meso-alkyl and meso-hydroxyalkyl substituted PAH.

Analysis of metabolites by Gas Chromatography and Mass Spectroscopy

Further analysis of metabolites was accomplished by GC/MS. Samples, dissolved in methylene chloride, were either directly analyzed by a Finnigan 4500 GC/MS or individual HPLC peaks were collected before analysis by GC/MS. The gas chromatographic column was a packed glass column packed with 3% OV-100 on chromosorb-WHP. The gas chromatographic oven was programmed at 200-270°C at 20°C/min increments. Mass spectral analysis of metabolites was accomplished by electron impact, operated at 70 eV. The instrument was equipped with an INCOS data system. Data obtained from each analysis included retention times on the gas chromatograph, and the mass spectral patterns. Individual spectra were compared to the GC/MS profiles of authentic standards of both meso-alkyl and meso-hydroxy-alkyl metabolites of the various PAH.

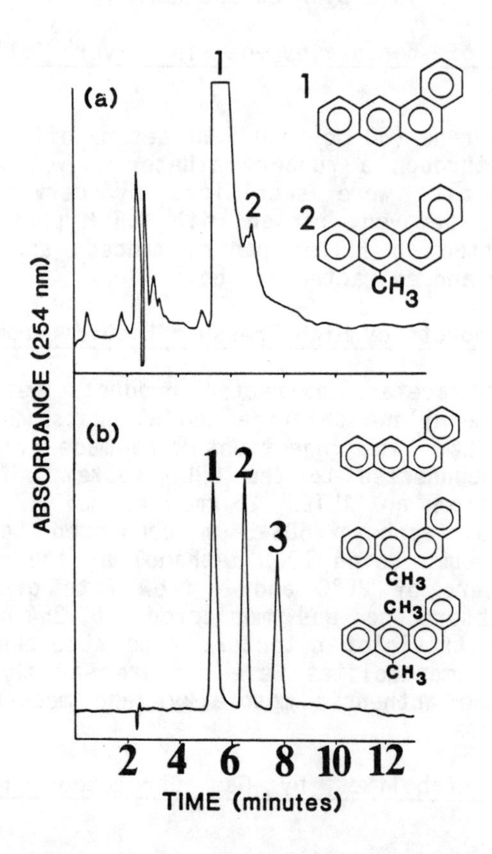

FIGURE 1. HPLC of the metabolites (a) of benz(a)anthracene after injection into the dorsal subcutaneous tissue. The profile reveals that benz(a)anthracene (peak 1) is bio-alkylated to form 7-methylbenz(a)anthracene (peak 2). Authentic standards of benz(a)anthracene (peak 1), 7-methyl-benz(a)anthracene (peak 2) and 7,12-dimethylbenz(a)anthracene (peak 3) are shown in (b).

RESULTS

Bioalkylation of benz(a)anthracene to 7-methylbenz(a)anthracene in subcutis tissue.

When benz(a)anthracene (BA) was administered by subcutaneous injection, and the tissue processed as described in Methods, a product was detected upon analysis by HPLC (Fig. 1(a)) that was indistinguishable from authentic 7-methylBA shown in Fig. 1b. Analysis of this material by single ion monitoring on GC showed that the substrate, BA, retention time 4 minutes, was bioalkylated to form 7-methylBA retention time of 4 minutes 50 seconds which was indistinguishable from authentic 7-methylBA. The mass spectrum of the metabolite identified by HPLC and GC as 7-methylBA yielded a parent molecular ion of 242 and other characteristic ions (Fig. 2) which were indistinguishable from authentic 7-methylBA.

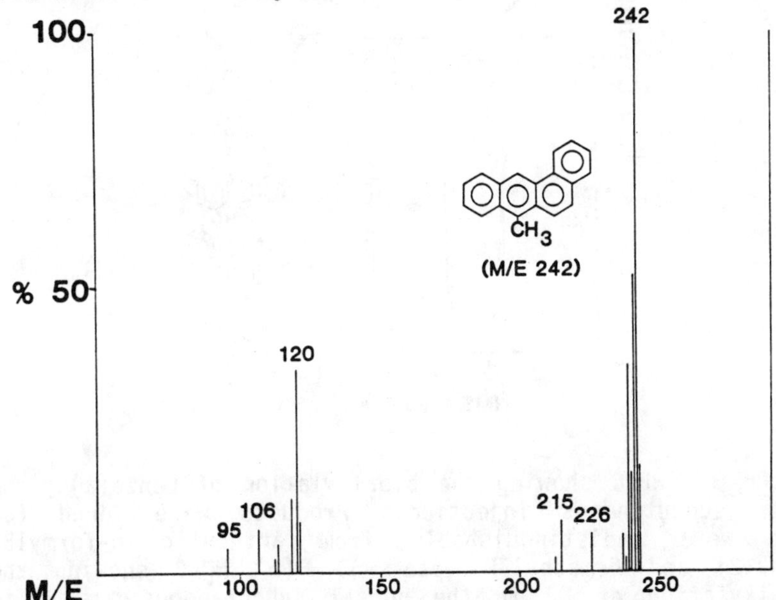

FIGURE 2. MS of the metabolite formed after subcutaneous administration of benz(a)anthracene. The metabolite yielded a molecular ion of 242 and other accompanying ions which were indistinguishable from authentic 7-methylbenz(a)anthracene.

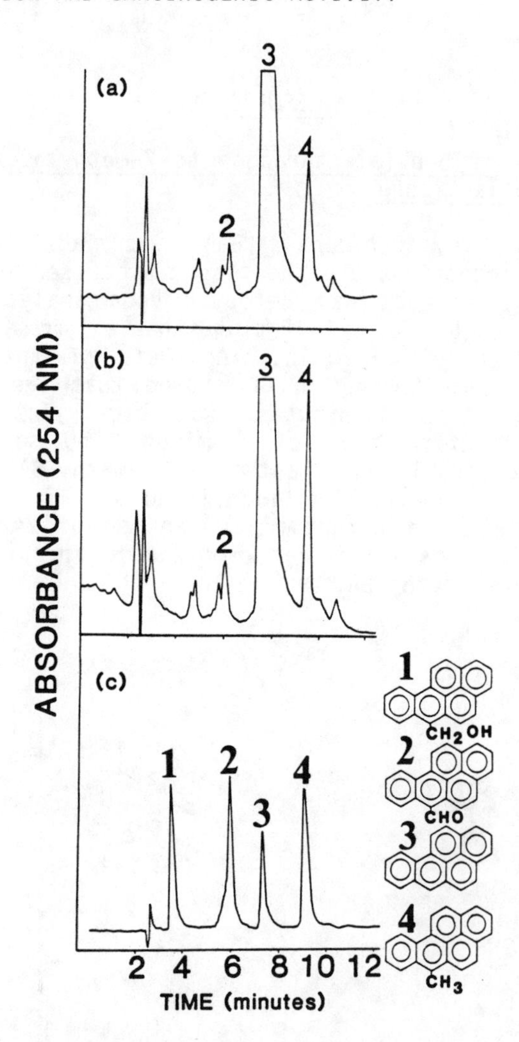

FIGURE 3. HPLC showing the bioalkylation of benzo(a)pyrene after subcutaneous injection. Products were found (a) that were indistinguishable from authentic 6-formylBP (peak 2) and 6-methylBP (peak 4). (b) HPLC showing the bioalkylation of BP in the dorsal subcutaneous tissue in the presence of exogenous S-adenosyl-L-methionine. Authentic standards of 6-hydroxymethylBP (peak 1), 6-formylBP (peak 2), BP (peak 3), and 6-methylBP (peak 4) are shown in (c).

Figure 3 shows the results obtained when the carcinogen benzo(a)pyrene (BP) (peak 3) was administered to the dorsal subcutaneous tissue of the rat as described in Methods, and the products extracted and analyzed by HPLC (Fig. 3a). Following analysis, a product was obtained that was indistinguishable from authentic 6-methylBP (peak 4) and 6-formylBP (peak 2).

Twenty-four hours after 0.4 μmoles of BP was injected subcutaneously along with 2.0 μmole of S-adenosyl-L-methionine, the products extracted and analyzed by HPLC 24 hrs

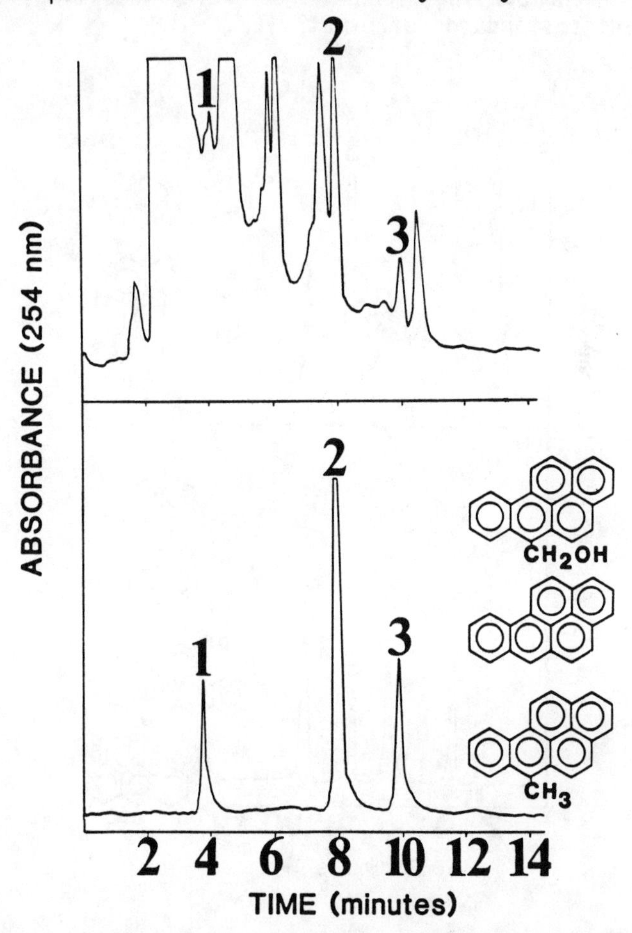

FIGURE 4. HPLC of the products obtained from liver (a) following the oral administration of 50 mgs of BP in 0.50 ml sesame oil. Products were found that were indistinguishable from the authentic standards shown in (b) of peak 1 6-hydroxymethylBP, peak 2 BP, and peak 3, 6-methylBP.

later (Fig. 3b), it was observed that the peak corresponding to 6-methylBP (peak 4) was enhanced. This proves that S-adenosyl-L-methionine is the carbon donor in the bioalkylation substitution reaction.

The identity of the metabolite was confirmed GC/MS. The product gave a characteristic molecular ion of 266 and other accompanying ions which were indistinguishable from authentic standards of 6-methylBP.

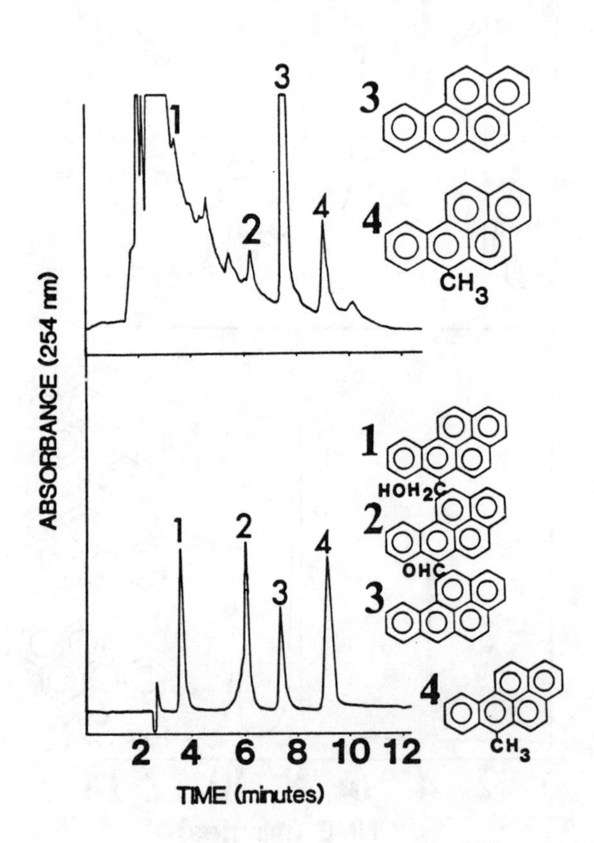

FIGURE 5. HPLC of the metabolites of BP following intraperitoneal injection of 5.0 mg of BP in 0.5 ml of DMSO. Metabolites (a) were found which were indistinguishable from authentic standards of 6-hydroxymethylBP (peak 1), 6-formylBP (peak 2) and 6-methylBP (peak 4) shown in (b).

Twenty-four hours after benzo(a)pyrene was administered orally to rats, products were isolated from liver which were indistinguishable (Fig. 4) from authentic standards of 6-hydroxymethylBP (peak 1), BP (peak 2), and 6-methylBP (peak 3). The identity of these metabolites was confirmed by GC/MS.

Two or 4 hours after the intraperitoneal administration of BP products were extracted from liver (Fig. 5a) which were indistinguishable from authentic standards of 6-hydroxymethylBP (peak 1), 6-formylBP (peak 2), BP (peak 3), and 6-methylBP (peak 3) shown in Fig. 5b. The material having a retention time of 2-4 minutes on HPLC is unidentified, but may be conjugates of the 6-hydroxymethylBP metabolite, since this metabolite has been demonstrated to be formed from 6-methylBP not only _in vitro_, but _in vivo_. The peak on HPLC corresponding to 6-methylBP was isolated and analyzed by mass spectral analysis. The product gave a molecular ion of 266 and other accompanying ions which were indistinguishable from authentic standards of 6-methylBP.

The metabolism of pyrene, a compound devoid of carcinogenic activity (18), was investigated following administration of 0.4 μmoles of pyrene in 200 μl sesame oil to the dorsal subcutaneous tissue of rats. Twenty-four hours later, the animals were sacrificed by cervical dislocation and the subcutaneous tissue in contact with the hydrocarbon isolated under UV light and the products of the reaction extracted with ethyl acetate and washed with water. Fig. 6(a) shows the results obtained following HPLC analysis of the metabolism of pyrene following injection into the dorsal subcutaneous tissue of the rat. A metabolite was identified that was found to be indistinguishable from authentic 1-methylpyrene shown in (b). The results demonstrate that in compounds that do not possess a reactive meso-anthracenic center, a center of reactivity may exist, which is sufficiently reactive toward a bioalkylation to confer weak carcinogenicity on the compound.

DISCUSSION

The present experiments have demonstrated that preprocarcinogens undergo a bioalkylation reaction that introduces an alkyl group into the molecule and that the reaction takes place at the reactive center or centers.

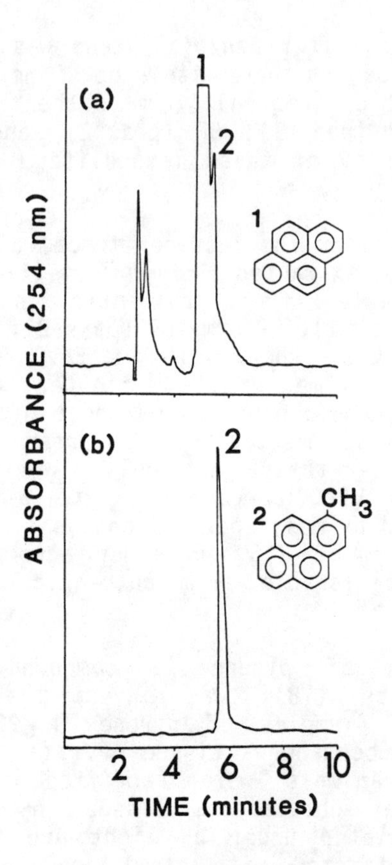

FIGURE 6. HPLC of the metabolite obtained following the administration of 0.4 µmole of pyrene in 200 µl sesame oil to the dorsal subcutaneous tissue of the rat (a). The substrate, pyrene (peak 1) is bioalkylated to form 1-methylpyrene (peak 2) which was found to be indistinguishable from authentic 1-methylpyrene shown in (b).

These results support the unified hypothesis that the chemical or biochemical introduction of an alkyl group, most favorably in the meso-anthracenic reactive centers of the molecule confers strong carcinogenic activity on the procarcinogen. Consequently, procarcinogens are more carcinogenic than preprocarcinogens. The hypothesis, that the meso-anthracenic reactive center of the molecule is most susceptible to substitution is consistent with the molecular orbital theory of M. J. S. Dewar (4) and the chemical studies of Professor L. F. Fieser (6). Some polynuclear aromatic hydrocarbons lacking meso-anthracenic reactive centers such as pyrene may nevertheless possess reactive centers that are sufficiently reactive toward bioalkylation to form both monomethyl and dimethyl derivatives. In the present experiments, however, only a monomethyl derivative was detected in subcutaneous tissue, *in vivo* and this was identified by HPLC and GC/MS as 1-methylpyrene. Whereas pyrene has been reported to be devoid of carcinogenic activity, 1-methylpyrene (29) has been shown to possess weak carcinogenic activity when administered by skin painting to mice. The unified hypothesis, which incorporates specific rules of molecular geometry, is able to account for the carcinogenic activity of most, if not all, of the polynuclear aromatic hydrocarbons that have been tested for carcinogenic activity by subcutaneous injection in mice and rats.

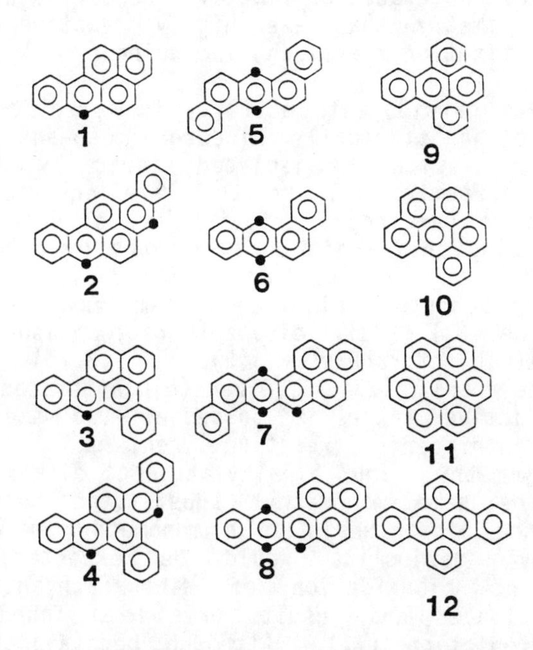

FIGURE 7. Structures of representative polycyclic aromatic hydrocarbons with and without meso-anthracenic reactive centers. The reactive meso-anthracenic center(s) are shown by the heavy dot.

The fundamental structural unit required for carcinogenic activity in polynuclear aromatic molecules is anthracene, which itself is carcinogenic (5). The 9,10 carbon atoms of anthracene are the reactive centers that form an L-region. Representative carcinogenic and non-carcinogenic hydrocarbons are shown in Figure 7.

RULE 1: Strong carcinogenic activity is correlated with the existence of a single sterically unhindered meso-anthracenic reactive center, or with two isolated sterically unhindered meso-anthracenic reactive centers that are equivalent by symmetry. Examples of strong carcinogens include benzo(a)pyrene (1) which possesses only a single meso-anthracenic reactive center since the other meso-anthracenic reactive center of the anthracene nucleus is abolished by ring fusion. The compound is a preprocarcinogen because the meso-anthracenic reactive center is highly reactive toward bioalkylation which forms the very powerful procarcinogen, 6-methylbenzo(a)pyrene (25). Benzo(rst)pentaphene (2), is a preprocarcinogen (19,20) because it possesses two isolated meso-anthracenic reactive centers that are equivalent by symmetry that are highly reactive toward all types of substitution reactions, including bioalkylation.

RULE 2: Moderate carcinogenic activity is correlated with the existence of a sterically hindered meso-anthracenic reactive center, or with two isolated sterically hindered meso-anthracenic reactive centers that are equivalent by symmetry. Moderate carcinogen activity is correlated with compounds possessing an L-region provided one or both meso-anthracenic reactive centers are sterically hindered by an angular benzo ring. Examples include dibenzo(a,e)pyrene (3) (22), dibenzo(h,rst)pentaphene (4) (23), dibenz(a,h)anthracene (5) (1,3,26), and benz(a)anthracene (6) (27). Dibenz(a,h)anthracene (5) is moderately active since it possesses two sterically hindered meso-anthracenic reactive centers that are equivalent by symmetry. Thus bioalkylation of either center would afford the same monomethyl substituted metabolite and therefore the biosynthesis of a monomethyl metabolite and a dimethyl metabolite would be expected. The bioalkylation and biooxidation of dibenz(a,h)anthracene has been investigated and results have been found which confirm this prediction (17). Although benz(a)anthracene is generally regarded as weakly carcinogenic under most conditions of testing, it has been shown to be a moderately active compound by subcutaneous injection in C57 black mice and to induce liver tumors in about half of the animals after oral administration to rats (27).

RULE 3: Weak carcinogenic activity is correlated with the existence of three reactive centers. Examples include phenanthro(2',3',3,4)pyrene (7) (21), and naptho(8,1,2-cde)naphthacene (8) (21), a compound in which two of the reactive centers are sterically hindered.

RULE 4: Carcinogenic activity is abolished when the reactive centers in the anthracene or naphthacene nucleus are eliminated by ring fusion or blocked by substituents which prevent the biochemical introduction of methyl groups. Carcinogenic activity is abolished in compound possessing two or more sterically unhindered L-regions or in compounds possessing a linear arrangement of benzene rings greater than three. Examples include benzo(e)pyrene (9) (18), benzo(ghi)perylene (10) (28), coronene (11) (28), and dibenzo(fg,op)naphthacene (12) (40). Carcinogenic activity is absent or weak in phenanthrene and its benzo ring derivatives chrysene and benz(c)phenanthrene. Furthermore, compounds with a phenanthrene nucleus have a tendency to undergo addition rather than substitution reactions. Compounds with unhindered L-regions have a marked tendency to undergo addition reactions that deactivate the compound as a carcinogen. According to the unified hypothesis, the conversion of a preprocarcinogen to a procarcinogen, *in vitro* and *in vivo*, is accomplished by a reaction that takes place in cytosol between the preprocarcinogen and S-adenosyl-L-methionine and that introduces an alkyl group most favorably in the meso-anthracenic reactive center(s) of the molecule.

In order that a strong procarcinogen be formed, bioalkylation of the preprocarcingen must take place at a sterically unhindered meso-anthracenic reactive center. For compounds with meso-anthracenic reactive centers arranged in such a way as to form an L-region, the bioalkylation reaction can take place only if the L-region is not too reactive toward addition reactions, since this type of reactivity does not lead to carcinogenesis, but leads to the inactivation of the hydrocarbon. The idea that there are certain points of correspondence between the chemical reactivity of hydrocarbons in their substitution reactions and their carcinogenic activity was strongly advocated by Fieser and his coworkers (6) who investigated several substitution reactions in seeking to establish a correlation between this type of reactivity and carcinogenic activity.

Although the rules of molecular geometry correctly predict the carcinogenic activity of most, if not all, unsubstituted polynuclear aromatic hydrocarbons, further studies to test the unified hypothesis and the rules of molecular geometry are required to determine their generality.

REFERENCES

1. Andervant, H.B. (1934): The production of dibenzanthracene tumors in pure strain mice. Public Health Report, 49, 620-624.
2. Badger, G.M. and Cook, J.W. (1939): Synthesis of growth inhibitory polycyclic compounds. J. Chem. Soc., 802-806, 1939.
3. Boyland, E. and Burrows, H. (1935): The experimental production of sarcoma in rats and mice by a colloidal aqueous solution of 1,2,5,6-dibenzanthracene. J. Path. Bact., 41:231-238.
4. Dewar, M.J.S. (1952): A molecular orbital theory of organic chemistry. VI. Aromatic substitution and addition. J. Am. Chem. Soc., 74:3357-3363.
5. Drückrey, H. and Schmähl, D. (1955): Carcinogene Wirkung von Anthracen. Naturwissenschaften 42:159-160.
6. Fieser, L.F. (1939): Carcinogenic activity, structure, and chemical reactivity of polynuclear aromatic hydrocarbons. Am. J. Cancer, 34:37-124.
7. Fieser, L.F. and Hershberg, E.B. (1937): The oxidation of methylcholanthrene and 3,4-benzpyrene with lead tetraacetate; Further derivatives of 3,4-benzpyrene. J. Am. Chem. Soc., 60:2542-2548.
8. Flesher, J.W., Soedigdo, S. and Kelley, D.R. (1967): Synthesis of metabolites of 7,12-dimethylbenz(a)anthracene, 4-hydroxy-7,12-dimethylbenz(a)anthracene, 7-hydroxy-12-methylbenz(a)anthracene, their methyl ethers, and acetoxy derivatives. J. Med. Chem., 10:932-936.
9. Flesher, J.W. and Sydnor, K.L. (1973): Possible role of 6-hydroxymethylbenzo(a)pyrene as a proximate carcinogen of benzo(a)pyrene and 6-methylbenzo(a)pyrene. Int. J. Cancer, 11:433-437.
10. Flesher, J.W., Stansbury, K.H. and Sydnor, K.L. (1982): S-adenosyl-L-methionine is a carbon donor in the conversion of benzo(a)pyrene to 6-hydroxymethylbenzo(a)pyrene by rat liver S-9. Cancer Letters, 16:91-94.

11. Flesher, J.W., Kadry, A.M., Chien, M., Stansbury, K., Gairola, C. and Sydnor, K.L. (1983): Metabolic activation of carcinogenic hydrocarbons in the meso-position (L-region). In: The 7th International Symposium of Polynuclear Aromatic Hydrocarbons, edited by W.M. Cooke and A.J. Dennis, pp. 505-515, Battelle Press.
12. Flesher, J.W., Stansbury, K.H., Kadry, A.M. and Myers, S.R. (1983): Bioalkylation of benzo(a)pyrene in rat lung and liver. In: Extrahepatic Drug Metabolism and Chemical Carcinogenesis, edited by J. Rydstrom, J. Montelius and M. Bengtsson, pp. 237-238, Elsevier Science, Amsterdam, New York.
13. Flesher, J.W. and Myers, S.R. (1985): Oxidative metabolism of 7-methylbenz(a)anthracene, 12-methylbenz-(a)anthracene, and 7,12-dimethylbenz(a)anthracene by rat liver cytosol. Cancer Letters, 26:83-88.
14. Flesher, J.W., Myers, S.R. and Blake, J.W. (1984): Biosynthesis of the potent carcinogen 7,12-dimethylbenz-(a)anthracene. Cancer Letters, 24:335-343.
15. Flesher, J.W., Myers, S.R. and Lovelace, G.E. (1984): Biotransformation of benzo(a)pyrene to 6-methylbenzo(a)-pyrene and related metabolites in rat liver and lung. In: The 8th International Symposium on Polynuclear Aromatic Hydrocarbons, edited by W.M. Cooke and A.J. Dennis, pp. 423-436, Battelle Press.
16. Flesher, J.W., Myers, S.R. and Blake, J.W.: Bioalkylation of polynuclear aromatic hydrocarbons: A predictor of carcinogenic activity. The 9th International Symposium on Polynuclear Aromatic Hydrocarbons, Battelle Press, In Press.
17. Flesher, J.W., Myers, S.R., Bergo, C.H. and Blake, J.W.: Bioalkylation of dibenz(a)anthracene in rat liver cytosol. Chem. Biol. Interacts, In Press.
18. Hartwell, J.L. (1951): Survey of compounds which have been tested for carcinogenic activity. U.S. Public Health Service, Publ. No. 149, Washington, D.C.
19. Lacassagne, A., Zajdela, F., Buu-Hoi, N.P. and Chalvet, H. (1957): Sur l'activite cancerogene du 3,4,9,10-dibenzopyrene et de quelques-uns de ses derives. Compt. Rend. Acad. Sci., 244:273-274.
20. Lacassagne, A., Buu-Hoi, N.P. and Zajdela, F. (1958): Relation entre structure moleculaire et activite cancerogene dans trois series d'hydrocarbures aromatiques hexacycliques. Compt. Rend. Acad. Sci., 246:1477-1480.

21. Lacassagne, A., Buu-Hoi, N.P. and Zajdela, F. (1960): Activite carcerogene d'hydrocarbures polycycliques derives du naphthacene. Compt. Rend. Acad. Sci., 250:3547-3548.
22. Lacassagne, A., Buu-Hoi, N.P., Zajdela, F. and Lavit-Lamy, D. (1963): Activite cancerogene elevee du 1,2,3,4-dibenzopyrene et 1,2,4,5-dibenzopyrene. Compt. Rend. Acad. Sci., 256:2728-2730.
23. Lacassagne, A., Buu-Hoi, N.P., Zazdela, F., and Lavit-Lamy, D. (1964): Activite cancerogene de derives substitutes du 1,2,3,4-dibenzopyrene, du 1,2,4,5-dibenzopyrene, et du 1,2,4,5,8,9-tribenzopyrene. Compt. Rend. Acad. Sci., 259:3899-3902.
24. Natarajan, R.K. and Flesher, J.W. (1973): Synthesis and carcinogenicity of compounds related to 6-hydroxymethylbenzo(a)pyrene. J. Med. Chem., 16:714-715.
25. Shear, M.J. and Leiter, J. (1940): Studies in carcinogenesis XIV. 3-Substituted and 10-substituted derivatives of 1,2-benzanthracene. J. Natl. Cancer Inst., 1:303-336.
26. Shimkin, M.B. and Bryan, W.R. (1943): Morphology and growth of subcutaneous tumors induced with carcinogenic hydrocarbons in strain C3H male mice. J. Natl. Cancer Inst., 4:25-35.
27. Steiner, P.E. and Edgecomb, J.H. (1952): Carcinogenicity of 1,2-benzanthracene. Cancer Res., 12:657-659.
28. Thompson, J.I. et al. Survey of compounds which have been tested for carcinogenic activity. DHEW Publ. No. (NIH) 73-75. Public Health Publ. No. 149, U.S. Dept. of Health, Education, and Welfare, Washington, D.C. (1961-1967 volume section II) 953-1012.
29. VanDuuren, B.L., Sivak, A., Segal, A., Orris, L. and Langseth, L. (1966): The tumor promoting agents of tobacco leaf and tobacco smoke condensates. J. Natl. Cancer Inst., 37:519-526.
30. White, F.R. and Eschenbrenner, A.B. (1945): Note on the occurrence of hepatomas in rats following the ingestion of 1,2-benzanthracene. J. Natl. Cancer Inst., 6:19-21.

DETERMINATION OF POLYCYCLIC AROMATIC HYDROCARBONS IN FLUE GASES FROM COAL-FIRED POWER PLANTS

WERNER FUNCKE*; JOHANN KÖNIG; ECKHARD BALFANZ
Gesellschaft für Arbeitsplatz- und Umweltanalytik GbR,
Nottulner Landweg 102, D-4400 Münster-Roxel,
Federal Republic of Germany.

INTRODUCTION

In the field of environmental pollution research, a steadily increasing number of investigations of polycyclic aromatic hydrocarbons (PAH) in different samples have been published in recent years, since these compounds have shown mutagenic and carcinogenic activity in bioassay methods. Incomplete combustion of fossil fuels is a main source of PAH. With regard to the high amounts of flue gases it is important to know the PAH emissions of power plants (e.g., 1,3,4,7).

In this paper PAH of molecular weights between 178 (phenanthrene) and 252 (benzo(a)pyrene) were analyzed in stack emissions of two coal-fired power plants. One of these plants (plant A) was merely equipped with an electrostatic precipitator (ESP), while the other one was additionally equipped with a desulphurisation system which allowed the clean-up of variable parts of the flue gases (up to 66 %). An all-glass Teflon sampling system is presented for a separate collection of particulate and vapor phase emissions (cf., 1,8,9). Additionally, particulate matter was taken from the ESP control units of both plants and from the desulphurisation system of plant B.

MATERIALS AND METHODS

Chemicals

The following PAH were used as reference substances: phenanthrene, pyrene, chrysene (Merck-Schuchard); fluoranthene (Carl Roth OHG); benzo(a)fluorene, benzo(b)fluorene, benzo(a) pyrene (EGA); benz(a)anthracene (Ferak); benzo(b)fluoranthene benzo(j)fluoranthene (BCR-Brussels); benzo(e)pyrene (Aldrich).

Prior to use the XAD-2 (Rohm and Haas) was washed with aqueous NaOH (4 %), deionized water, aqueous HCl (4 %), deionized water. A Soxhlet extraction with methanol, acetone, and dichloromethane for 24 h each followed. All solvents were distilled-in-glass quality before being used.

Sampling Procedure

The flue gases were isokinetically sampled in the horizontal duct between the ESP and the stack of plant A (sampling site A 1), between the ESP and the desulphurisation system of plant B (sampling site B 1), and in the stack of plant B in a height of about 54 m (sampling site B 2). Depending on the sulphur content of the coal, up to 66 % of the stack gases were cleaned up by the desulphurisation system at sampling site B 2. In all cases, the sampling was performed at a distance of at least 1 m from the duct or stack wall. The flue gas velocity was measured near to the nozzle during the whole sampling procedure.

Up to 2 kg particulate matter was collected from the ESP of plant A and B (sampling site A 2 and B3, respectively) and up to 2 kg calcium sulfate from the desulphurisation system of plant B (sampling site B 3). The mean concentration of the ESP particulates was in the range of about 10 g/m^3.

Further details about the combustion parameters and sampling procedures are given in Table 1.

Instrumentation

The all-glass Teflon sampling train consists of a glass probe liner with a Teflon nozzle for isokinetic sampling in the stack or in the horizontal ducts. The filter holder housing is heated up to the flue gas temperature; glass fiber filters are used for the collection of flue gas particulates. A glass cooler (Schott) cools the flue gas down from 105-131 ^0C to 15-20 ^0C at a flow rate of about 8 m^3/h. The condensate can be collected downstream the cooler. Gaseous substances are finally adsorbed by XAD-2 in a glass cartridge. While sampling, the whole train is protected against light.

For the clean-up procedures an HPLC instrument was used consisting of a high pressure pump Series 3 and an UV detector LC-55B (both Perkin Elmer) set at a wave length of 254 nm. A high pressure valve (Rheodyne 7105) equipped with a sample loop of 175 µl was utilized for sample injection. The stainless steel column (250 x 2.1 mm) was dry packed with Sephasorb HP (Deutsche Pharmacia) and kept at a controlled temperature of 50 ^0C. Propanol-2 was used as the solvent in all HPLC experiments.

The PAH fractions were analyzed by means of a Perkin Elmer gas chromatograph F 22 equipped with a FID and connected with the Hewlett Packard data system 3354. Using a 25 m fused

TABLE 1

COMBUSTION AND SAMPLING PARAMETERS (*MEAN VALUES, ** SAMPLING SITE B 2)

Power plant		A	B
Combustion parameters *			
- Load level	[MW]	312	720
- Combustion temperature	[^{o}C]	1,650	1,350
- Flue gas emissions	[m^3/h]	$4.61 \cdot 10^5$	$2.25 \cdot 10^6$ **
- O_2 content of the flue gas	[Vol %]	2.5	6 **
- CO content of the flue gas	[mg/m^3]	50	10 **
- NO_x content of the flue gas	[mg/m^3]	1,637	1,200 **
- SO_2 content of the flue gas	[mg/m^3]	2,100	1,200 **

Power plant		A	B	
Sampling site		A 1	B 1	B 2
Sampling parameters				
- Flue gas velocity	[m/s]	17	16	21
- Flue gas temperature at the glass fiber filter	[^{o}C]	131	125	105
- Flue gas temperature downstream glass cooler	[^{o}C]	20	15	17
Sampling amounts and rates of				
- Flue gases	[m^3]	48.2	41.2	40.9
	[m^3/h]	8.0	8.5	8.5
- Particulates	[g]	2.77	0.32	0.28
	[mg/m^3]	57.5	7.77	6.85
- Condensates	[g]	1,590	1,282	1,577
	[g/m^3]	330	31.1	38.6

silica column (i.d. = 0.32 mm) coated with 0.3 µm OV 73, the following chromatographic conditions were applied: oven temperature 4 min at 100 ^{o}C, 3 ^{o}C/min to 300 ^{o}C, hold at 300 ^{o}C; carrier gas helium at 1.2 bar.

GC-MS runs were performed by a Finnigan 4500 GC-MS system equipped with a fused silica column as described above or a 15 m fused silica column coated with 0.2 µm DB 5 which was linearly programmed from 70 to 270 ^{o}C at a rate of 2 ^{o}C/min and held at 270 ^{o}C; injector and interface temperature 300 ^{o}C; carrier gas helium at 0.6 bar. All mass spectrometric measurements were made in the EI mode at 70 eV with scanning from 50 to 400 amu every 1 s to obtain total ion chromatograms or with scanning selected masses to obtain mass chromatograms.

Clean-up Procedures

The filters loaded with particulates of the flue gases were extracted in a Soxhlet by benzene and the XAD-2 resins by dichloromethane for 24 h. The extracts of the filter, condensate combined with acetone from rinsing the glass cooler, and the XAD-2 resin were cleaned up separately using HPLC as described above and elsewhere (12). Prior to the HPLC clean-up procedure the XAD-2 extract was subjected to a silica gel chromatography.

RESULTS AND DISCUSSION

Three samples at different sampling sites as well at different sampling times were collected from two coal-fired power plants as described above. The PAH amounts in filter particulates, condensate, and XAD-2 were determined separately and summarized in Table 2 (expressed in ng/m^3). In all three cases, the PAH emission rates are to be found in the same order of magnitude and in the same range as observed by other authors, (1,3,7). Generally, the PAH concentrations of the flue gases are found to be rather low. Even in ambient air of highly industrialized areas of the Federal Republic of Germany, for instance, comparable PAH concentrations can be observed (5,10).

As shown in Table 2, additionally, more than 97 % of the total PAH amount was found in the gas phase at flue gas temperatures between 105 and 131 ^{o}C. Thus, merely sampling of particulates in hot flue gases is not sufficient for quantitative PAH analysis. Other authors obtained similar theoretical (11) and practical (e.g. 1) results.

Table 3 summarizes the PAH concentrations in the flue gas particulates and ESP particulates of plant A and B. In all cases, the PAH concentrations significantly decrease with their increasing molecular weight. With regard to the ESP particulates similar results have been obtained by other authors (6,13). The PAH concentrations of the particulates in the flue gases are observed to be up to a factor of 10^4 higher than in the ESP particulates. This result indicates that PAH are mainly adsorbed on very small particles which are not efficiently precipitated by the ESP control units. Furthermore, the PAH concentrations in the flue gas particulates of plant B are up to a factor of 20-50 higher than the corresponding values of plant A. An explanation for this 'enrichment' of the PAH in the particulates of plant B may be that a more efficient particle precipitation by the ESP of plant B (cf., Table 1) does not effect very small particles mainly containing the PAH.

TABLE 2

AMOUNTS OF PAH [NG/M³] IN THE FLUE GASES OF PLANT A AND B

Sampling site	PAH	Filter (α) [ng/m³]	Condensate (β) [ng/m³]	XAD-2 (γ) [ng/m³]	$\Sigma \alpha - \gamma$ [ng/m³]	$\frac{\beta + \gamma}{\Sigma \alpha - \gamma} \cdot 100$ [%]
A 1	1 Phenanthrene	0.4	0.3	28.8	29.5	99
	2 Fluoranthene	0.3	0.8	8.0	9.1	96
	3 Pyrene	0.1	1.6	2.4	4.1	98
	4 Benzo(a)fluorene	0.1	0.5	1.7	2.3	96
	5 Benzo(b)fluorene	0.1	0.5	1.6	2.2	95
	6 Benz(a)anthracene	<0.1	0.2	1.4	~1.6	>94
	7 Chrysene	0.2	1.2	14.1	15.5	99
	8 Benzofluoranthenes	0.4	0.9	23.5	24.8	98
	9 Benzo(e)pyrene	<0.1	<0.1	3.2	~3.2	~97
	10 Benzo(a)pyrene	<0.1	0.2	11.5	~11.7	>99
	Σ 1 - 10	~1.6	~6.2	96.2	~104.0	~98
B 1	1 Phenanthrene	0.6	0.5	53.7	54.8	99
	2 Fluoranthene	1.2	0.9	15.5	17.6	93
	3 Pyrene	0.3	0.3	13.2	13.8	98
	4 Benzo(a)fluorene	0.2	0.2	2.6	3.0	93
	5 Benzo(b)fluorene	0.5	0.2	2.6	3.3	85
	6 Benz(a)anthracene	<0.1	0.1	1.4	~1.5	>94
	7 Chrysene	0.2	0.4	4.4	5.0	96
	8 Benzofluoranthenes	0.5	0.3	11.6	12.4	96
	9 Benzo(e)pyrene	0.1	<0.1	3.7	~3.8	<97
	10 Benzo(a)pyrene	<0.1	<0.1	3.6	~3.6	~95
	Σ 1 - 10	~3.6	~2.9	112.3	~118.8	~97
B 2	1 Phenanthrene	1.2	1.1	80.0	82.3	99
	2 Fluoranthene	1.7	3.0	126.5	131.2	99
	3 Pyrene	1.7	1.0	95.2	97.9	98
	4 Benzo(a)fluorene	0.7	1.2	29.9	31.8	98
	5 Benzo(b)fluorene	1.0	0.8	40.1	41.9	98
	6 Benz(a)anthracene	0.4	2.0	6.8	9.2	96
	7 Chrysene	0.3	2.7	34.0	37.0	99
	8 Benzofluoranthenes	1.0	6.4	17.0	24.4	96
	9 Benzo(e)pyrene	0.6	2.5	6.3	9.4	94
	10 Benzo(a)pyrene	0.7	2.5	6.5	9.7	93
	Σ 1 - 10	9.3	23.2	442.3	474.8	98

Regarding the PAH amounts at comparable sampling sites of plant A and B and the PAH amounts in the ESP particulates in Table 4, the main part of the analyzed PAH passed the ESP control units (plant A: 95 % and plant B: 72 %).

Though the low PAH concentrations found in calcium sulfate of plant B indicate that the desulphurisation system contributes only to a minor degree to the precipitation of PAH, further investigations are necessary to get more detailed informations.

TABLE 3

AMOUNTS OF PAH [PBB] IN FLUE GAS PARTICULATES, ESP PARTICULATES, AND CALCIUM SULFATE FROM THE FLUE GAS DESULPHURISATION SYSTEM

	Flue gas particulates			ESP particulates		Calcium-sulfate
Sampling site	A 1	B 1	B 2	A 2	B 3	B 4
1 Phenanthrene	6.4	76	171	0.43	3.09	1.59
2 Fluoranthene	5.9	160	251	0.20	0.99	0.15
3 Pyrene	1.7	37	242	0.02	0.07	0.14
4 Benzo(a)fluorene	1.4	23	102	0.04	<0.01	0.01
5 Benzo(b)fluorene	1.7	63	141	0.06	<0.01	0.03
6 Benz(a)anthracene	<1.0	<10	53	<0.01	<0.01	0.03
7 Chrysene	2.6	27	47	0.02	0.15	0.08
8 Benzofluoranthenes	6.8	70	147	0.01	<0.01	0.04
9 Benzo(e)pyrene	<1.4	15	89	-0.01	<0.01	0.02
10 Benzo(a)pyrene	<1.4	<10	95	-0.01	<0.01	0.01
Σ 1 - 10	~26.5	~471	1,338	~0.78	~4.30	~2.10

TABLE 4

DISTRIBUTION [%] OF PAH IN THE PARTICULATES AND THE GAS PHASE OF FLUE GASES AS WELL AS IN ESP PARTICULATES (*DISTRIBUTION OF THE TOTAL PAH AMOUNT)

Sampling Site	A 1		A 2	B 1		B 3
PAH	Flue Gases		ESP	Flue Gases		ESP
	Particulates	Gas Phase	Particulates	Particulates	Gas Phase	Particulates
1 Phenanthrene	4	92	4	1	63	36
2 Fluoranthene	3	86	11	4	60	36
3 Pyrene	2	93	5	2	93	5
4 Benzo(a)fluorene	4	92	4	6	>91	<3
5 Benzo(b)fluorene	4	>92	<4	15	>82	<3
6 Benz(a)anthracene	<6	>88	<6	<6	>88	<6
7 Chrysene	1	>98	<1	4	>94	<2
8 Benzofluoranthenes	2	>97	<1	4	>96	<1
9 Benzo(e)pyrene	<3	>94	<3	3	>94	<3
10 Benzo(a)pyrene	<1	>98	<1	<3	>94	<3
Σ 1 - 10*	~2	~95	~3	~2	~72	~26

ACKNOWLEDGEMENT

The authors thank Thomas Romanowski for the GC-MS and HPLC analysis and Irmhild Schmitte for technical assistance.

APPOINTMENT

The investigations described were carried out at the Fraunhofer-Institut für Toxikologie und Aerosolforschung, formerly situated in D-4400 Münster-Roxel, Nottulner Landweg 102, Federal Republic of Germany.

REFERENCES

1. Alsberg, T. and Stenberg, U. (1979): Capillary GC-MS analysis of PAH emissions from combustion of peat and wood in a hot water boiler. Chemosphere, 7: 487-496.
2. Balfanz, E., König, J., Funcke, W., and Romanowski, T. (1981): Pittfalls of quantitative analysis of polycyclic aromatic hydrocarbons in airborne particulate matter by capillary gas chromatography. Fresenius Z. Anal. Chem., 306: 340-346.
3. Bennett, R.L., Knapp, K.T., Jones, P.W., Wilkerson, J.E., and Strup, P.E. (1979): Measurement of polynuclear aromatic hydrocarbons and other hazardous organic compounds in stack gases. In: Polynuclear Aromatic Hydrocarbons, edited by P.W. Jones and P. Leber, pp. 419-428, Ann Arbor Sci., Ann Arbor.
4. Bergström, J.G.T., Eklund, G., and Trzcinski, K. (1982): Characterization and comparison of organic emissions from coal, oil, and wood fired boilers. In: Polynuclear Aromatic Hydrocarbons, edited by M. Cooke, A.J. Dennis, and G.L. Fisher, pp. 109- 120, Battelle Press, Columbus, Ohio.
5. Funcke, W., König, J., Balfanz, E., Romanowski, T., and Großmann, I. (1982): Analytical investigations on the content of polycyclic aromatic hydrocarbons of airborne particulate matter from three cities of the Ruhr area and one rural town. Staub-Reinh. Luft, 42: 192-197.
6. Griest, W.H. and Guerin, M.R. (1979): Identification and quantification of polynuclear organic matter (POM) on particulates from a coal-fired power plant. EPRI EA 1092, Project 1057-1 DOE No. RTS 77-58.

7. Guggenberger, J., Krammer, G., and Lindenmüller, W. (1981): A contribution to the determination of emissions of polycyclic aromatic hydrocarbons from high-capacity furnaces. Staub-Reinh. Luft, 41: 339-344.
8. Johnson, L.D. and Merrill, R.G. (1983): Stack sampling for organic emissions. Toxicol. Environ. Chem., 6: 109-126.
9. Jones, P.W., Giammar, R.D., Strup, P.E., and Stanford, T.B. (1976): Efficient collection of polycyclic organic compounds from combustion effluents. Environ. Sci. Technol., 10: 506-510.
10. König, J., Funcke, W., Balfanz, E., Grosch, B., Romanowski, T., and Pott, F. (1981): Comparing investigations on samples of airborne particulate matter from 5 cities of the Federal Republic of Germany on their content of 135 polycyclic aromatic hydrocarbons. Staub-Reinh. Luft, 41: 73-78.
11. Natusch, D.S.F. and Tomkins, B.A. (1978): Theoretical consideration of the adsorption of PAH vapor onto fly ash in a coal-fired power plant. In: Carcinogenesis, Vol. 3, Polynuclear Aromatic Hydrocarbons, edited by P.W. Jones and R.I. Freudenthal, pp. 145-153, Raven Press, New York.
12. Romanowski, T., Funcke, W., König, J., and Balfanz, E. (1982): Isolation of polycyclic aromatic hydrocarbons in air particulate matter by liquid chromatography. Anal. Chem., 54: 1285-1287.
13. Tomkins, B.A., Reagan, R.R., Maskarinec, M.P., Harmon, S.H., Griest, W.H., and Caton, J.E. (1983): Analytical chemistry of polycyclic aromatic hydrocarbons present in coal-fired power plant fly ash. In: Polynuclear Aromatic Hydrocarbons: Formation, Metabolism, and Measurement, edited by M. Cooke and A.J. Dennis, pp. 1173-1187, Battelle Press, Columbus, Ohio.

SYNTHESIS AND BIOLOGICAL ACTIVITY OF NITRO-SUBSTITUTED CYCLOPENTA-FUSED PAH

JAY M. GOLDRING, LOUISE M. BALL, R. SANGAIAH, AND AVRAM GOLD
Department of Environmental Sciences and Engineering, School of Public Health, University of North Carolina at Chapel Hill, Chapel Hill, NC 27514

INTRODUCTION

Nitro-substituted polycyclic aromatic hydrocarbons have recently stimulated considerable interest because of their high mutagenic activity in the Ames assay and prevalence in the environment. They were first detected in photocopier toners (1) and have since been identified in coal fly ash particles (2), cigarette smoke condensates (3), emissions from diesel engines (4) and ambient air particulate matter (5).

Relatively little was known about the mechanism(s) by which these compounds exert their mutagenic activity. Rosenkranz and Mermelstein (6) recently initiated a series of computer-assisted studies aimed at identifying particular structural features associated with the mutagenic activity of nitroarenes. Klopman and Rosenkranz (7) concluded that nitroarenes require activation by the target cells before becoming biologically active; so-called "direct-acting" nitroaromatics are presumably converted by a family of nitroreductases to arylhydroxylamines that are mutagenic themselves or are conjugated to form hydroxamic acid esters that are non-enzymatically converted to arylnitrenium ions.

The present study attempts to extend the assessment of structure-activity relationships among the nitroarenes by synthesis and bioassay of nitro derivatives of benz(a)ace-anthrylene and cyclopenta(cd)pyrene (Fig. 1). We are interested in compounds substituted on the etheno bridge of the cyclopenta ring as the most likely products of atmo-

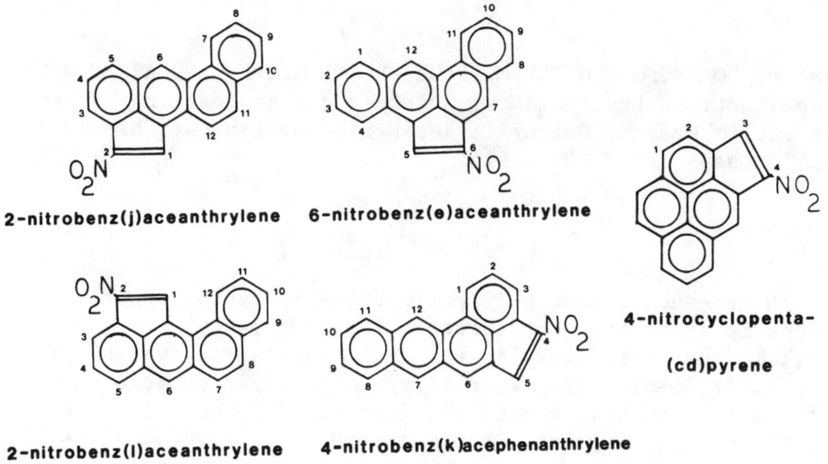

2-nitrobenz(j)aceanthrylene **6-nitrobenz(e)aceanthrylene** **4-nitrocyclopenta-(cd)pyrene**

2-nitrobenz(l)aceanthrylene **4-nitrobenz(k)acephenanthrylene**

FIGURE 1. Nitrated Cyclopenta-Fused Isomers of Benz(a)anthracene and Cyclopenta(cd)pyrene.

spheric nitration of the parent PAH. By synthesis and testing of a series of compounds in the same laboratory, the problem of interlaboratory variability can be avoided. The assay results for each compound can thus be directly compared for the series, and may also be compared with those of the parent hydrocarbons. This study will provide an opportunity to test several hypotheses concerning structural requirements for the activation of the nitroarenes.

We report here the preliminary findings for two compounds: 2-nitrobenz(l)aceanthrylene (2-nitroB(l)A) and 4-nitrocyclopenta(cd)pyrene (4-nitroCPP). The arylnitrenium ions which would be formed are expected to be highly stabilized, as illustrated in Fig. 2 for 2-NitroB(l)A. The positive charge developed on the nitrogen atom is an allylic extension of the resonance-stabilized benzylic carbonium ion fused to the *meso* carbon of the benz(a)anthracene substructure. The accessibility of the nitro group to enzymatic reduction is crucial to activation by the proposed route and hence both steric requirements for nitro reduction and the role of resonance stabilization of the putative nitrenium ion intermediate can be examined.

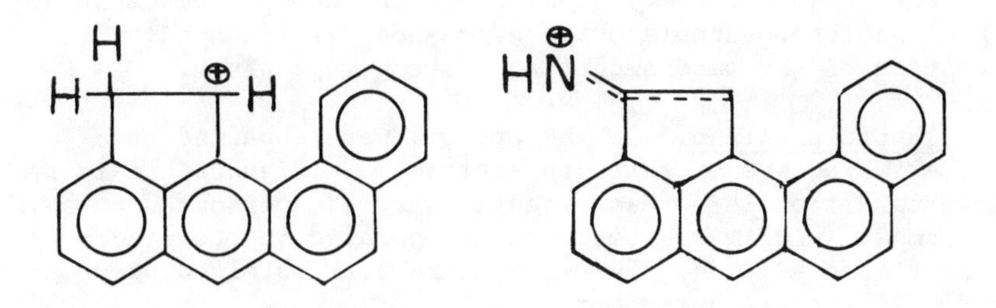

FIGURE 2. Formation of Arylnitrenium Ion in 2-Nitrobenz(1)aceanthrylene.

MATERIALS AND METHODS

Synthesis. The parent hydrocarbons were synthesized by published methods (8,9). The hydrocarbons were then nitrated by the method of Radner (10); 0.5 mL of N_2O_4, poured from a cooled tank (Matheson), was added to 20 mL of CH_2Cl_2. A stoichiometric amount of this solution was added to the parent hydrocarbon (dissolved in CH_2Cl_2; 5 mg/mL). The course of the reaction was monitored by UV-vis spectrophotometry; when the spectrum of the starting hydrocarbon disappeared, the reaction was terminated by evaporating the CH_2Cl_2 under a stream of nitrogen.

The crude mixture was then applied to a column of silica gel (Kieselgel 60, 230-400 mesh ASTM, EM Reagents) and chromatographed with CCl_4. The residual parent hydrocarbon eluted first, as a yellow band, followed by a deep red band of nitrated product. The nitro compounds were collected, dried under a stream of N_2, then recrystallized from acetone (melting point: 2-nitroB(1)A, 245°; 4-nitroCPP, 183°).

Bacterial Mutagenicity Assays. Bacterial mutagenicity assays were performed at the Health Effects Research Laboratory of the Environmental Protection Agency, Research Triangle Park, NC, according to the procedure of Ames et al. (11). All strains of **Salmonella typhimirium** and nutrient broths were prepared as described in the EPA standard protocol (12). True revertance to prototrophy was confirmed by plating small colonies onto histidine-free media and, in all cases, growth was observed.

Since HPLC analysis of DMSO solutions demonstrated that the nitro compounds are unstable in DMSO, the standard procedure was varied by dissolving the test compounds in 50 L of acetone. Spontaneous revertance was measured by addition of the same amount of acetone to a plate.

Instrumentation. ^1H NMR spectra were obtained on a Bruker WM250 spectrometer in acetone D_6. Signals are in ppm downfield from TMS. Mass spectrometry was performed on a VG Micromass 7070F mass spectrometer operated in the electron impact mode at 70eV. UV-vis spectra were obtained on a Cary model 15 spectrophotometer.

RESULTS

Structural Assignments. 2-nitroB(1)A. The ^1H NMR spectrum (Fig. 3A) shows the strongly deshielded bay region proton as a doublet at 9.1 ppm. This signal overlaps the singlet resonance from the proton at C6. The doublet from the proton on C5, which is adjacent to the strongly electron-withdrawing nitro group, appears at 8.9 ppm. The signal from the etheno-bridge proton, deshielded by the neighboring nitro group and the ring current within the "pseudo fjord" at 8.58 ppm, confirms the location of the nitro group on the etheno bridge. The mass spectrum (Fig. 3B) shows the expected M^+ peak at m/z=297, with major fragments at m/z=267 (M^+-NO) and m/z=250 (M^+-HNO_2). The UV-vis spectrum shows the broadened peaks characteristic of nitroarenes (Fig. 3C).

4 and 5-Nitrobenz(k)acephenanthrylene Mixture. The mass spectrum (Fig. 4A) demonstrates the presence of mononitrated benz(k)acephenanthrylene, with ions at m/z=297 (M^+), m/z=267 (M^+-NO) and m/z=250 (M^+-HNO_2). The ^1H NMR spectrum (Fig. 4B) suggests a mixture of two compounds, nitrated at either position on the cyclopenta ring. The presence of two compounds is confirmed by thin-layer chromatography (the appearance of two distinct spots on silica, solvent benzene/hexane). Preparative separation of the two products will be attempted by thick-layer chromatography on silica.

4-NitroCPP. The singlet for the proton of C5, which is in close proximity to the nitro group, appears at 9.2 ppm and is the most deshielded signal in the ^1H NMR spectrum (Fig. 5A). A singlet at 8.55 ppm is assigned to the proton on the etheno bridge and is also deshielded relative to the bridge proton resonances of the parent CPP. The mass spectrum (Fig.

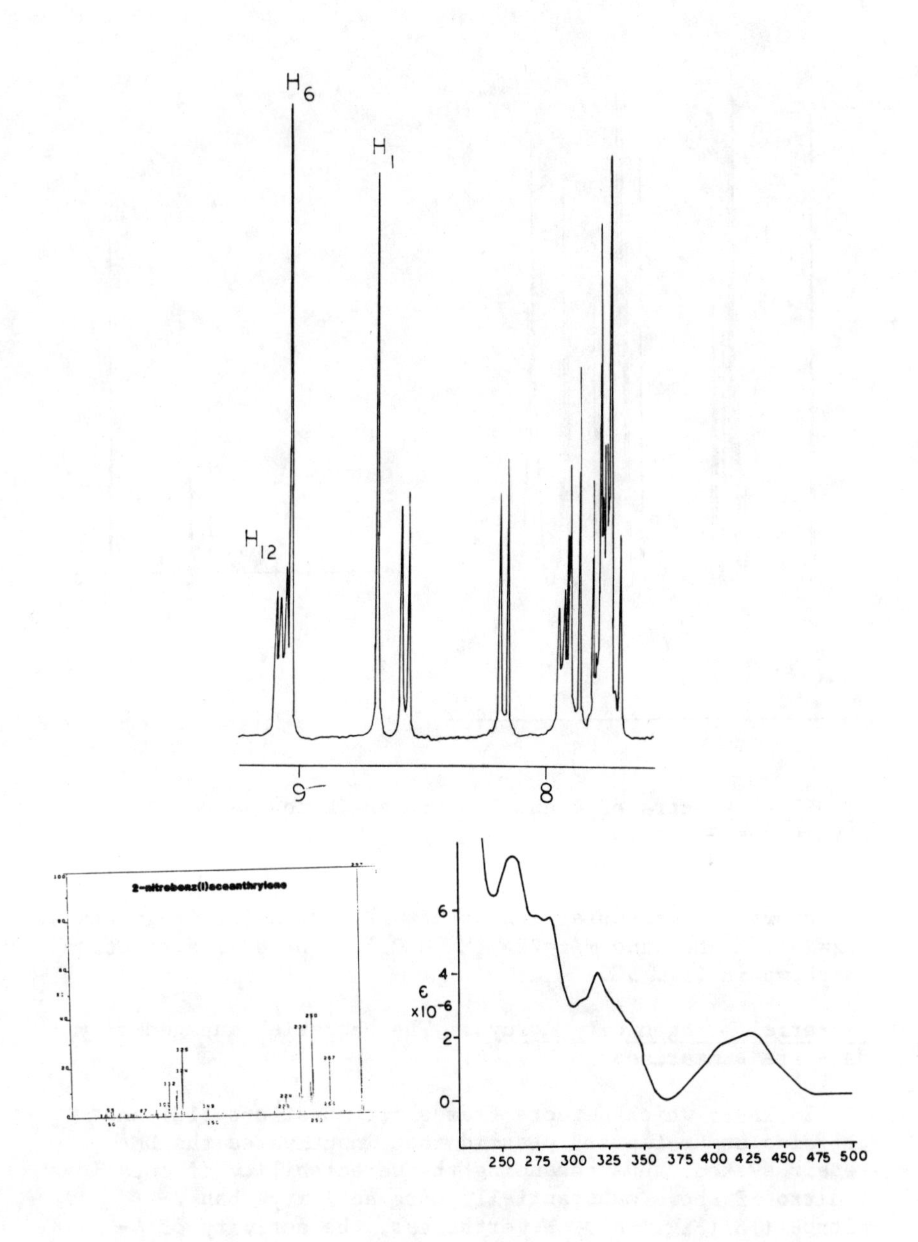

FIGURE 3. Spectra of 2-nitrobenz(l)aceanthrylene.
a. ^{1}H NMR b. Mass spectrum c. UV-vis

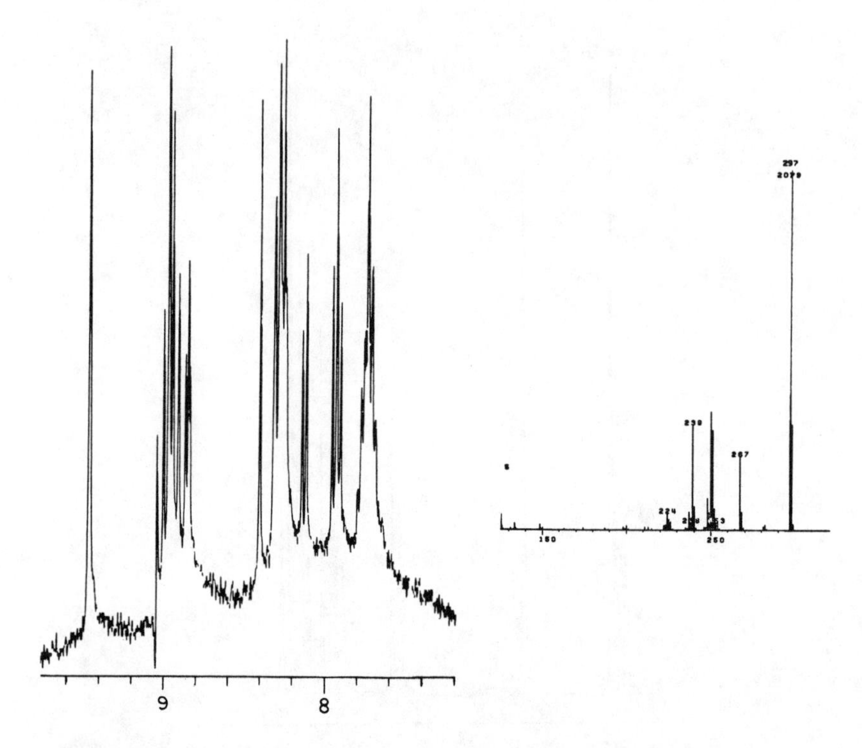

FIGURE 4. Spectra of 4 and 5 nitrobenz(k)acephenanthrylene.

5B) shows the molecular ion at m/z=271 and major fragments at m/z=241 (M^+-NO) and m/z=224 (M^+-HNO_2). The UV-vis spectrum is shown in Fig. 5C.

Bacterial Mutagenicity Assays. The bacterial mutagenicity data are summarized in Fig. 6.

In TA98, which detects frameshift mutations (12) and contains an "r-factor" plasmid that inactivates the DNA repair system, thus enhancing the detectability of mutations, 4-nitroCPP shows substantially more activity than 2-nitroB(1)A (Fig. 6A). Nevertheless, the activity of 4-

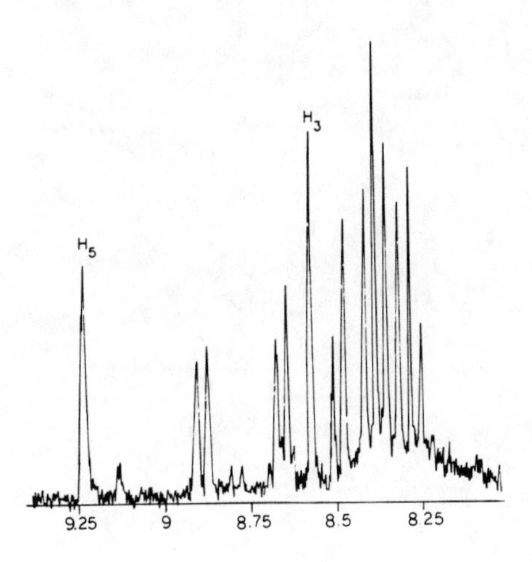

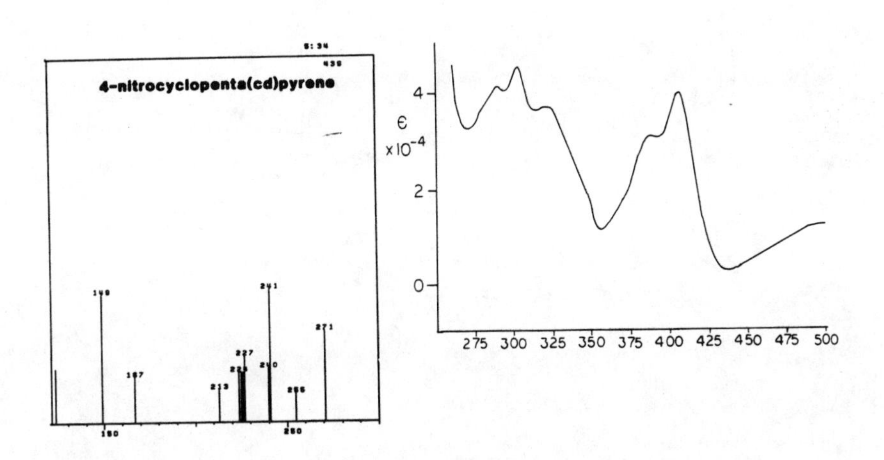

FIGURE 5. Spectra of 4-nitrocyclopenta(cd)pyrene.
a. ^{1}H NMR b. Mass Spectrum c. UV-Vis

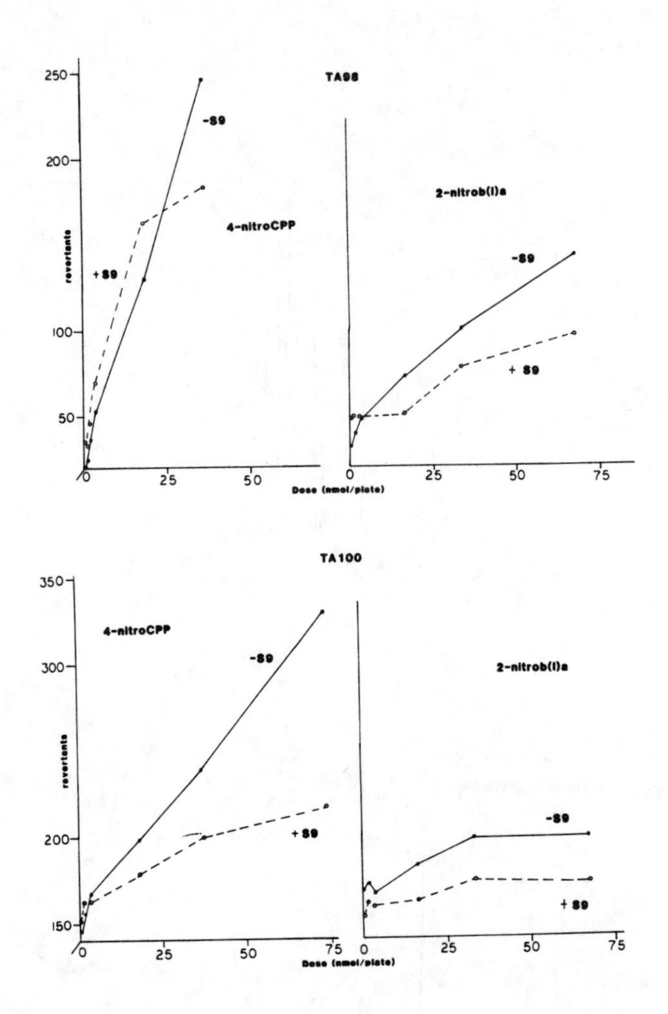

FIGURE 6. Bacterial mutagenicity assays in *S. typhimurium* for 4-nitrocyclopentapyrene and 2-nitrobenz($\overline{l}$)aceanthrylene. The results represent an average of two experiments performed on separate occasions with each dose repeated in triplicate. a. TA98 b. TA100 c. TA1535 d. TA1537 e. TA1538

TA1535

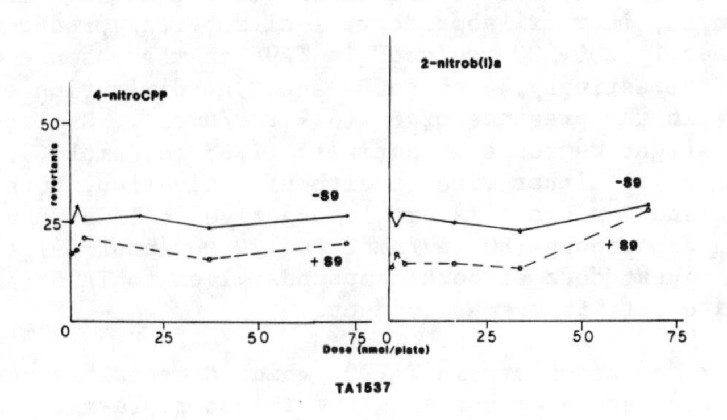

TA1537

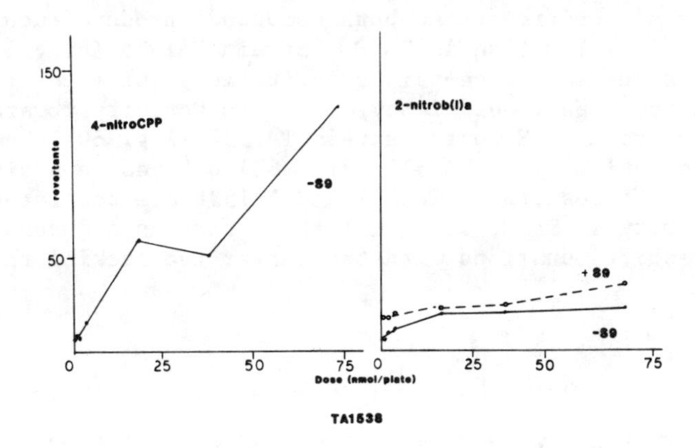

TA1538

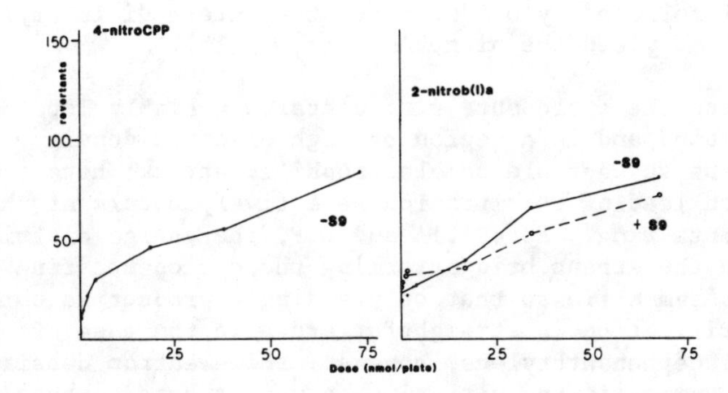

FIGURE 6. Bacterial mutagenicity assays in S. Typhimurium .... (continued)

NitroCPP in this strain (5.9 rev/nmol-S9) is low in comparison to the more potent known "direct-acting" mutagens; for example, in this laboratory, 1-nitropyrene produced approximately 250-300 rev/nmol in TA98 in the absence of S9 (13). Interestingly, 4-nitroCPP shows no diminution of activity in the presence of S9 (6.9 rev/nmol). 2-nitroB(1)A shows a slight decrease in activity (1.65 rev/nmol+S9, 0.7 rev/nmol-S9). Either with or without activation, both 4-nitroCPP and 2-nitroB(1)A are less active in TA98 than the parent hydrocarbons (80 rev/nmol and 20 rev/nmol+S9) (14). At the highest dose of both compounds given to TA98 (75 nmol/plate), toxicity was evident.

The response of strain TA100, which detects base-pair substitution and also contains the r-factor plasmid, is shown for the two nitro PAH in Figure 6B. Consistent with other direct-acting nitroarenes, both compounds produce much lower activity in TA100 than in TA98. Strain TA1535 (Fig. 6C), which also detects base-pair substitutions but lacks the plasmid for repair deficiency, shows no activity towards either nitroPAH. However, strain TA1537 (Fig. 6D) shows moderate activity and TA1538 (Fig. 6E) behaves in a similar manner. The results in TA1537 and TA1538 are consistent with the activity in TA98, since all three strains are sensitive to frameshift mutations with the former two lacking the plasmid.

DISCUSSION

In dry CH_2Cl_2, the reaction of PAH and nitrogen dioxide is presumed to involve electrophilic attack of the NO_2 radical followed by oxidation of the intermediate aryl radical to yield the nitroPAH (Fig. 7)(15).

Since the cyclopenta ring contains a highly localized double bond and is a region of high electron density, it should be susceptible to electrophilic attack; hence, the reaction leading to nitration selectively occurs at the cyclopenta ring. For B(1)A and CPP, the charge distribution on the etheno bridge forming the cyclopenta ring is highly asymmetric so that only a single product is obtained and purification is straightforward. In the case of benz(k)acephenanthrylene, however, the electron density is nearly symmetric and nitration occurs at both carbons of the etheno bridge.

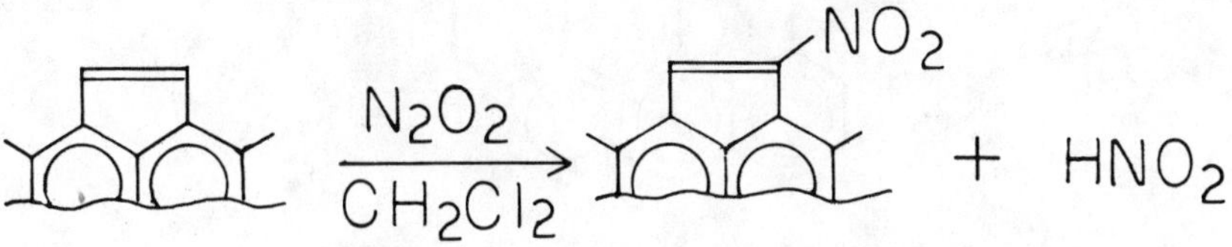

NO_2

$\xrightarrow[CH_2Cl_2]{N_2O_2}$ + HNO_2

FIGURE 7. Proposed reaction mechanism of formation of nitro cyclopenta-fused PAH

Biological Activity. The results of the Ames assay indicate both 4-nitroCPP and 2-nitroB(1)A are biologically active. Consistent with other nitroarenes previously tested, they do not require S9 activation, suggesting that they are "direct-acting" mutagens; in addition, they appear to induce primarily frameshift mutations.

However, these compounds are significantly less active than other nitroarenes of similar molecular weight, e.g., 1-nitropyrene and 3-nitrofluoranthene, nor are they nearly as active as the parent compounds (which require S9 activation). Moreover, preliminary results on 2-nitroB(1)A in the mammalian C3H10T1/2 morphological cell transformation assay indicate that this compound is highly toxic but induces little or no transformation (16). Possible explanations for this relative lack of activity are:

(i) These compounds may be poor substrates for the nitroreductase enzymes; little is known of the geometry of the active site. 2-nitroB(1)A is even less active than CPP perhaps as a result of the different stereochemical environments of the nitro group. In 2-nitroB(1)A, the nitro group is located at one end of a helical molecule and its access to the active site of the reductase enzymes or of the putative nitrenium ion to reactive sites on DNA may therefore be hindered. The nitro group of CPP is on

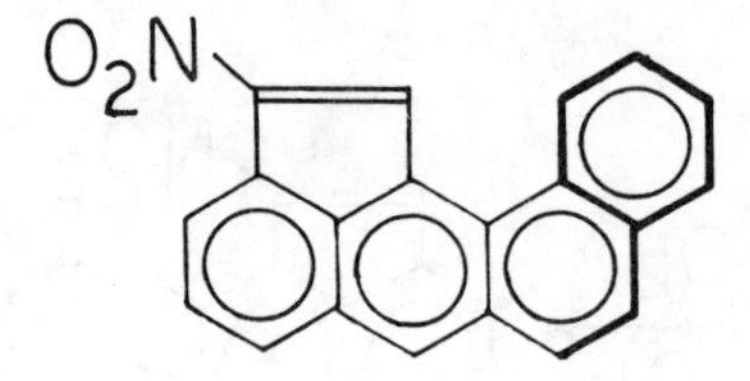

FIGURE 8. Formula and significant fragment of 2-nitroB(1)A. Heavy lines indicate deactivating fragments (7).

the periphery of a planar molecule which does not appear to impose any such steric constraints.

(ii) Klopman and Rosenkranz (7) have devised a system for predicting activities of nitroPAH and, according to this system, neither compound should be active. They categorize all the nitroarenes tested as active (greater than 1 rev/-nmol) or inactive and identified molecular fragments common to each group. Analysis of the structure of 2-nitroB(1)A according to this correlation scheme reveals a deactivating fragment (Fig. 8), which would predict a lack of mutagenicity. 4-NitroCPP possesses neither activating or deactivating fragments, a situation also considered to be predictive of little or no direct-acting mutagenicity. Although the behavior of 2-nitroB(1)A and 4-nitroCPP conforms to the predictions of this empirical structure-activity correlation, biochemical explanations for these predictions are completely lacking.

(iii) Bacterial nitroreductase enzymes may not be involved at all in the activation of nitro-substituted cyclopenta-fused compounds. This hypothesis is suggested by the case of 5-nitroacenaphthene (17), which is a very active mutagen and entirely S9-dependent. Perhaps the nitro group alters the electronic structure of the molecule so an as-yet unidentified active metabolite can be formed. Or maybe the nitro group must be removed prior to activation of these compounds; this step may in fact be rate-limiting.

Ames testing of the remainder of the compounds in the nitrobenzaceanthrylene series will reveal whether low activity is characteristic of this class of compounds or if one or more show greater activity. X-ray crystallography will also be performed on as many in the series as possible to determine their exact three-dimensional structure, and therefore evaluate if steric or electronic factors play a role.

Metabolism studies will also enable us to determine if the nitro group on the etheno bridge is reduced less readily than a nitro group on an aromatic ring. How the nitro group affects S9 activity also must be determined. Such studies in **S. typhimurium** will help elucidate whether and to what extent nitroreduction is hindered by structural constraints with the eventual goal of determining the steric requirements of nitroreductase active sites.

ACKNOWLEDGEMENTS

This work was supported by EPA Grant CR811817-01-0. This report has been reviewed by the Health Effects Research Laboratory, U.S. Environmental Protection Agency, and approved for publication. Approval does not signify that the contents necessarily reflect the views and policies of the Agency, nor does mention of trade names or commercial products constitute endorsement or recommendation for use.

REFERENCES

1. Rosenkranz, H.S., McCoy, E.C., Sanders, D.R., Butler, M., Kiriazades, D.K. and Mermelstein, R. (1980). Nitropyrenes; isolation, identification and reduction of mutagenic impurities in carbon black and toners, **Science 209**, 1039-1043.
2. Wei, C.I., Raabe, O.G. and Rosenblatt, L.S. (1982). Microbial detection of mutagenic nitroorganic compounds in filtrates of coal fly ash, **Envir. Mutagen. 4**, 382.
3. McCoy, E.C. and Rosenkranz, H.S. (1982). Cigarette smoking may yield nitroarenes, **Cancer Lett. 15**, 9-13.
4. Pederson, T.C. and Siak, J.S. (1981). The role of nitroaromatic compounds in the direct-acting mutagenicity of diesel particle extracts, **J. Appl. Toxicol. 4**, 54-60.
5. Wise, S.A., Chesler, S.N., Hilpert, L.R., May, W.E., Rebbert, R.E., Vogt, C.R., Nishioka, M.G., Austin, A. and Lewtas, J. (1985). Quantification of polycyclic aromatic hydrocarbons and nitro-substituted polycyclic aromatic hydrocarbons and mutagenicity testing for the characterization of amibent particulate matter, **Envr. Int. 11**, 147-160.
6. Rosenkranz, H.S. and Mermelstein, R. (1983). Mutagenicity and genotoxicity of nitroarenes: all nitro-containing chemicals were not created equal, **Mutat. Res. 114**, 217-267.
7. Klopman, G. and Rosenkranz, H.S. (1984). Structural requirements for the mutagenicity of environmental nitroarenes, **Mutat. Res. 126**, 227-238.
8. Sangaiah, R. and Gold, A. (1983). A short and convenient synthesis of cyclopenta(cd)pyrene and its oxygenated derivatives. In: **Polynuclear Aromatic Hydrocarbons: Mechanisms, Methods and Metabolism, Eighth International Symposium,** edited by M. W. Cooke and A. J. Dennis, pp. 1145-1150, Battelle Press, Columbus, OH.
9. Sangaiah, R., Gold, A. and Toney, G.E. (1983). Synthesis of a series of novel polycyclic aromatic systems: isomers of benz(a)anthracene containing a cyclopenta-fused ring, **J. Org. Chem. 48**, 1632-1638.
10. Radner, F. (1983). Nitration of polycyclic aromatic hydrocarbons with dinitrogen tetroxide. A simple and selective synthesis of mononitro derivatives, **Acta Chem. Scand. 37**, 65-67.
11. Ames, B.N., McCann, J. and Yamasaki, E. (1975). Methods for detecting carcinogens and mutagens with the salmonella/mammalian-nitrosome mutagenicity test. **Mutat. Res. 31**, 347-364.

12. Claxton, L.D., Kohan, M., Austin, A.C. and Evans, C. The Genetic Bioassay Branch Protocol for Bacterial Mutagenesis Including Safety and Quality Assurance Procedures, HERL-0323, U.S. Environmental Protection Agency, Research Triangle Park, NC, 147 pp.
13. Ball, L.M., Kohan, M.J., Claxton, L.D. and Lewtas, J. (1984). Mutagenicity of derivatives and metabolites of 1-nitropyrene: activation by rat liver S9 and bacterial enzymes, **Mutat. Res. 138**: 113-125.
14. Eisenstadt, E. and Gold, A. (1978). Cyclopenta(cd) pyrene: a highly mutagenic polycyclic aromatic hydrocarbon, **Proc. Nat. Acad. Sci. (USA) 75**, 1667-1669.
15. Eberson, L., Jonsson, L. and Radner, F. (1978). Nitration of aromatics via electron transfer; the revelancy of experiments involving generation of naphthalene radical cation in the presence of nitrogen dioxide, **Acta Chem. Scand. 32**, 749-753.
16. Mohapatra, N., Unpublished results.
17. McCoy, E.C., DeMarco, G., Rosenkranz, E.J., Anders, M. and Rosenkranz, H.S. (1983). 5-nitroacenaphthalene: a newly recognized role for the nitro function in mutagenicity, **Environ. Mutagen. 5**, 17-23.

ANALYSIS OF NITRATED POLYCYCLIC AROMATIC HYDROCARBONS, PAH-QUINONES AND RELATED COMPOUNDS IN AMBIENT AIR

ARTHUR GREENBERG*[1], FAYE DARACK[1], DEAN HAWTHORNE[1], DINA NATSIASHVILI[1], YALAN WANG[1], THOMAS B. ATHERHOLT[2], RONALD HARKOV[3], JUDITH B. LOUIS[3]
(1) Department of Chemical Engineering and Chemistry, New Jersey Institute of Technology, Newark, New Jersey 07102; (2) Department of Microbiology, Institute for Medical Research, Camden, New Jersey 08103; (3) Office of Science and Research, New Jersey Department of Environmental Protection, Trenton, New Jersey 08625.

INTRODUCTION

The search for major carcinogens associated with airborne particulate matter has evolved from investigations of polycyclic aromatic hydrocarbons (PAH) (1) through studies of nitro-PAH, PAH-quinones and other oxygenated derivatives to analysis of various mixed functionality compounds such as nitrohydroxy-PAH (2,3). It is clear that the PAH class itself is responsible for only a small fraction of the mutagenic activity of ambient airborne particulate matter. For example, the findings of the Airborne Toxic Elements and Organic Substances (ATEOS) project support this conclusion (4). In the series of three sequential extractions of airborne particulates, the first extract (non-polar, cyclohexane) exhibited about one third of the direct mutagenicity (rev/m^3) of the sum of the remaining two, more polar (dichloromethane and acetone) extracts (4). Even with S9 activation, the nonpolar cyclohexane extract, which contains at least 90% of the PAH (5), accounted for less than half of the total mutagenic activity. Furthermore, as we shall demonstrate later in this paper, the PAH are not responsible for most of the mutagenicity of the cyclohexane extract. Another study (6) indicated that approximately half of the mutagenic activity was attributed to nitrated compounds (6).

Although the absolute levels of nitro-PAH are low (7), as the result of their own instability or the unimportance of nitration as a PAH-disappearance pathway, their high mutagenicities remain a reason for concern. Oxidation of PAH appears to be a more important disappearance pathway. However, the ratio of BaP-quinones to BaP was found to increase from only 0.04 in the winter to only 0.09 in the summer (8). The high winter/summer ratios of the PAH themselves appear to be due to differences in seasonal output rather than to decomposition (9). Although the presence of nitro-PAH has been proposed by some to be, in part, an artifact of filter collection of particulates (10), it appears that gas-phase nitration (possibly with N_2O_5 does occur (11).

This paper reports an investigation wherein the sequential particulate extraction scheme (cyclohexane, dichloromethane, acetone) of the ATEOS project (12) has been studied in terms of 16 fractions. These fractions have been characterized by mass of extractable organic matter (EOM), mutagenicity and Fourier Transform Infrared (FTIR) Spectra. Furthermore, analytical techniques employed to identify and quantitate nitro-PAH, 1-nitropyrene in particular, and PAH-quinones will be described. The importance of determination of positional isomers of PAH derivatives will be briefly addressed.

MATERIALS AND METHODS

All solvents employed for extraction and chromatography are HPLC grade. 9- Nitroanthracene and 1-nitropyrene were obtained from Aldrich Chemical Company and used without further purification. Other nitro-PAH described here as well as the 7 benzo(a)pyrene (BaP) diones and one BeP dione were purchased from the National Cancer Institute Chemical Repository.

Twenty-four hour air samples were collected at the ATEOS Newark site (12) using an Anderson high volume sampler (10 micron) and pre-fired quartz filters. These were soxhlet extracted for 20-24 hours each using the sequence cyclohexane (CYC), dichloromethane (DCM), and acetone (ACE). Each extract was streaked onto a separate silica gel GF plate (Analabs, Inc.). The CYC extract was developed using 1:1 hexane/toluene and five fractions (#5, least polar, #1, most polar) were scraped from the plate and the adsorbent washed with tetrahydrofuran (freshly distilled over lithium aluminum hydride). The DCM and ACE plates were developed using DCM and the five DCM extract fractions (#5, least polar, #1, most polar) and the six ACE fractions (#5, least polar, #0, most polar) were obtained by scraping the adsorbent from the plate and washing with hot DCM and hot methanol sequentially and then combining washings. Solutions were clarified and 100 ul and 50 ul aliquots taken for EOM determination using a Cahn 26 automatic electrobalance (12). Quarterly composites of whole extracts and fractions were assayed using TA98. Relevant fractions were analyzed by HPLC following blowdown under high purity nitrogen. For PAH analysis, fraction 4 of the cyclohexane extract was employed. For nitro-PAH analysis, fraction 3 of the cyclohexane extract was used. Quinones appear to be present in fractions 1 and 2 of the cyclohexane extract and apparently in the more polar DCM extract. PAH are analyzed using gradient HPLC with the aid of Vydac 201TP54 columns, monitoring UV signals at 280 nm and 365 nm as well as fluorescence (Excit., 365 nm; Emiss., 440 nm). National Bureau of Standards Standard Reference Material (NBS SRM) 1647 and 1649 are routinely analyzed for quality assurance/quality control (5). Analysis of nitro-PAH employs essentially the same technique as that described by MacCrehan and May (13) except that modifications have been made on the zinc/silica reducer column. We have employed virtually the identical technique for analysis of PAH-quinones. NBS SRM 1587 and 1650 have been employed for quality assurance/quality control of nitro-PAH analysis. Analysis of 1-nitropyrene is described later in this chapter. FTIR spectra were obtained by M. Jaworsky and R. Mendelsohn of Rutgers University, Newark College of Arts and Sciences. Analysis of NO_3^-, NH_4^+, and SO_4^{--} was performed by L. Psota-Kelty and J.D. Sinclair of AT&T Laboratories.

RESULTS

Masses and Bacterial Mutagenicity Assays of Fractions

Masses and mutagenic activities of the 16 fractions obtained from the three extracts of a quarterly composite (October - December, 1984) are shown in Figure 1. Most of the mass is concentrated in the least polar fraction (cyclohexane # 5) and the most polar fraction (acetone #0). However, the greatest mutagenic activity, both on the basis of rev/mass of EOM and rev/m^3 is found in the most polar fraction (#1) of the DCM extract. (Those mutagenicity results indicated with question marks correspond to very small masses were the significance of the assays is in question). It is apparent that the only fraction to manifest a major increase in mutagenicity upon metabolic activation is cyclohexane fraction #4 which has at least 90% of the PAH (5). The acetone extract has 6 - 10% of its total mass due to NO_3^- and another 1% due to NH_4^+ with a negligible contribution from SO_4^{--}. These ions are concentrated in the acetone #0 fraction. It is likely that other inorganic substances also contribute to #0 which dominates the ACE EOM.

Fourier Transform Infrared Studies

FTIR results for cyclohexane fraction #5 indicate virtual absence of aromatic hydrocarbons (absence of 3000-3100 cm^{-1}

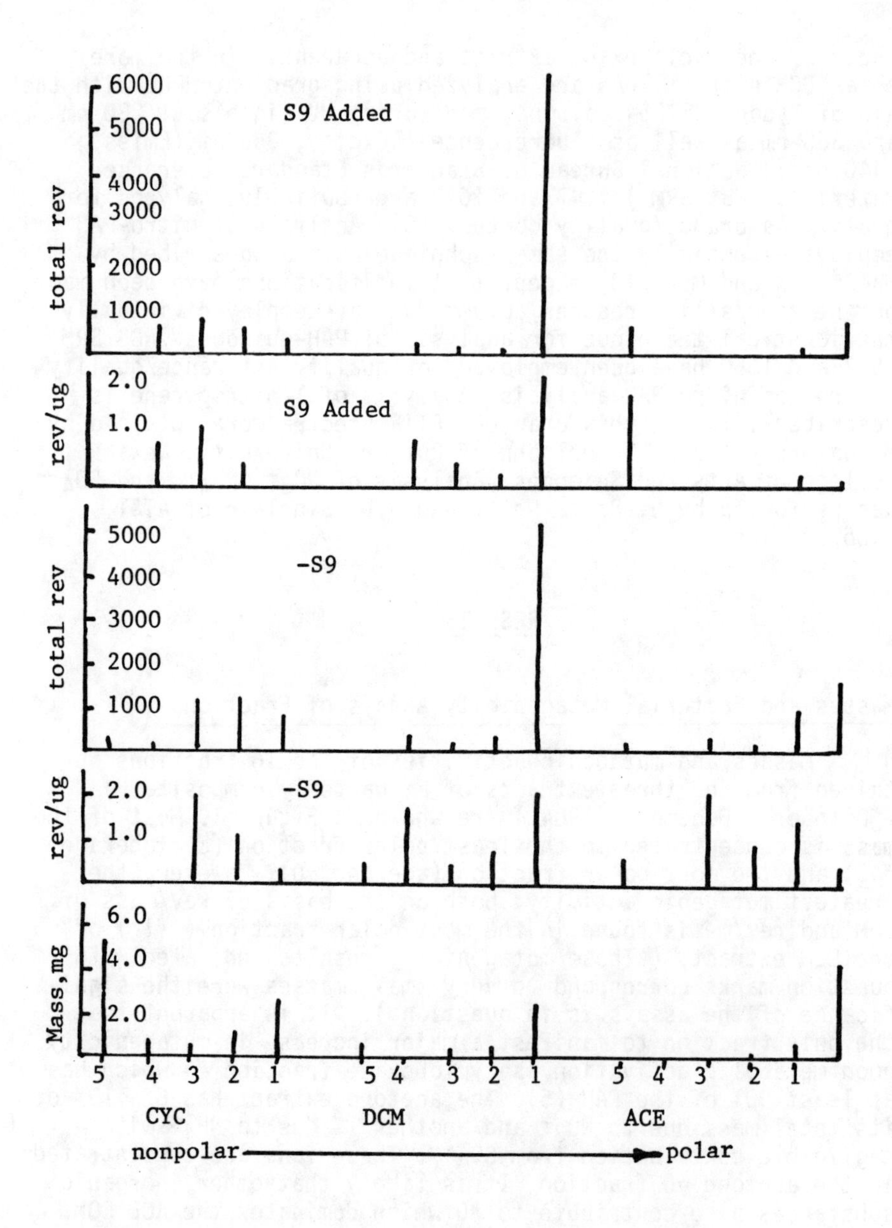

FIGURE 1. TA98 assays on 16 fractions from ambient airborne particulate.

bands) and are consistent with the presence of paraffins. Cyclohexane fraction#4 has a significant band at 3047 cm^{-1} attributable to PAH and a slight band at 1719 cm^{-1} in addition to paraffin and fingerprint bands. Cyclohexane fraction #1 is dominated by paraffin bands as well as a strong band at 1741 cm^{-1} and a weaker one at 1641 cm^{-1}. The next fraction in polarity, DCM #5, is rather similar to the previous one, thus indicating overlap in the extraction process. The rev/mass are also similar although the mass of DCM #5 is small thus limiting the significance of the results. The very potent DCM #1 has a small band at 3050 cm^{-1} and a large band at 1712 cm^{-1} in addition to a broad band centered at 3400 cm^{-1}. Acetone #5 looks similar to DCM #1 indicating overlap in the EOM of these two solvents. The revertants/mass of these two fractions have some similarity but the small ACE #5 mass limits the significance of these results. Acetone fraction #4 has a significant band at 3070 cm^{-1} and others at 1742 and 1295 cm^{-1} despite small mass. Acetone #3 is dominated by a band at 1120 cm^{-1}. Acetone #0 is qualitatively different from all other FTIR spectra in having very small paraffin bands; a band at 1724 cm^{-1} is present and a very strong band is found at 1382 cm^{-1}.

Analysis of Nitro-PAH

Use of nitro-PAH standards indicated that all of the mononitro-PAH are found in cyclohexane extract fraction #3. This was the most mutagenic of the cyclohexane fractions both with and without activation. Some dinitro-PAH may also be found in this fraction. The HPLC technique of MacCrehan and May (13) was employed with a modification of the reducer column. Figures 2a and 2b show HPLC chromatograms of NBS SRM 1587 without and with reduction. Quantitative conversion to highly fluorescent amino-PAH is achieved. Figures 3a and 3b show corresponding chromatograms (without reducer column and with reducer column) for cyclohexane fraction #3 of NBS SRM 1650 diesel particulate.

We are currently exploring the utility of 2-nitrotriptycene (1) as an internal standard for nitro-PAH analysis. As a

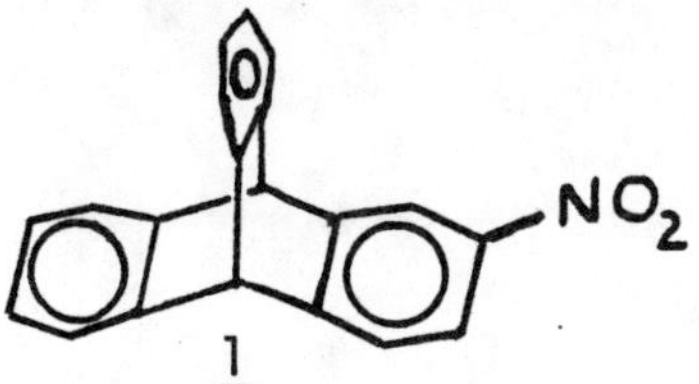

NO_2

1

nearly spherical molecule, it elutes much earlier than its

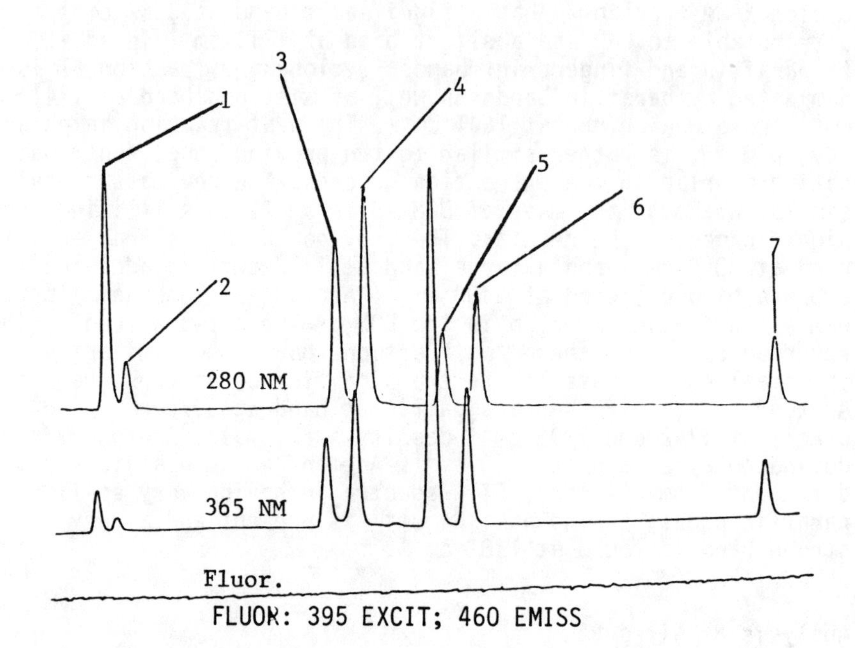

FIGURE 2A. HPLC of NBS SRM 1587 nitro-PAH without reducer column. Vydac 201 TP54, 60% aq CH_3CN to 100% CH_3CN, 35 min linear gradient.

1= 2-nitrofluorene
2= 9-nitroanthracene
3= 3-nitrofluorene
4= 1-nitropyrene
5= 7-nitro B(a)A
6= 6-nitrochrysene
7= 6-nitro B(a)P

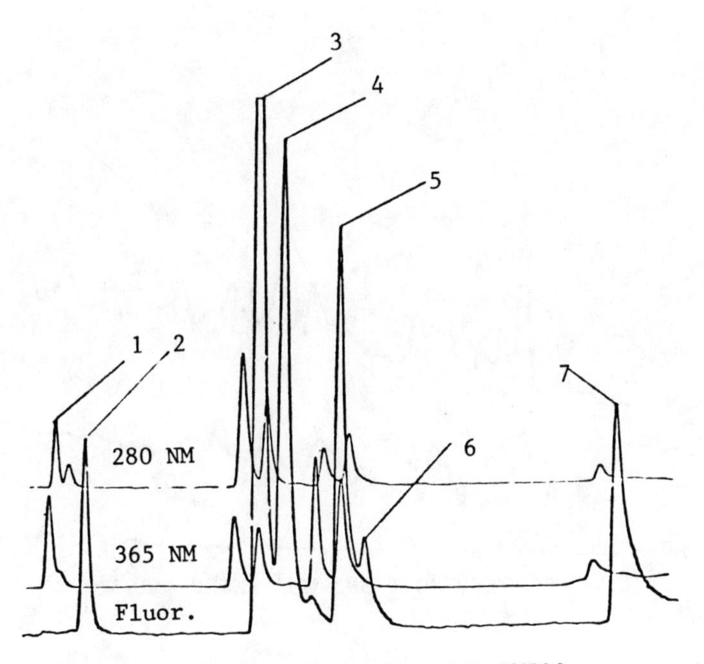

FIGURE 2B. HPLC of NBS SRM 1587 nitro-PAH with reducer column.

1= 2-nitrofluorene
2= 9-nitroanthracene
3= 3-nitrofluorene
4= 1-nitropyrene
5= 7-nitro B(a)A
6= 6-nitrochrysene
7= 6-nitro B(a)P

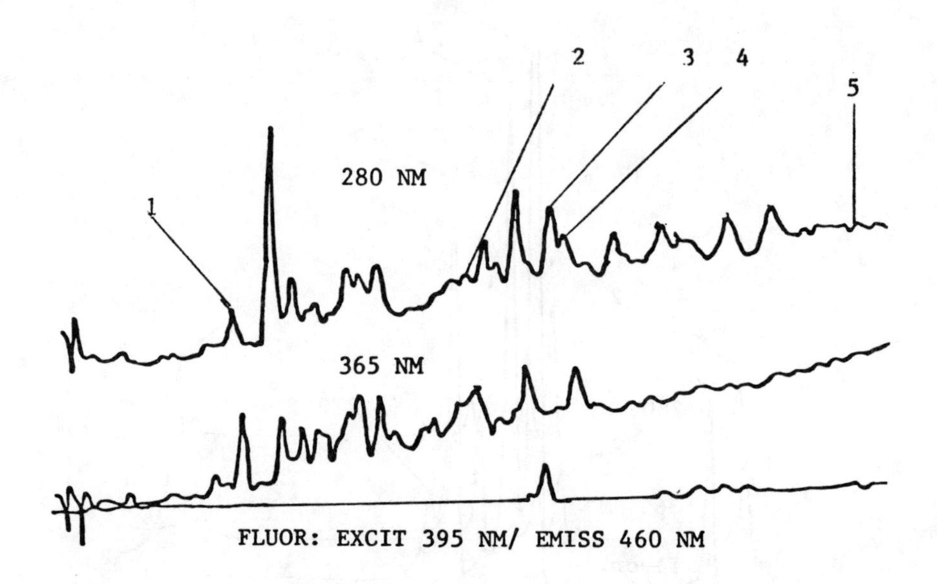

FIGURE 3A. HPLC of NBS SRM 1650 diesel particulate fraction #3 from TLC of cyclohexane extract Vydac column without reducer column.

1= 9-nitroanthracene
2= 1-nitropyrene
3= 7-nitro B(a)A
4= 6-nitrochrysene
5= 6-nitro B(a)P

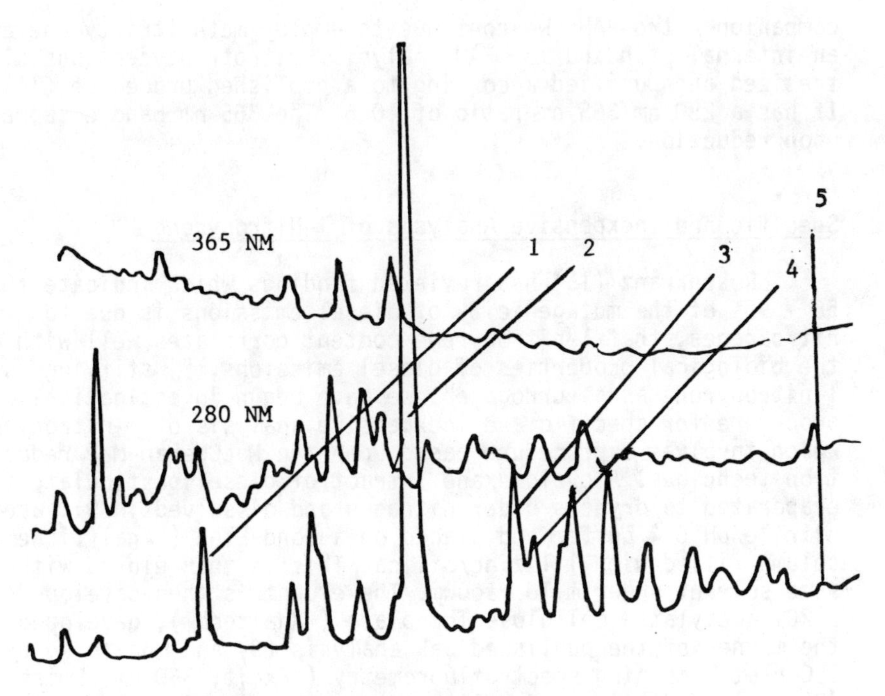

FIGURE 3B. HPLC of NMS SRM 1650 diesel particulate fraction #3 from TLC of cyclohexane extract Vydac column with reducer column.

1= 9-nitroanthracene
2= 1-nitropyrene
3= 7-nitro B(a)A
4= 6-nitrochrysene
5= 6-nitro B(a)P

companion nitro-PAH. We continue to employ methyltriptycene as an internal standard for PAH analysis. Nitrotriptycene was synthesized and purified according to a published procedure (14). It has a 280 nm/365 nm ratio of 10.6. The 365 nm band disappears upon reduction.

Specific and Inexpensive Analysis of 1-Nitropyrene

Rosenkranz (15) has reviewed findings which indicate that 50 - 90% of the mutagenicity of diesel emissions is due to nitroarenes, that 1-nitropyrene content correlates well with the biological properties of diesel emissions, "justifying using 1-nitropyrene as a surrogate". We have begun investigations of a procedure for specific and inexpensive analysis of 1-nitropyrene which involves a procedure based upon the MacCrehan-May reduction technique. A cyclohexane extract of diesel particulate is evaporated to dryness under nitrogen and dissolved in 9:1 acetonitrile/pH 5.4 buffer and placed on a Bond-Elut (Analytichem) column filled with 1:1 zinc/silica. This is then eluted with the same solvent under mild vacuum. The eluant is then developed on a 20% acetylated cellulose TLC plate (Analtech), developed in the manner of the published BaP analysis (9) and monitored by TLC plate scanning spectrofluorometry (Excit. 360 nm; Emiss. 430 nm). Reduction and collection is quantitative, 1-aminopyrene is stable on the plate and the technique works for diesel particulate although improvements in resolution must be made before ambient air particulates can be analyzed.

Analysis of PAH-Quinones

The MacCrehan-May chemical reduction technique was explor- for the PAH-quinones. These compounds are even more readily reduced than nitro-PAH. Figure 4a shows an HPLC chromatogram without reducer column for 7 BaP diones and 1 BeP dione using a Zorbax HPLC column. Figure 4b shows the result of reduction in producing dihydroquinones that strongly fluoresce under the conditions employed (EXCIT. 395 nm; EMISS, 460 nm). Reduction was done on another day and loss of resolution is not a necessary feature of the reducer column.

Exploratory Studies of 1-Nitropyrene-2-ol

The earlier-cited findings on nitrohydroxy-PAH prompted us to investigate one such compound, 1-nitropyrne-2-ol (2). This compound was synthesized photochemically from 1-nitro-

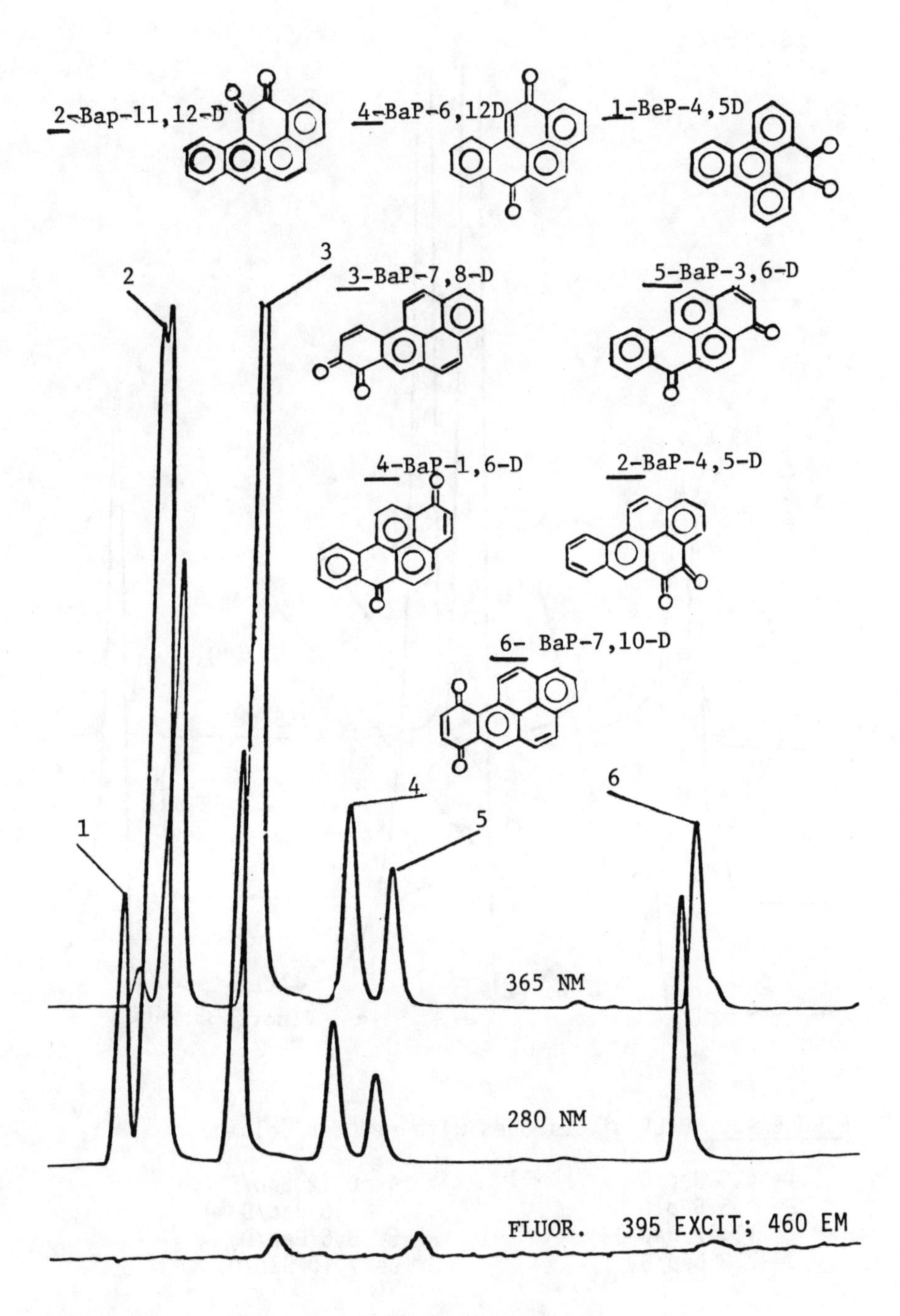

FIGURE 4A. HPLC of quinones without reducer column.

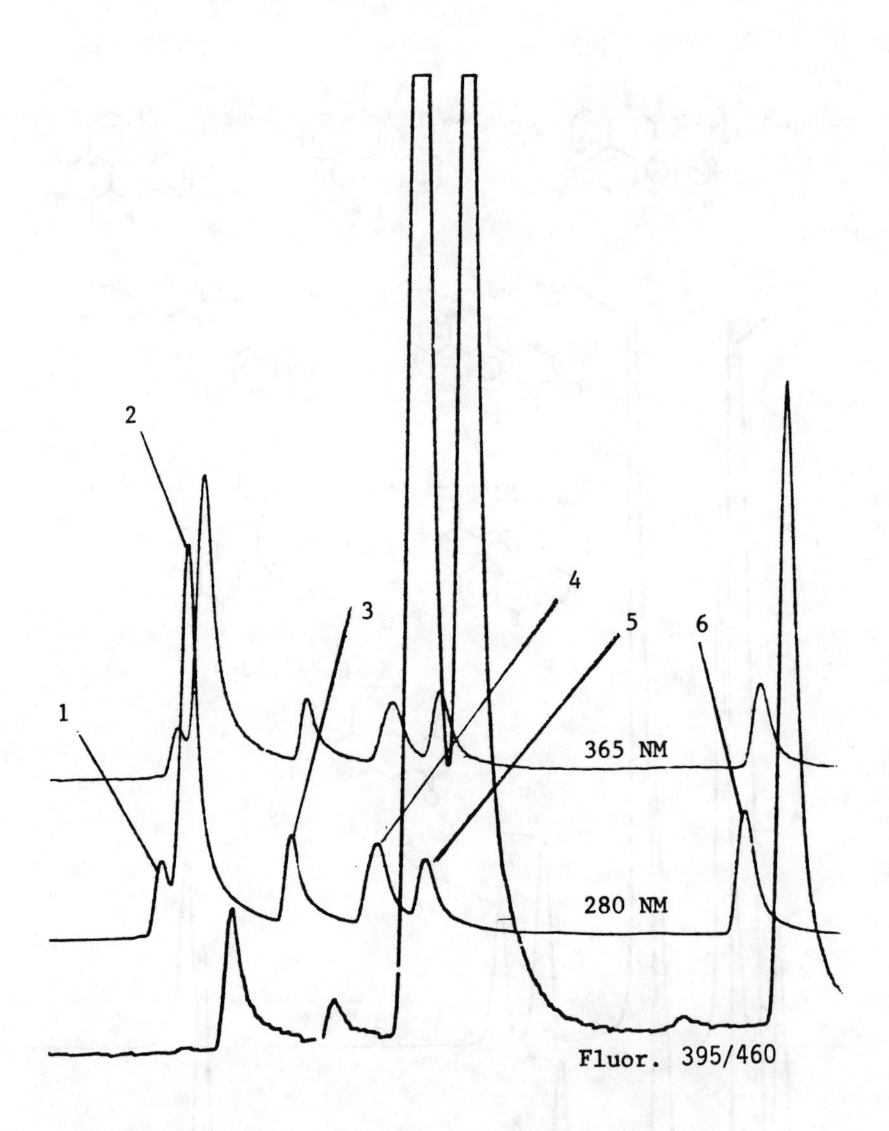

FIGURE 4B. HPLC of quinones with reducer column.

1= 4,5 Bep/D
2= 4,5 Bap/D
11,12 Bap/D
3= 7,8 Bap/D
4= 6,12 Bap/D
1,6 Bap/D
5= 3,6 Bap/D
6= 7,10 Bap/D

pyrene in the presence of air (16) in a manner highly suggestive of an ambient mechanism. This substance is nonmutagenic unless

MONITOR:
546 nm
365 nm

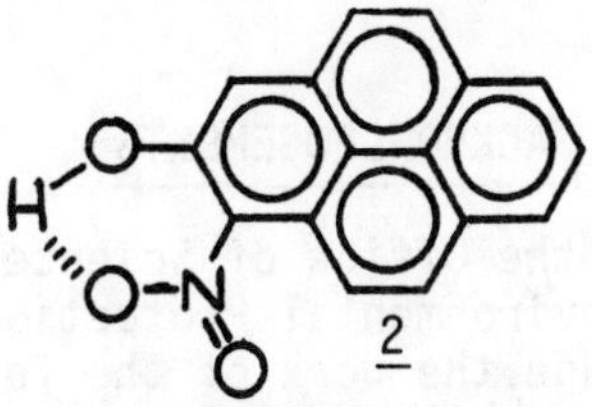

2

S9 is added when its activity is comparable to that of BaP(17). The absence of unactivated mutagenicity has been attributed to intramolecular hydrogen bonding which is probably the origin of its violet color and its relative nonpolarity (under DCM development it migrates toward the top of a silica gel plate indicating that if it is found in ambient air it should be present in cyclohexane fraction #3 or DCM #5. In fact, we have found 2 at levels of 1-3 ng/m^3 in urban air cyclohexane extract.

DISCUSSION AND CONCLUSIONS

The fact that most of the mass of the cyclohexane extract is due to nonmutagenic hydrocarbons explains the mediocre correlations between cyclohexane EOM and mutagenicity, and PAH and cyclohexane EOM observed in ATEOS (12). It should be noted that there are some differences between the distribution of extractable masses and mutagenicities between the composite sample described here and the ATEOS results. The earlier project found the greatest direct mutagenicity per mass in the DCM extract but greatest total potencies (rev/m^3) were frequently seen in the cyclohexane and acetone extracts. In the ATEOS study, acetone extracts usually had more than half the total mass.

The present study clearly concentrates on attempts to define surrogate compounds and classes for mutagenicity. Our work is now concentrating on further analysis using GC/MS of the more polar fractions, particularly DCM fraction #1.

Finally, we wish to note the importance of identification of positional isomers of selected PAH derivatives. Using simplified (oversimplified?) reasoning, one may be able to work backwards in order to understand the nature of the species which attacks a presumably preformed PAH. Thus, while attack on alternant hydrocarbons such as pyrene by radicals, Lewis acids and Lewis bases are all expected to occur at the same carbon, simple Huckel molecular orbital theory predicts that some nonalternant PAH are attacked at different sites by Lewis acids as opposed to radicals. Examples include fluoranthene, aceanthrylene (but not acephenanthrylene), benzo(ghi)fluoranthene among others (18).

ACKNOWLEDGEMENTS

We wish to thank the Office of Science and Research, New Jersey Department of Environmental Protection for support of this work. We acknowledge the work of the following students: Clint Brockway, Diane Dudacik, Paul Gross, Roman Pazdro, Julius Preston (American Chemical Society Project SEED student), and Orlando Rodriquez. FTIR spectra were obtained by Markian Jaworsky and Professor Richard Mendelsohn of Rutgers University Newark College of Arts and Sciences. Ion chromatographic results were obtained by Ms. Linda Psota-Kelly and Dr. J.D. Sinclair of AT&T Laboratories.

REFERENCES

1. National Academy of Sciences (1983): Polycyclic Aromatic Hydrocarbons: Evaluations of Sources and Effects, Washington, D.C.
2. Schuetzle, D., Riley, T., Prater, T.J., Harvey, T. M., Hunt, D.F. (1982): Analysis of nitrated polycyclic aromatic hydrocarbons in diesel particulates, Anal. Chem., 54:265-271.
3. Nishioka, M.G., Lucas, S.V., Lewtas, J. (1985): Identification and quantification of OH-NO_2-PAHs and NO_2-PAHs in an ambient air particulate extract by NCI HRGC/MS, Paper No. ANYL 12, 189th National Meeting of the American Chemical Society, April 28, 1985, Miami Beach, Florida.
4. Atherholt, T.B., McGarrity, G.J., Louis, J.B., McGeorge, L.J., Lioy, P.J., Daisey, J.M., Greenberg, A., and Darack, F. (1985): Mutagenicity studies of New Jersey ambient air particulate extracts. In: Short-Term Bioassays in the Analysis of Complex Environmental Mixtures IV, edited by M.D. Waters, S.S. Sandhu, J. Lewtas, L. Claxton, G. Strauss, and S. Nesnow, pp. 211-231, Plenum Pub.
5. Greenberg, A., Darack, F., Harkov, R., Lioy, P., and Daisey, J. (1985): Polycyclic aromatic hydrocarbons in New Jersey: A comparison of winter and summer concentrations over a two-year period, Atmos. Environ., 19: 1325-1339.

6. Siak, J., Chan, T.L., Gibson, T.L., and Wolff, G.T. (1985): Contribution to bacterial mutagenicity from nitro-PAH compounds in ambient aerosols, Atmos. Environ., 19:369-376.
7. Gibson, T.L. (1982): Nitro derivatives of polynuclear aromatic hydrocarbons in airborne and source particulate matter, Atmos. Environ., 16:2037-2040.
8. Pierce, R.C., and Katz, M. (1976): Chromatographic isolation and spectral analysis of polycyclic quinones: Application to Air Pollution Analysis, Environ. Sci. Technol., 10:45-51.
9. Harkov, R. and Greenberg, A. (1985): Benzo(a)pyrene in New Jersey: Results from a twenty-seven site study. J. Air Poll. Control Assoc., 35:238-243.
10. Risby, T.H. and Lestz, S.S. (1983): Is the direct mutagenic activity of diesel particulate matter a sampling artifact? Environ. Sci. Technol., 17:621-624.
11. Pitts, J.N., Jr., Zielinska, B., Sweeman, J.A., Atkinson, R., and Winer, A.M. (1985): Reactions of adsorbed pyrene and perylene with gaseous N_2O_5 under simulated atmospheric conditions, Atmos. Environ., 19:911-915.
12. Lioy, P.J., Daisey, J.M., Atherholt, T., Bozzelli, J., Darack, F., Fisher, R., Greenberg, A., Harkov, R., Kebbekus, B., Kneip, T.J., Louis, J., McGarrity, G., McGeorge, L., and Reiss, N.M. (1983): The New Jersey project on airborne toxic elements and organic substances (ATEOS): A summary of the 1981 summer and 1982 winter studies, J. Air Poll. Control Assoc., 33:649-657.
13. MacCrehan, W.A. and May, W.E. (1985): Determination of nitro-polynuclear aromatic hydrocarbons in diesel soot by liquid chromatography with fluorescence and electrochemical detection. In: Polynuclear Aromatic Hydrocarbons: Ninth International Symposium, edited by M. Cooke and A.J. Dennis, Battelle Press (in press).
14. Klanderman, B.H. and Perkin, W.C. (1968): Nitration of triptycene, J. Org. Chem., 34:630-633.
15. Rosenkranz, H.S. (1984): Mutagenic and carcinogenic nitroarenes in diesel emissions: risk identification, Mutation Res., 140:1-6.
16. Yasuhara, A. and Fuwa, K. (1983): Formation of 1-nitro-2-hydroxypyrene from 1-nitropyrene by photolysis, Chem. Lett., 347-348.
17. Lofroth, G., Nilsson, L., Agurell, E., and Yasuhara, A. (1984): Salmonella/microsome mutagenicity of 1-nitropyrene-2-ol, a nitropyrene phenol formed in the photolysis of 1-nitropyrene, Z. Naturforsch., 39c:193-195.

18. Greenberg, A. and Darack, F. (1986): Atmospheric reactions and reactivity indices of polycyclic aromatic hydrocarbons. In: Molecular Structure and Energetics, Vol. 4, edited by J.F. Liebman and A. Greenberg, VCH Publishers, Inc., Chapter 1 (in press).

EFFECT OF THE COCARCINOGEN FERRIC OXIDE ON BENZO(a)PYRENE METABOLISM BY HAMSTER ALVEOLAR MACROPHAGES.

ALICE GREIFE[1,2], RITA SCHOENY[2], DAVID WARSHAWSKY[2]
[1] National Institute for Occupational Safety and Health, Cincinnati, Ohio 45226 and [2] University of Cincinnati Medical Center, Department of Environmental Health, Cincinnati, Ohio 45267-0056.

INTRODUCTION

Alveolar macrophages (AM), the resident mononuclear phagocytes of the lung, defend the lower airways and alveolar spaces against inhaled foreign matter by the process of endocytosis. Endocytosis, which is accompanied by a respiration burst, is also associated with an increased production of oxygen radicals and/or hydrogen peroxide (1,2). Chemical carcinogens are one type of foreign matter which may be endocytized and metabolized by AM (3). Chemical carcinogens may also be adsorbed onto airborne particles and subsequently inhaled.

It has been demonstrated _in vitro_ that AM are able to metabolize the chemical carcinogen benzo(a)pyrene (BaP), and to phagocytize various particles such as ferric oxide (Fe_2O_3) (3,4). The effects of coadministration of particles and a carcinogen to AM _in vitro_, however, has not been fully evaluated. There is some evidence to suggest that _in vivo_ exposures of AM exposed to particles increases the phagocytic rate _in vitro_; however, a correlation between increased phagocytic rate and effect on metabolism of a carcinogen remains undetermined (4). It has been shown _in vivo_, that the presence of particles not only retards the clearance of a chemical carcinogen such as BaP, but also increases the tumor incidence and decreases the tumor latency associated with exposure to that carcinogen (5,6). An understanding of AM metabolism of the model carcinogen BaP under conditions of phagocytosis would, therefore, be important in understanding the effect that airborne particles may have on the defense mechanisms of the AM in the lung.

METHODS

AM were harvested by tracheal lavage, with cold, sterile saline, from 8-9 week old, 100-150 gm anesthetized male Syrian Hamsters. The AM were separated from the lavage fluid by centrifugation and cultured without serum in suspension in 2.5 ml Ca^{+2} Mg^{+2} - free Hanks balanced salt solution (HBSS) containing Hepes buffer, glucose, and Cytochrome C (pH 7.4) at 37°C. The AM were cultured (2.5 x 10^6 cells/vial) in vials which had been prewashed with NaOH and siliconized. Short-term culture viability was assessed by exclusion staining, and basal metabolic activity was examined by measuring reduction of cytochrome C by superoxide anions released by the AM. The cultures were divided into six experimental groups, (0.0, 1.0, 2.0, 3.0, 6.0, and 9.0 hours), and ^{14}C BaP (25uM) or ^{14}C BaP (25uM) adsorbed onto the selected dose of Fe_2O_3 (0.5, 1.0, 2.0 mg) was added to each culture.

Upon completion of each incubation period, the contents of each vial were centrifuged and the supernatant (media) was removed and assayed for the reduction of cytochrome C. The AM pellet was resuspended in 5 mls cold HBSS. The AM were separated from the particles by centrifuging the suspension with a subphase of Metrizamide*. All AM collected at the interface, while the cell-free particles formed a pellet. A sample of the AM was analyzed for viability. The remaining AM were washed twice to remove any radioactivity. The washes were combined with the media for quantitation. Both the AM and media were each extracted with hexane containing 1 ml .15 mM KOH in 85% DMSO. The hexane and hexane-non-extractable layers were counted in a liquid scintillation counter to quantify total BaP and total metabolites respectively. The AM pellet was also solubilized, decolorized and counted in a liquid scintillation counter.

RESULTS

Short Term Culture Viability

AM viability was greater than 95%, as determined by exclusion staining, in short-term suspension cultures in HBSS without serum for the first 6 hours regardless of

*(2-[3-acetamido-5-methyl-acetamido-2,4,6-triiodobenzamido]-2-deoxy-D-glucose)

treatment (Table 1). The viability of all cultures with or without particles declined over the next 3 hours. The degree of decline, to a maximum of 36% at the highest particle dose (2mg), appeared to be dependent on the concentration of administered particles.

TABLE 1

VIABILITY OF AM IN SERUM FREE MEDIUM[a]

Time in Culture HOURS	PERCENT VIABILITY Control[b]	.5 mg	1 mg	2 mg
0	98	99	99	98
1	97	97	99	96
2	98	98	97	97
3	97	96	95	95
6	96	95	94	94
9	52	48	40	36

[a] Determined from exclusion of erythrosin B
[b] Refers to added Fe_2O_3. Control = no added Fe_2O_3.

Cytochrome C Reduction

In control vials (no particles added) the amount of cytochrome C reduced increased from 0 to 6 hours, and appeared to reach a plateau after that time (Table 2). Addition of particles resulted in a significant increase ($p = .05$ when compared to controls) in the amount of cytochrome C reduced. This increase was a function of the concentration of Fe_2O_3 administered to the cultures. For all Fe_2O_3 concentrations, however, cytochrome C reduction reached a plateau at 6 hours. This was accompanied by a decrease in cell viability.

TABLE 2

CYTOCHROME C REDUCTION[a]

Time in Culture HOURS	ABSORBANCE UNITS Control[b]	.5 mg	1 mg	2 mg
0	0.0	0.0	0.0	0.0
1	0.09	0.12	0.13	0.14
2	0.10	0.20	0.25	0.28
3	0.14	0.30	0.39	0.42
6	0.32	0.41	0.54	0.62
9	0.38	0.48	0.62	0.70

[a] Measured by absorbance at 550 nm
[b] Refers to added Fe_2O_3. Control = no added Fe_2O_3.

ENDOCYTOSIS OF BAP

The amount of BaP endocytized by all cells increased for the first 3 hours and then declined. The highest noncytotoxic dose of BaP tolerated by the AM in this system was determined to be 25uM (data not shown). Quantitation of radiolabel in the hexane-extractable fraction by liquid scintillation indicated that the AM endocytized BaP (Figure 1). The largest concentration of endocytized BaP was found in the control cells (no particles, 8.0 nM BaP/10^6 cells). The cells exposed to particles may have endocytized more BaP overall than did control cells as evidenced by the fact that these cells produced more BaP metabolites than the control cells. The extent of BaP uptake in those cells exposed to particles appeared to be dependent on the dose of particles given.

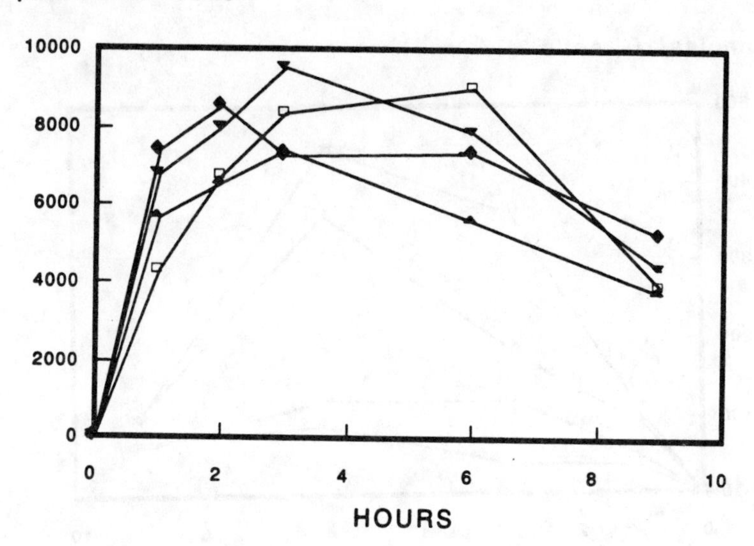

FIGURE 1. BaP Concentration in Alveolar Macrophages as a Function of Ferric Oxide Dose, Quantitation of hexane extractable radiolabel by liquid scintillation of the hexane non-extractable fraction of the medium shows that AM can metabolize BaP and subsequently release the metabolites into the medium. The increase observed in the amount of metabolites released between 6 and 9 hours is coupled with a decrease in AM viability and the amount of cytochrome C reduced. (◻) No Particles, (▲) 0.5mg Fe_2O_3 per flask, (◆) 1.0 mg, (▼) 2.0., (Mean of 3 experiments at each data point)

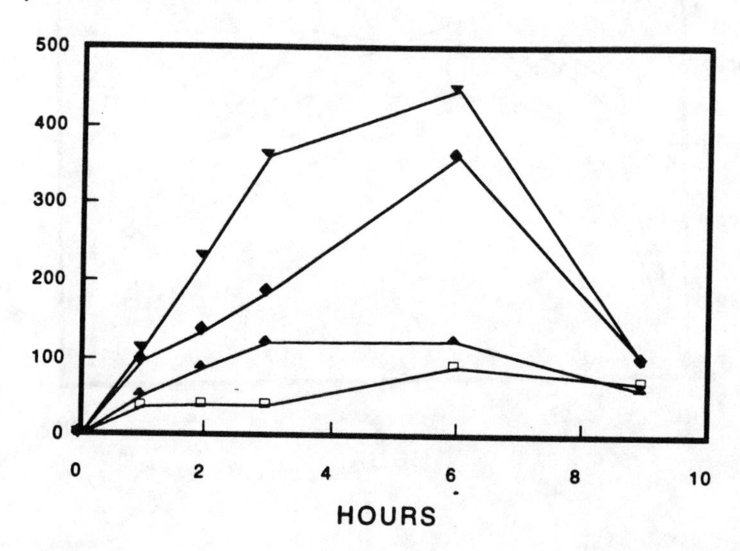

FIGURE 2. Intracellular BaP Metabolites Produced By Macrophages as a Function of Ferric Oxide Dose. Quantitation of intracellular radiolabel by liquid scintillation indicates that AM are able to metabolize the endocytized BaP. AM viability and reduction of cytochrome C increase up to 6 hours and then declines. This is indicative of AM quiescense or death. (□) No Particles, (▲) 0.5mg Fe_2O_3/ flask, (◆) 1.0 mg, (▼) 2.0, (Mean of 3 experiments at each data point).

METABOLISM OF BAP

Intracellular BaP metabolites, which were quantified by liquid scintillation of the hexane non-extractable fraction of the AM, indicated that AM metabolized the endocytized BaP (Figure 2). The total amount of intracellular metabolites increased in a near linear fashion for the first 6 hours and then declined. This decrease in metabolism was due to AM quiescence or death as indicated by a decrease in viability and reduction of cytochrome C. In all cases, more metabolites were associated with the cells exposed to particles than the control cells. This was significant ($p = .05$) for cells exposed to 2.0 mg Fe_2O_3 at all time points, excluding 9 hours, and for cells exposed to 1.0 mg Fe_2O_3 at 6 hours.

Extracellular BaP metabolites were quantified by liquid scintillation of the hexane non-extractable fraction of the medium. This indicated that the AM could metabolize BaP and subsequently release the metabolites into the medium (Figure 3). The total amount of metabolites released into the culture medium increased with the concentration of particles. This release was significant ($p = .05$) for all cells exposed to 1.0 mg and 2.0 mg Fe_2O_3 at all time points, excluding 6 hours, when compared to controls. Between 0 and 6 hours, the maximum amount of metabolites released was 2.04nM $/10^6$ cells at the highest dose of particles (2 mg). The increase observed in the amount of metabolites released between 6 and 9 hours was coupled with a decrease in AM viability and the amount of cytochrome C reduced.

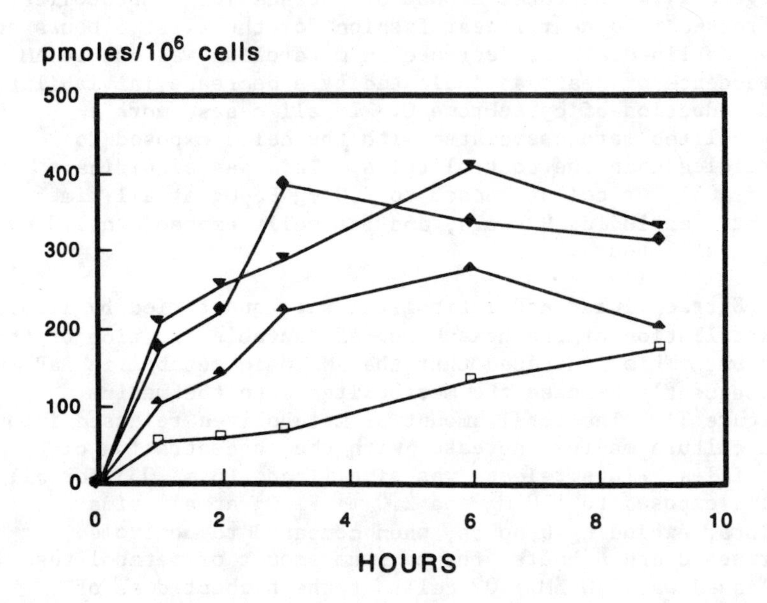

FIGURE 3. Extracellular BaP Metabolites Produced by AM as a Function of Ferric Oxide Dose. Quantitation by liquid scintillation of the hexane non-extractable fraction of the medium shows that AM can metabolize BaP and subsequently release metabolites. The increase observed in the amount of metabolites released between 6 and 9 hours is coupled with a decrease in AM viability and cytochrome C reducetion. (□) No Particles, (▲) 0.5mg Fe_2O_3/ flask, (◆) 1.0 mg, (▼) 2.0, (Mean of 3 experiments at each data point)

DISCUSSION

It has been demonstrated that AM release superoxide anions as a result of basal metabolic activity (8). Challenge of AM by foreign material, including particles and bacteria, results in phagocytosis of the foreign material. This phagocytosis is accompanied by an increase in the basal metabolic activity of the AM, and subsequent increase in the release of superoxide anions (8). It is thought that this release is associated with the defense system of the AM. There have been some in vitro studies examining the association between reduction of cytochrome C by superoxide anions released from AM and the presence of particles. One such study challenged rat AM with opsonized particles. This resulted in AM release of superoxide anions in a particle dose-dependent-manner. The anions released reduced cytochrome C (9).

The investigation presented here examined the release of superoxide anion from the AM cultured with and without particles (Fe_2O_3). It was observed that the presence of particles resulted in an increased release of superoxide anion by the cells and subsequent reduction of cytochrome C for the first six hours when compared to controls. This increase, which was significant at all dose levels, was indicative of phagocytosis. The observed decline in the amount of superoxide anion produced by the cells after six hours was indicative of AM quiescence or death.

The data presented in this investigation also show that AM in short term cultures can endocytize and metabolize BaP. Furthermore, the endocytosis and metabolism of the BAP is dependent on the dose of particles present. It has been demonstrated in vivo, that particles retard the clearance of intratracheally (IT) administered carcinogens. This carcinogen/particle interaction was demonstrated in a study of hamsters which were exposed IT to BaP or a coadministration of BaP and India ink (5). The hamsters treated with BaP and India ink cleared the BaP much more slowly from the lungs than those animals receiving only BaP. The presence of the particles thus increased the BaP residence time in the lung, by decreasing the clearance capacity of the pulmonary AM.

It has also been shown in vivo that particles such as Fe_2O_3 can enhance the activity of a coadministered carcinogen without eliciting a reaction itself. Hamsters were administered Fe_2O_3 intratracheally (IT) with and without BaP. The animals exposed to the BaP/Fe_2O_3 had a greater tumor incidence and decreased tumor latency developed more than the group exposed to BaP alone. Exposure to Fe_2O_3 alone did not cause any respiratory tumors in the animals. The enhancement of the action of the carcinogen appeared to be dependent on the presence of the particle (6). The tumors developed by these animals, at the tracheobronchial bifurcation, are characterized as broncogenic carcinomas. These tumors are very similar in morphology and histopathology to those developed in humans (7). Following IT administration of BaP coated particles, particle laden AM have been shown to migrate to and accumulate at this bifurcation (8).

There have been a few studies which examined the coadministration of particles and carcinogen to AM *in vitro*. For example, it was found that human AM obtained by lavage were capable of metabolizing BaP administered in culture (3). When the BaP was coadministered with Fe_2O_3, metabolism also occurred. In another study dog AM, obtained by lavage, were challenged with BaP alone or BaP adsorbed onto diesel particles (7). It was found that the presence of the diesel particles resulted in an increased production of metabolites, and that these metabolites were in general, in greater concentration in the culture medium than in the cells.

These results support the findings presented here. In all cases, the AM metabolized more BaP, at any given time up to 6 hours, in the presence of particles than without particles. In general, more metabolites were also associated with the culture medium than with the cells at any given time. The rapid increase in metabolites present in the media after 6 hours was probably due to deterioration of cell membranes or cell death. Increased production of BaP metabolites by the AM in the presence of particles may indicate that the AM endocytized more BaP in the presence of particles than without particles, and, therefore, had more BaP to metabolize. On the other hand, the AM may simply have metabolized the BaP which was endocytized in the presence of the particles faster than the BaP which was endocytized without particles. Elucidation of the individual metabolites produced by the AM *in vitro* and the effect that the particles may have on such production will be important in understanding the role of the AM in the development of respiratory disease.

ACKNOWLEDGMENTS

We thank the American Lung Association of Ohio for the grant which funded this research and Ms. Nancy Beck for assistance in preparation of the manuscript.

REFERENCES

1. Rossi, F., Romeo, D., and Patriarca, P. (1972): Mechanism of phagocytosis associated oxidative metabolism in polymorphonuclear leucocytes and macrophages, J. Reticuloendothel Soc., 12:127.
2. Karnovsky, M. (1962): Metabolic basis of phagocytic activity, Physiol. Rev., 42:143.
3. Autrup, H., Harris, C., and Stoner, G., et al. (1978): Metabolism of ^{3}H-benzo(a)pyrene by cultured human bronchus and cultured human pulmonary alveolar macrophages, Lab. Inv., 38:217.
4. Kavet, R., Brain, J., and Levens, D. (1978): Characteristics of pulmonary macrophages lavaged from hamsters exposed to iron oxide aerosols, Lab. Invest., 38:312.
5. Pylev, L., Roe, F., and Warwick, G. (1969): Elimination of radioactivity after intratracheal installation of tritiated 3,4-benzopyrene in hamsters, Br. J. Cancer, 23:103.
6. Stenback, F., Rowland, J., and Sellakumar, A. (1976): Carcinogenicity of benzo(a)pyrene and dusts in the hamster lung (instilled intratracheally with titanium oxide, aluminum oxide, carbon, and ferric oxide), Oncology, 33:29.
7. Saffiotti, U., Cefis, F., and Kolb, L. (1968): A method for the experimental induction of bronchogenic carcinoma, Cancer Res., 28:104.
8. Henry, M., Port, C., Kaufman, D. (1975): Importance of physical properties of benzo(a)pyrene-ferric oxide mixtures in lung tumor induction, Cancer Res., 35:207-217.
9. Bond, J., Butler, M., Medinsky, M., Muggenburg, B., McClellan, R. (1984): Dog pulmonary macrophage metabolism of free and particle-associated [^{14}C] benzo(a)pyrene, J. Tox. and Env. Health, 14:181.
10. Sweeny, T., Castranova, V., Bowman, L., and Miles, P. (1981): Factor which affect superoxide anion release from rat alveolar macrophages, Experimental Lung Research, 2:85.
11. Castranova, V., Bowman, L., Miles, P., and Reasor, M. (1980): Toxicity of metal ions to alveolar macrophages, Am. J. Ind. Med., 1:349.

PRODUCTION AND CHARACTERIZATION OF ANTIBODIES TO BENZO(A)PYRENE

G. D. GRIFFIN*[1], K. R. AMBROSE[1], R. N. THOMASON[2], C. M. MURCHISON[3], M. MCMANIS[4], P. G. R. ST. WECKER[3]; and T. VO-DINH[1]. (1) Advanced Monitoring Development Group, Health and Safety Research Division, Oak Ridge National Laboratory, P. O. Box X, Oak Ridge, Tennessee 37831; (2) High School Teacher Summer Research Program Participant; (3) Great Lakes College Association Science Semester Student; and (4) Summer Student, Department of Chemistry, The University of Tennessee, Knoxville, Tennessee 37916.

INTRODUCTION

Advances in analytical techniques have resulted in significant improvements in the ability to determine amounts of various polynuclear aromatic hydrocarbons (PAH) in environmental samples, such as air or water. These measurements can provide useful information about the extent of PAH contamination of an ambient environment, but cannot indicate differences in the internal dose of PAH received by different individuals within this PAH-contaminated ambient environment. Although measurement of the internal dose of PAH received by an individual is complicated by many factors such as route of intake, metabolism, etc., useful initial estimates may be made by measuring PAH (or metabolite) concentrations in various physiologic fluids. However, measurement of PAH concentrations in various physiological fluids which could be used for sampling is not an easy analytical task, due to the large number of different molecular species in most such fluids. Analytical techniques such as spectroscopy can be applied, but must cope with a large array of interfering substances. Chemical-specific antibodies could be potentially very useful analytical tools for detecting small quantities of various molecular species in complex mixtures. This is because antibodies can have high specificity for a particular molecular structure and can also exhibit high affinity for their particular antigen. If antibodies with suitable specificity and affinity could be developed for various PAH species of toxicological interest, they might serve as the basis of new analytical tools for PAH measurements in physiological fluids. We are currently in the process of developing such an analytical system (1). This report describes our initial work in producing an antibody to benzo[a]pyrene (BaP) in animals.

Synthesis of BaP-Protein Conjugate

Benzo[a]pyrene-6-isocyanate was prepared from BaP (98%, Aldrich Chemical Co.) by a 3-step synthesis, following an established procedure (2). The BaP-isocyanate was reacted with bovine serum albumin (BSA) (Sigma Chemical Co.) to produce the PAH-protein conjugate (BaP-BSA) (3), which was used for animal immunization.

Animal Immunization Procedures

New Zealand white rabbits or Fischer rats were initially immunized by subcutaneous or intramuscular injection using BaP-BSA in Freund's Complete Adjuvant (equal volumes of BaP-BSA in 0.0175 M phosphate buffer, pH 7.6 and Freund's Complete Adjuvant). The amount of antigen used was 0.3-0.5 mg/kg body weight in the case of the rabbits, and 1 mg/kg body weight for the rats. Succeeding doses (0.1-0.15 mg/kg) of BaP-BSA in Freund's Incomplete Adjuvant were administered by subcutaneous injection, at time intervals of 2-6 weeks, following the previous BaP-BSA administration. Blood was removed from the animals at various times following immunization and serum was prepared and tested for the presence of BaP antibodies, as described below. The injection schedule and time intervals of immunization for the antisera preparations was as follows. In the case of the immunized rabbits, rabbits 1 and 2 received two subsequent injections at two week intervals following the initial immunization, and serum was collected two weeks after the last injection. For rabbits 3 and 4, animals received a single injection six weeks after the initial antigen injection, and were bled 10 days following this second injection. The rats were reimmunized twice at two week intervals after the initial injection, and were bled one week following the last injection.

Immunological Assays

Sera from animals was routinely heated at 56°C for 30 min. to destroy complement, before being assayed. Sera from various animals were tested for BaP antibodies by one of four techniques: double immunodiffusion in agar (Ouchterlony) procedure; passive hemagglutination; enzyme-linked immunosorbent assay (ELISA); and radioimmunoassay. Ouchterlony plates were obtained from Cooper Biomedical Inc. Aliquots of serum samples in the center well were tested against varying dilutions of BaP-BSA or BSA in the peripheral wells.

The passive hemagglutination assay followed the procedure of Hirata and Brandiss (4). Formalinized sheep red blood cells were obtained from Cooper Biomedical, Inc. and coated with BaP-BSA or BSA according to the published procedure; serum samples were diluted as described (4). V-bottom, 96-well microtitration plates (Costar-Serocluster) were used for the hemagglutination assays.

ELISA assays were carried out in 96-well polystyrene plates (Nunc-Immuno Plate I), following, in general, usual procedures (5). Antigen (BaP-BSA or BSA) dilutions were made in phosphate-buffered saline (PBS - 0.01 M phosphate, pH 7.4, containing 0.137 M NaCl and 0.003 M KCl); antigen amounts from 10µg to 10 pg/well were used. Serum samples were diluted in PBS. Affinity purified alkaline phosphatase-conjugated anti-rabbit IgG antibody, heavy and light chain specific (Cooper Biomedical, Inc.) was used at a dilution of 1:1500 or 1:3000, the diluent being 0.03 M Tris, pH 8.0, 0.8% NaCl and 1% BSA. Ovalbumin (1% in PBS) was used as the blocking protein. The alkaline phosphatase substrate was p-nitrophenyl phosphate (Sigma Chemical Co.), used as 1 mg/ml solution in 0.2 M glycine buffer, pH 10.2. Incubation conditions for the ELISA assay were as follows: antigen - 3 hrs., 37°C, 18 hrs., 4°C; blocking protein - 30 min., 37°C; test antisera - 3 hrs., 37°C; alkaline phosphatase-conjugated anti-rabbit IgG antibody - 2 hrs., 25°C; enzyme substrate - 30 min., 37°C. Absorbances (405 nm) of individual wells were determined using a commercial ELISA reader (Model EL308, Bio-Tek Instruments, Inc.)

The radioimmunoassay used followed a procedure which employs dextran-coated charcoal to adsorb free antigen but reject antibody-bound antigen (6). ^{3}H- or ^{14}C- BaP was purchased from Amersham Corp. Details of the assay protocol followed the published procedure (6); the antigen-antibody incubation time was changed to 90 min. in the case of our experiments. Radioactivity was determined by liquid scintillation counting.

BAP ANTIBODIES

Immunoglobulin Purification and Assay

Immunoglobulin fractions were prepared from immune serum aliquots using: (1) $(NH_4)_2SO_4$ precipitation; and (2) DEAE-cellulose chromatography (7). Following pooling of the major protein fractions from the DEAE-cellulose column, the protein was collected by $(NH_4)_2SO_4$ precipitation, and subsequently dissolved in 0.0175 M phosphate buffer, pH 7.6. Immunoglobulin G concentrations in serum samples and in purified immunoglobulin fractions were determined by a radial immunodiffusion assay (8). The assay materials were purchased in kit form from Miles Scientific Co.

RESULTS

Preparation of BaP-BSA

The yields and melting points of the various compounds produced during the synthesis of the final product are given in the accompanying diagram.

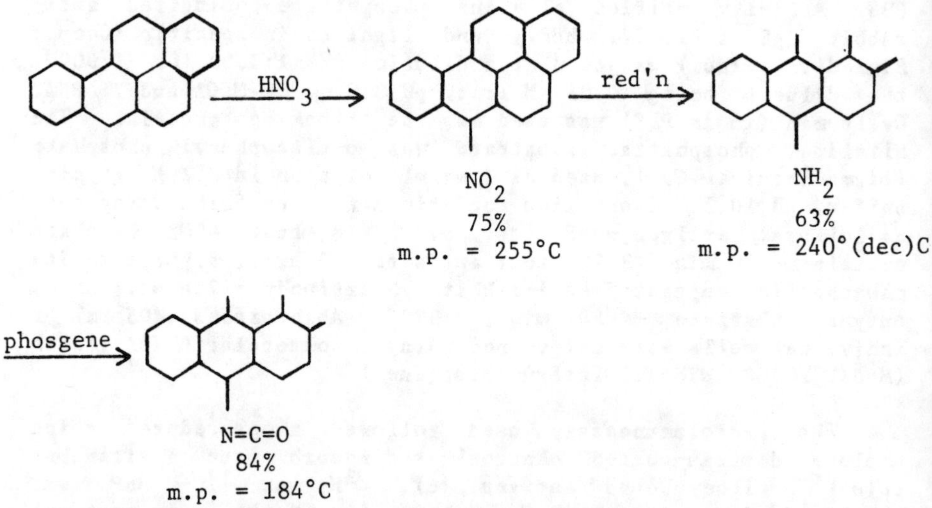

HNO_3 → NO_2 75% m.p. = 255°C; red'n → NH_2 63% m.p. = 240°(dec)C

phosgene → N=C=O 84% m.p. = 184°C

Four g of BSA was reacted with 800 mg of BaP-6-isocyanate; the final amount of BaP-BSA conjugate obtained was 2.4 g. Spectral analysis by the method of Creech and Jones (3) indicated 7-8 BaP residues were covalently attached per molecule of BSA.

Characterization of BaP Antibodies

Double immunodiffusion tests in Ouchterlony plates demonstrated that rabbit antisera preparations showed reactivity against BaP-BSA and BSA (Table 1).

TABLE 1

DOUBLE IMMUNODIFFUSION (OUCHTERLONY) TESTS OF RABBIT ANTISERA AGAINST VARIOUS BaP-BSA OR BSA CONCENTRATIONS

Antigen Concentration (mg/ml)	Animals Tested					
	Rabbit 1		Rabbit 3		Rabbit 4	
	BaP-BSA	BSA	BaP-BSA	BSA	BaP-BSA	BSA
6	+*	+	+	+	+	+
3	+	+	+	+	+	+
1	+	+	+	+	+	+
0.5	+	±	+	+	+	+
0.1	±	±	+	+	+	+
0.05	−	±	±	+	+	+
0.01			±	±	±	±
0.005			−		−	

*+ = definite precipitin line; ± = faint precipitin line;
− = no visible precipitin line.

Although some differences in the antisera preparations from different rabbits could be seen in terms of their reactivity toward lower concentrations of antigen, no particular discrimination in reactivity toward BaP-BSA as compared to BSA alone was observed. Rat antisera preparations were negative in Ouchterlony tests.

The results of passive hemagglutination assays of rabbit and rat antisera preparations are shown in Table 2.

TABLE 2

PASSIVE HEMAGGLUTINATION ASSAY OF VARIOUS ANTISERA PREPARATIONS, USING BaP-BSA OR BSA-COATED RED BLOOD CELLS

Antisera Source	Largest Dilution Producing a Positive Test	
	BaP-BSA	BSA
Rabbit 1	1:256	1:2
Rabbit 2	1:64	1:2
Rabbit 3	1:8	1:2
Rabbit 4	1:256	1:16
Rat 1	1:64	Neg.
Rat 2	1:32	Neg.
Rat 3	1:32	Neg.

There is a clear difference in reactivity of the various antisera preparations against BaP-BSA and BSA antigens; the antisera consistently shown higher reactivity against BaP-BSA-coated red blood cells. This result suggests that at least some of the antibodies present in the antisera are reacting against a BaP antigenic structural determinate, and not simply against some part of the BSA protein structure, when BaP-BSA coated cells are used in the assay. Attempts to demonstrate this directly in the hemagglutination assays by using BaP-coated red blood cells were unsuccessful, since we were unable to coat red blood cells with BaP, and still maintain the integrity of the cells.

The rabbit antisera preparations were also tested using an ELISA assay, and the results are shown in Table 3.

TABLE 3

ELISA ASSAY OF ANTISERA PREPARATIONS, USING BaP-BSA OR BSA AS THE ANTIGEN

Antisera Source	Largest Dilution Producing a Positive Test	
	BaP-BSA	BSA
Rabbit 1	1:4 x 10^5	1:3 x 10^3
Rabbit 2	1:2 x 10^4	1:125
Rabbit 3	1:2 x 10^6	1:3 x 10^3
Rabbit 4	1:10 x 10^6	1:600

The ELISA results are in agreement with the hemagglutination assay results, in that both show that the antisera preparations have different levels of reactivity against BaP-BSA and BSA. The ELISA assay is clearly far more sensitive than the passive hemagglutination assay in terms of ability to detect specific antibody. The reason for the apparent discrepancy in the comparative BaP-BSA antibody titer of the antisera from Rabbit 3 in the two assays is not apparent. The antisera from this animal showed a strong reactivity against BaP-BSA in the ELISA system, but was the weakest rabbit antisera in the hemagglutination assay. Attempts to demonstrate antisera recognition of BaP (not coupled to protein) directly in an ELISA assay were unsuccessful; BaP, due to insolubility in aqueous solution, could only be applied to the microwells in an organic solvent, but the organic solvent significantly affected the optical transparency of the wells.

The fourth assay applied to our antisera preparations was a radioimmunoassay, in which free BaP was used as the antigen. Thus in this assay system, positive results should unequivocally demonstrate the ability of the antisera to recognize BaP structural features, irrespective of whether BaP is part of some large proteinaceous structure. In this assay, a purified immunoglobulin fraction from Rabbit 1 antisera was used instead of whole antisera. The results of assays where differing amounts of rabbit immunoglobulin were tested for BaP binding using ^{3}H- or ^{14}C-labeled BaP are shown in Table 4.

TABLE 4

RADIOIMMUNOASSAY OF IMMUNOGLOBULIN FRACTION FROM AN IMMUNIZED RABBIT, USING BaP AS THE ANTIGEN

Amount of Immunoglobulin*(µg)	Amount of Antigen Bound, pmoles
100	0.74
200	2.02
300	2.64
400	3.49
900	6.05

*As determined by radial immunodiffusion assay.

The results shown above are consistent with the interpretation that the antibodies present in our antisera preparations are able to recognize and bind BaP. Increasing amounts of purified immunoglobulin fraction bound increasing amounts of the antigen. This behavior was not seen in control experiments when rabbit immunoglobulin at the same concentration was substituted in the assay system in place of the antigen-specific immunoglobulin fraction (i.e., no increase in BaP binding was observed in these assays, compared to tubes where no immunoglobulin was present).

Based upon the results from the radioimmunoassays, one can calculate the amount of BaP-specific antibody present in our antisera preparations. An average of 0.8×10^{-2} pmoles of BaP was bound per µmole of immunoglobulin. Thus 1 out of 10^{3} molecules in the immunoglobulin fraction of our antisera preparations binds BaP.

DISCUSSION

The production of antibodies which react against various PAH or PAH derivatives has been reported by a number of investigators (9,10,11,12,13,14,15). The studies of Curtis et al. (14), and Tompa et al. (15), are most directly comparable to our work. Tompa et al. (15), immunized rabbits with a BaP-horse serum albumin conjugate, and prepared a purified immunoglobulin fraction by $(NH_4)_2SO_4$ precipitation. They absorbed this preparation with horse serum albumin to remove antibodies to the horse serum albumin, and subsequently assayed for reactivity using the same passive hemagglutination procedure we employed. They found no reactivity against the horse serum albumin, but obtained titers of 1:512 against BaP-horse serum albumin-coated cells. We attempted to remove antibodies against BSA in our antisera preparations using the procedure described in Tompa et al. (15), but were unsuccessful, in that our antisera preparations showed the same titer toward BSA-coated red blood cells after absorption with BSA as before the absorption. The titer (1:512) reported by Tompa et al. (15), was somewhat higher than we had obtained, but this is probably to be expected, as they used immunoglobulin fractions of antisera, while we used the crude antisera. Curtis et al. (14), using the passive hemagglutination assay, report an antisera titer (1:256) from immunized rabbits which is the same we found for our most highly immune animals.

The differing levels of reactivity of our antisera preparations reflect, in general, the differing sensitivities of the different assay systems. Rabbits 3 and 4 produced the most reactive antisera toward BaP-BSA in the Ouchterlony and ELISA assays. This was also true of Rabbit 4 in the passive hemagglutination test, although Rabbit 3, as already noted, showed an anomalous lack of reactivity in this test. The immunized rats produced antisera which was much less reactive than the rabbit antisera, as judged from the results of the Ouchterlony and passive hemagglutination assays.

A highly significant result obtained in this present research is the demonstration that the antibody elicited with BaP-BSA could recognize and bind free BaP. Although this activity was assumed in the experiments of some other investigators (14,15), relatively few measurements of haptenic PAH binding to antisera or immunoglobulins derived from animals immunized with PAH-protein conjugates have been reported (12,16).

A major question remaining is in regard to the specificity of our antisera for the BaP structure, as compared to other PAH structures. This is obviously an important consideration in the final applicability of the antibody as a PAH analytical detector. We are just now beginning structural studies, using the radioimmunoassay. Previous reports in the literature provide us with some hope that the BaP antibody will show reasonable specificity for the BaP structure. Tompa et al. (15), found that their BaP antibody showed a much diminished reactivity against 7,12-dimethylbenz[a]anthracene as compared to BaP. Creech et al. (9,10,11), also demonstrated a reasonable degree of specificity of anti-PAH antibodies, although they cautioned that the specificity was not absolute (i.e., antisera prepared using one PAH-protein conjugate as an immunogen showed reactivity against other PAH-protein conjugates in which the PAH was different from the immunizing PAH). However, the PAH structural determinates which account for this cross-reactivity have not been investigated in a systematic fashion. The extent to which total fused ring size might be important in determining PAH antibody specificity (i.e., will antibodies produced against 5-ring PAH recognize 2- or 3-ring PAH?), has yet to be addressed. Furthermore, whether PAH-specific antibodies can also recognize and bind metabolites of these same PAH thus rendering such antibodies even more useful as analytical sensors of ingested dose of PAH, is still a question to be addressed.

Finally, monclonal antibody technology, with its potential for production of a mono-specific antibody against a single structural determinate, may be crucial in the successful development of a PAH-specific antibody of appropriate specificity and affinity. Our results from the radioimmunoassay suggest that the amount of BaP-specific antibody in polyclonal rabbit serum is a very small fraction of the total immunoglobulins. Application of the monoclonal antibody technique could entirely eliminate an otherwise forbiddingly difficult purification procedure.

ACKNOWLEDGEMENTS

We thank Miriam R. Griffin and Teresa B. Noe for assistance in preparation of the manuscript.

REFERENCES

1. Vo-Dinh, T., Griffin, G. D., Ambrose, K. R., Sepaniak, M., and Tromberg, B. J. (1985): Fiberoptics-based immunofluorescence spectroscopy for monitoring exposure to polynuclear aromatic compounds, 10th International Symposium on Polycyclic Aromatic Hydrocarbons, Battelle, Columbus, Ohio, October 21-24, 1985.
2. Creech, H. J. (1941): Isocyanates of 3,4-Benzpyrene and 10-methyl-1,2-Benzanthracene. J. Am. Chem. Soc. 63: 576-578.
3. Creech, H. J. and Jones, R. N. (1941): The conjugation of horse serum albumin with isocyanates of certain polynuclear aromatic hydrocarbons. J. Am. Chem. Soc. 63: 1661-1669.
4. Hirata, A. A. and Brandiss, M. W. (1968): Passive hemagglutination procedures for protein and polysaccharide antigens using erythrocytes stabilized by aldehydes. J. Immunol. 100: 641-464.
5. Monroe, D. (1984): Enzyme immunoassay. Anal. Chem. 56: 920-931.
6. Herbert, V., Lau, K-S., Gottlieb, C. W., and Bleicher, S. J. (1965): Coated charcoal immunoassay of insulin. J. Clin. Endocr. 25: 1375-1384.
7. Williams, C. A. and Chase, M. W., editors (1967): Methods in Immunology and Immunochemistry, Volume 1, Academic Press, New York and London, pp. 315-329.
8. Mancini, G., Carbonara, A. O., and Hereman, J. F. (1965): Immunochemical quantitation of antigens by single radial immunodiffusion. Immunochemistry 2: 235-254.
9. Creech, H. J., Oginsky, E. L., and Cheever, F. S. (1947): Immunological studies of hydrocarbon-protein conjugates. I. Precipitin reactions. Cancer Res. 7: 290-296.
10. Creech, H. J., Oginsky, E. L., and Allen, O. N. (1947): Immunological studies of hydrocarbon-protein conjugates. II. Quantitative results. Cancer Res. 7: 297-300.
11. Creech, H. J., Oginsky, E. L., and Tryon, M. (1947): Immunological studies of hydrocarbon-protein conjugates. III. Inhibition reactions. Cancer Res. 7: 301-304.
12. Moolten, F. L., Capparell, N. J., Boger, E., Mahathalang, P. (1978): Induction of antibodies against carcinogenic polycyclic aromatic hydrocarbons. Nature 272: 614-616.
13. Peck, R. M. and Peck, E. B. (1971): Inhibition of chemically induced neoplasia by immunization with an antihenic carcinogen-protein conjugate. Cancer Res. 31: 1550-1554.

14. Curtis, G. L., Ryan, W. L., and Stenback, F. (1978): Antibody stimulation of benzo(a)pyrene carcinogenesis. Cancer Letts. 4: 223-228.

15. Tompa, A., Curtis, G., Ryan, W., Kuszynski, C., and Langenbach, R. (1979): Benzo(a)pyrene antibody inhibition of benzo(a)pyrene-induced mutagenesis. Cancer Letts. 7: 163-169.

16. Moolten, F., Capparell, N., and Boger, E. (1978): Reduction of respiratory tract binding of benzo(a)pyrene in mice by immunzation. J. Natl. Cancer Inst. 61: 1347-1349.

EFFECT OF THE pH-VALUE OF DIESEL EXHAUST ON THE AMOUNT OF FILTER-COLLECTED NITRO-PAH

GERNOT GRIMMER, JÜRGEN JACOB, GERHARD DETTBARN, KLAUS-WERNER NAUJACK
Biochemical Institute for Environmental Carcinogens,
2070 Ahrensburg, Sieker Landstraße 19, Federal Republic of Germany.

INTRODUCTION

The objective of this investigation was to clarify the question whether nitro-arenes (NO_2-PAH) partially are formed during sampling of diesel exhaust by nitration of polycyclic aromatic compounds (PAC) already collected on the filter. Three different sources for the formation of NO_2-PAH may be taken into account: 1. the combustion chamber, 2. the tailpipe, and 3. the filter. As far as the combustion chamber and the tailpipe are concerned, Kittelson et al. (1) have demonstrated that most of the 1-nitropyrene is formed in the tailpipe. Pitts et al. (2), Gibson et al. (3), and Bradow et al. (4) reported on the formation of artificial nitro-PAH during sampling, the amount of which depends on the NO_2-concentration, on the temperature, as well as on the collecting time. More recently, Schuetzle (5) has investigated this problem using the dilution tunnel device and concluded that ´chemical conversion of PAH to nitro-PAH during dilution-tube sampling of particulates on Teflon filters and gases on XAD-2 resin is a minor problem (representing 10-20 %, on the average, of 1-nitropyrene found in extracts) at short sampling time (23 min), at low sampling temperatures (42 ^{o}C), and in diluted exhaust containing 3 ppm NO_2´.

Apart from dilution-tube sampling the collection of the total undiluted exhaust from passenger cars is common in Europe. In this case the exhaust gas is cooled to about 35 ^{o}C by a glass condenser before trapping the particles on a silicon-bound glass fiber filter. The concentration of NO_2 during the collection of the undiluted exhaust is much higher than in the dilution tube. As a consequence PAH previously precipitated on the filter might react with NO_2 forming additional nitro-PAH on the filter. We have tried to estimate this effect by preventing the nitration on the filter by admixing gaseous ammonia to the exhaust emitted from the tailpipe before it reaches the collection filter.

METHOD

Driving cycle: The car was driven on a chassis dynamometer following the ECE-reglement 15/04 which simulates the city traffic. This program lasts for 195 sec and the period is repeated four times (in total 780 min = 13 min). Figure 1 shows the scheme of the driving cycle according to the ECE reglement 15/04 (6).

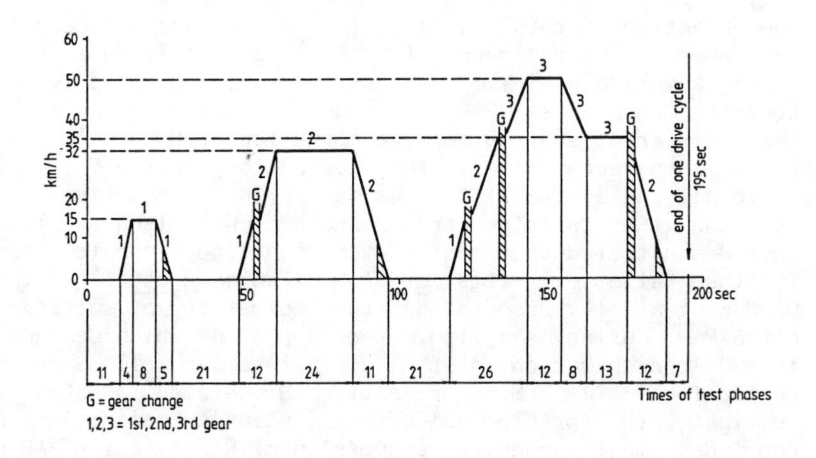

FIGURE 1. Scheme of the driving cycle according ECE reglement 15/04.

Determination of nitro-PAH: The scheme of enrichment of nitro-PAH is presented in Figure 2.

After an extraction of the silicon-bound glass fiber filter (area about 1 m^2) with toluene, the PAH are separated from the nitro-PAH by chromatography on silica gel (10 % water content) with cyclohexane and a mixture of cyclohexane-benzene as elutants. For further purification the nitro-PAH fraction is chromatographed on Sephadex LH 20. The resulting fraction is analyzed by gas chromatography using fused silica capillaries coated with SE 54 and a simultaneous detection by FID and a nitrogen-sensitive detector (NPD) as showed in Figure 3.

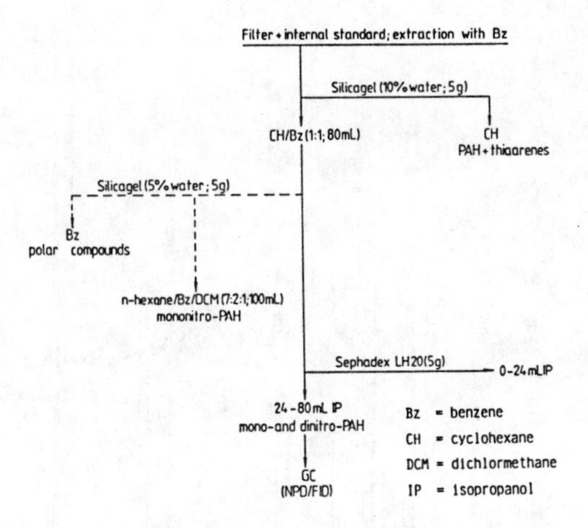

FIGURE 2. Scheme of enrichment of polycyclic aromatic compounds and nitro-PAH.

As an example, the evaluation of 1-nitropyrene by the comparison with the internal standard 6-nitrochrysene results in identical data with both detectors.

RESULTS AND DISCUSSION

To estimate the extent of chemical conversion of PAH to nitro-PAH during the sampling procedure of the particles from light-duty diesel vehicles on silicon-bound glass fibre filters, three series of experiments were carried out. In the first two series the undiluted exhaust was collected on filters directly connected with the tailpipe. In the last series the exhaust of various cars passed through a glass condenser before trapped on the filter.

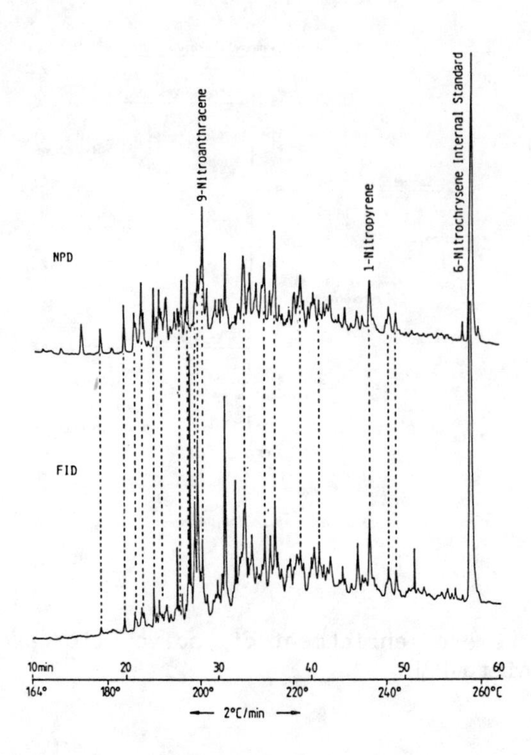

FIGURE 3. Nitro-PAH fraction of Diesel exhaust, simultaneously recorded by FID/NPD. Column: fused silica 25 m x 0.32 mm coated with SE 54, 0.25 um.

1. Collection of hot, undiluted exhaust with and without ammonia admixing

In this case the filter was directly connected with the tailpipe (Figure 4). A driving cycle of 195 sec was performed 4 times. At the end of each cycle the exhaust gas temperature increased continuously from 50 ^{o}C to 100 ^{o}C (50°, 75°, 90°, 100°). A complete ECE test took 13 min.

In total five ECE tests with and without ammonia admixing were driven. The resulting masses of 1-nitropyrene found on the filter as determined with FID as well as with NPD are given in Table 1.

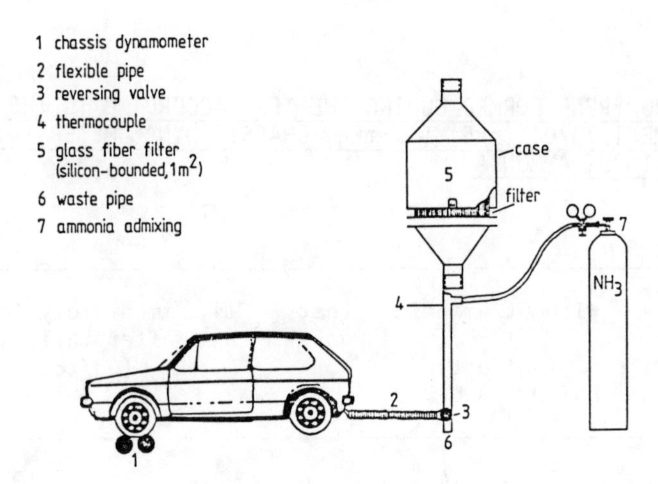

FIGURE 4. Collection of hot undiluted exhaust with and without ammonia admixing (maximum exhaust temperature 100 °C).

An average of 14.9 µg 1-nitropyrene/test (FID) was found on five filters collecting exhaust without neutralisation. The value of 15.2 µg obtained with the NPD lays within the margin of error. In contrast to the filters loaded with ammonia-admixed exhaust contained only about 50 % of this amount (7.1 µg/test). Although the deviation of data was markedly higher in this case, the difference to the above figures is significant. Assuming that admixing of ammonia to the acidic exhaust inhibits the nitration of the PAH on the filter, it may be concluded that about 50 % of the 1-nitropyrene is formed during the passage of the hot exhaust through the 90-100 °C hot filter.

2. Collection of undiluted exhaust particles on a filter subsequently exposed to a particle free, prefiltered hot exhaust.

The arrangement of two parallel filters directly behind the tailpipe is presented in Figure 5.

TABLE 1

1-NITROPYRENE FORMED DURING THE TEST ACCORDING TO THE ECE REGLEMENT 15/04 (n=5) (13 min, CHASSIS DYNAMOMETER) WITH AND WITHOUT AMMONIA

test no.	without ammonia (µg/test) FID	without ammonia (µg/test) NPD	test no.	NH_3 immediately injected after tailpipe (µg/test) FID	NH_3 immediately injected after tailpipe (µg/test) NPD
1	15.1	14.4	6	10.9	10.1
2	13.3	14.4	7	4.9	5.2
3	15.7	15.7	8	8.1	7.8
4	15.1	17.0	9	4.2	3.9
5	15.2	14.6	10	8.9	8.7
$\bar{x}$	14.9	15.2	$\bar{x}$	7.4	7.1
V	6.2 %	7.4 %	V	37.9 %	35.6 %

Note: Golf Diesel, NO_2 = 45 ± 3 ppm

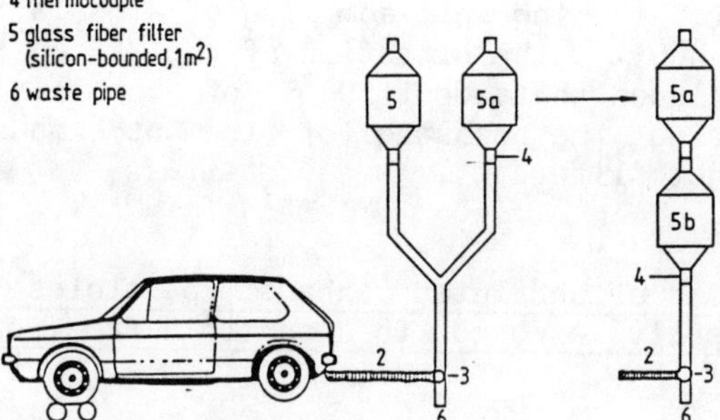

FIGURE 5. Collection of hot undiluted exhaust on parallel arranged filters (5 and 5a), subsequently one of the loaded filter was exposed a second time to the prefiltered exhaust gas (maximum exhaust gas temperature 100 °C).

As checked by a PAH-analysis both filters were equally loaded. One filter was analyzed for 1-nitropyrene. In a subsequent ECE test the second filter was placed behind a filter which retained the particle phase of the second ECE test. The gaseous phase of an ECE test then passed the second filter. The results of this experiment are presented in Table 2.

TABLE 2

1-NITROPYRENE FORMED DURING THE TEST ACCORDING ECE REGLEMENT 15/04. A LOADED FILTER HAS BEEN EXPOSED A SECOND TIME TO THE PREFILTERED PARTICLE-FREE HOT EXHAUST*

No.	4 ECE tests collected on 2 filters (μg/filter)	No.	loaded filter exposed to prefiltered exhaust (μg/filter)
1	4.9	1a	9.2
2	5.2	2a	7.8
3	5.4	3a	9.6
4	6.1	4a	11.2
$\bar{x}$	5.4	$\bar{x}$	9.5
V	9.4 %	V	14.8 %

*Maximum filter temperature 100 ^{o}C.
Note: One filter contains about 60 μg pyrene. Vehicle: Peugeot 205 GRD, NO_2 = 51 $\pm$ 3 ppm.

In total four ECE tests were driven. In the first test the filter is loaded with 4.9 μg 1-nitropyrene, in the second one with 5.2 μg etc. Since the exhaust passes two parallel filters it may be assumed that the second parallel filter is loaded with the same amount, <u>i.e.</u> that it contains also the average value of 5.4 µg nitropyrene. The particle-free exhaust of a second ECE test then passes through the second parallel and loaded filter. The amount of 1-nitropyrene on the so-treated filter has increased in all four ECE tests and was found to be 9.5 µg in average. From the results of this second series it may be also concluded that a part of the total pyrene on the filter (namely 60 µg) is nitrated by the 100 °C hot exhaust.

The question, which amount of 1-nitropyrene is formed on the filter by an exhaust cooled below 40 °C can be answered by the following series of experiments.

3. Collection of undiluted exhaust at 35 °C, with and without ammonia admixing

The common device for the collection of exhaust is given in Figure 6. The exhaust is cooled to 35 °C by passing a glass cooler.

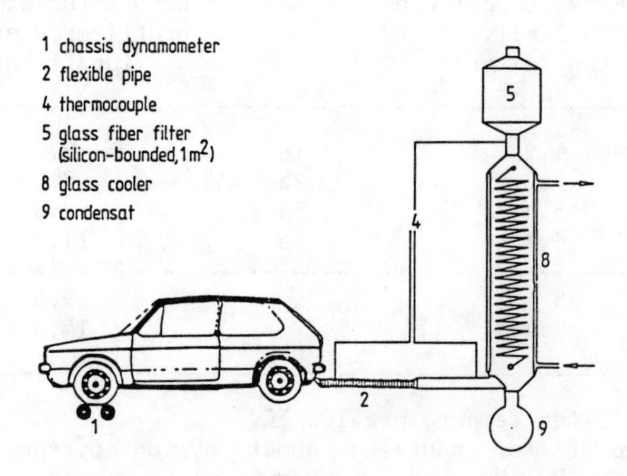

FIGURE 6. Collecting system for undiluted exhaust emitted during the test according to ECE reglement 15/04 (maximum exhaust gas temperature 100 °C).

Three different vehicles were investigated with this sampling system. They were driven three ECE tests each on a chassis dynamometer with and without ammonia admixing to the exhaust. The results of these experiments are presented in Figure 7.

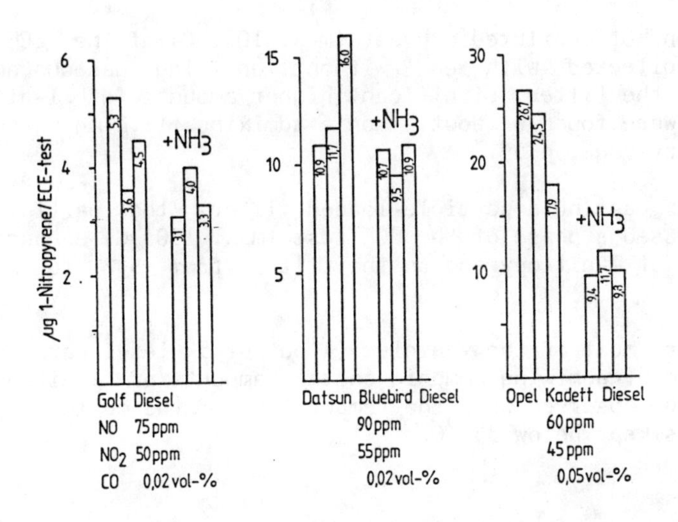

FIGURE 7. 1-Nitropyrene formed during the test according ECE reglemt 15/04 with and without ammonia admixing (maximum filter temperature 35 °C).

No influence of admixing ammonia on the amount of 1-nitropyrene found on the collecting filter could be observed in case of the Golf Diesel and the Datsun Diesel where identical values within the margin of error were found. In contrast to that were the results obtained with the third vehicle (Opel Kadett Diesel) in which admixing of ammonia to the exhaust resulted in significant decreases of the very high initial amounts of 1-nitropyrene.

In the meantime, we have investigated two further Diesel vehicles and these results confirm the data obtained with the Golf and Datsun Diesel: There is no difference of the 1-nitropyrene content on the collecting filter with or without ammonia admixing. The results of the Opel Kadett therefore seem to be an exception.

FORMATION OF NITRO-PAH ON COLLECTING FILTERS

SUMMARY

In order to estimate the extent of chemical conversion of PAH to nitro-PAH during the sampling procedure of particles from light-duty Diesel vehicles on silicon-bound glass fibre filters, three series of experiments were carried out:

1. When hot undiluted exhaust (max. 100 ^{o}C) of the ECE test was collected with and without admixing gaseous ammonia before the filter, significant higher amounts of 1-nitropyrene were found without ammonia admixing (15.2 µg versus 7.1 µg in average, n=5).

2. When passing a particle-loaded filter, the particle-free hot gaseous phase of the ECE test (max. 100 ^{o}C) enhances the amount of 1-nitropyrene on the filter from 5.2 to 9.5 µg (n=4).

3. For most of the vehicles (4 out of 5 Diesel cars) no influence of admixing ammonia on the amount of 1-nitropyrene could be observed when the temperature of the collecting filter was kept below 35 ^{o}C.

ACKNOWLEDGEMENTS

The authors would like to thank Dr.-Ing. W. Klank for having carried out the ECE tests. The present studies were carried out in accordance with the environmental plan of the Federal Environment Agency by order of the Federal Ministry of the Interior of the Federal Republic of Germany.

REFERENCES

1. Kittelson, D.B., Du, C.J., Bradow, R.L., Black, F., and Zweidinger, R.B. (1984): Society of Automotive Engineers Paper 840364.
2. Pitts Jr., J.N., van Cauwenberghe, K.A., Grosjean, D., Schmid, J.P., Fitz, D.R., Belser Jr., W.L., Knudson, G.B., and Hynds, P.M. (1978): Science 202:515-519.

3. Gibson, T.L., Ricci, A.I., and Williams, R.L. (1981): Measurements of PAH and their reactivity in diesel automobile exhaust. In: Polynuclear Aromatic Hydrocarbons, edited by M. Cooke and A.J. Dennis, pp. 707-717, Battelle Press, Columbus, Ohio.
4. Bradow, R.L., Zweidinger, R.B., Black, F.M., and Dietzmann, H.M. (1982): SAE Transactions 820182, pp. 13-21.
5. Schuetzle, D. (1983): Environ. Health Perspective 47:65-80.
6. Grimmer, G., Hildebrandt, A., and Böhnke, H. (1979): Europa test sampling procedure for vehicle exhaust gas. In: Environmental Carcinogens Selected Methods of Analysis, Vol. 3, edited by H. Egan, M. Castegnaro, P. Bogovski, H. Kunte, E.A. Walker, W. Davis, pp. 151-154, IARC Publ. No. 29.

REGIOSPECIFIC SILICON-MEDIATED ROUTE TO 7-METHOXY-1-INDANOLS AS MODELS FOR THE SYNTHESIS OF FLUORENE AND NITRO-FLUORENE PAH

W.R. HAHN, K. SHANKARAN, M.P. SIBI, V. SNIECKUS*
Guelph-Waterloo Centre for Graduate Work in Chemistry, University of Waterloo, Department of Chemistry, Waterloo ON N2L 3G1 Canada.

INTRODUCTION

Among environmental pollutants, polycyclic aromatic hydrocarbons (PAH) are demonstrably a widely dispersed class arising from incomplete combustion of fossil fuels and other organic matter (1). The strong inference that PAH are cancer causative agents in man (2) has led to extensive research on their formation, occurrence, detection, and metabolism (3). More recent work has been concerned with the corresponding aza-PAH (1b, 4, 5), methyl-PAH, a class being increasingly found co-occurring in PAH samples (2a, 6), and nitro-PAH, whose distribution in diesel exhaust (7), among other sources, has initiated thorough biological activity studies (7, 8).

Carcinogenic activity of PAH is expressed by prior metabolic activation via arene oxides and ultimately dihydrodiol epoxides which covalently modify DNA (2b, 9, 10). According to the Bay Region Theory of carcinogenesis, cancer growth is initiated via dihydrodiol epoxides of bay regions of PAH (9, 10, 11), although this theory cannot explain the effect of methyl substitution on activity (12).

Fundamental and practical studies in all areas of PAH research **(Figure 1)** require high analytical purity PAH stan-

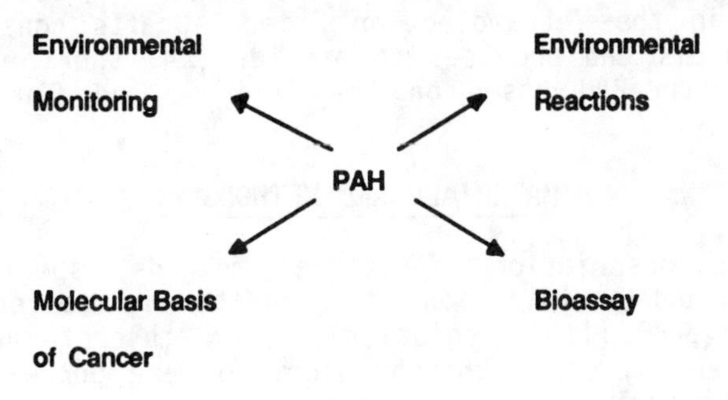

FIGURE 1.

dards for a) environmental monitoring of their fate and distribution; b) knowledge of their environmental reactions (e.g. with ozone, NO_x); c) understanding of their metabolic conversion to ultimate carcinogens and hence insight into the molecular basis of cancer; and d) bioassay of known and new PAH to establish mutagenicity/carcinogenicity and to warn of potential future environmental problems.

We have developed a new general methodology for the preparation of a variety of PAH and aza-PAH of environmental interest based on the aromatic directed metalation strategy (13). The original work (14) led to a one pot synthesis of a number of PAH-quinones and aza-PAH-quinones and this general approach was subsequently significantly improved in yield and ease of manipulation as a result of investigations concerned with the mechanism of the reaction (15, 16). Since the PAH-quinone to corresponding PAH conversion is well-documented (17), our overall route represents an efficient regimen which competes favorably with known classical procedures (18) for the synthesis of certain PAH as demonstrated in our laboratories (14).

In more recent work, we have connected the directed ortho metalation tactic (13) with the use of silicon functionality to protect reactive aromatic C-H and C-methyl sites and thereby devised new, unambiguous routes to peri-methyl-PAH (19, 20), peri-methyl-aza-PAH (19), and halo-PAH (21, 22, 23).

Herein we describe results which link the aromatic metalation methodology with an intramolecular ipso carbodesilylation process for the regiospecific synthesis of 1-indanols. This reaction constitutes a prototype of a new methodology which overrides the normal regioselectivity observed in the intramolecular Friedel-Crafts construction of 1-indanols and promises to provide (24) short routes to PAH and nitro-PAH based on the fluorene and fluoranthene nuclei.

MATERIALS AND METHODS

For a description of general methods, spectroscopic criteria, and purification of reagents and solvents, see (21). BH_3.THF (1.0 M solution), vinyl chloroformate, 18-crown-6, and triethyl phosphonoacetate were purchased from Aldrich Chemical Co.

N,N-Diethyl 3-methoxy-2-trimethylsilylbenzamide (4).

A solution of N,N-diethyl 3-methoxybenzamide (25) (15.0 g, 72.4 mmol) in anhydrous THF (25 mL) was added dropwise to a stirred solution of s-BuLi (69.8 mL, 89.6 mmol of a 1.14 M solution in cyclohexane) and TMEDA (9.25 g, 79.6 mmol) in THF (400 mL) at -78°C using syring-septum cap techniques. The mixture was stirred for 1 h, treated with TMSCl (39.3 g, 362 mmol) and allowed to warm to room temperature over an 8 h period. Standard work up followed by column chromatography (1:1 EtOAc-hexane eluent) and recrystallization (Et_2O-hexane) gave 18.0 g (88%) of product 4, mp 54-55°C; IR ($CHCl_3$) ν(max) 1630 cm^{-1}; NMR ($CDCl_3$) δ 0.25 (s, 9H), 1.06 (t, 3H, J = 7.0 Hz) 1.25 (t, 3H, J = 7.3 Hz), 3.14 (br s, 1H), 3.24 (br s, 1H), 3.41 (br s, 1H), 3.63 (br s, 1H), 3.80 (s, 3H), 6.76 (dd, 1H, J_{meta} = 1.0 Hz, J_{ortho} = 7.3 Hz), 6.82 (dd, 1H, J_{meta} = 1.0 Hz, J_{ortho} = 8.0 Hz), 7.3 (dd, 1H, J_{ortho} = 8.0 Hz, J_{ortho} = 7.3 Hz); Ms m/e (relative intensity) 279 (M^+, 3), 265 (61), 264 (100, 207 (28).

Anal. Calcd for $C_{15}H_{25}NO_2Si$: C, 64.47; H, 9.02; N, 5.01.
Found: C, 64.55; H, 9.31; N, 5.03.

N,N-Diethyl 3-Methoxy-2-trimethylsilylbenzylamine.

To a stirred solution of benzamide **4** (10.4 g, 37.2 mmol) in anhydrous THF (50 mL) was added BH_3•THF complex (185 mL of a 1 M solution) and the mixture was refluxed for 24 h. The solution was cooled, treated with water (200 mL), and the whole was evaporated to dryness in vacuo. The residue was extracted several times with CH_2Cl_2 and the combined extract was dried (Na_2SO_4) and evaporated to dryness to yield the crystalline amine BH_3 complex (IR ($CHCl_3$) ν(max) 2368 cm^{-1}) which was taken up in CH_2Cl_2 (40 mL), TMEDA (9.0 g, 75 mmol) was added, the solution was stirred at RT for 1 h, washed with water (3 x 10 mL) and saturated NH_4Cl (2 x 10 mL), and the organic phase was dried (Na_2SO_4) and evaporated yielding a two phase, solid-liquid system. To this was added hexane (100 mL) and the mixture was cooled on ice for 20 min. The white crystalline solid was removed by repeated filtration through a Celite bed to yield a clear colorless oil, which upon distillation gave 8.25 g (83.5%) of pure amine, bp 88-90°C (0.04 mm); IR(film) ν(max) 1588, 1565 cm^{-1}; NMR ($CDCl_3$) δ 0.42 (s, 9H), 1.06 (t, 6H), 2.58 (q, 4H, J = 7.4 Hz), 3.70 (s, 3H), 6.73 (d, 2H, J = 6.1 Hz), 6.76 (d, 1H, J = 5.6 Hz), 7.28 (t, 1H, J = 3.4 Hz); MS m/e (rel intensity) 265 (M^+, 2), 121 (100).

Anal. Calcd for $C_{15}H_{27}NOSi$: C, 67.87; H, 10.25; N, 5.28.
Found: C, 67.30; H, 10.11; N, 5.01.

3-Methoxy-2-trimethylsilylbenzyl chloride (5).

To a solution of the benzylamine (2.0 g, 7.5 mmol) obtained above in anhydrous 1,2-dichloroethane (15 mL) was added vinyl chloroformate (26) (109 mg, 10.2 mmol) and the mixture was refluxed for 6 h, cooled, and treated with water (15 mL). Dichloroethane was removed in vacuo and the remaining aqueous layer was extracted with CH_2Cl_2 (3 x 20 mL). The combined organic extract was dried (Na_2SO_4) and evaporated to dryness to give an oil which was chromatographed (hexane eluent) and distilled to furnish pure **5,** bp 86-88.5°C (0.25 mm); IR (film) ν(max) 1588, 1566, 1453 cm^{-1}; NMR ($CDCl_3$) δ 0.28 (s, 9H), 3.63 (s, 3H), 4.35 (s, 2H), 6.65 (m, 1H), 7.15 (m, 2H); MS m/e 230 (M^+ + 2, 2.8), 228 (M^+, 8.3), 119 (100).

Anal. Calcd for $C_{11}H_{17}ClOSi$: C, 57.75; H, 7.49; Cl, 15.49; Si, 12.23.
Found: C, 57.52; H, 7.40; Si, 12.09.

3-Methoxy-2-trimethylsilylbenzaldehyde (6).

A mixture of potassium phenylselenite (27) (3.7 g, 15.9 mmol), potassium hydrogen phosphate (2.2 g, 16.3 mmol), and 18-crown-6 (200 mg) was stirred in acetonitrile (10 mL) at 40°C until a homogeneous solution was achieved. To the warm solution was added the benzyl chloride **5** (3.13 g, 13.7 mmol) in acetonitrile (10 mL) and the mixture was refluxed for 6h. After quenching with water (25 mL) and cooling, the acetonitrile was removed in vacuo and the remaining aqueous solution was thoroughly extracted with CH_2Cl_2. The organic extract was washed successively with water and saturated NH_4Cl solution, dried, and evaporated. The residue was chromatographed (10% EtOAc-hexane) to give 2.2 g (91%) of **6,** mp 84-85°C (CH_2Cl_2-hexane); IR ($CHCl_3$) ν(max) 1723 cm^{-1}; NMR ($CDCl_3$) δ 0.42 (s, 9H), 3.92 (s, 3H), 6.65 (m, 1H), 7.15 (m, 2H), 11.85 (s, 1H); MS m/e 193 (M^+-15, 100).

Anal. Calcd for $C_{11}H_{16}O_2Si$: C, 63.42; H, 7.74; Si, 13.48.
Found: C, 63.54; H, 7.83; Si, 13.28.

Ethyl 3-(3-methoxy-2-trimethylsilylphenyl)propenoate.

To a suspension of sodium hydride (310 mg, 6.4 mmol, 50% suspension in mineral oil) freshly washed with Et_2O in

THF (60 mL) was added triethyl phosphonoacetate (1.43 g, 6.39 mmol) and the mixture was stirred for 0.5 h and dropwise treated with compound **6** (1.14 g, 5.48 mmol) in THF (40 mL). The reaction mixture was refluxed for 3 h, the heating bath was removed, and stirring was continued for 8 h. Water (20 mL) was added, the THF was removed by evaporation in vacuo, and the remaining aqueous solution was extracted with Et_2O. The organic extract was dried (Na_2SO_4) and evaporated to dryness to give an oil which was chromatographed (5% Et_2O-hexane eluent) to give 1.24 g (77%) of product, bp 121-123°C (0.6 mm); IR (film) ν(max) 1713 cm^{-1}; NMR ($CDCl_3$) δ 0.36 (s, 9H), 1.25 (t, 3H, J = 7 Hz), 3.53 (s, 3H), 4.01 (q, 2H, J = 7 Hz), 5.96 (d, 1H, J = 15.5 Hz), 6.56 (m, 1H), 7.15 (m, 2H), 7.93 (d, 1H, J = 15.5 Hz).

Anal. Calcd for $C_{15}H_{22}O_3Si$: C, 64.71; H, 7.96.
Found: C, 65.25; H, 8.23.

Ethyl 3-(3-methoxy-2-trimethylsilylphenyl)propionate (7, R = H).

A solution of the propenoate (480 mg, 2.14 mmol), prepared above, in absolute EtOH (50 mL) was hydrogenated in the presence of 10% Pd-C (95 mg) at 40 psi (Parr apparatus). Filtration through Celite followed by evaporation to dryness and distillation gave 465 mg (95%) of product, bp 100°C/0.04 mm; IR (film) ν(max) 1733 cm^{-1}; NMR ($CDCl_3$) δ 0.43 (s, 9H), 1.32 (t, 3H, J = 7.15 Hz), 2.52 (m, 2H), 3.01 (m, 2H), 3.73 (s, 3H), 4.12 (q, 2H, J = 7.15 Hz), 6.66 (d, 1H, J = 8.2 Hz), 6.77 (d, 1H, J = 7.6 Hz), 7.21 (t, 1H, 7.9 Hz); MS m/e (rel intensity) 265 (M^+-15, 100), 235 (24).

Anal. Calcd for $C_{15}H_{24}O_3Si$: C, 64.24; H, 8.63; Si, 10.02.
Found: C, 63.92; H, 8.41; Si, 9.67.

3-(3-Methoxy-2-trimethylsilylphenyl)propionaldehyde (8, R^1 = R^2 = H)

To a solution of the ester **7**, R = Me (290 mg, 1.03 mmol) in CH_2Cl_2 (15 mL) at -78°C under argon was added a solution of diisobutylaluminum hydride (0.98 mL, 1.52 mmol of a 1.52 M solution in toluene) and the mixture was allowed to stir for 1.5 h. Absolute MeOH (5 mL) was injected and the solution was allowed to warm to room temperature overnight, treated with saturated NaCl, and extracted with CH_2Cl_2. The organic extract was dried (Na_2SO_4) and evaporated to dryness to give an oil which upon preparative tlc (EtOAc:hexane = 30:70) gave 200 mg (81%) of product, IR (film) ν(max) 1724 cm^{-1}; NMR ($CDCl_3$) δ 0.3 (s, 9H), 2.5-3.2 (m, 4H), 3.75 (s, 3H), 6.6-6.85 (m, 2H), 7.1-7.35 (m, 1H), 9.8 (s, 1H), which was used directly in the next reaction.

7-Methoxy-1-Indanol(9, $R^1 = R^2 = H$).

To a suspension of anhydrous CsF (241 mg, 1.58 mmol) in anhydrous DMF (25 mL) was added aldehyde **8,** $R^1 = R^2 = H$ (250 mg, 1.05 mmol) and the resulting solution was stirred under nitrogen at 80°C for 24 h. After cooling, the solution was poured into water (50 mL) and the whole was extracted with methylene chloride (4 x 15 mL). The combined organic extract was dried (Na_2SO_4) and evaporated to dryness in vacuo. The residue was dissolved in ether (25 mL) and the solution was successively washed with water (2 x 10 mL) and saturated brine (1 x 10 mL), dried, and evaporated to give an oil which upon preparation tlc purification (36:3:1 hexane:CH_2Cl_2:ethanol eluent) and recrystallization (pet ether) gave 140 mg (76%) of product as colorless crystals, mp 47°C, lit (28) mp 47°C whose NMR spectrum was identical with that reported (28).

7-Methoxy-1-Indanone

The indanol **9,** $R^1 = R^2 = H$ was oxidized using pyridinium chlorochromate (29) to give 7-methoxy-1-indanone in quantitative yield, mp 103-104°C, lit (28) mp 104°C, NMR spectrum identical with that reported (28).

RESULTS AND DISCUSSION

Although the classical Friedel-Crafts reaction is a well-travelled route for annelation of 5-, 6-, and 7-membered ring ketones to aromatic substrates (30), its regiospecificity is dictated by the normal electrophilic substitution rules which preclude the preparation of certain substituted systems without resorting to protection divergencies and multi-step regimens. A case in point is the intramolecular Friedel-Crafts reaction of methoxy-substituted **1** **(Figure 2)** which invariably leads to para- **(2)** and not the desired ortho- **(3)** substituted product. As a result of our interest in connecting arylsilane derived methodology to the directed metalation reaction and the recent demonstration by Effenberger of fluoride-induced ipso carbodesilylation in reactions of arylsilanes with aromatic aldehydes (31), we sought to devise new routes to 7-methoxy-1-indanol **9,** $R^1 = R^2 = H$ **(Figure 3),** a synthon of demonstrated utility in a number of applications and of potential use for the preparation of fluorene and nitro-fluorene PAH.

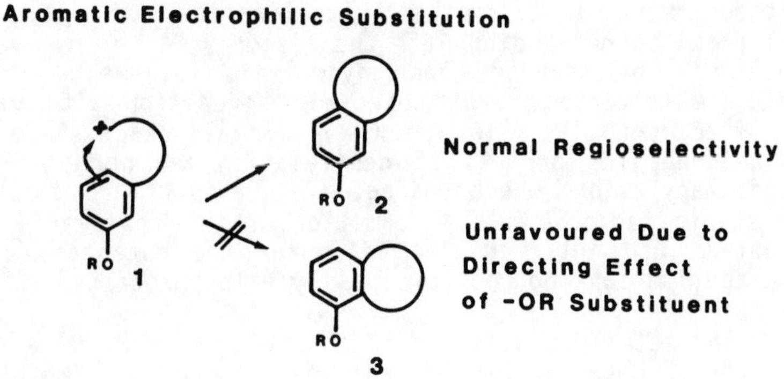
Aromatic Electrophilic Substitution
RO
1
RO
2
Normal Regioselectivity
Unfavoured Due to
Directing Effect
of -OR Substituent
RO
3

FIGURE 2 .

To derive the key intermediates **8**, $R^1 = R^2 = H$ **(Figure 3)**, the silylated meta-anisamide **4** was sequentially treated with diborane and vinyl chloroformate (26) to afford the unstable benzyl chloride **5** which, according to the excellent procedure of Syper and Mlockowski (27), was smoothly converted into the benzaldehyde **6**. Wadsworth-Emmons-Horner reaction followed by hydrogenation gave the phenyl propionate **7**, R = H, whose conversion into the corresponding aldehyde **8**, $R^1 = R^2 = H$ was achieved by standard diisobutylaluminum hydride (dibal) reduction. When the aldehyde **8**, $R^1 = R^2 = H$ was subjected to reaction with CsF in anhydrous DMF, the 1-indanol **9**, $R^1 = R^2 = H$ was obtained in 76% yield.

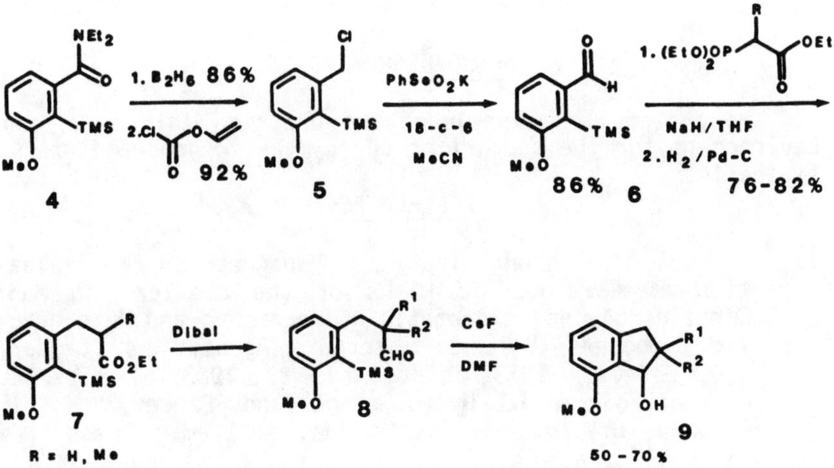
NEt2
TMS
MeO
4
1. B2H6 86%
2.Cl O
O 92%
Cl
TMS
MeO
5
PhSeO2K
18-c-6
MeCN
O
H
TMS
MeO
86% 6
R
1. (EtO)2OP OEt
O
NaH/THF
2. H2/Pd-C
76-82%
R
CO2Et
TMS
MeO 7
R = H, Me
Dibal
R1
R2
CHO
TMS
MeO 8
CsF
DMF
R1
R2
MeO OH
9
50 - 70 %

FIGURE 3.

To test the possibility of an intramolecular electrophile-induced ipso desilylation, a well-established method without sufficient synthetic application (32), the acid chloride corresponding to the ester **7,** R = H, readily obtained by standard base hydrolysis followed by oxalyl chloride treatment, was subjected to reaction with various Lewis acids ($AlCl_3$, $TiCl_3$, $SnCl_4$). In all cases, 5-methoxy-1-indanone, the product of desilylation and normal Friedel Crafts acylation, was obtained. This result, presumably due to steric factors in the formation of the Lewis-acid coordinated ipso intermediate, mitigates the potential use of the acylium ion-induced carbodesilylation process.

The methyl substituted aldehyde, **8,** R^1 = H, R^2 = Me was similarly obtained while the corresponding dimethylated derivative, **8,** R^1 = R^2 = Me was prepared by a different method (33). These aldehydes were converted under similar CsF/DMF conditions into the respective indanols **9,** R^1 = H, R^2 = Me and **9,** R^1 = R^2 = Me in 50-70% yields (33). Since the highest yields of indanols are obtained on **8,** R^1 = R^2 = H, the most likely system to undergo fluoride-catalyzed aldolization (34), the observed yields appear not to reflect such a process.

In summary, the fluoride-induced intramolecular carbodesilylation reaction, **8 → 9** is a useful procedure for the efficient access to 7-methoxy-1-indanols. The application of this method for the construction of fluorene- and nitrofluorene PAH is currently under investigation in our laboratories.

ACKNOWLEDGEMENT

We gratefully acknowledge the Ontario Ministry of the Environment for the support of our synthetic programs on PAH synthesis.

REFERENCES

1. a) World Health Organization. **Monograph on the Evaluation of Carcinogenic Risks of the Chemical to Man: Certain Polycyclic Aromatic Hydrocarbons and Heterocyclic Compounds,** Vol. 3, Internat. Agency Res. Cancer, WHO, Geneva, **1973;** ibid., Vol. 32, **1983;** b) Baum, E. J. in: **Polycyclic Hydrocarbons and Cancer,** Vol. 1, Gelboin, H.V., Ts'o, P.O.P. Eds., Academic Press, New York, **1978,** p. 45.
2. a) Dipple, A. in: **Chemical Carcinogens,** Searle, C.E. Ed., ACS Monograph 173, American Chem. Soc. Washington,

D.C., **1976,** p. 45; b) Harvey, R.G. **Amer. Scientist, 1982,** 70, 386.
3. Dipple, A.; Moschel, R.C.; Bigger, C.A.H. **Polynuclear Aromatic Hydrocarbons;** in: **Chemical Carcinogens,** Searle, C.E. Ed., 2nd Ed.; ACS Monograph 182; American Chemical Society: Washington, D.C. **1984,** p. 41. For reviews on chemical, physical, and analytical aspects, see Bjorseth, A., Ed. **Handbook of Polycyclic Aromatic Hydrocarbons,** Vols. 1 and 2, Dekker, New York, **1983, 1985.**
4. a) **Polynuclear Aromatic Hydrocarbons,** A Background Report Including Available Ontario Data, Ontario Ministry of the Environment, ARB-TDA-Report No. 58-79, September **1979;** b) Hoffman, D.; Wynder, E.L.; in: **Air Pollution,** Stern, A.C., Ed. Vol. 2, Academic Press, New York, **1977,** p. 361; c) Lee, M.L.; Novotny, M.; Bartle, K.D.: Gas chromatography/mass spectrometric and nuclear magnetic resonance spectrometric studies of carcinogenic polynuclear aromatic hydrocarbons in tobacco and marijuana smoke condensates, Anal. Chem. **1976,** 48, 405-416.
5. Blumer, M.; Dorsey, T.: Azaarenes in Recent Marine Sediments, **Science, 1977,** 195, 283-285; b) Hoffman, D.; Wynder, E.L. in: **Chemical Carcinogens,** Searle, C.E., Ed. American Chemical Society, Washington, D.C. **1976,** p. 324; c) Dipple, A. in: Ref. 5b), p. 245; Kosuge, T.; Zenda, H.; Nikaya, H.; Terada, A.; Okamoto, T.; Shudo, K.; Yamaguchi, K.; Iitka, Y.; Sugimura, T.; Nagao, M.; Wakabayashi, K.; Kosugi, A.; Saito, H. Isolation and structural determination of mutagenic substances in coal tar, **Chem. Pharm. Bull. Jpn. 1982,** 30, 1535-1538.
6. Hecht, S.S.; Bondinell, W.E.; Hoffmann, D.J. **Natl. Cancer Inst.** (US), **1974,** 53, 1121.
7. a) For recent reports, see **Proc. 10th Internat. Symp. on PAH, October, 1985, Battelle Press, Columbus OH, 1986,** pp. 18, 73-81; b) Quilliam, M.A.; Marr, J.; Gergeley, R.J. **Proc. Technology Transfer Conference No. 5,** Policy & Planning Branch, Ministry of the Environment, Ontario, Toronto, Canada, November, **1984,** Pt. 2, p. 903; c) Lewtas, J.; Nishioka, M.G.; Petersen, B.A. **ibid.** Pt. 1, p. 1.
8. For an excellent overview, see White, C.A., Ed. **Chromatographic Methods: Nitrated Polycyclic Aromatic Hydrocarbons,** Huthing Verlag, Heidelberg, **1985.**
9. Conney, A.H. **Cancer Res. 1982,** 42, 4875.
10. Harvey, R.G. **Acct. Chem. Res. 1981,** 14, 218.
11. Jerina, D.M.; Daly, J.W. in: **Drug Metabolism - from Microbe to Man,** Parke, D.V.; Smith, R.L. Eds.; Taylor

and Francis: London, **1977,** p. 13.

12. Hecht, S.S.; Melikian, A.A.; Amin, S. **Acct. Chem. Res. 1986,** 19, 174.
13. Snieckus, V. in: **Lectures in Heterocyclic Chemistry, J. Heterocyclic Chem. Suppl.** Castle, R.N. Ed., Heterocorp. Tampa, FL, **1984,** pp 95-106; Beak, P.; Snieckus, V. **Acct. Chem. Res. 1982,** 15, 306.
14. Watanabe, M.; Snieckus, V. **J. Am. Chem. Soc. 1980,** 102, 1457.
15. a) Doadt, E.G.; Iwao, M.; Reed, J.N.; Snieckus, V. in: **Polynuclear Aromatic Hydrocarbons,** Cooke, M.; Dennis, A.J., Eds. Battelle Press, Columbus OH, **1983,** p. 413.
16. Doadt, E.G.; Snieckus, V. unpublished results.
17. **Houben-Weyl, Methoden der Org. Chem.,** Vol. VII/3c, Georg Thieme Verlag, Stuttgart, **1979,** p. 282.
18. Clar, E. **Polycyclic Hydrocarbons,** Vol. 2, Academic Press, New York, **1964.**
19. Mills, R.J.; Snieckus, V. **J. Org. Chem. 1983,** 48, 1565-1568.
20. Mills, R.J.; Snieckus, V. in: **Polynuclear Aromatic Hydrocarbons,** Cooke, M.; Dennis, A.J., Eds. Battelle Press, Columbus OH, **1985,** p. 913.
21. Mills, R.J.; Snieckus, V. in: **Polycyclic Aromatic Hydrocarbons,** Cooke, M.; Dennis, A.J., Eds., Battelle Press, Columbus OH, **1986,** p. 579.
22. Mills, R.J., Ph.D. Thesis, University of Waterloo, **1985.**
23. Reed, J.N.; Snieckus, V. unpublished results. See also Beak, P.; Brown, R.A. **J. Org. Chem. 1982,** 47, 34.
24. Sharp, M.J.; Fu, J.-M.; Snieckus, V., work in progress.
25. Watanabe, M.; Snieckus, V. **J. Am. Chem. Soc., 1980,** 102, 1457.
26. Olofson, R.A.; Schnur, R.C.; Bunes, L.; Pepe, J.P. **Tetrahedron Lett. 1977,** 1563.
27. Syper, L.; Mlochowski, J. **Synthesis, 1984,** 747.
28. Tortai, J.-P.; Marechal, E. **Bull Soc. chim. Fr. 1971,** 2673; Loudon, J.D.; Razdan, R.K. **J. Chem. Soc. 1954,** 4299.
29. Corey, E.J.; Suggs, J.W. **Tetrahedron Lett. 1975,** 2647.
30. Olah, G.A. **Friedel-Crafts and Related Reactions,** Wiley--Interscience, New York, **1971,** p. 911.
31. Effenberger, F.; Spiegler, W. **Angew. Chem. Internat. Edit. Engl. 1981,** 20, 265.
32. Weber, W.P. **Silicon Reagents in Organic Synthesis,** Springer-Verlag, New York, **1983.**
33. Hahn, W.R.; Sibi, M.P.; Snieckus, V. manuscript in preparation.
34. Clark, J.H. **Chem. Rev. 1980,** 80, 429.

SYNTHESIS OF POTENTIALLY MUTAGENIC AND TUMORIGENIC CYCLOPENTA-FUSED POLYCYCLIC HYDROCARBONS

RONALD G. HARVEY, PASQUALE DI RADDO
Ben May Laboratory for Cancer Research, University of Chicago, Chicago, Illinois 60637, USA.

INTRODUCTION

Cyclopenta-fused polycyclic aromatic hydrocarbons (PAH) are a class of nonalternant PAH which contain a fused 5-membered ring with an unusual strained double bond which is more reactive than an aromatic bond or a normal olefinic bond. Examples of PAH of this class include aceanthrylene (1), acephenanthrylene (2), cyclopenta[c,d]pyrene (3), and cyclopenta[h,i]chrysene (4). Interest in the cyclopenta-PAH has recently been rekindled by reports of the activity of some members of this class as mutagens and carcinogens (1-3) and their occurrence as environmental pollutants (4-6).

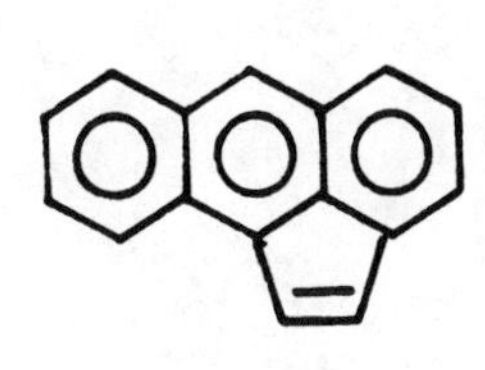

ACEANTHRYLENE

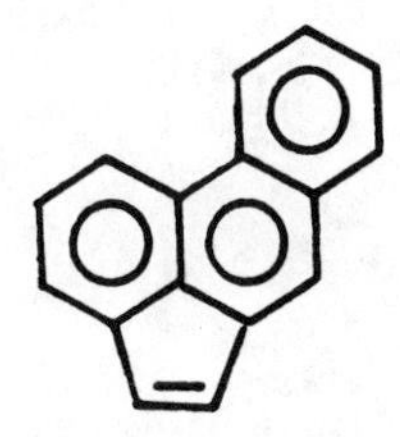

ACEPHENANTHRYLENE

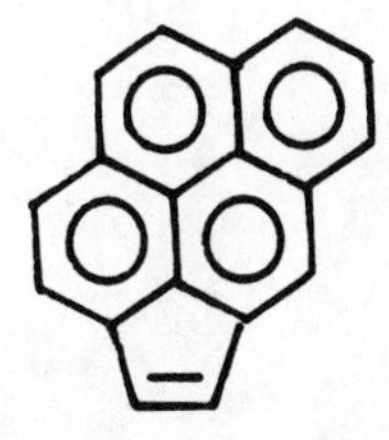

CYCLOPENTA [c,d] PYRENE

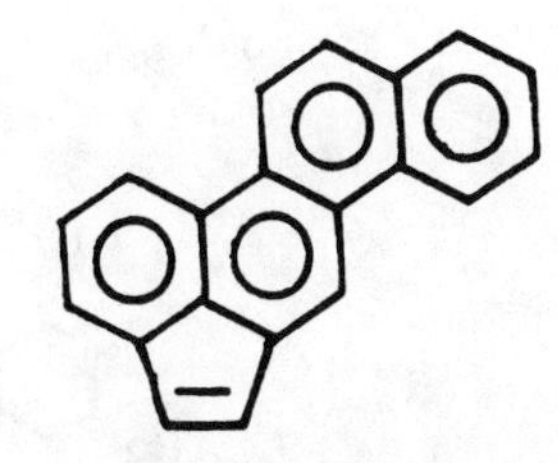

CYCLOPENTA [h,i] CHRYSENE

It is now generally accepted that the mechanism of carcinogenesis of alternant PAH, such as benzo[a]pyrene, involves metabolic activation to reactive diol epoxide intermediates which bind covalently to DNA (7). The mechanisms of action of nonalternant PAH are much less well understood. Studies on the metabolism of 3 by Gold (8) indicate that oxidative metabolism occurs predominantly in the electron rich ethylenic bridge via epoxidation. An arene oxide in this molecular region is the most likely active intermediate capable of alkylation of DNA. Molecular orbital calculations on arene oxides of this type show that the carbonium ions formed by ring opening possess large delocalization energies ($\Delta E_{deloc}/\beta$) (9). Therefore, these intermediates are predicted to be highly reactive electrophiles, and the bridge arene oxides are the most likely ultimate mutagenic and carcinogenic metabolites.

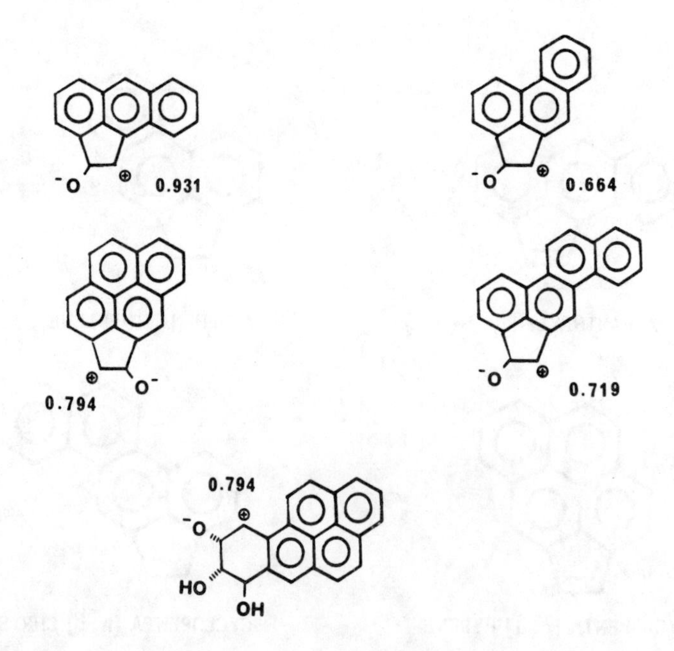

FIGURE 1. Calculated values of the delocalization energies ($\Delta E_{deloc}/\beta$) of the carbonium ions formed by opening of the epoxide ring of cyclopenta-PAH epoxides. The most stable of the two possible isomeric carbonium ions is shown.

Research on the cyclopenta-PAH has been restricted by their relative synthetic inaccessibility, except through complex multistep synthesis. For this reason, we have undertaken to develop more convenient synthetic approaches to molecules of this class. We now wish to report efficient syntheses of several cyclopenta-PAH directly from the parent alternant PAH in relatively few steps.

MATERIALS AND METHODS

Proton NMR spectra were recorded on a Varian EM360 spectrometer (60MHz) or on the University of Chicago 500 MHz NMR spectrometer. High resolution mass spectra were (electron impact mode) were taken on a VG 1021 instrument. Ultraviolet spectra were measured on a Perkin-Elmer Lambda 5 UV/VIS spectrometer in ethanol. The PAH starting materials were purified prior to use by chromatography on Florisil with 1% ethyl acetate in hexane as the eluent.

RESULTS

The first of the three general synthetic approaches which were investigated is illustrated for aceanthrylene (Figure 2). This method involves direct fusion of a 5-membered ring onto a PAH ring system by Friedel-Crafts reaction with oxalyl chloride and aluminum chloride. Reaction of anthracene with oxalyl chloride is reported in the literature (10) to afford aceanthraquinone in moderate yield (~50%); in our hands, slightly better yields (~70%) were obtainable. Reduction of this quinone with $NaBH_4$ in ethanol with oxygen bubbling through the solution afforded stereo-

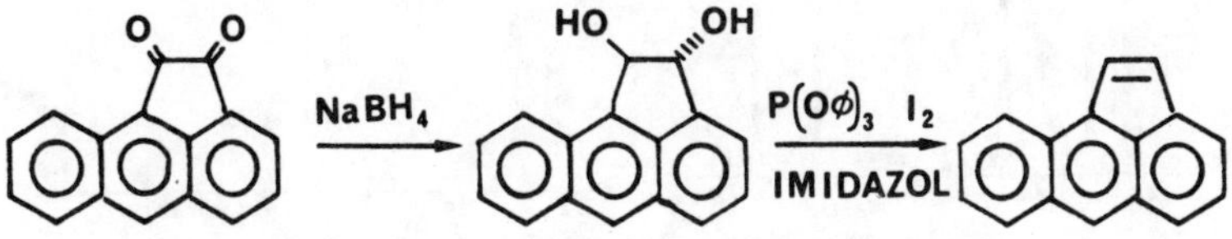

O O HO OH $NaBH_4$ $P(O\phi)_3$ I_2 IMIDAZOL

FIGURE 2. Synthesis of aceanthrylene

specifically the corresponding *trans*-dihydrodiol. The *trans* assignment is supported by NMR spectral analysis and by chemical evidence. It is also consistent with previous findings for the reduction of other polycyclic quinones with this reagent (11-13). The NMR spectrum of the dihydrodiol exhibited a small coupling constant ($J_{1,2} \leq 2$ Hz) characteristic of *trans*-dihydrodiols of this type (14). This dihydrodiol also failed to afford an acetonide derivative on treatment with $CuSO_4$ in refluxing acetone, conditions under which *cis*-dihydrodiols but not *trans*-dihydrodiols are readily converted to acetonides. The observed stereospecificity of this reduction contrasts with a recent report by Becker (15) that reduction of the same quinone with $NaBH_4$ without oxygen bubbling through the solution was less stereoselective, affording a substantial amount of the *cis* isomer as coproduct.

Conversion of the dihydrodiol to aceanthrylene was achieved in a single step on treatment with triphenyl

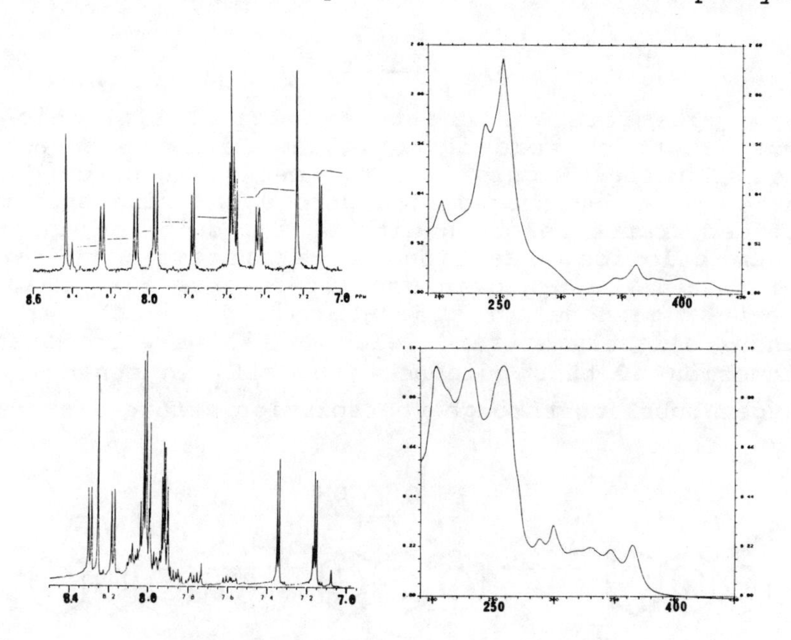

FIGURE 3. 500 MHZ NMR specta and UV spectra of aceanthrylene (top) and acephenanthrylene (bottom).

phosphite (4 Eq.), imidazole (4 Eq.), and iodine (3 Eq.) in benzene at 0°C. This method is a modification of a procedure developed earlier by Garegg and Samuelson for the deoxygenation of carbohydrate dihydrodiols (16). After purification, aceanthrylene was obtained in ~30% yield. The 500 MHz NMR spectrum and the UV spectrum (Figure 3) were in good agreement with this structural assignment. This three step sequence constitutes the shortest route yet for the synthesis of aceanthrylene (10,15,16). Another significant feature of this method from the viewpoint of metabolism studies is that the *trans*-dihydrodiol is also obtained and it can presumably be converted directly to the corresponding epoxide by established methods (17,18).

Similar reactions were conducted with acenaphthylene quinone to provide acenaphthylene in good overall yield (Figure 4). This method was also modified to prepare 1,2-dimethylacenaphthylene by reaction of the quinone with methyllithium to furnish the dimethyldiol which underwent deoxygenation with the same reagent. This would appear to be a good route to the dimethyl derivatives of cyclopenta-PAH. We plan to investigate its utility for this purpose in subsequent studies.

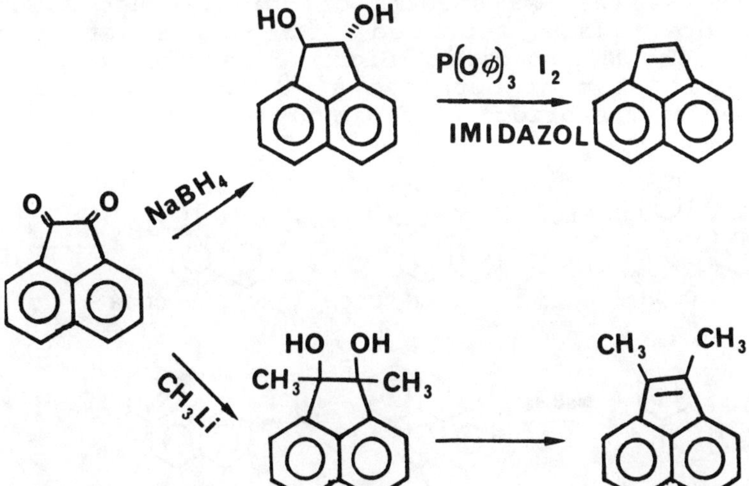

FIGURE 4. Synthesis of acenaphthylene and 1,2-dimethylacenaphthylene from acenaphthylene 1,2-dione

Extension of this synthetic approach to other cyclopenta-PAH has not proven successful. Thus, reaction of pyrene with oxalyl chloride under similar conditions gave two products, 1-pyrene carboxylic acid and a 2:1 adduct formed by reaction of one molecule of oxalyl chloride with two molecules of pyrene. Various modifications of this procedure were also investigated with equal lack of success. It appears that cyclization of the initially formed adduct is unfavorable relative to decomposition to the acid or reaction with a second molecule of pyrene.

In view of these difficulties, we investigated alternative methods which might potentially be more general in their applicability. One such method is outlined in Figure 5 for the synthesis of acephenanthrylene. This method takes advantage of the fact that acetylated derivatives of PAH are readily available by direct acetylation. 9-Acetylphenanthrene, which is commercially available, undergoes rearrangement with thallic (III) nitrate (19) smoothly to give phenanthrylacetic acid methyl ester. Cyclization of the acid in liquid HF furnished the cyclic ketone. Reduction of the latter with sodium borohydride followed by acid catalyzed dehydration afforded acephenanthrylene. The UV and NMR spectra of this product matched those previously reported for acephenanthrylene (5,20). The NMR spectrum (Figure 3) exhibited a characteristic pair of doublets at ∂ 7.17 and 7.35 ppm assigned to the bridge olefinic protons.

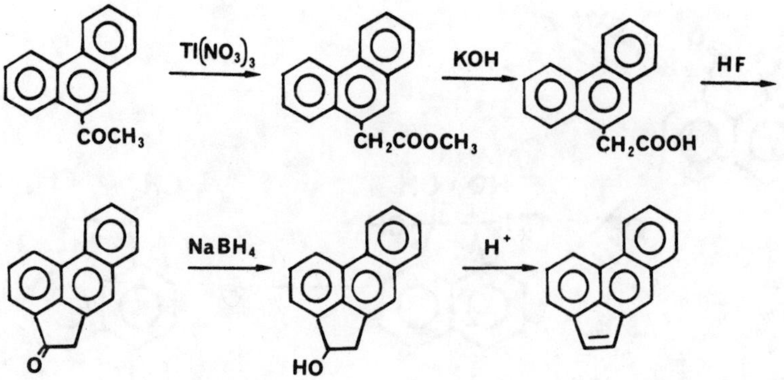

$COCH_3$ —$Tl(NO_3)_3$→ CH_2COOCH_3 —KOH→ CH_2COOH —HF→

O —$NaBH_4$→ HO —H^+→

FIGURE 5. Synthesis of acephenanthrylene.

A similar synthetic route was employed to prepare the pyrene analog, cyclopenta[c,d]pyrene (Figure 6). It was necessary to start with the acetyl group in the 4-position, since it was previously shown (21-23) that ring closure to form a 5-membered ring takes place readily to the 1-position, but not to the 4-position of pyrene. Acetylation of hexahydropyrene gave the 4-acetyl derivative which underwent rearrangement with thallium nitrate in methanol followed by dehydrogenation with DDQ to give 4-pyrenylacetic acid methyl ester in excellent yield. Basic hydrolysis gave the free acid which was cyclized in HF, then reduced, and dehydrated as previously described (21) to furnish cyclopenta[c,d]pyrene. The high resolution 500 MHz NMR spectrum showed the bridge protons as an AB quartet centered at ∂ 7.15 and 7.36 ppm. While this synthesis is similar to the reported methods (21-23), the use of the thallium-catalyzed rearrangement to prepare 4-pyrenylacetic acid represents a significant improvement.

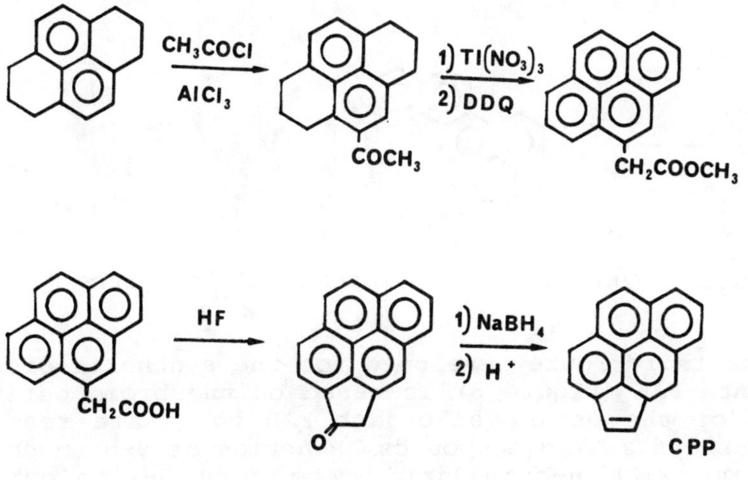

FIGURE 6. Synthesis of cyclopenta[c,d]pyrene

Synthesis of cyclopenta[h,i]chrysene has also been accomplished via an analogous method (Figure 7). 6-Acetylchrysene is readily available through direct acetylation of chrysene. The thallium-catalyzed rearrangement to the corresponding acetic acid derivative took place smoothly and in good yield. The main problem encountered was in the cyclization. Liquid HF, the acid employed in the previous examples, afforded only a very low yield (~5%) of the ketone. Polyphosphoric acid and methanesulfonic acid were even less satisfactory, providing no detectable ketone product. This was unexpected in view of the close structural similarity between the chrysene and phenanthrene examples. However, cyclization was achieved satisfactorily through conversion of the acid to its acid chloride by reaction with thionyl chloride followed by reaction with $AlCl_3$ or $SnCl_4$. Reduction and dehydration of the ketone gave the desired cyclopenta-PAH.

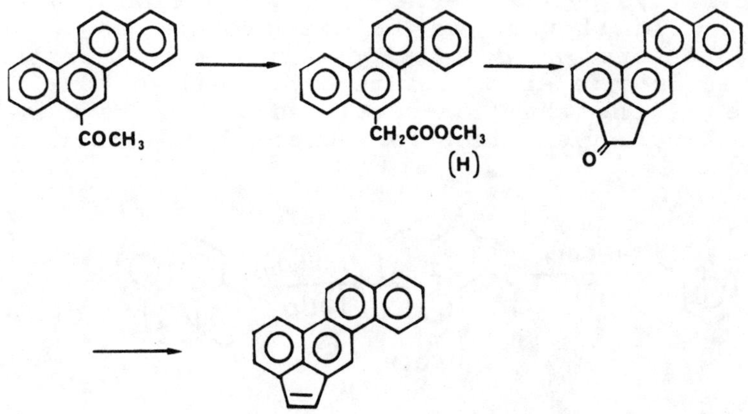

FIGURE 7. Synthesis of cyclopenta[h,i]chrysene

The third route developed for the synthesis of cyclopenta-PAH (Figure 8) is based on the bromo derivatives of the parent alternant PAH which are readily available starting compounds. Reaction of 9-bromophenanthrene with *n*-butyllithium to form 9-lithiophenanthrene followed by alkylation with ethylene oxide gave the ethyl alcohol derivative in good yield. Oxidation of the alcohol with pyridinium chlorochromate (PCC) furnished the corresponding aldehyde. Attempts to cyclize this aldehyde directly to acephenanthryene in polyphosphoric acid or other acids were unsuccessful. This is unfortunate, since it would have constituted a short and elegant synthesis of molecules of this type. Further oxidation with silver oxide furnished the acid whose physical properties were identical with those of the acid obtained via the thallium-catalyzed rearrangement route (Figure 5). This acid can be converted to acephenanthrylene by the sequence depicted in Figure 5.

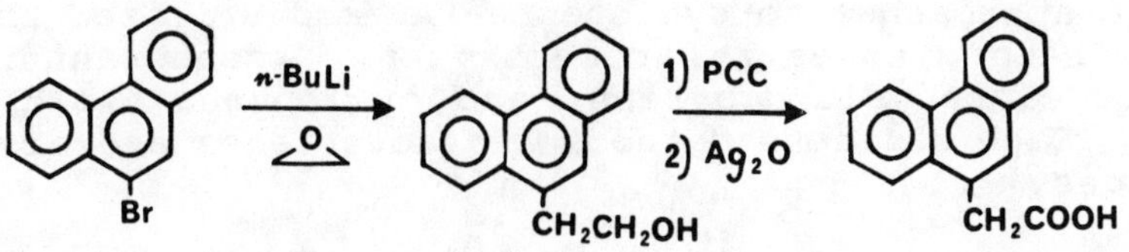

n-BuLi
1) PCC
2) Ag_2O
Br
CH_2CH_2OH
CH_2COOH

FIGURE 8. Alternative synthetic route to arylacetic acids.

Cyclopenta[c,d]pyrene can also be prepared from hexahydropyrene by a modification of the same procedure (Figure 9). However, this method involves more steps and does not offer any significant advantage in this case.

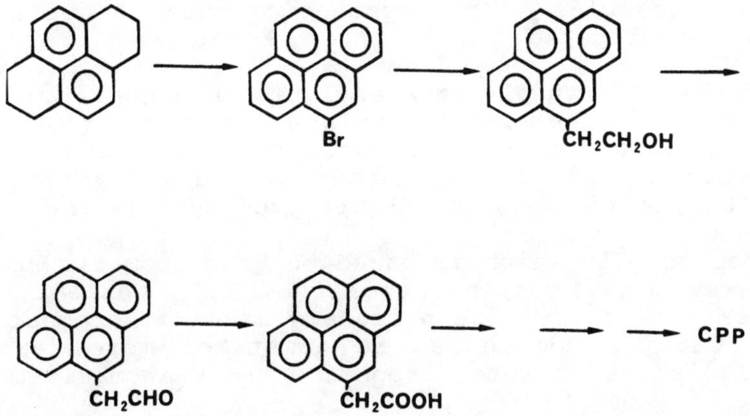

Br
CH_2CH_2OH
CH_2CHO
CH_2COOH
CPP

FIGURE 9. Synthesis of cyclopenta[c,d]pyrene

DISCUSSION

In summary, we have investigated a number of synthetic approaches to cyclopenta-PAH and utilized these methods to synthesize aceanthrylene, acephenanthrylene, cyclopenta[c,d]pyrene, and cyclopenta[h,i]chrysene. Each of these methods has advantages and disadvantages.

Method I involving reduction of the bridge quinones to the corresponding dihydrodiols followed by deoxygenation affords good yields in each of these steps and does not involve acid-catalyzed cyclization, as do the other methods. Since cyclopenta-PAH are quite sensitive to acid-catalyzed polymerization and decomposition, this is a significant advantage. The only problem with the quinone method is in obtaining the starting quinones. Their direct synthesis via reaction of the parent PAH with oxalyl chloride appears to be limited to a few hydrocarbons, such as anthracene and benz[a]anthracene. However, quinones of this type are potentially accessible by oxidation of the cyclic ketones obtained by cyclization of arylacetic acids. This synthetic route is currently being investigated.

Method II, involving acetylation and thallium catalyzed rearrangement to arylacetic acid derivatives appears to be of broad general utility. Its disadvantages are the larger number of steps involved and the low yields in the acid-catalyzed cyclization.

Method III, which is based on bromoaromatic starting compounds, complements Method II. While it requires more steps, the necessary isomeric bromoaromatic starting compounds are sometimes more readily available than the corresponding acetylaromatic molecules.

It is hoped that by combination of these methods, we can make available for environmental and biological studies substantial quantities of a large number of cyclopenta-PAH

ACKNOWLEDGEMENTS

These investigations were supported by grants CA 36097 and 14599 from the National Cancer Institute, DHHS. We also wish to acknowledge the contribution of Dr. Jayanta Ray who conducted some of the early studies on the synthesis of aceanthrylene.

REFERENCES

1. Eisenstadt, E. and Gold, A. (1978): Cyclopenta-[c,d]pyrene: a highly mutagenic polycyclic aromatic hydrocarbon. Proc. Nat'l Acad. Sci.USA, 75: 1667-1669.
2. Raveh, D., Slaga, T.J., and Huberman, E. (1982): Cell-mediated mutagenesis and tumor-initiating activity of the ubiquitous polycyclic hydrocarbon, cyclopenta[c,d]pyrene. Carcinogenesis, 3:763-766.
3. Nesnow, S., Leavitt, S., Easterling, R., Watts, R., Toney, S., Claxton, L., Sangaiah, R., Toney, G.E., Wiley, J., Fraher, P., and Gold, A. (1984): Mutagenicity of cyclopenta-fused isomers of benz-(a)anthracene in bacterial and rodent cells and identification of the major rat liver microsomal metabolites. Cancer Res., 44:4993-5003.
4. Wood, A.W., Levin, W., Chang, R.L., Huang, M-T., Ryan, D.E., Thomas, P.E., Lehr, R.E., Kumar, S., Koreeda, M., Akagi, H., Ittah, Y., Dansette, P., Yagi, H., Jerina, D.M., and Conney, A.H. (1980): Mutagenicity and tumor-initiating activity of cyclopenta[c,d]pyrene and structurally related compounds. Cancer Res., 40:642-649.
5. Kirshnan, F. and Hites, R.A. (1981): Identification of acephenanthrylene in combustion effluents. Anal.Chem., 53:342-343.
6. World Health Organization (1984): Evaluation of the Carcinogenic Risks of Chemicals to Humans: Polynuclear Aromatic Compounds, Vol. 32, IARC, Lyons, France.
7. Harvey, R. (1985): Polycyclic Hydrocarbons and Carcinogenesis, ACS Monograph No.283, American Chemical Society, Washington, D.C.
8. Gold, A. and Eisenstadt, E. (1980): Metabolic activation of cyclopenta[c,d]pyrene to 3,4-epoxycyclopenta[c,d]pyrene by rat liver microsomes. Cancer Res., 40:3940-3944.

9. Fu, P.P., Beland, F.A., and Yang, S.K. (1980): Cyclopenta-polycyclic aromatic hydrocarbons: Potential carcinogens and mutagens. Carcinogenesis, 1:725-727.
10. Plummer, B.F., Al-Saigh, Z.Y., and Arfan, M. (1984): Synthesis of aceanthrylene. J. Org. Chem., 49:2069-2071.
11. Harvey, R. (1985): Synthesis of dihydrodiol and diol epoxide metabolites of carcinogenic polycyclic hydrocarbons. In: Polycyclic Hydrocarbons and Carcinogenesis, ACS Monograph No. 283, American Chemical Society, Washington, D.C.
12. Jacobs, S., Cortez, C., and Harvey, R. (1983): Synthesis of potential proximate and ultimate carcinogenic metabolites of 3-methylcholanthrene. Carcinogenesis, 4:519-522.
13. Platt, K.L. and Oesch, F. (1983): Efficient synthesis of non-K-region *trans*-dihydro diols of polycyclic aromatic hydrocarbons from *o*-quinones and catechols. J. Org. Chem., 48:265-268.
14. Kinstle, T.H., Ihrig, P.J. (1970): Acenaphthylene oxide. J.Org. Chem.,35:257-258.
15. Becker, H-D., Hansen, L., and Andersson, K. (1985): Aceanthrylene. J. Org. Chem., 50:277-279.
16. Sangaiah, R. and Gold, A. (1985): A synthesis of aceanthrylene. Org. Prep. Proced. Internat., 17: 53-56.
17. Harvey, R.G., Goh, S.H., and Cortez, C. (1975): "K-Region" oxides and related oxidized metabolites of carcinogenic aromatic hydrocarbons. J. Amer. Chem. Soc., 97:3468-3479.
18. Cortez, C. and Harvey, R.G. (1978): Arene oxide synthesis: Phenanthrene 9,10-oxide. Org. Syn., 58: 12-16.
19. McKillop, A., Swann, E.C., Taylor, J. (1973): Thallium in organic synthesis. XXXIII. A one-step synthesis of methyl arylacetates from acetophenones using thallium (III) nitrate (TTN). J.Amer. Chem. Soc., 95:3340-3343.
20. Laarhoven, W.H. and Cuppen, T.J. (1976): Photodehydrocyclisations of stilbene-like compounds XVI. Photoreactions of a-(9-phenanthryl)stilbene and 1-(9-phenanthryl)-1-phenylethylene. Rec. Trav. Chim., 95:165-168.
21. Konieczny, M. and Harvey, R.G. (1979): Synthesis of cyclopenta[c,d]pyrene. J. Org. Chem., 44:2158-2160.

22. Gold, A., Schultz, J., and Eisenstadt, E. (1978): Relative reactivities of pyrene ring positions: Cyclopenta[c,d]pyrene via an intramolecular Friedel-Crafts acylation. Tetahedron Lett., 4491-4498.
23. Ittah, Y. and Jerina, D.M. (1978): Cyclopenta-[c,d]pyrene. Tetrahedron Lett., 4495-4498.

BIOMONITORING OF INDIVIDUALS EXPOSED TO HIGH LEVELS OF PAH IN THE WORK ENVIRONMENT

AAGE HAUGEN*, GEORG BECHER[1], CHRISTEL BENESTAD[2], KIRSI VAHAKANGAS[3], GLEENWOOD E. TRIVERS[3], MARK J. NEWMAN[4], CURTIS C. HARRIS[3]
*Institute of Occupational Health, P.O. Box 8149 Dep, 0033 OSLO 1, Norway, (1) National Institute of Public Health, 0462 OSLO 4, Norway, (2) Center for Industrial Research, P.O. Box 350, 0314 OSLO 3, Norway, (3) Laboratory of Human Carcinogenesis, National Cancer Institute, Bethesda, MD 20205, USA, and (4) Department of Pathology, Uniformed Services University of the Health Sciences, Bethesda, MD 20814, USA.

INTRODUCTION

Polycyclic aromatic hydrocarbons (PAH) are widely distributed in our environment and are implicated in cancer of the lung (14). The primary aluminum and the coke production are industrial processes leading to a high level of PAH exposure (5,6). The concentration of PAH in the work atmospheres of these industries frequently exceeds 1000-10000 times the concentration in urban atmospheres. Chemical analyses of dust samples from coke plants or aluminum industry have revealed the presence of over 100 different PAH, some of which are known carcinogens (5,6).

It is known that PAH (e.g. BaP) must be metabolically activated to electrophilic intermediates which bind covalently to DNA in order to exert their mutagenic or carcinogenic effects. The PAH intermediates react preferentially with the exocyclic nitrogens of the bases. Many studies have implicated a specific diol epoxide derivative of BP, 7β,8α-dihydroxy-(9α,10α)-epoxy-7β,8α-dihydroxy-(9α,10α)-epoxy-7,8,9,10 - tetrahydrobenzo(a)pyrene (BPDE-I) as the major carcinogenic metabolite involved in binding to DNA (11). Once metabolized, PAHs are eliminated through the urine and faeces (12).

In order to get a better understanding of the occupational hazard connected with PAH exposure, various biomonitoring methods for the measurement of PAH-exposure have been employed, *i.e.* quantitative measurements of urinary PAH metabolites, measurement of BP-DNA adducts in peripheral blood lymphocytes and determination of antibodies to the BPDE-DNA adducts in sera from coke oven workers.

MATERIALS AND METHODS

Subjects

Topside coke oven workers (22 smokers, 8 ex-smokers and 8 non-smokers) were investigated; 80% of the workers wore full-face mask. Urinary PAH analyses were performed on 15 smokers and 9 non-smokers. DNA from the 38 subjects were isolated from peripheral blood lymphocytes and analysed for BPDE-DNA adducts. Exposure levels of PAH were determined by personal sampling both inside and outside the respirator.

Analysis of Airborne PAH

The analytical method for the personal samples is described in detail elsewhere (8). Briefly, PAH on the filter and the gaseous PAH adsorbed on the XAD-2 resins were extracted in a Soxhlet apparatus. The PAH were cleaned up by liquid-liquid extraction followed by capillary gas chromatographic analysis.

Analysis of Urine Samples

Determination of PAH in urine was based on reduction of PAH-metabolites to the parent PAH compounds followed by HPLC with fluorescence detection (3).

Synchronous Fluorescence Spectrophotometry for Detection of BPDE-DNA Adducts

DNA was prepared from lymphocytes as previously described (18). The absorbance ratio (260 nm/280 nm) was over 1.7. DNA was acid-hydrolyzed in 0.9 M HCl at 90^0 C for 3 hrs and assayed by a Perkin-Elmer fluorescence spectrophotometer 650-40 by scanning excitation and emission synchronously with a constant wavelength difference of 34 nm (18).

Ultrasensitive Enzyme Radioimmunoassay for Detection of BPDE-DNA Adducts

The USERIA was performed in 96-well polystyrene microtiter plates precoated with 0.2 ng BPDE-DNA (10). Test DNA was heated to 90^{0} C for 15 minutes to obtain single stranded DNA. BPDE-modified calf thymus DNA was used as the standard reference DNA, an antiserum specific for BPDE-DNA as antibody (15), alkaline phosphatase-conjugated goat anti-rabbit IgG (Fab_2) as the secondary antibody, a mixture of tritiated paranitrophenyl phosphate (^{3}H-PNPP) and PNPP as the substrate and Econofluor 2 as the scintillate.

Detection of Antibody to BPDE-DNA Adducts

BPDE-modified or unmodified DNA were attached to polyvinylchloride microtiter plates. Serum samples were incubated with the antigen containing plates. The binding of immunoglobins was detected using goat anti-human immunoglobin reagents and the avidin-biotin horseradish-peroxidase system (ABC Vectastain kit, Vector laboratories, Burlingame, CA). A competitive ELISA was used to test sera that contained antibodies against the BPDE-modified DNA (7).

RESULTS

In table 1 a typical distribution of particulate and gaseous PAH is given. In table 2 the mean values for particulate and gaseous PAH and POM (polycyclic organic matter) of 4 air samples collected outside and inside the respirators are given. The respirators were able to remove 49-76% of the PAH.

TABLE 1
A TYPICAL DISTRIBUTION OF PARTICULATE AND GASEOUS PAH OUTSIDE THE RESPIRATORY DEVICE

	PAH ($\mu g/m^3$)	
	Particulate	Gaseous
Acenaphthylene	-	28.3
Acenaphthene	-	3.8
Fluorene	-	14.0
2-Methylfluorene	-	2.5
1-Methylfluorene	-	1.8
Phenanthrene	3.0	46.0
Anthracene	1.3	15.0
3-Metylphenanthrene	0.4	3.2
2-Metylphenanthrene	0.7	4.3
2-Methylanthracene	0.1	2.6
4,5-Metylenephenanthrene	0.8	4.2
4- and/or 9-Methylphenanthrene	0.1	2.8
1-Methylphenanthrene	0.4	3.0
Fluoranthene	8.0	13.6
Benz(e)acenaphthylene	1.5	2.1
Pyrene	7.1	9.7
Ethylmethylenephenanthrene	1.3	2.0
Benzo(a)fluorene	2.3	1.4
Benzo(b)fluorene	2.3	2.0
4-Methylpyrene		
2-Methylpyrene and/or Methylfluoranthrene	1.2	0.5
1-Methylpyrene	0.7	0.6
Benzo(ghi)fluoranthrene	1.0	0.6
Benzo(c)phenanthrene	0.9	0.5
Cyclopenteno(cd)pyrene	1.5	0.4
Benz(a)anthracene	5.8	1.7
Chrysene and Triphenylene	7.0	1.6
Benzo(b)fluoranthene	-	-
Benzo(j)fluoranthene	4.3	-
Benzo(k)fluoranthene	-	-
Benzo(e)pyrene	4.7	-
Benzo(a)pyrene	7.3	-
Perylene	1.8	-
Indeno(1,2,3-cd)pyrene	4.5	-
Dibenz(a,c and/or a,h) anthracenes	1.2	-
Benzo(ghi)perylene	4.4	-
Anthanthrene	2.4	-
Coronene	3.2	-
Sum identified PAH	81	168

TABLE 2

CONCENTRATION OF PAH AND POM ($\mu G/M^3$) OUTSIDE AND INSIDE THE RESPIRATOR

	OUTSIDE		INSIDE	
	Part.	Gaseous	Part	Gaseous
PAH	90	176	34	75
POM	2	15	1	7
PAH + POM	92	191	35	82
Total PAH	266		109	
Total PAH + POM	283		118	

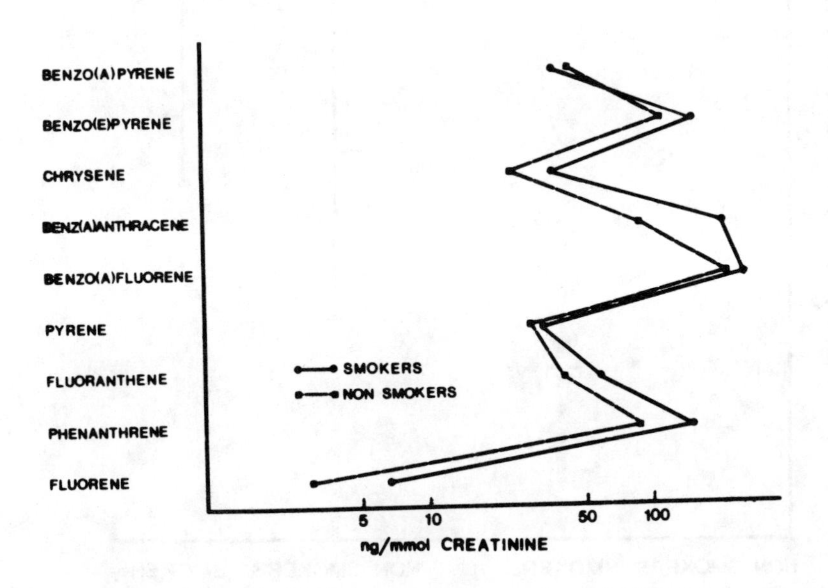

FIGURE 1. Mean values for concentrations of individual PAH in workers' urine.

The urinary PAH content in exposed smokers and non-smokers are shown in Figure 1-2. Figure 1 shows the level of individual PAH in the urine. The difference in PAH excretion between smokers and non-smokers is not statistically significant. The mean urinary excretion of PAH was relatively low; 0.57 µg/mmol creatinine for non-smokers and 0.87 µg/ mmol creatinine for smokers (Figure 2). However, in this study 80% of the coke oven workers wore full-face masks during work.

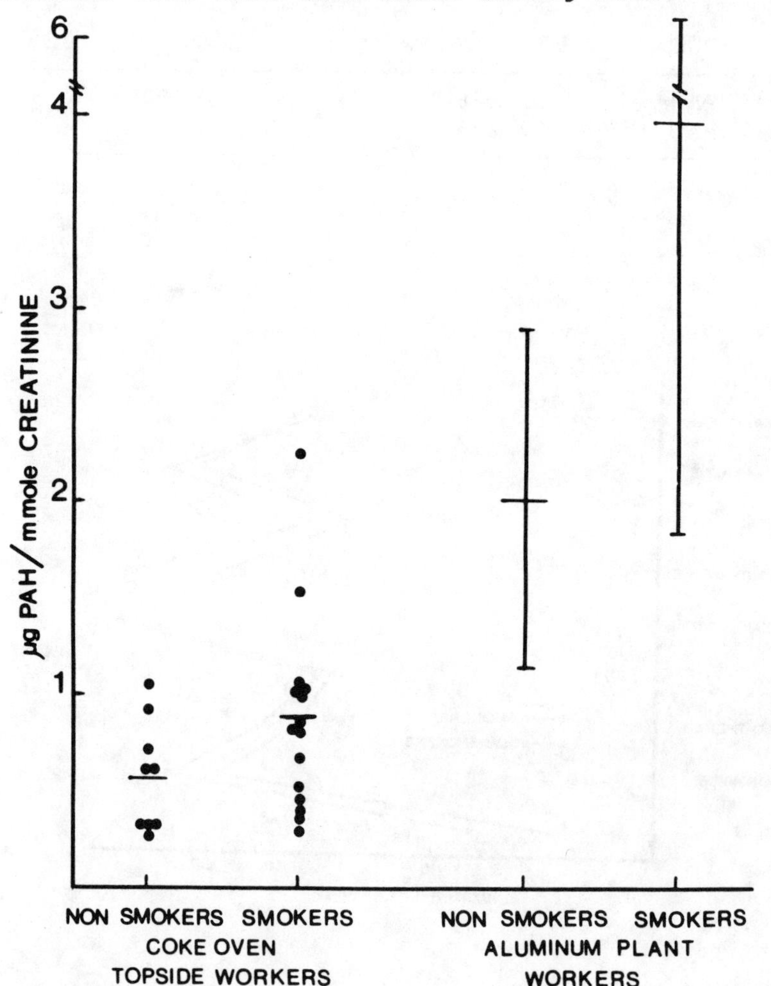

FIGURE 2. PAH levels of urine samples from exposed workers compared to results obtained for aluminum plant workers (3).

TABLE 3

DETECTION OF BPDE-DNA ADDUCTS IN LYMPHOCYTE DNA FROM COKE OVEN WORKERS USING BOTH COMPETITIVE USERIA AND SFS*

Sample	Smoking cig./day	USERIA fmole/μg DNA		SFS fmole/μg DNA	
		Work	Vacation**	Work	Vacation
1	Never	>13.7	0.15	2.2	ND
2	7	2.0	0.14	0.4	ND
3	3	1.8	NT	0.5	NT
4	Ex-s.	0.4	0.12	ND	ND
5	7	1.0	0.15	0.6	ND
6	Never	0.7	NT	ND	NT
7	Ex-s.	0.5	0.1	ND	ND
8	14	0.4	0.1	ND	ND
9	Ex-s.	0.6	NT	ND	NT
10	4	0.2	NT	ND	NT
11	Never	0.1	0.1	ND	ND
12	21	0.1	0.15	ND	ND
13	Never	0.4	NT	ND	NT

* ND, not detected; level of sensitivity: 0.2 fmole/μg DNA, SFS; 0.1 fmole/μg DNA, USERIA; NT, not tested.

**Samples taken 3 weeks after leaving workplace during vacation period.

BPDE-DNA adducts can be quantitatively detected in the peripheral lymphocyte DNA of exposed coke oven workers (table 3). Using SFS, 4 of 38 tested DNA samples isolated from peripheral lymphocytes were found to be positive, i.e. showed a peak at an emission wavelength of 379nm using a $\Delta\lambda$ of 34 nm. The adduct-level ranged from 2.2 fmole BPDE/μg DNA to 0.4 fmole BPDE/μg DNA. By USERIA, 13 out of 38 samples tested (34%) had detectable levels of BPDE-DNA antigenicity. The adduct level ranged from > 13.7 to 0.1 fmole BPDE/μg DNA. The 4 positive samples in SFS assay were also found to have the highest values in the USERIA. Thirty-seven per cent of the ex-smokers and 50% of the non-smokers were positive in the USERIA. The mean level of BPDE-DNA adducts were 0.41 ± 0.25 fmole/μg DNA for non-smokers, 0.49 ± 0.10 fmole/μg DNA for ex-smokers and 0.93 ± 0.75 fmole/μg DNA for smokers. After 3 weeks vacation 20 out of the 38 individuals were retested for BPDE-DNA adduct level (Table 3). By USERIA, the level of BPDE-DNA adducts for all but 2 of the individuals either remained the same or decreased. By SFS, all 20 samples were negative.

As shown in table 4, 12 of 38 (32%) serum samples contained antibodies to BPDE-DNA. Nine months later 13 of 38 (34%) serum samples were positive. During this period, no change was detected in 21 individuals, 8 lost titers and 9 gained titers.

TABLE 4

PRESENCE OF ANTIBODIES TO BPDE-DNA ADDUCTS IN SERA FROM COKE OVEN WORKERS

	Antibody Titer*	
No.	Sept. -83	May -84
Never smokers		
1	625	625
2	-	-
19	-	-
26	125	-
28	-	-
30	-	25
31	-	-
34	-	-
Ex-smokers		
4	125	-
5	125	-
15	-	-
10	25	25
13	-	-
16	-	-
18	25	-
22	125	125

Smokers		
3	-	-
6	-	25
7	-	-
8	-	125
9	125	125
11	-	-
12	-	125
14	-	-
17	-	125
20	125	-
21	-	-
23	-	125
24	625	-
25	-	-
27	25	-
29	125	-
32	-	25
33	-	-
35	-	125
36	-	-
37	-	125
38	-	-

*Antibody to BPDE-DNA was not detected in the sera when a number is not listed

DISCUSSION

In this study a battery of monitoring tests was used to determine individual exposure to PAH. Analysis of the samples from the work atmosphere demonstrated that the exposure to PAH is high. However, the masks reduced the average exposure 60%. The relatively low mean value of PAH excreted in urine from coke oven workers compared to aluminum workers could be explained by the fact that 80% of the coke oven workers wore full-face masks during work (3).

Many chemical carcinogens form covalent adducts at various sites in DNA. Persistent DNA adducts or adducts incorrectly repaired may give rise to mutations in the genome. Recently, quantitation of carcinogen-DNA adducts has become possible using specific antibodies as well as biophysical methods detecting specific alterations within the DNA at the fmole level (9,10,13,14,19). In the present study, 34% of the top side coke oven workers had detectable putative BPDE-DNA adducts in lymphocytes by USERIA and 10% of the DNA-samples had emission peaks at 379 nm. It is interesting that the level of BPDE-DNA adduct remained the same or decreased by 90% of the 20 workers retested after 3 weeks vacation period. By SFS, all 20 samples were negative.

Variation in exposure, the balance between metabolic activation and deactivation, and DNA repair capacity influence the adduct level. The bioavailability of the PAH compounds adsorbed to airborne particles is also an important factor. Although the exposure to PAH in an aluminum plant can be high the workers have only a slightly elevated cancer risk, normal number of SCE's in lymphocyte DNA and only 2% of the workers had detectable level of BPDE-DNA adducts (3,18). An explanation of these results could be that PAH in this industry is associated with strongly adsorbing particles and is therefore not readily bioavailable.

One-third of these workers also have antibodies against an epitope(s) on BPDE-DNA adducts in their sera. A previous study by Harris and coworkers among coke oven workers demonstrated that 28% of the workers had antibodies against this adduct (7). However, the persistence of the antibody titers and the structural component these antibodies react with needs to be investigated in more detail.

In conclusion, samples from individuals in such industries taken over a period of time and analysed with a battery of monitoring tests could provide means of assessing individuals exposure and give us a more comprehensive evaluation of genotoxic hazard in man.

ACKNOWLEDGEMENTS

We thank Dr. Erikstad for help with the sampling and Dr. Poirier for rabbit antibody to

BPDE-DNA. The secretarial aid of Anne Eide is appreciated.

REFERENCES

1. Andersen, Aa., Dahlberg, B.E., Magnus, K., and Wannag, A. (1982): Risk of cancer in the Norwegian aluminum industry. Int. J. Cancer, 29:295-298.
2. Becher, G., and Bjørseth, A. (1983): Determination of exposure to polycyclic aromatic hydrocarbons by analysis of human urine. Cancer Lett., 17: 301-311.
3. Becher, G., Haugen, A., and Bjørseth, A. (1983): Multimethod determination of occupational exposure to polycyclic aromatic hydrocarbons in an aluminum plant. Carcinogenesis, 5:647-651.
4. Bjørseth, A. (1977): Analysis of polycyclic aromatic hydrocarbons in particulate matter by glass capillary gas chromatography. Analyt. Chim. Acta, 94:21-27.
5. Bjørseth, A., Bjørseth, O., and Fjeldstad, P.E. (1978): Polycyclic aromatic hydrocarbons in the work atmosphere. I. Determination in an aluminum reduction plant. Scand. J. Work Environ. Health, 4:212-223.
6. Bjørseth, A., Bjørseth, O., and Fjeldstad, P.E. (1978): Polycyclic aromatic hydrocarbons in work atmospheres. II. Determination in a coke plant. Scand. J. Work Environ. Health, 4:224-236.
7. Harris, C.C., Vahakangas, K., Newman, M.J., Trivers, G.E., Shamsuddin, A., Sinopoli, N., Mann, D.L., and Wright, W.E. (1985): Detection of benzo(a)pyrene diol epoxide-DNA adducts in peripheral blood lymphocytes and antibodies to the adducts in sera from coke oven workers. Proc. Natl. Acad. Sci. USA, 82:6672-6676,1985.

8. Haugen, A., Becher, G. Benestad, C., Vahakangas, K.,Trivers, G.E.,Newman,M.J., and Harris, C.C.,(1986): Determination of polycyclic aromatic hydrocarbons (PAH) in the urine, benzo(a)pyrene diol epoxide-DNA adducts in lymphocyte DNA and antibodies to the adducts in sera from coke oven workers exposed to measured amounts of PAH in the work atmosphere. Cancer Res.,in press.
9. Haugen, A., Groopman, J.D., Hsu, I.C., Goodrich, G.R., Wogan, G.W., and Harris, C.C. (1981): Monoclonal antibody to aflatoxin B_1-modified DNA detected by enzyme immunoassay. Proc. Natl. Acad. Sci. USA, 78:4124-4127.
10. Hsu, I.-C., Poirier, M.C., Yuspa, S.H., Grunberger, D., Weinstein, I.B., Yolken, R.H., and Harris, C.C. (1981): Measurement of benzo(a)pyrene-DNA adducts by enzyme immunoassays and radioimmunoassay. Cancer Res., 41:1091-1095.
11. Jeffrey, A.M., Kinoshita, T., Santella, R.M., Grunberger, D., Katz, L., and Weinstein, I.B. (1980): The chemistry of polycyclic aromatic hydrocarbon-DNA adducts. Edited by B. Pullman, P.O. Tso and H.V. Gelboin. Carcinogenesis: Fundamental Mechanisms and Environmental Effects, pp. 565-579, D. Reidel Publ., Boston.
12. Kotin, P., Falk, H.L., and Busser, R. (1979): Distribution, retention and elimination of C^{14}-3,4-benzpyrene after administration to mice and rats. J. Natl. Cancer Inst., 23:541-555.
13. Müller, R., and Rajewsky, M.F. (1980): Immunological quantification by high-affinity antibodies of O^6-ethyldeoxyguanosine in DNA exposed to N-ethyl-N-nitrosourea. Cancer Res., 40:887896.
14. Perera, F. (1981): Carcinogenicity of airborne fine particulate benzo(a)pyrene: An appraisal of the evidence and the need for control Environ. Health Perspect., 42:163-185.
15. Poirier, M.C., Santella, R., Weinstein, I.B., Grunberger, D., and Yuspa, S.H. (1980): Quantification of benzo(a)pyrene-deoxyguanosine adducts by radioimmunoassay. Cancer Res., 40:412-416.

16. Reddy, M.V., Gupta, R.C., Randerath, E., and Randerath, K. (1984): ^{32}P-postlabelling test for covalent DNA binding of chemicals *in vivo*: application to a variety of aromatic carcinogens and methylating agents. *Carcinogenesis*, 5:231-243.
17. Shamsuddin, A.K.M., Sinopoli, N.T., Hemminki, K., Boesch, R., and Harris, C.C. (1985): Detection of benzo(a)pyrene: DNA adducts in human white blood cells. *Cancer Res.*, 45:66-68.
18. Vahakangas, K., Haugen, A., and Harris, C.C. (1985): An applied synchronous fluorescence spectrophotometric assay to study benzo(a)-pyrene-diolepoxide-DNA adducts. *Carcinogenesis*, 6:1109-1116.
19. Van der Laken, C.J., Hagenaars, A.M., Hermsen, G., Kriek, E., Kuipers, A.J., Nagel, J., Scherer, E., and Welling, M. (1982): Measurement of O^6-ethyldeoxyguanosine and N-(deoxyguanosin-8-yl)-N-acetyl-2-aminofluorene in DNA by high-sensitive enzyme immunoassays. *Carcinogenesis*, 3:569-572.

INHIBITORY EFFECTS OF PHENOLIC OR HYDROXY COMPOUNDS ON THE METABOLISM OF BENZO(A)PYRENE

Fu Z. HOU*, ZI P. BAO, and ZHI Y. WANG[1]
Institute of Environmental Science, Beijing Normal University and (1) Institute of Labor Protection, Ministry of Labor and Personnel, Beijing, CHINA

INTRODUCTION

Benzo(a)pyrene (BaP) is a carcinogenic polynucleararomatic hydrocarbon which has been thoroughly studied. Through the analysis of the enzymatic metabolites of BaP, S.K. Yang (1) proved 7,8-dihydrodiol-BaP is the proximate carcinogen which by further epoxidation forming 7,8-dihydrodiol-9,10-epoxide-BaP the electrophillic ultimate carcinogen binding to DNA or RNA. Eventually under carcinogenic promoting factors it results cancer.

Chemoprevention of carcinogenesis is recently developed research field. L.W.Wattenberg (2) has reported the antioxidants butylhydroxyanisole (BHA), butylhydroxytoluene (BHT), ethoxyquin (EQ) inhibited the carcinogenicity of benzo(a)pyrene on forestomach of mouse. They also inhibited mammary tumor formation produced by oral administration of dimethylbenzathracene (DMBA) to female rats. T.S. Slaga (3) has reported vitamin A, B_2, C, E, retinoids, flavonoids, and some steroids inhibit the effects of chemical carcinogens. Trace elements are not only essentials of life but also are important for effects on chemical carcinogenesis. Zinc and copper can inhibit the activity of monooxidase in cytochrome p-450 and the activity of epoxide hydrase, and reduce the formation of BaP metabolites (4).

In this paper we used the synthetic product p-ethoxy-phenol (PEP), 2,3,5-trimethylhydroquinone (TMHQ) the intermediate of vitamin E, cucurbitacin B (CU-B) from fruit of cucumis melo L, cucurbitaceae, paeonol (PL) from root of pycnostelma paniculatum (Bge) K. Schum and BHA (the last one was used for comparison) as inhibitors and observed their effects on the BaP metabolism by rat liver microsome <u>in vitro</u> incubation. The metabolites of BaP were analyzed by HPLC. The results proved that these five compounds could severely inhibit the formation of BaP metabolites in varying degrees.

MATERIALS AND METHODS

Chemicals

BaP Aldrich Chemical Co. 3-Methylcholanthracene (3-MC) Koch Light Laboratories Ltd. BaP metabolites were kindly supplied by National Cancer Institute U.S.A. CU-B a gift from Hunan Medicinal Industrial Research Institute. PEP was synthesized by ourself. TMHQ a gift from Beijing the Second Pharmaceutical Factory, recrystallized from water and benzene. PL a gift from Nanking Chinese Traditional Medical College Preparation Laboratories. BHA a gift from Dr. S.K. Yang (Uniformed service University of the Health Sciences, Bethesda, Md. U.S. A.). NADPH Sigma Chemical Co., Other chemicals are analytical pure grades.

Important Instruments

Super centrifuge VAC-602 Germany Demacratic Republic. HPLC model 244, from Water Associate U.S.A.

Preparation of Liver Microsomes

Male rats Beijing No.1 (weight 160-180 grams) after 40 hours of a single intraperitoneal injection of 5 mg of 3-MC (in 0.5 ml of corn oil) were sacrificed by decapitation. Before killing the diet was discontinued for 24 hours. 2 grams of liver lobe were cut into small chips with scissors and washed with ice cold 1.15% KC1 solution. It was then treated with 3 volumes of precold 0.25 M sucrose, 0.05 M Tris-HC1 buffer solution (PH 7.5) and was homogenized. The resulted liquid was filtered through cotton gauze. The filtrate was centrifuged at 1000 x g for 15 minutes. The upper layer of the liquid was again centrifuged at 105000 x g for 60 minutes. The temperature of centrifugal steps were kept at 4°C. The dark reddish brown microsome sediment was treated with precold 0.25 M sucrose and 0.05 M Tris-HC1 buffer solution (PH7.5). The resulting suspension was equivalent to 1 gram wet liver/ml. After the suspension was homogenized again, it was distributed into amples and kept in liquid nitrogen for use. The protein content of the liver microsome was determined according to the Lowry's method (5).

Vitro Incubation of BaP

Dr. S.K. Yang's method was adopted (6). To each 25 ml Erlenmeyer flask, 2 ml of 0.05 M Tris-HC1 diluted liver microsome, then BaP, PL, PEP, CU-B, TMHQ and BHA were added respectively. In each reaction liquid the final concentration of BaP was 25 uM, PL, PEP, CU-B, TMHQ and BHA were 25 uM, 50 uM, 100 uM respectively. The protein concentration of liver microsome in each reaction liquid was 0.15 mg/ml. In a mechanical shaker with a thermostat it was kept 2 minutes at 37°C. After 100 ul NADPH (final concentration in reaction liquid was 0.4 mM) was added, the reaction started and lasted for 30 minutes, and was terminated by adding 2 ml of cold acetone and followed by 4 ml of ethyl acetate. The reaction mixture was shaked for 10 minutes and was centrifuged at 3000 rpm for 5 minutes.

The organic solvent layer was separated and kept in a glass tube with well ground glass stopper. 1 gram of anhydrous Na_2SO_4 was added and dept at 4°C. It was filtered through thin filter with 0.5 micron pore diameter. The filtrate was evaporated to dryness under nitrogen with super pure quality. 200 ul acetone-methanol was added to redissolve the residue for HPLC Analysis. All manipulations were carried out under red or yellow light. The separation of BaP metabolites was performed with a Water Associates Model 244 HPLC Separation System equipped with a u-Bondapark C_{18} column (3.9x 300mm) and eluted with a curve NO.7 gradient of 60% methanol(Water) to 100% methanol in 70 minutes at room temperature, the solvent flow rate was 1.0 ml/min.

RESULTS

The effects of PL, PEP,CU-B and TMHQ on the metabolism of BaP were observed. The concentrations of these four compounds were 25 uM, 50 uM and 100 uM. The results were shown in the following figures 1,2,3,4.

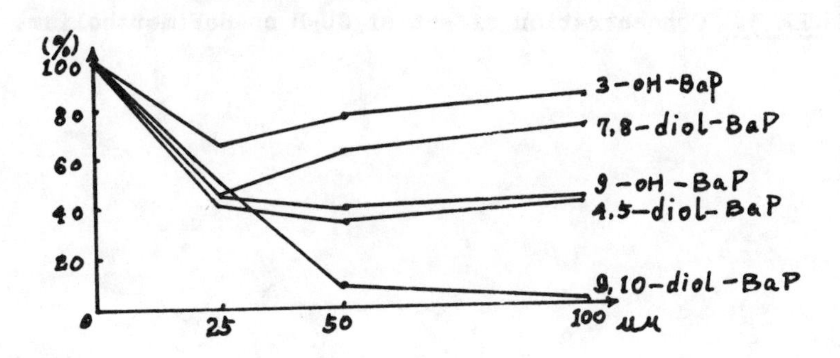

FIGURE 1. Concentration effect of PL on BaP metabolism.

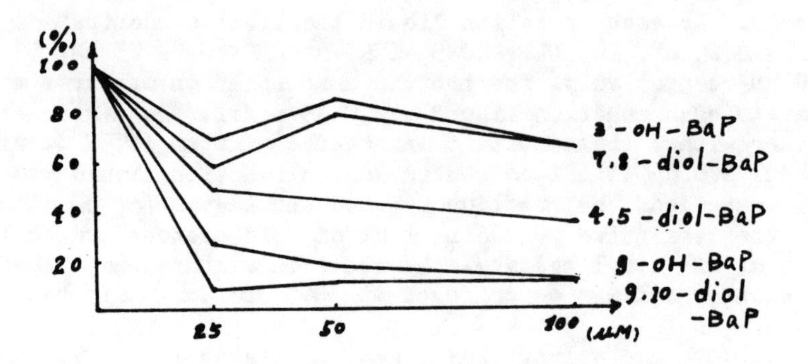

FIGURE 2. Concentration effect of PEP on BaP metabolism.

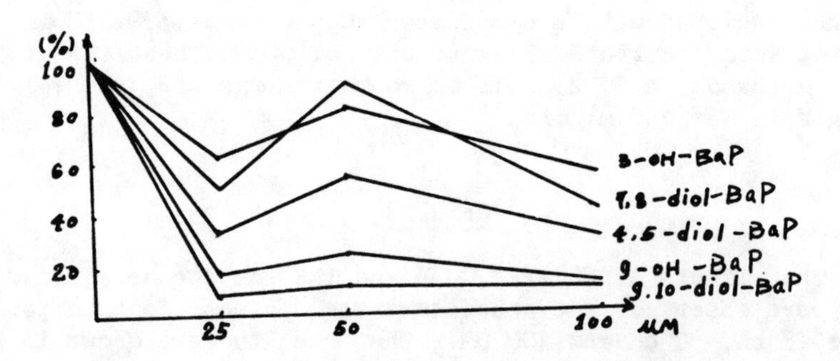

FIGURE 3. Concentration effect of CU-B on BaP metabolism.

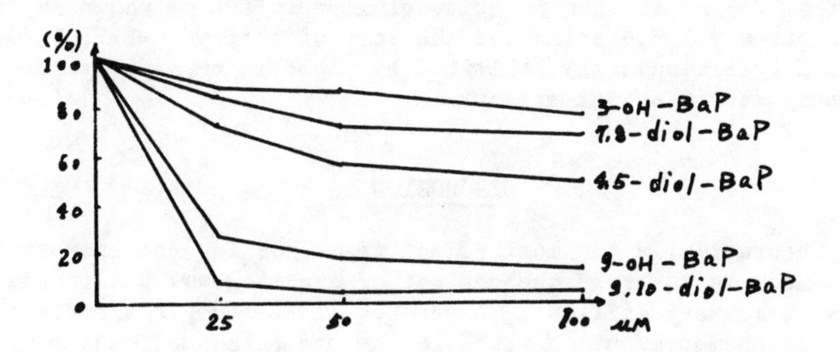

FIGURE 4. Concentration effect of TMHQ on BaP metabolism.

The comparison of the effects of PL, PEP, CU-B, TMHQ and BHA with the same concentration of 25 uM on BaP metabolism were indicated in figure 5.

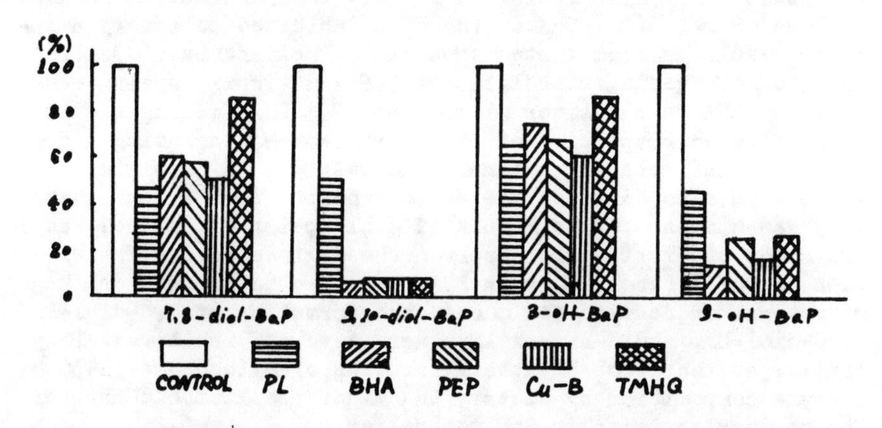

FIGURE 5. Comparison of the effects on BaP metabolism.

We have preliminarily found out that PL, PEP, Cu-B, TMHQ inhibiting the formation of the dihydrodiol and phenolic derivatives of BaP in varying degrees. Generally speaking the inhibition effect is positively related to the concentration of each compound used. From our experimental results the effect of BHA on the formation of the BaP metabolites was similar to the literature (7). However, the inhibition effects of PEP, CU-B on BaP metabolism is quite similar to BHA as shown in Fig. 5. Figures 1,2,3,4 indicated the formation of 9-OH-BaP, 9,10-diol-BaP were severely inhibited by phenolic or hydroxy compounds used in our experiment.

DISCUSSION

Theoretically the most effective method for the prevention of cancer is to avoid the contact of carcinogens. But in reality it is very difficult to pursue. Therefore, the research work of chemoprevention which is the prevention of cancer by adding anticarcinogenic agents to the diet to inhibit the chemical carcinogenic effect has been explored.

Antioxidant is a king of thoroughly studied anticarcinogenic agents which include BHA, BHT, vitamin C, A, E, EQ, and sodium selenite etc. They can inhibit the carcinogenicity of BaP or DMBA on mouse and rat. Among them BHA, BHT and EQ inhibited the cancer induced by BaP or DMBA on the stomach of the mouse. They also inhibited mammary tumor formation induced by DMBA on rats (2). BHA in the diet inhibited pulmonary neoplasia resulting from acute exposure to DMBA and BaP (8). The total carcinogenic metabolites of BaP was significantly reduced by BHA in rat liver microsomes (7). BHA feeding resulted in altered properties of liver microsomes, including a decrease in BaP metabolite binding to DNA (9). BHA applying to mouse skin also decrease the BaP metabolite to DNA, the quantity was similar to the result of J.L. Speier (10). 1981 Wasyl Jr. Sydor <u>et al.</u> (11,12) observed the regioselective inhibition of BaP metabolism, when BHA was added to hepatic or lung microsomal incubation fed mouse, the formation of 7,8-diol-BaP, 9,10-diol-Bap and 9-hydroxy-Bap were severely inhibited. They considered that probably the alteration of cytochrome p-450 isozyme composition by dietary BHA might result the change of the regioselectivity in BaP metabolism.

We considered that some natural occurring compounds with important physiological activity possessing antioxidizing property. For example, PL possesses antibacterial, antiinflammation and subsidence of swelling properties (13). CU-B a hydroxy-terpine like compound has been discovered recently to be effective for chronic hepatitis, causing blood protein return to normal and stimilating the function of cyto immunity and increasing CAMP/CGMP ratio of the blood plasma rabbit. This is beneficial for the curing and prognosis of hepatitis and it seems that the mechanism of the action of CU-B relates to the raising of the level of CAMP in the cell (14). L.W. Wattenberg (15) reported p-hydroxy anisole (PMP) possessed bigger inhibition effect on the BaP metabolism comparing to BHA. As PMP possesses bigger toxicity than BHA(16), in this paper we considered to find other inhibitors which might have similar effect but less toxicity than PMP. For this reason we have sythesized PEP and used TMHQ an intermediate of Vitamin E, which has more potent antioxidizing property but cheaper price.

The structures of the compounds we used as follows:

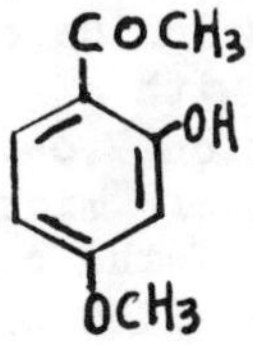

$COCH_3$ OH OCH_3

Paeonol (PL)

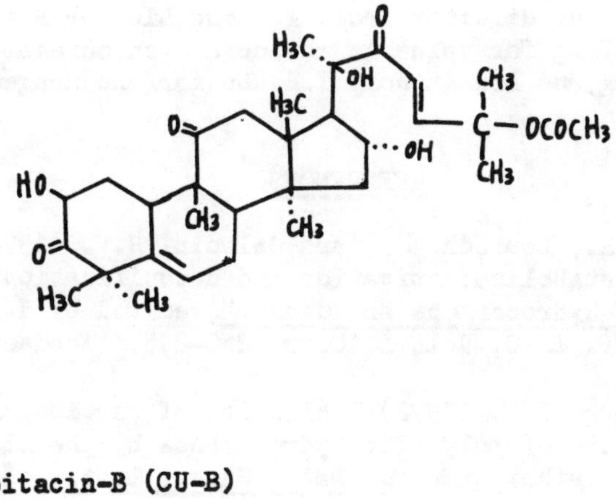

O H_3C OH CH_3 O H_3C C—$OCOCH_3$ OH CH_3 HO CH_3 CH_3 O H_3C CH_3

Cucurbitacin-B (CU-B)

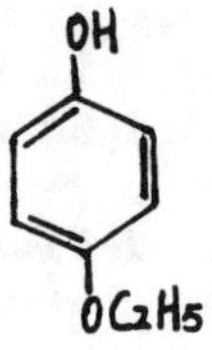

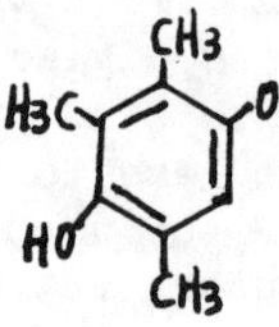

P-Ethoxyphenol (PEP) 2,3,5-Trimethylhydroquinone (TMHQ)

Our preliminary results indicated that the four compounds (PL, CU-B, PEP, TMHQ) each with a concentration of 25 uM could severely inhibit the formation 7,8-diol-BaP, 9,10-diol-Bap and 9-hydroxy-Bap by the rat liver microsomes. They were consistent with previous results of the regioselective inhibition of BaP metabolism when BHA was added to hepatic or lung microsomes as reported by Sydor et al. (11,12). It is valuable for us to continue the research work not only for aid in understanding the inhibitory effects of phenolic or hydroxy compounds on environmental chemical carcinogenesis, but also may open potential avenues to a rational approach for chemoprevention.

ACKNOWLEDEMENTS

We thank our director Prof. Peitong Liu for his support, Dr. Shen K. Yang for valuable advice and encouragement, Dr. Chung S. Yang and Dr. Anthony Y.H. Lu for encouragement.

REFERENCES

1. Yang, S.K., Deutsch, J., and Gelboin, H.V. (1978): Benzo(a) pyrene metabolism:Activation and detoxification. In: "Polycyclic Hydrocarbons and Cancer", edited by H.V. Gelboin, and P.O.P. TS'O, Vol. I 10,pp. 210-215. Academic Press, New York.
2. Wattenberg, L.W. (1972):Inhibition of carcinogenic and toxic effects of polycylic hydrocarbons by phenolic antioxidants and ethoxyquin. J. Natl. Cancer Inst., 48: 1425-1430

3. Slaga, T.J. (1980): Cancer: Etiology, Mechanisms, and Prevention- A Summary. In Carcinogenesis Vol. 5: Modifiers of Chemical Carcinogensis, edited by T.J. Slaga, p.243. Reven Press, New York.
4. Wang Zhiyuan, Bao Ziping, Xiao Xiulan, Liu Peitong, and Zhou Zongcan. (1982): Effect of Copper and Zinc ions on the metabolism and mutagenicity of benzo(a)pyrene, Acta Scientiae Circumstantiae,2:(1)28.
5. Lowry, O.H. et al. (1951): J. Biol. Chem., 193: 265.
6. Yang, S.K., Selkirk, J.K., Plotkin, E.V., and Gelboin, H. V. (1975): Kinetic analysis of the metabolism of benzo(a) pyrene to phenols, dihydrols and quinones by high-pressure chromatograph compared to analysis by arylhydrocarbons hydroxylase assay and the effect of enzyme induction. Cancer Res., 35:3642-3650.
7. Lam, L.K.T., Fladmoe, A.V., Hochalter, J.B., and Wattenberg, L.W. (1980):Short time interval effects of butylated hydroxyanisole on the metabolism of benzo(a)pyrene. Cancer Res., 40:2825.
8. Wattenberg. L.W. (1973): Inhibition of Chemical carcinogen-induced pulmonary neoplasia by butylated hydroxyanisole. J. Natl. Cancer Inst.,50:1541-1544.
9. Speier, J.L., and Wattenberg, L.W.(1975): Alterations in microsomal metabolism of benzo(a)pyrene in mice fed butylated hydroxyanisole. J. Natl. Cancer Inst., 55:470.
10. Slaga, T.J., and Bracken. W.H. (1977): The effects of antioxidants on skin tumor-initiation and arylhydrocarbon hydroxylase. Cancer Res., 37:1633.
11. Sydor, W. Jr., Chou, M.W., Yang, S.K. and Yang, C.S. (1983): Regioselective inhibition of Benzo(a)pyrene metabolism by butylated hydroxyanisole. Carcinogenesis (Lond.), 4:703-708.
12. Sydor, W. Jr., Lewis, K.F., and Yang, C.S. (1984): Effects of butylated hydroxyanisole on the metabolism of benzo(a) pyrene by mouse lung microsomes. Cancer Res., 44:134-138.
13. Nanking College of Traditional Chinese Medicine, Preparation Laboratories. (1973): Extraction of active principle and its therapeutic effect of the root of pycnostelma paniculatum (Bge) K. Schum, Zhong Cao Yiao Tongxun, No. 2:38.
14. Hunan pharmaceutical Industrial Research Institute and Shansi medical college hepatitis research groups (1982):Pharmacology research of the effects of cucurbitacin B,E on hepatitis. Chinese Traditional and Herbal Drugs. 13 (11): 505-506.

15. Wattenberg, L.W., Cocia, J.B., and Lam, L.K.T. (1980); Inhibitory effects of phenolic compounds on benzo(a)pyrene-induced neoplasia. Cancer Res., 40:2822.
16. Kaul, B.L. (1979): Cytogenetic activity of some common antioxidants and their interaction with X-rays, Mutation Res., 67:239-247.

2-CHLORO-PHENOTHIAZINE AND RESORUFIN INHIBIT BENZO(A)PYRENE MUTAGENESIS AND METABOLISM *IN VITRO*

JOSEPH E. JABLONSKI, PAUL D. SULLIVAN*
Department of Chemistry, Ohio University, Athens, Ohio 45701

INTRODUCTION

2-Chlorophenothiazine (2-Cl-PTH) and other derivatives of the neuroleptic drug phenothiazine are effective inhibitors of mutation in the Ames' test (1). Phenothiazines are metabolized by cytochrome P-450 mixed function oxygenases (2) suggesting that these drugs might interfere with benzo(a)pyrene (BaP) metabolism. Repetitive scan difference spectrophotometry was used to assess the degree of such inhibition in this paper.

The ethoxyresorufin de-ethylase assay has gained wide acceptance for the determination of cytochrome-P-448 activity since its introduction by Burke and Mayer (3). Pohl and Fouts (4) found that the presence of de-ethylated substrate, 7-hydroxy-phenoxazin-3-one, resorufin (Rr) inhibited de-ethylase activity significantly. Since the cytochrome P-448 or P-450c isozyme has been specifically implicated in the activation of PAH to ultimate mutagens and carcinogens (5), we thought this compound might be an effective inhibitor of mutation. Since Rr is not a type I substrate for cytochrome P-450 isozymes its mechanism of inhibition is unlike that of 2-Cl-PTH. This work shows the effect of Rr in the Ames' mutation assay and explores the mechanism of action through several biochemical assays.

MATERIALS AND METHODS

Chemicals

Benzo(a)pyrene, resorufin, and 2-chlorophenothiazine were obtained from the Aldrich Chemical Co. Dehydrogenases, substrates and cofactors for the NADPH regenerating systems were obtained from the Sigma Chemical Co.

Mutagenicity Assays

These were performed essentially as described by Ames *et al.* (6) with some slight modifications. 15 μl of DMSO containing 6 nmoles of BaP, 50 μl of inhibitor in DMSO, and

100 μl of an overnight culture of TA98 were added to 2.5 ml of top agar containing biotin and histidine. After these additions, 0.6 ml of S-9 mix containing 40 μl of 9000xg homogenate from the livers of beta-naphthoflavone-induced male 250g Sprague-Dawley rats, 6 μmoles of NADP and 5 μmoles of glucose-6-phosphate was quickly mixed with the top agar and poured over a minimal plate. The proper amounts of histidine and biotin were added to the melted solution of agar before it was autoclaved. The assays were run in duplicate using 5 plates for each inhibitor concentration with results expressed as percent of control reversion count.

In Vitro Metabolic Incubations

Male Sprague-Dawley rats (200-250 g) were injected twice with beta-naphthoflavone (90 mg/kg body weight) 48 and 24 hours before sacrifice. Liver microsomes were prepared by centrifugation of the 9000xg supernatant at 90,000xg for 90 minutes. Cytochrome P-450 concentration was determined according to Omura and Sato (7) and protein by the method of Bradford (8).

Difference spectroscopy to measure substrate/P-450 binding was performed as described by Jefcoate (9). Microsomes were suspended to 1 mg/ml protein concentration in 0.1 M KH_2PO_4 buffer, pH 7.4, 20% glycerol. K_s (spectral dissociation constant) and Amax (absorbance maximum) were determined from reciprocal plots using a weighted least squares program.

In vitro metabolism of BaP as affected by 2-Cl-PTH was measured using repetitive scan UV/visible spectroscopy as described by Prough (10). The diode-array spectrophotometer was programmed in BASIC to acquire spectra at selected intervals with a 1 second measurement per spectrum. Cuvettes contained 3ml of microsomal suspension at 0.5 mg/ml protein concentration and 80 μmolar BaP added in 12 μl of acetone. 0.3 μmol of NADPH was added to initiate metabolism sustained by a regenerating system consisting of isocitrate dehydrogenase (1 unit/ml), sodium isocitrate (5 mM), and $MgCl_2$ (5 mM).

In vitro metabolism of BaP as affected by Rr was measured using HPLC. Incubations were carried out in 6 ml volumes containing microsomal protein (1 mg/ml), glucose-6-phosphate dehydrogenase (1 unit/ml), glucose-6-phosphate (20 mM), $MgCl_2$ (5 mM) and NADP (2 mM). BaP was added in DMSO to a concentration of 40 μM. Rr was added in DMSO to provide concentrations of 40 and 400 μM. After 60 minutes of incubation with gentle shaking at 37°C the reactions were quenched with 6 ml of cold acetone. 12 ml of ethyl acetate was added with internal standards perylene or 6-methyl BaP to measure extraction efficiency. The solvent was evaporated and the sample reconstituted with 0.5 ml of methanol before injection.

NADPH oxidation was measured spectrophotometrically at 37°C using ε(340-400nm) of 6.22 mM^{-1} cm^{-1}. The reaction mixture contained 0.25mM NADP, 10 μM Rr, and 0.3 mg/ml microsomal or S-9 protein in a volume of 3 ml 0.05 M Tris buffer, pH 7.5, 0.15 M KCl.

Resorufin reduction was followed spectrophotometrically at 37°C using ε(572-700nm) of 73 mM^{-1} cm^{-1} for the oxidized form. The reaction mixture was essentially identical to the NADPH oxidation mixture except that .08 mg/ml of microsomal or S-9 protein was used.

Oxygen consumption was determined polarographically with a Clark-type electrode using 0.5 mg/ml microsomal protein, 0.25mM NADPH and 10 μM Rr in 3 ml of Tris .05 M, pH 7.4, 0.15 M KCl. Temperature was 30°C.

Instrumentation

Reversed-phase HPLC was performed with an ISCO system consisting of two model 2300 pumps, a model UA-5 absorbance detector with 254 nm filter, Valco C6W injector with a 20 μl sample loop. Solvent gradient formation and data acquisition were controlled by an Apple IIe computer via an Adalab interface and Chromatochart software from Interactive Microware. A Beckman Ultrasphere ODS column (4.6 x 250 mm), dp 5 μm was used with a gradient of 65% methanol/H_2O to 83% methanol/H_2O in 13 minutes, isocratic 13 minutes, to 100% methanol in 14 minutes.

UV-visible spectrophotometric experiments were conducted with a Hewlett-Packard 8451A diode array instrument with thermostatable cell attachment.

Oxygen consumption was measured on a Yellow Springs model 53 oxygen meter.

RESULTS

Effect of 2-Cl-Phenothiazine and Resorufin on Ames' Mutagenicity Assay

Figure 1 shows dose response curves for BaP induced

mutation in Salmonella strain TA 98. Both 2-Cl-PTH and Rr decrease the percent of control revertants to less than 50% at 5 x BaP concentration and reduce the number of mutant colonies to near-background levels at 25 x BaP and higher inhibitor concentrations. Some toxicity occurred with Rr at high concentrations seen as a negative percent of control reversion.

Spectral Binding Inhibition Studies

The inverse plot shown in Figure 2 indicates that 2-Cl-PTH acts as a non-competitive inhibitor of BaP/P-450 Type I binding shown by a decrease in Amax and unchanged K_s (spectral dissociation constant). Resorufin exerts no effect on BaP/P-450 binding. Since Rr produces only a weak modified Type II spectral response with P-450, its failure to inhibit BaP binding is not unexpected.

Repetitive Scan Spectroscopy of BaP Metabolism and 2-Cl-PTH

2-Cl-PTH exerts a dramatic effect on BaP metabolism as seen in Figure 3. At equimolar concentration it reduces the rate of BaP metabolism by a factor of 20, at only .25 x BaP a 50% reduction was observed. This rate is measured as nmoles 3-OH BaP formed/mg protein/minute using a molar absorptivity of 13,200 cm^{-1} M^{-1} for the absorbance difference 428nm - 454nm (10). Figure 4 summarizes aryl hydrocarbon hydroxylase (AHH) rates for the 10 minutes of measurement.

HPLC Study of Resorufin Effect on BaP Metabolism

Repetitive scan spectroscopy could not be used to measure BaP metabolism in the presence of Rr. As Rr is reduced in the presence of NADPH and DT-diaphorase a trough in the spectrum is produced at 572nm which alters absorbance at the 454nm isosbestic point. HPLC analysis is ideal in this case, because Rr is extremely water-soluble it does not significantly extract into the organic phase with the BaP metabolites. Figure 5 shows chromatograms from control (40 μM BaP) and 10 x (40 μM BaP and 400 μM Rr) incubations. BaP metabolism has been virtually eliminated by 400 μM Rr, Figure 5B. In the presence of 40 μM Rr, BaP metabolism was reduced 50%, data not shown. Chromatograms from incubations containing only Rr showed one peak other than the early-eluting peaks obtained from blank microsome incubations. When an organic extract from an Rr incubation was spiked with pure Rr this peak co-chromatographed with the pure Rr.

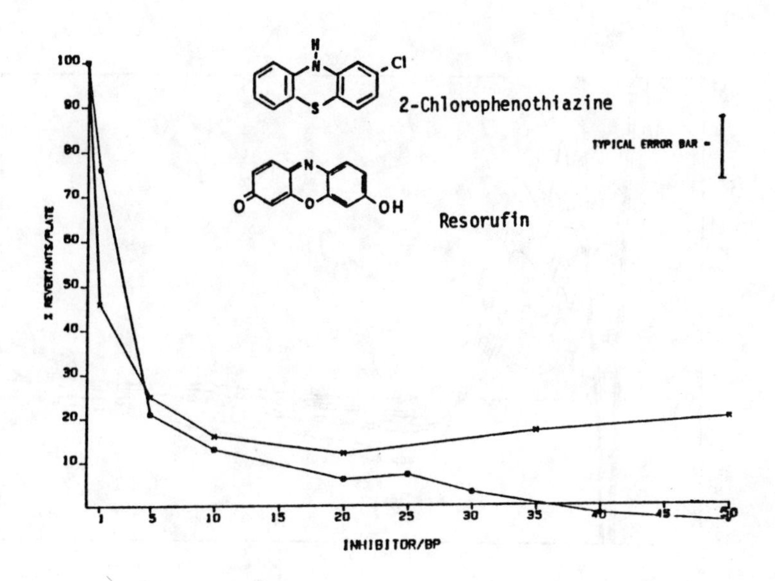

FIGURE 1. Dose response curves showing effect of resorufin(·) and 2-chlorophenothiazine (x) on BaP-induced mutation in Ames' test with tester strain TA98.

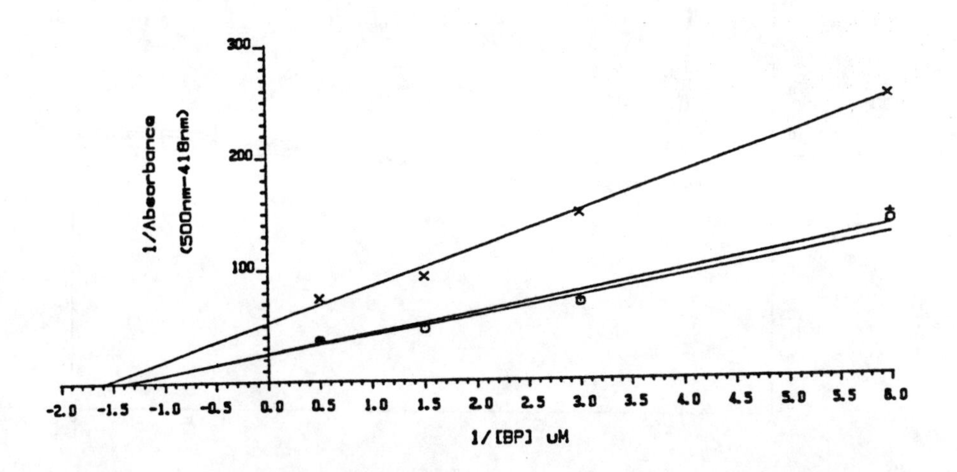

FIGURE 2. Inverse plot of BaP/cytochrome P-450 binding. Sample cuvette containing 1mg/ml of liver microsomes from betanaphthoflavone-induced rats was titrated with BaP in acetone. Reference cuvette received equal volumes of acetone.
(o) = BaP K_S = .695 μM ΔAmax = 0.041
(+) = BaP + 2 μM Rr K_S = .717 μM ΔAmax = 0.040
(x) = BaP + 2 μM 2-Cl-PTH K_S = .623 μM ΔAmax = 0.019

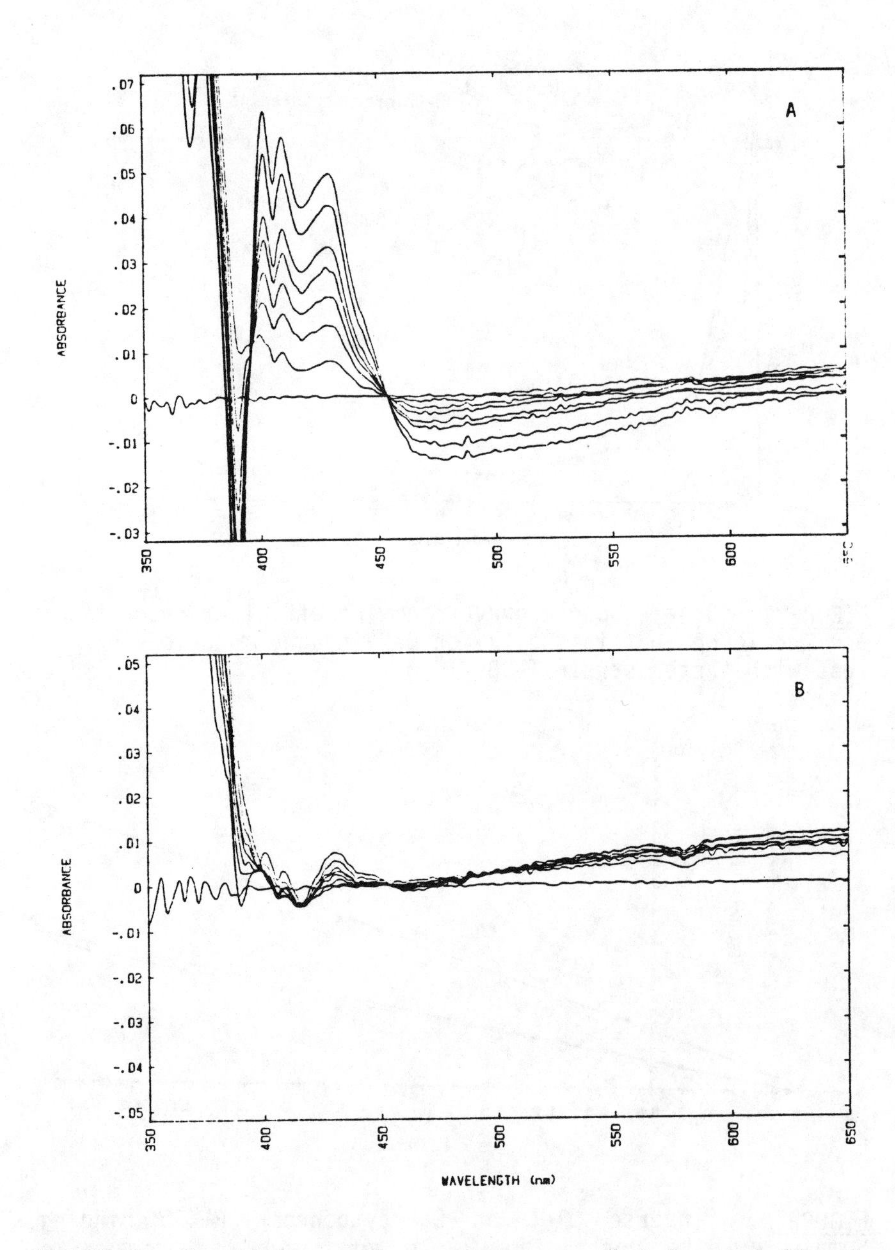

FIGURE 3. Repetitive scan spectroscopy of BaP metabolism in the absence (A) and presence of equimolar 2-Cl-PTH (B). Peaks at 402, 410 and 430 nm due to BaP phenol formation, isosbestic point at 454 nm. Control experiments with 2-Cl-PTH only, showed no spectral changes.

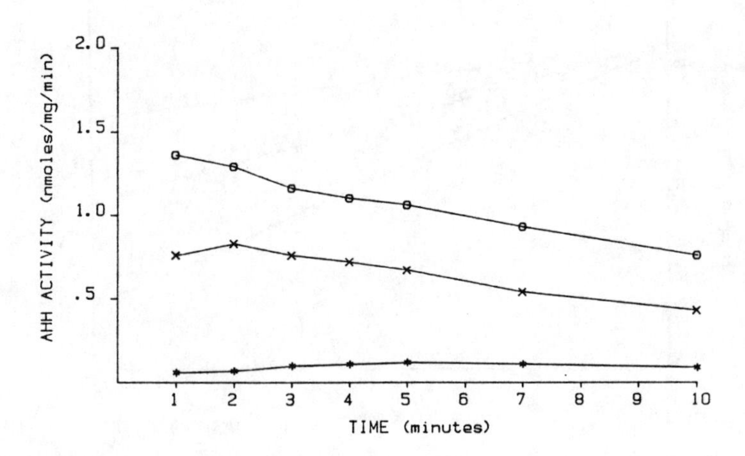

FIGURE 4. Aryl hydrocarbon hydroxylase activity calculated from rapid scan measurements of BaP metabolism.
(o) 80 μM BaP
(x) 80 μM BaP + 20 μM 2-Cl-PTH
(*) 80 μM BaP + 80 μM 2-Cl-PTH

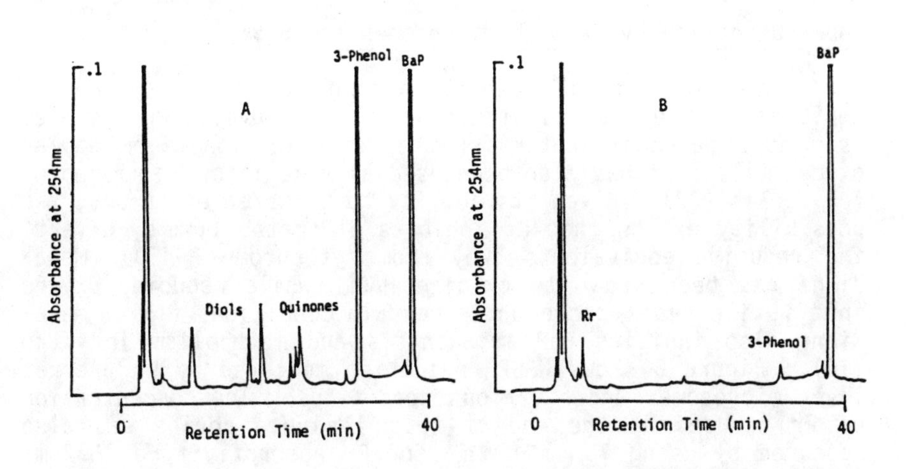

FIGURE 5. HPLC separation of BaP metabolites formed from incubations containing 40 μM BaP (A), and 40 μM BaP + 400 μM resorufin (B).

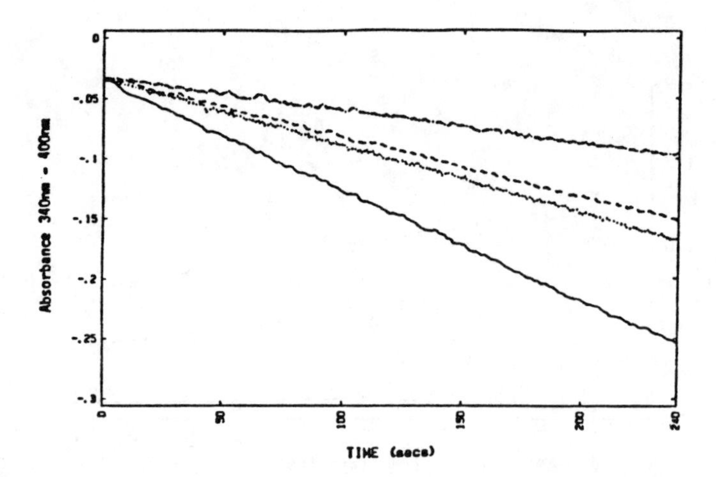

FIGURE 6. Resorufin and NADPH oxidation. To 3 ml of .05M Tris buffer, pH 7.4 containing .3 mg/ml microsomes and .15 mM NADPH the following additions were made.
——— = 10 μM Rr
------ = 10 μM Rr + 20 μM dicumarol
······ = 10 μM Rr + 2 mM $NADP^+$ or + N_2 purge
—·—·—· = 10 μM Rr + 20 μM dicumarol + 2 mM $NADP^+$

NADPH Oxidation by Rr with Microsomes and S-9

Since Rr did not inhibit binding of BaP to cytochrome P-450 it must exert its effect at another level. Rr has been used as a redox indicator for the determination of molecular oxygen (11). It has also been used as a mediator in microbiol fuel cells (12). Given its ability to transfer electrons, the possibility exists that Rr inhibits BaP metabolism by diverting reducing equivalents away from cytochrome P-450. Menadione has been shown to oxidize NADPH while reducing oxygen in a futile redox cycle under certain conditions (13). Menadione also inhibits BaP mutagenesis and metabolism *in vitro* (14). Figure 6 shows NADPH oxidation mediated by Rr and rat liver microsomes under various conditions. The concentration of Rr reduced at the end of this 4 minute period was also measured by using 73 cm^{-1} mM^{-1} for Rr absorptivity at 572nm. The solid line represents NADPH oxidation by 10 μmolar Rr, the ratio of NADPH oxidized/Rr reduced after 4 minutes was 5.9. The rate of NADPH oxidation in the absence of substrate was 20 nmoles/mg protein/minute. When the rate due to resorufin is corrected for this background, NADPH oxidation is 9.1

nmoles/mg protein/minute. Addition of 2 mM NADP or purging the cuvette with nitrogen reduced NADPH consumption levels to near-background, represented by dotted curve. Addition of both dicumarol and NADP reduced NADP oxidation to 8 nmoles/mg protein/minute. Table 1 summarizes NADPH oxidation assays with microsomes and S-9 under various conditions.

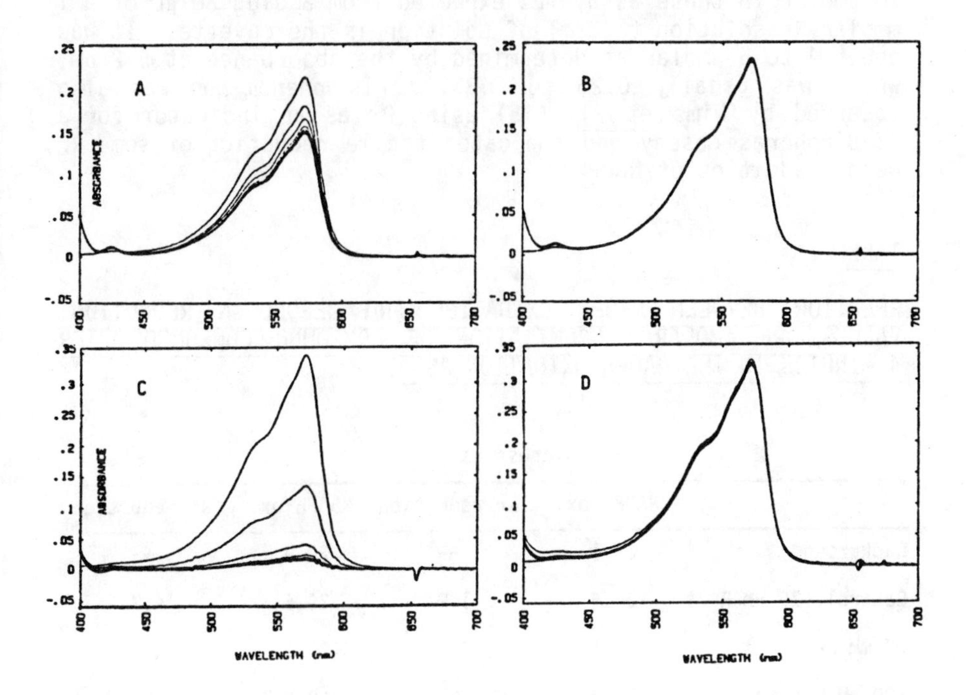

FIGURE 7. Spectrophotometric measurement of resorufin reduction. 10 μM Rr was added to cuvette and initial spectrum taken. .25 mM NADPH was then added and spectra were recorded at 30 second intervals. Absorbance decreases at 572 nm as Rr is reduced.
A) microsomes + Rr B) microsomes + Rr + 20 μM dicumarol
C) S-9 + Rr D) S-9 + Rr + 20 μM dicumarol

Resorufin Reduction by NADPH

Figure 7 shows resorufin reduction with microsomes and S-9 in the presence and absence of dicumarol. In the presence of S-9, reduction is nearly complete in 1.5 minutes, Figure 7C. Under identical conditions microsomes only partially reduce Rr, Figure 7A. Dicumarol completely inhibits Rr reduction by either system, Figures 7B and 7D. Microsomal reduction of Rr is probably due to traces of DT-Diaphorase, no Rr reduction occurs with microsomes which have been rinsed, resuspended and centrifuged a second time. It is interesting to note that the initial concentration of oxidized Rr is not 10 μmolar in these assays as expected from adding 30 μL of a 1 mmolar Rr solution to 3 ml of solution in the cuvette. It was about 4 to 5 μmolar as determined by the absorbance at 572 nm, which was usually 0.25 to 0.35. This phenomenon was also observed by Nims *et al.* (15) using Rr as an indicator for a DT-diaphorase assay and indicates facile reduction of some Rr before addition of NADPH.

TABLE 1

RELATION BETWEEN NADPH OXIDATION AND RESORUFIN REDUCTION. VALUES ARE μMOLAR CONCENTRATION OF COMPOUND CONSUMED AFTER 4 MINUTES IN THE NADPH OXIDATION ASSAY.

	Microsomes		S-9	
	NADPH ox.	Rr reduction	NADPH ox.	Rr reduction
Background	24.1	--	20.9	--
Control (20 μM Rr)	35.4	1.9	29.5	4.2
+2 mM NADP	22.5	2.3	27.3	4.4
+20 μM Dicumarol	19.3	0	8.8	0
+2 mM NADP +20 μM Dicumarol	9.6	0	--	--
N_2 purge	25.7	2.6	22.5	4.2
N_2 purge +20 μM Dicumarol	--	--	4.8	0

Oxygen Consumption Measurements

Figure 8 shows that Rr does not increase the consumption of oxygen above background levels in the presence of microsomes. As a control experiment, the same assay was run with 1,4 naphthoquinone, which did consume oxygen (16).

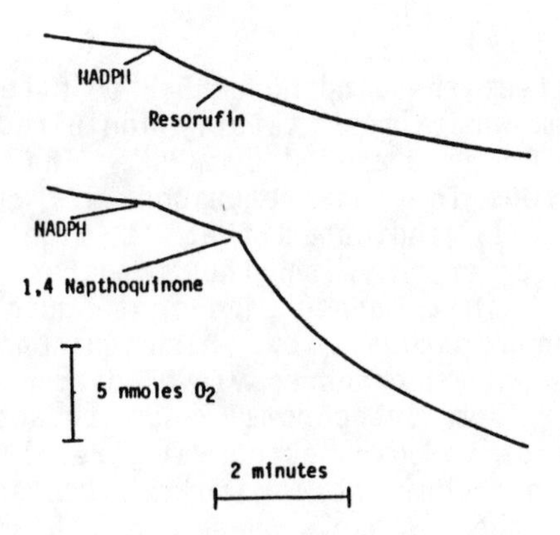

FIGURE 8. Oxygen consumption mediated by resorufin and 1,4 napthoquinone with microsomes.

DISCUSSION

2-Chloro-phenothiazine Inhibition

From the spectral binding data, Figure 2, and repetitive scan metabolism spectra, Figure 3 it is apparent that 2-Cl-PTH inhibits BaP metabolism by interfering with cytochrome P-450/BaP binding. The non-competitive nature of this inhibition suggests that bound 2-Cl-PTH is oriented differently than BaP at the P-450 substrate binding site, even though both compounds give type I binding spectra. The binding of 2-Cl-PTH to cytochrome P-450 translates into a large decrease in aryl hydrocarbon hydroxylase metabolism of BaP, Figure 4, and subsequent formation of reactive metabolites which cause mutation. However, the results are not so clear cut as to establish that this is the only contributing mechanism of inhibition for 2-Cl-PTH. Other possible contributions could be due to the known intercalating properties of 2-Cl-PTH with DNA (17) which might interfere with the binding of the ultimate mutagenic metabolites of the substrates. Alternatively a major metabolite of 2-Cl-PTH is 3-hydroxy-2-chlorophenothiazine (2), this metabolite is capable of redox cycling in much the same way as Rr, which could also contribute to the inhibition of BaP metabolism. Further work is necessary to fully elucidate these possibilities.

Resorufin Inhibition

Rr did not affect the binding of BaP to cytochrome P-450, yet BaP metabolism was almost totally inhibited by 10 x BaP concentrations of Rr as measured by HPLC. The ability of Rr to transfer electrons in redox reactions (12) and react with molecular oxygen (11) indicates that it might inhibit BaP metabolism by interfering with the transfer of electrons to cytochrome P-450 in a manner similar to quinones (13,18). There are two known pathways by which quinones can accept electrons, a two-electron transfer via DT-diaphorase or a one-electron transfer from cytochrome P-450 reductase (13,18). Two electron transfer to quinones normally results in the formation of stable hydroquinones which can be conjugated, and occurs mainly in the cytosol where most DT-diaphorase is located (13). One-electron transfer occurs in the microsomal fraction and results in the formation of unstable semiquinone intermediates which can reduce oxygen to superoxide radicals. In this case, the net result is a futile redox cycle with NADPH oxidation catalyzed by the quinone and molecular oxygen consumed. Similar one electron and two electron pathways may occur for quinoneimines (i.e. Rr), see Figure 9, below.

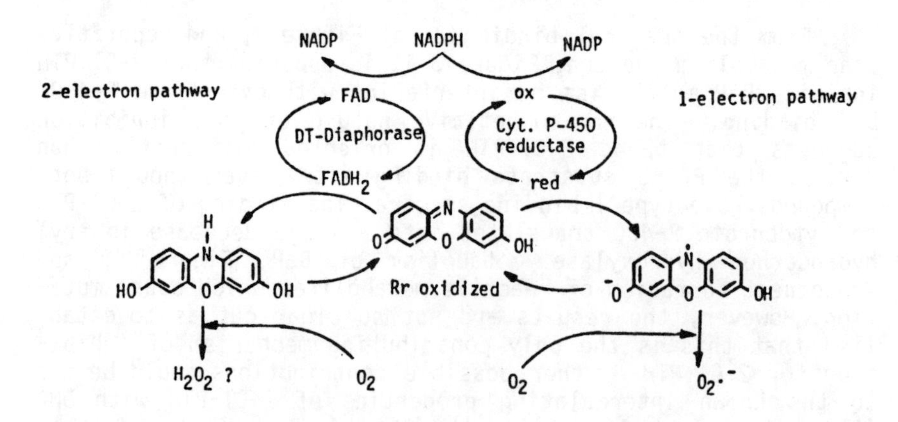

FIGURE 9. Redox pathways by which resorufin may cycle and transfer electrons from NADPH to molecular oxygen.

Table 1 shows that the quantity of NADPH oxidized is greater than that of resorufin reduced with microsomes or S-9. The NADPH concentration used in the HPLC incubations was 2mM, the glucose-6-phosphate concentration was 20mM. At 10 x BaP the Rr concentration was only 400 μM. If only the two-electron pathway with no cycling was operable then 400 μM of reducing equivalents would be consumed by Rr which is only a 2% reduction in the total electron pool available in the form of glucose-6-phosphate. The results indicate that cycling of Rr between oxidized and reduced states occurs while reducing equivalents from NADPH are transferred to an acceptor molecule. This non-stoichiometric consumption of NADPH reducing equivalents by Rr could explain the large decrease in BaP metabolites formed as observed by HPLC. Figure 6 and Table 1 show that NADPH oxidation with Rr is sensitive to the cytochrome P-450 reductase inhibitor, NADP, and the DT-Diaphorase inhibitor, dicumarol. This suggests that both the one-electron and two-electron pathways participate in the reduction of Rr. Under anaerobic conditions, NADPH oxidation was reduced to near background levels indicating that molecular oxygen serves as an electron acceptor in this redox cycle. Rr is different from quinones tested as redox inhibitors, such as menadione, in that the reduced form of Rr is readily oxidized by molecular oxygen (15), enabling it to cycle reducing equivalents through the two-electron, DT-Diaphorase pathway as well as the one-electron pathway via cytochrome P-450 reductase.

Despite the observed effect of anaerobicity on the NADPH oxidation assay with Rr, no direct oxygen consumption was observed polarographically with the oxygen-sensitive electrode. Oxygen was consumed upon addition of NADPH to microsomes, and enhanced when 1,4 naphthoquinone was added as a control (16). Considering the well-known property of Rr to react with oxygen, the absence of oxygen consumption is puzzling. A possible explanation might involve the ability of Rr to effect electrode processes as previously mentioned (12). Another question concerns the high background rate of NADPH oxidation using microsomes and S-9. In the absence of microsomes, NADPH is oxidized at 4.0 nmoles/mg protein/minute, addition of microsomes or S-9 increases the rate to 24.1 and 17.5 nmoles/mg protein/minutes, respectively. This endogenous consumption rate may be due to lipid peroxidation initiated by superoxide anions (19).

CONCLUSIONS

The data presented indicate that 2-chlorophenothiazine inhibits BaP metabolism and mutagenesis by interfering with the initial binding of BaP to cytochrome P-450. Resorufin also inhibits BaP mutagenesis and metabolism _in vitro_, but through a mechanism involving cytochrome P-450 reductase and

DT-Diaphorase. Reducing equivalents and molecular oxygen are coupled in a redox cycle mediated by Rr thus depriving cytochrome P-450 mixed function oxygenase of cofactors needed to convert BaP to reactive metabolites.

ACKNOWLEDGEMENTS

This work was supported, in part, by Grant # CA-34966 awarded by the National Cancer Institute, DHEW to P.D.S. We would also like to thank P. Espino and F. Fernandes for technical assistance with some of the experiments.

REFERENCES

1. Kittle, J.D. Jr., Calle, L.M., Sullivan, P.D. (1981): The effect of substituted phenothiazines on the mutagenicity of benzo(a)pyrene. Mutation Res., 80:259-264.
2. Beckett, A.H., and Navas, G.E. (1978): Aromatic oxidation of some phenothiazines. Xenobiotica, 8:721-736.
3. Burke, M.D., and Mayer, R.T. (1974): Ethoxyresorufin: Direct fluorometric assay of a microsomal O-dealkylation which is preferentially inducible by 3-methylcholanthrene Drug Metab. Disp., 2:583-588.
4. Pohl, R.J., Fouts, J.R. (1980): A rapid method for assaying the metabolism of 7-ethoxyresorufin by microsomal subcellular fractions. Anal. Biochem., 107:150-155.
5. Levin, W., Wood, A.Y., Lu, A.Y.H., Ryan, D., West, S., Conney, A.H., Thakker, D.R., Yagi, H., and Jerina, D.M. (1977). Role of purified cytochrome P-448 and epoxide hydrase in the activation and detoxication of benzo(a)pyrene. Drug Metabolism Concepts, (D.M. Jerina, ed.), ACS Symposium Series, No. 44:99-126. American Chemical Society, Washington, D.C.
6. Ames, B.N., McCann, J., and Yamasaki, E. (1975): Methods for detecting carcinogens and mutagens with the Salmonella mammalian-microsome mutagenicity test. Mutation Res., 31: 347-364.
7. Omura, T., and Sato (1964): The carbon monoxide-binding pigment of liver microsomes. J. Biol. Chem. 239:2370-2385.
8. Bradford, M.M. (1976): A rapid and sensitive method for the quantitation of microgram quantities of protein utilizing the principle of protein-dye binding. Anal. Biochem. 72: 248-254.
9. Jefcoate, C.R. (1978): Measurement of substrate and inhibitor binding to microsomal cytochrome P-450 by optical difference spectroscopy. In: Methods in Enzymology Vol. 52 part C:258-279. Academic Press, New York.

10. Prough, R.A., Patrizi, V.W., Estabrook, R.W. (1976): The direct spectrophotometric observation of Benzo(a)pyrene phenol formation by liver microsomes. Cancer Research 36:4439-4443.
11. Teijin Ltd., Jpn. Kokai Tokkyo Koho 81,108,955. Reagents for oxygen detection (Cl.GOIN31/22) Patent.
12. Delany, G.M., Bennetto, H.P., Mason, S.D., Stirlington, J.L., Thurston, C.F. (1984): Electron-transfer coupling in microbiol fuel cells. 2. Performance of fuel cells containing selected microorganism mediator-substrate combinations. J. Chem. Tech. Biotechnol. 34B:13-27.
13. Lind, C., Hochstein, P., Ernster, L. (1982): DT-Diaphorase as a quinone reductase: a cellular control device against semiquinone and superoxide radical formation. Arch. Biochem. Biophys. 216:178-185.
14. Israels, L.G., Walls, G.A., Ollman, D.J., Friesen, E., Israels, E.D. (1983): Vitamin K as a regulator of benzo(a)pyrene metabolism, mutagenesis and carcinogenesis. J. Clin. Invest., 71:1130-1140.
15. Nims, R.M., Prough, R.A., Lubet, R.A. (1984): Cytosol-mediated reduction of resorufin: a method for measuring quinone oxidoreductase. Arch. Biochem. Biophys., 229:459-465.
16. Thornally, P.J., d'Arcy Doherty, M., Smith, M.T., Bannister, J.F., and Cohen, G.M. (1984): The formation of active oxygen species following activation of 1-naphthol, 1,2- and 1,4-naphthoquinone by rat liver microsomes Chem. Biol. Interactions, 48:195-206.
17. deMol, N.J., and Maanders, J.P.A.C.M. (1983): Non-covalent binding of some phenothiazine drugs to DNA. Int. J. Pharmaceutics, 16:153-162.
18. Chesis, P.L., Levin, D.E., Smith, M.T., Ernster, L., Ames, B.N. (1984): Mutagenicity of quinones: pathways of metabolic activation and detoxification. Proc. Natl. Acad. Sci. USA, 81:1696-1700.
19. Wills, E.D. (1969): Lipid Peroxide formation in microsomes. Biochem. J., 113:315-24.

RAT-LIVER MICROSOMAL OXIDATION OF SULFUR-CONTAINING POLYCYC-CYCLIC AROMATIC HYDROCARBONS (THIAARENES)

JÜRGEN JACOB[1], ACHIM SCHMOLDT[2], GERNOT GRIMMER[1]
(1)Biochemical Institute for Environmental Carcinogens, 2070 Ahrensburg, FRG; (2) Institute for Forensic Medicine, University Hamburg, 2000 Hamburg 54, FRG.

INTRODUCTION

Sulfur-containing polycyclic aromatic compounds (thiaarenes, S-PAC) have been detected in various environmental matter such as crude oil (1), fresh and used engine oil (2, 3) and as a consequence also in vehicle exhaust (4). They occur in considerable concentrations in the emission of brown-coal (5) and of hard- coal combustion (6). Some thiaarenes have been shown to be potent carcinogens (7) and in a more recent paper Croisy *et al*. (8) stated that the S-analogues of chrysene, benzo(c)chrysene, dibenz(a,h)anthracene, and dibenz-(a,j)anthracene were even more potent carcinogens than the parent PAH. Evidence for the mutagenic activities of various thiaarenes in the Ames *Salmonella typhimurium* test system has also been presented (9, 10). Despite of these biological activities the metabolism of thiaarenes in mammalian systems has not been analyzed in detail. The present investigation therefore was undertaken to study the microsomal oxidation of four thiaarenes (benzo(b)naphtho(1,2-d)thiophene; benzo-(b)naphtho(2,1-d)thiophene; benzo(b)naphtho(2,3-d)thiophene; triphenyleno(4,5-bcd)thiophene; for structure see Figure 1) in rat liver.

MATERIAL AND METHODS

Materials

Benzo(b)naphtho(1,2-d)-, -(2,1-d)-, and -(2,3-d)thiophene were obtained from the Commission of the European Community, Community Bureau of Reference , BCR, Rue de la Loi 200, B-1049 Brussels in a purity better than 99.0%. Triphenyleno-(4,5-bcd)thiophene was synthesized by insertion of sulfur into triphenylene following Klemm and Lawrence (11). The

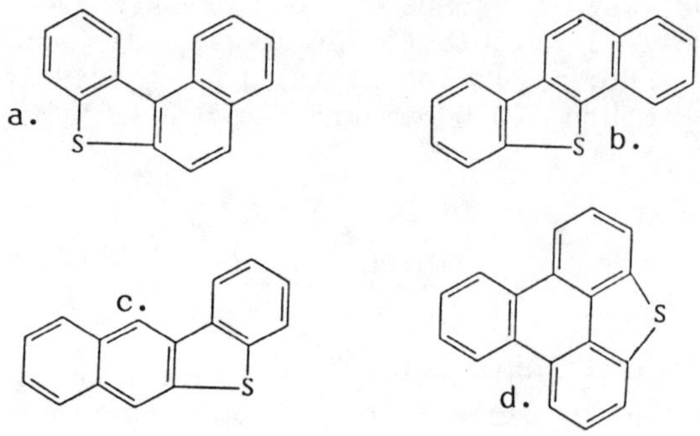

FIGURE 1. Structure of a. benzo(b)naphtho(1,2-d)thiophene, b. benzo(b)naphtho(2,1-d)thiophene, c. benzo(b)naphtho-(2,3-d)thiophene, and d. triphenyleno(4,5-bcd)thiophene.

crude product was distilled in vacuo and repeatedly recrystallized to a purity better than 99.0 %. Sulfones of all four thiaarenes were prepared by oxidation with H_2O_2 in acetic acid (11). The sulfoxides were obtained by treating the thiaarenes with iodobenzene diacetate (12). Sulfones and sulfoxides were recrystallized from acetic acid and from ethanol/water resulting in products of purities better than 99.0%. Benzo(b)naphtho(2,3-d)thiophene-6,11-quinone was synthesized from benzo(b)naphtho(2,3-d)thiophene by oxidation with CrO_3 in boiling acetic acid (0.3 g thiaarene + 0.3 g CrO_3 in 20 ml acetic acid, 1 hr) and purified by silica chromatography and repeated recrystallization from propanol-2.

Animals

Male Wistar rats (body weight 150-160 g) were used for the experiments. For induction of the monooxygenases the animals were pretreated by a threefold application of 40 mg inducer/ kg body weight, dissolved in 2 ml arachis oil, over three consecutive days.

Microsomal preparation and incubation

Microsomal preparation (13), determination of microsomal proteins (14) and of cytochrome P450 content (15) were performed according to the literature.

Incubations were carried out at 37 °C in a total volume of 2 mL containing 50 μmol/L thiaarene (dissolved in 20 μL acetone, 50 mmol/L tris HCl, pH 7.45, 0.15 mol/L KCl, 5 mmol/L $MgCl_2$, 8 mmol isocitrate and 1 mg/L microsomal protein), 20 μg isocitratedehydrogenase, and 0.5 mmol/L NADPH, and stopped after 15 min by adding 10 ml acetone. The mixtures were stored at -20 °C under N_2-atmosphere. Control incubations were obtained by omitting NADPH and the NADPH-regenerating system.

After evaporating acetone, 11.5 μg indeno(1,2,3-cd)fluoranthene in case of the benzonaphthothiophenes or 15.84 μg benzo(c)phenanthrene in case of the triphenylenothiophene incubations was added as internal standard and the pH-value adjusted to 3 by addition of acetic acid. The extracts obtained by double extraction with 30 ml ethylacetate were cautiously evaporated to dryness, dissolved in 2 ml isopropanol and chromatographed on 10 g Sephadex LH 20 columns. Isopropanol was used as elutant. The first 45 ml were discarded. The following volume (45-200 ml) contains unreacted substrate and the metabolites.

Analysis

Metabolites were analyzed as such as well as in derivatized form. To this end the metabolites were dissolved in 8 uL toluene to which 16 μL Trisil (Pierce) and 2,2,2-trifluoro-N,O-bis(trimethylsilyl)acetamide (1:1; v/v) were added. Allowing for a 45 min reaction, specimens were gas chromatographed on a Perkin-Elmer (Sigma 2B, FID) instrument. A 25 m glass capillary column with 0.37 mm i.d. and an impregnated layer (CP sil 5) thickness of 0.8 μm was used for the separation. FID-signals were registered with a Spectra-Physics SP 4100-02 integrator. Injections were carried out at a 85 °C inlet temperature, with a temperature program after a 3 min delay of 85-180 °C with a heating rate of 25 °C/min followed by a second program of 180-270 °C with 2 °C/min and a approximate carrier gas flow rate of 1 ml helium/min.

Analyses were repeated three times.

For the identification the metabolites formed, a combination of gas chromatography (Perkin-Elmer F 22) and mass spectrometry (MAT Varian 112S with 70eV ionisation potential) was used. Samples were injected at 85 ^{o}C and the solvent eluted over 10 min with closed split. After opening the split, the column was rapidly heated (30 $^{o}C/min$) to 160 ^{o}C followed by a temperature program from 160-270 ^{o}C with 1.5 $^{o}C/min$. The injection port and interface temperature was 270 ^{o}C and the temperature of the ion source was 220 ^{o}C.

RESULTS

Benzo(b)naphtho(1,2-d)thiophene

With liver microsomes of untreated Wistar rats this thiaarene was metabolized to the corresponding sulfone. The mass spectroscopy of the sulfone GC peak (RRT = 1.38) indicated that minor amounts of the sulfoxide were hidden under this peak. The mass spectrum showed an intense molecular ion M^+=266 and the typical fragments (M-29), (M-32), (M-45), (M-48), (M-66), and (M-77), corresponding to eliminations of CHO, S, CHS, SO, H_2SO_2, and $CHSO_2$, respectively. The presence of the sulfoxide could be deduced from the molecular ion M^+=250 in the ascent of the sulfone peak. In addition only one further metabolite was detected which could be identified to be a sulfone phenol. Its TMS derivative showed intense signals at m/z 354 (= M^+) and m/z 339 (M-15), however, no intense (M-73) was observed, which may indicate that the OTMS-group is located in the neighbourhood of the sulfone group (possibly at C-6 or C-8).

DDT treatment of the rats resulted in a 50 % increase of the sulfone formation and stimulated the phenol formation by a factor of 6. After treatment with the potent cytochrome P_{MC}-inducer benzo(k)fluoranthene a three-fold increase of the sulfone formation was observed, but no phenol was detected. In this incubation, however, the formation of a dihydrodiol was observed in which the dihydroxy-groups are located in a non-K-region as can be deduced from the mass spectrum of its TMS-derivative since m/z 191 is more intense than m/z 147 (further key ions are M^+=412; M-15; M-90). The quantitative composition of the metabolites formed are presented in Table

1, their structures are given in Figure 2.

TABLE 1

BENZO(b)NAPHTHO(1,2-d)THIOPHENE METABOLITES FORMED BY LIVER MICROSOMES OF UNTREATED AND DDT- OR BENZO(k)FLUORANTHENE-PRETREATED WISTAR RATS

pretreatment	metabolites formed (nmol/mg microsomal protein)		
	Sulfone (1.38)*	Sulfonephenol (1.03)	non-K-Dihydrodiol (1.16)
none	9.43	2.07	-
DDT	15.57	12.03	-
BkF	30.03	-	0.44

*GC retention time relative to benzo(b)naphtho(1,2-d)thiophene

Sulfoxide

Sulfone

Sulfonephenol

non-K-region Dihydrodiol

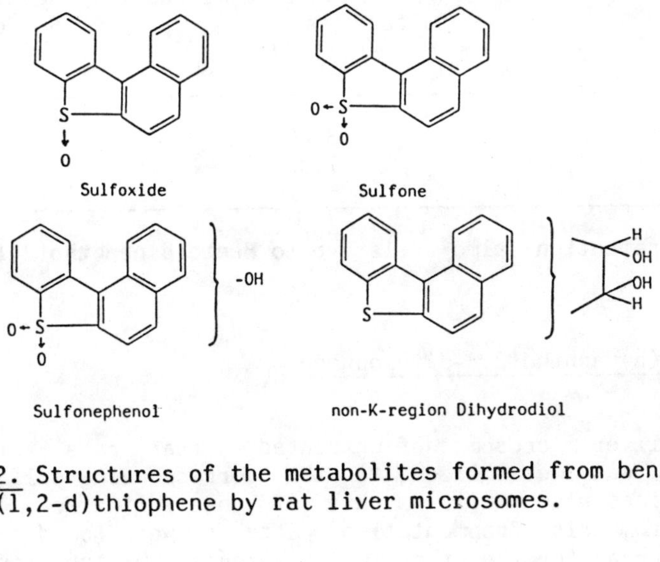

FIGURE 2. Structures of the metabolites formed from benzo(b)-naphtho(1,2-d)thiophene by rat liver microsomes.

Benzo(b)naphtho(2,1-d)thiophene

Benzo(b)naphtho(2,1-d)thiophene was converted to the corresponding sulfone by liver microsomes of untreated Wistar rats as indicated by GC/MS (RRT = 1.36; main fragments were m/z 266 (=M^+), 237, 234, 221, 218, and 189 corresponding to CHO-, S-, CHS-, SO-, and $CHSO_2$-elimination). As in case of benzo(b)naphtho(1,2-d)thiophene small amounts of the sulfoxide were also found. DDT treatment did not give rise to the sulfone formation but stimulated the formation of a sulfonephenol (M^+=354; intense m/z 339 (= M-15) for the TMS-ether). The quantitative composition of the metabolites are listed in Table 2, their structure given in Figure 3.

TABLE 2

BENZO(b)NAPHTHO(2,1-d)THIOPHENE METABOLITES FORMED BY LIVER MICROSOMES OF UNTREATED AND DDT-TREATED WISTAR RATS

pretreatment	metabolites formed (nmol/mg microsomal protein)	
	Sulfone (1.36)*	Sulfonephenol (1.07)
none	8.00	-
DDT	7.39	4.15

*GC retention time relative to Benzo(b)naphtho(2,1-d)thiophene.

Benzo(b)naphtho(2,3-d)thiophene

With liver microsomes of untreated Wistar rats this thiaarene was metabolized to the corresponding sulfone which contained also small amounts of the sulfoxide. The mass spectrometric fragmentation patterns were found to be very similar as for the other two thiaarenes. DDT treatment of

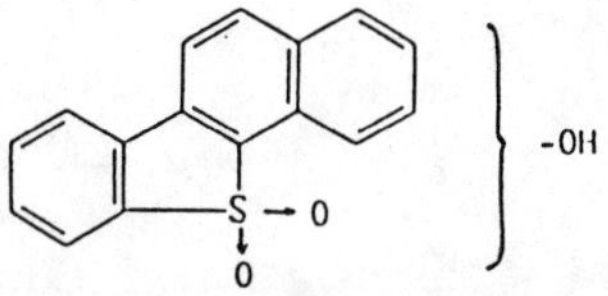

Sulfonephenol

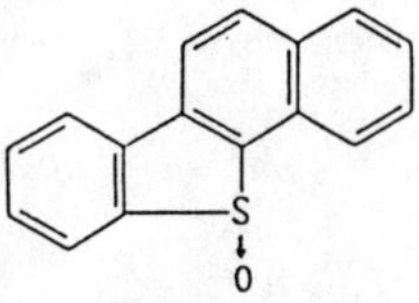

Sulfoxide

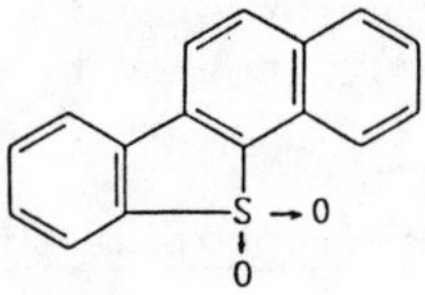

Sulfone

FIGURE 3. Structures of the metabolites formed from benzo(b)-naphtho(2,1-d)thiophene by rat liver microsomes.

the rats did not enhance the sulfone formation; however, the formation of a 6,11-quinone was observed after this pretreatment (Table 3 and Figure 4).

TABLE 3

BENZO(b)NAPHTHO(2,3-d)THIOPHENE METABOLITES FORMED BY LIVER MICROSOMES OF UNTREATED AND DDT-TREATED WISTAR RATS

pretreatment	metabolites formed (nmol/mg microsomal protein)	
	sulfone (1.36)*	6,11-quinone (1.12)
none	7.17	-
DDT	6.43	10.49

*GC retention time relative to benzo(b)naphtho(2,3-d)thiophene

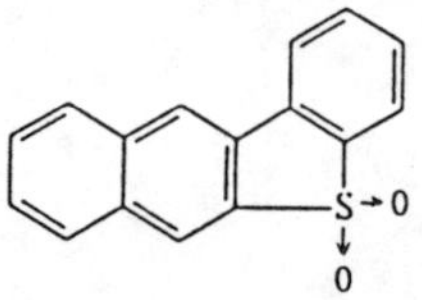

Sulfone

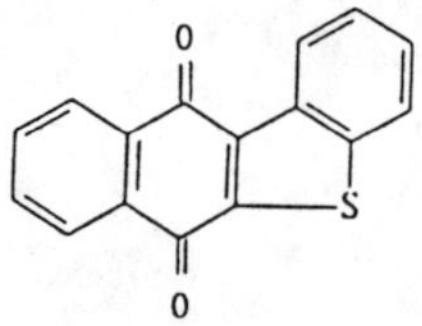

6,11-Quinone

FIGURE 4. Structures of the metabolites formed from benzo(b)-naphtho(2,3-d)thiophene by rat liver microsomes.

Triphenyleno(4,5-bcd)thiophene

The only metabolite of this thiaarene after incubation with liver microsomes of either untreated or DDT-pretreated Wistar rats was the sulfone (Figure 5) which contained also minor amounts of the sulfoxide which could not be separated by GC. In this and the aforementioned cases, however, the sulfoxide concentration was less than 10 % of that of the sulfone as indicated by mass spectrometry. DDT treatment did not stimulate the sulfone formation significantly as can be seen from Table 4.

TABLE 4

TRIPHENYLENO(4,5-bcd)THIOPHENE METABOLITES FORMED BY LIVER MICROSOMES OF UNTREATED AND DDT-TREATED WISTAR RATS.

pretreatment	metabolites formed (nmol/mg microsomal protein) sulfone (1.29)*
none	5.16
DDT	6.25

*GC retention time relative to triphenyleno(4,5-bcd)thiophene.

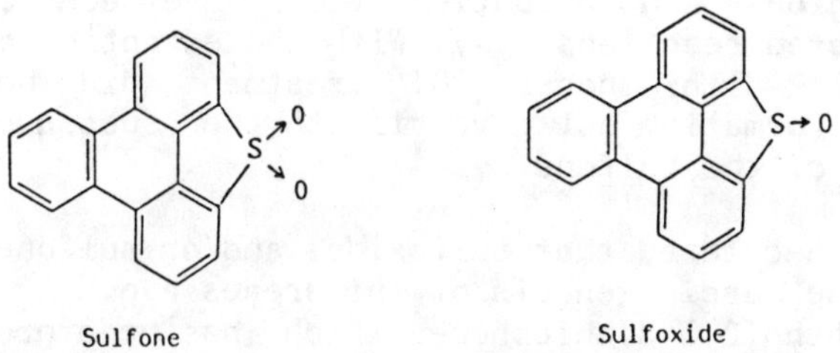

Sulfone Sulfoxide

FIGURE 5. Structure of the metabolite formed from triphenyleno(4,5-bcd)thiophene by rat liver microsomes.

DISCUSSION

Although the physico-chemical properties of thiaarenes, since depending on the π-electron system, very much resemble those of the corresponding parent PAH, their metabolic behaviour is dictated by the sensitivity of the sulfur atom against oxidation. This reaction results in the formation of sulfoxides and sulfones and predominates in all incubation experiments which have been carried out with liver microsomes of untreated rats. In case of benzo(b)naphtho(1,2-d)thiophene, which is isoster to benzo(c)phenanthrene subsequent ring epoxidation yielding a sulphonephenol was observed. However, the metabolism of benzo(c)phenanthrene, which is preferentially metabolized at the K-region (16) does not resemble at all that of benzo(b)naphtho(1,2-d)thiophene. Ring oxidation of the latter, can be stimulated by treatment with the potent cytochrome P_{MC}-inducer benzo(k)fluoranthene. The same holds true for benzo(b)naphtho(2,1-d)thiophene, the S-isoster of chrysene. No similarities between their metabolites profiles can be noticed. The thiaarene is a better substrate for rat liver microsomes, and the chrysene specific metabolic activations (formation of 3,4- and 1,2-dihydrodiols (17)) cannot be observed for the sulfur analog.

In case of benzo(b)naphtho(2,3-d)thiophene being isoster to benz(a)anthracene the S-oxidation predominates again, but surprisingly the formation of a p-quinone can be observed after DDT treatment - a reaction which does not take place under the same circumstances with benz(a)anthracene (18). S-oxidation was the exclusive reaction with triphenyleno(4,5-bcd)thiophene, the S-isoster of benzo(e)pyrene for which K-region- and oxidation at C-1 have been reported to be the prefered reactions (19). With the exception of benzo-(b)naphtho(1,2-d)thiophene, DDT treatment did not enhance the sulfone formation but gave rise to the subsequent arene epoxidation of the sulfone.

It may be speculated that sulfoxides and/or sulfones are involved in the carcinogenesis of thiaarenes e.g. in case of benzo(b)naphtho(2,1-d)thiophene which has been proven to be carcinogenic (8) and which is exclusively S-oxidized by liver microsomes of untreated rats. We are presently testing various sulfones and sulfoxides to check this hypothesis.

ACKNOWLEDGEMENTS

The authours like to thank Mr.A.Raab and Mr. A.Hildebrandt for technical assistance. The present studies were carried out in accordance with the environmental plan of the Federal Ministry of the Interior of the Federal Republic of Germany.

REFERENCES

1. Grimmer, G., Jacob, J., and Naujack, K.-W. (1983): Polycyclic aromatic compounds from crude oils. Part 3. Inventory by GCGC/MS. PAH in environmental materials. Fres. Z. Anal. Chem., 314:26-36.
2. Grimmer, G., Jacob, J., and Naujack, K.-W. (1981): Profile of polycyclic aromatic hydrocarbons from lubricating oils. Inventory by GCGC/MS - PAH in environmental materials. Part 1. Fres. Z. Anal. Chem., 306:347-355.
3. Grimmer, G., Jacob, J., Naujack, K.-W., and Dettbarn, G. (1981): Profile of the polycyclic aromatic hydrocarbons from used engine oil - Inventory by GCGC/MS - PAH in environmental Materials Part 2. Fres. Z. Anal. Chem., 309: 13-19.
4. Jacob, J. (1983): Carcinogenic impact from automobile exhaust condensate and the dependence of the PAH-profile on various parameters. In: Mobile Source Emissions Including Polycyclic Organic Species, edited by D. Rondia, M. Cooke, R.K. Haroz, pp. 110-114, D. Reidel Publ. Co.

5. Grimmer, G., Jacob, J., Naujack, K.-W., and Dettbarn, G. (1983): Determination of polycyclic aromatic compounds emitted from brown-coal-fired residential stoves by gas chromatography/mass spectrometry. Anal. Chem., 55:892-900.
6. Grimmer, G., Jacob, J., Dettbarn, G., and Naujack, K.-W. (1985): Determination of polycyclic aromatic hydrocarbons, azaarenes, and thiaarenes emitted from coal-fired residential furnaces by gas chromatography/mass spectrometry. Fres. Z. Anal. Chem., in press.
7. Tilak, B.D. (1960): Carcinogenicity by thiophene isosters of polycyclic hydrocarbons. Synthesis of condensed thiophenes. Tetrahedron, 9:76-95.
8. Croisy, A., Mispelter, J., Lhoste, J.M., Zajdela, F., and Jacquignon, P. (1984): Thiophene analogues of carcinogenic polycyclic hydrocarbons. Elbs pyrolysis of various aroylmethylbenzo(b)thiophenes. J. Heterocyclic. Chem., 21:353-359.
9. Karcher, W., Nelen, A., Depaus, R., van Eijk, J, Glaude, P., and Jacob, J. (1981): New results in the detection, identification and mutagenic testing of heterocyclic polycyclic aromatic hydrocarbons. In: Polynuclear Aromatic Hydrocarbons: Chemical Analysis and Biological Fate, edited by M. Cooke and A.J. Dennis, pp. 317-327, Battelle Press, Columbus/Ohio.
10. Pelroy, R.A., Stewart, D.L., Tominaga, Y., Iwao, M., Castle, R.N., and Lee, M.L. (1983): Microbial mutagenicity of 3- and 4-ring polycyclic aromatic sulfur heterocycles. Mutation Research, 117:31-40.
11. Klemm, L.H. and Lawrence, R.F. (1979): The insertion and extrusion of heterosulfur bridges. X. Conversions in the triphenylene-triphenylo(4,5-bcd)thiophene system (1). J. Heterocycl. Chem., 16:599-601.
12. Lucas, H.J. and Kennedy, E.R. (1955): Org. Synth., Coll. Vol. III, 482.
13. Kutt, H., and Fouts, J.R. (1971): Diphenylhydantoin metabolism by rat liver preparations. J. Pharmacol. Exp. Ther., 176: 11-26.
14. Lowry, G.L., Rosebrough, N.J., Farr, A.L., and Randall, R.J. (1951): Protein measurement with Folin phenol reagent. J. Biol. Chem., 193:265-275.
15. Omura, T. and Sato, R. (1964): The carbon-monoxide binding pigment of rat liver microsomes. J. Biol. Chem., 239:2370-2378.

16. Ittah, Y., Thakker, D.R., Levin, W., Croisy-Delcey, M., Ryan, D.E., Thomas, P.E., Conney, A.H., and Jerina, D.M. (1983): Metabolism of benzo(c)phenanthrene by rat liver microsomes and by a purified monooxygenase system reconstituted with different isozymes of cytochrome P-450. Chem. Biol. Interactions, 45:15-28.
17. Jacob, J., Schmoldt, A., and Grimmer, G. (1982): Formation of carcinogenic and inactive chrysene metabolites by rat liver microsomes of various monooxygenase activities. Arch. Toxicol., 51: 255-265.
18. Jacob, J., Grimmer, G., and Schmoldt, A. (1981): The influence of polycyclic aromatic hydrocarbons as inducers of monooxygenases on the metabolite profile of benz(a)-anthracene in rat liver microsomes. Cancer Lett., 14:175-185.
19. Jacob, J., Schmoldt, A., and Grimmer, G. (1983): Benzo-(e)pyrene metabolism in rat liver microsomes: dependence of the metabolite profile on the pretreatment of rats with various monooxygenase inducers. Carcinogenesis, 4:905-910.

THE INFLUENCE OF TEMPERATURE ON THE DAYTIME PAH DECAY ON ATMOSPHERIC SOOT PARTICLES

RICHARD KAMENS*, ZHISHI GUO, JAMES FULCHER, AND DOUGLAS BELL
Department of Environmental Sciences and Engineering, School of Public Health, University of North Carolina, Chapel Hill, N.C, USA 27514

INTRODUCTION

Polyaromatic hydrocarbons (PAH) are formed during the combustion of fossil or biomass fuels and many of these compounds are carcinogenic. PAH can contribute as much as 10-25% of the indirect acting TA98 bacterial mutagenicity to freshly emitted wood soot particles (1,2). We have previously shown that atmospheric wood soot PAH degrade quickly at moderate ambient temperatures when exposed to natural midday sunlight. This loss rate is approximately four times faster than that produced by reacting wood soot PAH with 0.2 ppm O_3 or 0.6 ppm NO_2 (3). The loss of PAH in the daylight (NO_x <0.07 ppm) is also accompanied by a 50 to 70 percent decline in indirect acting mutagenicity of the particulate extracts as measured with Salmonella typhimurium (TA98 -S9). Direct acting mutagenicity did not appreciably change under these conditions (4). Similar experiments with internal combustion gasoline soot particles exposed to natural sunlight also exhibited a dramatic reduction in indirect acting bacterial TA98-S9 mutagenicity and a lower but still significant reduction in direct acting TA98 mutagenicity (Figure 1).

More recently we have observed (5) that the daytime decay of wood soot PAH is much slower at cooler temperatures than at more moderate temperatures (Figure 2). In this paper we will attempt to semi-quantify the loss of PAH on atmospheric soot particles with respect to temperature and solar radiation effects. Wood soot particles will be used to represent a model soot system. We expect that similar relationships can be developed for PAH on particles generated from the combustion of other selected biomass or fossil fuels.

ATMOSPHERIC DAYTIME PAH DECAY

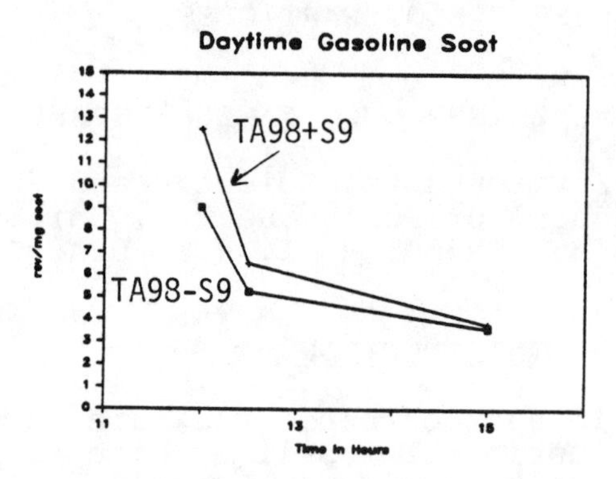

FIGURE 1. The mutagenic decay of airborne gasoline soot particles exposed to mid-summer, mid-day sunlight in the UNC 25 m^3 outdoor Teflon chambers. Particles generated from a 1979, 225 cubic inch Volare sedan.

EXPERIMENTAL

Wood smoke emissions were added directly from the chimney of a double door Buck wood stove (Smoky Mountain Enterprises, Asheville, NC, USA) to one or both 25 m^3 outdoor Teflon film chambers. Split pine logs (6 cm x 10 cm x 50 cm) were used as a fuel and these were burned at a rate of 2 to 6 kg/hr. The resulting dilute wood smoke chamber systems, which had initial particle concentrations of 400 to 2000 ug/m^3, were then exposed and aged in the presence of midday sun for two to three hours. Wood smoke injections into the chambers took place during the middle part of the solar day and during a period when the solar radiation was not rapidly increasing or decreasing.

Preliminary comparison experiments were also performed with gasoline internal combustion soot. A 4.5 horsepower, Briggs and Straton, single cylinder, internal combustion engine, and a 1979 225 cubic inch Volare Sedan engine were used to add soot particles to the chambers. The single cylinder engine was operated with the needle valve adjusted to produce a rich gasoline to air mixture, and the Volare particle injections were made immediately after a cold start.

Solar radiation measurements were taken with an Eppley black and white pyrometer (Newport, RI, USA) which was placed on the floor of the chamber.

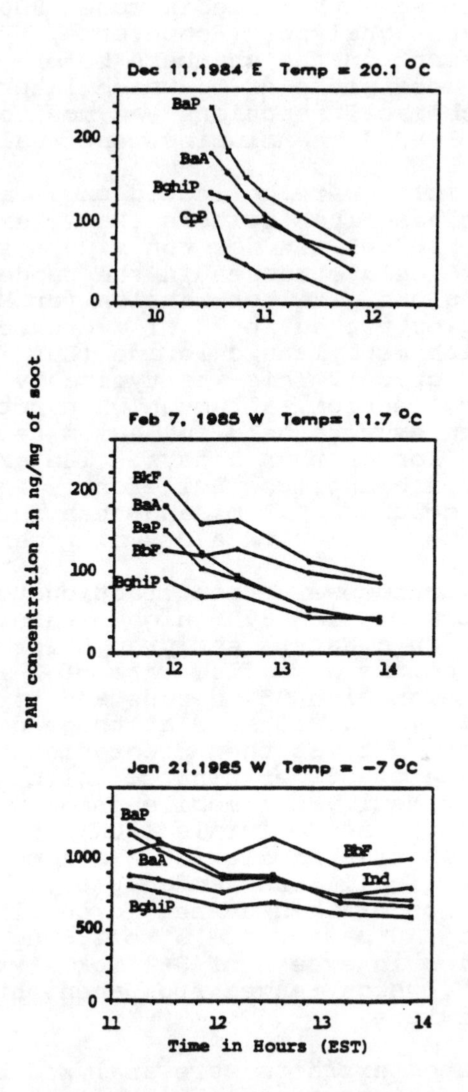

FIGURE 2. The daytime decay of PAH on airborne wood soot particles at three different temperatures.

Legend: BaP, benzo(a)pyrene; BaA, benz(a)anthracene; BghiP, benzo(ghi)perylene; CpP, cyclopenta(cd)pyrene; BbF, benzo(b)fluoranthene; BkF, benzo(j&k)fluoranthenes; Ind, indeno(1,2,3-cd)pyrene.

NO_x was monitored with a Bendix model 8002 chemiluminescent analyzer (Ronceverte, WV 24970, USA). O_3 present in the chambers before the injection of wood smoke or O_3 which later appeared due to photochemical reactions was monitored with a Bendix model 8002 chemiluminescent analyzer.

Soot particles were collected on 47 mm teflon impregnated glass fiber filters (Pallflex T60A20) at a flow of 0.07 m^3/min for ten minute sampling periods. Particulate masses in the range of 0.2 to 2.5 mg resulted. Filter samples for PAH analysis were extracted in 25 ml micro soxhlet extractors with methylene chloride (Burdick and Jackson, HPLC grade). This was typically done 3-4 hours after collection, although when not possible, samples were held in sealed test tubes at 6^oC for no longer than 3 days. The extract was concentrated with Supelco (Bellefonte, PA, USA) micro Snyder columns to 3 ml and then further concentrated to 50 ul with a dry nitrogen stream.

Filter extracts were then fractionated with a ternary Spectra Physics 8700 high pressure liquid chromatograph on a Waters semiprep u Porasil, 300 x 7.8 mm ID, column at a flow rate of 2 ml/min. The program began with 95% hexane and 5% methylene chloride ($MeCl_2$) and was held at these conditions for 12 minutes. It was then programmed to 100% $MeCl_2$ at a rate of 5% per minute. After one minute at 100% $MeCl_2$, the mobile phase was rapidly ramped up to 100% acetonitrile (ACN) and held for 25 minutes. This is similar to a scheme reported by Alfheim <u>et al.</u> (1) and Ramdahl (6). A detailed characterization of this system in our laboratory with injected PAH masses of 15 to 250 ng has shown mean recoveries in excess of 86% for sixteen different PAH, which ranged from acenaphthene to benzo(ghi)perylene.

Fractionated extracts were analyzed by capillary gas chromatography with flame ionization detection on a model 4130 Carlo Erba gas chromatograph. A thirty meter J & W DB-5 fused silica column was used with hydrogen as a carrier. One microliter samples were injected into a 300^oC split/splitless injector, while the chromatographic oven was held at 40^oC. The oven was then quickly ramped to 130^oC and temperature programmed at 6^oC/min to 290^oC.

In some experiments unfractionated extracts were analyzed directly by reverse phase high pressure liquid chromatography. A Spectra Physics 8700 ternary gradient pump and programmer were used with a Supelco 5u 4.6 x 150 mm reverse phase C18 PAH analytical column (Supelco Inc., Bellefonte, PA, USA). The initial condition of the mobile phase was a 50:50 mixture of ACN and water at a flow of 1.8 ml/min. This was programmed at the same flow to a 90:10 mixture in ten minutes after sample injection, and held at these conditions for 8 minutes. A Shimadzu model RF-5306 fluorescence detector was used with emission and excitation wave lengths of 305 nm and 430 nm. Monitored species were benzo(b)fluoranthene (BbF), benzo(k)fluoranthene (BkF), benzo(a)pyrene (BaP), and benzo(ghi)perylene. It should be noted that reported BkF concentrations by GC include benzo(j)fluoranthene (BjF), while HPLC measurements of BkF do not include BjF. Chrysene (Chry/Tri) GC measurements include the compound triphenylene.

DISCUSSION

Dark Stability of PAH on Soot Particles

The dark stability of PAH varies depending upon the substrate or sorbant on which the PAH is adsorbed or bound. In the dark, PAH concentrations on wood soot particles do not change with time (2,3). The same is true for anthracene and pyrene adsorbed onto a high surface carbon black (7), and for gasoline soot particles, aged over a period of three hours in our outdoor chambers (Figure 3). Unlike these substrates, flyash particles apparently promote the rapid dark decay of the PAH, fluorene (8). We have also observed a rapid loss of the anthracene that was coated onto silica particles when these particles were aged in the dark in our outdoor chambers (7). We could not however, distinguish between losses due to vaporization of anthracene from the particles during sampling and/or dark oxidation reactions of anthracene promoted by the silica surface.

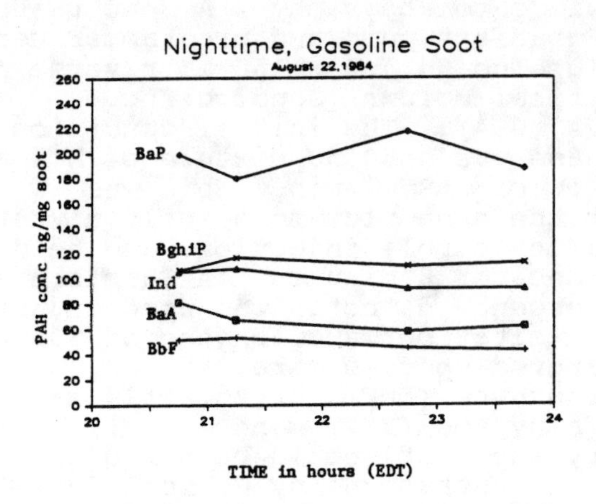

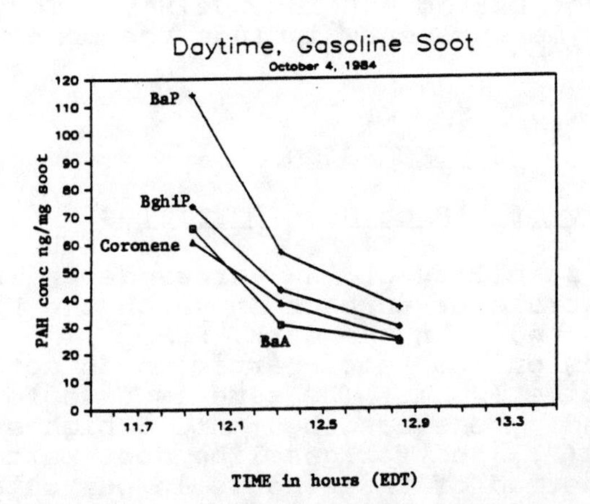

FIGURE 3. Stability of gasoline soot PAH in the presence and absence of sunlight. Nightime particles came from a single cylinder 4 stroke gasoline engine; daytime particles from a 1979 Volare sedan. The average chamber temperature for the nighttime run was 20.3°C and the daytime run, 30.0°C. Coronene and BghiP concentrations have been divided by 10.

Decay of PAH in Sunlight

Studies which have addressed the photo-induced decay of PAH on filters, silica particles, or airborne wood soot particles have suggested that the apparent loss of PAH may be approximated by an expression which is first order in PAH concentration (3,5,9). Preliminary data reported by Cimberle et al. (9) and Blau and Güsten (10), and Kamens et al. (3) have also indicated that the apparent PAH decay rate constant is directly proportional to the absorbed light energy or solar intensity.

The dramatic effect of sunlight on gasoline and wood soot particle PAH decay is illustrated in Figure 4. We have also observed a loss of PAH from diesel combustion soot and China coal soot particles (11). The gasoline soot experiment in Figure 4 took place over a period of thirty hours, and one can see the effect of sunlight during the first daylight period, the general stability of PAH during the following hours of darkness, and the subsequent decay of PAH decay after the sun came up on the second day.

The last sample taken on the first day occurred three hours before the sun went down. An estimate of the PAH decay that occurred between this sample and sundown was made from the observed afternoon decay and the available sunlight measured during this period. We performed similar estimates of the PAH decay for the second day between the time of sunrise and the first sample at 9:34 am (33.6 hr in Figure 4).

The data in Figure 4 were obtained from direct HPLC analysis of unfractionated particulate extracts. A comparison from the first day of BaP and BghiP concentrations obtained from wood soot extract fractionation, and GC analysis vs direct HPLC analysis of unfractionated extracts, is shown in Figure 5. As can be seen, both methods yield similar concentrations and PAH behavior. We have therefore concluded that unfractionated HPLC PAH data can be used to illustrate the pattern of PAH decay.

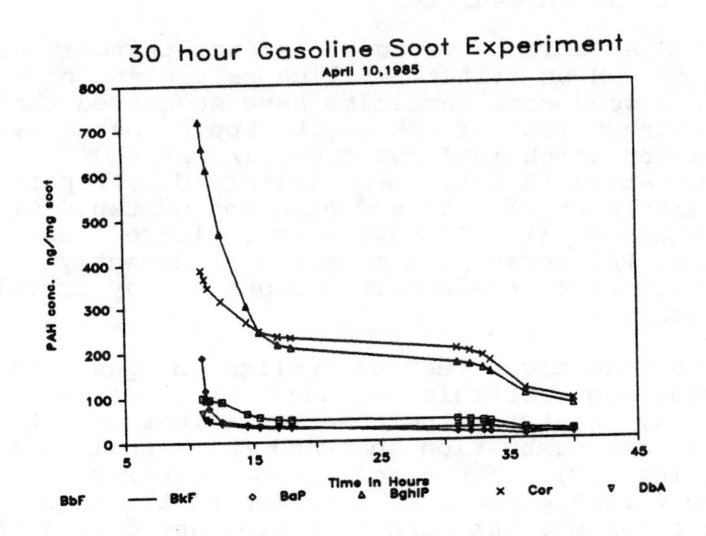

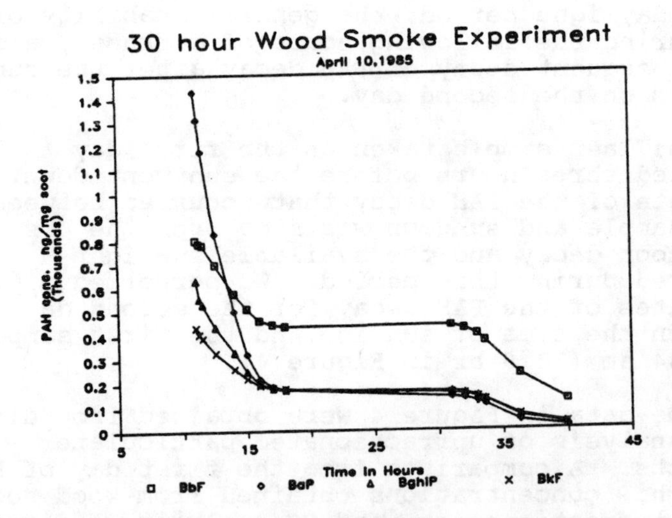

FIGURE 4. Decay of wood and gasoline soot PAH over a 30 hour period of outdoor sunlight and darkness. Data points at 16, 17,18,31,32 & 33 hrs. were predicted from estimated decay during this period. All other data points were derived from HPLC analysis of unfractionated extracts.

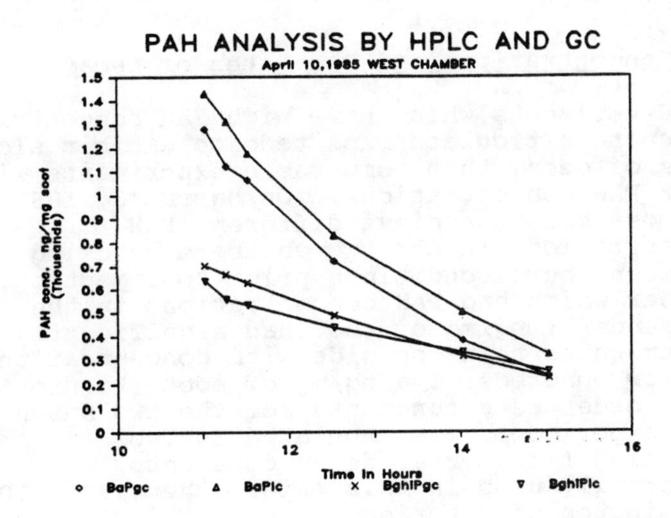

FIGURE 5. PAH decay on wood soot particles as measured by direct HPLC analysis of wood soot extracts or fractionation and GC analysis.

TABLE 1

FIRST ORDER RATE CONSTANTS FOR PAH DAYTIME DECAY AT DIFFERENT PAH CONCENTRATIONS*

	March 13,1985		Feb 7,1985		Apr 10, 1985
PAH conc	High min-1	low min-1	low min-1	low min-1	High min-1
BaA	0.0125	0.0250	0.0128	0.0147	0.0057
Chry/Tri	0.0056	0.0090	0.0027	0.0050	0.0022
BbF	0.0065	0.0090	0.0034	0.0046	0.0023
Bk&jF	0.0057	0.0153	0.0065	0.0063	0.0031
BaP	0.0090	0.0211	0.0110	0.0116	0.0062
Ind	0.0054	0.0125	0.0066	0.0087	0.0035
Bghip	0.0077	0.0116	0.0064	0.0100	0.0042
ave temp	23.3 °C	24.1 °C	11.7 °C	11.7 °C	14.6 °C
ave solar intensity	0.93 Ly	0.94 Ly	0.67 Ly	0.67 Ly	0.97 Ly

* high PAH conc. = 700-2000 ng/mg soot
low PAH conc. = 50 - 350 ng/mg soot

** 1 Ly = 1 cal min-1 cm-2

PAH Concentrations and PAH Rates of Decay

Experiments which have high PAH concentrations per unit particulate mass tend to exhibit slower rates of decay than comparable experiments with lower PAH concentrations. On March 13, 1985 it was possible to achieve different PAH concentrations in the two chambers by using different burn conditions prior to injection. The chamber which had PAH concentrations in the thousands of ng/mg of soot had significantly slower rates than the side with concentrations in the low hundreds of a ng/mg of soot (Figure 6). First order rate constants for the two March 13, 1985 experiments are tabulated in Table 1. In computing first order decay constants, we arbitrarily used in this paper, the first 110 to 150 minutes of reaction.

On the February 7, 1985 the date of one of the experiments described in Figure 2 in this paper, there was also a companion experiment in the other chamber. Since the injection of wood soot into both chambers occurred within minutes of one another, similar PAH concentrations resulted. This run had PAH concentrations in the low hundreds of a ng/mg soot. The PAH decay profiles from both experiments for most of the monitored PAH were very similar (see computed first order rate constants in Table 1 and Figure 7).

As with the high PAH concentration March 13 experiment, the thirty hour April 10, 1985 wood soot experiment (Figure 4) also had relatively high PAH particle concentrations. The temperature on this day was 14.6 ^{o}C, almost 3 degrees higher than the February 7, 1985 experiment. As would be expected the April 10 experiment had much lower first order rate constants (Table 1) than the February 7 experiments, even though April 10, had a higher average solar intensity and temperature.

At colder temperatures we have made similar comparisons, and have shown (5) that as temperatures cool, the rates of PAH decay become very slow and the influence of PAH concentrations on rates of decay becomes less apparent.

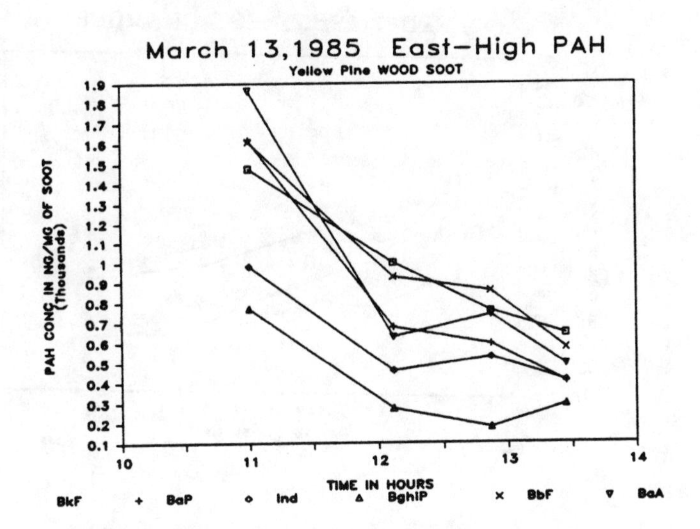

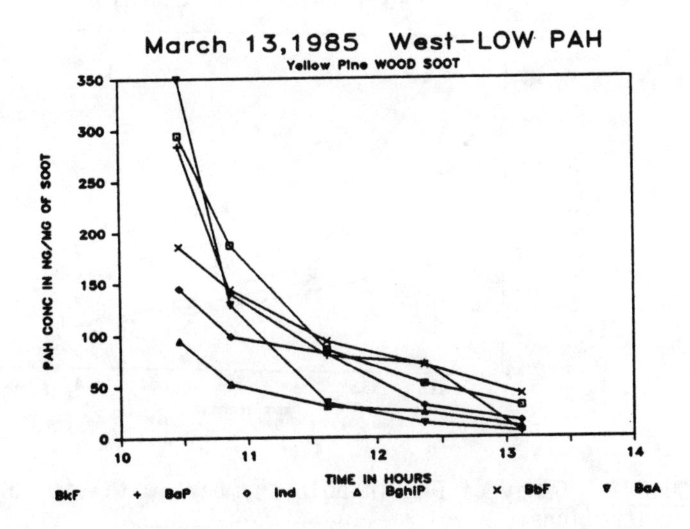

FIGURE 6. The effect of high and low PAH concentration on the daytime decay of PAH.

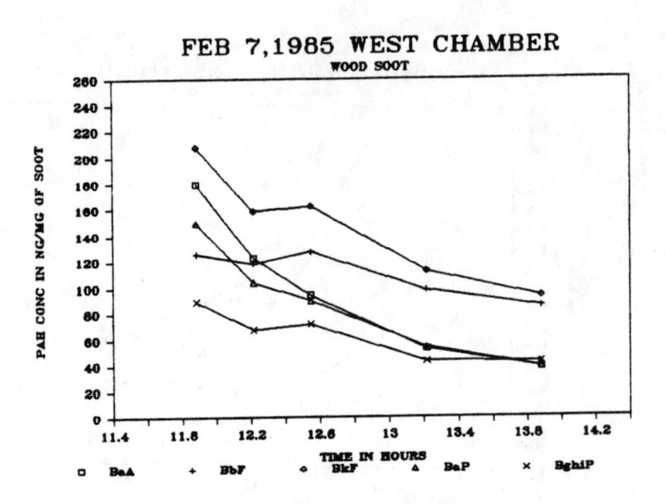

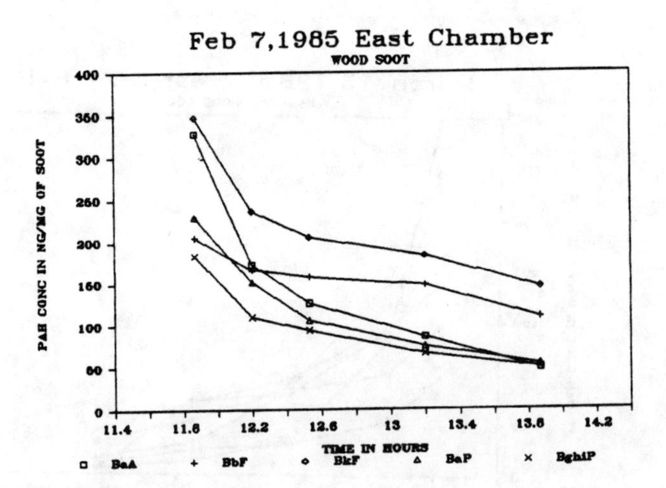

FIGURE 7. Decay of PAH in both chambers with similar PAH concentrations.

Isolating Sunlight and Temperature Effects

Since the concentration of PAH on airborne soot particles appears to be associated with observed PAH rates of decay, we have used experiments in this preliminary analysis which had PAH initial concentrations between 350 and 30 ng/mg. The only exception to this was the compound retene,(1-methyl-7-isopropyl-phenanthrene). Its concentrations were generally in the 1000 to 2000 ng/mg of soot, while other in PAH from the same soot particles were in the low hundreds of a ng/mg. This compound has been proposed as a possible tracer for wood smoke from the residential burning of soft wood (12), although more recently it has been identified in other sources.

In order to isolate sunlight from temperature effects on PAH decay, a number of experiments were conducted under similar outdoor temperatures and different solar intensities. The relationship between the first order decay rate and available sunlight for BaP is illustrated in Figure 8. We have assumed that similar rate vs solar intensity lines exist for all temperatures. This has permitted us to scale all of the experimentally determined rates, at different temperatures and solar intensities, to rate constants which would have resulted if the solar intensity had been one cal cm^{-2} min^{-1}. The scaled or normalized rate constants were then plotted vs the inverse of temperature to determine if the data conformed to an Arrhenius expression (equation 1).

$$k = Ae^{-Ea/RT} \qquad \text{eq 1.}$$

As illustrated in Figure 8, this approach can provide a reasonable approximation to the data and makes it possible to estimate "activation energies", Ea/R, and pre-exponential factors, A, for different PAH (Table 2). With these values we have calculated the half-lives for a number of PAH on soot particles at different temperatures and solar intensities (Table 3). At warm ambient temperatures and an average midday solar intensity of one cal cm^{-2} min^{-1}, PAH half-lives were of the order of one hour. At very cool temperatures and

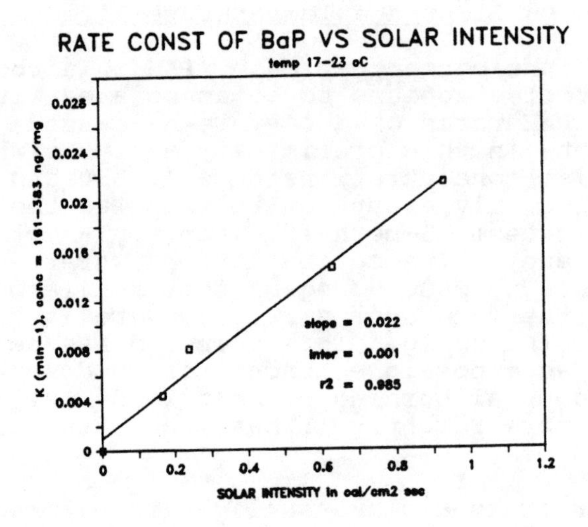

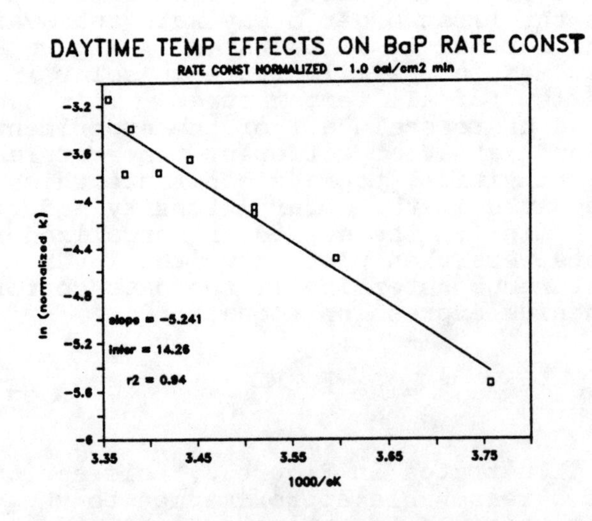

FIGURE 8. The relationship between Temperature, Solar Intensity, and the first order daylight rate constant for BaP.

TABLE 2

ACTIVATION ENERGIES AND PRE-EXPONENTIAL FACTORS FOR PAH PHOTO-INDUCED FIRST ORDER RATE CONSTANTS*

COMPOUND	A	Ea/R	r2	estimated rate constants "k" in(min-1)
Chry/Tri	3.36E+07	6405	0.88	0.0022
BbF	9.29E+05	5386	0.84	0.0025
BkF/BjF	6.16E+08	7112	0.91	0.0030
Ind	2.49E+10	8173	0.93	0.0025
BghiP	2.25E+06	5479	0.89	0.0044
BaP	1.56E+06	5241	0.94	0.0072
BaA	2.22E+06	5273	0.94	0.0092
Retene	3.28E+05	4981	0.90	0.0039

* at zero ^{0}C and an average solar intensity of 1 cal min-1 cm-2 calculated from linear regression of ln(k) vs 1000/^{0}K to give the form: k = A*exp(Ea/RT) ; R = 1.9872 cal/deg mole

TABLE 3

PAH HALF-LIVES IN THE ATMOSPHERE AT DIFFERENT SOLAR INTENSITIES AND TEMPERATURES

	(HOURS OF DAYLIGHT)			
COMPOUND	+ 20 oC (1 Ly)	0 ^{0}C (0.8 Ly)	- 10 ^{0}C (0.6 Ly)	- 20 ^{0}C (0.4 Ly)
Chry/Tri	1.1	6.6	55.7	74.8
BbF	1.2	5.7	36.0	48.4
BkF/BjF	0.6	4.8	49.6	66.6
Ind	0.6	5.7	81.1	109.0
BghiP	0.7	3.3	21.5	28.8
BaP	0.4	2.0	12.1	16.3
BaA	0.3	1.6	9.6	12.9
Retene	0.8	3.7	20.6	27.7

* 1 Ly = 1 cal min-1 cm-2

low angle sunlight PAH daytime half-lives on airborne soot particles increased for many compounds to a period of days. These estimates assumed constant sunlight and no periods of darkness. Hence, if one were to include these factors, atmospheric half-lives for PAH under very cold conditions, would be much longer.

In similar experiments with gasoline soot particles at 30°C and 14.6°C, we have also observed PAH behavior that is similar to wood soot particles. One may thus reasonably speculate, that under much cooler conditions, gasoline soot PAH would also show the kinds of temperature effects that have been observed with wood soot PAH.

PAH Decay And Humidity

As we have previously reported (5) that the the very cold experiments that were used to illustrate a temperature effect on PAH rates of decay also had dew points which were much lower than those present during the warmer experiments. It would thus appear that lower water vapor concentrations could also be associated with the lower rates of PAH decay. Very recently we have simultaneously performed wet and dry soot experiments in the two chambers. Preliminary analysis suggests that the lower dew point experiments had slower rates of PAH decay.

In computing the PAH activation energies described above we did not segregate humidity (or other possible factors with the exception of sunlight), from temperature. Thus the above reported "activation energies" mathematically include (or are confounded by) other factors like humidity. They therefore represent a composite or lumping of the different parameters, other than solar radiation, which influence PAH decay. The ways that temperature and humidity affect the mechanism of PAH photo-induced decay on atmospheric wood soot particles is not understood. The reactions which involve singlet delta molecular oxygen oxidation or OH attack are possibilities. In our current work we are attempting to isolate the impact of humidity from temperature. This will be the subject of a future manuscript.

SUMMARY AND CONCLUSIONS

Based on the results of these studies it is anticipated that during the winter months, regions which are in the extreme northern or southern latitudes, will not experience significant reductions in wood soot PAH due to sunlight as these particles are aged and transported in the atmosphere. This will occur as a result of extremely low ambient temperatures and the low angle of the sun during this season. On warmer winter days in moderate climates, PAH will show some deterioration, depending on the temperature and the available sunlight.

Preliminary indications also suggest that PAH on soot from gasoline exhaust are stable at night and decay at moderate temperatures in the presence of sunlight. Since this is the type of behavior we have seen with wood soot PAH, one is led to speculate that the decay of gasoline soot may also be temperature dependent. This would further explain why PAH concentrations (per mass of particulate matter) in urban atmospheres tend to be generally higher in the winter rather than in the summer months.

ACKNOWLEDGEMENTS

This work was sponsored by a grant, #R812256, from the US EPA to the University of North Carolina. This paper has not under gone US EPA review for publication and the contents do not necessarily reflect EPA views or policy. Mention of trade names or commercial products also does not constitute EPA endorsement or recommendation for use.

REFERENCES

1. Alfheim, I., Becher, G., Hongslo, J.K., and Ramdahl, T. (1984): Mutagenicity testing of high performance liquid chromatograph, fractions from wood stove emission samples using modified salmonella assay requiring smaller sample volumes, Environmental Mutagenesis, 6:91-102.

2. Kamens, R.M., Bell, D.A., Dietrich, A., Perry, J.M., Goodman, R.G., and Claxton, L.D. (1985a): Mutagenic transformations of dilute wood smoke systems in the presence of ozone and nitrogen dioxide. Analysis of selected high-pressure liquid chromatography fractions from wood smoke particle extracts, Environmental Science & Technology, 19:63-69.
3. Kamens, R.M., Perry, J.M., Saucy, D.A., Bell, D.A., Newton, D.L., and Brand, B. (1985b): Factors which influence polynuclear aromatic hydrocarbon decomposition on wood smoke particles, Environment International, 11:131-136.
4. Bell, D.A. and Kamens, R.M. (1986): Photodegradation of wood smoke mutagens under low NO_x conditions, Atmospheric Environment, 20:2, 317-322.
5. Kamens, R.M., Fulcher, J.N., and Guo, Z. (1986): Effects of temperature on wood soot PAH decay in atmospheres with sunlight and low NO_x, Atmospheric Environment (in press).
6. Ramdahl, T. (1983): Polycyclic aromatic ketones in environmental sample. Environmental Science & Technology, 17:666-670.
7. Saucy, D.A., Kamens, R.M., and Linton, R.W. (1986): Reactivity of PAH and nitro-PAH adsorbed on particulate substances, Polyaromatic Hydrocarbon, Ninth International Symposium, edited by Marcus Cooke), Battelle Press, Columbus (in press).
8. Korfmacher, W., Mamanotov, G., Wehry, E., Natusch, D, and Mauney, T. (1981): Nonphotochemical decomposition of fluorene vapor-adsorbed on coal fly ash, Environmental Science and Technology, 5(11):1370-1375.
9. Cimberle, M.R., Bottino, P., and Valerio, F. (1983): Decomposition of benzo(a)pyrene deposited on glass fiber filters and exposed to sunlight, Chemosphere, 12:317-324.
10. Blau, L. and Güsten, H. (1982): Quantum yields of the photodecomposition of polynuclear aromatic hydrocarbons absorbed on silica gel. Polynuclear Aromatic Hydrocarbons. Sixth International Symposium, edited by Marcus Cooke, Anthony Dennis and Gerald Fisher, pp. 135-144, Battelle Press, Columbus.

11. Kamens, R.M., Guo, K., Fulcher, J.N., and Wenxing, W. (1985c): The Reaction of PAH on Combustion Soot Particles from a Chinese Stove in an Atmosphere with Sunlight and low NO_x, A Preliminary Study, Final Report to the US EPA Contract #5D2458NAEX to the Univ. of North Carolina. William Wilson, Project Officer, U.S. Environmental, Research Triangle Park, NC.
12. Ramdahl, T. (1983): Retene-A molecular marker of wood combustion in ambient air, Nature, 306:580-582.

DIFFERENTIAL INDUCTION OF THE MONOOXYGENASE ISOENZYMES IN MOUSE LIVER MICROSOMES BY POLYCYCLIC AROMATIC HYDROCARBONS.

ANNETTE KEMENA* , KLAUS H. NORPOTH, JÜRGEN JACOB[1]
Institute of Hygiene and Occupational Medicine, University Medical Center, Essen University, Hufelandstr.55, D-4300 Essen, FRG and (1)Biochemical Institute of Environmental Carcinogens, Sieker Landstr.19, D-2070 Ahrensburg, FRG

INTRODUCTION

Polycyclic aromatic hydrocarbons (PAHs) are regarded as a class of compounds capable to induce specifically the cytochrome (cyt) P-448 dependent microsomal monooxygenases. Most of the induction experiments reported have been carried out with 3-methylcholanthrene (3-MC) or benzo[a]pyrene B(a)P, taken as representatives of the PAHs (1-15). This investigation was performed in order to evaluate, if selective induction of the cyt P-448 dependent monooxygenases is a general characteristic of all PAHs, and how the patterns of enzymatic activity vary with time. For this purpose we chose several PAHs known as poor or potent inducers from previous experiments on mutagenicity (16), and compared them with 3-MC and phenobarbital, which are regarded as typical representatives of cyt P-448 type inducers and the cyt P-450 type inducers, respectively. For discrimination of the two isoenzyme groups, the two specific inhibitors α-naphthoflavone (for cyt P-448) and metyrapone (for cyt P-450) were employed.

MATERIALS AND METHODS

Chemicals

Benzo[j]fluoranthene (BjF), benzo[k]fluoranthene (BkF) and dibenz[ah]anthracene (DBahA) were obtained from the Bureau of Reference of the European Community (BCR, Brussels, Belgium). Pyrene (Pyr) was purchased from commercial sources and was sublimated and recrystallized in order to reach 99% purity. Benzo[e]pyrene (BeP) and indeno[1,2,3-cd]pyrene (IP) were synthesized according to the methods described elsewhere (17,18). The 99% purity of all compounds was checked by capillary gas chromatography, HPLC, UV spectroscopy and high-resolution mass spectrometry (70 eV). 3-MC and umbelliferone were purchased from Serva (Heidelberg), phenobarbital and deoxycholic acid from Merck (Darmstadt), isocitrate dehydrogenase, NADP and 7-ethoxycoumarin from Boehringer (Mannheim), isocitrate, sodium-dithionite and bovine serum albumin from Sigma (St. Louis).

Treatment of Animals

3 female B6C3F1 mice (25-30 g, inhouse breeding) were each submitted to a single i.p. injection of 40 mg/kg body weight of a PAH or phenobarbital suspended in 0.1 ml corn oil by ultrasonification. This amount was the maximal tolerable dose for IP. The animals were killed one to five days later for liver extirpation. Controls remained untreated. Induction experiments were carried out once with each PAH and phenobarbital, except with IP where two identical experiments were performed.

Liver Extracts

Livers of three similarly treated animals were pooled, minced, homogenated in 5 volumes of 0.5 M Tris/HCl buffer pH 7.4 containing 0.25 M sucrose, centrifuged at 11,000 x g and 4°C for 20 min. The supernatant (S11 fraction) was used for the tests.

7-Ethoxycoumarin O-Deethylase Activity

The enzymatic activity was determined fluorimetrically according to the methods described by Ullrich and Weber (19) with slight modifications. The assay mixture contained in a total volume of 1 ml: 0.1 M Tris/HCl buffer pH 7.4, 0.5 mM NADP, 5 mM isocitrate, 5 mM $MgCl_2$, 5 μM $MnCl_2$, 1 U isocitrate dehydrogenase and 0.2-1 mg S11 protein. Before addition of liver extract the assay mixture was incubated for 20 min at 37°C for NADPH generation. The reaction was started by addition of 0.1 mM 7-ethoxycoumarin (dissolved in 0.1 M Tris/HCl buffer pH 7.4), and umbelliferone generation was monitored for 5 min in a Perkin Elmer fluorimeter 650-10S. Subsequently 25 μl of 0.1 mM umbelliferone (dissolved in 0.1 M Tris/HCl buffer pH 7.4) was added for standardization. In inhibition experiments metyrapone (0.1 mM) and α-naphthoflavone (0.5 mM) were added to the preincubated assay mixture before starting the reaction with 7-ethoxycoumarin. Measurements were carried out in duplicate (deviation of the mean value: 15%).

Cyt P-450 Content

The total content of cyt P-450 dependent monooxygenases in the S11 fraction was determined as the CO difference spectrum at 450 and 500 nm according to the method described by Omura and Sato (20). Measurements were carried out in duplicate in a double beam Kontron spectrophotometer Uvicon 810 (deviation of the mean value: 10%).

Protein content

Protein contents were determined by the Biuret method according to Gornall et al. (21). Bovine serum albumin was used as standard. After decolourization by addition of a small amount of KCN, extinctions were determined at the same wavelength in order to correct the data for turbidity caused by microsomal particles. Measurements were carried out in duplicate (deviation of the mean value: 8%).

RESULTS

The 7-ethoxycoumarin O-deethylase activity as well as the total content of cyt P-450 in S11 fractions of induced mouse liver was monitored over a time period of 5 days. The PAHs used for induction of the monooxygenase isoenzymes were chosen according to their induction potency determined in previous mutagenicity tests (16). Pyr and BeP were employed as poor inducers, BkF as an inducer of medium potency, and IP, DBahA and BjF as potent inducers.

According to the enzyme pattern induced, the PAHs could be classified into 3 different groups:
- the "P-448 type", where α-naphthoflavone was almost exclusively inhibitory, while metyrapone did not or just slightly affect the enzymatic activity: DBahA, BkF (Figure 1);
- the "mixed type", where both α-naphthoflavone and metyrapone inhibited the 7-ethoxycoumarin O-deethylase activity to different extents: Pyr, BjF (Figure 2), 3-MC, BeP (Figure 3);
- a "special P-448 type", where α-naphthoflavone inhibited and metyrapone stimulated the activity of the 7-ethoxycoumarin O-deethylation: IP (Figure 4).

The time courses of the enzymatic activities were specific for each PAH, and in the case of the "mixed type" inducers both activities responsive to inhibition by α-naphthoflavone and metyrapone, respectively, did not always parallel each other (Figures 2 and 3).

Induction with 3-MC which is generally referred to as specific for cyt P-448, in our experiments turned out to be of a "mixed type" (Figure 3A). The most selective inducer of the cyt P-448 isoenzyme group was DBahA, which thereby exactly opposites the phenobarbital mediated specific cyt P-450 induction (Figures 1A and 5).

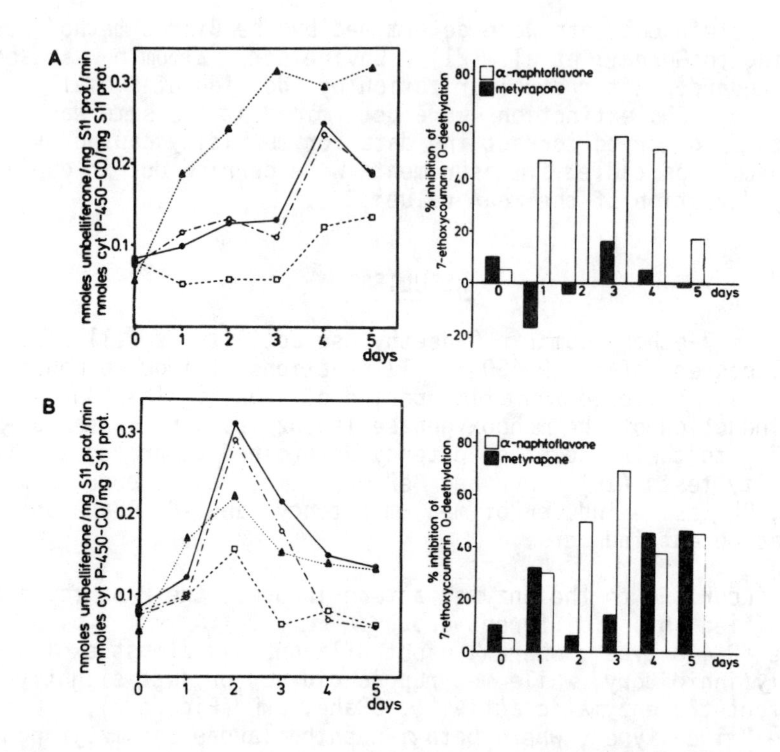

FIGURE 1. Induction potency of DBahA (A) and BkF (B) on the microsomal 7-ethoxycoumarin O-deethylase activity in mouse liver (●) and its inhibition by α-naphthoflavone (□) and metyrapone (○). 7-ethoxycoumarin O-deethylase activity is given as absolute activity in the absence and presence of ligand in nmoles umbelliferone/ mg S11 protein/minute, and as relative inhibition by either ligand expressed as % of the activity in the absence of ligand. Total content of cyt P-450 dependent monooxygenases is given as nmoles cyt P-450-CO/mg S11 protein (▲).

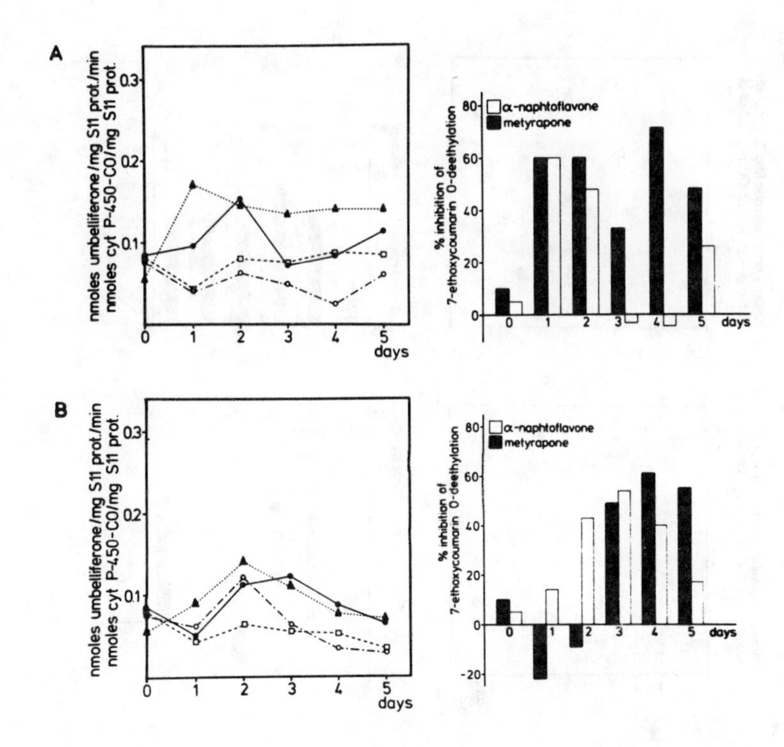

FIGURE 2. Induction potency of Pyr (A) and BjF (B) on the microsomal 7-ethoxycoumarin O-deethylase activity in mouse liver (●) and its inhibition by α-naphthoflavone (□) and metyrapone (○). 7-ethoxycoumarin O-deethylase activity is given as absolute activity in the absence and presence of ligand in nmoles umbelliferone/mg S11 protein/minute, and as relative inhibition by either ligand expressed as % of the activity in the absence of ligand. Total content of cyt P-450 dependent monooxygenases is given as nmoles cyt P-450-CO/mg S11 protein (▲).

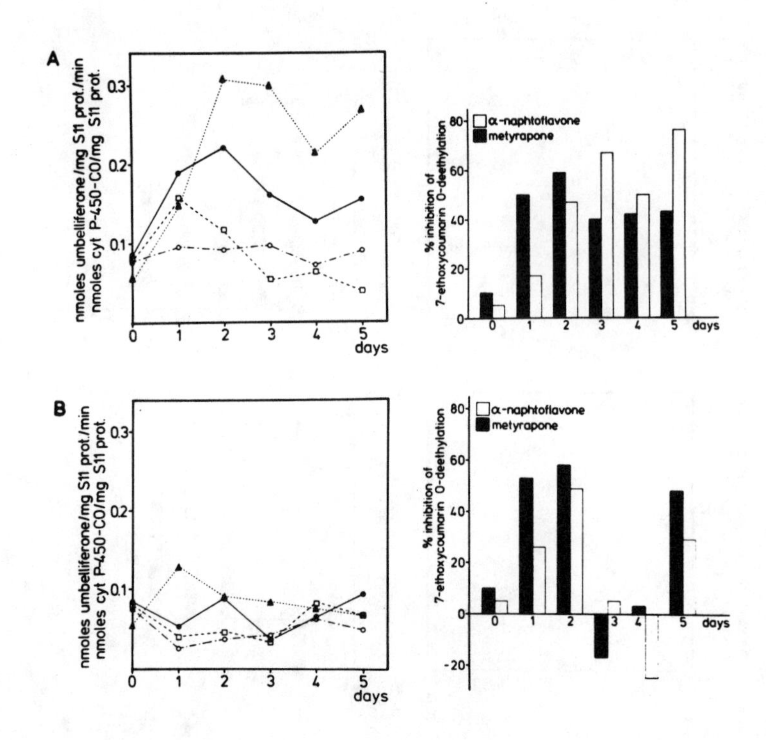

FIGURE 3. Induction potency of 3-MC (A) and BeP (B) on the microsomal 7-ethoxycoumarin O-deethylase activity in mouse liver (●) and its inhibition by α-naphthoflavone (□) and metyrapone (○). 7-ethoxycoumarin O-deethylase activity is given as absolute activity in the absence and presence of ligand in nmoles umbelliferone/mg S11 protein/minute, and as relative inhibition by either ligand expressed as % of the activity in the absence of ligand. Total content of cyt P-450 dependent monooxygenases is given as nmoles cyt P-450-CO/mg S11 protein (▲).

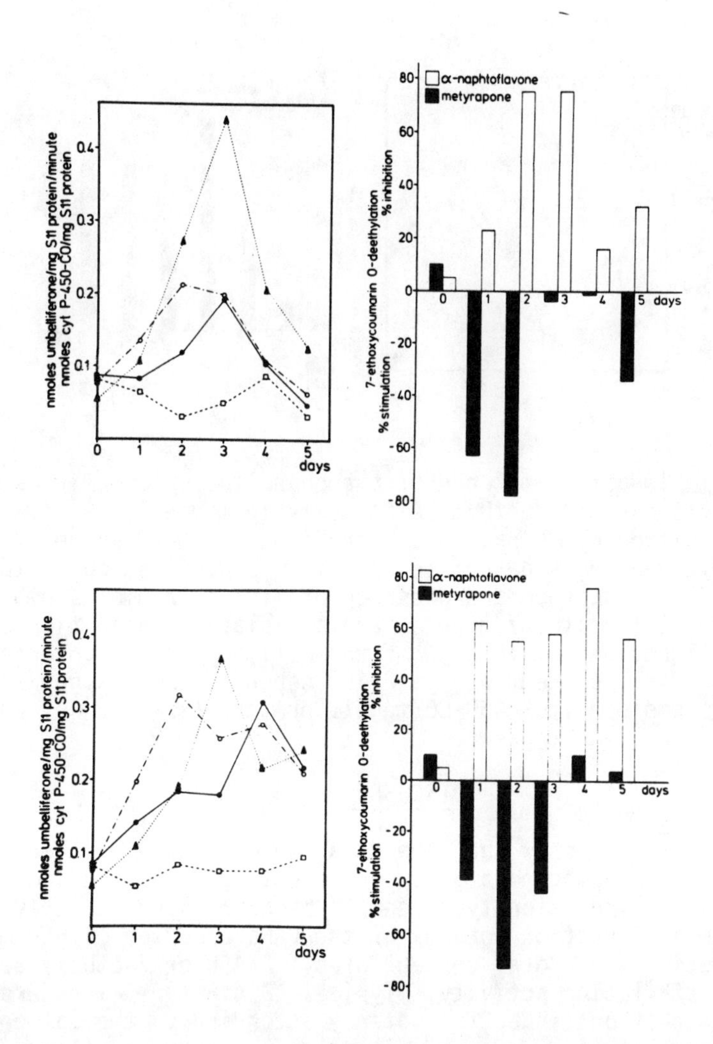

FIGURE 4. Induction potency of IP (two identical experiments) on the microsomal 7-ethoxycoumarin O-deethylase activity in mouse liver (●) and its inhibition by α-naphthoflavone (□) and metyrapone (○). 7-ethoxycoumarin O-deethylase activity is given as absolute activity in the absence and presence of ligand in nmoles umbelliferone/mg S11 protein/minute, and as relative inhibition by either ligand expressed as % of the activity in the absence of ligand. Total content of cyt P-450 dependent monooxygenases is given as nmoles cyt P-450-CO/ mg S11 protein (▲).

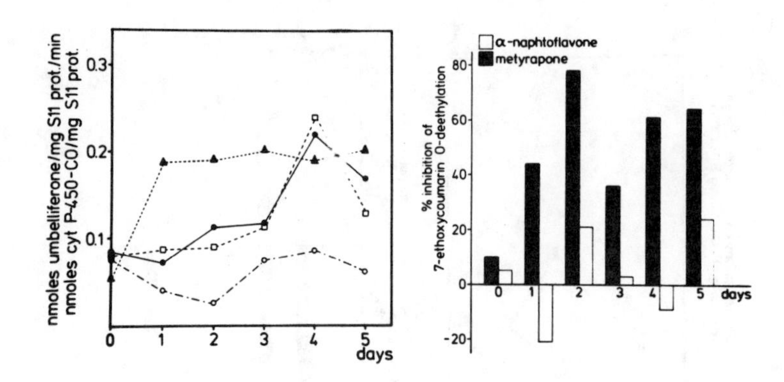

FIGURE 5. Induction potency of phenobarbital on the microsomal 7-ethoxycoumarin O-deethylase activity in mouse liver (●) and its inhibition by α-naphthoflavone (□) and metyrapone (○). 7-ethoxycoumarin O-deethylase activity is given as absolute activity in the absence and presence of ligand in nmoles umbelliferone/mg S11 protein/ minute, and as relative inhibition by either ligand expressed as % of the activity in the absence of ligand. Total content of cyt P-450 dependent monooxygenases is given as nmoles cyt P-450-CO/mg S11 protein (▲).

DISCUSSION

Our results show that the PAHs can obviously neither be regarded as a homogenous group of selective cyt P-448 inducers, nor does the induction type seem to be correlated strictly with the general induction potency of the PAHs based on either mutagenic activity or total content of cyt P-450 or 7-ethoxycoumarin O-deethylation activity. Typical "P-448 type" inducers are DBahA with strong induction potency according to the Salmonella oxygenase test (16), and BkF with medium induction potency in the Salmonella oxygenase test (16), but with a high 7-ethoxycoumarin O-deethylation activity. The group of the "mixed type" inducers consists of PAHs with poor inducing activity in the mutagenicity test, like Pyr and BeP (16), and also of strong inducers like 3-MC (22) and BjF (16), the latter showing only a medium 7-ethoxycoumarin O-deethylase activity. As variability of the enzyme pattern with time is marked after induction with various PAHs, a "mixed type" inducer may show transitory selectivity for cyt P-448 (see BjF day 2; Figure 2B) or for cyt P-450 (see Pyr days 3-4; Figure 2A). Neither the enzyme patterns nor the days of maximum enzymatic activities correlate with general induction potency of the PAH.

In the case of adult female B6C3F1 mice, DBahA and not 3-MC turned out to exactly oppose the specific cyt P-450 inducer PB. Selectivity of 3-MC for cyt P-448 induction has been shown with the aid of several different methods, e.g. specific enzymatic activities in microsomal preparations (15,23), catalytic activities of purified enzymes (24) and specific antibodies (2,5, 10,11,25). Apart from these results there are also some reports of "mixed" induction by 3-MC. In the O-deethylation reaction of 7-ethoxycoumarin 10 and 20% of the total cyt P-450 content of 3-MC induced female rats were susceptible to inhibition by metyrapone (7,26). When tested for several enzymatic activities, two cytochrome species purified after induction of rats with 3-MC both exhibited a substantial part of catalytic activities known to be selective for the cyt P-450 isoenzymes (27). An immunochemical quantitation of different microsomal monooxygenases of 3-MC pretreated male rats showed that besides the predominant 3-MC inducible isoenzymes a minor amount of cyt P-450s occurred which were also induced to appreciable extents by phenobarbital treatment (12). Furthermore, one form of immunochemically quantitated two major 3-MC inducible cyt P-448s, called P-450 d or βNF/ISF-G (12) or P-448$_{HCB}$ (13) was shown to be susceptible to metyrapone binding (29). On the basis of these findings the "mixed induction" we found after pretreatment of female mice with 3-MC and also with other PAHs could be attributed either to simultaneous induction of isoenzyme species belonging to the minor phenobarbital inducible forms (12) or to a certain form of 3-MC inducible cyt P-448 susceptible to metyrapone (9). Induction of this latter form, called P-450 d, has been shown to vary greatly among different rodent species.

Among the PAHs tested induction with IP resulted in a unique effect of metyrapone on the 7-ethoxycoumarin O-deethylation reaction: it stimulated the enzymatic activity, most pronounced during the first 2-3 days after induction (Figure 4) without showing any significant inhibitory effect during the final days of the period of time tested. α-naphthoflavone on the other hand was strongly inhibitory, reflecting high cyt P-448 activity. Enhancement of enzymatic activity by metyrapone has been reported (30,31,32). A substrate dependent increase in enzymatic activity in the presence of metyrapone (maximal for acetanilide hydroxylation) was shown to reflect the stimulation of the cyt P-450 PB-1 isoenzyme(s). This effect only occurred when the metyrapone repressible cyt P-450 PB-4 did not contribute significantly to the total microsomal oxygenation activity (32).

The 7-ethoxycoumarin O-deethylation reaction employed in the present investigation is less specific for certain monooxygenase isoenzymes than acetanilide hydroxylation, although both of them show highest enzymatic activity with cyt P-450$_{\beta NF-B}$ (33). A possible explanation for the stimulating effect of metyrapone we found after pretreatment of mice with IP could be a selective induction of cyt P-450 PB-1 next to the P-448 isoenzymes without stimulation of the typical metyrapone repressible enzymes.

ACKNOWLEDGEMENT

We thank Miss C.Schulz for her technical assistance.

REFERENCES

1. Conney, A.H. (1982): Induction of microsomal enzymes by foreign chemicals and carcinogenesis by polycyclic aromatic hydrocarbons: G.H.A. Clowes Memorial Lecture. Cancer Res., 42, 4875-4917.
2. Harada, N., Omura, T. (1983): Phenobarbital-and 3-methylcholanthrene- induced synthesis of two different molecular species of microsomal cytochrome P-450 in rat liver. J. Biochem., 93: 1361-1373.
3. Levin, W., Thakker, D.R., Wood, A.W., Chang, R.L., Lehr, R.E. Jerina, D.M., Conney, A.H. (1978): Evidence that benzo[a]-anthracene 3,4-diol-1,2-epoxide is an ultimate carcinogen on mouse skin. Cancer Res., 38: 1705-1710.
4. Ioannides, C., Lum, P.Y., Parke, D.V. (1984): Cytochrome P-448 and the activation of toxic chemicals and carcinogens. Xenobiotica, 14: 119-137.
5. Kawajiri, K., Yonekawa, H., Gotoh, O., Watanabe, J., Igarashi, S., Tagashira, Y. (1983): Contributions of two inducible forms of cytochrome P-450 in rat liver microsomes to the metabolic activation of various chemical carcinogens. Cancer Res., 43: 819-823.
6. Mishin, V.M., Gulyaeva, L-F., Mishina, D.V. (1984): Immunochemical studies of cytochrome P-450 identity in mice and rat liver tissue after induction by means of 3-methylcholanthrene. Vop. Med. Khim. SSR, 30: 98-101.
7. Ullrich, V., Frommer, U., Weber, P. (1973): Characterization of cytochrome P-450 species in rat liver microsomes, I. Differences in the O-dealkylation of 7-ethoxycoumarin after pretreatment with phenobarbital and 3-methylcholanthrene. Hoppe Seyler's Z. Physiol. Chem., 354: 514-520.
8. Thorgeirsson, S.S., Atlas, S.A., Boobis, A.R., Felton, J.S. (1979): Species differences in the substrate specifity of

hepatic cytochrome P-448 from polycyclic hydrocarbon-treated animals. Biochem. Pharmacol., 28: 217-226.

9. Thomas, P.E., Reidy, J., Reik, L.M. (1984): Use of monoclonal antibody probes against rat hepatic cytochrome P-450 and P-450d to detect immunochemically related isoenzymes in liver microsomes from different species. Arch. Biochem. Biophys., 235: 239-253.
10. Kuwahara, S., Harada, N., Yoshioka, H., Miyata, T., Omura, T. (1984): Purification and characterization of four forms of cytochrome P-450 from liver microsomes of phenobarbital-treated and 3-methylcholanthrene-treated rats. J. Biochem., 95: 703-714.
11. Ryan, D.E., Thomas, P.E., Levin, W. (1982): Purification and characterization of a minor form of hepatic microsomal cytochrome P-450 from rats treated with polychlorinated biphenyls. Arch. Biochem. Biophys., 216: 272-288.
12. Dannan, G.A., Guengerich, F.P., Kamminsky, L.S., Aust, S.D. (1983): Regulation of cytochrome P-450. J. Biol. Chem., 258: 1282-1288.
13. Luster, M.I., Lawson, L.D., Linko, P., Goldstein, J.A. (1983): Immunochemical evidence for two 3-methylcholanthrene inducible forms of cytochrome P-448 in rat liver microsomes using a double-antibody radioimmunoassay procedure. Molecular Pharmacol., 23: 252-257.
14. Greenlee, W.F., Poland, A. (1978): An improved assay of 7-ethoxycoumarin O-deethylase activity: induction of hepatic enzyme activity in C57 Bl/6J and DBA/2J mice by phenobarbital, 3-methylcholanthrene and 2,3,7,8-tetrachloro. J.Pharm. Exper. Therap., 205: 596-605.
15. Phillipson, C.E., Godden, P.M.M., Lum, P.Y., Ioannides, C., Parke, D.V. (1984): Determination of cytochrome P-448 activity in biological tissues. Biochem. J., 221: 81-88.
16. Norpoth, K., Kemena, A., Jacob, J., Schümann, C. (1984): The influence of 18 environmentally relevant polycyclic aromatic hydrocarbons and Clophen A50, as liver monooxygenase inducers, on the mutagenic activity of benz[a]anthracene in the Ames test. Carcinogenesis, 5: 747-752.
17. Studt, P. (1978): Notiz über die Synthese von Benzo[e]pyrene. Liebigs Ann. Chem., 1: 530-531.
18. Studt, P. (1978): Notiz über die Synthese von Indeno[1,2,3-cd]pyren. Liebigs Ann. Chem., 1: 528-529.
19. Ullrich, U., Weber, P. (1972): The O-dealkylation of 7-ethoxycoumarin by liver microsomes. Hoppe Seyler's Z. Physiol. Chem., 353: 1171-1177.
20. Omura, T., Sato, R. (1964): The carbon monoxide-binding pigment of liver microsomes. J. Biol. Chem., 239: 2370-2378.
21. Gornall, A.G., Bordawill, C.J., David, M.M. (1949): Determination of serum proteins by means of the Biuret reaction. J. Biol. Chem., 177: 751-766.

22. Ames, B.N., McCann, J., Yamasaki, E. (1975): Methods for detecting carcinogens and mutagens with the Salmonella/mammalian-microsome mutagenicity test. Mutation Res., 31: 347-364.
23. Phillipson, C.E., Ioannides, C., Barrett, D.C.A., Parke, D.V. (1985): The homogeneity of rat liver microsomal cytochrome P-448 activity and its role in the activation of benzo[a]pyrene to mutagens. Int. J. Biochem., 17: 37-42.
24. Ryan, D.E., Thomas, P.E., Korzeniowski, D., Levin, W. (1979): Separation and characterization of highly purified forms of liver microsomal cytochrome P-450 from rats treated with polychlorinated biphenyls, phenobarbital, and 3-methylcholanthrene. J. Biol. Chem., 254: 1365-1374.
25. Thomas, P.E., Reik, L.M., Ryan, D.E., Levin, W. (1981): Regulation of three forms of cytochrome P-450 and epoxide hydrolase in rat liver microsomes. J. Biol. Chem., 256: 1044-1052.
26. Dent, J.G., Elcombe, C.R., Netter, K.J., Gibson, J.E. (1978): Rat hepatic microsomal cytochrome(s) P-450 induced by polybrominated biphenyls. Drug Metabolism & Disposition Amer. Soc. Pharmacol. & Exper. Therap., 6: 96-101.
27. Ohmori, S., Motohashi, K., Kitada, M. (1984): Purification and properties of cytochrome P-450 from untreated monkey liver microsomes. Biochem. Biophys. Res. Commun., 125: 1089-1096.
28. Thomas, P.E., Reik, L.M., Ryan, D.E., Levin, W. (1983): Induction of two immunochemically related rat liver cytochrome P-450 isoenzymes, cytochromes P-450c and P-450d, by structurally diverse xenobiotics. J. Biol Chem., 258: 4590-4598.
29. Steward, A.R., Dannan, G.A., Guzelian, P.S.(1985): Changes in the concentration of seven forms of cytochrome P-450 in primary cultures of adult rat hepatocytes. Mol. Pharmacol., 27: 125-132.
30. Leibman, K.C. (1969): Effects of metyrapone on liver microsomal drug oxidations. Mol. Pharmacol., 5: 1-9.
31. Leibman, K.C., Ortiz, E. (1973): Metyrapone and other modifiers of microsomal drug metabolism. Drug Metabolism & Disposition, 1: 184-190.
32. Waxman, D.J., Walsh, C. (1983): Cytochrome P-450 isozyme 1 from phenobarbital-induced rat liver: purification, characterization, and interactions with metyrapone and cytochrome b_5. Biochemistry, 22: 4846-4855.
33. Guengerich, F.P., Dannan, G.A., Wright, S.T., Martin, M.V., Kaminsky, L.S. (1982): Purification and characterization of microsomal cytochrome P-450s. Xenobiotica, 12: 701-716.

IMPACTS OF AIRTIGHT AND NONAIRTIGHT WOOD HEATERS ON INDOOR LEVELS OF POLYNUCLEAR AROMATIC HYDROCARBONS IN A WEATHERIZED RESIDENTIAL HOME

C. V. KNIGHT,* M. P. HUMPHREYS[1]
(1) Energy Use Test Facility, Tennessee Valley Authority, Chattanooga, Tennessee 37405; *Under Contract to TVA from the University of Tennessee at Chattanooga, School of Engineering, Chattanooga, Tennessee 37402.

INTRODUCTION

Polynuclear Aromatic Hydrocarbons (PAH) are a group of organic compounds known to be produced by residential wood combustion. The concentrations of such PAH pollutant compounds in the indoor environment directly attributed to the use of wood heaters have only recently been investigated by GEOMET (1,2), Lawrence Berkeley Laboratory (3), TVA (4), and others. The benzo(a)pyrene (BaP) component of total PAH is of special interest since it is known to be a carcinogen.

Studies (1,2,3) conducted during the past five years have investigated indoor air quality conditions associated with conventional airtight wood heaters and non-weatherized homes having air infiltration rates typically higher than one air exchange per hour. This study was developed to specifically determine the impacts of airtight and non-airtight wood heaters on indoor air quality. Both conventional and catalytic-type airtight as well as box and freestanding fireplaces, Franklin-type non-airtight wood heaters were operated in the test home. The test program was conducted in a modular home having an infiltration rate of approximately 0.4 air changes per hour, a rate that is typical of current homes having weatherized features.

The study was conducted during the winter months of 1983, 1984, and 1985, with TVA receiving joint funding from the Bonneville Power Administration for the first and third phases of the study and funding from the Consumer Products Safety Commission for the second phase of testing. The test program was conducted by the TVA Energy Use Test Facility staff at Chattanooga, Tennessee.

MATERIALS AND METHODS

Test System

The test home was a relatively tight, unoccupied residential home of approximately 337 cubic meters interior volume. The wood heater to be tested was located in the living room area of the home and was vented into either the 6-inch masonry chimney or an 8-inch metal stack installed for those heaters requiring a larger flue gas system. The wood heater was placed on an electronic weight scale for wood fuel burn rate determination. Although gaseous pollutants, as well as particulate matter were sampled from indoor and outdoor locations throughout each test, only PAH results will be reported in this paper. Air samples for PAH analysis were collected from both the living room area and the outdoor environment of the test home. The transient air exchange rate of the test home was determined by the tracer gas dilution method.

The wood heaters tested in the three phases of the study included (1) conventional and catalytic airtight wood heaters and (2) conventional non-airtight wood heaters, such as the freestanding fireplace, Franklin-type. The airtight wood heaters were tested during phases 1 and 3, while the non-airtight wood heaters were tested during phase 2 of the study.

Each wood heater was tested at low and high burn rates, while some heaters were also tested at medium burn rates. Replicate testing was conducted for selected tests that appeared to be of special importance relative to impact on indoor air quality. Each phase 1 test was conducted over a 24-hr period of time, while each of the phases 2 and 3 tests were conducted for a 12-hr period. Two wood fuel load sizes were burned during testing to determine both the interdependence of damper setting (controlling wood fuel burn rate) and wood loading size on indoor pollutant generation.

The test fuel consisted primarily of medium moisture (21 percent, wet basis), split, red-oak hardwood for all three phases of testing. Medium-moisture, split, field pine cordwood was also used during phase 3 testing.

Sampling Methods

Sampling protocols for the collection of airborne particulates have traditionally used high-volume samplers containing glass fiber particulate filters. Recent investigations, however, have shown that the collection of such particulates under high-volume conditions may result in significant losses of PAH by either chemical reactions while sorbed on the filter media or direct volatilization from particulate matter during sampling (5,6,7,8,9). Keller and

Bidleman (10) reported losses from a high-volume glass fiber filter to be over 80 percent for phenanthrene and pyrene and also that the majority of three- and four-ring PAHs were not retained by the filter under high-volume sampling conditions. Similarly, high losses of PAH recovery were found in two European studies (11,12). Improved collection efficiencies have been reported for collectors using Tenax, XAD-2, and polyurethane foam (13,14,15).

In view of these findings, an integrated sample of PAH was collected by drawing a metered volume of air through a sorbent resin trap. The trap was filled with purified XAD-2 resin. A sufficient sample flow rate was maintained over the 24- or 12-hr test period to provide for recovery of 10 cubic meters of either indoor or outdoor air.

Indoor and outdoor XAD-2 resin sampling systems were operated simultaneously for the full duration of each test. The indoor sampling system was located at the center of the living room, four feet above the floor. The outdoor sampling system, located approximately 200 feet to the north of the test home, was used for collection of ambient PAH.

Indoor-Outdoor Air Source → XAD-2 Sorbent → Soxhlet Extraction for 24 (72) Hours →

Kuderna Danish Concentration → 1.0 ml Aliquot for Spec. Compound Analysis → HPLC/UV Fluorescence

or Capillary Column GCMS → Report PAH

After completion of each test, the resin traps were removed, capped, and transported to the chemical laboratory for analysis. As shown above, the first step in the analysis procedure involved continuous extraction with methylene chloride for 24 hours (phases 1 and 2) or for 72 hours (phase 3). The methylene chloride extract was then dried, reduced in volume by Kuderna Danish evaporation, and followed by identification and quantification of the PAH compounds in the extract by HPLC with UV and fluorescence detection (phase 1) or capillary column GC/MS (phases 2 and 3) techniques. The PAH results were corrected to account for whatever background contamination existed for the XAD-2 resin by the use of blank traps.

DISCUSSION OF RESULTS

Tables 1, 2, and 3 present a summary of PAH and BaP results for each phase of the study conducted during the winters of 1983, 1984, and 1985, respectively. Each table presents indoor and outdoor PAH and BaP results, as well as indoor-to-outdoor ratios and average air changes per hour existing during each test. The air exchange per hour data is included so that a source strength may be computed by considering an overall mass balance for the home. The source strength (μg/hr) directly attributed to the wood heater represents a more valid criteria of comparison than a simple indoor-to-outdoor ratio, even though the ratio has commonly been used for pollutant source evaluations. This is clearly shown by results presented for tests 1 and 3 in Table 1 and many other tests in Tables 2 and 3. Whereas test 3 had a much higher indoor-to-outdoor ratio, it had a much lower source strength. A discussion of the source strength model is presented later. The indoor-to-outdoor ratio does not represent the dynamic situation. The evidence of an indoor source is signified by the difference between indoor and outdoor concentrations, not the absolute level of either the indoor or outdoor concentrations or the ratio of the two.

The source strength values for PAH and BaP presented in Tables 1, 2, and 3 were developed by considering the following mass balance:

$$\dot{S} = V\left[\frac{dC_{in}}{dt} + RC_{in} + \alpha C_{in} - \alpha f C_{out}\right]$$

where dC_{in}/dt represents the rate of change of the indoor concentration, which is equal to zero for integrated test sampling,

$\dot{S}$ = the average source strength (μg/hr) attributed to the device being used to sustain the elevated indoor concentrations,

α = the average air changes per hour for the test,

f = wall transmission factor for PAH compounds in outdoor air,

C_{in}, C_{out} = indoor and outdoor PAH and BaP average concentrations (μg/m³),

TABLE I

SUMMARY OF 24-HR POLYNUCLEAR AROMATIC HYDROCARBONS (PAH) AND BENZO(A)PYRENE (BaP) CONCENTRATIONS (ng/m^3) AND SOURCE STRENGTHS (µg/hr) FOR 1983 TVA/BPA STUDY

		PAH				BaP			
Test No.	ACPH	Indoor	Outdoor	Indoor/ Outdoor	SS	Indoor	Outdoor	Indoor/ Outdoor	SS
1	0.37	3187	1206	2.64	1375	*	15.1	0.07	-0.8
2	0.34	1585	576	2.75	181	*	7.6	0.13	-0.4
3	0.37	618	178	3.47	76	13.2	34.6	0.38	-0.4
4	0.38	5353	3555	1.51	548	25.7	5.7	4.51	3.4
5	0.44	3363	3260	1.03	314	14.3	10.6	1.35	1.6
6	0.28	2164	2490	0.87	123	2.2	11.5	0.19	1.2
7	0.62	3128	–	–	–	*	–	–	–
8	0.33	4057	3612	1.12	319	51.6	10.3	5.01	6.0
9	0.35	7555	3459	2.18	814	*	9.9	0.10	-5.0
10	0.41	**	–	–	–	*	–	–	–
11	0.31	1655	**	16.6	196	40.5	*	40.5	4.9
12	0.38	2240	256	11.2	308	3.9	2.8	1.39	0.4
13	0.48	1535	**	15.4	266	*	1.9	0.53	0.2
14	0.51	2037	**	20.4	356	5.3	*	5.30	1.0
15	0.45	2318	520	4.46	351	35.2	11.0	3.86	5.1
16	0.31	1076	360	2.99	112	37.4	*	37.4	4.5
17	0.37	4264	**	42.6	597	40.3	*	40.3	5.7
18	0.37	2369	**	23.7	329	50.7	*	50.7	7.1
19	0.38	1204	**	12.0	168	35.3	*	35.3	5.1
20	0.41	228	**	2.28	28	15.7	8.0	1.96	1.9
21	0.38	1150	**	11.5	160	8.5	*	8.50	1.2
22	0.37	785	152	5.16	102	17.4	*	17.4	2.4
23	0.40	1097	519	2.11	131	23.2	19.3	1.20	2.2
Avg.		2309	997			18.5	7.4		
St. Dev.		1957	1354			17.5	8.8		
95% Con.		841	584			7.5	3.8		
A			3260				52.0		

* – Below BaP detection limit of 1 ng/m^3.
** – Below total PAH detection limit of 100 ng/m^3.
A – Average for Petersville Study 16.
BaP – Represents sum of BaP and BeP because the two were not separated.

TABLE 2

SUMMARY OF 12-HR POLYNUCLEAR AROMATIC HYDROCARBONS (PAH) AND BENZO(A)PYRENE (BaP) CONCENTRATIONS (ng/m^3) AND SOURCE STRENGTHS (μg/hr) FOR 1984 TVA/CPSC STUDY

		PAH				BaP		
Test No.	ACPH	Indoor (1)	Outdoor (4)	Indoor/ Outdoor	SS	Indoor (1)	Outdoor (4)	SS
7	0.55	576	**	7.2	109	*	*	a
8	0.40	1195	163	7.3	170	*	*	a
9	0.43	1582	769	2.1	200	*	*	a
BL 10	0.43	-	-	-	-	-	-	-
11	0.44	1862	845	2.2	244	*	*	a
12	0.33	3073	900	3.4	343	*	*	a
13	0.33	3421	879	3.9	389	8.2	*	1.0
14	0.58	270	209	1.3	37	*	*	a
15	0.53	1991	-	-	-	-	-	-
16	0.37	2552	332	8.3	341	*	*	a
BL 17	0.30	-	-	-	-	-	-	-
18	0.41	533	148	3.6	72	*	*	a
19	0.53	1422	739	1.9	213	*	*	a
20	0.41	981	267	3.7	157	*	*	a
21	0.36	1395	608	2.3	156	*	*	a
22	0.40	973	820	1.2	92	*	*	a
23	0.33	1925	499	3.9	219	5.2	*	0.6
BL 24	0.31	636	-	-	-	-	-	-
25	0.40	602	**	7.5	86	*	*	a
26	0.30	839	630	1.3	67	*	*	a
27	0.31	1093	634	1.7	100	*	*	a
28	0.51	689	159	4.3	116	*	*	a
29	0.51	876	327	2.9	137	*	*	a
30	0.30	2623	1488	1.8	234	9.8	*	1.1
BL 31	0.32	1096	-	-	-	-	-	-
32	0.53	754	266	2.8	124	*	*	a
33	0.46	538	349	1.5	66	*	*	a
34	0.49	617	269	2.3	70	*	*	a
Avg.		1357	498	2.7				
St. Dev.		859	34					

* - Below BaP detection limit of 2 ng/m^3.
** - Below total PAH detection limit of 80 ng/m^3.
a - Below detection.
BL - Baseline tests without wood heater use.

TABLE 3

SUMMARY OF 12-HR POLYNUCLEAR AROMATIC HYDROCARBONS (PAH) AND BENZO(A)PYRENE (BaP) CONCENTRATIONS (ng/m^3) AND SOURCE STRENGTHS (μg/hr) FOR 1985 TVA/BPA STUDY

		PAH				BaP		
Test No.	ACPH	Indoor (1)	Outdoor (4)	Indoor/ Outdoor	SS	Indoor (1)	Outdoor (4)	SS
BL 1	0.54	–	–	–	–	–	–	–
2 LO	0.57	120	650	0.2	-37	*	*	a
3 LO	0.46	1659	406	4.1	254	3.6	*	0.5
4 LP	0.49	829	236	3.5	103	*	*	a
BL 5	0.41	–	–	–	–	–	–	–
6 HO	0.68	2088	1092	1.9	388	*	*	a
7 HO	0.59	98	64	1.5	15	*	*	a
8 HP	0.60	519	433	1.2	70	*	*	a
9	0.36	**	–	–	–	**	–	–
10 LO	0.57	543	115	4.7	102	*	*	a
11 LO	0.39	593	478	1.2	57	*	*	a
BL 12	0.37	–	–	–	–	–	–	–
13 LP	0.45	285	6	48.3	48	3.7	*	0.6
14 LP	0.44	455	166	2.7	63	*	*	a
15 LO	0.45	92	89	1.0	9	*	*	a
16 LO	0.43	418	322	1.3	94	*	*	a
17 LP	0.47	368	280	1.3	84	*	*	a
18 LO	0.49	154	758	0.2	-34	*	2.2	a
19 LO	0.37	**	298	–	-15	**	*	a
20 LP	0.36	2482	6519	0.4	-52	*	*	a
BL 21	0.30	–	–	–	–	–	–	–
22 HO	0.38	154	2105	0.1	-112	*	*	a
23 HO	0.44	93	616	0.2	-71	*	*	a
24 HP	0.50	599	2810	0.2	-126	*	*	a
25 LO	0.39	972	356	2.7	121	*	*	a
26 LO	0.46	1233	497	2.5	151	*	*	a
27 LP	0.18	1107	515	2.2	70	*	*	a
28 HO	0.31	1133	728	1.6	99	86.7	*	10.4
29 HO	0.34	1004	483	2.1	100	24.2	*	3.1
BL 30	0.27	–	–	–	–	–	–	–
31 HP	0.29	675	359	1.9	60	4.1	*	0.4
32 LO	0.26	422	459	0.9	24	7.5	*	0.7
33 LO	0.30	545	299	1.8	49	26.2	*	3.0
34 LP	0.28	481	189	2.5	44	7.6	*	0.8
BL 35	0.24	184	444	0.4	0	4.0	*	0.4
36 HO	0.37	555	285	1.9	61	5.1	*	0.7
37 HP	0.50	401	74	5.4	58	7.6	*	1.3

LO – Low burn rate oak fuel   HO – High burn rate oak fuel
LP – Low burn rate pine fuel   HP – High burn rate pine fuel
* – Below BaP detection 2 ng/m^3 ** – Below PAH detection 80 ng/m^3

R = deposition, plate-out factor for indoor PAH and BaP (hr^{-1}), and

V = the test home volume (337 m^3).

A wall transmission factor of 0.5 and a deposition factor of 0.05 were used in evaluating the source strengths. The transmission value of 0.5 was also used throughout each final report for each phase of the study to generate the average total suspended particulate source strength for each test. The same factor was used for PAH and particulates since most PAH compounds are associated with particulates. The particulate source strengths, along with in-depth consideration of other pollutant findings for each phase of the study, are presented in the three final reports (17,18,19).

The PAH results presented in Figure 1 were determined using packed column HPLC with ultraviolet and fluorescence detections with both BaP and BeP being reported as BaP. The PAH results presented in Figures 2 and 3 were determined using a capillary column GC/MS system with BaP and BeP being separated. Therefore, the reported BaP must be corrected, since other studies (21,22,23) found BaP and BeP concentrations to be generally of similar magnitudes in ambient and wood heater stack gases.

As shown in Figure 1, airtight wood heaters 1 through 3 were found to generate highly variable source strengths, with low burn rate tests (designated by (a)) commonly having the highest PAH and BaP source strengths. Catalytic wood heater 2 generated a higher PAH (814 μg/hr) and BaP (6 μg/hr) source strength than conventional airtight wood heaters 1 and 3. This was rather unexpected, because catalytic heater 2 contributed less to indoor carbon monoxide and suspended particulate pollution than conventional wood heaters 1 or 3. Catalytic wood heater 4 was responsible for consistently higher levels of PAH for all tests. This heater also generated high levels of all types of indoor pollutants associated with wood heater flue gases. The front door seal area of catalytic wood heater 4, as well as some welded joints in the primary combustion area, were found to discharge flue gases rich in products of incomplete combustion. The same wood heater was tested again in the third phase (wood heater 3) of the study, with a significant improvement in its performance.

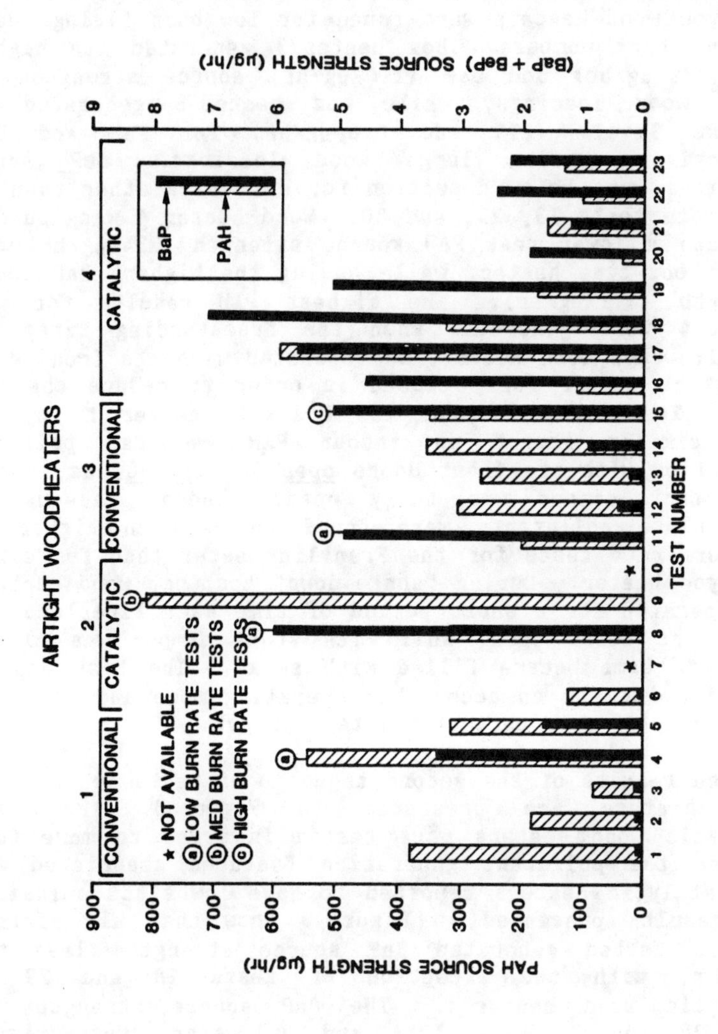

FIGURE 1. Average (24 hr) PAH and (BaP + BeP) source strengths for four airtight wood heaters (Winter 1983).

Non-airtight heater results for winter 1984 are presented in Figure 2. The peak PAH and BaP source strengths found for the three non-airtight heaters were less than half of the peak levels found for the phase 1 (winter 1983) study. Highest PAH and BaP source strengths for the box-type wood heaters were found for low burn firings using common stack dampers. Box heater 3 generated its highest PAH (389 μg/hr) and BaP (1.0 μg/hr) source strengths for small wood loadings, while box heater 5 generated its highest levels (219 and 0.6 μg/hr) for PAH and BaP, respectively, using large wood loadings. BaP source strengths were below detection for all tests other than low burn rate tests 13, 25, and 30. Wood heater 4 was found to generate a lower peak PAH source strength (234 μg/hr) than either box-type heater, while having the highest BaP source strength (1.1 μg/hr). The highest PAH results for wood heater 4 were generated when the freestanding fireplace, Franklin-type wood heater, was operated with its front doors and stack damper fully closed in order to reduce the burn rate. Test results found the Franklin heater to operate with similar impact on indoor PAH and BaP pollutant generation with the front doors open and closed, as long as the stack damper was fully open. Indoor gaseous and particulate pollutants were found to be much higher for low-burn-rate tests for the Franklin heater than for either box-type heater. During "shake-down" testing, wood heater 4 was operated for a short period of time as a fireplace with its front doors open, while its stack damper was closed. The test home became filled with smoke. The test was not considered to be an acceptable operating mode and hence was not conducted as a part of the test program.

The results of the second sequence of tests for airtight wood heaters are presented in Figure 3. Primarily, catalytic wood heaters were tested in order to more fully explore the pollutant generation features associated with new catalytic heaters reported to have low stack emissions. The results presented in Figure 3 show that all airtight heaters tested generated BaP source strengths less than 1 μg/hr, with the exception of tests 28 and 29 for catalytic wood heater 4. The BaP source strengths for tests 28 and 29 were 10.4 and 3.1 μg/hr, respectively. The mechanism associated with the high BaP source strengths for the two tests has not been revealed. The total PAH source strengths for the same two tests were not elevated over those commonly found for other tests. The results also show that the highest PAH source strengths for the catalytic wood heaters were found most often for high burn rate

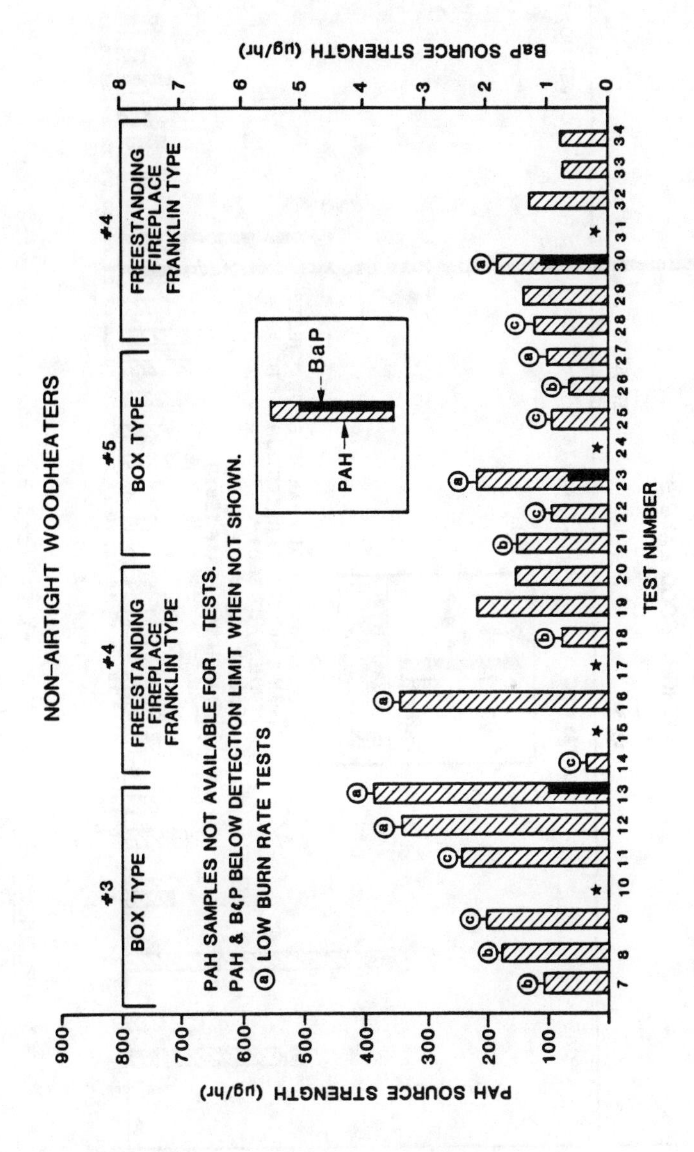

FIGURE 2. Average (12 hr) PAH and BaP source strengths for three non-airtight wood heaters (Winter 1984).

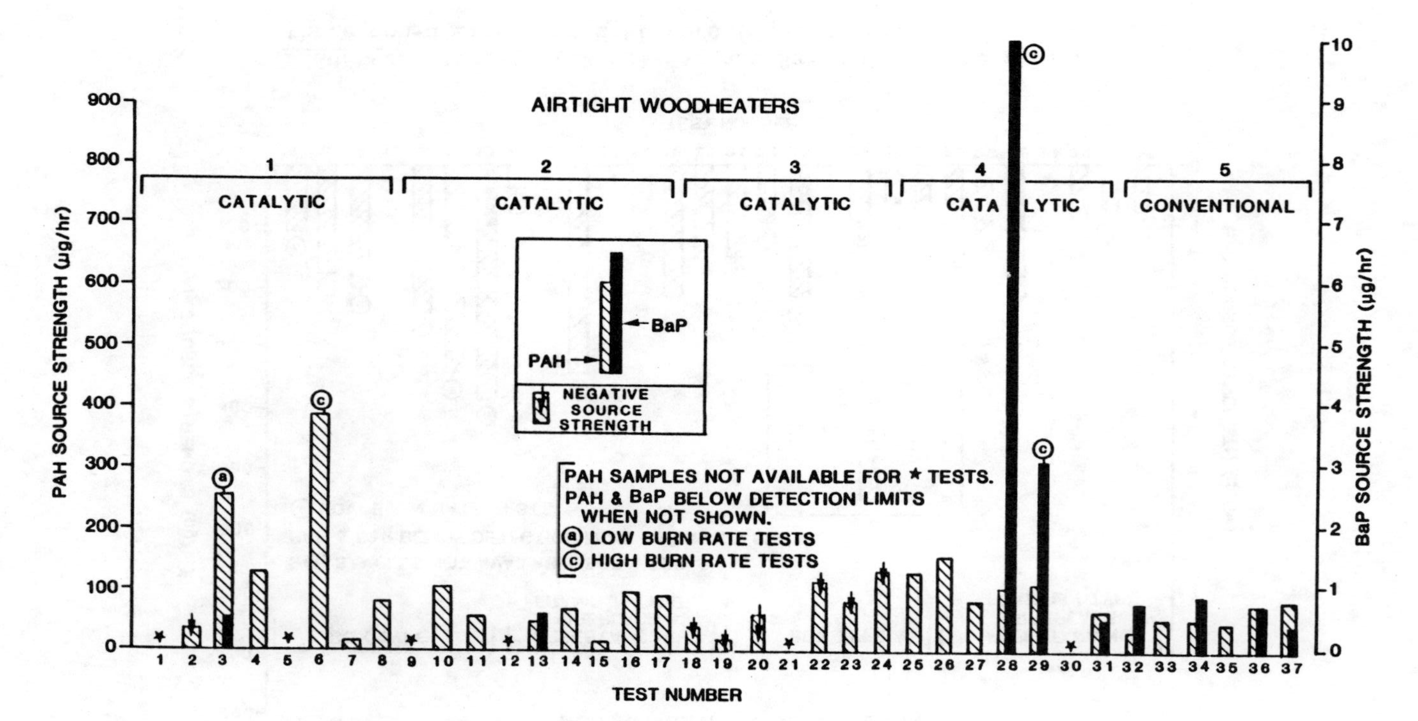

FIGURE 3. Average (12 hr) PAH and BaP source strengths for five airtight wood heaters (Winter 1985).

tests. This represents a significant finding in that non-airtight wood heaters were found to have higher source strengths for low burn rate tests.

Tests 3 and 7 represented an effort at repeating tests 2 and 6 for low and high burn rate firings of catalytic wood heater 1. Similarly, many other tests represent an effort at replicate testing. Results presented in Figure 3 show that such tests were indeed not replicate tests as the PAH and BaP results varied substantially for such tests. Gaseous and suspended particulate pollutant source strengths also varied substantially for tests intended to be replicates. This was not surprising in view of earlier stack emissions testing conducted by TVA (20) having found that replicate cordwood testing using cordwood fuel was not possible due to irregular sizing and/or stacking features associated with combustion in the fire zone.

PAH and BaP source strengths for worst conditions for the airtight wood heaters in Figure 3 were found to be generally lower than for the non-airtight wood heaters found in Figure 2, with the exception being tests 28 and 29 in Figure 3. Higher PAH and BaP peak source strengths were found for airtight wood heaters presented in Figure 1 than for similar airtight wood heaters presented in Figure 3. It is thought that the major portion of these differences can be attributed to analytical methods associated with using HPLC in the winter 1983 study versus GCMS quantification and detection for the winter 1985 study.

The results presented in Tables 1, 2, and 3 show that the modeling effort provided small negative PAH and/or BaP source strenghts for several tests. Two isolated tests (9 in 1983 and 2 in 1985) had significant negative PAH and/or BaP source strengths that were incorrectly attributed to use of the wood heater in the home. During the 1985 phase of testing, all PAH source strengths for catalytic wood heater 3 were found to be negative, due to high levels of outdoor PAH, primarily napthalene. The source of the high concentration of napthalene in the outdoor PAH sample was not conclusively found. The napthalene could have been associated with wood heater stack gases as an earlier study conducted by TVA (20) found napthalene to be by far the largest component of PAH in the flue gases.

Test 35 and five other tests in the 1985 study were conducted to determine indoor-to-outdoor relationships for no-burn or baseline (BL) periods. Table 3 shows that the

PAH/BaP mass balance presented earlier correctly yielded a zero source strength for PAH and a small BaP source strength for test 35. PAH sampling was not performed on all baseline tests because of cost considerations.

CONCLUSIONS

• Polynuclear Aromatic Hydrocarbons (PAH)

All wood heaters were found to be an indoor source of PAH under some operating conditions. The non-airtight wood heaters were found to generate highest PAH source strengths for low burn rate firings, while the catalytic airtight wood heaters were usually found to generate highest PAH source strengths for high burn rate firings.

PAH source strengths for airtight wood heaters were found to be generally lower than for non-airtight wood heaters.

The freestanding, Franklin-type wood heater generated PAH source strengths lower than for other box-type non-airtight heaters, even though it represented the most significant source of gaseous and particulate indoor pollution for any heater tested.

• Benzo(a)pyrene (BaP)

Source strengths attributed to each wood heater were found to be highly variable with the non-airtight heaters having highest BaP source strengths for low burn rate firings.

REFERENCES

1. Moschandreas, D. J., Zabransky, J. Jr., and Rector, H. E. (1980): The effect of woodburning on the indoor residential air quality, Environmental International, Vol. 4:463-468.
2. Moschandreas, D. J., Zabransky, J. Jr., and Pelton, D. J. Comparison of Indoor-Outdoor Air Quality. Electric Power Research Institute, EPRI EA-1733, March 1981.
3. Berk, J. V., Hollowell, C. D., Lin, C., and Pepper, J. H. Design of a Mobile Laboratory for Ventilation Studies and Indoor Air Pollution Monitoring. Lawrence Berkeley Laboratory, University of California, Berkeley, California, LBL-7817, April 1978.

4. Knight, C. V. and Humphreys, M. H. Indoor Air Quality Associated with Conventional and New Technology Wood Heaters in a Weatherized Home. Proc. of 77th Annual APCA Meeting, San Francisco, California, June 1984.
5. Moschandreas, D. J., McFadden, J. E., and Stark, J.W.C. Indoor Air Pollution in the Residential Environment, Volume II - Field Monitoring Protocol, Indoor Episcondic Pollutant Release, Experiments and Numerical Analysis. GEOMET Report No. EF-688, GEOMET, Inc., Gaithersburg, Maryland, EPA Report No. EPA 600/7-78-2296.
6. Strup, P. E., Giammar, R. D., Stanford, T. B., and Jones, P. W. Improved measurement techniques for polycyclic hydrocarbons from combustion effluents. In: Carcinogenesis--A Comprehensive Survey, Polynuclear Aromatic Hydrocarbons - Chemistry, Metabolism, and Carcinogenesis, Vol. 1, Raven Press, New York.
7. Jones, P. W. (1980): Measurement and environmental impact of PAH - some closing remarks. In: Polynuclear Aromatic Hydrocarbons - Fourth International Symposium on Chemistry, Biology, Carcinogenesis, and Mutagenesis, edited by P. W. Jones and P. Leber, Ann Arbor Science, Ann Arbor, MI.
8. Lao, R. C. and Thomas R. S. (1980): The volatility of PAH and possible losses in ambient sampling. In: Polynuclear Aromatic Hydrocarbons - Fourth International Symposium on Chemistry, Biology, Carcinogenesis, and Mutagenesis, edited by P. W. Jones and P. Leber, Ann Arbor Science, Ann Arbor, MI.
9. Davis, C. S., Caton, R. B., Guerin, S. G., and Tam, W. C. (1983): Effects of ozone on the collection of selected PAH in airborne particulate matter. In: Polynuclear Aromatic Hydrocarbons: Mechanisms, Methods, and Metabolism, Eighth International Symposium, Battelle Press, Columbus, Ohio.
10. Keller, C. and Bidleman, T. F. (1982): Collection of vapor phase polycyclic aromatic hydrocarbons in ambient air. Proceedings of Annual American Chemical Society Meeting, Kansas City, Missouri.
11. Cautreels, W. and Van Cauwenberghe, K. (1979): Experiments on the distribution of organic pollutants between airborne particulate matter and the corresponding gas phase. Atmospheric Environment, 12:1133-1141.
12. Thrane, K. E. and Mikalsen, A. (1981): High volume sampling of airborne polycyclic aromatic hydrocarbons using glass fibre filters and polyurethane foam. Atmospheric Environment, 15(6):909-918.
13. Lindgren, J. L., Krauss, H. J., and Fox, M. A. (1978): A comparison of two techniques for the collection and analysis of polynuclear aromatic hydrocarbons in ambient air. APCA Annual Meeting, Houston, Texas.

14. Hanson, R. L., Clark, C. R., Carpenter, R. L., and Hobbs, C. H. (1981): Evaluation of Tenax-GC and XAD-2 as polymer adsorbents for sampling fossil fuel combustion products containing nitrogen oxides. Environmental Science and Technology, 15:701.
15. Pellizzari, E. D. Ambient Air Carcinogenic Vapours: Improved Sampling and Analytical Techniques and Field Studies. EPA 600/2-79-081, US-EPA, Environmental Sciences Research Laboratory, Research Triangle Park, North Carolina.
16. Imhoff, R. E., Manning, J. A., Cooke, W. M., and Hayes, T. L. Final Report on a Study of the Ambient Impact of Residential Wood Combustion in Petersville, Alabama. Proceedings of Air Pollution Control Association Specialty Meeting on Residential Wood and Coal Combustion, Louisville, Kentucky, March 1982.
17. Tennessee Valley Authority - Bonneville Power Administration Indoor Air Quality Study. Tennessee Valley Authority, Division of Conservation and Energy Management, TVA/OP/C&EM-84/84, Chattanooga, Tennessee, 1984.
18. Consumer Products Safety Commission Indoor Air Quality Study. Tennessee Valley Authority, Division of Conservation and Energy Management, Chattanooga, Tennessee (in press, 1985).
19. Tennessee Valley Authority - Bonneville Power Administration Indoor Air Quality Study, Phase II. Tennessee Valley Authority, Division of Conservation and Energy Management, Chattanooga, Tennessee (in press, 1985).
20. Residential Wood Heater Test Report. Tennessee Valley Authority, Division of Energy Conservation and Rates, TVA/OP/ECR-83/69, Chattanooga, Tennessee, 1983.
21. Grimmer, G., Naujack, K. W., and Schneider, D. (1980): Changes in PAH profiles in different areas of a city during the year. In: Fourth International Symposium on Polynuclear Aromatic Hydrocarbons, Battelle Press, Columbus, Ohio.
22. Grimmer, G., Naujack, K. W., and Schneider, D. (1981): Comparison of profiles of PAH in different areas of a city by glass-capillary-gas chromatography in the nanogram range. International Journal of Environmental Analytical Chemistry, 10:265-276.
23. Truesdale, R. S. and Cleland, J. B. Residential Stove Emissions from Coal and Other Alternative Fuels Combustion. Proceedings of Residential Wood and Coal Combustion Specialty Conference, Air Pollution Control Association, Pittsburgh, Pennsylvania, March 1982.

A PERSONAL COMPUTER DATABASE FOR THE CHEMICAL, PHYSICAL AND THERMODYNAMIC PROPERTIES OF POLYCYCLIC AROMATIC HYDROCARBONS

DOUGLAS A. LANE, DONNA M.A. McCURVIN
Environment Canada, Atmospheric Environment Service, 4905 Dufferin Street, Downsview, Ontario, M3H 5T4, Canada.

INTRODUCTION

Polycyclic Aromatic Hydrocarbons (PAH) arise from the incomplete combustion of fossil fuels, industrial waste, wood smoke and tobacco smoke, and are associated predominantly with the respirable (less than 10 μm) particulate matter. Once thought to be stable in the atmosphere for days to weeks (1), PAH have since been shown to be quite reactive to atmospheric oxidants and to sunlight (2-8), the reactivity depending on the substrate on which they are adsorbed (9-12). As a consequence, many decomposition products of the PAH are being isolated from various environmental samples and, in some cases, the products exhibit mutagenic potentials far greater than those of the parent PAH from which they were derived.

In order to assess the potential chemical and biological threat of the PAH and their derivatives to the environment, to follow their migration through the biosphere, to perform effective analytical determinations and to characterize their behaviour in natural and synthetic fuels, it is necessary that the chemical, physical, thermodynamic, spectral and biological properties of the PAH be available and readily accessible. Unfortunately, there is a great paucity of such information in the literature and, that which does exist, is widely scattered through journals, textbooks, and government reports (many of which are difficult, if not impossible, to obtain). In addition, much of the data are rather old and, as a consequence, may be of questionable quality.

Since the data were not readily accessible and were needed frequently in our laboratory, a compilation of the available data on the chemical, physical and thermodynamic properties of the PAH, their alkyl, amino, nitro and oxygenated derivatives, in the most easily accessible and updatable format possible was initiated.

HARDWARE

From the outset, it was felt that the most effective and versatile vehicle for generating the database would be the microcomputer. In our laboratory, there were several makes and models from which to choose, however, since it was our desire to make the database as accessible as possible, we decided that the IBM PC-XT personal computer should be used. The IBM PC-XT comprises 256 Kbytes of RAM memory, a 10 Mbyte fixed disk and a 5 1/4 inch floppy disk drive and an Epson FX-80 printer. Each double sided, double density floppy diskette could hold 362 Kbytes of data.

SOFTWARE

The database was established using the SCIMATE software produced and distributed by the Institute for Scientific Information (ISI). Installation of the software simply required a customized disk drive assignment. The help files and the program files were placed on the hard disk (reserved as the system disk) and the user files, work files, template files and report files were placed on the floppy disk.

The SCIMATE software comprises two separate software packages -- the Personal Data Manager (PDM) and the On-Line Data Search (OLS). On entering the SCIMATE software, the user is given the option of entering the PDM, the OLS (one need not obtain the OLS in order to use the PDM) or the Help files. On selecting the PDM, the user is presented with a variety of choices amongst which are to perform a data search, to enter a new record into an established database, to select or to change the current user file, to create a new template and to generate a columnar data report. Other choices are available, however, those mentioned above are the ones most often used. After making a selection, the user is well guided by the menu-driven software. The software can easily be mastered in less than an hour.

To place data in a data file, it is necessary to name the user file and to set up an appropriate template to overlay the data. Each template may contain from one to twenty separate fields of data. The individual fields within a record (or accession number as it is called in the software) may be any length, however, each record is limited to a maximum of 1894 characters of data. If a particular record is anticipated to extend beyond 1894 characters, then, at the first entry, two records may be linked together. Linking of records cannot occur after the first edit. Once the template and user file are created, data may be entered as desired.

The data may be searched in one of two modes: by accession number and by text. When searching the text, the Boolean operators AND and OR are available to assist in the search. In later versions of the SCIMATE software, searches may be restricted to certain user defined ranges of accession number - a distinct advantage when there are several hundred or a thousand records in a user file. The Boolean operators ANDNOT and ORNOT have been added to the search routines in order to allow the user to be more restrictive in the nature of the search. It is also possible to carry out field directed searches in which data from a specified field is searched instead of the entire record. This greatly enhances the usefulness of the search routines.

Data may be printed out in a columnar format. The format is user defined but suffers from a lack of wrap-around capabilities in printing out the data. As a result, in columnar data presentations, words are split at odd positions. This is not a serious problem but is at times frustrating.

THE DATABASE

It became apparent very quickly that, if one data file were used for all the PAH and their derivatives, a single diskette would not adequately store the data. The database was, therefore, subdivided into smaller databases each of which was placed on a separate diskette. There are 8 separate databases which include: the parent PAH (109 chemicals); the alkyl substituted PAH (355 chemicals); the amino substituted PAH (30 chemicals); the PAH ketones (62 chemicals); the PAH quinones (81 chemicals); the nitro substituted PAH (92 chemicals); the aza-arenes (222 chemicals); and the literature references (399 entries).

For each compound in the database, three templates were utilized. One template (Table 1) contained the basic chemical, physical and some spectroscopic data while another template (Table 2) contained, principally, the thermodynamic data. The third template contained the literature reference data.

To demonstrate the nature of the output from the data-

TABLE 1

INFORMATION CONTAINED IN THE "BASIC" TEMPLATE

Field No.	Name	Field No.	Name
1	Chemical Name	11	Proton Affinity
2	Synonyms	12	Aqueous Solubility
3	Physical Appearance	13	Henry's Law Constant
4	Chemical Formula	14	Partition Coefficient
5	Molecular Weight	15	Fluorescence Wavelengths
6	Melting Point	16	UV Maximum
7	Boiling Point	17	Phosphorescence Spectra
8	Density	18	Half-Life
9	Vapor Pressure	19	Thermal Reactivity
10	Ionization Potential	20	Carcinogenic Potential

TABLE 2

INFORMATION CONTAINED IN THE "THERMODYNAMIC" TEMPLATE

Field No.	Name	Field No.	Name
1	Chemical Name	11	Free Energy of Solubilization
2	Molecular Weight	12	Heat Capacity of Solubilization
3	Chemical Abstracts No.	13	Entropy of Solubilization
4	Molecular Connectivity Index	14	Entropy at 298 K
5	Heat of Formation	15	Singlet Energy
6	Heat of Ionization	16	Hydrogen Affinity
7	Heat of Vaporization	17	Saturation Vapor Concentration
8	Heat of Sublimation	18	Equilibrium Vapor Concentration
9	Heat of Fusion	19	Organic Solubility
10	Heat of Solubilization	20	Heat of Combustion

```
USER File Accession Number 61

NAME      PYRENE (1)
SYNONYMS
PHYS.APP Pale yellow platelets with slight blue fluoresence  [21]
         (Yellow colour originates from a trace of Tetracene)  [44]

CHEM.FOR C16H10

MOLECWT  202.26 gm/mol

MELT.PT  156 C (corrected)  [4]; 150 C  [2]; 149-151 C  [22];
         153-155 C  [148]; 145-147 C  [286]; 148.0 C  [288]; 151 C  [331]

BOIL PT  393 C  [2]; 385 C  [144]; 260 C (60 mmHg)  [2]; 360 C  [1];
         404 C  [4]; 399 C  [44]; 384 C (predicted)  [284]; 394 C  [333]

DENSITY  In gm/ml  (T(K),torr)
         1.271 (23,4)  [2]; 1.272(20,-)      [290]; 1.264(20,-)   [286];
         1.25(20,-)  [286]; 1.1012(18.8,-)  [286]; 1.1052(17,-)   [286]

VAP.PRES In atm at 25 C (experimental technique and/or temperature)
         8.75E-09 (extrap. from solid V.P.)  [51]; 8.98E-09 (298 K)  [20];
         8.69E-09 (Effusion method - extrap. from 72-85 C)  [51];
         3.16E-09 (solid - 20 C)  [24]; 7.37E-08 (subcooled liq - 20 C)  [24];
         8.98E-10  [4]; 5.92E-09 (HPLC - method)  [14]; 9.01E-10  [21]
         2.66E-09 (lit. ave.)  [92]

         for equation log P=A-B/(t+C), where t=Temp.;  (A,B,C)
         (4.75902,1127.529,16.020)  [156]

ION.POT  In eV (CTS=Charge Transfer Spectrum, EI=Electron Impact Spectroscopy,
               (-)=not stated)

         7.58 (CTS)    [78]; 7.56 (-)      [90]; 7.31 (CTS)   [111];
         7.48 (CTS)  [109]; 7.53 (CTS)  [116]; 7.55 (CTS)   [112];
         7.70 (CTS)  [110]; 7.72 (CTS)  [113]; 7.41 (-)      [79];
         7.72+/-0.3 (EI)  [118]; 7.45 (-)  [377]; 7.43 (-) [388]

PROT.AFF 208.5 Kcal/mol  [27]; 206.1 Kcal/mol  [386]; 862 KJ/mol  [386]

AQ. SOL. In mg/L at 25 C unless otherwise stated
         0.148  [53]; 0.135  [10]; 0.132  [55]; 0.175  [60]; 0.171  [56];
         0.129  [54]; 0.165 (29 C)  [59]; 0.090 (at 35% salinity)   [73]

         In umol/L at 25 C unless otherwise stated
         0.653  [55]; 0.668  [10]; 0.733  [53]; 0.817 (27 C)  [59];
         0.64   [73]; 0.72  [363]; 0.77   [60]; 0.16 (24 C)  [364];
         1.0 (20 C)  [76]; 0.8 (22 C)  [365]; 0.52 (20 C)    [366];

         -log S; where S=aqueous solubility(mol/L) at 25 C
         6.176  [12]; 6.185  [12]; 6.192  [12]; 6.18  [13]

HENRY'S
PART. CO
FLUOR.
UV.MAX
PHOS
HALFLIFE
THERMRX
CARC.POT
```

FIGURE 1. The above table shows the chemical and physical data for pyrene located in accession number 61. The numeral one in brackets after the name indicates that this record is the first of two or more records. The remainder of these data may be found in Figure 2. The numbers in square brackets are the literature references.

```
USER File Accession Number 62

NAME      PYRENE (2)
SYNONYMS
PHYS.APP
CHEM.FOR
MOLECWT
MELT.PT
BOIL PT
DENSITY
VAP.PRES
ION.POT
PROT.AFF
AQ. SOL.
HENRY'S   In KPa.m3/mol (c=calculated, e=experimental, r=recommended) at 25 C
          0.00121 (c)  [1]; 0.00133 (c)  [1]; 0.00136 (c)  [1];
          0.00102 (c)  [1]; 0.00105 (c)  [1]; 0.0011 (e)  [57];
          0.0012+/-0.002 (r)  [1]
PART. CO  C=concentration
          C(cyclohexane)/C(methanol-water)=150  [25];
          C(nitromethane)/C(cycohexane)=4.40  [25];
          C(DMSO)/C(n-pentane)=9.0  [101]; C(DMSO)/C(n-heptane)=5.1  [101];
          C(DMSO)/C(isooctane)=8.7  [101]; C(cyclohexane) 3.3  [104];
          C(DMF-water)/C(cyclohexane)=3.4  [103]
FLUOR.    In nm (excitation:emission)
          (302:338)  [15]; (305:335)  [38]; (320:335)  [38]
          (313:373, 380, 385, 390, 395, 414 & 397(ave.) - in cyclo.)  [169]
          (340:385, 392, 430 & 445)  [196]; (325:398)  [204];
          (332:373, 378, 363, 387 & 392)  [196]; (332:392)  [205];
          (328:389 - in diethyl ether/cyclohexane)  [183];
          (330:382 - in pentane)  [200]; (305:>370)  [387]
UV.MAX    In nm
          230.5, 241, 251, 261.5, 272, 292, 305, 318, 333.5, 351.5, 356, 362
          and 371.5 (in methanol/ethanol)  [44];
          305, 320 and 335 ( in cyclohexane)  [169]; 371  [196]
PHOS      In nm (excitation,emission - experimental technique)
          (322,590 - Paper(AgNO3))  [161]; (346,597 - Paper(NaI))  [158];
          (330,600 - Paper(AgNO3))  [163]; (330,600 - Paper(TiNO3))  [162];
          (340,600 - Paper(AgNO3))  [162]; (340,600 - Chalk(TiNO3))  [162];
          (343,595 - Paper(Pb(CH3COO)2))  [160]; (350,600)  [45];
          (343,595 - Paper(TiCH3COO))  [164];
          (342,597 - Cyclodextrin induced)  [39];
          (330,600 - H3BO3/T-7Clay/NaOH(AgNO3)  [162];
          (330,600 - CaHPO4/T-7Clay/NaOH(TiNO3)  [162];
          (343,595 - Sodium acetate treated paper(TiCH3COO))  [159]
HALFLIFE  5.7 days (under nitrating conditions in the dark)  [16]
          Under Photolytic Conditions  [385]
          21 hrs (on Silca Gel); 31 hrs (on Alumina);
          46 hrs (on Fly Ash); >1000 hrs (on Carbon Black)
THERMRX   unreactive - no carbonaceous observed at 750 C
CARC.POT  0 - non-carcinogenic [3]
          (-) non-carcinogenic (calc. & exp.)  [396]
          Co-carcinogen  [397]
```

FIGURE 2. This figure shows the information available in accession number 62. This data completes the chemical and physical data file for pyrene. Again, the numbers in square brackets refer to the literature references in the literature data file and do not refer to the references at the end of this paper.

```
USER File Accession Number 167

NAME       PYRENE
MOLEC.WT 202.26 gm/mol

CAS#       129-00-0

MOL.CONN 5.559  [282, 283]

H.FORM     216 KJ/mol  [386]; 52 Kcal/mol  [386]

H.ION.     173.0 Kcal/mol  [27]

H.VAP.     15.49 Kcal/mol  [284]; 14.57 Kcal/mol (predicted)  [284];
           76.6 cal/gm  [291]

H.SUB.     In KJ/mol (temperature range measured)
           91.2+/-0.5 (10-50 C)  [14]; 94.1 (10-50 C)  [14];
           95.2 (10-50 C)  [14]; 80.1 (10-50 C)  [14]; 100.4  [136];
           100  [137]
H.FUS.
H.SOL.
G.SOL.
C.SOL.
S.SOL.
S298
SING.ENG 322 KJ/mol  [38]

HYD.AFF. 68 Kcal/mol  [27]

SATVCONC 7.4E04 ng/m3  [4]

EQ.VCONC In ng/m3  (Temp.)
           5.8E02 (30 C)     [4]; 1.4E05 (-10 C)  [4]; 7.4E04 (25 C)  [21];
           7.6E04 (25 C)    [20]; 9.0E05 (50 C)  [20]; 6.3E07 (93 C)  [20];
           9.4E08 (130 C)   [20]

SOLUB.     Soluble in alcohol, ether, benzene and carbon disulfide  [21]

H.COMB.    1872.97 Kcal/mol  [288];
           9261.12 cal/gm  [295]; 9261 cal/gm  [291]
```

FIGURE 3. The complete datafile for the thermodynamic data for pyrene available in accession number 167 is reproduced above. The numbers in square brackets refer to the literature citations.

base, the information for pyrene is displayed in Figures 1 to 3. Figures 1 and 2 show two linked records (accession numbers 61 and 62) for the chemical and physical data while Figure 3 displays the thermodynamic data from accession number 167. The numbers in square brackets are the literature references from which the data was obtained.

The PAH database currently contains information on 729 PAH and their substituted derivatives, and 222 aza-arenes. If the current database were to be printed out in its entirety, it would represent a compendium of about 2500 pages. From this endeavour, it has become obvious that here is, in general, very little chemical, physical and thermodynamic data on the PAH in the available literature and government reports.

To illustrate the lack of data available, a search was carried out through the data base to look for the presence of data in the melting point, density, vapor pressure and aqueous solubility fields. This data is, of course, fundamental data required for almost any environmental assessment exercise. The results of this survey are shown in Table 3 in which the number of compounds for which data was found is tabulated as a function of database and parameter. Of 109 parent PAH in the database, for example, melting points were found for only 75 compounds, densities were found for only 14 compounds, vapor pressures for 14 and aqueous solubilities for 26. For the amino-, keto-, quinone- and nitro-PAH, no information at all was available for either vapor pressure or aqueous solubility. Considering the database as a whole, densities were found for 52 of 729 compounds, aqueous solubilities for 47 of 729 and vapor pressures for only 16 of 729 compounds,

TABLE 3

NUMBER OF COMPOUNDS IN EACH CLASS FOR WHICH DATA IS AVAILABLE

Database	Compounds in Database	Melting Point	Density	Vapor Pressure	Aqueous Solubility
Parent-PAH	109	75	14	14	26
Alkyl-PAH	355	300	26	2	21
Amino-PAH	30	30	6	0	0
Keto-PAH	62	5	1	0	0
Quinone-PAH	81	60	4	0	0
Nitro-PAH	92	78	1	0	0
Total	729	548	52	16	47

Inconsistencies have been noted in the literature data. Consider first, for example, the data for pyrene which is shown in Figure 1. Seven separate melting point determinations varying from 145 to 156°C were found. Seven boiling points (at atmospheric pressure) were also found for pyrene and they varied from 360 to 404°C. Six determinations (at 25°C) of the vapor pressure ranged from

8.98E-10 to 8.98E-09 atmospheres.

For perylene, four references were found for the boiling point. One reference reported a value of 500°C, another a value of 497°C, another claimed 460°C while the fourth stated that perylene sublimes between 350 and 400°C.

Another example shows 11 melting points for chrysene between 248 and 256°C with most values in the 250 to 252°C range. However, the boiling point data revealed 5 values between 436°C and 488°C and a predicted value of 437°C.

Probably the worst case of disagreement is in the vapor pressure data for fluoranthene. One value (1.77E-05 atm) was extrapolated from liquid state data, another (2.51E-06 atm) was calculated by extrapolating a vapor pressure measurement incorporating a fugacity ratio correction, while the third (1.21-08 atm) was derived from a High Pressure Liquid Chromatographic method.

It must be pointed out that no attempt was made to assess the correctness of the data included in the database. All information which could be found in the literature were included and referenced.

On the other hand, the melting point data for benzo(k)-fluoranthene fared somewhat better. Four references were found and all four were within the range of 215.5°C to 217.4°C. Boiling points (three were found) ranged from 280 to 282°C. Vapor pressures (two were found), however, varied from 1.28E-13 to 1.24E-12 atmospheres.

As one might expect, there was considerable data retrieved for benzo(a)pyrene. Ten melting point determinations ranged from 174 to 180.2°C. Boiling points (three at atmospheric pressure) ranged from 475 to 496°C. Only three vapor pressure measurements were found. They, however, showed remarkable agreement, ranging from 7.19E-12 to 7.22E-12 atmosphere.

Significant variations could be found in the reported values for other parameters as well. For example, the aqueous solubility data for benzo(a)pyrene has been reported in at least nine articles and varies from 0.0001 mg/L at 25°C all the way up to 0.0061 mg/L. One report even quotes a value for the aqueous solubility of 0.012 mg/L at 27°C.

SUMMARY

Data have been compiled from the available literature for 38 different chemical, physical, thermodynamic and spectral characteristics for 729 PAH and their alkyl, amino, oxygenated and nitro derivatives and have been entered into a database on a personal computer. The compilation of the data, apart from collecting considerable data in a single repository, has revealed a number of important findings:

- very little data exist for most of the PAH and their derivatives
- the data which does exist are often inconsistent and, in some data fields, very large differences occur
- different measurement techniques can give widely varying results
- there is a need to evaluate the validity of the various methodologies and to standardize techniques for the determination of basic chemical and physical data.

Since very little measured data does exist, and since the cost of determining the data for all the PAH would be prohibitive, predictive methods such as those developed by White (13) should prove invaluable in providing the data where, at present, none exists.

REFERENCES

1. National Academy of Sciences. (1972): Particulate Polycyclic Organic Matter. 361 pp. National Academy of Sciences, Washington, D.C.
2. Lane, D.A. and Katz, M. (1977): The Photomodification of Benzo(a)pyrene, Benzo(b)fluoranthene and Benzo(k)-fluoranthene Under Simulated Atmospheric Conditions. Adv. Environ. Sci. Technol., 8(2): 137-154 edited by I.H. Suffet, Wiley Interscience, New York.
3. Rajagopalan, R., Vohra, K.G. and Mohan Rao, A.M. (1983): Studies on Oxidation of Benzo(a)pyrene by Sunlight and Ozone. Sci. Total Environ., 27: 33-42.
4. Pitts, J.N., Jr. (1983): Formation and Fate of Gaseous and Particulate Mutagens and Carcinogens in Real and Simulated Atmospheres. Environ. Health Perspect., 47: 115-140.
5. Grosjean, D., Fung, K. and Harrison J. (1983): Interactions of Polycyclic Aromatic Hydrocarbons with

Atmospheric Pollutants. Environ. Sci. Technol., 17: 673-679.

6. Nielson, T. (1984): Reactivity of Polycyclic Aromatic Hydrocarbons toward Nitrating Species. Environ. Sci. Technol. 18: 157-163.
7. Eisenberg, W.C., Taylor, K., Cunningham, D.L.B. and Murray, R.W. (1985): Atmospheric Fate of Polycyclic Organic Material. In: Polynuclear Aromatic Hydrocarbons: Mechanisms, Methods and Metabolism, edited by M. Cooke and A.J. Dennis, pp 395-410, Battelle Press, Columbus, OH.
8. Pitts, J.N., Jr., Zielinska, B., Sweetman, J.A., Atkinson, R. and Winer, A.M. (1985): Reactions of Adsorbed Pyrene and Perylene with Gaseous N_2O_5 Under Simulated Atmospheric Conditions. Atmos. Environ., 19(6): 911-915.
9. Daisey, J.M., Lewandowski, C.G. and Zorz, M. (1982): A Photoreactor for Investigations of the Degradation of Particle-Bound Polycyclic Aromatic Hydrocarbons Under Simulated Atmospheric Conditions. Environ. Sci. Technol., 16(12): 857-861.
10. Daisey, J.M., Low, M.J.D. and Tascon, J.M.D. (1984): The Nature of the Surface Interactions of Adsorbed Pyrene on Several Types of Particles. In: Polynuclear Aromatic Hydrocarbons: Mechanisms, Methods and Metabolism, edited by M. Cooke and A.J. Dennis, pp 307-315, Battelle Press, Columbus, OH.
11. Taskar, P.K., Solomon, J.J. and Daisey, J.M. (1984): Rates and Products of Reaction of Pyrene Adsorbed on Carbon, Silica and Alumina. In: Polynuclear Aromatic Hydrocarbons: Mechanisms, Methods and Metabolism, edited by M. Cooke and A.J. Dennis, pp 1285-1298, Battelle Press, Columbus, OH.
12. Dlugi, R. and Gusten, H. (1983): The Catalytic and Photocatalytic Activity of Coal Fly Ashes. Atmos. Environ. 17(9): 1765-1771.
13. White, C.M. (1985): Prediction of the Boiling Point, Heat of Vaporization, and Vapor Pressure at Various Temperatures for Polycyclic Aromatic Hydrocarbons. J. Chem. Eng. Data, (submitted for publication).

QUANTITATIVE ASSESSMENT OF THE DESORPTION OF PAH FROM PARTICULATE MATTER IN A RESONANT MICROWAVE CAVITY.

DOUGLAS A. LANE, STEPHEN W.D. JENKINS
Environment Canada, Atmospheric Environment Service, 4905 Dufferin Street, Downsview, Ontario, M3H 5T4, Canada.

INTRODUCTION

Polycyclic Aromatic Hydrocarbons (PAH) are released to the atmosphere through the combustion of fossil fuels, wood, tobacco and refuse. In general, PAH are adsorbed onto the surface and in the interstitial cavities of soot particles generated during the combustion process. It has been observed that about 80 percent of the PAH are adsorbed on particles less than 5 μm in mass median diameter (1). When particulate matter is collected by filtration on a glass fiber filter, the problem of how to remove the PAH from the particulate matter is one which is not easily solved.

In general, solvent elution and thermal desorption are the two basic methods by which organics have been removed from particulate matter. Thermal desorption is a simpler, more direct route, however, it has been less successful than solvent elution because of the inability to achieve quantitative removal of the organics under infra-red (IR) irradiation and the charring of the compounds which can occur under IR irradiation. The use of microwave irradiation has been described by Sevcik (2) for the removal of organics from charcoal adsorption tubes, and by Lane (3) for the thermal desorption of organics from the surface of particulate matter. Microwave thermal desorption has shown promise of being an efficient and effective method for the thermal desorption of organic contaminants from the surface of various substrates.

Since our presentation last year (3), we have made a number of modifications to the microwave desorption system, and as a result, have improved the desorption efficiency for many of the compounds of interest.

Background

When a plane wave of microwave (300 MHz to 300 GHz) radiation is incident upon a molecule of a substance, the absorption of that radiation by the molecule is dependent upon the total molecular polarizability of the molecule which comprises the sum of the distortion polarizability (related to the ability of the molecule to be polarized) and the orientation polarizability (related to the permanent dipole moment of the molecule). The total molar polarizability of a substance is also related to the dielectric constant (ε) as shown by the Debye equation:

$$P_m = \frac{4}{3}\pi N \left(\alpha + \frac{\mu^2}{3kT}\right) = \frac{(\varepsilon - 1)}{(\varepsilon + 2)}\frac{M}{\rho} \qquad [1]$$

where N is Avagadro's Number, α is the molecular polarizability, μ is the permanent dipole moment, k is the Boltzman constant, M is the molecular weight and ρ is the density.

The major drawback to heating with infra-red radiation is that radiation at those frequencies will penetrate a substance, such as particulate matter, only to a depth of a fraction of a micron. Infra-red heating is, therefore, a surface heating phenomenon. Sub-surface material is heated through the process of conduction. In addition, IR frequencies are very close to bond breaking energies and, as a result, molecular changes can occur during desorption with IR radiation.

Microwave radiation, on the other hand, will pass through particles and heat the entire volume of the particle uniformly as well as the molecules adsorbed on the surface. The degree of heating will depend upon the absorptive properties of the particle and the adsorbed molecules. Also, microwave radiation does not have sufficient energy to cause bond disruption.

The depth at which microwave radiation is decreased to 1/e of its incident power has been shown by von Hipple (4) to be related to the wavelength of the incident radiation (λ), the relative dielectric constant at that frequency (k') and the loss tangent ($\tan\delta$) of the substance by the expression:

$$D = \frac{\lambda}{2\pi}\left(\frac{2}{k'\left(\sqrt{1 + \tan^2\delta} - 1\right)}\right)^{1/2} \qquad [2]$$

There are three general cases for this expression and those occur when $\tan^2\delta << 1$, $\tan^2\delta \approx 1$, and $\tan^2\delta >> 1$. If $\tan^2\delta << 1$, the value of the quantity $(\sqrt{1 + \tan^2\delta} - 1) \approx 1/2(\tan^2\delta)$ and the depth of penetration may be simplified to:

$$D = \frac{\lambda}{\pi\sqrt{k'}\tan\delta} \qquad [3]$$

If $\tan^2\delta \approx 1$, then equation [2] must be used. However, if $\tan^2\delta >> 1$, then the value of the quantity $(\sqrt{1 + \tan^2\delta} - 1) \approx \tan\delta$ and, hence, the expression may be written as:

$$D = \frac{\lambda}{2\pi}\left(\frac{2}{k'\tan\delta}\right)^{1/2} \qquad [4]$$

The attenuation of electromagnetic radiation through a particle is sometimes expressed in terms of a loss in decibels per meter. The relevant expression derived by von Hipple is:

$$\text{loss}\left(\frac{db}{m}\right) = \frac{8.686\ \pi\ 2}{\lambda}\left[\frac{k'}{2}\left(\sqrt{1 + \tan^2\delta} - 1\right)\right]^{1/2} \qquad [5]$$

This information may now be applied to the current example in which atmospheric particulate matter on a glass fiber filter is to be irradiated by microwave energy. Since glass is virtually transparent to microwave radiation, the radiation will encounter essentially just the particulate matter in the desorption cavity. A search through the literature failed to identify any references to the electrical properties of particulate matter, and in particular, the dielectric constant or the loss tangent. However, since the structure of the soot upon which the PAH are adsorbed (extended coronene structure) is similar to that of coal, it is suggested that, to a first approximation, the electrical properties of soot be considered to be somewhat similar to those of coal.

Again, a search of the literature failed to reveal the dielectric constant or the loss tangent for coal so that a calculation of the depth of penetration could not be performed.

Recently, however, Rush (5) determined that the attenuation of microwave radiation through high organic content, dried coal was less than 0.1 db/cm. This implies that the microwave radiation will pass through the particulate matter without appreciable attenuation and result in an even, volumetric heating of the particles.

MATERIALS AND METHODS

The desorption system (shown in Figure 1.) comprised a Microtron 200 Mk III continuous wave, variable power generator, a resonant transverse magnetic (Tm_{010}) cavity, a metered flow of nitrogen, and a high pressure liquid chromatography pump for introducing a low flow of water and/or other microwave absorbing fluid into the desorption

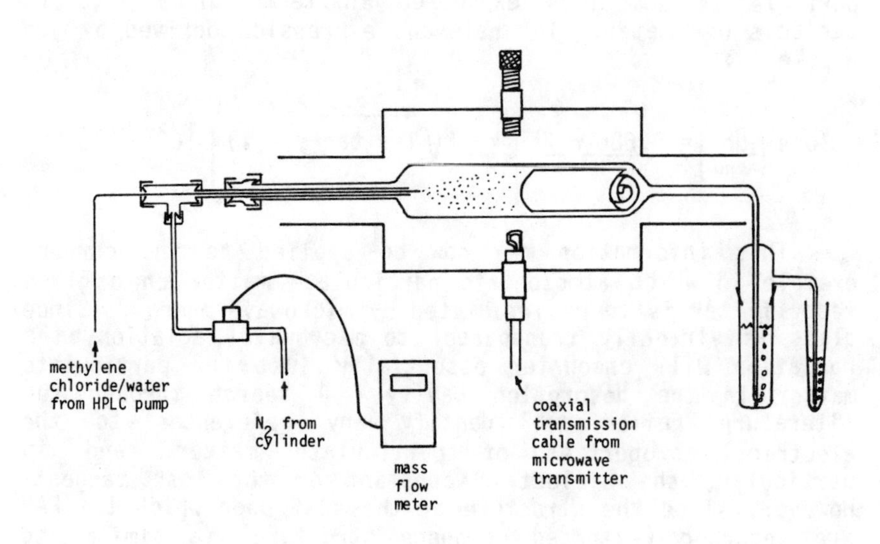

FIGURE 1. Schematic diagram (not to scale) of the resonant cavity, microwave desorption system.

zone of the cavity. The cavity was fitted with 3/4 wavelength traps on each end to close the cavity electrically to microwave radiation. The cavity has been designed so that any low-loss material tube such as quartz, glass or teflon with an outer diameter up to 4.4 cm can be placed coaxially through the cavity without allowing microwave radiation to escape.

When transmitting electromagnetic power into any antenna or cavity, the impedances of the load, transmission line and transmitter must all be properly matched. If a mismatch occurs, power will be reflected back along the transmission line and will be dissipated there as heat. Depending upon how well the final output circuits of the transmitter are designed, high reflected powers may cause serious damage to the final amplifier stage of the transmitter (in this case an expensive magnetron tube). The transmitter was equipped with a reflected power (RP) module and, as per the manufacturer's recommendations, the RP was not permitted to exceed 50 watts.

To desorb the organics from atmospheric particulate matter, one quarter of a standard 8 inch by 10 inch glass fiber filter was coiled with the particulate matter on the inside and placed in the desorption tube. The microwave generator was turned on to low power (20 watts input) and the cavity was tuned for maximum forward power and least reflected power. At this point, the HPLC pump was started and the aerosol spray of water/methylene chloride mixture was started. The nitrogen flow through the desorption tube was then adjusted to 100 ml/min. The input power was raised to 200 watts, care being taken not to exceed 50 watts RP. The hot gases evolved during the desorption process passed through a methylene chloride bubbler where the desorbed organics were trapped.

After the desorption had been completed, the methylene chloride was separated from the water in the bubbler. The filter was removed from the desorption tube and the desorption tube was rinsed with methylene chloride and the rinse was added to the eluate from the bubbler. The methylene chloride fraction was reduced in volume in a rotary flash evaporator and then solvent exchanged into toluene for injection onto the Finnigan 4023 GC/MS/DS. The water phase was not analysed. For calibration purposes, D_{10}-phenanthrene was added to each solution prior to injection, as an internal standard.

RESULTS AND DISCUSSION

In general, there was not sufficient mass of particulate matter (generally about 10 to 30 mg) on the quarter filter to effect a proper impedance match between the cavity and the transmission line. The most effective way to load the cavity was to add a microwave absorbing fluid which would quickly evaporate and also assist the desorption process by vapour phase stripping the organics off the particulate surface. When water alone was used as the microwave absorber, the high molecular weight PAH (the 5-ring and above compounds) were not desorbed from the particulate filter when the maximum input power of 200 watts was applied. An organic solvent with the highest possible flash point and auto-ignition temperatures and the largest possible dielectric constant was sought. Methylene chloride was selected and used in place of the water. The desorption using methylene chloride succeeded in removing the high molecular weight compounds, however, the temperature in the cavity rose rather quickly to the point where the particulate matter on the filter ignited and subsequently destroyed the filter. This happened even when the power was reduced to 100 watts. To eliminate this problem, a mixture of methylene chloride and water (60:40) was pumped into the cavity. The appropriate mixing ratio was set on the HPLC pump and the flow set at 1.0 ml/min. It was then possible to perform the desorptions without ignition of the particulate matter.

The methylene chloride water mixture was injected directly into the desorber in the microwave cavity in the form of a fine aerosol. The mixture was pumped through a 0.37 mm ID capillary which, in turn, passed through another capillary tube of 0.50 mm ID through which also passed a flow of nitrogen at 100 ml/min (see Figure 1.). The flow of gas around the inner capillary tube resulted in the production of a fine aerosol mist which vaporized upon encountering the high density of microwave energy in the cavity.

A solution of 11 PAH and 2 chlorinated biphenyls was prepared and analysed on the gc/ms. The reconstructed ion chromatogram (RIC) from the gc/ms analysis is shown in Figure 2. The numbers on the chromatogram refer to the compounds as listed in Table 1. To obtain the relative response factors for each compound relative to that of D_{10}-phenanthrene, the parent molecular ion response relative to the parent molecular ion of D_{10}-phenanthrene

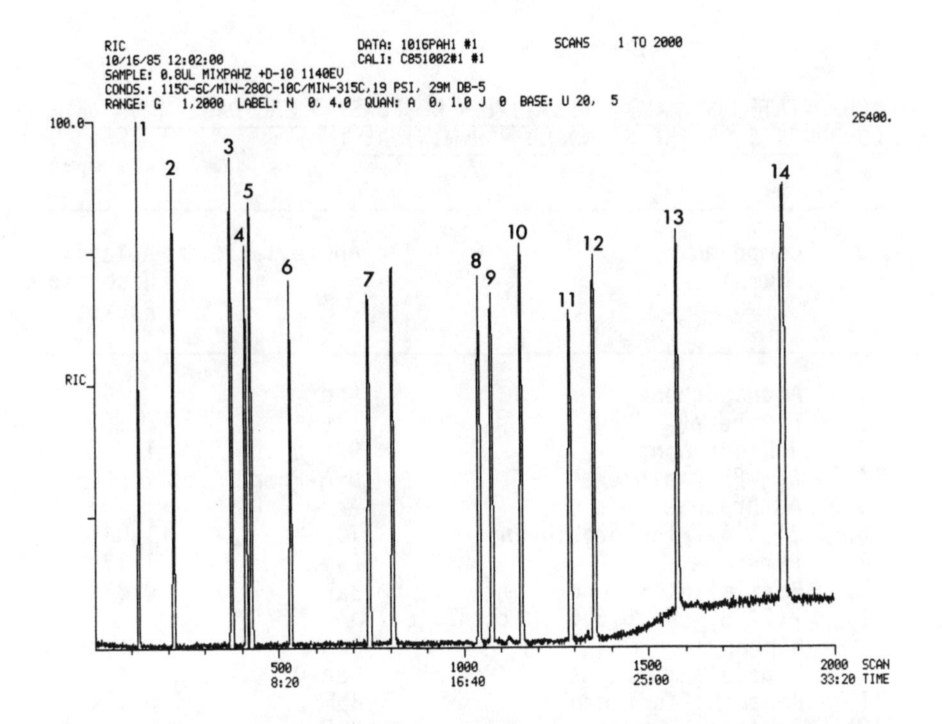

FIGURE 2. The reconstructed ion chromatogram of the standard solution. The numbers refer to the compounds as listed in Table 1.

was determined. For the chlorinated biphenyls, the sum of the parent molecular ion isotopes was used. The values obtained are shown in Table 1.

A blank glass fiber filter was doped with a mixture of PAH, placed in the cavity and desorbed in the usual manner. The reconstructed ion chromatogram (RIC) obtained from the gc/ms showed that the high molecular weight organics (5-ring compounds and higher) were not removed. In addition, a higher than normal RP dictated a maximum input power of 150 watts. This was anticipated since there was no absorbing medium in the cavity other than the PAH molecules and the cavity was, consequently, not adequately loaded.

When desorbing the PAH from a glass surface, only the PAH were heated by the absorption of the microwave energy. The glass filter remained cool. The polarizable PAH were

TABLE 1

ABBREVIATIONS AND RELATIVE RESPONSE FACTORS FOR THE COMPONENTS OF THE STANDARD PAH MIXTURE

No.	Compound Name	Abbreviation	Relative Response Factor
1.	Acenaphthene	Ace	0.746
2.	Fluorene	Flu	0.880
3.	9-Fluorenone	9-Flu	1.206
4.	D_{10}-Phenanthrene	D_{10}-Phen	1.000
5.	Anthracene	Anth	1.265
6.	2',3,4-Trichlorobiphenyl	TCB	1.094
7.	Pyrene	Pyr	1.178
8.	Benz(a)anthracene	BaA	0.834
9.	2,2',3,4,4',5,6,6'-Octachlorobiphenyl	OCB	1.152
10.	Benz(a)anthracene-7,12-dione	BaA-D	0.411
11.	Benzo(b)fluoranthene	BbF	0.794
12.	Benzo(a)pyrene	BaP	0.809
13.	Benzo(ghi)perylene	BghiP	0.710
14.	Coronene	Cor	0.655

thermally excited as evidenced by the removal of the PAH up to, and including, the 4-ring membered species, however, there was not sufficient heating to remove the higher PAH. It appears that higher power would be required to effect a complete removal of the PAH from the glass fiber filter. By contrast, all of the PAH were desorbed from the surface of the particulate matter. The thermal energy from the absorption of the microwave radiation by the particulate matter was imparted to the molecules adsorbed on the particle surface, causing them to leave the particle. It is apparent that atmospheric particulate matter is a relatively lossy material and, therefore, plays a significant role in the desorption process.

Air particulate samples for these experiments were obtained by high volume filtration at our location on the northern boundary of Metropolitan Toronto. The filter samples used for the desorption tests, although very light in loading (20 $\mu g/m^3$), were typical of the loadings in this area.

A filter containing atmospheric particulate matter was divided into four equal parts. One section was desorbed in the cavity and the eluted organics which partitioned into the methylene chloride were quantitatively analysed. The analysis revealed the presence of 12 ng of 9-Flu, 260 ng of Anth, 170 ng of Pyr and 13 ng of BaA. No other compounds (corresponding to the standard components) were detected in the sample.

A second filter section was doped with the standard solution. The amount of each component in the standard deposited on the filter is shown in Table 2. It is realized

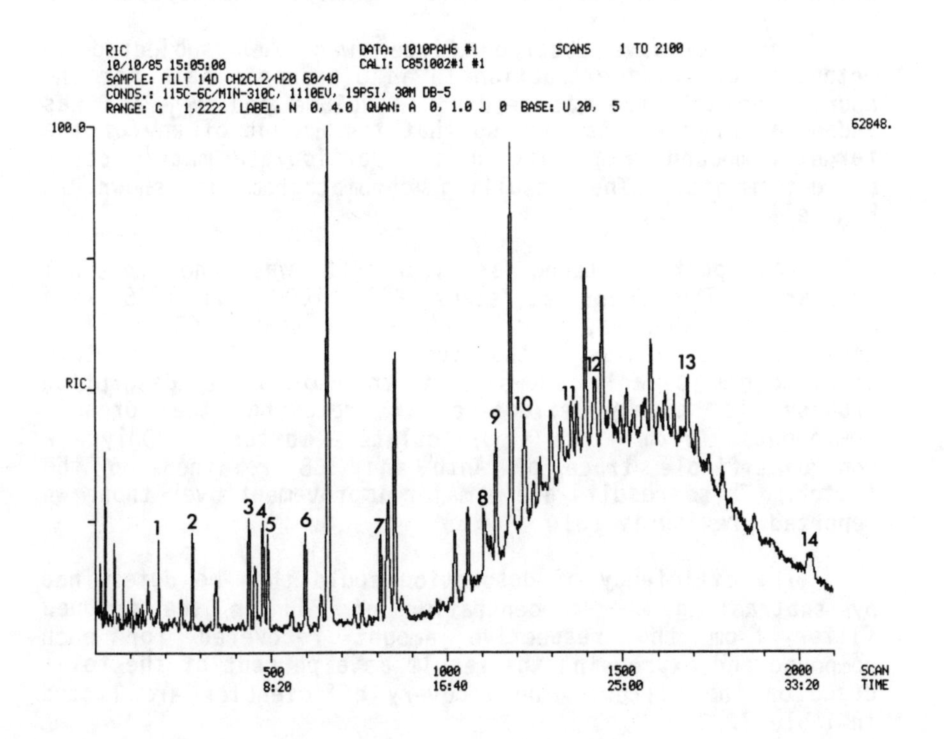

FIGURE 3. Reconstructed ion chromatogram showing the recovery of sample from a doped particulate filter.

that solution coating was not the ideal way in which to dope the filter and it is appreciated that the PAH which are solution doped onto a filter may not exist on the particle surface in the same physical or chemical state as the PAH which were originally on the particle surface. However, this was considered to be the best available strategy to use. The filter was then microwave desorbed and the methylene chloride extract was analysed. The RIC of this desorption is reproduced in Figure 3.

In Figure 3, all of the doped species are clearly identified. The large baseline hump seen in this trace is typical of the chromatograms obtained when the sample has not been cleaned up prior to analysis. The drift is due to unresolved components and not due to column bleed. The amount of each compound recovered was determined and entered in Table 2. The water phase of this desorption was brownish-yellow in colour, however, it was not analysed.

The microwave desorbed filter was then subjected to methylene chloride extraction in an ultrasonic bath for one hour. The solution was reduced in volume and D_{10}-Phen was added as internal standard so that the amount of any of the target compounds remaining on the particulate matter could be determined. The resulting chromatogram is shown in Figure 4.

The peak centered at scan 455 was the internal standard. The peaks at scans 270, 1000 and 1375 were unidentified whereas, those centered at scans 643 and 1163 were the ubiquitous phthalates. The baseline of this chromatogram clearly shows that the microwave desorption process is highly effective at removing the organic components from the particulate matter. Only a non-quantifiable trace of Anth and TCB remained on the filter. These results are a major improvement over those we reported previously (3).

The efficiency of desorption could then be determined by subtracting the concentration of PAH on the undoped filter from the respective amount recovered for each compound and expressing the result as a percent of the total doped on the filter. The recovery efficiencies are listed in Table 2.

The poor efficiency of recovery for Ace and Flu may be explained by their relatively high volatility. It is considered likely that the loss of these compounds took

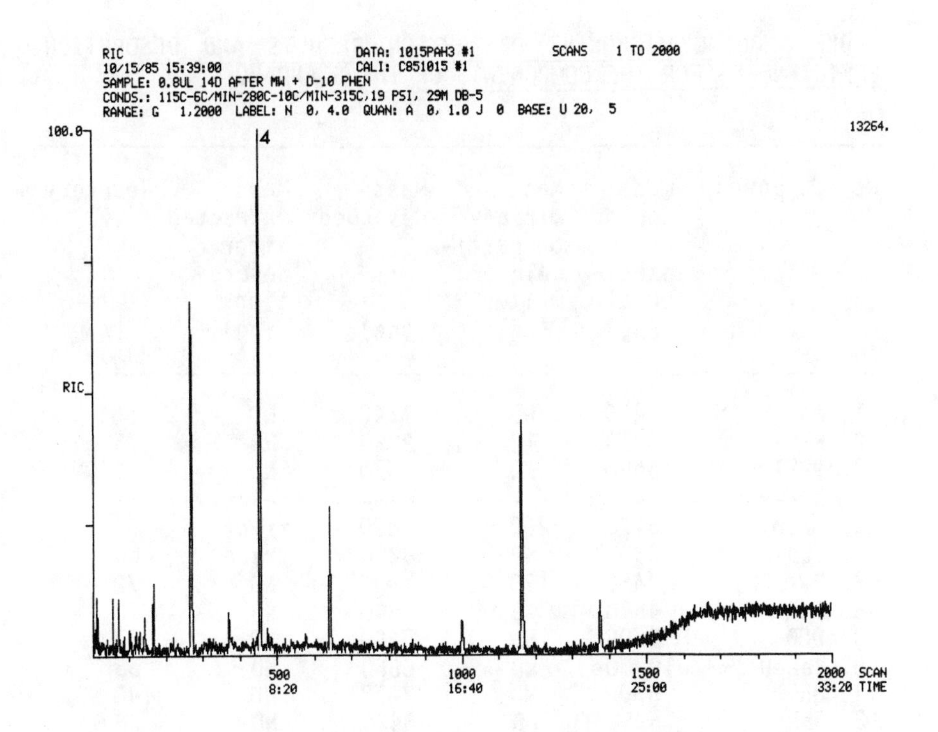

FIGURE 4. Reconstructed ion chromatogram showing the species remaining on the doped filter after microwave desorption.

place during the solvent reduction stage of the analysis. This may also explain the lack of recovery of the other compounds. Further tests will have to be performed to prove or to disprove this contention. Some of the loss may also be due to problems in quantifying the species when there may be interfering components in the sample. This is likely the reason for the apparent recovery of 140 percent for the BbF.

CONCLUSIONS

A system for desorbing organic compounds from atmospheric particulate matter collected on glass fiber filters has been described. The desorption is rapid,

TABLE 2

TABLE SHOWING MICROWAVE DESORPTION RESULTS AND DESORPTION EFFICIENCIES FOR THE COMPONENTS OF THE STANDARD

No.	Compound	Mass doped on particulate (ng)	Mass already on particulate matter (ng)	Mass desorbed (ng)	Mass extracted after desorption (ng)	Recovery (%)
1.	Ace	2420	ND	1340	ND	55
2.	Flu	3020	12	2220	ND	73
3.	9-Flu	5690	ND	4670	ND	82
4.	D_{10}-Phen	----	---	----	--	--
5.	Anth	3170	260	2830	trace	81
6.	TCB	10300	ND	8220	trace	80
7.	Pyr	2450	170	1930	ND	72
8.	BaA	4840	13	3540	ND	73
9.	OCB	9600	ND	5490	ND	57
10.	BaA-D	10200	ND	6660	ND	65
11.	BbF	5300	ND	7473	ND	140
12.	BaP	5950	ND	3420	ND	58
13.	BghiP	4900	ND	3023	ND	62
14.	Cor	18800	ND	19000	ND	100

requiring only about 15 minutes for the complete desorption. The attempt at a mass balance has shown that more care must be taken in the solvent reduction stage of the process and errors in the gc/ms quantitation stage must be assessed. The method does appear, however, to be very effective for the removal of most of the organic compounds adsorbed on the particulate matter. In view of the definite brownish colour of the water phase, it is clear that there are many water soluble organic compounds on the particulate matter. This aspect of the procedure needs to be investigated.

REFERENCES

1. Rierce, R.C. and Katz, M. (1975): Dependency of Polynuclear Aromatic Hydrocarbon Content on Size Distribution of Atmospheric Aerosols. Environ. Sci. Technol. 9(4) 347-353.
2. Sevcik J. (1984): Thermal Desorption of Environmental Samples. Amer. Lab. 16(7) 48-57.
3. Lane, D.A. and Jenkins, S.W.D. (1985): Microwave Desorption of Organic Compounds from Particulate Matter. In: Polynuclear Aromatic Hydrocarbons: Chemistry and Carcinogenesis. (in press).
4. von Hipple, A.R. (1954): Dielectrics and Waves. John Wiley and Sons Inc., New York, 284 pp.
5. Rush, W.F. and Onischak, M. (1983): A Criterion for Selective Microwave Heating of Sulphur in Coal. Fuel 62 459.

QUINOLINES AND BENZOQUINOLINES: STUDIES RELATED TO THEIR METABOLISM, MUTAGENICITY, TUMOR-INITIATING ACTIVITY, AND CARCINOGENICITY

EDMOND J. LaVOIE, AKEMI SHIGEMATSU, ELIZABETH ANN ADAMS, NORA G. GEDDIE, AND JOSEPH E. RICE
Naylor Dana Institute for Disease Prevention, American Health Foundation, Valhalla, NY 10595.

INTRODUCTION

Quinolines and benzoquinolines are among the major aza-arenes which have been detected in the environment. In addition to being major industrial chemicals, quinoline and isoquinoline have been detected as components of urban air pollution, cigarette smoke, shale oil, and coal liquefaction products (1-11). The various isomeric benzoquinolines have also been shown to be major components in the basic portion of automobile exhaust, urban air particulates, and cigarette smoke (2-14).

Quinoline is a hepatocarcinogen in both rats and mice (15,16). Recently, we have shown that quinoline and both 4- and 8-methylquinoline are active as tumor initiators on Sencar mouse skin (17). Quinoline is mutagenic in *S. typhimurium* TA100 in the presence of rat liver homogenate (18-21) and induces unscheduled DNA synthesis in primary cultures of rat hepatocytes (22). In contrast, isoquinoline is inactive as a genotoxic agent in assays with *Salmonella typhimurium* and in the hepatocyte primary culture (HPC)/DNA repair test. On the basis of these differences in biological activity, we have compared the metabolites of quinoline and isoquinoline as formed *in vitro* (23).

There are no convincing data on the potential carcinogenic activity of benzoquinolines. In a comparative study on the mutagenic potential of all five isomeric benzoquinolines in *S. typhimurium* TA98 and TA100, benzo(f)quinoline, benzo(h)quinoline, and phenanthridine were active as mutagens (24). Both benz(f)isoquinoline and benz(h)isoquinoline were inactive under these assay conditions. In view of these results, we have studied the metabolism of benzo(f)-quinoline, benzo(h)quinoline, and phenanthridine *in vitro* using a similar rat liver preparation as employed in these mutagenicity assays (18,24-26).

We have, in the present study, attempted to:

(1) Investigate the possible mechanisms by which quinoline is ultimately activated to a mutagen;

(2) Determine the activity of each isomeric benzoquinoline in the HPC/DNA repair test;

(3) Evaluate the tumor-initiating activity of benzo(f)-quinoline, benzo(h)quinoline, phenanthridine and select derivatives in mouse skin; and

(4) Investigate the possible mechanism(s) by which benzo(f)quinoline, benzo(h)quinoline, and phenanthridine exert their mutagenic activity in *S. typhimurium*.

MATERIALS AND METHODS

Chemicals

Quinoline, quinoline-N-oxide, benzo(f)quinoline, benzo-(h)quinoline, and phenanthridine were purchased from Aldrich Chemical Co. Benz(f)isoquinoline and benz(h)isoquinoline were prepared as previously described (24). The synthesis of the 5,6-oxide and the 7,8-oxide of quinoline are outlined in Scheme 1 and Scheme 2. 2-Fluoroquinoline was synthesized as previously described (27). 3- and 4-Fluoroquinoline were prepared by decomposition of the fluoroborate derivatives of their respective diazonium salts (28). Each of these fluoroquinolines gave PMR and MS spectra consistent with their structure. The purity of these fluoroquinolines was >99% as determined by capillary GC. The 5,6-oxides of benzo(h)quinoline and benzo(f)quinoline were prepared by singlet oxygen oxidation as previously described (29). The 5,6-dihydrodiols of benzo(f)quinoline and benzo(h)quinoline were prepared by hydrolysis of their corresponding oxides in 1.0 N NaOH. The syntheses of 4-methylbenzo(h)quinoline, 1-methylbenzo(f)quinoline, and 1,3-dimethylbenzo(f)quinoline were performed using established literature procedures (30). The preparation of ^{14}C-phenanthridine has been described (26). The preparation of ^{14}C-labeled benzo(f)quinoline and benzo(h)quinoline was accomplished using 2- and 1-naphthylamine, respectively, and radiolabeled glycerol in the Skraup synthesis in a similar manner to that used previously to prepare ^{14}C-labeled quinoline (7).

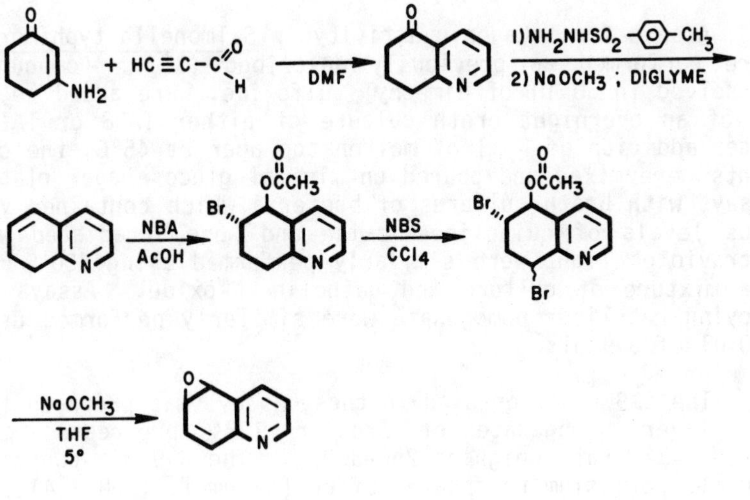

SCHEME 1. Synthesis of 5,6-epoxy-5,6-dihydroquinoline.

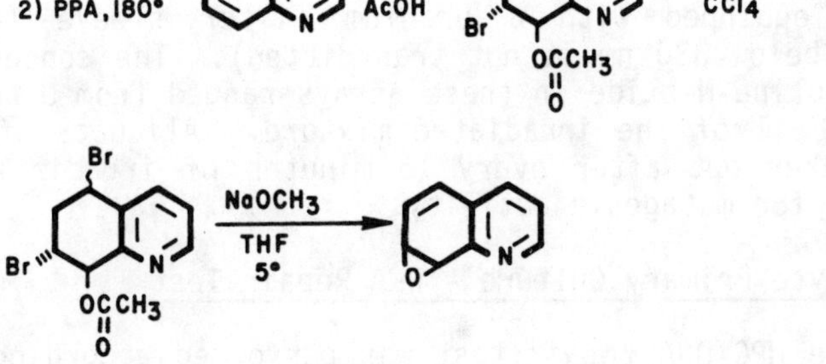

SCHEME 2. Synthesis of 7,8-epoxy-7,8-dihydroquinoline.

Mutagenicity Assays

Assays for mutagenic activity in *Salmonella typhimurium* were performed as previously described (31). Compounds, dissolved in 50 µl of dimethyl sulfoxide, were added to 0.1 ml of an overnight broth culture of either TA98 or TA100. After addition of 2 ml of molten top agar at 45°C, the contents were mixed and poured on minimal glucose agar plates. Assays with broth cultures of bacteria which contained various levels of quinoline-N-oxide and were irradiated with ultraviolet light were similarly performed using 400 µl of the mixture of culture and quinoline-N-oxide. Assays employing rat liver homogenate were similarly performed using 200 µl of S-9 mix.

The S-9 fraction used in these assays was prepared from the liver homogenate of Aroclor 1254 pretreated male Fischer-344 rats weighing 250-300 g. The S-9 mix contained per ml: potassium phosphate buffer (50 µmoles, pH 7.4), 8.0 µmoles $MgCl_2$, 33 µmoles KCl, 5.0 µmoles glucose 6-phosphate, and 4.0 µmoles $NADP^+$ and 0.5 ml of S-9 fraction.

Exposure Of Bacterial Broth Cultures To Ultraviolet Irradiation

Assays for mutagenic activity were also performed using an overnight broth culture of TA100 which was exposed to UV irradiation in the presence of quinoline-N-oxide. In these assays, 10 ml solutions containing 5 ml of an overnight broth culture of TA100 and quinoline-N-oxide in 5 ml of distilled water were irradiated using a 450 watt Conrad-Hanovia immersion lamp in an Ace Glass photochemical reaction assembly equipped with a uranium filter sleeve (radiated energy below 330 nm is not transmitted). The concentration of quinoline-N-oxide in these assays ranged from 0 to 100 µg per 100 µl of the irradiated mixture. Aliquots of 400 µl were taken out after every 15 minutes of irradiation, and assayed for mutagenic activity.

Hepatocyte Primary Culture - DNA Repair Test

The HPC/DNA repair test was performed according to established protocols (32,33). Two hours after inoculation of freshly prepared hepatocytes from adult male F-344 rats, triplicate coverslips of HPCs are exposed to the test compound and 10 µCi/ml tritiated thymidine ([^{3}H]-TdR). After 18-20 hours of exposure to the test compound and [^{3}H]-TdR in

Williams' Medium E, the cultures are washed and the nuclei swelled and in Na citrate (10-15 minutes) and are fixed in three 30 minute changes of 3:1 ethanol-glacial acetic acid. The cultures are air dried and mounted, cell surface up, on glass slides with Permount (Fischer Scientific). Autoradiographs are prepared by dipping slides in NTB-2 emulsion. The slides are dried (1-4 hours) and stored for 7 days at 4°C. The slides are developed in D19 (Eastman Kodak). Results are then quantified by determining the net increase in grains/nucleus induced by a chemical. Grain counts are performed on a Arlet Model 880 electronic counter with microscopic attachment. Net nuclear grain counts of repair synthesis are calculated by subtracting the cytoplasmic area count from the nuclear area count.

Tumor Initiation Bioassays

Compounds tested as tumor initiators were applied to the skin of CD-1 mice or Sencar mice. When the animals were in the second telogen phase of the hair cycle, the compound to be tested was applied in an acetone solution (100 µl) ten times on alternate days to the shaved backs of mice. Ten days after the last initiation dose, promotion was begun by application thrice weekly of 2.5 µg of tetradecanoylphorbol acetate in acetone (100 µl) for several weeks as outlined in Table 4.

RESULTS AND DISCUSSION

On the Metabolic Activation of Quinoline to a Mutagen

The ethyl acetate extractable metabolites of quinoline as formed with rat liver homogenate have been identified as 5,6-dihydroxy-5,6-dihydroquinoline, 2-hydroxyquinoline, 3-hydroxyquinoline, and quinoline-N-oxide (23). None of these metabolites were mutagenic in *S. typhimurium* in the presence or absence of rat liver homogenate. It was initially hypothesized that the epoxide precursor to the major metabolite, 5,6-dihydroxy-5,6-dihydroquinoline, was responsible for the mutagenic activity of quinoline. To evaluate this hypothesis, 5,6-epoxy-5,6-dihydroquinoline was synthesized as outlined in Scheme 1. We also prepared 7,8-epoxy-7,8-dihydroquinoline as illustrated in Scheme 2. We evaluated both of these arene oxides for mutagenic activity in *S. typhimurium* TA100 in the presence and absence of rat liver homogenate. The results of these assays as shown in Table 1 demonstrate that neither of these oxides were mutagenic under these

TABLE 1

BIOASSAYS ON QUINOLINE, 5,6-EPOXY-5,6-DIHYDROQUINOLINE, AND 7,8-EPOXY-7,8-DIHYDROQUINOLINE FOR MUTAGENIC ACTIVITY IN S. TYPHIMURIUM AND UNSCHEDULED DNA SYNTHESIS IN RAT HEPATOCYTES

Compounds	S. Typhimurium TA100 Dose (μg)	His⁺/Plate	HPC/DNA Repair Assay Results	Grains/ Nucleus	Dose M
Quinoline	200	936 *		33.5±8.3	$5x10^{-4}$
	100	784	(+)	14.4±5.7	$1x10^{-4}$
	50	517		0.3±0.9	$5x10^{-5}$
	0	133		0.1±0.3	$1x10^{-5}$
5,6-Epoxy-5,6-dihydroquinoline	200	142 **		Toxic	$5x10^{-5}$
	100	159	(−)	0.5±1.2	$1x10^{-5}$
	50	137		0.1±0.3	$5x10^{-6}$
	0	155		0.4±1.4	$1x10^{-6}$
7,8-Epoxy-7,8-dihydroquinoline	50	Toxic **	(−)	Toxic	$5x10^{-5}$
	25	117		3.1±4.9	$1x10^{-5}$
	12	175		0.1±0.5	$5x10^{-6}$
	0	146		2.1±3.0	$1x10^{-6}$

* + S-9
** - S-9

assays conditions. In contrast to quinoline, which is active in the HPC/DNA repair test, neither of these epoxides were found to induce unscheduled DNA synthesis, see Table 1. Mutagenicity assays performed on both the 5,6- and 7,8-dihydrodiol derivatives of these oxides with and without metabolic activation also indicated that these dihydrodiols did not contribute to the mutagenic activity of quinoline.

These data prompted us to examine the role of positions 1-4 in the ultimate activation of quinoline to a mutagen. While 4-methylquinoline is a potent mutagenic analog of quinoline, it is known that both 2- and 3-methylquinoline are much less active than quinoline (23). It is also known that 2-chloroquinoline is inactive as a mutagen in S. typhimurium TA100 and as a hepatocarcinogen in rats (14,34). To more effectively probe the possible involvement of positions 2-4 in the activation of quinoline to a mutagen, we prepared 2-, 3-, and 4-fluoroquinoline. Substitution of a halogen atom at these positions will alter the electronic properties of quinoline. In the case of either fluoro- or chloro-substituted quinolines, metabolism would be expected to be inhibited at the carbon atom to which it is attached. In contrast to a chlorine atom which has approximately the steric bulk of a methyl substituent, the fluorine atom is sterically similar to a hydrogen atom. Thus, these fluorinated derivatives more closely approximate isosteric analogs of quinoline. When evaluated in S. typhimurium TA100 in the presence of rat liver homogenate, each of these fluoro-derivatives was inactive. These data suggest that the overall electronic effects of halogen substitution on the quinoline nucleus rather than inhibition of metabolism at any one specific position has a pronounced impact on mutagenic activity.

The steric effects of methyl substitution on an aromatic hydrocarbon generally results in inhibition of metabolism at the carbon to which it is attached, as well as oxidation at an adjacent carbon atom. The mutagenic potency of 4-methylquinoline and its tumorigenic activity on mouse skin suggest that formation of an electrophilic oxide at the 3,4 position is not likely to be involved in the ultimate activation of quinoline. The metabolic formation of an oxaziridine at the 1 and 2 positions of quinoline could ultimately be responsible for the mutagenic activity of quinoline in the presence of rat liver homogenate.

It has been suggested that during the photolysis of quinoline-N-oxide this oxaziridine is formed as a transient intermediate, see Figure 1 (35,36). We have shown that ir-

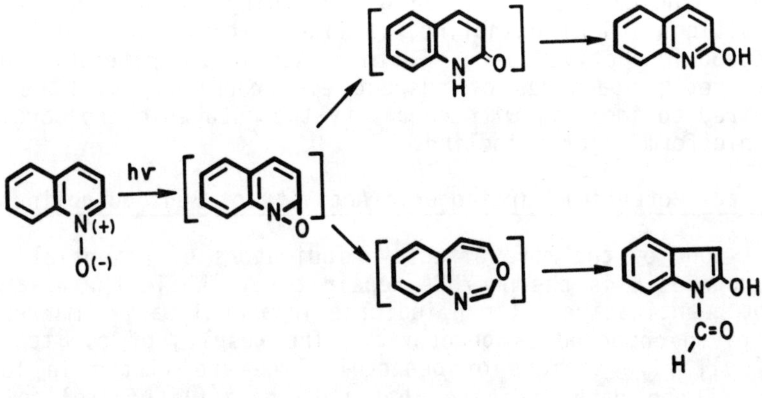

FIGURE 1. Suggested involvement of an intermediate oxaziridine in the formation of photolysis products from quinoline-N-oxide.

radiation of nutrient broth cultures of S. typhimurium under the conditions outlined in Materials and Methods had no effect upon the spontaneous mutation frequency. In studies in which quinoline-N-oxide was added to the nutrient broth culture prior to irradiation, a significant increase in mutagenic activity was observed. These results are shown in Table 2.

TABLE 2

MUTAGENIC ACTIVITY OF QUINOLINE-N-OXIDE IN S. TYPHIMURIUM TA100 AFTER UV IRRADIATION

Concentration of Quinoline-N-Oxide (μg/400 μl)	His$^+$ Revertants Per Plate After Exposure to UV Irradiation (min)				
	0	15	30	45	60
400	151	455	454	410	271
300	166	454	439	426	350
200	162	391	389	430	348
100	153	297	331	375	356
50	152	269	297	329	328
25	145	252	258	300	311
0	155	221	200	241	248

These data suggest that the formation of this oxaziridine may be related to the ultimate activation of quinoline to a mutagen. These studies are consistent with our data which indicate that the ring containing the nitrogen atom, positions 1-4 of quinoline, is likely to be involved in its metabolic activation. Further studies, preferably with synthetic standards of suspect electrophiles, will be required to identify with certainty the mutagenic and carcinogenic form(s) of quinoline.

On the Potential Carcinogenic Activity of Benzoquinolines

One of the more effective indicators of potential carcinogenesis is the HPC/DNA repair test. While the assay is not quantitative, it can indicate in a qualitative manner if a given compound is genotoxic. The results of our testing of all five isomers of benzoquinoline are listed in Table 3. These data indicate that both benz(f)isoquinoline and benz(h)isoquinoline do not induce unscheduled DNA synthesis. The results obtained for benzo(f)quinoline, benzo(h)-quinoline, and phenanthridine were equivocal. The toxicity of these benzoquinolines to hepatocytes did not permit either a clear confirmation of a positive response or a definitive negative response. The limited data obtained from this bioassay, however, did focus our further studies on assessing the genotoxicity of these three specific isomers of benzoquinoline.

TABLE 3

THE RESPONSE OF ALL FIVE ISOMERIC BENZOQUINOLINES IN THE HPC/ DNA REPAIR ASSAY

Chemical	Concentration (M)	UDS (grains/nucleus)[a]
1. Benzo(f)quinoline[b]	5×10^{-3}	Cytotoxic
	1×10^{-3}	5.7 ± 6.7
	5×10^{-4}	-
2. Benzo(h)quinoline[b]	5×10^{-3}	Cytotoxic
	1×10^{-3}	7.7 ± 1.9
	5×10^{-4}	-
3. Phenanthridine[b]	5×10^{-3}	Cytotoxic
	1×10^{-3}	9.8 ±12.3
	5×10^{-4}	8.7 ±13.1
4. Benz(f)isoquinoline	1×10^{-3}	Cytotoxic
	5×10^{-4}	Cytotoxic
	1×10^{-4}	--
5. Benz(h)isoquinoline	1×10^{-3}	Cytotoxic
	5×10^{-4}	Cytotoxic
	1×10^{-4}	--

[a] Mean gain counts greater than 6 is considered positive. At lower concentrations, all compounds failed to produce a significant increase in UDS activity.

[b] Replicate assays on these compounds failed to confirm the weak UDS activity observed in this assay.

Quinoline, 4-methylquinoline, and 8-methylquinoline exhibited significant tumor-initiating activity when assayed on the skin of Sencar mice. We evaluated the tumor-initiating activity of benzo(f)quinoline, benzo(h)quinoline, and phenanthridine at two dose levels as outlined in Table 4. We also evaluated the tumor-initiating activity of 4-methylbenzo(h)quinoline on the skin of Sencar mice at a total initiator dose of 1.0 mg per mouse. None of these benzoquinolines were active as tumor initiators under these assay conditions. These results paralleled similar data obtained in

TABLE 4

TUMOR INITIATION BIOASSAYS OF BENZOQUINOLINES AND SUBSTITUTED BENZOQUINOLINES PERFORMED ON MOUSE SKIN

Compound	Total Initiator Dose (mg)	Strain	Number of Mice	% Tumor-Bearing Animals	Tumors/ Animal
1. Benzo(f)quinoline	1.0	Sencar[a]	20	20.0	0.2
	10.7	Sencar[b]	30	0	0
2. Benzo(h)quinoline	1.0	Sencar[a]	20	0	0
	10.7	Sencar[b]	30	16.7	0.23
3. Phenanthridine	1.0	Sencar[a]	20	15.0	0
	10.7	Sencar[b]	30	0	0
4. 1-Methylbenzo(f)-quinoline	1.0	CD-1[b]	20	10.0	0.10
5. 1,3-Dimethylbenzo-(f)quinoline	1.0	CD-1[b]	20	0	0
6. 4-Methylbenzo(h)-quinoline	1.0	Sencar[a]	20	5.0	0.05
	1.0	CD-1[b]	20	0	0
7. Acetone	---	Sencar[a]	20	15.0	0.15
	---	Sencar[b]	40	7.5	0.08
	---	CD-1[b]	20	10.0	0.10
8. Benzo(a)pyrene	.03	Sencar[a]	20	65	2.0
	.03	Sencar[b]	20	90	2.1
	.03	CD-1[b]	20	45	1.2

[a] 18 weeks of promotion
[b] 20 weeks of promotion

our laboratories on the tumor-initiating activity of 1-methylbenzo(f)quinoline, 4-methylbenzo(h)quinoline, and 1,3-dimethylbenzo(f)quinoline as evaluated on the skin of CD-1 mice, see Table 4.

The mutagenic activity of benzo(f)quinoline, benzo(h)-quinoline, and phenanthridine is well established. These data on mouse skin, as well as the results of other earlier studies, do not provide definitive data on the potential of these aza-arenes to be carcinogenic. Further studies are in progress in our laboratories to evaluate their tumorigenic activity in newborn mice and rats.

On the Metabolic Activation of Benzo(f)quinoline, Benzo(h)-quinoline, and Phenanthridine to Mutagens

The major ethyl acetate extractable metabolites of benzo(f)quinoline, benzo(h)quinoline, and phenanthridine formed using a similar rat liver homogenate preparation as employed in assays on their mutagenic activity have been identified. While the relative amounts of each of the metabolites of phenanthridine was reported (26), similar data is not available for either benzo(f)quinoline or benzo-(h)quinoline. We synthesized ^{14}C-labeled benzo(h)quinoline and benzo(f)quinoline by employing 1- and 2-naphthylamine and ^{14}C-labeled glycerol in the Skraup reaction. The relative amounts of each of the major ethyl acetate extractable metabolites of these three benzoquinolines as formed *in vitro* with rat liver homogenate was determined. The results are listed in Table 5.

Previous studies have shown that phenanthridone, 7,8-dihydroxy-7,8-dihydrobenzo(f)quinoline, 9,10-dihydroxy-9,10-dihydrobenzo(f)quinoline, and 7,8-dihydroxybenzo(f)quinoline can contribute to the observed mutagenic activity of phenanthridine and benzo(f)quinoline (24,37).

One of the major metabolites of benzo(h)quinoline is 5,6-dihydroxy-5,6-dihydrobenzo(h)quinoline. To more fully determine the mutagenic potency of the metabolites of benzo-(h)quinoline as formed *in vitro*, we synthesized this 5,6-dihydrodiol of benzo(h)quinoline and its epoxide precursor, 5,6-epoxy-5,6-dihydrobenzo(h)quinoline. We evaluated the mutagenic activity of these two oxygenated metabolites of benzo(h)quinoline with the 5,6-dihydrodiol and 5,6-epoxide of benzo(f)quinoline. Neither of these dihydrodiols were mutagenic in the presence or absence of rat liver homo-

TABLE 5

RELATIVE DISTRIBUTION OF THE ETHYL ACETATE EXTRACTABLE METABOLITES OF ^{14}C-LABELED BENZO(f)QUINOLINE, BENZO(h)QUINOLINE, AND PHENANTHRIDINE

	Percent of Ethyl Acetate-Extractable Metabolites		
Metabolite	Benzo(f)-quinoline	Benzo(h)-quinoline	Phenanthridine
1,2-Dihydrodiol	--	--	29.9
5,6-Dihydrodiol	--	40.5	--
7,8-Dihydrodiol	35.3	59.5	--
9,10-Dihydrodiol	44.3	--	22.3
N-Oxide	7.5	--	35.8
2-Hydroxy	--	--	1.3
7-Hydroxy	4.9	--	--
Phenanthridone	--	--	7.9
Unknown	8.0	--	2.9
	45.7*	35.7*	27.2*

*Percentage of the Total Amount of Radioactivity Added to the Incubation Mixture Which Were Ethyl-Acetate Extractable Metabolites.

genate. Both of these dihydrodiols, however, were found to be toxic to the bacteria at doses of 20 µg per plate. The epoxides did have significant mutagenic activity as illustrated in Figure 2. The 5,6-epoxide of benzo(h)quinoline was a more potent mutagen than the 5,6-epoxide of benzo(f)-quinoline. While neither the 5,6-diol nor the 5,6-epoxide of benzo(f)quinoline were identified as metabolites, metabolism studies on benzo(h)quinoline clearly indicate that both its 5,6-diol and 5,6-epoxide are formed. The mutagenic activity of benzo(h)quinoline may be explained, in part, by the formation of this K-region epoxide. The exceptional toxicity of benzo(h)quinoline at the higher doses used in these mutagenicity assays may be related to the concentration of 5,6-dihydrodiol formed _in vitro_ under these experimental conditions.

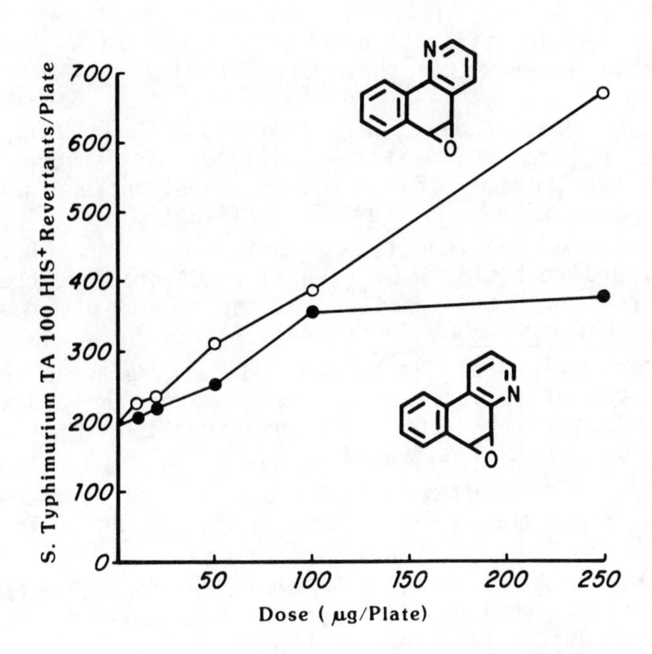

FIGURE 2. Mutagenic activity of the 5,6-epoxide of benzo(h)-quinoline and benzo(f)quinoline.

ACKNOWLEDGEMENTS

This study was supported by the National Institute of Environmental Health Sciences #ES02338. We also thank Ms. Marianne DeFloria for her technical assistance in the mutagenicity assays, Ms. Tomiko Shimada for performing the HPC/DNA repair assay, and Ms. Debbie Conroy for her assistance in the preparation of this manuscript.

REFERENCES

1. Adams, J.D., LaVoie, E.J., Shigematsu, A., Owens, P. and Hoffmann, D. (1983): Quinoline and methylquinolines in cigarette smoke: comparative data and the effect of filtration. J. Anal. Toxicol. 7: 293-296.
2. Dong, M.W., Locke, D.C. and Hoffmann, D. (1977): Characterization of aza-arenes in basic organic portion of suspended particulate matter. Environ. Sci. Technol. 11: 612-618.

3. Shue, F.-F. and Yen, T.F. (1981): Concentration and selective identification of nitrogen- and oxygen-containing compounds in shale oil. Anal. Chem. 53: 2081-2084.
4. Ciupek, J.D., Zakett, D., Cooks. R.G. and Wood, K.V. (1982): High and low energy collision mass spectrometry/mass spectrometry of aza and amino polynuclear aromatic hydrocarbons. Anal. Chem. 54: 2215-2219.
5. Royer, R.E., Mitchell, C.E., Hanson, H.L., Dutcher, J.S. and Bechtold, W.D. (1983): Fractionation, chemical analysis, and mutagenicity testing of low-btu gasifier tar. Environ. Res. 31: 460-471.
6. Pelroy, R.A. and Wilson, B.W. (1981): Relative concentrations of polyaromatic primary amines and aza-arenes in mutagenically active nitrogen fractions from a coal liquid. Mutat. Res. 90: 321-335.
7. Dong, M., Schmeltz, I., Jacob, E. and Hoffmann, D. (1978): Aza-arenes in tobacco smoke. J. Anal. Toxicol. 2: 21-25.
8. Snook, M.E., Fortson, P.J. and Chortyk, O.T. (1981): Isolation and identification of aza-arenes of tobacco smoke. Beitr. Tabakforsch. 11; 67-78.
9. Schmitter, J.M., Ignatiadis, I. and Guiochon, G. (1982): Capillary gas chromatography of aza-arenes. II. Application to petroleum nitrogen bases. J. Chromatogr. 248: 203-216.
10. Buchanan, M.V., Ho, C.-h., Guerin, M.R. and Clark, B.R. (1981): Chemical Characterization of Mutagenic Nitrogen Containing Polycyclic Aromatic Hydrocarbons in Fossil Fuels. In: Polynuclear Aromatic Hydrocarbons, edited by M. Cooke and A.J. Dennis, pp. 133-144, Battelle Press, Columbus, OH.
11. Hoffmann, D. and Wynder, E.L. (1977): Organic Particulate Pollutants: Chemical Analysis and Bioassay for Carcinogenicity. In: Air Pollution, Ed. 3, Vol. 2, edited by A.C. Stern, pp. 361-455, Academic Press, New York, NY.
12. Sawicki, E., Mecker, J.E. and Morgan, C. (1965): Polynuclear aza compounds in automotive exhaust. Arch. Environ. Health 11: 773-775.
13. Sawicki, E., Meeker, J.E. and Morgan, M.J. (1965): The quantitative composition of air pollution: source of effluents in terms of aza heterocyclic compounds and polynuclear aromatic hydrocarbons. Int. J. Air Water Pollut. 9: 291-298.

14. Blumer, M. and Dorsey, T. (1977): Aza-arenes in recent marine sediments. Science 195: 282-285.

15. Hirao, K., Shinohara, Y., Tsuda, H., Fukushima, S., Takahashi, M. and Ito, N. (1976): Carcinogenic activity of quinoline on rat liver. Cancer Res. 36: 329-335.

16. Shinohara, Y., Ogiso, T., Hananouchi, M., Nakanishi, K., Yoshimura, T. and Ito, N. (1977): Effect of various factors on the induction of liver tumors in mice by quinoline. GANN 68: 785-796.

17. LaVoie, E.J., Shigematsu, A., Adams, E.A., Rigotty, J. and Hoffmann, D. (1984): Tumor-initiating activity of quinoline and methylated quinolines on the skin of Sencar mice. Cancer Lett. 22: 269-273.

18. Dong, D., Schmeltz, I., LaVoie, E. and Hoffmann, D. (1978): Aza-Arenes in the Respiratory Environment: Analysis and Assays for Mutagenicity. In: Carcinogenesis, Vol. 3, Polynuclear Aromatic Hydrocarbons, edited by P.W. Jones and R.I. Freudenthal, pp. 97-108, Raven Press, New York, NY.

19. Matsumoto, T., Yoshida, D., Mizusaki, H., Tomita, H. and Koshimizu, K. (1978): Structural requirements for mutagenic activities of N-heterocyclic bases in the Salmonella test system. Agric. Biol. Chem. 42: 861-864.

20. Nagao, M., Yahagi, T., Seino, Y., Sugimura, T. and Ito, N. (1977): Mutagenenesis of quinoline and its derivatives. Mutat. Res. 42: 335-342.

21. Hollstein, M., Talcott, R. and Wei, E. (1978): Quinoline: conversion to a mutagen by human and rodent liver. J. Natl. Cancer Inst. 60: 405-410.

22. Williams, G.M. (1981): The detection of genotoxic chemicals in the hepatocyte primary culture/DNA repair test. Gann 27: 45-55.

23. LaVoie, E.J., Adams, E.A., Shigematsu, A. and Hoffmann, D. (1983): On the metabolism of quinoline and isoquinoline: possible molecular basis for differences in biological activities. Carcinogenesis 4: 1169-1173.

24. Adams, E.A., LaVoie, E.J. and Hoffmann, D. (1983). Mutagenicity and Metabolism of Azaphenanthrenes. In: Polynuclear Aromatic Hydrocarbons: Formation, Metabolism and Measurement, eds. M. Cooke and A.J. Dennis, pp. 73-87, Battelle Press, Columbus, OH.

25. LaVoie, E.J., Adams, E.A. and Hoffmann, D. (1983): Identification of the metabolites of benzo(f)quinoline and benzo(h)quinoline formed by rat liver homogenate. Carcinogenesis 4: 1133-1138.

26. LaVoie, E.J., Adams, E.A., Shigematsu, A. and Hoffmann, D. (1985): Metabolites of phenanthridine formed by rat liver homogenate. Drug. Metab. Disp. 13: 71-75.
27. Hamer, J., Link, W.J., Jurjevich, A. and Vigo, T.L. (1962): Preparation of 2-fluoroquinoline and 2,6-difluoropyridine by halogen exchange. Rec. Trav. chim. 81: 1058-1060.
28. Roe, A. and Hawkins, G.F. (1949): The preparation of heterocyclic fluorine compounds by the Schiemann reaction. II. The monofluoroquinolines. J. Amer. Chem. Soc. 71: 1785-1786.
29. Krishnan, S., Kuhn, D.G. and Hamilton, G.A. (1977): Direct oxidation in high yields of some polycyclic aromatic compounds to arene oxides using hypochlorite and phase transfer catalysts. J. Amer. Chem. Soc. 99: 8121-8122.
30. Kaslow, C.E. and Sommer, N.B. (1946): Substituted lepidines. J. Amer. Chem. Soc. 68: 644-647.
31. Maron, D.M. and Ames, B.N. (1983): Revised methods for the Salmonella mutagenicity test. Mutat. Res. 113: 173-215.
32. Williams, G.M. (1977): Detection of chemical carcinogens by unscheduled DNA synthesis in rat primary cell cultures. Cancer Res. 37: 1845-1851.
33. Williams, G.M. (1980): The Detection of Chemical Mutagens/Carcinogens by DNA Repair and Mutagenesis in Liver Culture. In: Chemical Mutagens, Vol. 6, A. Hollaender, ed., pp. 61-79, Plenum Press, NY.
34. Fukushima, S., Ishihara, Y., Nishio, O., Ogiso, T., Shirai, T. and Ito, N. (1981): Carcinogenicities of quinoline derivatives in F344 rats. Cancer Lett. 14: 115-123.
35. Buchardt, O. (1967): Photochemical studies. The formation of benz[d]-1,3-oxazepines in the photolysis of quinoline N-oxides in solution. Acta. Chem. Scand. 21: 1841-1854.
36. Buchardt, O., Kumler, P.L. and Lohse, C. (1969): Photochemical studies. The liquid phase photolysis of quinoline N-oxides unsubstituted in the 2-position. Acta. Chem. Scand. 23: 159-170.

CONTROLLED OXIDATION STUDIES OF BENZO(A)PYRENE

EDWARD LEE-RUFF*, HIRA KAZARIANS-MOGHADDAM, MORRIS KATZ
Department of Chemistry, York University, Downsview, Ontario
M3J 1P3, Canada

INTRODUCTION

The presence of environmental polynuclear aromatic hydrocarbons (PAH) has been of great concern due to the toxicities as carcinogenic initiators exhibited by a number of these derivatives (1). Various oxidized species have been implicated as being responsible in the molecular binding with vital biopolymers such as DNA, proteins and polysaccharides (2). Benzo(a)pyrene (BaP) (1) is a classical example of such PAHs. It has been well established that epoxides and diol epoxides, metabolites of this derivative, bind with DNA to form stable adducts and are responsible for the mutagenic activity. Certain peroxidase systems are also known to catalyze the oxygenation of BaP to a mixture of quinones. Whereas most of the metabolic studies on BaP have focussed on the 7,8-opoxide and 7,8-diol-9,10-epoxide, interest in the oxidation to the 1,6, 3,6 and 6,12-quinones has recently picked up (3) as the result of the observation that these quinones themselves exhibit moderate mutagenic activity (4). Recently, Tso (5) has shown that all three quinones caused degradation of DNA and that this process is oxygen dependent. Esr studies have shown that the semiquinones of these derivatives covalently bind to DNA and polyglutamic acid (6). Katz has shown that the three benzo(a)pyrene quinones are also produced by oxidation of BaP adsorbed on cellulose acetate plates and on glass under simulated environmental conditions (7,8) involving a high intensity visible light source. Recently, Eisenberg *et al*. (9) obtained evidence that certain PAHs sensitize production of singlet oxygen under environmental conditions and that the oxidation of these PAHs, including BaP, involves the self-sensitized singlet oxygenation of these compounds (10). However, to our knowledge, no systematic study of the singlet oxygen reaction of BAP has been reported. We undertook a study of the oxidation of benzo(a)pyrene under photosensitized singlet oxygen production as well as the use of one-electron oxidations in order to establish whether singlet oxygen is responsible for the photooxidation process. Furthermore, we report that the photooxidation and photosensitized singlet oxygenation of BaP produces a major product which has not been previously reported and which suggests a possible mechanism for quinone production by singlet oxygen reaction. Heretofore, the structures of the three isomeric 1,6, 3,6, and 6,12 benzo(a)-pyrene diones 2-4 were all assigned by comparison of uv and mass spectra with 'authentic' material obtained from chromium trioxide oxidation of BaP based on an early literature report (11) which predates nmr spectroscopy. Neither uv nor mass spectrometry represent absolute molecular structure characterization methods and in order to confirm the original structural assignments, we carried out a detailed structural analysis of the abovementioned three quinones as well as 4,5 BaP quinone based on ^{1}H 2-D COSY and selective NOE nmr studies.

RESULTS

Confirmation of Benzo(a)pyrene Quinone Structures by High-Field NMR Studies

The mixture of 6,12, 1,6 and 3,6-benzo(a)pyrene diones (2-4) were prepared according to an early literature procedure (11). The mixture was separated on a dry column of

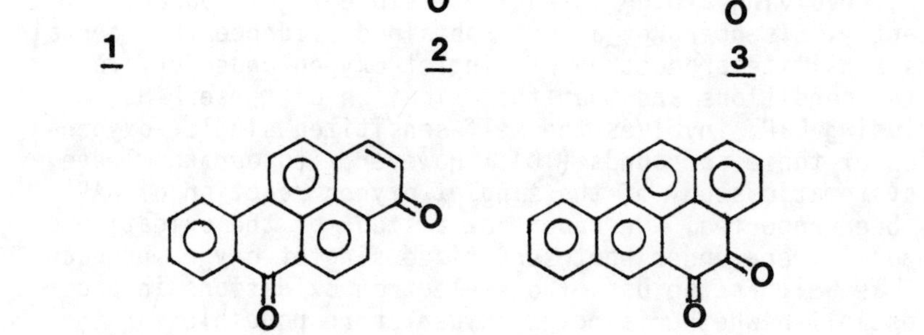

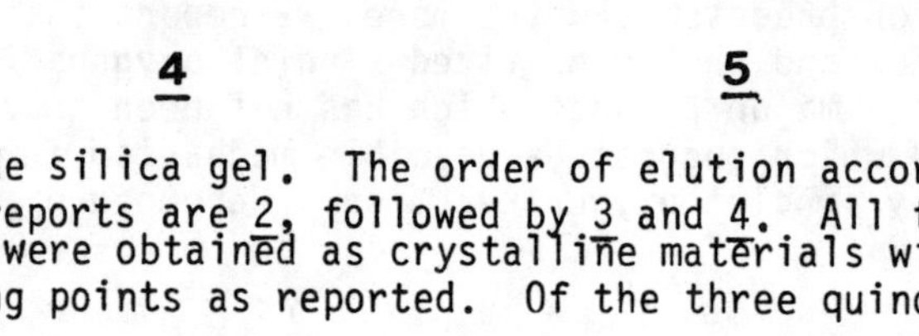

flash-grade silica gel. The order of elution according to previous reports are 2, followed by 3 and 4. All three compounds were obtained as crystalline materials with identical melting points as reported. Of the three quinones,

TABLE 1

^{1}H CHEMICAL SHIFT ASSIGNMENTS OF BaP OXIDATION PRODUCTS

Compound	H_1	H_2	H_3	H_4	H_5	H_6	H_7	H_8	H_9	H_{10}	H_{11}	H_{12}
BaP* (1)	8.26	8.00	8.11	7.94	8.01	8.54	8.31	7.79	7.84	9.06	9.07	8.35
4,5-dione (5)	8.26	7.76	8.55	-	-	8.88	8.10	7.71	7.86	8.69	8.67	8.07
6,12-dione (2)	8.63	7.85	8.22	8.17	8.47	-	8.46	7.68	7.77	8.19	7.58	-
3,6-dione (4)	7.73	6.71	-	8.82	8.72	-	8.45	7.59	7.77	8.28	8.37	7.80
1,6-dione (3)	-	6.76	7.74	7.91	8.68	-	8.47	7.79	7.63	8.35	8.57	8.67
seco BaP (6)	7.25	7.70	7.86	7.77	6.51	9.72	8.00	7.57	7.70	8.13	8.26	7.55

*Daub et a. JACS 105, 773(1983)

only the 6,12-dione 2 had been prepared by independent synthesis (12). We have repeated this synthesis and the product obtained was identical in all respects with the faster eluting component in the dione mixture. Benzo(a)pyrene-4,5-dione 5 was also prepared by literature methods (13,14) for comparison with products obtained in the oxidation studies of BaP. The confirmation of structures of all four benzo(a)pyrene diones 2-5 was established by the mapping of all vicinally coupled and proximal protons by a combination of the 2-D COSY and selective NOE difference spectroscopy methods (15) and checking for self consistency. In this manner, all of the proton nmr chemical shifts for each of the isomers 2-5 have been unequivocally established and listed in Table 1 and the original structural assignments have now been confirmed. One feature of note is that bay region hydrogen signals (H_{11},H_{10}) which appear at lowest field in benzo(a)-pyrene 1 are shifted to higher fields by about 0.5-0.7ppm in the four quinones. This can be rationalized in terms of decreasing ring current effects brought about by the decrease in π-electron density in the quinones. The lowest-field signals in the nmr spectra appear to be for those protons at the peri position relative to the carbonyl groups. The sensitivity of the high field nmr method now permits analysis of trace photooxidation products of PAHs by an absolute method.

Oxidation Studies of Benzo(a)pyrene

It has been suggested that the environmental oxidation of PAHs proceed via self-sensitized singlet oxygen reaction (16). Evidence for gas-phase singlet oxygen formation using heterogeneous gas-phase photosensitization involving PAHs has been obtained from trapping experiments with known singlet oxygen scavengers such as 9,10-diphenylanthracene (9). In order to establish that singlet oxygen is involved in the direct photooxidation reaction and to elucidate the mechanism of quinone formation, we analysed and compared the product distribution of the direct photooxidation with those derived from photosensitized singlet oxygen reaction using the common sensitizers, tetraphenylporphyrin (TPP) and methylene blue. Furthermore, it has been suggested that certain reactions involving singlet oxygen proceed via electron transfer processes involving superoxide O_2^-. In order to establish whether such a mechanism operates in the photooxidation of benzo(a)pyrene, we prepared the benzo(a)pyrene radical cation by reaction of BaP with tris (4-bromophenyl) aminiumhexachloroantimonate and subjected the radical cation to reaction with potassium superoxide. The product distribution was compared with those of the direct photoxidation and photo-sensitized singlet oxygen reaction.

The direct photooxidations of BaP were carried out in one of two ways: aerated benzene solutions of 1 (10^{-3}M) were irradiated with a uv source (450W medium pressure Hg lamp) using a pyrex filter ($\lambda >$ 300 nm). Alternatively, aerated solutions of 1 (10^{-3}M) in CHCl were irradiated using a visible light source (500W) with a pyrex filter. The only products in the uv irradiations are the diones 2-4 with a small amount of product having the identical HPLC retention time as the 4,5-dione (5). Ir and nmr spectal comparisons of this minor product with authentic 4,5-dione 5 showed that the two were not identical. The uv data of this minor product were also not compatible with those of a 11,12-benzo(a)pyrene dione (17) and no further attempts at identifying this compound were made. The product mixture obtained from irradiation with visible light consisted of the three quinones 2-4 and major product which constituted 48% of this mixture.

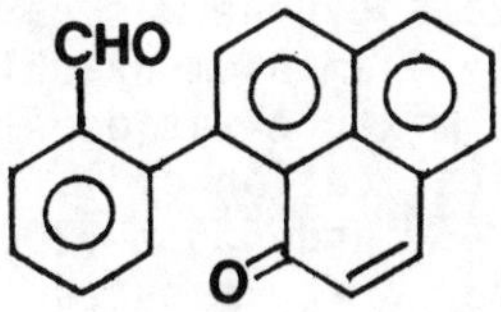

CHO

O

6

$(4\text{-BrC}_6\text{H}_4)_3\ \dot{N}^+\ \text{SbCl}_6$

7

This product was identified as the 6-secobenzo(a)pyrene derivative 6 and has not been previously reported in benzo(a)pyrene photooxidations. This assignment was based on high-field nmr 2-D COSY correlations which indicate a vicinal proton coupling system very similar to benzo(a)pyrene; however, a low field proton at 9.7ppm assigned to the aldehydic C-6 proton was found to be long-range coupled to the C-7 proton. Furthermore, the unusually shielded C-5 proton at 6.5ppm is characteristic of the α-proton (next to carbonyl) in phenalenones (18). The uv spectrum of this compound was similar to that of phenalenone. Further confirmation of this assignment was obtained from comparison of spectral data with those reported for the same compound obtained in 0.03% yield from the oxidative γ-radiolysis of benzo(a)pyrene (19).

Up to 65% quenching of the oxidation occurred in the presence of 1 eq of DABCO implying presence of $^1O_2$ as the reactive species. The photosensitized oxidation was carried out using either TPP or methylene blue. A filter had to be used for selective excitation of the senitizer. The filter chosen was a chemical filter solution (21) consisting of $CuSO_4$, $NaNO_3$ in ammonium hydroxide with a transparent window between 405 and 520mm. Benzo(a)pyrene 1 showed a weak tail-end absorption in this region. Using a 2cm path width of filter solution the direct photooxidation of 1 proceeds at 1/20 of the rate in the absence of the filter. This was used as the control for the photosensitized oxygenation experiments. TPP exhibits strong absorption at 416mm so that using the filter assembly, the sensitizer (TPP) absorbs greater than 95% of the light. Methylene blue on the other hand has a maximum absorption at 610mm; however, this band extends into the filter window. In both instances the photosensitized oxygenation rate was at least twice as fast as that in the absence of photosensitizer. The product mixture consisted of the non-polar fraction obtained in the direct photooxidation, a mixture of the diones 2-4 and the seco-product 6. The relative yields of the products are listed in Table 2.

The one-electron oxidation of benzo(a)pyrene was carried out using the oxidant tris (4-bromophenyl) aminiumhexachloroantimonate 7, an instant rose-coloured mixture was developed. This species was attributed to the radical cation of 1. The uv/visible spectrum exhibited peaks at 450 and 560mm (20). In air, the coloured solution faded after a few minutes. The coloured species was instantly quenched with water. It was also quenched by addition of powdered potassium superoxide. In each case, the reaction mixture consisted of the three quinones with no detectable trace of the seco-product 6. The product distributions of the quinone mixture obtained in all of the oxidations are listed in Table 2.

DISCUSSION

The possibility that PAHs can sensitize singlet oxygen formation has been aptly demonstrated by Eisenberg (9). On energetic grounds (see Figure 1) both the first singlet as well as the triplet states of benzo(a)pyrene 1 can result in singlet oxygen formation by energy transfer by mechanisms shown in Scheme 1. The lifetime of both excited states of 1 permit bimolecular collisions with oxygen to take place for energy transfer and it has been shown for other PAHs that fluorescence and phosphorescence can be quenched by oxygen.

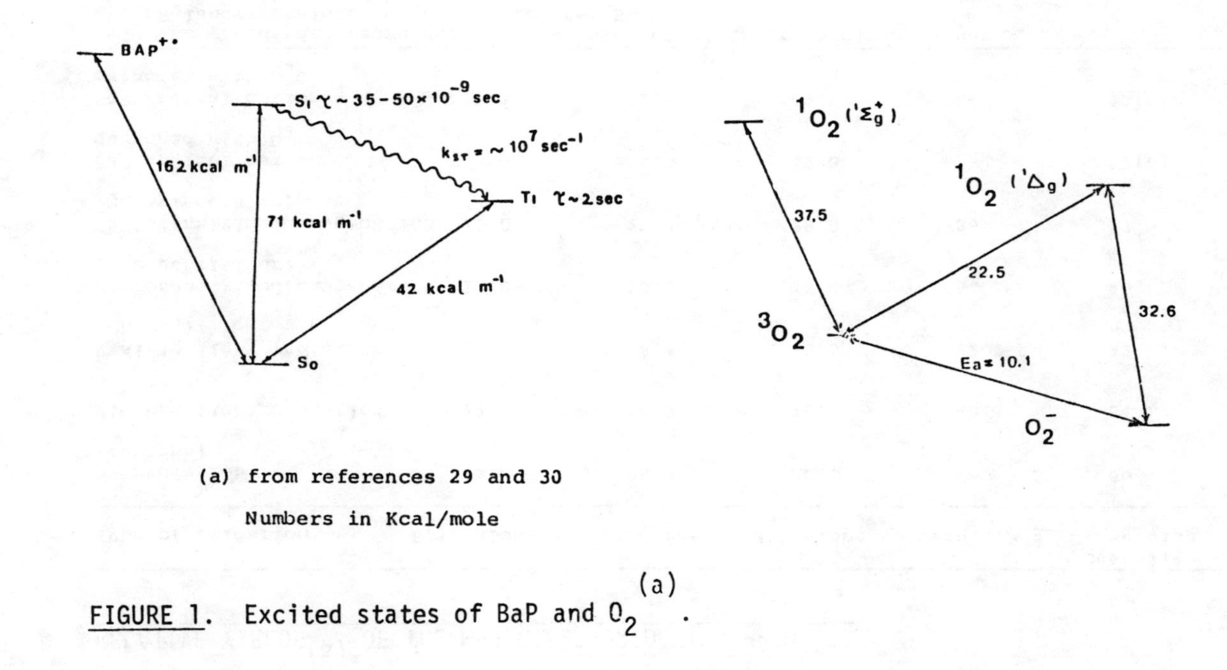

(a) from references 29 and 30

Numbers in Kcal/mole

FIGURE 1. Excited states of BaP and O_2 [(a)].

TABLE 2

RELATIVE YIELDS[a] OF THE MAJOR BaP OXIDATION PRODUCTS

Type of Oxidation	6,12-dione 2	6,1-dione 3	3,6-dione 4	seco BaP 6	Overall Conversion
UV radiation (benzene)	43.3	26.7	30.0	-	60
Visible light radiation ($CHCl_3$)	22.0	18.8	11.0	48.1	64
Visible light radiation (CH_2Cl_2)	25.9	26.4	25.5	22.2	27
Photosensitized oxygenation (TPP sensitizer)	13.8	10.3	11.9	54	20
Photosensitized oxygenation (Methylene Blue)	13.0	27.3	23.9	36	17
BaP radical cation quenched with O_2^-	3.4	54.3	42.3	-	57(b)
BaP radical cation quenched with H_2O	13.1	37.6	49.3	-	13(b)

(a) The yields are reproducible to within 10% of their absolute values.
(b) Balance consisted of only unreacted BaP.

Scheme 1

$BaP \xrightarrow{h\nu} BaP^1$

$BaP^1 + {}^3O_2 \longrightarrow BaP^3 + {}^1O_2$

$BaP^1 \longrightarrow BaP^3$

$BaP^3 + {}^3O_2 \longrightarrow {}^1O_2 + BaP^1$

$BaP + {}^1O_2 \longrightarrow$ reaction products.

In other readily reducing aromatic derivatives, it has been suggested that singlet oxygen reactions occur by way of single electron transfer processes involving PAH radical cation formation (22). Electron transfer from BaP to either singlet (Δg) or ground state oxygen would be endoergonic a and not likely to occur on the basis of the high ionization potential of BaP (162Kcal/mole) (23). Similar consideration even for the excited states of BaP would preclude electron transfer to oxygen. In order to test this, we have shown that the reaction of BaP radical cation, generated independently, reacts with superoxide or water to give a significantly different product distribution (see Table 2). The 6,12-dione in this case is formed only as a minor product. These relative yields are very similar to ones reported for the electrochemical oxidation of BaP (24). Greenstock has reported that neutral BaP reacts with superoxide generated by pulse radiolysis (25). In our hands, mixtures of BaP with KO_2 were found to be stable indefinitely. It is likely that under conditions of pulse radiolysis, ionization of 1 to the radical cation takes place and that the decomposition of BaP with O_2 takes place by way of this radical cation. The direct photooxidation of BaP 1 produces a mixture which is similar to one obtained from the TPP sensitized oxidations in which the 6,12-dione 2 and seco-product 6 predominate. Evidence that the direct photooxidation proceeds via singlet oxygen was shown from DABCO quenching and the trapping of diphenylanthracene-9,10 peroxide when diphenylanthracene, a known singlet oxygen quencher, was added to the photomixture.

The different product distributions observed in the one-electron and photooxidations of BaP suggest that two fundamentally different mechanisms operate in quinone formation in the two reactions. Adams has suggested (24) that quinone formation by way of electrochemical oxidation proceeds through the intermediacy of the radical cation 8 which, in aqueous solution, is trapped by water leading to 6-hydroxy BaP 9. Further oxidation by way of the corresponding oxyradical leads to the mixture of quinones (Scheme 2). Formation of the quinones from reaction of the radical cation with superoxide could proceed by nucleophilic trapping at position 6 and formation of dihydroperoxides. These dihydroperoxides, under basic conditions, could conceivably rearrange to the diones.

Scheme 2 : $BaP^{\cdot+}$ (9) $\xrightarrow[-H^+,\,-H^\cdot]{H_2O}$ 6-HO-BaP $\xrightarrow{^3O_2}$ 2 - 4

Scheme 3 :

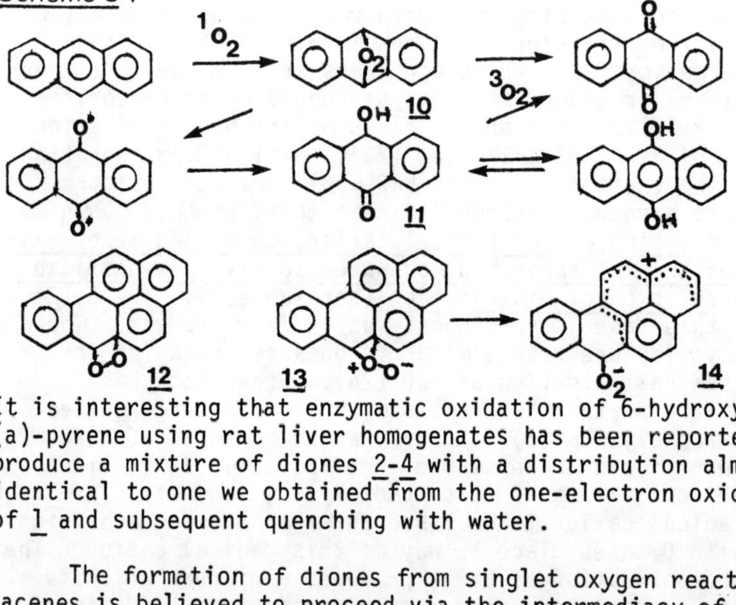

$^1O_2$ 10 $^3O_2$ OH O 11 OH 12 13 14 O_2^-

It is interesting that enzymatic oxidation of 6-hydroxybenzo-(a)-pyrene using rat liver homogenates has been reported to produce a mixture of diones 2-4 with a distribution almost identical to one we obtained from the one-electron oxidation of 1 and subsequent quenching with water.

The formation of diones from singlet oxygen reaction of acenes is believed to proceed via the intermediacy of meso-endoperoxides which represent Diels-Alder adducts of singlet oxygen (26). This is illustrated by the example of anthracene which undergoes cyclo-addition with singlet oxygen to produce the meso-peroxide 10. Formation of 9,10-anthraquinone from the peroxide could involve homolytic O-O bond cleavage and subsequent rearrangement to the half oxidized hydroquinone 11. Further oxidation to the dione could be initiated either by ground state or singlet oxygen. The formation of diones 2-4 from benzo(a)pyrene is unlikely to involve a mechanism as described above. It would be unlikely for a peroxo-bridge to span the 6,12 1,6 and 3,6 centers. The formation of the seco-derivative 6 as the major product in the singlet oxygen reaction of benzo(a)pyrene suggests that a dioxetan 12 or perepoxide 13 is involved in the formation of this product. Such an intermediate may also be responsible for quinone formation. It is conceivable that these intermediates could lead to the zwitterion 13. Subsequent trapping of this species by water and further oxidation could lead to the observed quinones.

CONCLUSION

Structures of the three major isomeric diones 2-4 formed from benzo(a)pyrene 1 oxidation have been confirmed by 2D-COSY and selective NOE nmr techniques. The relative distribution of diones is markedly dependent on the nature of the oxidation. One -electron oxidation of benzo(a)pyrene using tris(4-bromophenyl)-amminium hexachloroantimonate as a one-electron oxidant produces exclusively a mixture of diones 2-4 with 3 predominating and 2 present as a minor component. On the other hand, direct photooxidation or photosensitized oxygenation of 1 produces a mixture of diones with dione 2 as the major product. Another major product identified as the seco-derivative 6 was isolated in these reactions. The presence of 6 suggests a mechanism involving a dioxetan or perepoxide as the probably precursor to diones 2-4 under conditions of singlet oxygen reaction. It is interesting to note that biochemical systems in which singlet oxygen has been implicated catalyze the oxidation of BaP to give a similar mixture of diones as reported by us for the direct or photosensitized oxidations. Marnett (27) has reported that 1 is oxidized by sheep seminal vesicals arichodonate dependent system to a mixture of diones 2-4 with a relative distribution of 45:25:35. On the other hand, oxidation of 1 by rat liver homogenate (28) (cytochrome dependent) produces a dione mixture 2-4 with relative yields (5:41:44) which is similar to our observed product distribution from the one-electron oxidation. Oxidized cytochromes contain complexed ferric ions which would initiate one-electron oxidation processes with aromatic derivatives with sufficiently low oxidation potentials.

The environmental analysis of benzo(a)pyrene is complicated by its oxidation to the diones 2-4 during sampling. In our study we find that a new major product 6 is also present in simulated environmental oxidations. Thus a complete analysis of environmental benzo(a)pyrene would have to include analysis of diones 2-4 as well as aldehyde 6.

ACKNOWLEDGMENTS

We would like to thank NSERC for their generous support of a strategic grant in environmental toxicology. One of us (E. Lee-Ruff) would like to acknowledge the support of an Environment Canada AES grant.

MATERIALS AND METHODS

Benzo(a)pyrene and tris (4-bromophenyl)aminiumhexachloroantimonate were obtained from Aldrich.

Dry column chromatography was performed with silica gel, Merck Kieselgel 60, 230-400 mesh ASTM deactivated with water (15% w/w). High performance liquid chromatography (HPLC) was carried out with a Waters Associate Chromatograph equipped with a Model 6000A pump, model U6K injector and a model 440 absorbance detector set at 254 nm. A Waters straight phase μ-porasil column (3.9 mmID x 30 cm/10) was used with a mixture of hexane and chloroform as the mobile phase. The diones 2-4 were prepared according to the literature and their separation was carried out using dry column chromatography with chloroform as the eluant. The 6,12-dione 2 was first to elute followed by the 1,6 and 3,6-isomers 3 and 4 respectively. The purity of the individual fractions was tested by HPLC and their retention times are: 6,12-dione 2: 8.78 min; 1,6-dione 3: 13:43 min; 3,6-dione: 15.21 min.*

Benzo(a)pyrene-4,5-dione (5) was synthesized according to literature methods (13,14). The uv and mass spectral data of 2-5 were in good agreement with those reported in the literature.

The nmr acquisitions were performed on a Bruker AM-300 spectrometer.

Selective NOE and 2-D COSY Acquisitions

The Bruker pulse sequence 90^{o}_{PH1}-t_1-90^{o}_{PH2}-FID(t_2) was used for the 2-D COSY experiment with a 1Kx1K array of data points. The sweepwidth was restricted to the aromatic region 900Hz. The 2-D plot allows the correlation of vicinally coupled protons. The selective NOE experiments

---
* The response of all four diones were identical to within 2% when standard solutions were analysed.

were carried out using a decoupler power of 45-50L and a decoupler pulse approximately equal to the longest proton T_1 (7 sec) to ensure maximum saturation and polarization transfer. NOE enhancements ranged from 3-15% with bay region protons exhibiting the largest effects. In all cases a marker proton was established (H_{11} in 2, H_2 in 3, H_2 in 4 and H_6 in 5) and bay-region as well as peri relationships of protons determined by this method.

Photooxidation Reactions

General: All photooxidations were performed in glass-distilled solvents in pyrex tubes equipped with a gas purger.

For the uv radiations, the tube was strapped around a quartz cooling well and the assembly was placed in an ice water bath. A pyrex inner filter was used and a Hanovia 450W medium pressure lamp served as the light source. Dry oxygen was bubbled through the solution during the photolysis.

For the visible light irradiation, the sample tube was placed in the inner well of the cooling vessel. A GE 500W quartzline lamp was used as the light source and placed exterior to the photolysis assembly. Irradiations with a chemical filter solution were run using a concentric tube placed in the interior of the cooling well which was filled with a Cu^{+2} ammonia complex solution (21).

Triethylene Diamine (DABCO) Quenching Studies

A solution of 10mg of 1 in 80ml of $CHCl_3$ containing 10mg of DABCO was irradiated with the visible light source for 48 hours. The usual work-up yielded 2.7 mg of photooxidized products and HPLC analysis showed a reduction of 50% of each of the quinones 2-4 and 60% of the seco-product 6.

9,10-Diphenylanthracene Trapping Experiment

A solution of 20mg of diphenylanthracene and 5mg of 1 in 80ml of $CHCl_3$ was irradiated for 24 hours using the visible light source and the inorganic filter solution. In addition to quinone 2-4 formation, a new product identified as the 9,10-diphenylanthracene peroxide was isolated and its structure confirmed by comparison of its ir spectrum with that of an authentic sample. A control experiment without benzo(a)pyrene present was carried out under identical conditions and, in this case no detectable amounts of the peroxide was obtained. The photomixture consisted only of unreacted starting material by t.l.c. analysis.

Photosensitized Oxidation of Benzo(a)pyrene (1)

A solution of 10mg of 1 and 5 mg of methylene blue in 80ml of $CHCl_3$ was irradiated with the visible light source for 48 hours using the inorganic filter solution. After separating unreacted 1 from the photomixture, continued elution with $CHCl_3$ on a silica gel column gave a mixture which was analysed by HPLC (60/40 hexane/$CHCl_3$; 1ml/min.).

The above experiment was repeated with tetraphenylporphyrin as the photosensitizer under identical conditions. A 20% conversion to photooxygenated products was observed by HPLC analysis.

Oxidation of Benzo(a)pyrene (1) with Tris(4-bromophenyl) aminium-hexachloroantimonate and Quenching with O_2^-

To a solution of 20mg of benzo(a)pyrene in 50ml of dry CH_2Cl_2 was added slowly a solution of 65.2mg of tris(4-bromophenyl)aminiumhexachloroantimonate in 50ml of dry CH_2Cl_2. The solution turned to a rose purple colour. To this solution was added a suspension of KO_2 in dry CH_2Cl_2 at which point the colour changed to a yellow solution. The solution was stirred for three minutes and poured over crushed ice. The organic layer was separated and washed with 3x30ml of 10% HCl, followed by 3x30ml of water and dried over magnesium sulphate. The solvent was removed under reduced pressure and the residue was applied to a dry silica gel column. The unreacted benzo(a)pyrene (5mg) was eluted with 50/50 hexane/benzene mixture. Following elution with $CHCl_3$ the quinone fraction 2-4 (15mg) was obtained and analysed by HPLC.

In a control experiment, no reaction was observed between KO_2 and benzo(a)pyrene in the absence of tris(4-bromophenyl)aminiumhexachloroantimonate.

Repetition of the above experiment using water instead of KO_2 as the quenching agent, led to 2.85mg (12.6%) conversion to quinones 2-4. The balance of the mixture was unreacted 1.

REFERENCES

1. Cavalieri, E., Rogan, E. and Roth, R. (1982): In: Free Radicals and Cancer, edited by R.A. Floyd, pp. 117-158, Dekker, New York.
2. Wood, A.W., Levin, W., Chang, R.L., Yagi, H., Thakker, D.R., Lehr, R.E., Jerina, D.M., and Conney, A.H. (1979): In: Polynuclear Aromatic Hydrocarbons: Third International Symposium on Chemistry and Biology-Carcinogenesis and Mutagenesis, edited by P.W. Jones and P. Leber, pp. 531-551, Ann Arbor Science Publishers, Ann Arbor, Michigan.
3. Morgenstern, R., Guthenberg, C., Mannervik, B., de Pierre, J.W., and Ernster, L. (1982): Cancer Res., 42:4215.
4. Salamone, M.F., Heddle, J.A., and Katz, M. (1979): Env. Int., 2:37.
5. Lesko, S.A., Lorentzen, R.J., and Tso, P.O.P. (1978): In: Polycyclic Hydrocarbons and Cancer, edited by P.O.P. Tso and H.V. Gelboin, pp. 261-269, Academic Press, NY.
6. Nagata, C., Kodama, M, and Ioki, Y. (1978). In: Polycyclic Hydrocarbons and Cancer, edited by P.O.P. Tso and H.V. Gelboin, pp. 247-260, Academic Press, NY.
7. Lane, D.A. and Katz, M. (1978): In: Advances in Environmental Science and Technology, Fate of Pollutants in Air and Water Environments, Vol. 9, edited by I.H. Suffet, pp. 137-154, Wiley Interscience, NY.
8. Katz, M., Chan, C., Tosine, H. and Sakuma, T. (1980): In: Polynuclear Aromatic Hydrocarbons, edited by P.W. Jones and P. Leber, pp. 171-189, Ann Arbor Science Publishers, Ann Arbor, MI.
9. Eisenberg, W.C., Taylor, K., Cunningham, D.L.B., and Murray, R.W., (1984): Preprint to appear in: Eighth International Symposium on Polynuclear Aromatic Hydrocarbons, Battelle Press, Columbus, OH.
10. Eisenberg, W.C., Taylor, K., and Murray, R.W., (1984): Carcinogenesis, 5:1095.
11. Vollmann, H., Becker, H., Corell, M., and Streeck, H. (1937): Ann., 531:2.
12. Schroeder, H.E., Stilmar, F.B., and Palmer, F.S. (1956): J. Am. Chem. Soc., 78:446.
13. Harvey, R.G., Goh, S.G., and Cortez, C. (1975): J. Am. Chem. Soc., 97:3468.
14. Raha, C.R., Keefer, L.K., and Loo, J. (1973): J. Chem. Eng. Data, 18:332.
15. Bax, A. and Freeman, R. (1981): J. Magn. Res., 43:259. See also Bruker publication (1982) "Two Dimensional NMR, Aspect 2000, 3000".

16. Zander, M. (1980): In: Environmental Chemistry: Anthropogenic Compounds, Vol. 3, Pt. A., pp. 121, Springer Verlag.
17. Bodine, R.S., Hylarides, M., Daub, G., and VanderJagt, D.L. (1978): J. Org. Chem., 43:4025.
18. Prinzbach, H., Freudenberger, V., and Scheidegger, V. (1971): Helv. Chim. Acta., 50:1087.
19. Gibson, T.L. and Smith, L.L. (1979): J. Org. Chem., 44:1842.
20. Caspary, W., Cohen, B., Lesko, S., and Tso, P.O.P. (1973): Biochemistry, 12:2649.
21. Gollnick, K. and Schnatterer, A. (1984): Tet Letters, 185.
22. Schaap, A.P., Zaklikr, K.A., Kashar, B., and Fung, L.W.H. (1980): J. Am. Chem. Soc., 102-389.
23. Cavalieri, E.L., Rogan, E.G., Roth, R.W., Saugier, R.K., and Hakam, A. (1983): Chem.-Biol. Inter., 47:87.
24. Jeftic, L. and Adams, R.N. (1970): J. Am. Chem. Soc., 92:1332.
25. Greenstock, C.L. and Ruddock, G.W. (1978): Photochem. Photobiol., 28:887.
26. Frimer, A.A. (1983): Singlet oxygen in peroxide chemistry. In: The Chemistry of Functional Groups, edited by S. Patai, John Wiley and Sons, Ltd., NY.
27. Marnett, L.J., Reed, G.A., and Johnson, J.T. (1977): Biochem. Biophys. Res. Commun., 79:569.
28. Lesko, S., Caspary, W., and Lorentzen, R. (1978): Biochem., 14:3978.
29. Birks, J.B. (1976): In: Photophysics of Aromatic Molecules, Wiley Interscience, NY.
30. Frimer, A.A. (1984): The organic chemistry of superoxide anion radical. In: Superoxide Dismutase, edited by L.W. Oberley, CRC Press, Boca Raton. (See also reference 26.)

NOMENCLATURE OF METABOLIC PRODUCTS OF PAH

KURT L. LOENING AND JOY E. MERRITT
Nomenclature Division, Chemical Abstracts Service,
P.O. Box 3012, Columbus, Ohio 43210

INTRODUCTION

In previous years, we have discussed such topics as the nomenclature of polycyclic hydrocarbons and the nitrogen, oxygen, and sulfur heterocyclic analogs as well as the nomenclature of functional derivatives containing these elements (1,2,3). This year, we will focus on the topic of metabolic products of PAH. To keep the discussion simple, we will concentrate on the benz[a]anthracene derivatives shown in Figure 1 (4).

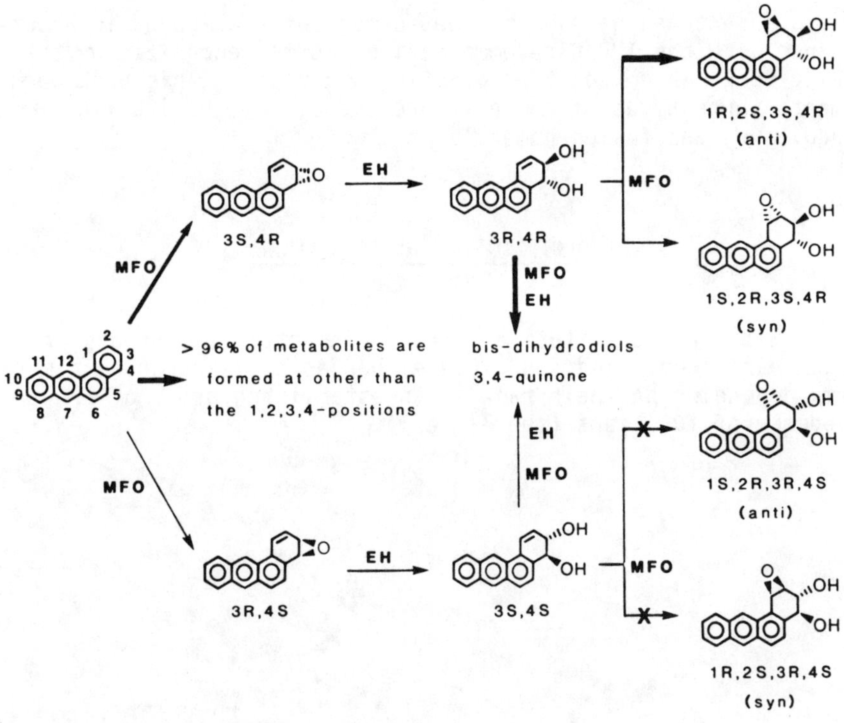

FIGURE 1. Reprinted from the ACS Symposium Series 283. Metabolic activation pathways of BA. MFO abbreviates for the cytochrome P-450-containing mixed-function oxidases. The absolute configurations of the metabolites are as shown. (4)

The nonstereochemical aspects of the nomenclature of these derivatives will be discussed first, followed by a discussion of the stereochemical descriptors.

The nomenclature recommendations of the International Union of Pure and Applied Chemistry (IUPAC) are codified in several books most readily identified by their colors. The Blue Book outlines the nomenclature of organic chemistry, while the Red Book codifies that of inorganic chemistry (5,6). A compendium of recommendations in the field of analytical chemistry is contained in the Orange Book, and the Green Book functions as a manual for physicochemical symbols, units, and terminology (7,8). The White Book, commonly referred to as The Compendium, contains the nomenclature recommendations of the IUPAC and the International Union of Biochemistry (IUB), Joint Commission on Biochemical Nomenclature (IUPAC-IUB/JCBN) (9). And finally, the book "Enzyme Nomenclature" is offered by the IUB (10).

More than one name may be formed for a compound in accordance with the IUPAC recommendations as is recognized by the Blue Book, which codifies existing practices. This book documents three types of nomenclature systems: substitutive, additive, and fusion nomenclature.

SUBSTITUTIVE NOMENCLATURE

With the substitutive method, characteristic groups or atoms are identified and cited either as prefixes or as a suffix depending on their rank in an established order of precedence of functions (see Table 1).

TABLE 1

SUFFIXES AND PREFIXES OF CHARACTERISTIC GROUPS IN THE ORDER OF DECREASING PRIORITY FOR CITATION AS PRINCIPAL GROUP

Class	Formula[a]	Prefix	Suffix
Cations		-onio -onia	-onium
Carboxylic acid	-COOH	Carboxy	-carboxylic acid
	-(C)OOH	-	-oic acid
Sulfonic acid	$-SO_3H$	Sulfo	-sulfonic acid
Salts	-COOM	-	Metal...carboxylate
	-(C)OOM	-	Metal...oate
Esters	-COOR	R-oxycarbonyl	R...carboxylate
	-(C)OOR	-	R...oate
Acid halides	-CO-Halogen	Haloformyl	-carbonyl halide
	-(C)O-Halogen	-	-oyl halide
Amides	$-CO-NH_2$	Carbamoyl	-carboxamide
	$-(C)O-NH_2$	-	-amide
Amidines	$-C(=NH)-NH_2$	Amidino	-carboxamidine
	$-(C)(=NH)-NH_2$	-	-amidine
Nitriles	-C≡N	Cyano	-carbonitrile
	-(C)≡N	-	-nitrile
Aldehydes	-CHO	Formyl	-carbaldehyde
	-(C)HO	Oxo	-al
Ketones	>(C)=O	Oxo	-one
Alcohols	-OH	Hydroxy	-ol
Phenols	-OH	Hydroxy	-ol
Thiols	-SH	Mercapto	-thiol
Hydroperoxides	-O-OH	Hydroperoxy	-
Amines	$-NH_2$	Amino	-amine
Imines	=NH	Imino	-imine
Ethers	-OR	R-oxy	-
Sulfides	-SR	R-thio	-
Peroxides	-O-OR	R-dioxy	-

[a]Carbon atoms enclosed in parentheses are included in the name of the parent compound and not in the suffix or prefix (IUPAC, Rules C11.11, C11.31, etc.)

When cited as prefixes, these groups are arranged in alphabetical order among themselves. Hydro prefixes are cited either within this order, or they may be cited directly in front of the name of the parent compound in accordance with IUPAC Rule C-16.33:

> "Hydro prefixes may be treated as detachable and arranged in alphabetical order among other detachable prefixes or they may be treated as non-detachable from the name of the parent compound or radical."

Example:

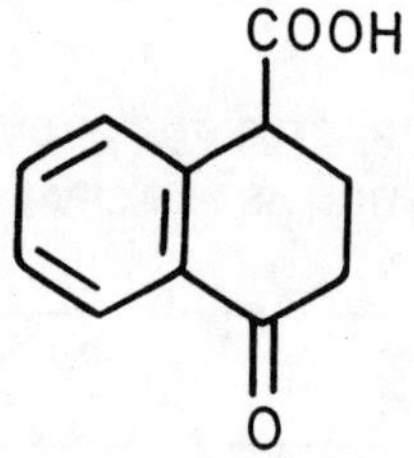

Detachable: 1,2,3,4-Tetrahydro-4-oxo-1-naphthoic acid

Non-detachable: 4-Oxo-1,2,3,4-tetrahydro-1-naphthoic acid

When a characteristic group is cited as a prefix, it is formally considered to substitute for one or more hydrogen atoms of a conceptual parent structure. This is in contrast to the additive method of nomenclature in which atoms are simply added to the conceptual parent structure, leaving the total number of hydrogen atoms in the compound unchanged.

We will use the epoxide or oxide derivatives of benz[a]anthracene to illustrate the differences between the substitutive and additive nomenclature methods. With the substitutive method, IUPAC Rule C-212.2 states:

> "An oxygen atom directly attached to two carbon atoms already forming part of a ring system or to two carbon atoms of a chain may be indicated by the prefix 'epoxy-', particularly when it is desired to preserve the name of a specific complex structure, as, for example, steroids or carotenoids."

Example:

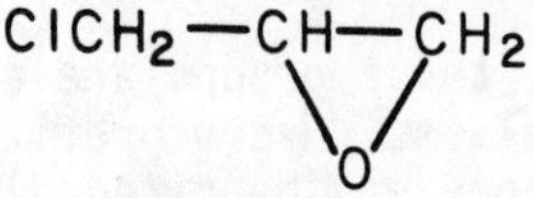

$ClCH_2—CH—CH_2$ (with O bridging CH and CH_2)

Substitutive name: 1-Chloro-2,3-epoxypropane

Thus, metabolic products of benz[a]anthracene are named by the substitutive nomenclature method as shown in Figure 2.

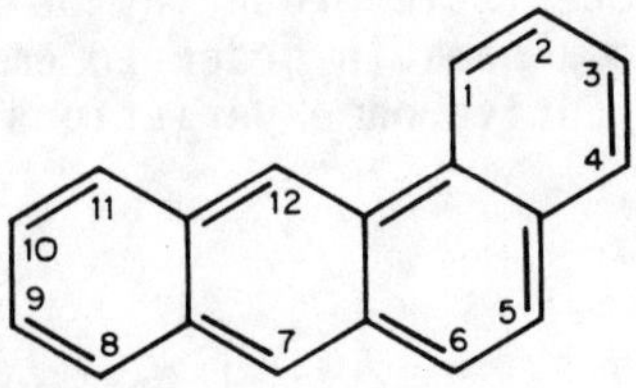

Parent: Benz[a]anthracene

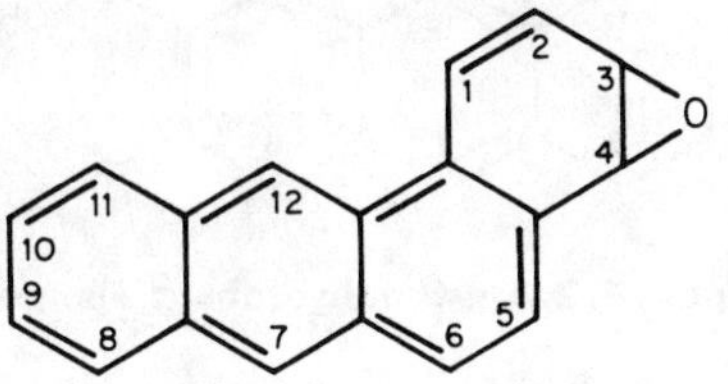

3, 4-Epoxy-3, 4-dihydrobenz[a]anthracene

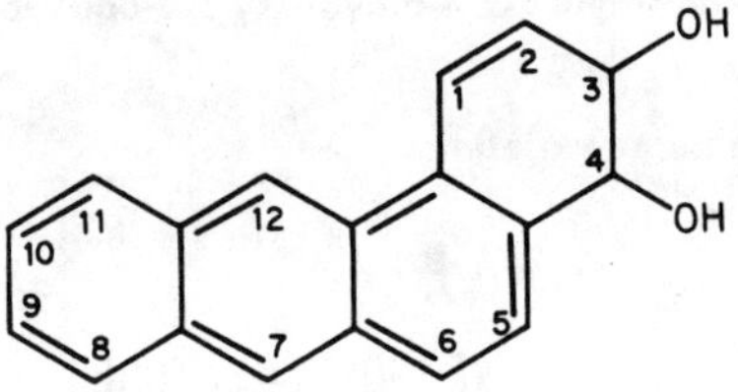

3, 4-Dihydrobenz[a]anthracene-3, 4-diol

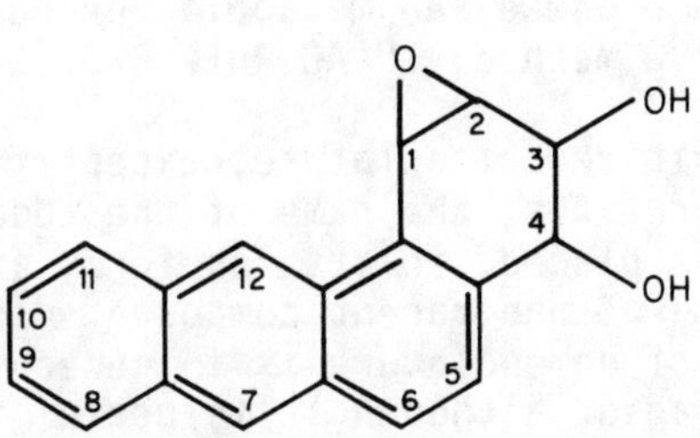

1, 2-Epoxy-1, 2, 3, 4-tetrahydrobenz[a]anthracene-3, 4-diol

FIGURE 2. Substitutive nomenclature.

In addition to placement of the hydro prefixes in a name, other variations within a name are permissible. For example, the principal characteristic group may be overstepped and cited as a prefix in a name in order to emphasize the ring itself. Some substitutive name variations are illustrated in Figure 3.

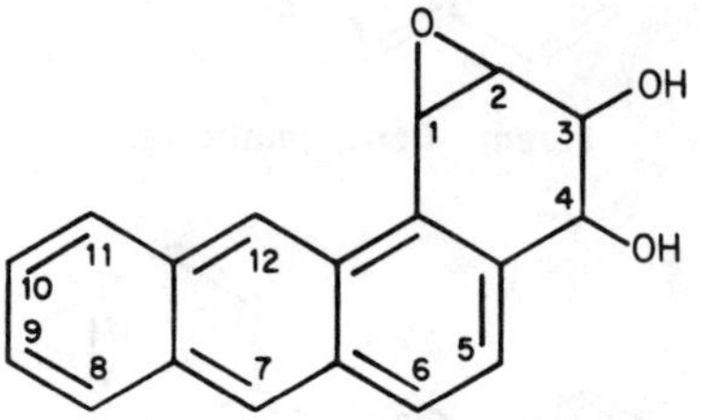

O
OH
OH
1 2 3 4 5 6 7 8 9 10 11 12

PREFERRED: 1, 2-Epoxy-1, 2, 3, 4-tetrahydrobenz[a]anthracene-3, 4-diol

OTHER: 1, 2-Epoxy-1, 2, 3, 4-tetrahydro-3, 4-dihydroxybenz[a]anthracene
1, 2-Epoxy-3, 4-dihydroxy-1, 2, 3, 4-tetrahydrobenz[a]anthracene

WRONG: 1, 2, 3, 4-Tetrahydro-3, 4-dihydroxy-1, 2-epoxybenz[a]anthracene

FIGURE 3. Substitutive name variations.

ADDITIVE NOMENCLATURE

Another type of nomenclature system, the additive method, may be used to name these PAH metabolic products. In contrast to the substitutive method, IUPAC Rule C-33.1 states

> "In additive nomenclature, except for hydro prefixes, the name of the added atoms is placed, in its ion form, after the name of the parent compound. This method of nomenclature is in general discouraged, although it is useful in a few specialized cases, particularly when the stereochemistry of addition is unknown or when the systematic name of the adduct might obscure the relation to the parent compound;"

Example:

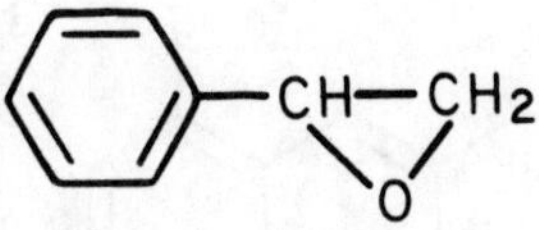

$$C_6H_5\text{-}CH\text{-}CH_2 \text{ (with O bridging, epoxide)}$$

Additive name: Styrene oxide

or

Styrene epoxide

Substitutive names: 2-Phenyloxirane

or

(1,2-Epoxyethyl)benzene

Thus, the epoxides of benz[a]anthracene are named by additive nomenclature as shown in Figure 4.

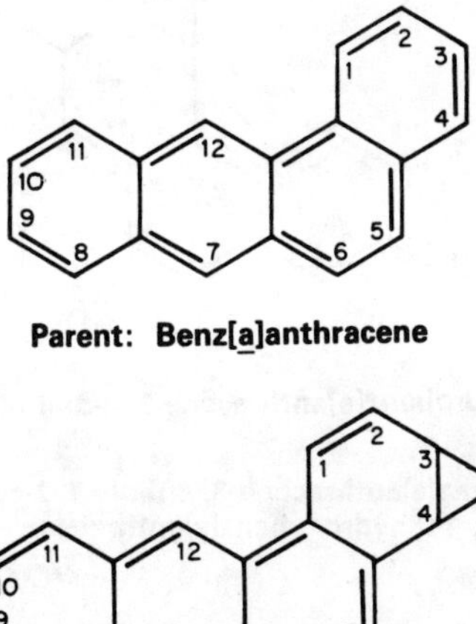

Parent: Benz[a]anthracene

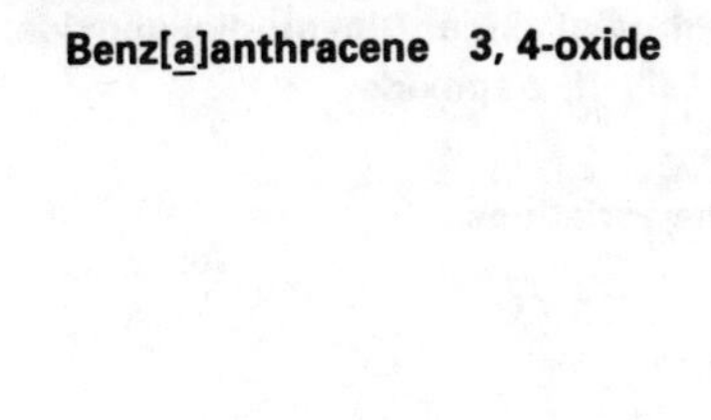

Benz[a]anthracene 3, 4-oxide

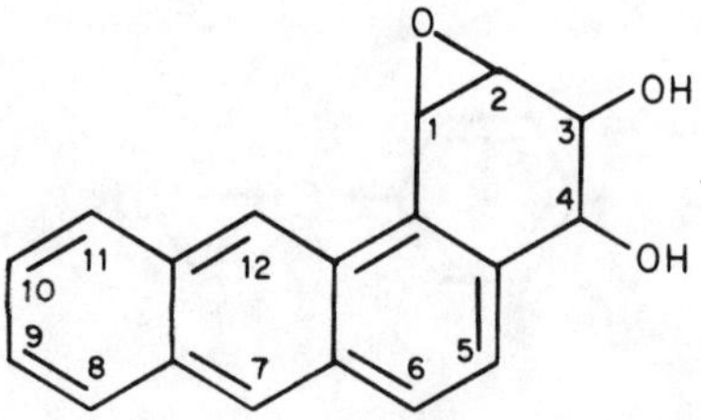

3, 4-Dihydrobenz[a]anthracene-3, 4-diol 1, 2-oxide

FIGURE 4. Additive nomenclature.

With this method, the term epoxide is a permissible variation of the term oxide. Some additive name variations are illustrated in Figure 5.

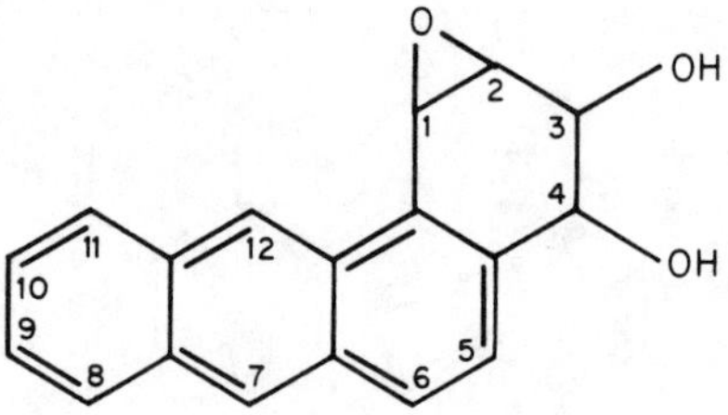

PREFERRED: 3, 4-Dihydrobenz[a]anthracene-3, 4-diol 1, 2-oxide

OTHER: 3, 4-Dihydrobenz[a]anthracene-3, 4-diol 1, 2-epoxide
3, 4-Dihydro-3, 4-dihydroxybenz[a]anthracene 1, 2-epoxide

WRONG STOICHIOMETRY: 1, 2, 3, 4-Tetrahydrobenz[a]anthracene-3, 4-diol 1, 2-epoxide

CLASS NAMES: Dihydrodiol oxide; Dihydrodiol epoxide;
3, 4-Dihydrodiol 1, 2-epoxide

FIGURE 5. Additive name variations.

FUSION NOMENCLATURE

Finally, a third method may be used to name benz[a]anthracene epoxide derivatives: the fusion method. This type of nomenclature is documented in IUPAC Rule A-21.3 for hydrocarbon rings and in IUPAC Rule B-3 for heterocyclic rings (5). This method was also extensively discussed and presented in the proceedings of previous PAH symposia (1,2,3). It is important to recall that in fusion nomenclature, rings containing the maximum number of non-cumulative double bonds are fused, and the completed system is then oriented and renumbered. Figure 6 illustrates the naming of benz[a]anthracene epoxide derivatives by fusion nomenclature.

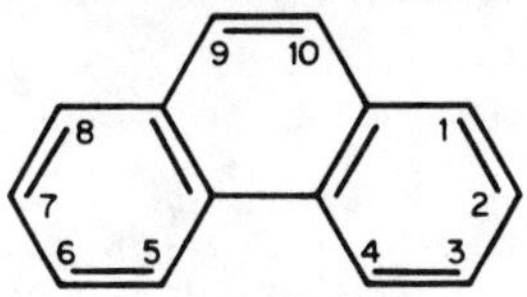

Phenanthrene

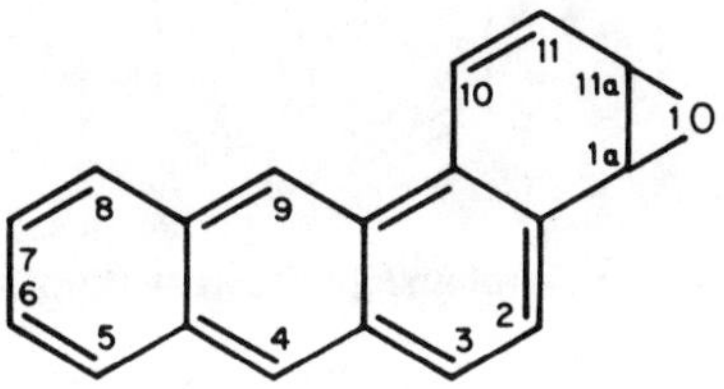

1a, 11a-Dihydrobenzo[6, 7]phenanthro[1, 2-b]oxirene

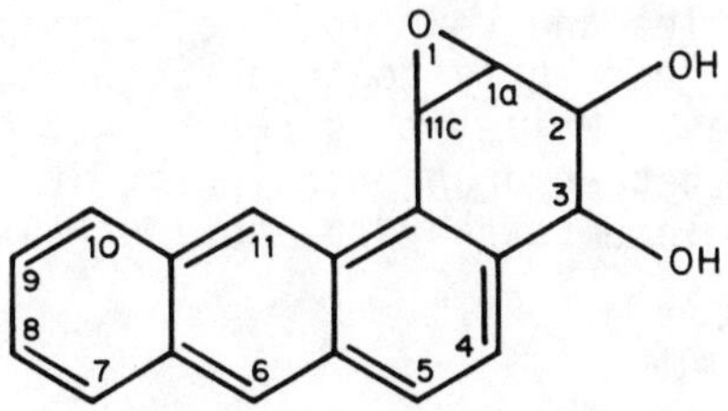

1a, 2, 3, 11c-Tetrahydrobenzo[6, 7]phenanthro [3, 4-b]oxirene-2, 3-diol

FIGURE 6. Fusion nomenclature.

For ease of comparison, three types of names derived by substitutive, additive, and fusion methods respectively are illustrated for a single compound in Figure 7.

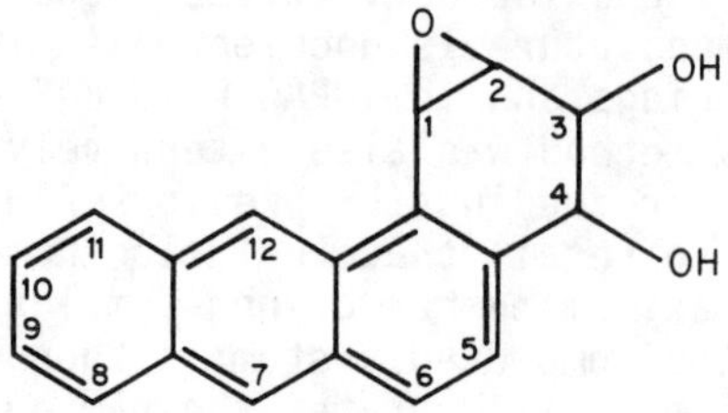

SUBSTITUTIVE: **1, 2-Epoxy-1, 2, 3, 4-tetrahydrobenz[a]anthracene-3, 4-diol**

ADDITIVE: **3, 4-Dihydrobenz[a]anthracene-3, 4-diol 1, 2-oxide**

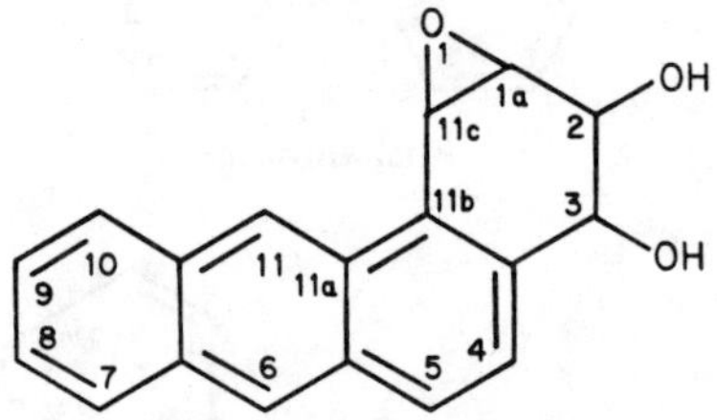

FUSION: **1a, 2, 3, 11c-Tetrahydrobenzo[6, 7]phenanthro[3, 4-b]oxirene-2, 3-diol**

FIGURE 7. IUPAC preferred nomenclature.

Nomenclature errors occur frequently in the literature, especially in titles and text where class names are used. The stoichiometry is frequently confused between the substitutive and additive names. While it is correct to call the compounds in Figure 8 epoxytetrahydronaphthalenes, it is incorrect to call them naphthalene tetrahydro epoxides (11).

H O H H O H

SUBSTITUTIVE CLASS NAME: Epoxytetrahydronaphthalenes

ADDITIVE CLASS NAME: Dihydronaphthalene oxides or Dihydronaphthalene epoxides

FUSION CLASS NAME: Tetrahydronaphthoxirenes

FIGURE 8. Class names.

Other errors commonly encountered include disregard for the alphabetical order. Even in combination with abbreviations, such as DMBA, alphabetical order can be maintained. Use of the name 3,4-Dihydroxy-1,2-epoxy-1,2,3,4-tetrahydro-DMBA is incorrect (12), but 1,2-Epoxy-1,2,3,4-tetrahydro-3,4-dihydroxy-DMBA is better. Still another error occurs when authors cite a characteristic group both as a prefix and as a suffix, and when they cite the hydro groups with the suffix as in 1-Hydroxycholanthrene-9,10-dihydrodiol-7,8-epoxide (13). The correct name for this compound is 9,10-Dihydrocholanthrene-1,9,10-triol 7,8-epoxide, or 9,10-Dihydro-1,9,10-trihydroxycholanthrene 7,8-epoxide. Hyphenation also causes some problems in the literature. For example, epoxide should be separated by a space from the name it modifies.

STEREOCHEMISTRY

IUPAC has carefully developed recommendations for the definitions and use of stereochemical descriptors. These are documented as Section E in the IUPAC Blue Book (5). To illustrate how the stereochemical configurations of the PAH are described in the IUPAC system, we shall continue using the example of the benz[a]anthracenes. The absolute configurations of four 3,4-dihydrodiol 1,2-oxides of benz[a]anthracene have been reported (see Figure 1). Our discussion addresses both relative and absolute stereodescriptors.

Syn and Anti

The terms syn and anti are used often in the PAH field to describe the relative configurations of dihydrodiol epoxides (14); however, these terms have no official IUPAC definition at present. These descriptors have been used in the literature to describe the geometry of oximes, but IUPAC recommends their abandonment for such compounds (5, Rule E-2.2.2). Their use by Chemical Abstracts Service (CAS) is restricted to describing only the senior substituent on the Z bridge of a bicyclo[X.Y.Z]ane (see Figure 9) (15).

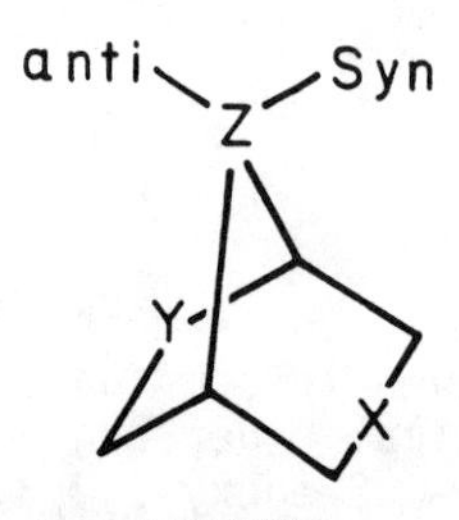

FIGURE 9. Use of syn and anti in Chemical Abstracts.

PAH chemists' use of syn and anti is unique to their field. Therefore, these terms may not be adequate for unambiguous communication to non-PAH chemists.

The Cahn-Ingold-Prelog Sequence Rule Procedure

If syn and anti cannot be used for interdisciplinary communication, what should the PAH chemist use in their place? Consider the absolute stereodescriptors R and S for describing absolute configurations (5, IUPAC Section E, 16). Assignment of the absolute terms R and S depends upon the priority ranking of atoms or groups attached to the stereochemical element whose chirality ("handedness") is to be determined by the Cahn-Ingold-Prelog Sequence Rule. This ranking depends first upon the descending order of atomic number of the atoms directly attached to the chiral center; thus, for bromochlorofluoroiodomethane the order is I, Br, Cl, F. In Figure 10 these atoms are represented by a, b, c, and d, respectively.

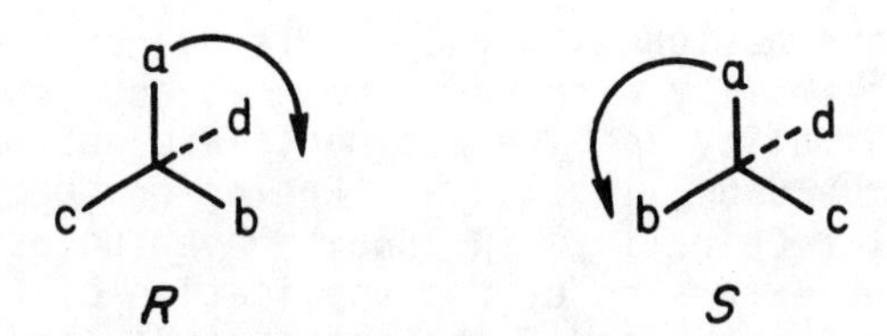

FIGURE 10. **Assignment of the stereodescriptors *R* and *S*. The symbol *R* is assigned to a clockwise (right-handed) sequence of a, b, c, while *S* denotes a counterclockwise (anticlockwise) sequence.**

The least preferred atom or group, d, is represented by a dotted line to indicate it is to be considered to lie below the plane of the paper, while a, b, and c are to be imagined to project toward the viewer at an angle. The analogy of an automobile steering wheel with three radial bars is useful to consider when visualizing this concept. The descriptor R is assigned to a clockwise (right-handed) sequence of a, b, c, while S denotes a counterclockwise (anticlockwise) sequence. This system is extremely comprehensive and can be used to describe almost all chiral centers. However R and S describe absolute configurations only. Syn and anti are relative descriptors, and therefore R and S cannot be used in their place.

The IUPAC Stereodescriptors R* and S*

To bridge this gap in the use of the Sequence Rule procedure, the IUPAC rules introduce the use of two new symbols, R* and S* for describing relative configurations (see Figure 11).

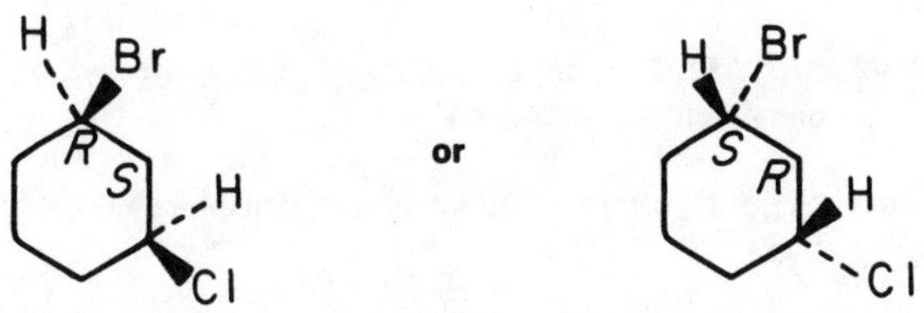

FIGURE 11. **Assignment of stereodescriptors *R** and *S** to denote relative configurations.**

These symbols are assigned by the same sequence procedure, but the center of chirality with the lowest locant is arbitrarily assigned the chirality *R*. These symbols are not part of the original paper by Cahn, Ingold, and Prelog on the "Specification of Molecular Chirality" in 1966. Nevertheless, they are a very important extension of the application of the Sequence Rule procedure. Like *R* and *S* the scope of *R** and *S** is very comprehensive, and thus they are excellent for use in interdisciplinary communication. Therefore, we recommend these IUPAC symbols be used in place of *syn* and *anti* for the PAH dihydrodiol oxides. Names that denote the relative configuration of a dihydrobenz[*a*]anthracenediol oxide derivative using *R** and *S** are shown in Figure 12.

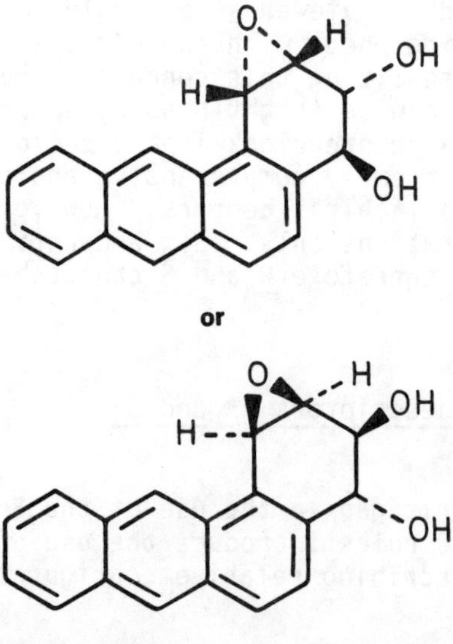

or

SUBSTITUTIVE: (1*R, 2*S**, 3*S**, 4*R**)-1, 2-Epoxy-1, 2, 3, 4-tetrahydrobenz[*a*]anthracene-3, 4-diol**

ADDITIVE: (1*R, 2*S**, 3*S**, 4*R**)-3, 4-Dihydrobenz[*a*]anthracene-3, 4-diol 1, 2-oxide**

FUSION: (1a*R, 2*R**, 3*S**, 11c*S**)-1a, 2, 3, 11c-Tetrahydrobenzo[6, 7]phenanthro[3, 4-*b*]oxirene-2, 3-diol**

FIGURE 12. Names that denote the relative configuration using *R and *S**.**

The R*/S* symbols may be cited together as a group in front of the entire name or placed within the name nearest the part to which they refer. These descriptors may also be used with either the substitutive or additive type names. However if used in the name in which the oxirene ring is expressed by fusion, the locants will change.

Racemates

Racemates are described either by placing the symbol ($\pm$) before the R*,S* relative stereodescription, or by the pairs RS or SR according to whether the center is R or S when the first is R (see Figure 13).

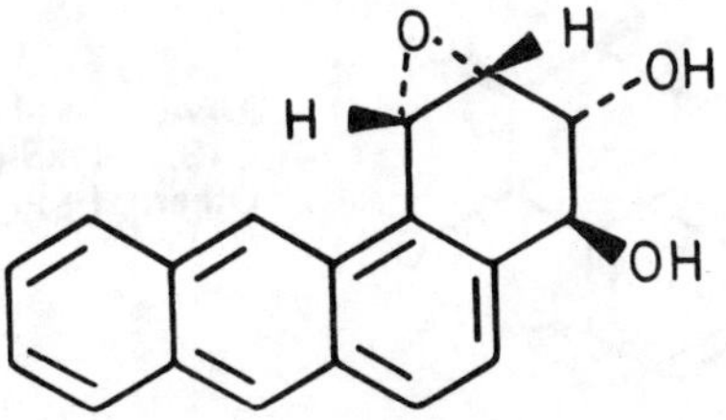

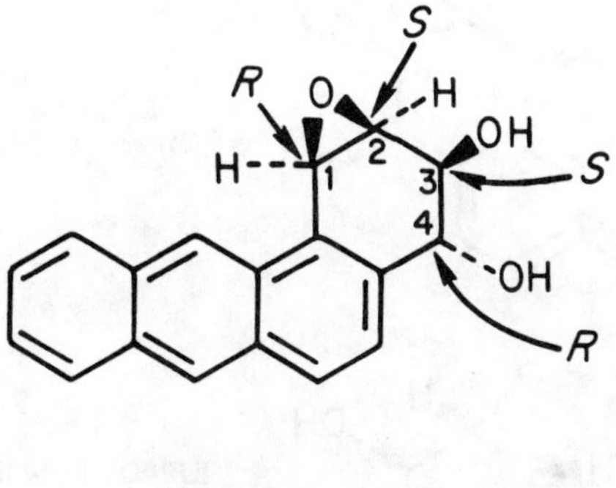

(±)-(1*R**, 2*S**, 3*S**, 4*R**)-
or
(1*RS*, 2*SR*, 3*SR*, 4*RS*)-

FIGURE 13. IUPAC Stereodescription of a racemate.

The Absolute Stereodescriptors R and S

The absolute R and S stereodescriptors for the (+)-syn, (-)-syn, and similarly for the (+)- and the (-)-anti isomers are shown in Figure 14.

IUPAC: (+)-(1*R*, 2*S*, 3*R*, 4*S*)-
CAS: [1a*S*-(1aα, 2α, 3β, 11cα)]-
Other: (+)-syn-

IUPAC: (+)-(1*R*, 2*S*, 3*S*, 4*R*)-
CAS: [1a*S*-(1aα, 2β, 3α, 11cα)]-
Other: (+)-anti-

IUPAC: (−)-(1*S*, 2*R*, 3*S*, 4*R*)-
CAS: [1a*R*-(1aα, 2α, 3β, 11cα)]-
Other: (−)-syn-

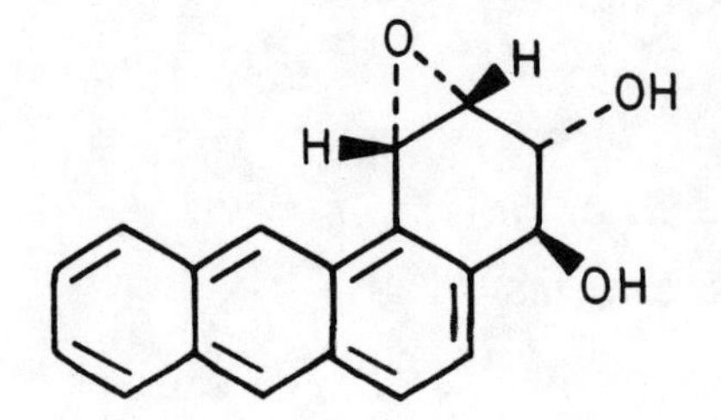

IUPAC: (−)-(1*S*, 2*R*, 3*R*, 4*S*)-
CAS: [1a*R*-(1aα, 2β, 3α, 11cα)]-
Other: (−)-anti-

FIGURE 14. **Stereodescriptors for the absolute configurations.**

In the actual assignment of R and S to these compounds it is easy to make an incorrect assignment at position 2 of 1,2-epoxy-1,2,3,4-tetrahydrobenz[a]anthracene-3,4-diol. This position is analogous to position 9 of 9,10- epoxy-7,8,9,10-tetrahydrobenzo[a]pyrene-7,8-diol (see Figure 15).

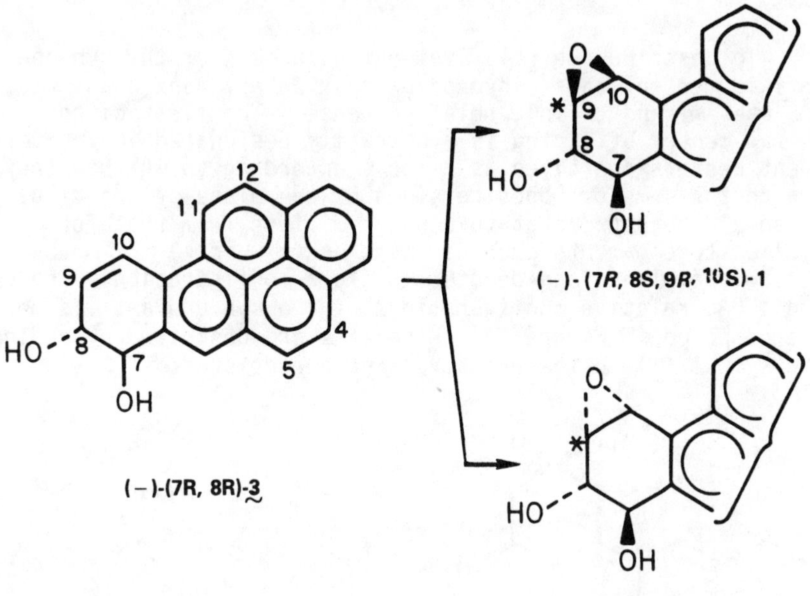

(–)-(7R, 8R)-3

(–)-(7R, 8S, 9R, 10S)-1

(+)-7R, 8S, 9S, 10R)-2

FIGURE 15. Metabolic conversion of (–)-(7R, 8R)-dihydrodiol ((–)-3) into the diastereomeric 7, 8-diol 9, 10-epoxides ((–)-1 and (+)-2).
Reprinted from the *J. Am. Chem. Soc.* 1977, 99(7) 2358.

If in the second level of exploration, one is not careful to explore the connections of the preferred pathway first before proceeding to explore the less preferred pathway, an incorrect assignment of R or S will result (16, 17).

CAS Stereodescriptors

The CAS system for describing stereogenic compounds is documented in Appendix IV of the Chemical Abstracts Index Guide (15). Stereodescriptors for systematically named compounds are covered in paragraph 203 of Appendix IV. This paragraph is also coincidentally a part of a section labelled "Section E," but should not be confused with Section E of the IUPAC Blue Book.

CAS Relative Stereodescriptors and Description of Racemates

To describe the relative configurations of the _syn_ and _anti_ isomers of the "dihydrodiol epoxide" of benz[_a_]anthracene CAS uses α and β in a relative sense. The first cited chiral center of a ring is arbitrarily designated α . Subsequent centers are cited as α or β according to whether they are on the same or opposite sides of the plane of the ring. It should be noted that this usage differs from that for cyclic stereoparents such as steroids in which " α " means "below the plane" and denotes absolute configuration. The CAS and IUPAC relative configurations of the _syn_ and _anti_ isomers are compared in Figure 16. Racemates are described by adding the symbol ($\pm$) to the relative stereodescriptor.

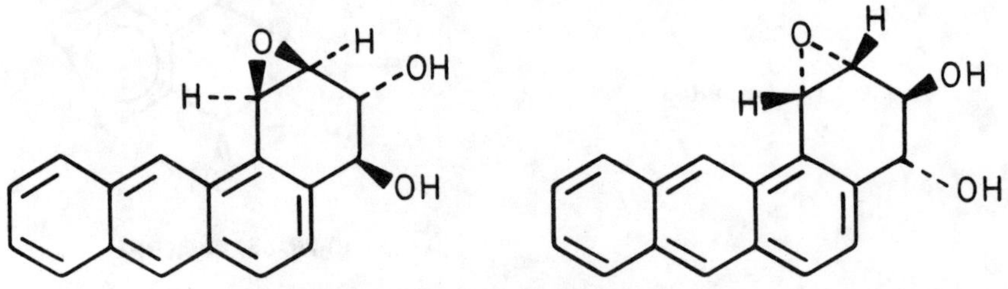

IUPAC: (1*R, 2*S**, 3*R**, 4*S**)-**
CAS: (1aα, 2α, 3β, 11cα)-
Other: _syn_-

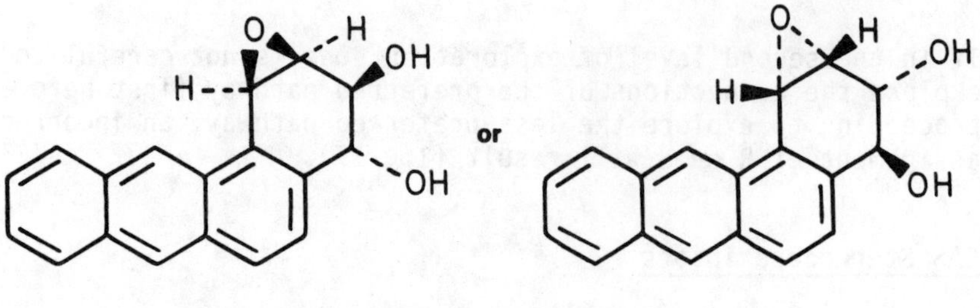

or

IUPAC: (1*R, 2*S**, 3*S**, 4*R*)-**
CAS: (1aα, 2β, 3α, 11cα)-
Other: _anti_-

FIGURE 16. Stereodescriptors for the relative configurations.

CAS Absolute Stereodescriptors

To describe the absolute configurations of systematically named compounds, the CAS system is based on analogy with the carbohydrate system for describing stereoisomers. A chiral reference center is defined and the remaining chiral centers are related to it by a relative descriptor. For example in the carbohydrate descriptor D-gluco, D describes the reference center and gluco describes the relative configurations of the chiral centers. Analogously, among systematic stereochemical descriptors the absolute terms R and S are employed for a single "reference" center, and the following relative terms are cited under various conditions: cis, trans, exo, endo, syn, anti, α, β, R*, S*, E, and Z.

The absolute configurations of the (+)- and (-)-syn, and the (+)- and (-)-anti isomers as described in the CAS system are compared in Figure 14. The reference center for the CAS system is the first cited chiral center, which is position "1a" of the ring, numbered according to the fusion name 1a,2,3,11c-tetrahydrobenzo[6,7]phenanthro[3,4-b]oxirene-2,3-diol. The relative configuration is described by α and β.

CONCLUSION

Each of the three nomenclature systems documented in the IUPAC rules for organic nomenclature (substitutive, additive, and fusion) can be used to describe the IUPAC metabolic products of PAH, such as the benz[a]anthracene metabolites. One must exercise care in forming these names so that the correct stoichiometry is expressed. Class names and abbreviations should also reflect correct nomenclature practices.

The relative stereodescriptors syn and anti have no official definition at present. For interdisciplinary communication, the IUPAC stereodescriptors R* and S* are used to describe relative configurations, while R and S are used to describe absolute configurations. The CAS system for describing stereoisomers is based on an analogy to carbohydrate stereodescriptors: a single reference center is described absolutely, which is then added to the complete relative descriptor.

REFERENCES

1. Loening, K. L. and Merritt, J. E. (1983): Some aids for naming polycyclic aromatic hydrocarbons and their heterocyclic analogs. In: Polynuclear Aromatic Hydrocarbons: Formation, Metabolism, and Measurement, edited by M. Cooke and A.J. Dennis, pp. 819-843, Battelle Press, Columbus, Ohio.
2. Jacob, J. and Loening, K. L. (1983): The nomenclature of PAH. Presented at the 8th International Symposium on Polynuclear Aromatic Hydrocarbons, Columbus, Ohio, 27 October 1983.
3. Later, D. W., Wright, C. W., Loening, K. L., and Merritt, J. E. (1986): Systematic nomenclature of the nitrogen, oxygen, and sulfur functional polycyclic aromatic compounds. In: Polynuclear Aromatic Hydrocarbons: Chemistry and Carcinogenesis, edited by M. Cooke and A. J. Dennis, Battelle Press, Columbus, Ohio.
4. Yang, S. K., Mushtaq, M. and Chiu, P. L. (1985): Stereoselective metabolism and activations of polycyclic aromatic hydrocarbons. In: Polycyclic Hydrocarbons and Carcinogenesis (ACS Symposium Series No. 283), edited by R. G. Harvey, p. 26, American Chemical Society, Washington, D.C.
5. IUPAC (1979): Nomenclature of organic chemistry, Sections A,B,C,D, E, F and H. Pergamon, Oxford. 559 pp. (The Blue Book)
6. IUPAC (1971): Nomenclature of inorganic chemistry. 2nd ed., Butterworths, London. 110 pp. (The Red Book)
7. IUPAC (1978): Compendium of analytical nomenclature. Pergamon, Oxford, 223 pp. (The Orange Book)
8. IUPAC (1979): Manual of symbols and terminology for physicochemical quantities and units. Revised ed. Pergamon, Oxford. 41 pp. (The Green Book)
9. IUPAC-IUB (1978): Biochemical nomenclature and related documents. The Biochemical Society, London. 223 pp. (The Compendium).
10. IUB (1984): Enzyme nomenclature 1984. Academic Press, Orlando, Florida. 646 pp.
11. Gillilan, R. E. (1982): Substituent effects on rates and product distributions in the hydrolysis reactions of naphthalene tetrahydro epoxides. Models for tetrahydro epoxides of polycyclic aromatic hydrocarbons, J. Am. Chem. Soc., 104:4481-4482.
12. LeBreton, P. R. (1985): The intercalation of benzo[a]pyrene and 7,12- dimethylbenz[a]anthracene metabolites and metabolite model compounds into DNA. In:

Polycyclic Hydrocarbons and Carcinogenesis (ACS Symposium Series No. 283), edited by R. G. Harvey, pp. 213, American Chemical Society, Washington, D.C..
13. Cooke, M. and Dennis, A. J. (1983): Key word index. In: Polynuclear Aromatic Hydrocarbons: Formation, Metabolism, and Measurement, Battelle Press, Columbus, Ohio.
14. Beland, F. A. and Harvey, R.G. (1976): The isomeric 9,10-oxides of trans-7,8-dihydroxy-7,8-dihydrobenzo[a]pyrene. J. Chem. Soc, Chem. Commun., 84-85.
15. Chemical Abstracts Index Guide (1985): Appendix IV, Section 203.
16. Cahn, R. S., Ingold, C. and Prelog, V. (1966): Specification of molecular chirality. Angew. Chem., Int. Ed. Engl. 5(4), 385-415.
17. Yagi, H., Akagi, H., Thakker, D. R., Mah, H. D., Koreeda, M. and Jerina, D. M. (1977): Absolute stereochemistry of the highly mutagenic 7,8-diol 9,10-epoxides derived from the potent carcinogen trans-7,8-dihydroxy-7,8-dihydrobenzo[a]pyrene. J. Am. Chem. Soc, 99(7) 2358.

MUTAGENICITY TESTING AND DETERMINATION OF POLYNUCLEAR AROMATIC HYDROCARBONS AND NITROAROMATICS IN PHOTOCOPIER TONERS, EXPOSED COPIES AND AMBIENT AIR

M. MALAIYANDI*, B.S. DAS**, D.J. KOWBEL AND E.R. NESTMANN

Environmental Health Directorate HPB, Health and Welfare Canada, Ottawa, Canada K1A 0L2; **Applied Chemistry Division, Ontario Research Foundation, Mississauga, Ontario Canada L5K 1B3. * To whom correspondence should be addressed.

INTRODUCTION

Recent studies have indicated that photocopier toners and exposed copies from photocopiers may contain traces of polycyclic aromatic hydrocarbons (PAH) and nitroaromatics (NA), such as nitropyrenes.[1-3] Several PAH are either known or suspected carcinogens, and the NA have been reported to be direct-acting mutagens in the Ames *Salmonella* assay.[1,2] For example, the extracts of selected xerographic copies and toners are known to produce positive responses in *Salmonella*. This mutagenic activity was traced to various nitropyrenes found as trace impurities in the carbon black used as the toner colorant. More recently, presence of 2,4,7-trinitro-9-fluorenone, (2,4,7-TNF) a nitro PAH derivative, has been reported in ambient air near a photocopier using toners that contained 2,4,7-TNF[4,5]

No work has been reported regarding worker exposure to PAH and NA from airborne toner and constituents of exposed photocopies during the operation and maintenance of photocopying machines. This study describes chemical analysis of toners, exposed copies, and ambient air near photocopying machines, for PAH and NA compounds, to determine whether the widespread use of such machines may pose a potential hazard to human health. The analytical method presented here uses reverse phase high performance liquid chromatography (HPLC), with fluorescence detection for the analysis of PAH, and with UV detection for NA analysis and capillary gas chromatographic technique. Biological studies to screen extracts isolated from toner and from exposed paper for mutagenicity using the Ames *Salmonella* assay also are reported.

MATERIALS AND METHODS

Preparation of Standards

Samples of PAH and NA compounds used as references in the development of HPLC analytical methods were obtained from the following sources and were used without further purification: fluoranthene, (Fl) benz[a]anthracene (B[a]A), benzo[k]fluoranthene (B[k]Fl), and benzo[a]pyrene (B[a]P), from the repository of the laboratory of the Air Pollution Control Directorate, Department of the Environment, Ottawa; dibenz[a,h]anthracene, (DB[a,h]A), benzo[g,h,i] perylene (B[g,h,i]Per), 3-methyl cholanthrene (3-MeCho), coronene (Cor), 1-nitronaphthalene (1-NO_2-Naph), 2-nitrofluorene (2-NO_2-F), 2,4,7-trinitro-9-fluorenone (2,4,7-TNF) and 9-nitroanthracene (9-NO_2-A) from Aldrich Chemical Company Inc., Milwaukee, Wis.; and 1-nitropyrene (1-NO_2-P), from Pfaltz and Bauer Inc., Stamford, Conn. Stock solutions of the standards were prepared by dissolving them in cyclohexane at mg/mL concentration. Appropriate dilutions of standards with cyclohexane, followed by replacement of cyclohexane with methanol, gave concentrations of either low μg/mL or ng/mL range prior to chromatographic analysis.

High Performance Liquid Chromatography(HPLC)

Analysis of PAH. Analyses of the PAH were performed with a Beckman Altex Model 322 HPLC instrument using a variable wavelength fluorescence detector (Model FS 970, Schoeffel Instrument Corp.) operating at λex 280 nm/λem > 389 nm and a Beckman Altex Ultrasphere ODS column (250 mm x 4.6 mm). Two elution conditions were used: a) Gradient elution. Solvent A - 1% Methanol + 99% H_2O + 0.1% acetic acid, Solvent B - 90% Methanol + 10% H_2O + 0.1% acetic acid; programmed at 0% B - 100% B in 30 min; hold at 100% B for 30 min; 100% B - 0% B in 10 min at a flow rate of 1.5 mL/min with an injection volume of 10 μL. This condition was used in the analysis of PAH in toners and copied paper extracts. (See chromatogram of PAH in FIGURE 1A). b) Isocratic: 82% acetonitrile at a flow rate of 0.80 mL/min with an injection volume of 50 μL. Extracts of particulates and vapor phase PAH in ambient air near a photocopier were analyzed using the above condition. (See chromatogram of PAH standard in FIGURE 1B).

Analysis of NA. Analyses of the NA in extracts of toner, exposed paper, ambient air particulates and XAD-2 adsorbent, were made using a Beckman Altex Model 345 HPLC

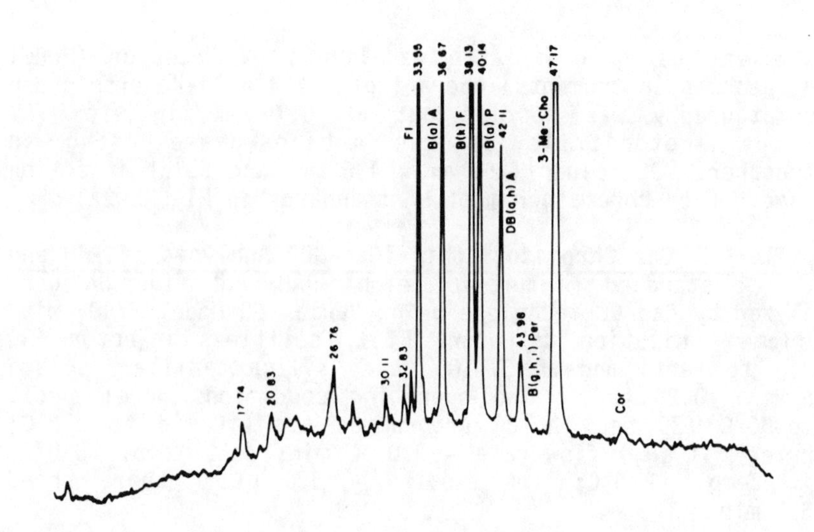

FIGURE 1A. HPLC-fluorescence chromatogram of eight PAH standards. (Analytical conditions as in A. See text.)

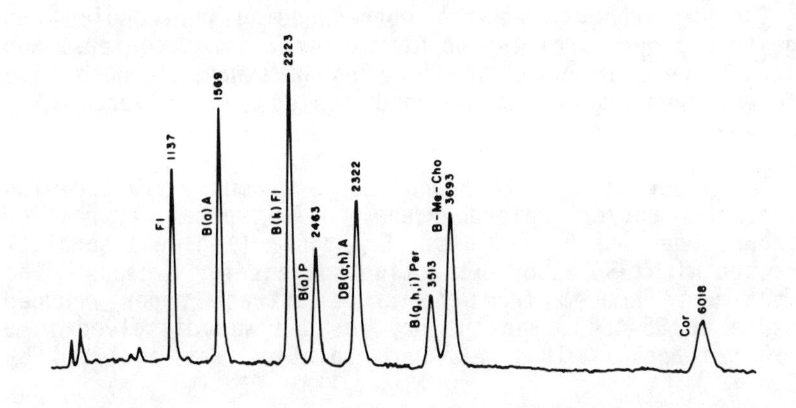

FIGURE 1B. HPLC-fluorescence chromatogram of eight PAH standards. (Analytical conditions as in B. See text.)

instrument using a variable wavelength UV detector (Model 153, Beckman Instruments) operating at 254 nm. Reverse phase chromatography was carried out at 0.70 mL/min with 70% aqueous acetonitrile as the mobile phase using an Ultrasphere ODS column (250 mm x 4.6 mm) and 50 μL injection volume. (See chromatogram of NA standards in FIGURE 2A).

Capillary - Gas Chromatographic (Cap-GC) Analysis of PAH and NA. A standard mixture of eight PAH and five NA were analyzed by Cap-GC technique using Varian GC Model 3700, with a flame ionization detector (FID), splitless injector (30 sec. to vent) and an OV-101 fused silica capillary column (25 m x 0.25 mm). The operating conditions were: col. temp-50°C (30 sec. hold) programmed at 10°C/min to 270°C; carrier gas (N_2) flow rate - 2.0 mL/min; det. temp. -350°C; inj. temp. -270°C; sample size - 5.0 μL. Chart speed 0.33"/min.

The chromatogram of Cap-GC analysis of a combined mixture of eight standard PAH and five NA is shown in FIGURE 2B.

Selection and Extraction of Toners and Copied Papers

The dry toners and the corresponding photocopied and image-free papers from two popular photocopiers, copier 1 and copier 2, were reported in the present study although five different photocopier toners and copied papers were fully investigated.

The toner samples from photocopiers were extracted with cyclohexane and/or chlorobenzene. A typical extraction procedure used was as follows. Dry toner (1 g) was soxhlet-extracted with 350 mL of extracting solvent for 8 hours. The solvent was removed from the crude extract under reduced pressure at 35-45°C, and the dry residue was dissolved in 5 mL of methanol. This methanol solution was analyzed by HPLC.

For preparation of photocopied and image-free paper extracts (image-free paper samples used as a control blank), 100 sheets of paper 8" x 11" size of each type were cut into small pieces, and then soxhlet-extracted (three sheets per thimble) with cyclohexane for 8 hours. After removing the solvents from the crude extract the dry residue was dissolved in methanol, and this resulting solution was analyzed by HPLC. The cyclohexane and chlorobenzene extracts of the

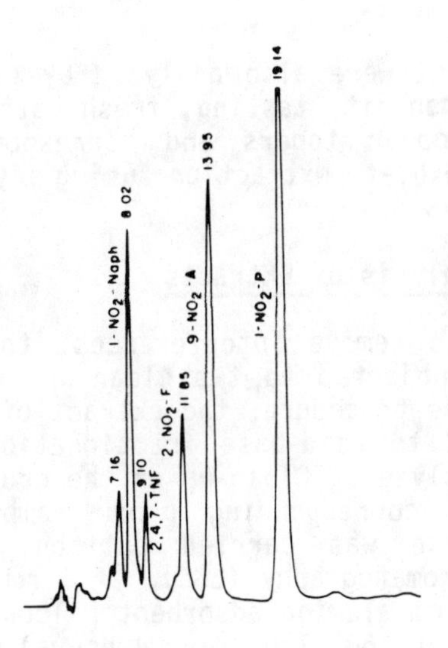

FIGURE 2A. HPLC-UV chromatogram of five NA standards. (Analytical conditions as in C. See text.)

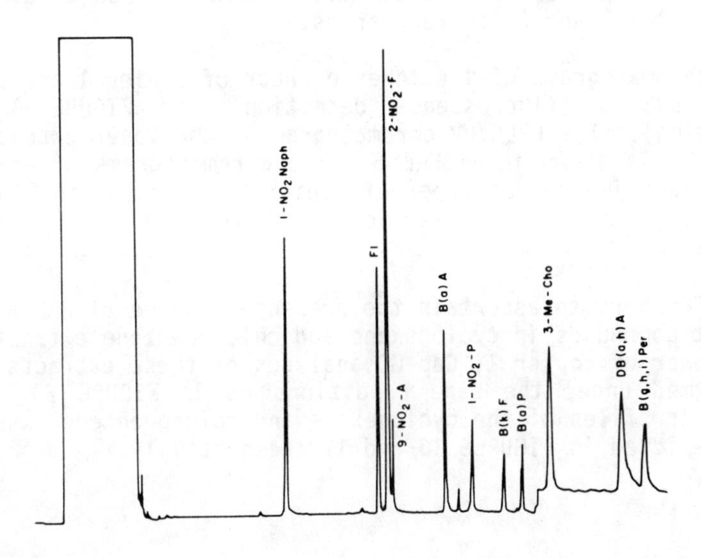

FIGURE 2B. Cap-GC analysis of standard mixture of 5 NA and 8 PAH. (Conditions same as in Cap-GC analysis described in Materials and Methods.)

toner of copier 1 were also analyzed by Cap-GC method. For *Salmonella* mutagenicity testing, fresh batches of extracts of the two photocopier toners and corresponding papers were obtained by soxhlet extraction using cyclohexane. (See *Salmonella* Assays)

Clean-Up and Analysis of Extracts

In order to remove interferences, the toner and paper extracts were subjected to two clean-up procedures prior to analysis. In one procedure, the extract of toner from copier 1 was subjected to acid-base fractionation and the neutral fraction was analyzed. Clean-up of the crude extracts of the toners and the corresponding paper samples from the two photocopiers also was carried out on activated alumina liquid-solid chromatography (LSC). A 5 mm I.D. glass column packed with 5.0 cm alumina adsorbent (Alcoa Type F-20, 80-200 mesh, activated at 160°-170° for 24 hours) was used. Eluents used, were cyclohexane, benzene and 5% MeOH in benzene, (50 mL each) in that order, at a flow rate of 0.4 mL/min. The eluates were concentrated to 1 mL volume. Recovery studies of PAH and NA standards also were studied by LSC on alumina (See Tables 1 and 2 for recoveries.)

HPLC chromatograms of the toner extract of copier 1 are shown in FIGURE 3 (fluorescence detection) and FIGURE 4 (UV detection). The HPLC/UV chromatogram of the toner extract of copier 2 is shown in FIGURE 5. The chromatograms of exposed copies and image-free paper of copier 1 are shown in FIGURES 6 and 7 (fluorescence detection for PAH) and FIGURES 8 and 9 (UV detection for NA).

Further, to ascertain the presence/absence of PAH and NA target compounds in cyclohexane and chlorobenzene extracts of the toner of copier 1, Cap-GC analyses of these extracts were performed under the same conditions as in FIGURE 2B. The Cap-GC profiles of the cyclohexane and chlorobenzene extracts are depicted in FIGURES 10 and 11, respectively.

TABLE 1

PAH RECOVERY FROM ALUMINA COLUMN CHROMATOGRAPHY

PAH	CYCLOHEXANE ELUATE %	BENZENE ELUATE %	MeOH-Bz ELUATE %
Fl	0.00	88.1	0.00
P	0.00	90.0	0.00
B[a]A	0.00	85.9	0.00
B[k]Fl	0.00	85.1	0.00
B[a]P	0.00	87.4	0.00
DB(a,h)A	0.00	63.7	0.00
B(g,h,i)Per+ 3-Me-Cho	0.00	68.5	0.00

N.B. a) B(g,h,i) Per and 3-Me-Cho peaks overlapped in this analysis.
b) The amount of PAH used ranged from 1.1 to 2.28 μg.

TABLE 2

RECOVERY OF NA COMPOUNDS FROM ALUMINA COLUMN CHROMATOGRAPHY

NA	CYCLOHEXANE ELUATE %	BENZENE(Bz) ELUATE %	MeOH-Bz ELUATE %
1-NO_2-Naph	0.00	79.3	0.00
2,4,7-TNF	0.00	292.0(%)	82.2 (?)
2-NO_2-F	0.00	82.2	17.7
9-NO_2-A	0.00	100.0	0.00
1-NO_2-P	0.00	57.8	35.1

N.B. The amount of NA used ranged from 1.0 to 3.98 μg.

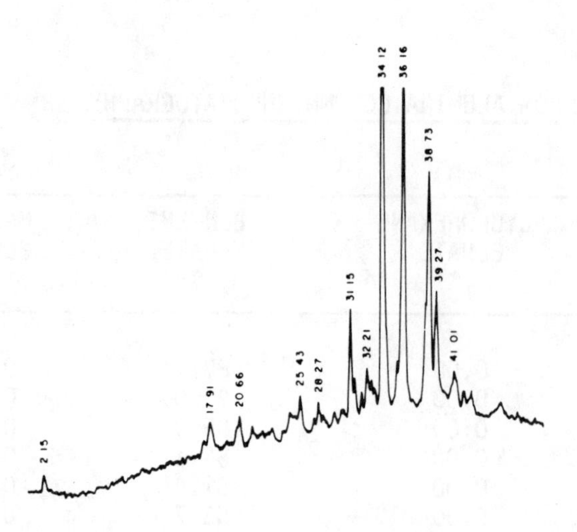

FIGURE 3. HPLC-fluorescence chromatogram of the toner extract (dilution 1:10) of copier 1. (Conditions same as in Figure 1A for PAH analysis.)

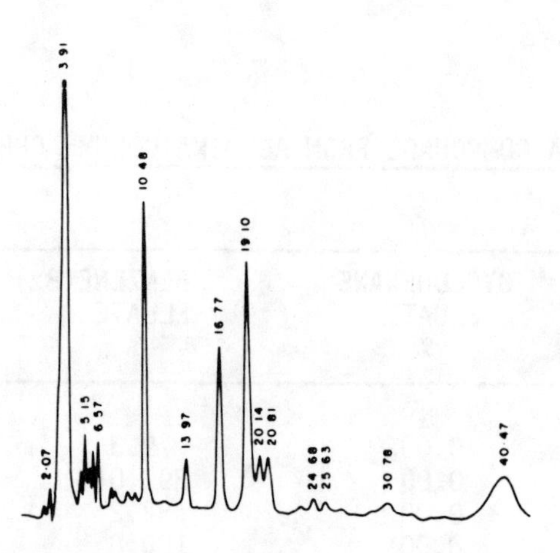

FIGURE 4. HPLC-UV chromatogram of the toner extract (dilution 1:10) of copier 1. (Conditions same as in Figure 2A NA analysis.)

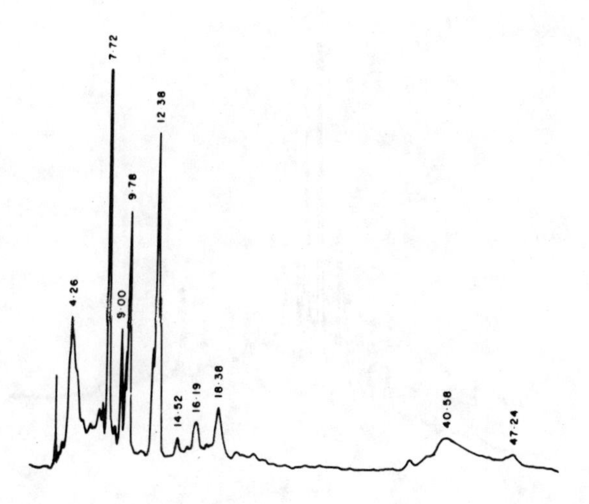

FIGURE 5. HPLC-UV chromatogram of toner extract of copier 2. (Conditions same as in Figure 2A for NA analysis.)

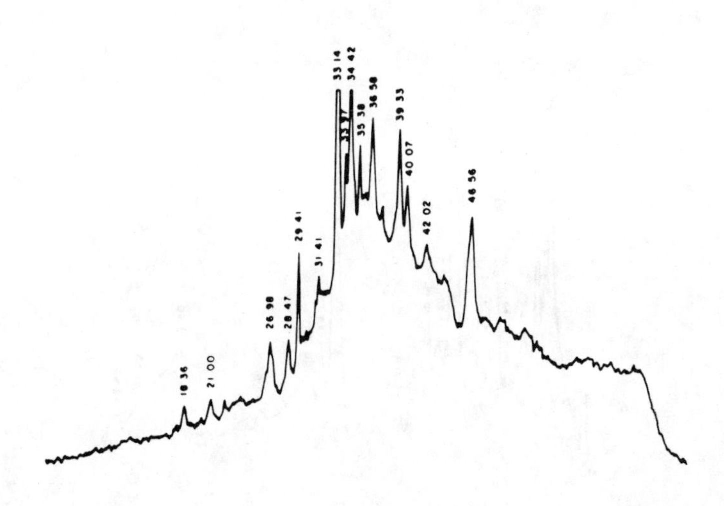

FIGURE 6. HPLC-fluorescence chromatogram of image-free paper extract (Control blank) from copier 1. (Conditions same as in Figure 1A for PAH analysis.)

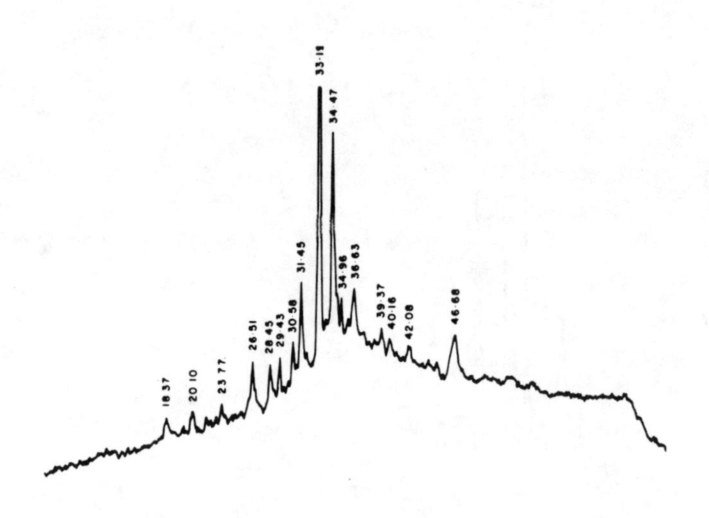

FIGURE 7. HPLC-fluorescence chromatogram of extract of exposed paper from copier 1. (Conditions same as in Figure 1A for PAH analysis.)

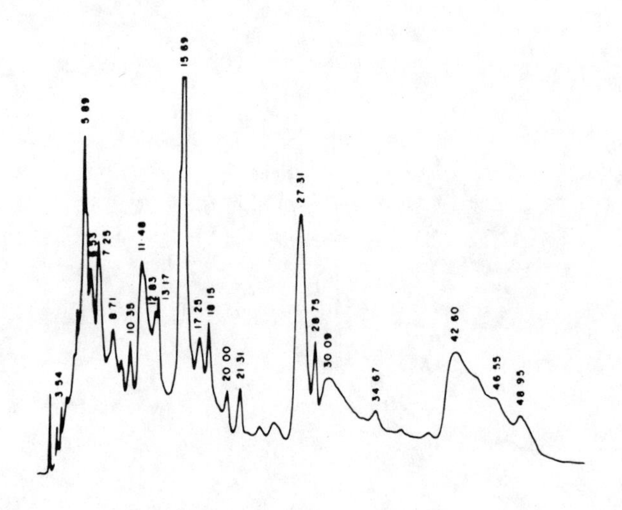

FIGURE 8. HPLC-UV chromatogram of image-free paper extract (Control blank) from copier 1. (Conditions same as in Figure 2A for NA analysis.)

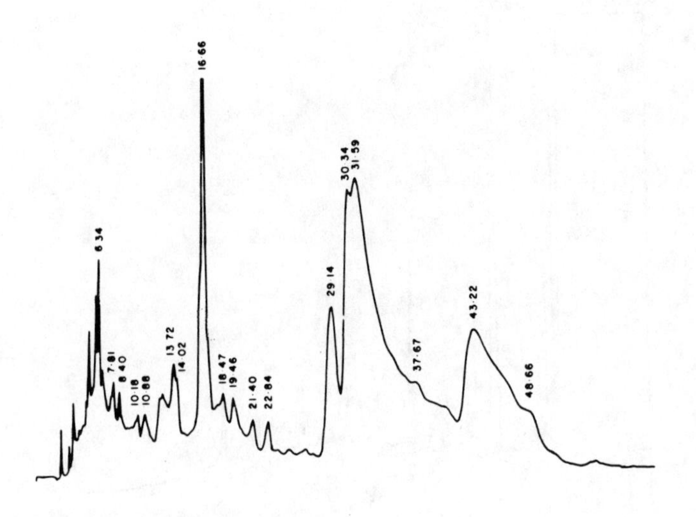

FIGURE 9. HPLC-UV chromatogram of the extract of exposed paper from copier 1. (Conditions same as in Figure 2A for NA analysis.)

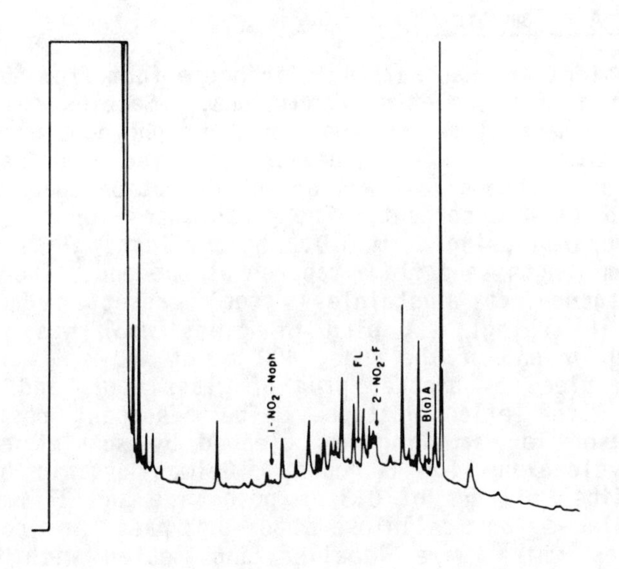

FIGURE 10. Cap-GC analysis of cyclohexane extract of the toner of copier 1. (Conditions same as in Figure 2B.)

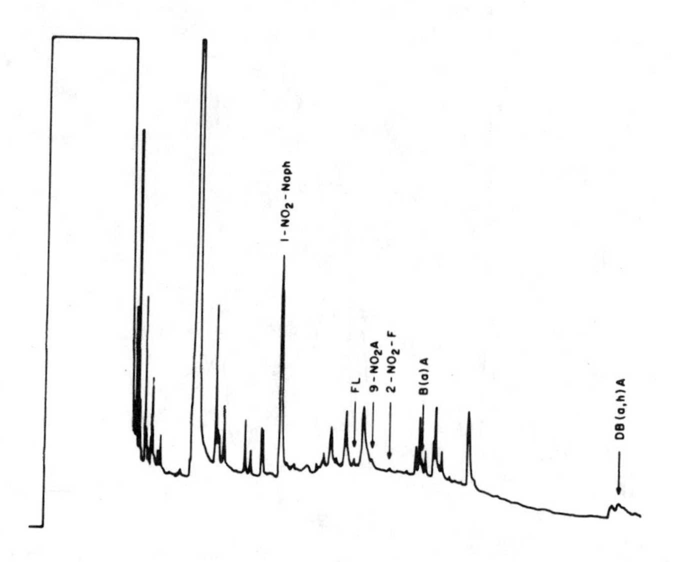

FIGURE 11. Cap-GC analysis of chlorobenzene extract of the toner of copier 1. (Conditions same as in Figure 2B.)

Ambient Air Sampling

Ambient air was sampled within one foot from photocopier 2 under normal operating conditions. Samples collected in corridors several meters away from the photocopier were used as controls. The sampling device consisted of Gelman binderless glass filters followed by an adsorption tube containing purified XAD-2 adsorbent. The glass tubes for the adsorbent were prepared using 8 mm O.D. borosilicate glass tubes cut to 80 mm lengths and flame tapered at one end. The other end was attached to a stainless steel cassette adaptor with Teflon(R) tubing. A plug of glass wool was placed in the adaptor end of the tube, 400 mg of XAD-2 was added, and kept in place by another plug of glass wool, and both ends sealed with Teflon(R) tape. The adsorbent resin (60-80 mesh) used for sampling was cleaned by soxhlet extraction with cyclohexane for 8 hours. Gelman organic binderless glass fiber filters of 0.3 μm pore size and 37 mm diameter were placed on cellulose back-up pads in polystyrene cassettes which were labelled and sealed with cellulose shrink bands. Sampling was performed with a Bendix BDX 44 battery operated pump. The total volume of air sampled was about 0.4 m^3 in 8.5 hours in each sampling run.

Analysis of Filters and Adsorbent Samples

The filters and the adsorbent, after air sampling, were soxhlet-extracted with cyclohexane, and the extracts concentrated to dryness at 30°-35°C under reduced pressure in a rotary evaporator. The residue was dissolved in methanol prior to analysis. The HPLC- fluorescence profiles of the extracts of filter and XAD-2 adsorbent are shown in FIGURES 12 and 13, respectively. The NA analyses of the same extracts of particulates and XAD-2 adsorbents are seen in HPLC-UV profiles (FIGURE 14 and 15, respectively).

Salmonella Assays

The *Salmonella* plate incorporation assay was performed as described previously[6]. Strains TA1537, TA97[8], TA98 and TA100 were used in the present study. Aroclor 1254-induced rat liver S9 was prepared and used for metabolic activation as described previously[7]. Additional details are provided in the legend of TABLE 4.

RESULTS AND DISCUSSION

Analytical Methods

C_{18} reverse phase (RP) HPLC-fluorescence detection techniques in both isocratic and gradient elution modes were used for the analysis of PAH following the procedures reported in the literature [9,10] For NA analysis, methods based on C_{18} RP HPLC-UV detection have been developed. Use of a similar technique for the analysis of 2,4,7-TNF in toner and ambient air in the vicinity of the photocopier has been reported in the past[5]. The compound 2,4,7-TNF was included in the study because of its reported presence in some toners. The other four NA compounds used in this study were 1-NO_2-Naph, 2-NO_2-F, 2-NO_2-A, and 1-NO_2P. The eight PAH frequently studied and often cited in the literature were also selected as target compounds for this study. The PAH were: Fl, B[a]A, B[k]Fl, B[a]P, DB[a,h]A, B[g,h,i]Per, 3-Me-Cho and Cor. Fl is included in the study because of its ubiquitous occurrence in most environmental samples.

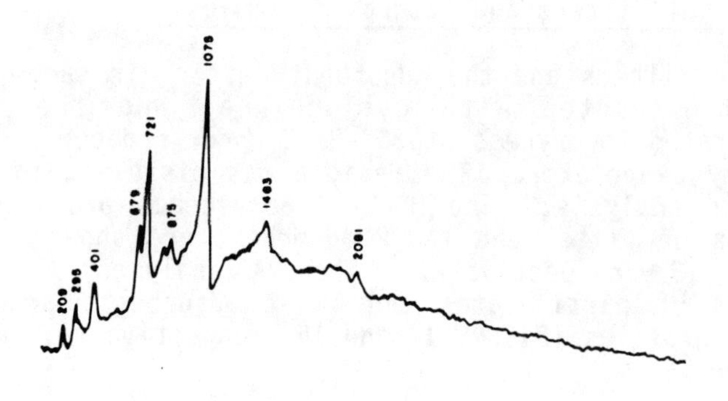

FIGURE 12. HPLC-fluorescence chromatogram of extract of filter particulates from ambient air sample near copier 1. (Conditions same as in Figure 1B for PAH analysis.)

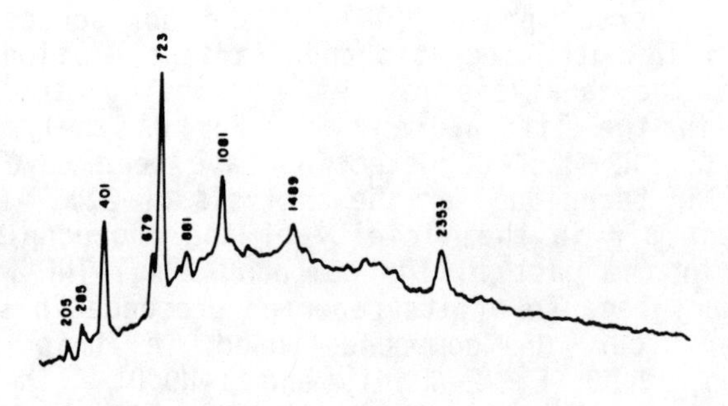

FIGURE 13. HPLC-fluorescence chromatogram of vapor phase PAH in XAD-2 cartridge from ambient air sample near copier 1. (Conditions same as in Figure 1B for PAH analysis.)

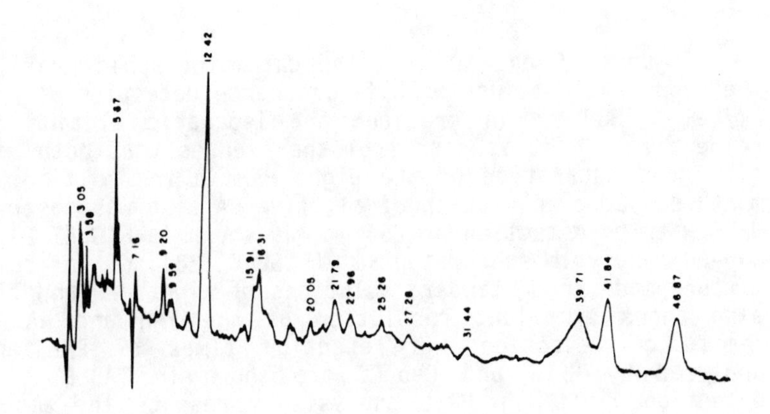

FIGURE 14. HPLC-UV chromatogram of the extract of filter particulates of ambient air sample near copier 1. (Conditions same as in Figure 2A for NA analysis.)

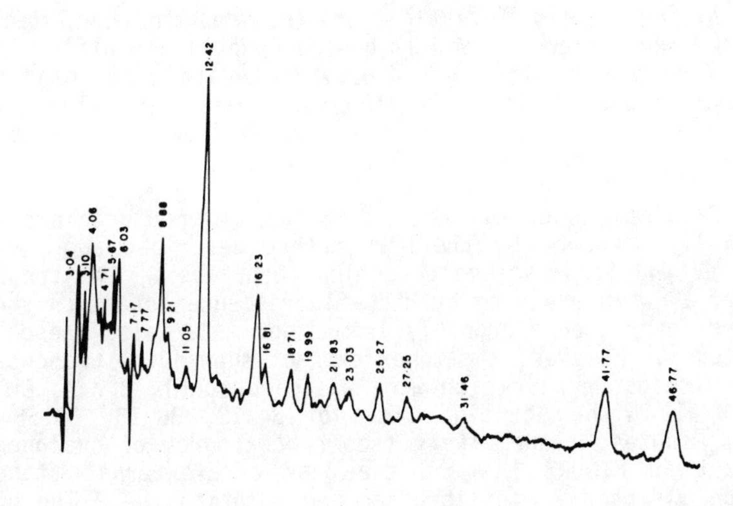

FIGURE 15. HPLC-UV chromatogram of vapor phase compounds in XAD-2 cartridge extract from ambient air sample near copier 1. (Conditions same as in Figure 2A for NA analysis.)

Figures 1A and 1B show the chromatographic profiles of the eight PAH obtained with fluorescence detection at λex 280 nm/λem > 389 nm in gradient and isocratic elution modes, respectively. It is seen from the figures that both methods give good separation of the eight PAH in the multicomponent mixture. Good resolution of all five NA also was observed in HPLC with UV detection at 254 nm, as shown in FIGURE 2A. The Cap-GC analysis chromatogram (FIGURE 2B) of a combined mixture made from standard solutions of eight PAH and five NA also shows excellent resolution of the PAH and NA. The limits of detection and retention times of PAH and NA analyzed by HPLC and Cap-GC are shown in TABLE 3. The detection limit for HPLC analysis represents the amount of standard material required to give a response which is roughly twice the background noise. In the case of Cap-GC method, under the experimental conditions, the amounts of standards used gave practically a noiseless baseline and these levels were considered as detection limit in this study. The actual sensitivity of the method would however depend on the amount of sample material injected, the interferences and the recovery throughout the method.

Analysis of Toners and Copied Papers

Although this investigation included the analyses of toners, exposed copies and image-free papers and plain papers from five photocopiers, the analytical data pertaining only to two popular photocopier toners, exposed and image-free papers are described, since the chromatograms of the photocopier toners and papers were similar.

Original toner extracts from the two photocopiers were initially screened by the HPLC method described above using both UV and fluorescence detection techniques. The toner of copier 1 when analyzed by HPLC-fluorescence detection showed strong interference probably from other materials present in the toner. However, the toner extract showed the presence of four distinct peaks which were assigned to Fl, B[a]A, B[k]Fl and B[a]P with retention times of 34.12, 36.16, 38.58 and 39.87 minutes, respectively (see chromatogram of cyclohexane extract in FIGURE 1A) and the peak area-external standard method was used for calibration and quantitation. The other target PAH were either not present or masked in the interference peaks. The presence of some PAH compounds also has been reported in carbon black generally used as the main

TABLE 3

RETENTION TIMES AND DETECTION LIMITS FOR PAH AND NA

NO	NAME OF COMPOUND	HPLC METHOD A		HPLC METHOD C		CAP-GC METHOD	
		RET. TIME (min)	DET. LIMIT (pg)	RET. TIME (min)	DET. LIMIT (ng)	RET. TIME (min)	DET. LIMIT (ng)
	NA						
1	1-NO_2-Naph	-	-	7.0	3.2	14.8	0.73
2	2,4-7-TNF	-	-	7.9	2.1	x	x
3	2-NO_2-F	-	-	10.0	3.2	20.4	1.5
4	9-NO_2-A	-	-	11.2	0.6	20.8	1.36
5	1-NO_2-P	-	-	15.5	1.9	24.8	2.58
	PAH						
1	Fl	11.4	23	ND	ND	19.8	0.40
2	B(a)A	15.7	16	ND	ND	23.6	0.48
3	B(k)Fl	22.2	25	ND	ND	26.8	2.30
4	B(a)P	24.6	52	ND	ND	27.6	1.90
5	DB(a,h)A	28.3	52	ND	ND	33.0	4.56
6	B(g,h,i)Per	35.1	83	ND	ND	34.4	4.92
7	Me-Cho	37.0	55	ND	ND	29.4	3.45
8	Cor	60.2	147	ND	ND	x	x

\- - Not detectable by Fluorescence Method
ND - Not determined
x - Could not be eluted from the column under the GC conditions used.

ingredient in toner formulation.[3] The four PAH were found present in the toner at low levels (ppb), e.g. Fl (17.41), B[a]A (1.60), B[k]Fl (1.79), and B[a]P (0.66). It is uncertain that the presence of such low levels of PAH in toner would pose any potential health hazard.

The extracts of toners of both copier 1 and 2 when analyzed for NA compounds by the HPLC-UV detection technique showed the presence of a large number of peaks with retention times very close (± 0.2 min) to those of the five NA standards. The profiles are shown in FIGURES 4 and 5 for toner extracts of copier 1 and 2, respectively. However, no peaks identitical in retention times to those of the five NA target compounds (see NA standard chromatogram in FIGURE 2A) were observed. However, mass spectrometric analysis may be necessary to precisely characterize the compound responsible for those peaks.

Strong interferences were also noted in the HPLC analysis of exposed copies and the corresponding image-free paper (used as control blank) extracts of both copiers 1 and 2, respectively. However, it is very difficult to ascertain the presence of PAH and NA compounds in the copied paper extracts, since extracts of control blank and exposed copies displayed identical HPLC profiles, as demonstrated in FIGURES 6-7 and FIGURES 8-9. It is however probable that some target NA may be present and masked by the peaks with similar retention times.

Further, the cyclohexane and chlorobenzene extracts of toner of copier 1 were analyzed by Cap-GC and the profiles are shown in FIGURES 10 and 11, respectively. It would appear that, on the basis of retention times (FIGURES 10), traces of 1-NO_2-Naph, Fl, 2-NO_2-A and B(a)A were detected in the cyclohexane extract. Similarly, in FIGURE 11 one can observe traces of 9-NO_2-A, 2-NO_2-F and B(a)A but considerable amount (ppb) of 1-NO_2-Naph was found. Further confirmation is under investigation.

In order to remove interferences (which probably originated from paper additives and coatings rather than from the toners during photocopying) two clean-up methods were studied with the toner and copied paper extracts. Initially, an acid-base clean-up was performed and the neutral fraction was isolated and analyzed. However, desired purification of the extract was not achieved by this procedure. The second clean-up method studied consisted of liquid-solid chromatography (LSC) using activated alumina and gradient elution. Initial LSC studies with the eight PAH and five NA target compounds showed high recoveries of the compounds in the benzene eluate (see TABLES 1 and 2. However, the interfering impurities in the toner and paper extracts although diminished in concentration, also were recovered in the benzene eluate fraction under identical LSC conditions. This suggests that the development of a suitable clean-up/fractionation method is necessary in order to conclude unequivocally that the NA compounds were present in the toners and copied papers.

Analysis of Ambient Air Samples

To determine whether the widespread use of photocopiers poses any potential hazard to human health, the ambient air in the vicinity of the copier 1 was analyzed for both PAH and NA compounds collected on filters and XAD-2 adsorbents, respectively. After sampling with a Bendix BDX-44-battery operated sampling pump, the filters and the adsorbent cartridges were separately extracted and the extracts analyzed by methods similar to those used for toners and copied papers as described previously. PAH were analyzed by HPLC-fluorescence detection and NA compounds by the HPLC/UV method. Samples collected in corridors several meters away from the location of copier 1 were used as controls.

HPLC-fluorescence chromatograms of the filter and adsorbent extracts (see chromatograms in FIGURES 12 and 13, respectively) and the corresponding control chromatograms (not shown) showed identical profiles. This indicated the ambient air in the vicinity of the photocopier did not show any detectable level of the target PAH, e.g. 6.5 ng/m^3 for B[a]P. Identical chromatographic profiles also were observed with the filter and adsorbent extracts and their corresponding control extracts in the HPLC/UV analysis for NA compounds. The two HPLC/UV chromatograms of the filter and adsorbent extracts of the ambient air samples are shown in FIGURES 14 and 15, respectively. It would appear that, similar to the observation made with PAH analysis above, the extract showed some peaks very close to some of the target NA compounds. Further confirmation would be necessary to characterize the compounds under these peaks.

Salmonella Assay of Extracts for Mutagenicity

None of the extracts tested for mutagenicity was active in the Salmonella/mammalian-microsome assay (TABLE 4).

TABLE 4

RESULTS OF SALMONELLA TESTS ON EXTRACTS OF TONERS, COPIED AND PLAIN SHEETS OF PAPER

EXTRACT OF	DOSE PER PLATE mg(ug)[a]	REVERTANTS PER PLATE[b]							
		TA100		TA97		TA98		TA1537	
		-S9	S9	-S9	S9	-S9	S9	-S9	S9
TONER #1	0	147	131	104	160	12	34	6	5
	0.01	114	129	113	159	16	40	5	7
	0.1	106	132	129	138	14	43	2	9
	1.0	126	113	105	113	9	14	5	5
	10.	102[c]	74[c]	105[c]	106[c]	6[c]	17[c]	1[c]	1[c]
TONER #2	0	127	89	139	145	17	31	3	9
	0.01	193	112	160	188	25	35	6	8
	0.1	144	123	118	179	13	30	10	12
	1.0	NT	NT	NT	NT	25	46	NT	NT
PAPER #1[d] (COPIED)	0	138	125	136	162	20	31	8	9
	0.01	143	126	131	170	19	15	5	3
	0.1	156	127	130	184	13	21	4	6
	1.0	140	128	150	148	17	24	6	7
	10.	169	128	128	181	16	27	10	8
PAPER #1[d] (PLAIN)	0	138	125	136	162	20	31	8	9
	0.01	131	117	123	150	14	17	4	5
	0.1	119	141	131	193	12	23	5	2
	1.0	170	155	129	178	10	18	3	9
	10.	86	119	48	103	19	35	2	7

TABLE 4 (Cont'd)

RESULTS OF SALMONELLA TESTS ON EXTRACTS OF TONERS, COPIED AND PLAIN SHEETS OF PAPER

EXTRACT OF	DOSE PER PLATE mg(ug)[a]	REVERTANTS PER PLATE[b]							
		TA100		TA97		TA98		TA1537	
		-S9	S9	-S9	S9	-S9	S9	-S9	S9
PAPER #2[d] (COPIED)	0	138	125	136	162	20	31	8	9
	0.01	156	139	137	158	16	33	5	4
	0.1	128	127	127	158	30	29	3	7
	1.0	145	135	130	178	28	44	5	3
	10.	164	137	92	141	28	35	7	1
PAPER #2[d] (PLAIN)	0	127	89	139	145	17	31	3	9
	0.01	174	151	125	165	24	34	11	10
	0.1	142	139	147	159	19	31	9	10
	1.0	CS	CS	CS	CS	CS	CS	NT	NT
POSITIVE CONTROLS[e]									
1-Nitropyrene[f]	0					25			
	(0.094)					107			
	(0.187)					242			
	(0.374)					351			
MNNG	(5.)	2710							
$2\text{-}NO_2\text{-}F$	(20.)					1800			
$9\text{-}NH_2\text{-}Ac$	(75.)			1311					
	0.15							1459	
$2\text{-}NH_2\text{-}A$	0.15	200	1962	217	1469	54	1834	34	230

[a] The solvent was cyclohexane except for the experiment with $1\text{-}No_2\text{-}P$ which was dissolved in dimethyl sulfoxide. The volume added to each plate was 100 uL for all doses.

[b] Average of duplicate plates. S9, supernatant of rat liver homogenate used for metabolic activation; NT, not tested; CS, couldn't score (due to heavy precipitate of test sample).

[c] The test material partly precipitated out of solution.

[d] Extracts were prepared of 100 sheets of paper.

[e] Positive control results are averages of 4 experiments that were involved in the screening of these samples (except for $1\text{-}NO_2\text{-}P$). MNNG, N-methyl-N'-nitro-N-nitrosoguanidine; $2\text{-}NO_2\text{-}F$, 2-nitrofluorene; $9\text{-}NH_2\text{-}Ac$, 9-aminoacridine; $2\text{-}NH_2\text{-}A$, 2-aminoanthracene.

The samples were tested over a wide range of concentrations, both in the presence and absence of a mammalian metabolizing fraction from rat liver (S9). In some cases, material precipitated form solution when the highest test concentrations were added to the aqueous top agar overlay in the plate incorporation method. Certain extracts induced lethality in the tester strains, as shown by reduction in the numbers of revertants at the highest doses tested. The frequencies of spontaneous revertants in the negative (solvent) controls were in the normal ranges for all four strains [6,7]. Results with the positive controls confirmed that the strains were characteristically sensitive to specific mutagens (including PAH and NA) and that the S9 preparation was active. Moreover, results with 1-NO_2-P, a mutagen that previously was found to be partially responsible for the mutagenic activity of photocopying toners that since have been withdrawn from use[2], were positive. This finding shows that the test as performed was capable of detecting nitropyrene-induced activity if mutagenic nitro- pyrenes had been present.

In 1980, some photocopier toners were found to be mutagenic in the Salmonella test [1,2] and this mutagenicity appeared to be due to nitropyrene contamination of carbon black[2] which has been used as a colorant. Since then, conditions in the industrial process for the production of carbon black colorant were changed and mutagenic ingredients were removed from the photocopier toners[2]. The present results in Salmonella for extracts of currently available photocopier toners confirm that the residues from these commerial products are non-mutagenic and are in agreement with the findings reported by Rosenkranz et al.[2]. The crude extracts from exposed copies and the image-free papers also gave negative responses in Salmonella indicating that the extract in its entirety is non-mutagenic.

SUMMARY

Low levels of some target PAH and NA compounds were detected in the toners and exposed copies used in the present study. The methods described indicate that a suitable clean-up procedure for the extracts of toners and exposed copies is necessary to unequivocally ascertain the presence/absence of the target PAH and NA. The present results on mutagenicity showed no positive responses of the extracts from the toners and exposed copies confirming the results reported earlier[2].

ACKNOWLEDGEMENTS

The authors would like to thank L. Dias, C.M. Smith for their skilled technical assistance, R. Otson and G. LeBel for critically reviewing the manuscript, A. Cservik for word processing and D. Bryant, McMaster University for providing purified 1-nitropyrene.

REFERENCES

1. Löfroth, G., Hefner, E., Alfheim, I., and Moller, M.,(1980) Mutagenic activity in photocopies, Science, 209: 1037-1039.
2. Rosenkranz, H.S., McCoy, E.C., Sanders, D.R. , Butler, M., Kiriazides, D.K., and Mermelstein, R., (1980): Nitropyrenes: Isolation, identification and reduction of mutagenic impurities in carbon black and toners, Science, 209: 1039-43
3. Sanders, D.R. (1981): The isolation of trace mutagenic impurities from the toluene extract of an after-treated carbon black in chemical analysis and biological fate, Polynuclear Aromatic Hydrocarbons, 5th Int. Symp. Eds. M. Cooke, and A.J. Dennis p. 145.
4. Seymour, M.J. (1982): Determination 2,4,7-trinitro-9-fluorenone in workplace environmental samples using high performance liquid chromatography, J. Chromatogr., 236, 530-534.
5. Bagon, D.A. and Purnell, C.J. (1981): Determination of 2,4,7-trinitro-9-fluorenone in air by high performance liquid chromatography, J. High Resolut. Chromatogr. Chromatogr. Comm., 4: 586-588.
6. Maron, D.M. and Ames, B.N. (1983): Revised methods for the Salmonella mutagenicity test, Mutat. Res., 113: 173-215.
7. Ames, B.N. McCann, J., and Yamasaki, E. (1975): Methods for detecting carcinogens and mutagens with the Salmonella/mammalian-microsome mutagenicity test, Mutat. Res., 31: 347-364.
8. Levin, D.E., Yamasaki, E., and Ames, B.N. (1982): A new Salmonella tester strain, TA97, for the detection of frameshift mutagens. A run of cytosines as a mutational hot-spot, Mutat. Res., 94: 315-330.
9. Malaiyandi, M. , Benedek, A., Holks, A.P., and Bansci, J.J. (1981): Measurement of Potentially

hazardous polynuclear aromatic hydrocarbons from occupational exposure during roofing and paving operations, Polynuclear Aromatic Hydrocarbon, 6th Int. Symp. Eds. M. Cooke, A.J. Dennis and G.L. Fisher, p 471-489.

10. Das, B.S., and Thomas, G.H., (1978): The use of a fluorescence detector in high performance liquid chromatographic routine analysis of polycyclic aromatic hydrocarbons in environmental pollution and occupational health studies, Natl. Bur. Stds. special Publ. 519. Trace org. Anal: A new frontier in anal. chem., proc. of the 9th Materials Res. Symp. April 10-13.
11. Crathorne, B., Watts, C.D., and Fielding, M. (1979): Analysis of nonvolatile organic compounds in water by high performance liquid chromatography J. Chromatogr., 185: 671-690.

NITROARENES ARE INDUCERS OF CUTANEOUS AND HEPATIC MONOOXYGENASES IN NEONATAL RATS: COMPARISON WITH THE PARENT ARENES.

HASAN MUKHTAR*[1,2], PARTHASARATHY ASOKAN[1], MUKUL DAS[1], DANIEL P. BIK[1], PAUL C. HOWARD[2], G.D. McCOY[2], HERBERT S. ROSENKRANZ[2] AND DAVID R. BICKERS[1,2].
(1)Department of Dermatology, University Hospitals of Cleveland, Case Western Reserve University and the VA Medical Center, and (2)Department of Environmental Health Sciences, School of Medicine, Case Western Reserve University, Cleveland, Ohio, 44106, USA.

INTRODUCTION

Nitrated derivatives of polynuclear aromatic hydrocarbons (nitroarenes, NA) have recently been recognized as a group of potential environmental pollutants. Nitroarenes are produced by the reaction of the parent arenes with low concentrations of NO_2 in the presence of traces of HNO_3 (1,2) and during the combustion of coal (3), and diesel (4,5). Only very few nitroarenes have been chemically identified (4,6). Several nitroarenes have recently been shown to be highly mutagenic and carcinogenic to rodents (2,7,8).

Many chemical carcinogens require metabolic activation for conversion into reactive electrophilic metabolites which ultimately bind to cellular macromolecules to exert their carcinogenic effects. The active metabolites are formed by microsomal cytochrome P-450 dependent monooxygenases that have been shown to be induced in experimental animals after exposure to complex mixtures of chemicals such as diesel exhaust, coal gas condensate, and PAHs present in coal tar and cigarette smoke (9-12).

In a recent study (13), we have shown that a single topical application of a mixture of anthropogenic nitrated pyrenes induces certain microsomal cytochrome P-450 dependent monooxygenase activities in skin and liver. We have now extended these studies to several pairs of nitroarenes and their parent arenes. We have evaluated the effect of topical application of these compounds on skin and liver monooxygenases and on the metabolism of benzo(a)pyrene (BP). Our results suggest that topically applied nitroarenes are inducers of cutaneous and hepatic monooxygenases in neonatal rats and that they resemble 3-methylcholanthrene (3-MC) type of inducers in this regard.

MATERIALS AND METHODS

Chemicals

7-Ethoxycoumarin, pyrene, 3-MC, fluoranthene, perylene, benzo(ghi) perylene and resorufin were purchased from Aldrich Chemical Co., (Milwaukee, WI). BP, bovine serum albumin, NADP and NADPH were purchased from Sigma Chemical Co. (St. Louis, MO). 7-Ethoxyresorufin was a product of Pierce Chemicals. Benzphetamine HCl was a gift from the Upjohn Company, Kalamazoo, MI. [7,10-^{14}C]BP (specific activity, 58.5 mCi/mmol) was purchased from Amersham Searle (Chicago, IL). Prior to use, radiolabeled BP was purified on a silica gel (Partisil 10 μ; Waters Associates) column with hexane as the eluting solvent and subsequently by reverse-phase HPLC using a DuPont Zorbax ODS column (6.2 mm x 25 cm) eluted with methanol:water (19:1, v/v). The purity of BP was greater than 99% as judged by HPLC. BP reference standards were provided by the Cancer Research Program, Division of Cancer Cause and Prevention, National Cancer Institute (Bethesda, MD).

Preparation of Nitroarenes

The nitroarene mixtures were prepared by overnight nitration in 250-fold (mole/mole) nitric acid at 22°C. The nitroarenes were collected following precipitation in water. Gas chromatographic analysis of nitrated pyrenes revealed: pyrene - 43%, 1-nitropyrene - 7%, dinitropyrenes - 50%. GC analysis of the nitrated fluoranthenes revealed: fluoranthene - 1%, 1-7-nitrofluoranthene - $<$0.1%, 3-nitrofluoranthene - 5%, 8-nitrofluoranthene - 3%, n-(poly)nitrofluoranthenes - 91%. Reversed-phase HPLC analysis of the nitroperylene and nitrobenzo(ghi)perylene mixtures indicated single peaks in addition to the parent arene. 1-Nitropyrene ($>$98%) was a generous gift from Dr. F.A. Beland. 3-Nitrofluoranthene ($>$99.5%) was synthesized as described by Kloetzel *et al.*(14).

Animals and Treatment

Pregnant Sprague-Dawley rats were obtained from Holtzman Rat Farm, Madison, WI and were housed in individual plastic cages. The room was maintained at 21°C with a 12 hr light and dark cycle. Newborn rats were withdrawn from their mothers on day 4 after birth. The advantages of using neonatal rodents for studies on cutaneous drug and carcinogen metabolism have been described earlier (15). All parent arenes (pyrene, perylene, benzo(ghi)perylene, and fluoranthene) and nitroarenes (1-nitropyrene, nitropyrene mixture, 3-nitrofluoranthene, nitrofluoranthene mixture, nitrobenzo(ghi)-

perylene mixture, and nitroperylene mixture) were dissolved in dime thyl sulfoxide and diluted with THF (1:1) and 100 µl of this mixture was applied topically to deliver a dose of 10 mg kg body wt. 3-MC was also applied topically to a group of animals to serve as a standard reference inducer. Animals receiving an identical volume of the vehicle served as controls.

Preparation of Liver and Skin Microsomes

Twenty-four hours after a single topical application of the test compounds, the animals were killed and skin and livers removed and microsomal fractions prepared according to procedures established in this laboratory (14,15). The microsomal pellets were suspended in buffer A (100 mM potassium phosphate, pH 7.4 containing 10 mM dithiothreitol, 10 mM EDTA, and 20% v/v glycerol) and frozen at -170°C under nitrogen.

Estimation of Microsomal Enzymes

For the determination of microsomal enzyme activities and BP metabolism, the frozen pellets were slowly thawed in an ice bucket (within 4-6 days of tissue preparation) and used as the enzyme source. Enzyme activities were stable for at least 3 weeks under these storage conditions. Aryl hydrocarbon hydroxylase (AHH), 7-ethoxyresorufin O-de-ethylase (ERD), 7-ethoxycoumarin O-de-ethylase (ECD) and benzphetamine N-demethylase (BPD) activities were determined as described earlier (13). NADPH-cytochrome P-450 reductase and NADH-ferricyanide reductase activities were estimated by the method of Comai and Gaylor (16) Cytochrome P-450 and cytochrome b_5 levels were determined according to Omura and Sato (17). Protein was estimated according to the method of Lowry _et al._ (18) using bovine serum alumin as the reference standard.

In Vitro BP Metabolism by Liver and Skin Microsomes

The incubation mixture in a final volume of 1.0 ml contained 0.5 mg of skin or liver microsomal protein, 0.10 mmol of phosphate buffer, pH 7.4, 3 µmol of $MgCl_2$, and 1.3 µmol of NADPH. The reaction was initiated by the addition of 80 nmol of (^{14}C)BP in 40 µl of acetone. The samples were incubated for 10 min (liver) or 30 min (skin) in the dark at 37°C in a Dubnoff metabolic shaker. The reaction was terminated by the addition of 1 ml of cold acetone followed by 2 ml of ethyl acetate. The mixture was vortexed for 1 min to extract BP metabolites as well as unreacted BP into the organic phase. The organic and aqueous phases were separated by centrifugation at 1500 rpm for 10 min. The organic phase was dried under a stream of nitrogen and dissolved in 100 ul of methanol for HPLC analysis. All operations were performed under yellow subdued light.

HPLC Analysis of Formation of BP Metabolites

A Waters Associates model 204 liquid chromatograph, fitted with a Waters Associates radial compression system equipped with a C_{18}-Radial Pak column was used for the analysis of radiolabeled metabolite mixtures of BP. Identification of metabolites was based on reference standards. For the separation of BP metabolites the column was eluted at ambient temperature with 25 min linear gradient of 65 to 100% methanol:water at a solvent flow rate of 0.6 ml per min. The eluates were monitored at 254 nm, fractions of approximately 0.2 ml were collected dropwise, and the radioactivity of each fraction was determined on a Packard TriCarb 460 CD liquid scintillation spectrometer.

RESULTS

Effect of Arenes and Nitroarenes on Cutaneous Monooxygenase

The effect of a single topical application of several pairs of arenes and nitroarenes to neonatal rats on skin microsomal cytochrome P-450 dependent AHH, ECD and ERD activities is shown in Table 1. Among the arenes tested, fluoranthene and benzo(ghi)perylene were found to be moderately effective in inducing the activities of AHH (1.7 and 2.7-fold respectively), ERD (1.5 and 3.1-fold respectively), and ECD (1.4 and 2.4-fold respectively). With the exception of 1-nitropyrene, a single topical application of all studied nitroarenes resulted in a highly significant induction of AHH (3.2-14.6-fold), ERD (4.6-7.9-fold) and ECD (3.9-9.7-fold) activities. Induction by 1-nitropyrene was modest (1.5-1.6-fold) but was significant as compared to control or pyrene treatment. The induction of each enzyme by nitroarenes was highly significant when compared to control or their corresponding parent arenes. The inducibility of cutaneous monooxygenase activities after topical application to animals of different nitroarenes was in the following order: Nitrofluoranthene mixture > nitroperylene mixture > 3-nitrofluoranthene > nitrobenzo(ghi)perylene mixture > nitropyrenes mixture > 1-nitropyrene. Cutaneous NADPH-cytochrome P-450 reductase and NADH-ferricyanide reductase activities were not altered by treatment of animals with either arenes or nitroarenes (data not shown).

TABLE 1

EFFECTS OF TOPICAL APPLICATION OF NITROARENES (NA) AND THEIR PARENT ARENES (PA) ON SKIN MONOOXYGENASES

Test Compound	SKIN MICROSOMAL MONOOXYGENASE		
	AHH	ERD % of control values*	ECD
Control	100 ± 9	100 ± 4	100 ± 5
Pyrene (PA)	107 ± 17	118 ± 5	131 ± 7
1-Nitropyrene (NA)	150 ± 9[a,b]	155 ± 9[a,b]	162 ± 6[a,b]
Nitropyrenes mixture (NA)	342 ± 28[a,b]	459 ± 32[a,b]	390 ± 23[a,b]
Benzo(ghi)perylene (PA)	175 ± 9[a]	155 ± 9[a]	140 ± 9[a]
Nitrobenzo(ghi)perylene mixture (NA)	700 ± 32[a,b]	522 ± 46[a,b]	409 ± 27[a,b]
Fluoranthene (PA)	275 ± 29[a]	315 ± 26[a]	236 ± 20[a]
3-Nitrofluoranthene (NA)	942 ± 48[a,b]	636 ± 52[a,b]	721 ± 49[a,b]
Nitrofluoranthene mixture (NA)	1458 ± 126[a,b]	713 ± 48[a,b]	971 ± 49[a,b]
Perylene (PA)	158 ± 17[a]	117 ± 5	125 ± 7
Nitroperylene mixture (NA)	1208 ± 96[a,b]	791 ± 54[a,b]	868 ± 49[a,b]
3-MC	1525 ± 126[a]	354 ± 26[a]	1136 ± 65[a]

*Data represent mean ± S.E. of 4 determinations with control values expressed as 100. Control enzyme activities are AHH, 1.2; ERD, 0.64; ECD 2.10 pmol product/min/mg protein. For each determination 4 neonatal rats (4 day old) were treated with a single topical application of PA or NA (10 mg/kg) dissolved in THF:DMSO (1:1). The treatment of vehicle alone had no detectable effect on enzyme activites. Animals were killed 24 hr after the treatment, microsomal fractions prepared and enzyme activities determined.

[a]Statistically significant from control $p < 0.05$.

[b]Statistically significant from the corresponding PA, $p < 0.05$.

TABLE 2

EFFECTS OF TOPICAL APPLICATION OF NITROARENES (NA) AND THEIR PARENT ARENES (PA) ON LIVER MONOOXYGENASES

Test compound	LIVER MICROSOMAL MONOOXYGENASE			
	AHH	ERD	ECD	BPD
	% control values*			
Control	100 ± 9	100 ± 4	100 ± 4	100 ± 6
Pyrene (PA)	119 ± 10	307 ± 18[a]	120 ± 6	106 ± 10
1-Nitropyrene (NA)	143 ± 19[a]	480 ± 24[a,b]	125 ± 5[a]	89 ± 8
Nitropyrene mixture (NA)	390 ± 33[a,b]	1793 ± 45[a,b]	145 ± 8[a]	98 ± 10
Benzo(ghi)perylene (PA)	281 ± 28[a]	270 ± 6[a]	104 ± 4	90 ± 8
Nitrobenzo(ghi)-perylene mixture (NA)	909 ± 81[a,b]	2709 ± 140[a,b]	170 ± 6[a,b]	108 ± 6
Fluoranthene (PA)	289 ± 28[a]	1117 ± 86[a]	140 ± 8[a]	101 ± 8
3-Nitrofluoranthene (NA)	814 ± 86[a,b]	4167 ± 270[a,b]	243 ± 12[a,b]	109 ± 8
Nitrofluoranthene mixture (NA)	924 ± 76[a,b]	4193 ± 317[a,b]	226 ± 9[a,b]	102 ± 5
Perylene (PA)	119 ± 10	118 ± 8	100 ± 8	102 ± 5
Nitroperylene mixture (NA)	486 ± 43[a,b]	3271 ± 296[a,b]	171 ± 11[a,b]	99 ± 5
3-MC	1271 ± 108[a]	3825 ± 316[a]	273 ± 20[a]	108 ± 8

*Values represent mean ± S.E. of 4 determinations with control values expressed as 100. Control enzyme activities are AHH, 21.0; ERD, 8.3; ECD, 524; BPD, 0.90 pmol product/min/mg protein. For treatment schedule, and other details see Table 1.
[a]Statistically significant from control p 0.05 as compared by paired student's 't' test.
[b]Statistically significant from the corresponding PA, $p < 0.05$.

TABLE 3

EFFECTS OF TOPICAL APPLICATION OF NITROARENES (NA) AND THEIR PARENT ARENES (PA) ON SKIN MICROSOMAL METABOLISM OF BENZO(A)PYRENE

Test Compound	Metabolites (pmol/min/mg protein)				
	Total Diol	Total Phenols	Total Quinones	Total Metabolites	Percent of Control
Control	0.18	0.61	1.06	1.85	100
Pyrene (PA)	0.21	0.65	0.95	1.81	98
1-Nitropyrene (NA)	0.28[a,b]	0.75[a]	1.42[a]	2.55[a]	138[a]
Nitropyrenes mixture (NA)	0.64[a,b]	2.48[a,b]	2.34[a,b]	5.46[a,b]	295[a,b]
Benzo(ghi)-perylene (PA)	0.28[a]	1.13[a]	1.48[a]	2.89[a]	156[a]
Nitrobenzo(ghi)-perylene mixture (PA)	1.43[a,b]	5.12[a,b]	5.08[a,b]	11.63[a,b]	628[a,b]
Fluoranthene (PA)	0.44[a]	1.66[a]	2.12[a]	4.22[a]	228[a]
3-Nitrofluoranthene (NA)	1.57[a,b]	5.50[a,b]	7.24[a,b]	14.31[a,b]	773[a,b]
Nitrofluoranthene mixture (NA)	2.79[a,b]	7.44[a,b]	11.86[a,b]	22.09[a,b]	1194[a,b]
Perylene (PA)	0.27[a]	0.75[a]	1.21	2.23	121
Nitroperylene mixture (NA)	2.30[a,b]	7.35[a,b]	9.20[a,b]	18.85[a,b]	1019[a,b]
3-MC	3.01[a]	8.15[a]	13.46[a]	24.61[a]	1330[a]

Eighty nmol [^{14}C]BP (specific activity = 58.5 mCi/mmol) was incubated with skin microsomes (0.5 mg protein) in a final volume of 1.0 ml. Numbers shown are average values of triplicate samples which agree within 10% of the indicated values. For treatment and other details see Table 1.

[a]Statistically significant from control $p<0.05$.

[b]Statistically significant from PA, $p<0.05$.

TABLE 4

EFFECTS OF TOPICAL APPLICATION OF NITROARENES (NA) AND THEIR PARENT ARENES (PA) ON HEPATIC MICROSOMAL METABOLISM OF BENZO(A)PYRENE

Test Compound	Metabolites (pmol/min/mg protein)				
	Total Diol	Total Phenols	Total Quinones	Total Metabolites	Percent of Control
Control	8.7	22.3	8.0	39.0	100
Pyrene (PA)	9.4	23.4	8.4	41.2	106
1-Nitropyrene (NA)	11.5[a,b]	28.3[a,b]	9.5[a,b]	49.3[a,b]	126[a,b]
Nitropyrenes mixture (NA)	23.8[a,b]	77.6[a,b]	19.8[a,b]	121.2[a,b]	311[a,b]
Benzo(ghi)-perylene (PA)	25.0[a]	53.3[a]	14.5[a]	92.8[a]	238[a]
Nitrobenzo(ghi)-perylene mixture (PA)	69.4[a,b]	169.8[a,b]	64.4[a,b]	303.6[a,b]	778[a,b]
Fluoranthene (PA)	26.7[a]	60.1[a]	20.1[a]	106.9[a]	274[a]
3-Nitrofluoranthene (NA)	60.2[a,b]	159.1[a,b]	54.2[a,b]	273.5[a,b]	701[a,b]
Nitrofluoranthene mixture (NA)	68.7[a,b]	173.7[a,b]	61.9[a,b]	304.3[a,b]	780[a,b]
Perylene (PA)	8.9	23.1	7.8	39.8	102
Nitroperylene mixture (NA)	30.4[a,b]	89.2[a,b]	34.4[a,b]	154.0[a,b]	395[a,b]
3-MC	111.9[a]	246.1[a]	85.7[a]	423.7[a]	1086[a]

Note: For footnotes see Table 3.

Effect of Arenes and Nitroarenes on Hepatic Monooxygenases

The effect of a single topical application of several pairs of arenes and nitroarenes to animals on hepatic AHH, ERD, ECD, and BPD activities is shown in Table 2. Among the parent arenes fluoranthene and benzo(ghi)perylene were found to be effective in inducing the activities of AHH (2.7 to 2.8 fold) and ERD (2.7-11.2 fold). A single topical treatment of all nitroarenes resulted in significant induction of AHH (1.4-9.2 fold), ERD (4.8-41.9 fold), and ECD (1.3-2.4 fold) activities as compared to control or pyrene treatment. Consistent with the effects on skin enzyme activities a single topical application of nitroarenes resulted in significantly increased enzyme activities as compared to control or their corresponding parent arene. The inducibility of hepatic monooxygenases by different nitroarenes was in the following order: nitrofluoranthene mixture > 3-nitrofluoranthene > nitroperylene mixture > nitrobenzo(ghi)perylene mixture > nitropyrenes mixture > 1-nitropyrene. However, the activities of hepatic BPD, NADPH-cytochrome P-450 reductase, (data not shown), and NADH-ferricyanide reductase (data not shown), remained unchanged following the treatment of animals with either arenes or nitroarenes.

Effect of Arenes and Nitroarenes on Cytochrome P-450 and b_5

The effect of a single topical application of different pairs of arenes and nitroarenes to animals on hepatic cytochrome P-450 levels and on the absorption maximum of the cytochrome was studied. Cytochrome P-450 levels were not affected by any treatment. However, a shift of approximately 1 nm to the blue region in the absorption maximum of cytochrome P-450 was observed in animals treated with nitroarenes. This shift was not evident in the case of 1-nitropyrene (data not shown). Hepatic and cutaneous cytochrome b_5 levels remained unchanged following the treatment with arenes and nitroarenes (data not shown).

Effect of arenes and nitroarenes on in vitro BP metabolism

The metabolic profile of BP by cutaneous and hepatic microsomes obtained from animals treated with arenes and nitroarenes is shown in Tables 3 and 4. Consistent with the effects on AHH activity shown in Tables 1 and 2 treatment of animals with each nitroarene resulted in greater metabolism of BP as compared to controls or the parent arenes. Among the nitroarenes, 1-nitropyrene, was the weakest and nitrofluoranthene mixture was the most potent inducer of cutaneous and hepatic BP metabolism.

The induction potency of arenes and nitroarenes

The relative potency of various nitroarenes and their parent arenes as inducers of AHH in the skin and liver is summarized in table 5. Among the arenes, fluoranthene was moderately active in inducing enzyme activity in both skin and liver. With the exception of 1-nitropyrene all the nitroarenes tested were potent inducers of cutaneous and hepatic monooxygenase activities in neonatal rats.

TABLE 5

RELATIVE INDUCTION EFFECTS OF NITROARENES AND THEIR PARENT ARENES ON SKIN AND LIVER AHH ACTIVITY.

% Increase	Skin	Liver
Low (<100%)	Pyrene Perylene 1-Nitropyrene Benzo(ghi)perylene	Pyrene Perylene 1-Nitropyrene
Moderate (100-200%)	Fluoranthene	Benzo(ghi)perylene Fluoranthene
High (>200%)	Nitropyrene mixture Nitrobenzo(ghi)perylene mixture 3-Nitrofluoranthene Nitrofluoranthene mixture Nitroperylene mixture	Nitropyrene mixture Nitrobenzo(ghi)perylene mixture 3-Nitrofluoranthene Nitrofluoranthene mixture Nitroperylene mixture

DISCUSSION

Nitrated polycyclic aromatic hydrocarbons have recently been recognized as potent toxic and carcinogenic chemicals. Long term exposure to nitropyrene may have adverse effects on human health (1,19). In an effort to define the mode of toxicity of the nitropyrenes, we have recently studied the effects of topical application of a mixture of nitropyrenes, in the proportion of their occurrence in the environment, on hepatic and cutaneous monooxygenase activities (13). The effects were compared with the non-nitrated parent compound, pyrene, which is nonmutagenic and non-carcinogenic (20). Our studies have shown that a single topical application of the nitropyrenes mixture to rats resulted in a several-fold induction of hepatic and cutaneous AHH, ERD and ECD activities. We have now extended these studies to several pairs of nitrated arenes and their parent arenes. In each case the nitrated arenes had higher inducing effects than the corresponding parent arenes.

The deethylation of ethoxyresorufin (or ERD) is highly specific for cytochrome P-448 (21) whereas BP hydroxylation (or AHH) is predominately catalyzed by cytochrome P-448 and deethylation of 7-ethoxycoumarin (or ECD) is metabolized by both cytochrome P-448 and cytochrome P-450 (22,23). Cytochrome P-448 is selectively induced by 3-MC type of chemical inducers (23). The N-demethylation of aminopyrine or benzphetamine (or BPD) is catalyzed by cytochrome P-450 which is induced by phenobarbital type of chemical inducers (24). Our results demonstrating high inducibility of hepatic and cutaneous AHH and ERD activities and BP metabolism without any effects on cytochrome b_5 levels, NADPH cytochrome P-450 reductase and BPD activities following treatment with nitrated arenes suggest that nitropyrenes behave like 3-MC type of inducers of monooxygenase activities. This is further supported by the lack of effect on hepatic epoxide hydrolase activity (data not shown). Prior studies have shown that 3-MC has no effect whereas phenobarbital results in the induction of hepatic epoxide hydrolase activity in rodents (25).

The absorption maximum of the CO-difference spectra from control neonatal rat liver was at 452 nm which is consistent with that reported previously (26). In the present study a shift of 1 nm to the blue region was observed in the absorption maximum of the hepatic hemeprotein following treatment with the nitroarenes (with the exception of 1-nitropyrene). Since the compounds were applied to animals topically for only 24 hours, it is possible that prolonged application or parenteral administration may result in greater shifts in the wavelength absorption maximum and increased levels of the hemoprotein as is observed after prolonged administration of 3-MC. These studies have not been performed.

Another interesting finding of our studies is the observation that topically applied nitroarenes can produce effects on hepatic monooxygenase activities. This suggests that nitropyrenes present in the ambient environment can penetrate the cutaneous barrier and gain entry into the circulation to produce effects on hepatic drug metabolizing enzyme activities and thereby may produce toxicity including carcinogenicity in cutaneous as well as extracutaneous tissues.

In summary our results indicate that nitrated arenes are inducers of the 3-MC type and they possess higher inducing potential as compared to parent arenes. Of special interest is the effect of mixture of nitropyrenes. This mixture consisted of dinitropyrenes and 1-nitropyrene with a ratio of 7:1. Since 1-nitropyrene is a weak inducer by itself the observed enzyme induction effects of the mixture of nitropyrene suggest that probably dinitropyrenes are inducers of enzyme activities. We prepared another mixture of nitropyrenes which was found to possess tri and tetranitropyrenes. This mixture was found to be ineffective in inducing the enzyme activities in our system (data not shown). Further studies are needed to define the relative inducibility of individual anthropogenic nitropyrenes known to be present in mixtures of nitropyrenes. Our findings further emphasize that airborne nitrated polynuclear aromatic hydrocarbons may be potent environmental toxins and that the skin is a potential target tissue for their effects.

ACKNOWLEDGMENTS

Supported by NIH Grants ES-1900, CA 38028 and AM 34368 and funds from the Veterans Administration. Thanks are due to Ms. Sandra Evans for preparing the manuscript.

REFERENCES

1. Pitts, J.N. Jr., Van Cauwenberghe, K.A., Grosjean, D., Schmid, J.P., Fitz, D.R., Belser, W.L. Jr., Knudson, G.B., and Hynds, P.M. (1978): Atmospheric reactions of polycyclic aromatic hydrocarbons: Facile formation of mutagenic nitro derivatives. Science, 202:515-519.
2. Tokiwa, H., Nakagawa, R., Morita, K., and Ohnishi, Y. (1981): Mutagenicity of nitro derivatives induced by exposure to aromatic compounds to nitrogen dioxide. Mutation Res., 85: 195-205.
3. Hanson, R.L., Henderson, T.R., Hobbs, C.H., Clark, C.R., Carptenter, R.L., and Dutcher, J.S. (1983): Detection of nitroaromatic compounds on coal combustion particles. J. Toxicol. Environ. Hlth., 11: 971-980.

4. Schuetzle, D., Riley, J.L., Prater, T.J., Harvey, T.M., and Hunt, D.F. (1982): Analysis of nitrated polycyclic aromatic hydrocarbons in diesel particulates. Anal. Chem., 54, 265-271.
5. McCoy, E.C., and Rosenkranz, H.S. (1982): Cigarette smoking may yield nitroarenes. Cancer Lett., 15: 9-13.
6. Yu, W.C., Fine, D.H., Chiu, K.S., and Biermann, K. (1984): Determination of nitrated polycyclic aromatic hydrocarbons in diesel particulates by gas-chromatography with chemiluminescent detection. Anal. Chem., 56: 1158-1162.
7. Ohgaki, H., Matsukura, N., Morino, K., Kawachi, T., Sugimura, T., Morita, K., Tokiwa, H., and Hirota, T. (1982): Carcinogenicity in rats of the mutagenic compounds 1-nitropyrene and 3-nitrofluoranthene. Cancer Lett., 15: 1-7.
8. El-Bayoumy, K., Hecht, S., Sackl, T., and Stoner, G.D. (1984): Tumorigenicity and metabolism of 1-nitropyrene in A/J mice. Carcinogenesis, 5: 1449-1452.
9. Lee, I.P., Suzuki, K., Lee, S.D., and Dixon, R.L. (1980): Aryl hydrocarbon hydroxylase induction in rat lung, liver, and male reproductive organs following inhalation exposure to diesel emission. Toxicol. Appl. Pharmacol., 52: 181-184.
10. Mukhtar, H., Link, C.M., Cherniack, E., Kushner, D.M., and Bickers, D.R. (1982): Effect of topical application of defined constituents of coal tar on skin and liver aryl hydrocarbon hydroxylase and 7-ethoxycoumarin deethylase activities. Toxicol. Appl. Pharmacol., 64 : 541-549.
11. Mukhtar, H., and Bickers, D.R. (1983): Age-related changes in the benzo(a)pyrene metaoblism and epoxide metabolizing enzyme activities in rat skin. Drug Metab. Dispos., 11: 562-567.
12. Uotila, P., Pelkonen, O., and Cohen, G.M. (1977): The effects of cigarette smoke on the metabolism of ^{3}H benzo(a)pyrene by rat lung microsomes. Cancer Res., 37: 2156-2161.
13. Asokan, P., Das, M., Rosenkranz, H.S., Bickers, D.R., and Mukhtar, H. (1985): Topically applied nitropyrenes are potent inducers of cutaneous and hepatic monooxygenases. Biochem. Biophys. Res. Commun., 129: 134-140.
14. Kloetzel, M.C., King, W., and Menkeys, T.J. (1956): Fluranthene derivative. III. 2-Nitrofluranthene and 2-amino-fluranthene. J. Amer. Chem. Soc., 78 : 1165-1168.
15. Mukhtar, H., and Bickers, D.R. (1981): Drug metabolism in skin. Comparative activity of the mixed-function oxidases, epoxide hydratase and glutathione-S-transferase in liver and skin of neonatal rat. Drug Metab. Dispos., 9 : 311-314.

16. Comai, K., and Gaylor, J.L. (1973): Existence and separation of three forms of cytochrome P-450 from rat liver microsomes. J. Biol. Chem., 248:4947-4955.
17. Omura, T., and Sato, R. (1964): The carbon monoxide binding pigment of liver microsomes. Evidence for its hemoprotein nature. J. Biol. Chem., 239: 2370-2378.
18. Lowry, O.H., Rosebrough, N.J., Farr, A.L., and Randall, R.J. (1951): Protein measurement with Folin phenol reagent. J. Biol. Chem., 193:265-275.
19. Rosenkranz, H.S., McCoy, E.C., Sanders, D.R., Butler, M., Kiriazides, D.K., and Mermelstein, R. (1980): Nitropyrenes: Isolation, identification, and reduction of mutagenic impurities in carbon black and toners. Science, 209: 1039-1043.
20. Wood, A.W., Levin, W., Chang, R.L., Huang, M.T., Ryan, D.E., Thomas, P.E., Lehr, R.E., Kumar, S., Koreeda, M., Akagi, H., Ittah, Y., Dansette, P., Yagi, H., Jerina, D.M., and Conney, A.H. (1980): Mutagenicity and tumor-initiating activity of cyclopenta(c,d)pyrene and structurally related compounds. Cancer Res., 40: 642-649.
21. Burke, M.D., and Mayer, R.T. (1975): Inherent specificities of purified cytochromes P-450 and P-448 towards biphenyl hydroxylation and ethoxyresorufin deethylation. Drug Metab. Dispos., 3: 245-253.
22. Ioannides, C., Lum, P.Y., and Parke, D.V. (1984): Cytochrome P-448 and the activation of toxic chemicals and carcinogens. Xenobiotica, 14: 119-137.
23. Weibel, F.J., Selkirk, J.K., Gelboin, H.V., Haugen, D.A., Van der Hoeven, T.A., and Coon, M.J. (1975): Position-specific oxygenation of benzo(a)pyrene by different forms of purified cytochrome P-450 from rabbit liver. Proc. Natl. Acad. Sci. USA, 72: 3917-3920.
24. Ryan, D.E., Iida, S., Wood, A.W., Thomas, P.E., Liebert, C.S., and Levin, W. (1984): Characterization of three highly purified cytochrome P-450 from hepatic microsomes of adult male rats. J. Biol. Chem., 259: 1239-1250.
25. Bresnick, E., Mukhtar, H., Stoming, T.A., Dansette, P.M., and Jerina, D.M. (1977): Effect of phenobarbital and 3-methycholanthrene administration on epoxide hydrolase levels in liver microsomes. Biochem. Pharmacol., 26: 891-892.
26. Mukhtar, H., and Bickers, D.R. (1982): Evidence that coal tar is a mixed inducer of microsomal drug metabolizing enzymes. Tox. Lett., 11: 211-227.

DIOXIRANES 3. ACTIVATION OF POLYCYCLIC AROMATIC HYDROCARBONS BY REACTION WITH DIMETHYLDIOXIRANE[1]

ROBERT W. MURRAY*, RAMASUBBU JEYARAMAN
Department of Chemistry, University of Missouri-St. Louis,
St. Louis, Missouri 63121, USA

INTRODUCTION

Polycyclic aromatic hydrocarbon (PAH)-caused carcinogenesis is known to require metabolic activation, *i.e.*, oxidation, of the PAH (2-7). The metabolites include arene oxides, phenols, diols, and diol epoxides among others. Of these metabolites the arene oxide functionality appears to be the chemical moiety which is essential for binding of the activated PAH to important biopolymers including DNA and RNA (8).

There have now been a large number of studies describing the PAH content of various atmospheres (9). It is also the case that airborn PAH undergo a variety of chemical transformations and, in particular, oxidation reactions (10,11). Of great significance to environmental health is the observation that oxygenated fractions of atmospheric samples can be carcinogenic (12). Indeed particulate-absorbed PAH are believed to be major contributors to the observed higher death rates caused by lung cancer in urban areas as compared to rural areas (13).

We have been attempting to identify mechanisms by which PAH can be converted to oxidation products in the atmosphere. In particular we are focusing on those oxidation processes which might 'activate' PAH, that is, produce oxidation products similar to those known to be involved in the *in vivo* metabolic activation to carcinogens. Of interest in this connection is the report by Pitts and coworkers that benzo[a]pyrene can be converted to its K-region oxide by ozone (14). Our work has concentrated on two reactive species known to be produced in ozone-olefin reactions, namely, carbonyl oxide, 1, and dioxiranes, 2. We have been able to show that a carbonyl oxide, produced *via* ozonolysis of an olefin-bearing PAH absorbed on silica gel as a model particulate, can react with the PAH to give the K-region oxide (15). We had earlier described the ability of carbonyl oxides to form arene oxides in solution processes (16,17).

The involvement of carbonyl oxides in the ozonolysis process has led to a considerable literature describing

their chemistry (18). Dioxiranes, on the other hand, are relatively rare. The parent dioxirane has been shown to be formed by gas phase ozonolysis of ethylene (19-21). The patent literature contains a description of the synthesis of two halogen-substituted dioxiranes (22). Dioxiranes have also been invoked as intermediates in certain solution phase reactions of ketones with peracids (23-29). We have recently described a method of preparation of solutions of pure methyldioxiranes (30). This achievement has permitted us to study the spectroscopic properties and chemistry of dimethyldioxirane in particular. We had earlier described the reaction of <u>in situ</u> generated dimethyldioxirane with PAH (29). In most cases these reactions produce arene oxides.

In the current work we describe the reaction of dimethyldioxirane in acetone solution with eight PAH. All of the PAH chosen for study have been found to be present in airborne particulate matter (9). In addition 5 of the 8 have been found to be at least weakly carcinogenic in small animals (31). The structure types employed also give us an opportunity to determine the relative reactivity of defined regions of the PAH.

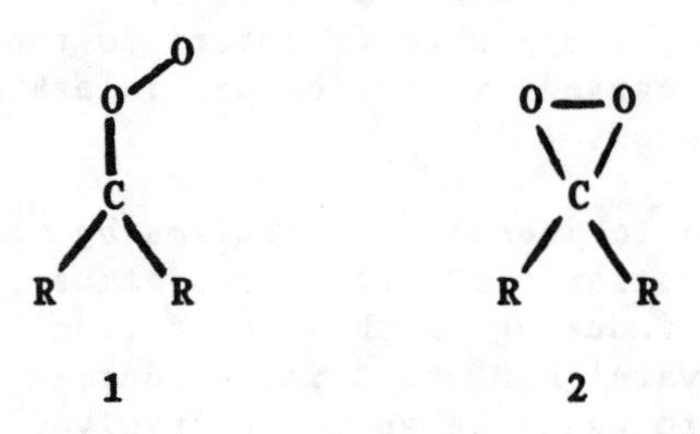

MATERIALS AND METHODS

Materials

Oxone (DuPont Co., $2KHSO_5 . KHSO_4.K_2SO_4$), acetone (reagent grade), pyrene, chrysene, 4H-cyclopenta[*def*]-phenanthrene, dibenz[*ah*]- anthracene, and benz[e]pyrene were purchased from Aldrich Chemical Co., Milwaukee, Wisconsin. Methylene Chloride, methanol, hexanes, $MgSO_4$, and anhydrous Na_2SO_4 were obtained from Fisher Scientific Co., Fairlawn, N.J. All solvents were distilled prior to use. Benzo-[b]chrysene and benz[a]pyrene were obtained from Columbia

Organic Chemicals Co., Columbia, So. Carolina. Benz[a]-anthracene was purchased from Eastman Organic Chemical, Rochester, N.Y. Preparative TLC was done on Analtech 1000 micron, pre-coated, silica gel plates. HPLC analyses were performed using a cyanosilica column (CN-10) (50 cm x 8 mm) obtained from Varian aerograph, Palo Alto, California.

Instrumentation

NMR spectra were recorded on Varian T-60 (^{1}H) or JEOL XL-100 (^{1}H and ^{13}C) NMR spectrometers in a 1:1 solution of $CDCl_3$ and acetone-d_6. Chemical shifts are reported relative to internal $(CH_3)_4Si$. Mass spectra were recorded at 70 eV ionizing voltage on an AEI MS 1201-B spectrometer. Melting points were recorded on a Thomas-Hoover capillary melting point apparatus and are uncorrected. HPLC analyses were done on a Varian Model 5000 gradient liquid chromatograph using a CN-10 (8 mm x 50 cm) column. The solvent composition was 50% CH_2Cl_2:50% hexane (5 min) and then to 100% CH_2Cl_2 in 10 min. flow 4 ml/min. A variable wavelength UV detector set at 254 nm was used for detection.

Methods

Oxidation of Polycyclic Aromatic Hydrocarbons

Pyrene. Pyrene (0.10 g, 0.5 mmol) was dissolved in a previously dried (Na_2SO_4) solution of dimethyldioxirane (30) in acetone (50 ml, 0.07 M) and CH_2Cl_2 (50 ml). The solution was stirred at room temperature for two days. The progress of the oxidation was followed by TLC. The solvent was removed on the rotary evaporator and the residue dissolved in CH_2Cl_2 (3-5ml) and then subjected to TLC using CH_2Cl_2-CH_3OH (99:1) as eluent. The band at R_f 0.55 was extracted with CH_2Cl_2-CH_3OH (95:5) (50 ml x 3) and the solvent evaporated (cold) to give solid pyrene-4,5-oxide in 35% yield. ^{1}H NMR ($CDCl_3$) δ4.5 (s, 2H), δ7.3-8.2 (m). This solid was recrystallized from CH_2Cl_2-hexane to give crystals with MS 218 (parent) and m.p. 190-195°C, lit (32) m.p. 192-200°C.

Chrysene. Chrysene (0.05 g; 0.21 mmol) was dissolved in an acetone solution of dimethyldioxirane (50 ml, 0.08 M) and the solution stirred at room temperature while protected from light. The progress of the reaction was followed by TLC. After 18 hr the solvent was evaporated and the residue analyzed by TLC using CH_2Cl_2. The band corresponding to chrysene oxide was removed. This material had ^{1}H NMR ($CDCl_3$) δ4.1 (d), 4.7 (d), 7.3-8.4 (m). Yield was 29%. HPLC retention time (10% CH_2Cl_2, 90% hexane, flow 1 ml/min on a 30 cm

IBM CN-10 column) was 20 min. This chromatographic behavior was the same as that of an authentic sample of oxide prepared by the method of Hamilton et al (33).

4H-Cyclopenta[def]phenanthrene. A solution of 4H-cyclopenta[def]phenanthrene (0.20 g; 1.05 mmol) in CH_2Cl_2 (20 ml) was added to a solution of dimethyldioxirane in acetone (25 ml; ca. 0.06 M) and CH_2Cl_2 (25 ml). The reaction was stirred at room temperature for 24 hr and in the absence of light. The solvent was then evaporated and the residue analyzed by TLC using CH_2Cl_2. Four fractions were isolated. The major fraction, the 8,9-dione, was obtained as orange-red crystals (ca. 50% yield) m.p. 257-259°C, lit m.p. 256°C (34). This material had ^1H NMR ($CDCl_3$) δ 4.1 (s), 7.1-8.0 (m). The fraction with R_f = 0.8 using CH_2Cl_2:CH_3OH (98.2) was pale yellow in color and provided a few crystals from CH_2Cl_2-hexane (0°C) with m.p. 166-169°C, lit m.p. 169.5-170°C (34) for the 4-one compound. This material had ^1H NMR ($CDCl_3$) δ7.2-8.4 (m). The third TLC fraction had R_f = 0.5 and is believed to be the K-region oxide. This material had ^1H NMR ($CDCl_3$) δ4.4 (sharp singlet, oxirane protons) and δ4.2 (CH_2). The compound had M.S. 206 (parent of oxide). This material is rapidly converted to a brown substance which was also present as the fourth fraction in the TLC analysis. The brown material was not further characterized.

Benz[a]anthracene. A freshly-prepared solution of dimethyldioxirane in acetone (30 ml, ca. 0.08 M) was mixed with 100 ml of CH_2Cl_2, dried with $MgSO_4$, filtered and then combined with a solution of benz[a]anthracene (0.10 g; 0.44 mmol) in CH_2Cl_2 (30 ml). The solution was kept at room temperature for 4 hr at which time TLC and HPLC analysis indicated that products in addition to the K-region oxide were beginning to form. The solvent was evaporated and analyzed by TLC. The major product (65% yield) was found to be benz[a]anthracene-5,6-oxide. This material had m.p. 128-131°C, lit m.p. 130-132°C (32). The material had ^1H NMR ($CDCl_3$) δ4.44 (d), 4.58 (d) and 7.2-8.4 (m). A second yellow-colored product was isolated in 25% yield. This material had m.p. 162-165°C and ^1H NMR ($CDCl_3$) δ 7.2-8.7 (m). It is tentatively assigned the structure of benz[a]anthracene-7,12-dione, lit. m.p. 168°C (35) and yellow-colored. It is noted that benz[a]anthracene-5,6-dione has lit. m.p. 262-263°C (36) and is bright red in color.

Benzo[a]pyrene. A freshly-prepared solution of dimethyldioxirane in acetone (10 ml, ca. 0.05 M) was added to a solution of benz[a]pyrene 90.05 g; 0.20 mmol) in acetone (5 ml) at 0°C. An immediate, rapid reaction took place leading to color changes first to pale yellow, then brown, and then dark brown within a few minutes. TLC analysis of the reaction mixture indicated the presence of several components. Attempts to extract and analyze the individual fractions did not yield any crystalline materials. Instead only black pasty masses or greenish-violet intractable materials were obtained. TLC analysis performed at different stages of the reaction suggested that a material was formed, perhaps the K-region oxide, which was rapidly transformed to other materials.

Benzo[e]pyrene. Benzo[e]pyrene (0.08 g; 0.32 mmol) was dissolved in hexane (50 ml). To this solution was added a solution of dimethyldioxirane in acetone (30 ml., ca. 0.05 M diluted with 30 ml of hexane and dried). The reaction mixture was stirred at room temperature for 6 hr and the solvent evaporated. TLC analysis of the reaction mixture, using CH_2Cl_2; hexane (1:1) and also containing 1% isopropyl alcohol, indicated the presence of polar compounds (R_f = 0.11) and an arene oxide (R_f = 0.53). This latter band is assigned to an arene oxide on the basis of R_f comparison with a number of other arene oxides. Extraction of this TLC band and work-up led to the isolation of the rearrangement product, 4-hydroxybenz[e]pyrene (20% yield). This material had ^1H NMR ($CDCl_3$) 7.2-8.6 (m) and MS 268 (parent) and had m.p. 236-239°C, lit. m.p. 242-243°C (37).

Dibenz[a,h]anthracene. A fresh solution of dimethyldioxirane in acetone (30 ml, ca. 0.07 M) was mixed with CH_2Cl_2 (50 ml) and dried with Na_2SO_4. A solution of dibenz[a,h]anthracene (0.10 g; 0.36 mmol) in CH_2Cl_2 (40 ml) was then added to the dried diemthyldioxirane solution and the combined solution stirred at room temperature for 20 hr. The solvent was then evaporated and the residue analyzed by TLC using CH_2Cl_2:CH_3OH (95:5) for development. The major product was found to be dibenz[a,h]anthracene-5,6-oxide in 51% yield, m.p. 254-257°C, lit m.p. 255-258°C (38). This material had MS 294 (parent) and ^{13}C NMR (Acetone-d_6 and $CDCl_3$, dil. soln.) 56.8 and 57.0 (oxirane C's) and several absorptions at 122.6-135.1. The reaction mixture was also found to contain a compound with R_f = 0.75 which was deep yellow-colored and a second compound which was a brown, polar material and was not further characterized. The yellow-colored material is tentatively assigned the structure of the 7,14-dione.

Benzo[b]chrysene. Benzo[b]chrysene (0.045g; 0.18 mmol) was dissolved in methyl acetate (20 ml) and this solution was then added to a solution of dimethyldioxirane in acetone (20 ml, ca, 0.04 M). The combined solution was dried rapidly with Na_2SO_4 and filtered. The reaction solution was allowed to stand at room temperature for 24 hr and the solvent evaporated. The residue was analyzed by TLC, using CH_2Cl_2 for development, to give as the major product (70% yield) benzo[b]chrysene-7,12-dione which had R_f = 0.55 and m.p. 265-269°C, lit m.p. 269-270°C (39) and MS 308 (parent). The TLC analysis also indicated the presence of an arene oxide (R_f = 0.30, 20% yield) and several unidentified polar compounds (R_f = 0.01 - 0.08).

RESULTS AND DISCUSSION

Eight polycyclic aromatic hydrocarbons, 3-10, have been reacted with dimethyldioxirane at room temperature or below. The results of the oxidations are shown in Scheme I. Our intention at this phase of our studies was to determine whether oxidation of the various PAH would occur. Thus yields and, in some cases, relative distribution of products are preliminary and probably not optimal.

Pyrene, 3, is converted to its 4,5-oxide (K-region oxide) in 35% yield. Likewise chrysene, 4, is converted to the corresponding K-region oxide (5,6-oxide) in 29% yield. These results are the same as those obtained in the earlier in situ use of dimethyldioxirane (29). Oxidation of 4-H-cyclopenta[def]phenanthrene, 5, led to a number of products although the 8,9-dione (50% yield) was the major product. Traces of the 4-one, the compound resulting from oxidation of the methylene group, and the K-region oxide were also obtained. A fourth TLC fraction, a brown, intractable material was not identified further. It was observed that workup and handling of the K-region oxide appeared to convert it to this same brown material. The results do suggest that the K-region is more reactive than the methylene group toward oxidation by dimethyldioxirane.

Benz[a]anthracene, 6 is converted to the K-region oxide (5,6-oxide) in good yield (65%). A quinone was also formed (25%) which is assigned the structure of the 7,12-dione primarily on the basis of melting point (162-165°C). It is noted that the 5,6-dione is also known, but has a considerably high melting point (262-263°) (36) and is bright red in color as opposed to the yellow color of the material

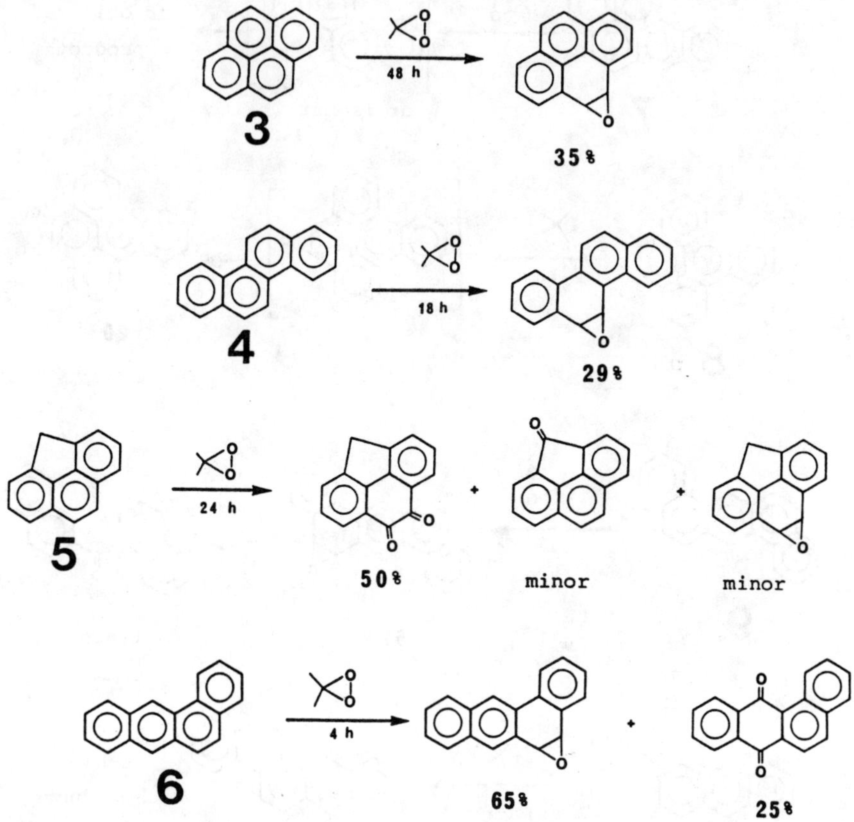

3

48 h

35%

4

18 h

29%

5

24 h

50%

minor

minor

6

4 h

65%

25%

SCHEME I

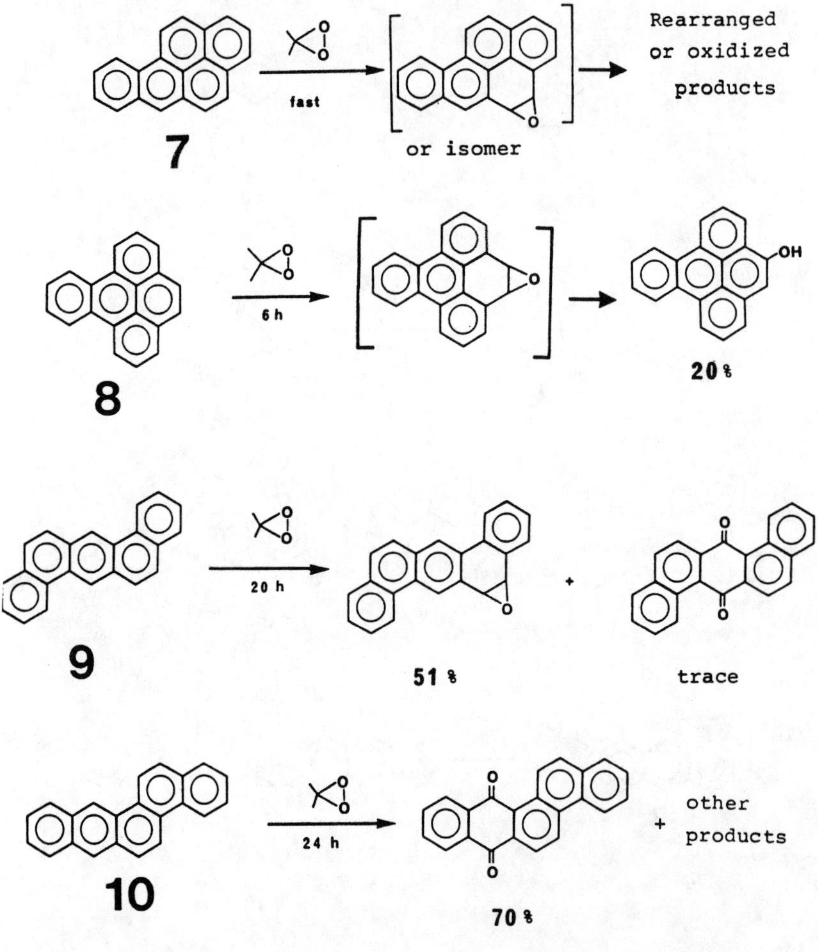

Rearranged or oxidized products

fast

7

or isomer

6 h

8

OH

20 %

20 h

9

51 %

trace

24 h

10

70 %

+ other products

SCHEME I

produced in this oxidation. Here again the K-region is found to be more reactive than other possible reaction sites, in particular the L-region. Benz[a]pyrene, 7 was found to be extremely reactive toward dimethyldioxirane. The reaction was accompanied by a series of color changes suggesting that initially produced materials are further transformed under the conditions used. No identifiable materials could be isolated from the reaction mixture. It appears that it will be necessary to run this reaction at lower temperature and with a reduced amount of dimethyldioxirane. Benzo[e]pyrene, 8, was oxidized at room temperature and this led to the isolation of a product determined to be the K-region alcohol 4-hydroxy-benzo[e]pyrene. TLC analysis of the reaction mixture suggests that this alcohol may be formed from the K-region oxide upon workup, that is, the analysis indicates the presence of a material with an R_f value associated with a number of K-region oxides. Dibenz[a,h]anthracene, 9, is conveniently converted in 51% yield, to the K-region oxide. The reaction mixture also contained a yellow-colored material which is believed to be the 7,14-dione. Finally, benzo[b]chrysene, 10 was oxidized in high yield (70%) to the 7,12-dione. The reaction mixture also contained smaller amounts of several other compounds perhaps including an arene oxide. The oxidation of 10 stands in sharp contract to that of the related compounds, 6 and 9 where the K-region oxide is the major product and illustrates the influence of subtle structural differences on the course of the oxidation.

The oxidation of PAH by dimethyldioxirane is significant in several ways. These reactions may be of synthetic utility particularly where K-region oxides and/or diones are desired products. Secondly, the fact that dioxirane has been shown to be a product of gas phase ozonolysis of ethylene (19-21) combined with the observation that polluted atmospheres frequently contain high concentrations of ozone and particulate - absorbed PAH suggests that similar oxidations could occur in such atmospheres. This chemistry could explain in part, the observation that PAH are partially 'activated' in polluted atmospheres in the absence of metabolic processes (10-13).

ACKNOWLEDGEMENTS

We gratefully acknowledge support of this work by the National Institutes of Health through grant no. R01 ESO 1984.

REFERENCES

1. Paper 2 in this series is reference 30.
2. Harvey, R. G. (1981): Activated metabolities of carcinogenic hydrocarbons. Acc. Chem. Res., 14:218-226.
3. Harvey, R. G. (1982): Polycyclic hydrocarbons and Cancer. Am. Scientist, 70:386-393.
4. Gelboin, H. V. (1980): Benzo[a]pyrene metabolism, activation and carbinogenesis: Role and regulation of the mixed function oxidases and related enzymes. Physiol. Rev., 60:1107-1166.
5. Grover, P. L., Hewer, A., and Sims, P. (1971): Epoxides as microsomal metabolites of polycyclic hydrocarbons. Fed. Eur. Biochem. Soc. Lett., 18:76-80.
6. Jerina, D. M. and Daly, J. W. (1974): Arene oxides: A new aspect of drug metabolism. Science: 185:573-582.
7. Boyland, E. (1950): The biological significance of metabolism of polycyclic compounds. Biochem. Soc. Symp. 5:40-54.
8. Weinstein, I. B., Jeffrey, A. M. Jenette, K. W., Blotstein, J. H., Harvey, R. G., Harris, C., Autrup, H., Kasai, H., and Nakanishi, K. (1976): Benzo[a]pyrene diol epoxides as intermediates in nucleic acid binding in vitro and in vivo Science: 193:592-595.
9. For a summary see: Lee, M. L., Novotny, M. V., and Bartle, K. D. (1981): Analytical Chemistry of Polycyclic Aromatic Hydrocarbons, Academic Press, New York, 462 pp
10. (1972): Biologic Effects of Atmospheric Pollutants: Particulate Polycyclic Organic Matter, Nat. Acad. Sci., Washington, D.C., 528 pp.
11. Sawicki, E. (1967): Airborne carcinogens and allied compounds. Arch. Environ. Health, 14:46-53.
12. Epstein, S. S. (1965): Photoactivation of polynuclear hydrocarbons. Arch. Environ. Health, 10:233-239.
13. Lave, L. B. and Seskin, E. P. (1970): Air pollution and human health, Science, 169: 723-733.
14. Pitts, J. N., Jr., Lokensgard, D. M. Ripley, P. S., Cauwenberghe, K.A.V. Luk, V. V., Shaffer, S. D., Thill, A. J. and Belser, W. L. Jr. (1980): Atmospheric epoxidation of benzo[a]pyrene by ozone: Formation of the metabolite benzo[a]pyrene-4,5-oxide. Science, 210:1347-1349.

15. Murray, R. W. and Banavali, R. (1983): Formation of a K-region arene oxide by intramolecular O atom transfer. Tetrahedron Lett., 24:2327-2330.

16. Murray, R. W. and Kumar, S. (1982): Oxidation of phenanthrene with a carbonyl oxide. In Polynuclear Aromatic Hydrocarbons: Physical and Biological Chemistry, edited by M. Cooke, A. J. Dennis and G. L. Fisher, pp 575-584, Battelle Press, Columbus, Ohio.

17. Kumar, S. and Murray, R. W. (1984): Carbonyl oxide chemistry: The NIH shift. J. Am. Chem. Soc., 106: 1040-1045.

18. Bailey, P. S. (1978): Ozonation in Organic Chemistry, Vol. I, Academic Press, New York, 372 pp

19. Lovas, F. J. and Suenram, R. D. (1977): Indentification of dioxirane (H_2COO) in ozone-olefin reactions via microwave spectroscopy. Chem. Phys. Lett. 51: 453-456.

20. Suenram, R. D. and Lovas, F. J. (1978): Dioxirane: Its synthesis, microwave spectra, structure, and dipole moment, J. Am. Chem. Soc. 100:5117-5122.

21. Martinez, R. I., Huie, R. E., and Herron, J. T. (1977): Mass Spectrometric detection of dioxirane, H_2COO, and its decomposition products, H_2 and CO, from the reaction of ozone with ethylene. Chem. Phys. Lett. 51:457-459.

22. Talbott, R. I. and Thompson, P. G., U.S. Patent 3 632 606, 1972.

23. Edwards, J. O., Pater, R. H., Curci, R., and DeFuria, F. (1979): On the formation and reactivity of dioxirane intermediate in the reaction of peroxoanions with organic substrates. Photochem. Photobiol. 30:63-70.

24. Montgomery, R. E. (1974): Catalysis of peroxymonosulfate reactions by ketones. J. Am. Chem. Soc., 96:7820-7821.

25. Gallopo. A. R. and Edwards, J. O. (1981): Kinetics and mechanism of the oxidation of pyridine by Caro's acid catalyzed by ketones. J. Org. Chem., 46:1684-1688.

26. Cicala, G., Curci, R., Fiorentiono, M., and Laricchiuta, O. (1982): Stereo- and regioselectivities in the epoxidation of some allylic alchols by the dioxirane intermediate generated in the reaction of potassium caroate with acetone. J. Org. Chem. 47:2670-2673.

27. Curci, R., Fiorentino, M., Troisi, L., Edwards, J. O., and Pater, R. H. (1980): Epoxidation of alkenes by dioxiranes intermediates generated in the reaction of potassium caroate with ketones, J. Org. Chem. 45:4758-4760.
28. Murray, R. W. and Ramachandran, V. (1979): Oxidation of acetone under Bayer-Villiger conditions; Evidence for formation of acetone carbonyl oxide or its isomeric dioxirane. Photochem. Photobiol. 30: 187-189.
29. Jeyaraman, R. and Murray, R. W. (1984): Production of arene oxides by the caroate-acetone system (dimethyldioxirane). J. Am. Chem. Soc. 106:2462-2463.
30. Murray, R. W. and Jeyaraman, R. (1985): Dioxiranes; Synthesis and reactions of methyldioxiranes, J. Org. Chem., 50:2847-2853.
31. Reference 9, Appendix 5.
32. Harvey, R. G., Groh, S. H. and Cortez, C. (1975): K-region xoides and related oxidized metabolites of carcinogenic aromatic hydrocarbons. J. Am. Chem. Soc., 97:3468-3479.
33. Krishnan, S., Kuhn, D. G., and Hamilton, G. A. (1977): Direct oxidation in high yield of polycyclic aromatic compounds to arene oxide using hypochlorite and phase transfer catalysts. J. Am. Chem. Soc., 99:8121-8123.
34. Yashida, M., Kadokura, A. and Minabe, M. (1980); Some reactions on 4-H-cyclopenta[def]phenanthrene-8,9-dione. Bull. Chem. Soc. Japan, 53:1179-1180.
35. Newman, M. S. and Lilje, K. C. (1979): Synthesis of 7-fluorobenz[a]anthracene, J. Org. Chem., 44:1347-1348.
36. Fieser, L. F. and Dietz, E. M. (1929): 1,2-Benz-3,4-anthraquinone, J. Am. Chem. Soc., 51:3141-3148.
37. Lee, H., Shyamasundar, N. and Harvey, R.G. (1981): Isomeric Phenols of benzo[e]pyrene, J. Org. Chem., 46:2889-2895.
38. McClausland, D. J., Fischer, D. L., Kolwyck, K. C., Duncan, W. P., Wiley, J. C., Jr., Menon, C. S., Engel, J. F., Selkirk, J. K., and Roller, P. P. (1976): Polynuclear aromatic hydrocarbon derivatives; Synthesis and physiochemical characterization. In Carcinogenesis, Vol. 1 edited by R. I. Freudenthal and R. W. Jones, p. 349-411, Raven Press, New York.

39. **Clar, E. (1929); Zur kenntnis mehrkerniger aromatischer kohlenwasserstoffe und ihrer abkommlinge, IV, Mitteil: Naphthophenanthrene und ihre Chinone, Chem. Ber., 62:1574-1582.**

A COMPARISON OF THE SKIN CARCINOGENICITY OF CONDENSED ROOFING ASPHALT AND COAL TAR PITCH FUMES

R. W. NIEMEIER*, P. S. THAYER(1), K. T. MENZIES(1),
P. VON THUNA(1), C. E. MOSS, J. BURG
Department of Health and Human Services, U. S. Public Health Service, Centers for Disease Control, National Institute for Occupational Safety and Health, Division of Biomedical and Behavioral Science, Experimental Toxicology Branch, 4676 Columbia Parkway, Cincinnati, Ohio 45226
and (1)Arthur D. Little, Inc., Acorn Park, Cambridge, Massachusetts 02140

INTRODUCTION

Asphalt and coal tar pitch are bituminous materials which are similar in appearance and have been used interchangeably for roofing, paving and in other industrial applications (1,2). Whether exposure to asphalt represents significant health risk has been debated for years. This debate is further clouded by the confusion in the definition and usage of the terms asphalt, pitch, coal tar, bitumen, petroleum pitch, and coal tar pitch. Bingham and coworkers, (2) in a critical review of the literature described petroleum technology and included a glossary of terms, which should serve as a guide in elucidating this sometimes perplexing terminology.

Coal tar pitch, a residue in the partial evaporation or fractional distillation of coal tar produced by the destructive distillation of bituminous coal, was among the first substances detected as a human carcinogen and successfully tested for its carcinogenic effects in animals (3). Asphalt is a bitumen which occurs in nature or is obtained in the refining of crude petroleum. Several attempts (2,4,5,6,7,8,9,10,11,12) to produce tumors in animals exposed to raw, solvent diluted or aerosolized asphalt were reviewed by NIOSH which concluded that "reliable reports associating malignant tumors with exposure to asphalt fumes have not been found in the literature" and "although available information has not clearly demonstrated that a direct carcinogenic hazard is associated with asphalt fumes, NIOSH is concerned that future investigations may suggest a greater occupational hazard from asphalt fumes than is currently documented in the literature."(1)

Wallcave et al. (11) suggested that differences in bioassay results with coal tar pitches and asphalts were related to polynuclear aromatic hydrocarbon (PAH or PNA) content, reflecting their thermal histories. Asphalt production temperatures range from 350° to 400° C and those of coal tar pitches exceed 1000° C. PAH content of coal tar pitches may be several orders of magnitude greater than that of petroleum asphalts, with coal tar pitch containing more higher-molecular weight PAH (13). Wallcave's suggestion is useful for the raw products, but under field conditions thermal histories may be much different from those of the raw materials. For example, Thomas and Mukai (14) reported that heating in roofing kettle operations is poorly controlled and commonly the materials are heated as high as 638° C (1000° F), well above their recommended kettle temperatures of 204 to 273° C. This results in pyrolysis and increased production of PAH. Some field operations such as the asphalt hot mix operation for road paving do maintain moderate temperatures, generally lower than 160° C. (15)

In a study of workers with at least 5 years and an average of 15.1 years exposure to asphalt in 25 oil refineries, Baylor and Weaver (16) found a few cases of bronchitis, asthma, and emphysema but no other significant health effects compared to a control population. They cited further information on workers employed in highway construction, roofing manufacture and truck driving over asphalt highways that exposure to petroleum asphalt constituted no health hazard. Zeglio in 1950 (1), reporting on Italian workers who used natural asphalt, possibly adulterated with coal tar pitch to insulate electrical cables and telegraph and telephone lines, concluded that characterizing bitumen vapors only as irritating did not recognize the rhinitis, oropharyngitis, laryngitis, bronchitis, and X-ray and respiratory changes from rales to emphysema that they caused.

More recently Hammond et al. (17) found that occupational exposures of roofing workers to asphalt and coal tar pitch fumes for more than 20 years were associated with increased mortality from cancer and other pulmonary diseases including emphysema, chronic bronchitis and asthma. Leukemia and cancer of the lung, upper respiratory tract, stomach, bladder, and skin had excess mortality. Smoking habits were not considered, and the causative factors could not be identified because of mixed exposure to coal tar pitch and asphalt fumes.

Lawther (18) studied mean 8-hour shift exposures of roofing and gas retort workers to benzo[a]pyrene (BaP), a ubiquitous carcinogen often mentioned as an indicator of PAH exposure. (19,20) He found for roofers, exposures of less than 1 and 13 μg for asphalt and hot pitch, respectively and for gas retort workers exposures averaged 27 μg. Bingham et al. (2) noted that a single carcinogen, such as BaP, can serve as a guide to carcinogenic potency, but the presence or absence of BaP does not always account for the observed potency, and roles of other carcinogens, cocarcinogens and inhibitors of carcinogenesis must be considered. The International Agency for Research on Cancer (21) stated that predictions of human cancer risks cannot be made from simple knowledge of PAH levels.

In addition to the risks associated with exposure to the coal tar pitch and petroleum asphalt fumes in the roofers environment, the added risk of sunlight exposure in the out-of-doors environment must be considered. Various investigators working with experimental animals have found that ultraviolet and visible light augment the carcinogenicity of PAH exposure (22,23).

NIOSH estimates that 12,000 roofing contractors employ over 116,000 workers in the U.S. (24) Since these workers have combined exposures to asphalt, coal tar pitch and sunlight, an experiment was designed to study the relative importance of each and combinations of them. The purposes were to assess: (a) the carcinogenic potentials of condensed volatiles from two commonly used roofing asphalts and two commonly used coal tar pitches collected from fumes generated at recommended application temperatures; (b) the carcinogenic potentials of the condensed volatiles collected from fumes generated at temperatures in excess of their recommended application temperatures; (c) the effects of simulated sunlight on the carcinogenic outcome of the above materials; and (d) the responses in pigmented and nonpigmented mice.

MATERIALS AND METHODS

Sample Identity, Collection, and Characterization

There are four types of asphalt and three types of coal tar pitch used on roofs. Two types of each, chosen for this study on the basis of common use and extremes of classification, were Type I and Type III asphalt, referred to in the industry as "dead level" and "steep," respectively, and Type I and Type III coal tar pitch often referred to as "regular roofing" and "low fuming" or "low burn" pitch, respectively. They are produced by several manufacturers to physical specifications recommended by the American Society of Testing and Materials (25).

Approximately 270 kg of each of the materials were purchased: Type I and Type III asphalts, purchased from a distributor of Exxon, Inc., Roofing Products (Beacon Sales, Inc., Somerville, MA), were produced by distillation and air blowing of Arabian crude; Type I coal tar pitch, obtained from Reilly Tar and Chemical Corp. (Cleveland, OH), and Type III coal tar pitch, provided by Koppers Company, Inc. (Monroeville, PA). All materials were manufactured to ASTM specifications and were shipped from available inventory. Fumes were generated from an easily controlled glass generation system and the condensed material was collected in a glass cryogenic system. Laboratory generation and collection was used instead of a roofing kettle to avoid problems of inadequate sample mixing, generation temperature extremes, exposure to sunlight and difficulty in collecting condensed field samples at a subambient temperature. The roofing material was made into small pieces, approximately 10L placed into a weighed 12L round bottom reaction flask, and warmed in a forced air oven at 150° C. When the the material was soft it was stirred with a stainless steel multifinned stirring rod at 250 to 300 rpm. The rod was driven by a compressed air motor and inserted through the neck of the flask lubricated with a teflon stirring gland. An electric heating mantle was used to attain the desired generation temperatures (± 5° C) of 232° C, the approximate recommended kettle temperature, and 316° C, just below the flashpoint of the materials.

The fume collection system consisted of 20 mm O.D. glass transfer tubes and 500 mL glass impingers placed in three cryotraps, one containing ice (0° C) and two dry ice/isopropanol (-77° C). An impinger containing a 50/50 mixture of cyclohexane/acetone was used after the cryotraps to provide additional collection through dissolution. Air, precleaned and dried using a high efficiency filter, silica gel and granular activated charcoal for removal of particulates, water, and organic vapors, respectively, and heated to 100° C was pulled through the system at a rate of 10 Lpm. The laboratory light was filtered through yellow cellulose acetate-butyrate filters to reduce ultraviolet light induced photooxidation and decomposition during generation and collection.

After the fumes were condensed and collected, the individual impingers and transfer lines were weighed and the material was quantitatively transferred from them with an excess of 50/50 cyclohexane/acetone solvent mixture to a large flask. The solvent was then removed at reduced pressure at 50° C, and analyzed by gas chromatography/mass spectrometry (GC/MS) to confirm that significant amounts of compounds of interest were not present. The collected condensates were weighed and dissolved in a 50/50 mixture of cyclohexane/acetone. One to two generations were needed to obtain the necessary amount of material with the coal tar pitches at either temperature; with the asphalts 34 to 59 generations were required at 232° C and 6 to 7 at 316° C. The materials collected from each generation were combined to produce about 4L each of eight skin painting solutions (4 materials x 2 temperatures).

The collections at 316° C compared to the collections at 232° C, yielded 9 to 16 times more volatile material from asphalts but only 2 to 7 times more from coal tar pitch.

For the final concentration of the skin painting solutions, the coal tar pitch condensates were diluted to attain a BaP concentration of 0.01 percent in order to produce an estimated 100 percent tumor incidence at approximately one year after the start of the experiment (26). The final asphalt condensate solutions contained 50 percent total solids, which is a commonly used practice for assaying complex mixtures with low PAH concentrations. Concentrations of selected PAH in skin painting solutions were determined by combined GC/MS (Finnigan Model 4023) analysis using a high resolution glass capillary column (25-m coated with SP-2250) for separation and quantitation of closely related isomers. A compound was considered as identified when the retention time and the mass spectrum of the PAH of interest matched those of a standard PAH. Table 1 shows the concentrations of 18 identified chemicals and total concentrations of condensed fumes in the 8 skin painting solutions.

Benzo[a]pyrene (BaP) for the positive controls was obtained from the NCI Chemical Repository (IITRI, Chicago, IL). Cyclohexane and acetone were both HPLC grade (Fisher Scientific, Inc.). Cyclohexane/acetone was chosen as the solvent system for collection, rinsing and preparation of skin painting solutions because: (a) the mixture was a better solvent of the material than either alone; (b) each solvent had a low boiling point and was compatable with PAH

TABLE 1

MEAN CONCENTRATION (μg/mL) OF PAHS AND FINAL CONCENTRATION OF CONDENSED FUMES IN SKIN-PAINTING SOLUTIONS

PAH	Analytical GC/MS Ion	Asphalt Type I 232°C	Asphalt Type I 316°C	Asphalt Type III 232°C	Asphalt Type III 316°C	Pitch Type I 232°C	Pitch Type I 316°C	Pitch Type III 232°C	Pitch Type III 316°C
Naphthalene	128	22	4	17	49	>1800	1770	288	620
Fluorene	166	36	22	39	28	--	740	--	--
Carbazole	167	20	1	6	--	1980	1450	540	1400
Anthracene/Phenanthrene	178	180	53	300	69	> 960	2960	>2580	>5200
Fluoranthene	202	86	10	97	7	>2940	2350	> 960	>2800
Pyrene	202	70	9	63	8	>2070	1790	> 720	>2300
Benz[a]anthracene	228	11	10	8	6	570	330	330	800
Chrysene/Triphenylene	228	25	19	13	14	460	300	290	710
Benzofluoranthenes	252	2	4	5	--	230	230	250	250
Benzo[e]pyrene	252	6	8	4	1	42	51	45	46
Benzo[a]pyrene	252	2	2	3	--	96	85	102	90
Indeno[c,d]pyrene	276	3	3	2	--	33	2	11	7
Benzo[g,h,i]perylene	276	1	2	1	--	28	2	7	1
Dibenzanthracenes	278	2	--	2	--	12	--	4	--
Coronene	300	--	--	--	--	--	--	--	--
Dibenzopyrenes	302	--	--	--	--	--	--	--	--
Final concentration of condensed fumes (mg/mL)		500	500	500	500	78	55	84	30

stability; (c) the solvents did not absorb light in the simulated sunlight range; and (d) both solvents were inactive with respect to mouse skin bioassay.

Solar Simulator

The solar simulator used a 15-cm Atlas 6.5 kW Xenon arc, water cooled, quartz enveloped burner located about 53-cm above and midway between two turntables. Two additional turntables, completely shielded from light, were contained in the unit for sham irradiation. Wavelengths shorter than 290 nm were excluded by metal oxide coated, 3 mm Tempax glass, No. 114 Schott filters that also reflected infrared radiation (>790 nm) away from the mice. Light measurements were made to validate the geometric concept of the turntable arrangement using a silicon photodiodetector (EG&G) equipped with narrow-band optical interference filters (309 to 790 mm). An automatic integrating system adjusted exposure time for fluctuations and changes in intensity of the arc due to aging to maintain a constant exposure. The solar simulator was contained in an aluminum housing.

Stainless steel (3 x 3 mesh) cage units, specially designed so that two of them fit on each of the four round turntables, contained 50 4 x 9 x 3 cm cells to individually house all mice of a group for simultaneous exposure, and permitted simultaneous light exposure of four groups and sham exposure of four groups. During exposures the turntables were rotated at one rpm, so that on the average each mouse was exposed at the same distance from the source. Cooling air, was delivered to maintain the chamber temperature at about 24 to 28° C.

The goal of the experiment was to produce skin tumor onset in a light exposed control group at approximately 50 weeks. From Bingham and Nord (27) and Burns (28) it was estimated that the total UVB (280-320 nm) dose required was approximately 2 x 10^5 Wsec/m^2 (2 X 10^8 ergs/cm^2). Since the mice were theoretically to be exposed twice weekly for a total of 100 exposure sessions, and the wattage at the exposure site was 0.68 W/m^2, the required duration of exposure per session was approximately 49 minutes. More accurate spectral data revealed that the original integrated irradiance between 280 and 320 nm was too high and thus the actual exposure per session was 1480 Wsec/m^2 with the desired total dose delivered in 135 sessions instead of 2000 Wsec/m^2 delivered in 100 sessions.

CARCINOGENICITY OF CONDENSED ROOFING FUMES

Skin Carcinogenesis Tests

Nonpigmented Swiss CD-1 (Charles River) and pigmented C3H/HeJ (Jackson Laboratories) male mice were used. Upon arrival, at six weeks of age, they were quarantined for 6 to 9 weeks. They were housed individually in suspended stainless steel cages and provided food and water ad libitum except during the chemical and light exposures. Each of the 48 experimental groups consisted of 50 randomized mice of one strain, and each mouse was individually identified. The 48 experimental groups were:

a. Thirty-two groups for the primary factorial experiment, i.e., 2 strains x 4 materials x 2 generation temperatures x 2 light exposure conditions (presence or absence of simulated sunlight); each animal dosed twice weekly with 50 µL of the appropriate test material.

b. Four groups for the solvent control, i.e., 2 strains x 2 light exposure conditions; each animal dosed twice weekly with 50 µL of the cyclohexane/acetone (1:1) vehicle.

c. Two groups for cage control, i.e., 2 strains, dosed twice weekly with 50 µL solvent and not sham irradiated but always maintained in their individual cages.

d. Four groups for the positive control, i.e., 2 strains x 2 light exposure conditions dosed twice weekly with 50 µL of the solution of 0.01 percent BaP in cyclohexane:acetone (1:1).

e. Four groups for a combination treatment of asphalt and coal tar pitch fume condensate, i.e., 2 strains x 2 light exposure conditions; each animal dosed twice weekly with the high temperature condensate from Type III asphalt and Type I coal tar pitch, on alternate weeks.

f. Two groups, i.e., 2 strains, for light exposure twice weekly with no skin painting treatment.

These groups are indicated in Table 2 where the effective number of mice are presented.

TABLE 2

EFFECTIVE NUMBER OF MICE PER GROUP AND WEEK OF FIRST PAPILLOMA*

Temperature of Generation	232° C				316° C			
Light Condition	Light		No Light		Light		No Light	
Mouse Strain	P	N	P	N	P	N	P	N
Asphalt I	49(33*)	10(56)	49(30)	46(24)	47(18)	10(50)	46(23)	45(24)
Asphalt III	44(34)	48(17)	47(21)	48(31)	49(16)	46(41)	46(24)	46(41)
Coal Tar Pitch I	46(37)	45(31)	47(33)	46(22)	45(37)	46(35)	47(26)	45(31)
Coal Tar Pitch III	48(17)	45(40)	43(34)	49(14)	42(39)	48(28)	46(28)	48(28)
Combination (Type III Asphalt and Type I Pitch)	--	--	--	--	44(21)	41(21)	48(25)	45(30)
No Temperature Variations								
BaP	45(40)	49(28)	42(38)	49(28)				
Solvent Controls	33(40)	-(-)	-(-)	-(-)				
Solvent Cage Controls	-(-)	-(-)	-(-)	48(41)				
Untreated Controls	21(80)	-(-)	--	--				

P = pigmented C3H/HeJ
N = nonpigmented CD-1
(-) = no tumors
-- = not tested
* = week of first grossly observed papilloma

Cage racks were regularly rotated within the animal quarters, and to minimize effects of fluorescent lights, lamps were enclosed in clear filter tubes (Crown Plastics Corp.). The light cycle was 12 hours light and 12 hours dark. Each mouse was weighed prior to each first weekly application for 6 weeks, and biweekly thereafter.

Hair was clipped as needed from the interscapular region, using a separate No. 40 clipper head for each test material, and the test material was applied with disposable-tip automatic pipettes in a ventilated hood. After chemical treatment, each mouse was placed in a cell of a clean solar exposure cage unit and kept under a hood. At 30 minutes following the application to the last mouse in an experimental group, the group was exposed to the light, or sham exposed. To insure uniformity of exposure, the mice were rotated weekly into different cells of the solar cage unit by a formal procedure. The treatments continued for 78 weeks (18 months).

Mice were observed daily for systemic toxicity, and gross appearance of tumors. Mice found dead were necropsied. Those that were moribund were killed and necropsied and when groups were terminated at 78 weeks, those remaining were necropsied. Tissues were examined and preserved in 10 percent buffered formalin. Skin lesions were excised and fixed in buffered 10 percent formalin and prepared for microscopic examination.

Additional details of the experimental methods are given elsewhere (29).

Statistical Methods

The survival curves were estimated using the product limit (Kaplan Meier) estimate (30) with day as the interval of time. The null hypothesis that the groups of interest had the same survival distributions was tested using the Breslow (generalized Wilcoxson) statistic. It should be noted that the Breslow test gives greater weight to the early events and is less sensitive to the events of interest that occur later in time when few animals remain in study. Animals sacrificed at termination of the study were considered censored, others as dying naturally. The mean, standard error of the mean, and 25, 50 and 75 quantiles were calculated. It should be noted that if the longest surviving animal is censored (as is usually the case in this study), the calculated mean value underestimates the true mean.

The time-to-tumor curves were calculated in the manner described previously for survival only the time of observation for the animal's first tumor (a tumor that was subsequently confirmed histopathologically) was used as the event rather than time of death.

To test for simple differences in the number of tumor bearing animals at risk, the Fisher-Irwin Exact Test was used. For tumor incidence, the heterogeneity of groups was tested, after allowing for differences in longevity, utilizing the "onset rate" analysis as suggested by Peto _et al_. (31) This method assumes that the appearance of skin tumors are mortality-independent events. Only animals that had tumors that were observed and histopathologically confirmed were considered tumor bearing animals. The time period used in calculations was one day and the number of animals at risk for a given group on a given day are those animals alive and without any previously observed tumors. The number of tumor bearing animals were compared, although there may have been multiple tumors per animal.

For all analyses, the significance level selected was $p=0.01$. It was recognized that a given group was included in multiple comparisons and the probability level of a given test may have been greater than that calculated; therefore, $p=0.01$ level was used to insure conservative conclusions for the study.

The following group comparisons were made: (a) each group versus appropriate solvent group; (b) each group versus appropriate benzo[_a_]pyrene group; (c) light versus no light groups; (d) high temperature versus low temperature groups; and (e) C_3H/HeJ species versus CD-1 species. The following comparisons were made for the groups: (a) survival distribution; (b) time to tumor distribution; (c) number of tumor bearing animals; and (d) onset rate analysis combining the above factors.

RESULTS

General Health and Survival

Body weights reflected no effect of the treatments. Weight gains occurred in the first 2 months; after that there were only minor oscillations. Killing moribund or cachectic mice was a factor in these oscillations.

Mean survival times in the treated C_3H/HeJ groups ranged from 44.3 to 68.7 weeks. Mean survival times (± standard error of the mean) for the solvent controls were 65.6 ± 3.0 and 73.9 ± 2.2 weeks for non-solar and solar groups, respectively. In the CD-1 groups, mean survival for the treated groups ranged from 52.6 to 67.8 weeks with the mean ± S.E. for solvent controls for non-solar and solar groups, 63.9 ± 2.6 and 67.8 ± 2.1 weeks, respectively. The control groups had expected survival rates, although there were more deaths in the CD-1 mice due to endemic urinary tract infections.

Tumor Incidence and Latency

Tumors resulted from administration of condensed fumes from both types of asphalt and coal tar pitches. The percentages of effective total mice bearing tumors are shown in Figure 1 for benign (papillomas, kerato- acanthomas, fibromas and unclassified benign epitheliomas), malignant (squamous cell carcinoma and fibrosarcomas) and total tumors. Most of the benign tumors were papillomas and most malignant tumors were squamous cell carcinomas. The incidence of all malignant tumors was much lower in the CD-1 mice, about 5 percent versus about 60 in the C_3H/HeJ strain. Fibrosarcomas were seen more often in the C3H/HeJ mice than in the CD-1 mice. The average latent period (Figure 2) ranged from 39.5 to 56.1 weeks among the C3H/HeJ groups, and from 47.4 to 76.5 weeks among the CD-1 groups treated with the condensed fumes of the roofing materials.

Comparisons to Solvent Controls

The solvent control groups did not develop tumors with the exception of one benign tumor found in the C_3H/HeJ strain exposed to light. Therefore all groups had significantly more tumors than the comparable solvent control group.

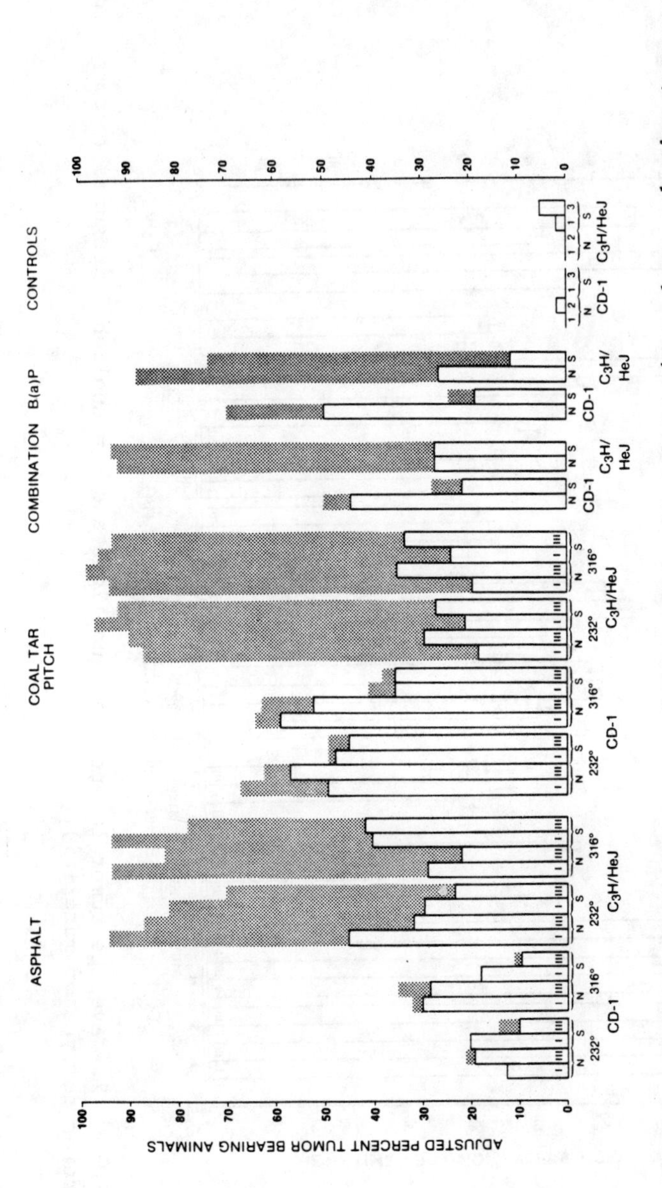

FIGURE 1. Tumorigenic response to condensed roofing fumes (open bars = benign tumors; shaded bars = malignant tumors; N = no light and S = sunlight exposures; roman numerals refer to type of material)

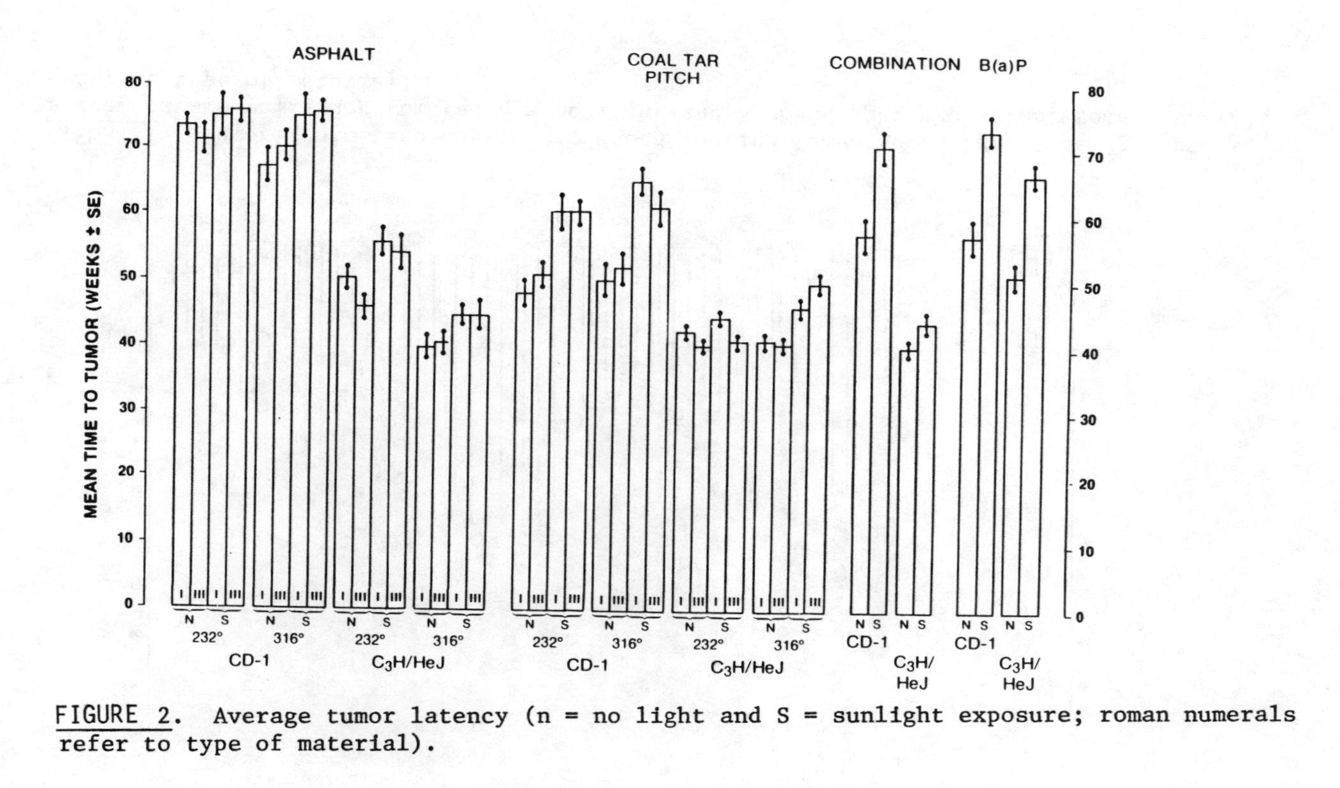

FIGURE 2. Average tumor latency (n = no light and S = sunlight exposure; roman numerals refer to type of material).

All C_3H/HeJ groups with two exceptions had significantly different survival curves from the appropriate solvent control groups. The two exceptions were Type I asphalt (232° C) with and without light exposure. In all cases the treated mice died earlier than the solvent control groups. Survival of CD-1 groups differed significantly from the appropriate solvent control groups with the following six exceptions: both asphalt types at high and low temperature of generation without light exposure; Type I pitch (316° C) with light exposure; and Type III pitch (316°C) without light exposure.

In the negative cage control groups only one CD-1 mouse developed a histologically confirmed benign tumor when treated with solvent only; the C_3H/HeJ strain remained tumor free. Exposure to the light source without solvent treatment caused development of one malignant tumor in the C_3H/HeJ strain but no tumors were observed in the CD-1 group. Regressed tumors not confirmed by histopathology, were observed in almost every group (Table 3).

Mouse Strain Comparisons

In comparing strains of mice, except for the solvent control groups, the survival time was in general shorter for the C_3H/HeJ strain. This difference was significant for: Type I asphalt (232° C) with light; Type III asphalt (316° C) without light; Type I pitch (232° C) with and without light; Type I pitch (316° C) without light; Type III pitch (232 and 316° C) without light; both combination exposure groups; and the C_3H/HeJ survived longer than CD-1 solvent control with light.

Comparing mouse strain differences in response to the various treatments, all C_3H/HeJ groups had a significantly higher incidence of tumor bearing animals (TBA). With two exceptions, i.e., Type I pitch (232°) and BaP groups without light exposure, the time to tumor curves were significantly different with the C_3H/HeJ groups developing tumors earlier.

It should be emphasized that a significantly large number of grossly observed tumors regressed in the CD-1 strain as compared to the C_3H/HeJ strain (Table 3). These data were not incorporated into the current analyses since histopathological confirmation was not available.

TABLE 3

PERCENT OF ORIGINAL MICE WITH REGRESSED TUMORS

Material	Temperature(^{o}C)	Light Exposure	C3H/HeJ	CD1
Asphalt I	232	-	2	14
		+	6	28
	316	-	4	16
		+	0	20
Asphalt III	232	-	2	18
		+	8	14
	316	-	4	14
		+	0	12
Pitch I	232	-	10	26
		+	0	26
	316	-	2	28
		+	0	26
Pitch III	232	-	2	28
		+	4	20
	316	-	0	26
		+	4	34
Combination	316	-	0	22
	316	+	0	26
Solvent		-	8	2
		+	12	4
BaP		-	6	14
		+	10	22
Cage Control		-	2	2
Untreated		+	2	0

Effects of Fume Generation Temperature

For both asphalts the mean latency period was reduced an average of 9.5 ± 2.5 weeks (mean ± S.D.) when C_3H/HeJ mice were treated with the higher temperature material compared to the response from the lower temperature condensed fumes (Table 4). Similarly, mean survival was reduced 10.0 ± 3.0 weeks which corresponds to the reduced latency period. Groups not exposed to light showed no differences in percent TBA and those exposed to light were found to have ~9.5 percent increased, but statistically insignificant, tumor incidence by Fisher's Exact Test. The onset rate analysis determined that all groups exposed to the higher temperature materials were significantly more susceptible.

Those C_3H/HeJ mice exposed to coal tar pitch fumes exhibited very little change in latency or survival due to increased temperature of generation, except for the Type III groups exposed to light, where an inhibitory effect was noted on time to tumor development, but there was no effect on tumor incidence. Onset rate analysis revealed a significant difference in the light exposed Type III pitch groups which appears to be an effect based solely on latency differences. The groups not exposed to light got more tumors (~9 percent) with increased temperature of generation but these increases were not significant. Even though the BaP concentrations were normalized for the pitch groups, changes were observed in the measured responses.

The CD-1 groups exposed to asphalts without light generally exhibited a non-significant increase in percent TBA with increasing temperature, but no changes in survival or latency were observed. The CD-1 groups exposed to condensed pitch fumes generally responded less to the higher temperature materials but no significant differences were noted for any individual measured parameters, or using the onset rate analysis.

Effects of Light

The C_3H/HeJ asphalt groups exposed to light generally responded by developing less tumors (Table 5). This was particularly noted with the low temperature groups. Latency and survival increased with this decreased tumor response. Onset rate analysis showed significant differences for both low temperature materials. The pitch groups of this mouse strain showed little change in percent TBA by altering light

TABLE 4

THE EFFECTS OF INCREASED FUME GENERATION TEMPERATURE ON NET CHANGE IN RESPONSE

Type (± light)		C3H/HeJ				CD-1			
		Corrected Percent TBA	Mean Latency (Wks)	Mean Survival Time (Wks)	Onset Rate Analysis	Corrected Percent TBA	Mean Latency (Wks)	Mean Survival Time (Wks)	Onset Rate Analysis
Asphalt	I-	- 0.4	-10.9[s]	-13.1[s]	s	18.3*	-5.6	-3.5	ns
	I+	12.0	-11.8[s]	-11.8[s]	s	10.0	0.0	-0.8	ns
	III-	- 4.6	- 6.1	- 6.5	s	14.0	-1.7	4.7	ns
	III+	7.1	- 9.1[s]	- 8.7[s]	s	- 3.7	3.5	-2.1	ns
Pitch	I-	8.5	- 0.4	1.3	ns	- 3.0	2.6	-1.3	ns
	I+	- 2.2	0.9	0.0	ns	- 8.7	4.8	1.7	ns
	III-	9.4	0.4	4.4	ns	1.3	0.5	3.0	ns
	III+	- 1.2	8.7[s]	3.5[s]	s	-11.4	0.9	3.9	ns

* significant at $p<0.05$ by Fisher-Irwin Exact Test
s significant at $p<0.01$
ns not significant

TABLE 5

THE EFFECTS OF LIGHT ON NET CHANGE IN RESPONSE

Type	°C	C3H/HeJ Corrected Percent TBA	C3H/HeJ Mean Latency (Wks)	C3H/HeJ Mean Survival Time (Wks)	C3H/HeJ Onset Rate Analysis	CD-1 Corrected Percent TBA	CD-1 Mean Latency (Wks)	CD-1 Mean Survival Time (Wks)	CD-1 Onset Rate Analysis
Asphalt	I 232	-12.3	5.7	2.2	s	7.0	0.9	-9.2s	ns
	I 316	0.1	4.8s	3.5	ns	- 1.1	6.5s	-6.5	ns
	III 232	-16.7*	6.5s	2.2	s	- 6.2	0.9	0.8	ns
	III 316	- 5.0	3.5	0.0	ns	-23.9s	6.1	-6.0	ns
Pitch	I 232	12.7*	2.2	5.2s	s	-17.4	13.4s	2.2	s
	I 316	- 2.0	3.5s	3.9s	s	-23.1*	15.6s	5.2	s
	III 232	3.3	0.8	6.5s	s	-12.3	8.7s	-2.6	ns
	III 316	- 4.9	9.1s	5.6s	s	-25.0s	9.1s	-1.7	s
BaP		-17.0*	24.8s	9.1	s	-44.9s	16.5s	2.2	s
Combination		1.5	3.4s	1.3s	s	-26.7s	14.3s	0.4	s

* significant at $p<0.05$ by Fisher-Irwin Exact Test
s significant at $p<0.01$
ns not significantly different

exposure. The most prominent change in TBA ($p<0.05$) was observed in the Type I pitch (232° C). Increased latency and survival were the trends. Onset rate analysis showed all four materials to affect mice differently in the presence of light. Light inhibited ($p<0.05$) tumor formation in the BaP group with a corresponding increase in latency time and survival. Onset rate analysis showed a significant difference. The combination group responded in a fashion similar to the pitch group, with the overall effect being significantly different with light exposure.

The CD-1 mice exposed to asphalt fumes reacted erratically. The overall response in the higher temperature groups appeared to be decreased with light exposure, but these were not significant using the onset rate analysis method. There was an increased time to tumor noted in the higher temperature groups; however, these same groups showed decreased mean survival. The results of exposing the pitch fume groups to light were very dramatic; tumor development in all groups was inhibited by an average of 19.5 ± 5.8 percent. There was a corresponding significant increase in time to tumor development, an average of 11.7 ± 3.4 weeks. The survival time was not affected. Onset rate analysis showed significant differences in three of the four groups. As with the C_3/HeJ strain, tumor development caused by BaP in the CD-1 strain was severely inhibited by light exposure with a corresponding increase in latency, but there was no effect on survival. The combination group, as before, appeared to mimic the coal tar pitch fume effects. Both comparisons were statistically significant using an onset rate analysis.

Comparison to the BaP Response

Reviewing the responses in the C_3H/HeJ strain, fumes collected from Type I asphalt appeared to be more carcinogenic than the 0.01 percent BaP increasing the average percent TBA by approximately 11 ± 8 percent and decreasing the latency period by an average of 11.5 ± 8.5 weeks (Table 6). However, there was no consistent effect on survival; the largest decrease occurred with the Type 1 (316° C) receiving concommitant light exposure. The onset rate analysis showed significance for only the 316° C groups. The Type III asphalt fumes did not induce consistent change when compared with the corrected percent TBA from the BaP group; although, mean time to tumor was again significantly decreased an average of 12.8 ± 7.2 weeks and survival was generally decreased. As with the Type I

TABLE 6

GROUP COMPARISONS TO BaP RESPONSES

Type (± light)	°C	C3H/HeJ Corrected Percent TBA	C3H/HeJ Mean Latency (Wks)	C3H/HeJ Mean Survival Time (Wks)	C3H/HeJ Onset Rate Analysis	CD-1 Corrected Percent TBA	CD-1 Mean Latency (Wks)	CD-1 Mean Survival Time (Wks)	CD-1 Onset Rate Analysis
Asphalt I-	232	5.8	- 1.3	8.7	ns	-56.4^{s}	17.4^{s}	- 3.0	s
	316	5.4	-12.2^{s}	4.4	s	-38.3^{s}	11.8^{s}	- 6.5	s
II+	232	10.5	-10.4^{s}	- 1.8	ns	- 4.5	1.8	-14.4^{s}	ns
	316	22.5^{s}	-22.2^{s}	-10.0	s	5.5	1.8	-15.2^{s}	ns
III-	232	- 0.9	- 4.8	1.7	ns	-48.6^{s}	15.2^{s}	- 4.3	s
	316	- 5.5	-10.9^{s}	- 4.8	s	-34.6^{s}	13.5^{s}	0.4	s
III+	232	- 0.6	-13.1^{s}	- 5.2	ns	- 9.9	- 0.4	- 5.7^{s}	ns
	316	6.5	-22.2^{s}	-13.9^{s}	s	-13.6	3.1	- 7.8^{s}	ns
Pitch I-	232	- 3.0	- 9.6^{s}	- 9.1^{s}	s	- 2.0	- 9.5	- 8.7^{s}	ns
	316	5.5	-10.0^{s}	- 7.8^{s}	s	- 5.0	- 6.9	-10.0^{s}	ns
I+	232	26.7^{s}	-22.2^{s}	-13.0^{s}	s	25.5^{s}	-12.6^{s}	- 8.7^{s}	s
	316	24.5^{s}	-21.3^{s}	-13.0^{s}	s	16.8	- 7.8	- 7.0^{s}	s
III-	232	0.3	-11.7^{s}	-13.5^{s}	s	- 8.2	- 4.8	- 6.9^{s}	ns
	316	9.7	-11.3^{s}	- 9.1^{s}	s	- 6.9	- 4.3	- 3.9	ns
III+	232	20.6^{s}	-25.7^{s}	-16.1^{s}	s	24.4^{s}	-12.6^{s}	-11.9^{s}	s
	316	21.8^{s}	-17.0^{s}	-12.6^{s}	s	13.0	-11.7^{s}	- 7.8^{s}	s
Combination -		3.6	-11.7^{s}	- 7.4^{s}	s	-16.1	0.5	- 8.2	ns
+		22.1^{s}	-23.1^{s}	-15.2^{s}	s	2.3	- 1.7	-10.0^{s}	ns

s significant at $p<0.01$
ns not significant
positive number indicates increased response compared to BaP

asphalt, the Type III (316° C) group exposed to light showed the most change and only the 316° C groups were significantly different by the onset rate analysis.

Consistent trends were found in the C_3H/HeJ mice exposed to condensed pitch fumes as compared to BaP. The mice receiving the low temperature material without light showed no change in percent TBA but significant decreases in latency and survival. When the temperature of generation was increased, again with no light exposure, increases in percent TBA were observed with similar changes in latency and survival as noted with the low temperature material. Mice exposed to light and pitch fumes from both temperatures experienced significant increases in tumor development and decreases in latency when compared to the BaP group. Decreased mean survival times paralleled the time to tumor changes. All pitch groups were found to be significantly different by the onset rate method from the corresponding BaP group. The groups receiving the combination treatment paralleled the effects noted above with the individual member component groups.

The CD-1 strain, in comparing their reaction to both types of condensed asphalt fumes and 0.01 percent BaP, had significantly decreased tumor incidences when exposed to the fumes and sham-irradiated. The average difference in reduced tumor response was 44.5 ± 9.9 percent TBA, with the largest reduction in the lower temperature groups. A corresponding significant increase in time to tumor was observed, i.e., approximately 14.5 ± 2.4 weeks with little reduction in survival as compared to the BaP group. All four groups were significantly different from the non-irradiated BaP group by the onset rate analysis method. The light exposed CD-1 mice as a subset of asphalt exposed groups showed a general slight but non-significant reduction in percent TBA, a significant decreased survival time and no change in mean latency time when compared to the BaP group. The overall effects analyzed by the onset rate method were not different from the BaP group.

The effects were reversed in the pitch exposed CD-1 groups. Slight reductions in percent TBA, significantly decreased survival and non-significant reductions in latency were observed in the groups not exposed to light as compared to the appropriate BaP group. These sham irradiated CD-1 pitch groups were not different from BaP by the onset rate method. Light exposed CD-1 mice, concurrently receiving condensed pitch fumes, had increased incidences of percent

TBA, averaging 19.9 ± 6.0. Here as before with the non-solar asphalt fume groups, the lower temperature materials induced more dramatic results. Survival was significantly reduced an average of 8.9 ± 2.1 weeks in these groups compared to the BaP group and latency time was generally decreased (11.2 ± 2.3 weeks) to a significant degree. Onset rate analysis showed significant differences for all light exposed pitch groups of CD-1 mice. The combination group reacted as anticipated and appeared to be more affected by the pitch component.

Combination Treatment

In comparing the combination group to their respective controls, the C_3H/HeJ strain showed no differences in latency, but the combination group showed a significantly increased response by the onset rate analysis method as compared to the asphalt group (Type III--316° C) without light. The responses in CD-1 mice were similar in that the non-solar exposed combination group had an significantly increased tumor response by the onset rate method than the asphalt group. They also died earlier and developed tumors at a significantly earlier time than the asphalt group. When the combination groups were compared to the responses of the pitch groups, onset rate analysis showed that the sham irradiated pitch group (Type I--316° C) had a significantly greater tumor response than the combination group. No other differences were noted.

To summarize the results, it is obvious from an examination of the data that: the asphalt and coal tar pitch fumes were highly carcinogenic; the condensed coal tar pitch fumes had more carcinogenic activity than the asphalts; the two types of each material had very similar activities; the nonpigmented CD-1 strain was less responsive to the carcinogenic activity of the materials than the pigmented C_3H/HeJ strain; increased temperature significantly increased the tumorigenic response of C_3H/Hej mice to the condensed asphalt fumes (the only pitch group to show a similar increased response was the Type III with concomitant exposure to light); no significant changes in the tumorigenic response of the CD-1 mice to increased temperature of fume generation were noted; simulated sunlight significantly inhibited tumorigenic response in C_3H/HeJ mice to both types of asphalt fumes collected at the lower temperature and no overall significant affects were noted in the CD-1 strain; both strains reacted similarly when exposed to condensed coal tar pitch fumes, BaP or the combination of pitch and asphalt in the presence of simulated sunlight, where the general response was a significant inhibition of tumorigenesis; in comparison to the C_3H/HeJ strain treated with 0.01 percent BaP, all coal tar pitch groups, the combination groups and those exposed the high temperature asphalt fumes showed an increased tumorigenic response; and, in the CD-1 strain, only non-solar asphalt fume exposed and solar exposed pitch fume groups showed this same significantly increased tumorigenic response when compared to the positive control.

DISCUSSION

Condensed volatiles from heated asphalt roofing materials were strikingly more tumorigenic and carcinogenic to C3H/HeJ mice than was expected from previous studies (1,2); their activity being nearly equal to that of the condensed volatiles from heated coal tar pitch roofing materials diluted to contain approximately 0.01 percent BaP. The CD-1 strain was more refractory to the asphalt fumes and the results were more in line with expectations based on historical observations, showing high tumorigenic activity with the coal tar pitch material and lower activity with the asphalts. These observations raise questions about the unique qualities of the two strains of mice used, e.g., levels of inducible microsomal enzymes and metabolic profile differences in handling PAHs. Pelkonen et al. (32), have found that additional genes besides the Ah locus may cause a particular mouse strain to be more sensitive or resistant to BaP initiated tumors than would be expected on the basis of metabolic induction capabilities. They found the C_3H inbred strain to be 5 to 15 times more responsive to subcutaneous fibrosarcoma initiation than other strains. They did not study the CD-1 strain.

It is interesting to note that Bingham et al. (2) reported very low or no activity with neat roofing asphalts in toluene (1:1 by weight) using the C3H mouse strain with essentially an identical protocol to the one used in the current study. Their results with solvent diluted roofing pitch derived from coal tar produced similar tumor incidences but with considerably shorter latent periods than observed in this study. The decreased latency is presumed to be related to the higher dosage of pitch used in the former investigation.

A significantly increased tumorigenic response was observed in the C_3H/HeJ strain using higher temperatures to generate the asphalt fumes. Most of this increased response was attributed to a decrease in time to tumor development rather than an increase in tumor bearing animals. This effect with asphalt fumes was not observed in the CD-1 non-pigmented strain. The pigmented strain, but not the CD-1 mice, also exhibited a significantly increased tumor response to higher temperature Type III coal tar pitch fumes in the presence of simulated sunlight, in spite of concentration adjustments to standardize the BaP content. It is inferred that the higher temperature coal tar pitch fumes may have higher specific activity in both strains when undiluted.

There appeared to be virtually no differences in tumorigenic activity in the same animal strain as a result of exposure to condensates of the two different types of asphalt or coal tar pitch materials generated at the same temperature. Therefore, it can be concluded that there is not a large change in relative carcinogenic risk with the type of asphalt or coal tar pitch material used. The objective to show differences in carcinogenic activity of the different ASTM classifications of materials and hence relative safety of the material could not be demonstrated in this study. However, Lowe et al. (33), noting that many investigators found differences in efficacy using tar to treat dermatoses, demonstrated that pharmacologic action varied significantly among sublots of the same crude coal tar collection. There is no reason to expect that asphalt fume activity would not also vary with different sources of crude oil.

Simulated sunlight, as used in this investigation, clearly caused an inhibition of the tumorigenic response. This is somewhat surprising and quite the opposite effect of that expected. There is some evidence suggesting long wavelength ultraviolet (280-350 nm) may act cooperatively way with chemicals to induce skin tumors in mice under certain laboratory conditions. (22,23) Further, Bingham and Nord (27) reported that exposure of mice to light at wavelengths above 350 nm coupled with exposure to normal alkanes results in tumor induction, whereas exposure to light alone or alkane alone yields few, if any, tumors. However, inhibition of the carcinogenic effects of UV light and dimethylbenzanthracene (DMBA) on the skin of Swiss mice was observed by Stenback and Shubik. (34) This was demonstrated when both UV light and DMBA were administered twice weekly for 4 weeks, with each UV treatment following DMBA treatment by 1 hour. Grossly observed tumors produced by DMBA alone numbered 36, whereas only 21 occurred with the combined treatments. In another schedule, a single UV exposure 1 hour before the first DMBA treatment, followed by 21 weeks of twice-weekly treatment of DMBA alone, resulted in an increased number of tumors as compared to the DMBA controls (52 vs. 30). Although a full comparison of these studies is not possible because of the different experimental conditions, i.e., strains of mice, materials and light source, it is clear that careful consideration must be given to the designs of future experiments involving interaction of light and chemicals. Attention must be given to similarities between the design and actual occupational exposure experience, such as with roofing workers.

Is this inhibition of carcinogenic response due to photooxidation/photodestruction of the carcinogens in these materials or to the modification of the skin, lessening its responsiveness, or to an actual cytotoxic effect on precancerous cells? Zepp and Schlotzhauer (35) as well as Mill et al. (36), determined half-lives of various photo-irradiated PAH in water to range from less than one hour to several hours. BaP and other carcinogenic PAH were among those most easily photooxidized in less than one hour. Katz et al. (37) found the half-lives of many of these same chemicals in simulated sunlight atmospheric conditions to be about the same order of magnitude. The presence of 0.2 ppm ozone markedly decreased the observed half-lives. The relative contribution of this observation to the general inhibition phenomenon observed in the current study remains to be determined. Fisher and Kripke (38) using C_3H/HeN(MTV-) mice concluded that suppressor T lymphocytes induced in mice by UV radiation play a decisive role in skin carcinogenesis by reducing the latent period in photocarcinogenesis. This mechanism does not appear to apply to the current study where the general response was an inhibition of the carcinogenic process.

Preliminary chemical characterization according to EPA-RTP Level 1 procedures (39,40) of a direct extract of coal tar pitch used in these studies and of a pitch fume sample generated from a full-scale roofing kettle (which was not the identical fume material used in these carcinogenicity studies) suggests that heterocyclic sulfur and nitrogen compounds and organic nitriles are present in the pitch samples at levels of 1-10 percent of the PAH content. El-Bayoumy and coworkers (41) tested tumor initiating activity of four nitro PNAs and their parent hydrocarbons and found that only nitroperylene induced higher activity than the parent compound; yet, this activity was still much less than that induced by BaP. Pelroy and coworkers (42,43) used microbial mutagenicity methods to assess the activity of sulfur heterocyclics. None of the compounds tested exhibited activity greater than BaP or the neutral fractions from which they were isolated. Thus, species of these types may contribute slightly to the carcinogenic effect of the roofing pitch fume materials; however, it remains clear based on the data in Table 7 and from a previous report (20) that BaP is a reasonable indicator substance of coal tar pitch fume exposure. From Table 7 it can be calculated that BaP accounts for approximately 75 percent of the total carcinogenic activity of the coal tar pitch fumes.

Since the PAH content of the asphalt volatiles is very low compared to that of the coal tar pitch preparations, what other chemical components do these asphalts fumes contain which contribute to the observed carcinogenicity? Are these promoters, cocarcinogens, or even other carcinogens not of the PAH chemical class? This point is further emphasized by the analysis presented in Table 7. The amount of total test material applied by the time of 50 percent tumor incidence (C3H mice) is very much larger for the asphalts compared to the coal tar pitch fume. Approximately 10 fold more asphalt fume is required to obtain the same biological effect. The amounts of total PAH analyzed or BaP found in asphalt fumes are 30 to 40 fold less than those amounts found to be effective in eliciting the same response with condensed coal tar pitch fumes. This suggests that the asphalt preparations contain other types of chemical components which may augment the activity of the very low concentrations of measured PAH found in this study. It is likely that alkylated PAH derivatives and/or aliphatic hydrocarbons may account for some of this activity. Although complete chemical characterization data of the asphalt and pitch fume samples are not available at this time, the PAH analyses have been supplemented by nuclear magnetic resonance (NMR) analysis (39). The latter results imply that the asphalt fume material is <1 percent aromatic, >99 percent aliphatic, with straight-chain, unbranched materials predominating. In contrast, NMR analysis of the pitch fume indicated >90 percent aromatic content. These analyses are reinforced by earlier work. Puzinauskas and Corbet (44) characterized the carbon types found in roofing materials. They found roofing pitch was composed of 79, 18 and 3 percent while roofing asphalt was

TABLE 7

TOTAL DOSE OF TEST MATERIAL APPLIED UP TO TIME OF 50 PERCENT TUMOR INCIDENCE (C3H MICE--NON SOLAR)

Test Material	Time to 50 percent T.I.* (weeks)	Amount of Material per application			Total Dose Applied for ED50**		
		Total Solids mg	PAH µg	BaP µg	Total Solids mg	PAH mg	BaP µg
Asphalts							
Type I - 232°	50.4	25	23.2	0.110	2520	2.34	11.1
Type I - 316°	39.5	25	7.4	0.095	1975	0.58	7.5
Type III - 232°	46.9	25	28.0	0.145	2345	2.63	13.6
Type III - 316°	40.8	25	9.1	<0.025	2040	0.74	<2.0
Coal Tar Pitches							
Type I - 232°	42.1	3.90	>560	4.80	328	>47.2	404
Type I - 316°	41.7	2.75	603	4.25	229	50.3	354
Type III - 232°	40.0	4.20	>306	5.10	336	>24.5	408
Type III - 316°	40.4	1.50	>711	4.50	121	>57.4	364
BaP	51.7	--	5.0	5.00	0.5	0.5	517

* Time in weeks to produce 50 percent tumor incidence (histologically confirmed) where number of applications = weeks X 2.

** No. of applications x amount per application = Effective Dose for 50 percent Tumor Incidence characterized as 37, 23 and 40 percent aromatic, naphthene (cycloparaffins and cyclohexanes) and paraffinic carbon, respectively. The high aliphatic content of the asphalt fume sample is interesting in light of earlier observations (45,46,47) that some alkyl hydrocarbons exhibit cocarcinogenic activity. Further characterization, fractionation and testing of these materials are now underway which will emphasize the indentification of indicator compound(s) for asphalt fume occupational exposure.

In order to more fully judge the comparative carcinogenicity of exposure to roofing material fumes, Figure 3 was compiled to illustrate relative emission rates, dosage of fumes to elicit an effective 50 percent tumor incidence (ED_{50}) and the number of minutes of generation needed to produce sufficient fumes for an ED_{50}. This was composed for asphalts and coal tar pitches at the suggested kettle operation temperature and at the temperature found during an overheat condition.

Theoretically, an asphalt kettle being maintained at the recommended temperature would emit approximately 37 and 11.5 times less fume than an overheated coal tar pitch or asphalt kettle, respectively. A similar comparison using a high temperature asphalt kettle could theoretically emit as much fume as a properly operated coal tar pitch kettle and approximately one third that of an overheated coal tar pitch kettle. When the time of generation needed to collect sufficient material to produce 50 percent tumor incidence in C3H/HeJ mice is calculated, the ratio of low temperature asphalt fumes to high temperature coal tar pitch fumes represents a greater than 500 fold difference; however, this difference is much less at the lower suggested temperatures of operation for the pitch. This analysis points out the need to maintain kettle temperatures as low as possible in order to reduce the carcinogenic risk.

Penalva and coworkers (48) studied mutagenic activity of aerosols and vapors emitted by road coating tar at various temperatures and found increasing activity as the temperature of the tar was increased from 250° to 550° C. Similar effects were noted by Mahlum (49) in his assessment of distillation cuts from coal liquids. This same trend was determined by Tye et al. (50) in assessing the carcinogenic potency of a catalytically cracked petroleum. He found that tumor initiating activity increased as the boiling point of the fraction increased. These results agree well with the observations from the

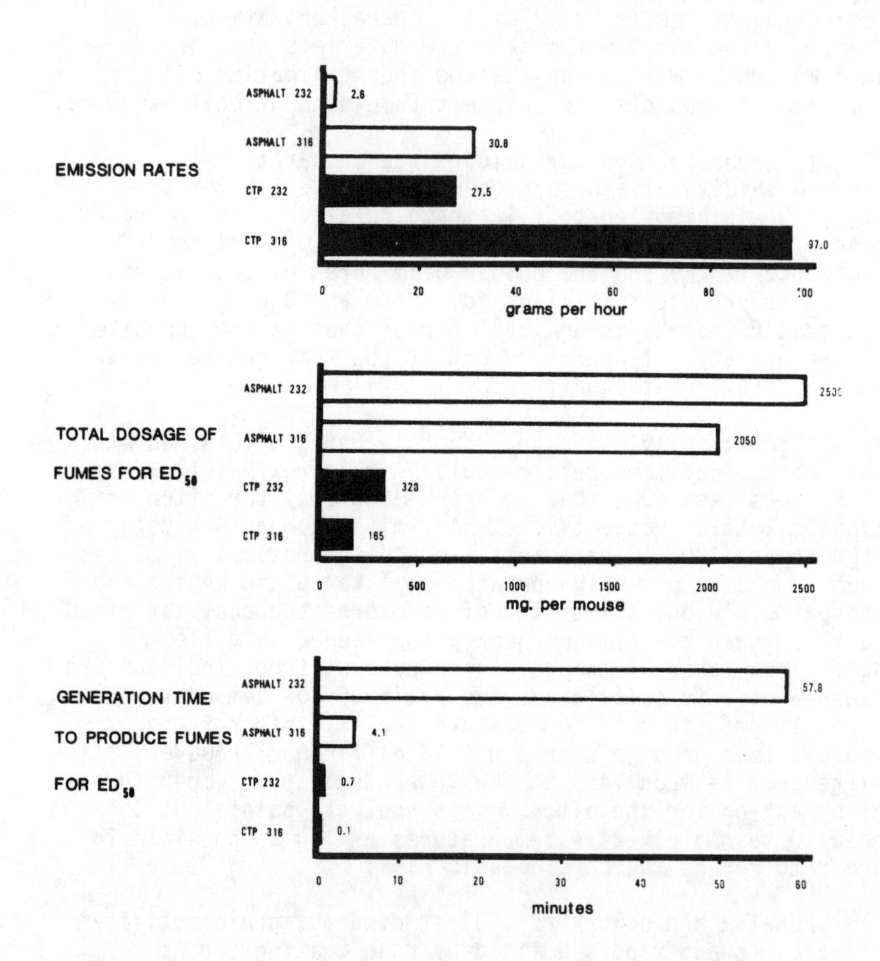

FIGURE 3. Comparison of generation times necessary to produce sufficient roofing material fumes for an ED_{50} tumorigenic response in C3H/HeJ mice

current study, that higher temperature materials are more hazardous.

Table 8 lists theoretical air concentrations of total condensed fume, PAH and BaP from asphalt and coal tar pitch fumes when generated in the laboratory apparatus at 232 and 316° C. Also listed are field concentrations of PAH and BaP measured by Malaizandi and coworkers (51) at roofing sites using either asphalt or coal tar pitch materials. The absolute numbers are orders of magnitude higher for the laboratory data as compared to field data, but this is expected since field concentrations should reflect significant dilution by ambient air. However, there is reasonable agreement among laboratory and field data when the ratios of PAH/BaP concentrations are calculated. The ratios for laboratory generation of asphalt fumes is 157 and 122 for low and high temperatures, respectively, and the field collected ratio is 330. For coal tar pitch fumes the laboratory ratio is 88 and 153 for low and high temperature generations, respectively, where the field collected ratio is 159. We conclude from these observations that the relative carcinogenic activity of fumes generated in the field should be similar to those demonstrated in the laboratory setting.

CONCLUSIONS

The data from these experiments vindicate the prior concern of NIOSH that there is a greater occupational hazard from exposure to asphalt fumes than was documented in the literature. This investigation clearly demonstrates that exposure to asphalt fumes significantly increases the potential carcinogenic risk, and as expected, higher use temperatures further increase this potential risk.

The carcinogenic activity of roofing coal tar pitch fume materials may be understandable, to a first order approximation, in terms of their BaP content. This ubiquitous carcinogen is therefore a satisfactory indicator substance in estimating relative coal tar pitch fume exposures in roofing operations.

The carcinogenic activity of the asphalt fume materials, on the other hand, cannot even approximately be explained on the basis of their BaP content. It can be hypothesized, although it has not been demonstrated, that cocarcinogenic effects of aliphatic hydrocarbons, which may be major components of the asphalt fumes generated in this

TABLE 8

ASPHALT AND COAL TAR PITCH THEORETICAL LABORATORY[a] VS. FIELD SURVEY[b] AIR CONCENTRATIONS

	Total Condensed Fumes (g/m^3)	PAH ($\mu g/m^3$)	BaP ($\mu g/m^3$)	Ratio of PAH/BaP
Asphalt				
232°	4.3	3300	21	157
316°	51.3	15000	122	122
IH[b]	--	33.1(±38.7)	0.119(±0.152)	330
Coal Tar Pitch				
232°	45.8	5.0 X 10^6	56.6 X 10^3	88
316°	161.7	56.1 X 10^6	366.7 X 10^3	153
IH[b]	--	476(±260)	3.1(±4.3)	159

a from this study based on emission rates at 10 Lpm
b Malaizandi et al.[51] field study mean (± standard deviation)

study, are probably responsible for this enhanced activity. A recent Danish report (52) suggests that emissions from an asphalt production plant may have augmented the lung cancer risk in smokers living and working in immediate vicinity. It is evident that further research is necessary before the active components of asphalt fumes are identified. Evidence suggests that emphasis should be placed on the promoting/ cocarcinogenic activity of asphalt fumes and that an approach different from that used in studying coal tar pitch must be devised. A protocol using PAH-free asphalt fume condensate administered simultaneously or following subcarcinogenic doses of BaP, similar to that used by Bingham and Falk (46), could provide useful information in the future characterization of this material.

In the interim reasonable measures should be taken to limit exposure as much as feasible to both coal tar pitch and asphalt fumes using available engineering and work practice controls. One major work practice that can be instituted immediately is the maintenance of proper heating conditions in the kettle thereby reducing potential risk possibly to a substantial degree, i.e., approximately a 7 to 14 fold theoretical decrease for coal tar pitch and asphalt fumes, respectively.

Even though it was not demonstrated in this study, it is quite possible that exposure to both asphalt and coal tar pitch fumes may be more hazardous than exposure to either alone. For if our suspicions prove to be correct, that the activity of asphalt fumes is primarily attributable to its cocarcinogenic activity, then reduction of carcinogen levels to as low as one thousandth of the original concentration may still lead to carcinogenic expression above the threshold response (46). This is partially supported in the simulated sunlight experiments where photooxidation probably destroys many carcinogens, but not the cocarcinogenic moieties. One must be very careful in assuming large reductions in risk due to relatively lower exposures to carcinogens and cocarcinogens. As Bingham and Falk (46) demonstrated, cocarcinogens of the aliphatic hydrocarbon type are most noticeable in their effect when the concentration of the carcinogen is low. They also demonstrated that low levels of cocarcinogens (less than 20 to 30 percent) appear to have the most marked effect on tumor outcome at lower levels of carcinogen exposure. This may be the condition most frequently encountered in the working population and therefore its importance cannot be overemphasized in attempts to determine the hazards to man.

It is interesting that sunlight may offer some protection against the carcinogenic activity of these materials; however, this will be extremely difficult to demonstrate in the working population. The future needs of extensive epidemiological studies, increased medical surveillance and detailed industrial hygiene surveillance in this hazardous work environment are quite evident.

ACKNOWLEDGMENT

The fume generation, PAH analyses, and skin-painting carcinogenicity studies reported here were conducted under NIOSH Contract 210-78-0035 through an interagency agreement with the National Cancer Institute. The supplementary Level 1 analyses of pitch samples and NMR analyses of pitch and asphalt fumes were supported by EPA Contract 68-02-3111. We also acknowledge the assistance of Dr. J. C. Harris from A. D. Little and the invaluable assistance of the following NIOSH personnel: Ms. Sandra Clark in typing this manuscript, and Dr. Robert Mason for his critique and guidance in composing the manuscript.

DISCLAIMER

Mention of companies or commercial products in this paper does not indicate endorsement by NIOSH.

REFERENCES

1. NIOSH (1977): Criteria for a Recommended Standard--Occupational Exposure to Asphalt Fumes. DHEW (NIOSH) Publication No. 78-106.
2. Bingham, E., Trosset, R.P., Warshawsky, D.: Carcinogenic potential of petroleum hydrocarbons. J Environ Path Toxicol 3:483-563, 1980. (See also Trosset R.P., Warshawsky D., Menefee C.L., Bingham E.: Investigation of selected potential environmental contaminants: asphalt and coal tar pitch. Final Report EPA 560/2-77-005, September 1978.)
3. NIOSH: Criteria for a Recommended Standard: Occupational Exposure to Coal Tar Products. DHEW(NIOSH) Pub. No. 78-107, 1977.
4. Simmers, M.H., Podolak, E., Kinosita, R. (1959): Carcinogenic effects of petroleum asphalt. Proc Soc Exper Biol Med 101:266-268.
5. Simmers, M.H. (1964): Petroleum asphalt inhalation by mice. Arch Environ Health 9:727-734.
6. Simmers, M.H. (1965): Cancer from air-refined and steam-refined asphalt. Ind Med Surg 34:255-261.
7. Simmers, M.H. (1965): Cancers in mice from asphalt fractions. Ind Med Surg 34:573-577.
8. Simmers, M.H. (1966): Tumors from asphalt fractions injected into mice. Ind Med Surg 35:889-894.
9. Hueper, W.C., Payne, W.W. (1960): Carcinogenic studies on petroleum asphalt, cooling oil, and coal tar. AMA Arch Path 70:372-384.
10. Hueper, W.C. (1965): Blown asphalt not carcinogenic. Am Ind Hyg Assoc J 26:95.
11. Wallcave, L., Garcia, H., Feldman, R., et al.(1971): Skin tumorigenesis in mice by petroleum asphalts and coal-tar pitches of known polynuclear aromatic hydrocarbon content. Toxicol Appl Pharmacol 18:41-52.
12. Kireeva, I.S. (1968): (Carcinogenic properties of coal-tar pitch and petroleum asphalts used as binders for coal briquettes). Hyg San 33:180-186. English Translation of Gig Sanit 33:35-40.
13. Hervin, R.L., Emmett, E.A. (1976): Sellers and Marquis Roofing Company, AJ Shirk Roofing Company, Western Roofing Company and the Quality Roofing Company--A Joint Venture--Kansas City, Missouri, Health Hazard Evaluation Determination Report No. 75-102-304. Cincinnati, U.S. Department of Health, Education, and Welfare, Center for Disease Control, National Institute for Occupational Safety and Health Hazard Evaluation and Technical Assistance Branch, June 1976, 34 pp.

14. Thomas, J.F., Mukai, M. (1975): Evaluation of emissions from asphalt roofing kettles with respect to air pollution. College Park, MD: The Asphalt Institute Research Report No. 75-2, 21 pp.
15. Puzinauskas, V.P., Corbett, L.W. (1975): Report on emissions from asphalt hot mixes. College Park, MD: The Asphalt Institute Research Report No. RR-75-1A.
16. Baylor, C.H., Weaver, N.K. (1968): A health survey of petroleum asphalt workers. Arch Environ Health 17:210-214.
17. Hammond, E.C., Selikoff, I.J., Lawther, P.L., et al. (1976): Inhalation of benzpyrene and cancer in man. Ann. N.Y. Acad. Aci 271:116-124.
18. Lawther, P.J. (1971): Pollution at work in relation to general pollution. Royal Soc Health J 91:250-253.
19. Shabad, L.M., Yanysheva, N.Y. (1981): The significance of the polycyclic aromatic hydrocarbons as carcinogenic agents for man. Gig Trudy Prof Zabol 25:29-33.
20. Bingham, E.: Report of the standards advisory committee on coke oven emissions to the Assistance Secretary of Labor, OSHA. May 24, 1975.
21. International Agency for Research on Cancer. Monographs on the evaluation of carcinogenic risk of chemicals to man Vol 3. Certain polycyclic aromatic hydrocarbons and heterocyclic compounds. IARC, Lyon, France, 271 pp, 1973.
22. Santamaria, L., Giordano, G.G., Alfici, M., et al. (1966): Effects of light on 3,4-benzpyrene carcinogenesis. Nature 210:824-825.
23. Urbach, F. (1959): Modification of ultraviolet carcinogenesis by photoactive agents. J Invest Derm 32:373-378.
24. NIOSH: National Occupational Hazard Survey Vol 1 Survey Manual DHEW(NIOSH) Pub. No. 74-127, 1974. Vol 2 Data Editing and Data Base Development DHEW(NIOSH) Pub. No. 77-213, 1977. Vol 3 Survey Analysis and Supplemental Tables DHEW(NIOSH) Pub. No. 78-114, 1977.
25. American Society for Testing Materials: 1978 Annual Book of Standards, Part 15, Roads and paving materials; Bituminous materials for highway construction, waterproofing and roofing, and pipe; Skid Resistance. 1916 Race Street, Philadelphia, PA 19103.
27. Bingham, E., Nord, P.G. (1977): Cocarcinogenic effects of n-alkanes and ultraviolet light on mice. J Natl Cancer Inst 58:1099-1101.
28. Burns, F. (1978): NYU Medical Center--Personal communication.

29. Thayer, P.S., Menzies, K.T., von Tnuna, P.C. (1981): Roofing asphalts, pitch and UVL carcinogenesis. Final report on NIOSH Contract 210-78-0035, Cincinnati, OH.
30. Miller, R.G. (1981): Survival Analysis. John Wiley and Son USA.
31. Peto, R., Pike, M.C., Day, N.E., et al. (1980): Guideline for simple, sensitive significance tests for carcinogenic effects in long-term animal experiments, Chapt. in Long-Term and Short-Term Screening Assays for Carcinogens: A critical Appraisal. IARC Monograph, Lyon.
32. Pelkonen, D., Boobis, A.R., Levitt, R.C., et al. (1979): Genetic differences in the metabolic activation of benzo[a]pyrene in mice. Pharmacology 18:281-293.
33. Lowe, N.J., Breeding, J., Wortzman, M.S. (1979): The pharmacological variability of crude coal tar. Brit. J. Dermatol. 107:475-480.
34. Stenback, F., Shubik, P. (1973): Carcinogen-induced skin tumorigenesis in mice: enhancement and inhibition by ultraviolet light. Z Krebsforsch 79:234-240.
35. Zepp, R.G., Schlotzhauer, P.F. (1979): Photoreactivity of selected aromatic hydrocarbons in water. In Polynuclear Aromatic Hydrocarbons, Third International Symposium. Jones, P.W., Leber, P (eds) Ann Arbor Science Publishers, Inc., Ann Arbor, MI. pp 141-158.
36. Mill, T., Mabey, W.R., Lan, B.Y. et al (1981): Photolysis of polycyclic aromatic hydrocarbons in water. Chemosphere 10:1281-1290.
37. Katz, M., Chan, C., Tosine, H. et al. (1979): Relative rates of photochemical and biological oxidation (in vitro) of polynuclear aromatic hydrocarbons. In Polynuclear Aromatic Hydrocarbons, Third International Symposium. Jones PW, Leber, P (eds) Ann Arbor Science Publishers, Inc., Ann Arbor, MI. pp 171-189.
38. Fisher, M.S., Kripke, M.L., Suppressor, T. (1982): Lymphocytes control the development of primary skin cancers in ultraviolet-irradiated mice. Science, 216:1133-1134.
39. Thayer, P.S., Harris, J.C., Menzies, K.T., et al. (1983): Integrated chemical and biological analysis of asphalt and pitch fumes. In Short-Term Bioassays in the Analysis of Complex Environmental Mixtures III. Waters, M.D., Sandhu, S.S., Lewtas, J., Claxton, L., Chernoff, N., and Nesnow, P. Plenum Publishing Corp., pp. 351-366.

40. Lentzen, D.E., Wagoner, D.E., Estes, E.D., et al. (1978): IERL-RTP Procedures Manual: Level 1 Environmental Assessment (Second Edition). EPA-600/7-78-201.
41. El-Baryoumy, K., Hecht, S.S., Hoffmann, D. (1982): Comparative tumor initiating activity on mouse skin of 6-introbenzo[a]pyrene, 6-nitrochrysene, 3-nitropyrene, 1-nitropyrene and their parent hydrocarbons. Cancer Letters, 16:333-337.
42. Pelroy, R.A., Stewart, D.L., Tominaga, Y., Iwao, M., Castle, R.N., Lee, M.L. (1983): Microbial mutagenicity of 3- and 4-ring polycyclic aromatic heterocycles. Mutation Research, 117:31-40.
43. Willey, C., Pelroy, R.A., Stewart, D.L. (1982): Comparative analysis of polycyclic aromatic sulfur heterocycles isolated from four shale oils. In Polynuclear Aromatic Hydrocarbons: Physical and Biological Chemistry--Sixth International Symposium, Cooke M, Dennis AJ, and Fisher GL editors, Battelle Press/Springer-Verlag pp 907-917.
44. Puninanskas, V.P., Corbet, L.W. (1978): Differences between petroleum asphalt, coal-tar pitch and road tar. The Asphalt Institute, Research Report 78-1 (RR-78-1), 31 p.
45. Horton, A.W., Denman, D.T., Trosset, R.P. (1957): Carcinogenesis of the skin. The accelerating properties of aliphatic and related hydrocarbons. Cancer Res 17:758-766.
46. Horton, A.W., Burton, M.J., Tye, R., et al.(1963): Composition vs. carcinogenicity of distillate oils. ACS Div. of Petroleum Chem. Preprints 8, No. 4C:59-65.
47. Bingham, E., Falk, H.L. (1969): Environmental carcinogens. The modifying effect of cocarcinogens on the threshold response. Arch Environ Health 19:779-783.
48. Penalva, J.M., Chalabreysse, J., Archimbaud, M., Bourgineau, G. (1983): Determining the mutagenic activity of a tar, its vapors and aerosols. Mutation Research, 117:93-104.
49. Mahlum, D.D. (1983): Skin-tumor initiation activity of coal liquids with different boiling point ranges: J. Applied Toxicol, 3:254-258.
50. Tye, R., Burton, M.J., Bingham, E., Bell, Z., Horton, A.W. (1966): Carcinogens in cracked petroleum residium. Arch. Environ. Health 13:202-207.

51. Malaiyandi, M., Benedek, A., Holko, A.P., Bancsi, J.J. (1982): Measurement of potentially hazardous polynuclear aromatic hydrocarbons from occupational exposure during roofing and paving operations. In Polynuclear Aromatic Hydrocarbons, Sixth International Symposium, Cooke M, Dennis AJ, Fisher GL, editors, Battelle Press/Springer-Verlag pp 471-489.
52. Wilson, K. (1984): Asphalt production and lung cancer in a Danish village. Lancet (Aug 11):354.

MUTAGENICITY AND TOXICITY OF PAH CONTAMINATED MARINE SEDIMENT

R. B. ODENSE[1], M. S. HUTCHESON[1], J. D. POPHAM[2], B. F. FOWLER[2]
(1)Seakem Oceanography Ltd., 46 Fielding Ave., Dartmouth, Nova Scotia B3B 1E4; (2)Seakem Oceanography Ltd., 2045 Mills Road, Sidney, British Columbia V8L 3S1.

INTRODUCTION

The sediments of Sydney Harbour, Nova Scotia have been contaminated by polynuclear aromatic hydrocarbon (PAH) containing discharges from the Sydney Steel Mill for over 80 years (1). Air and water contamination by the steel mill may in part be responsible for the increased cancer incidence in the local area.

Sediment samples were collected from the area of the harbour adjacent to the steel mill. Mutagencity tests of the sediments were performed in conjuction with a program to develop a microorganism test of dredge spoil toxicity. Chemical analysis of the sediments included PAH, chlorophenol, phthalate ester and chlorinated hydrocarbon determinations.

MATERIALS AND METHODS

Sampling Sites

Sediment samples were collected from twenty sites along Muggah Creek and stored frozen at -20°C. See Figure 1. The character of the sediment ranged from coarse sand at the mouth of the creek to black oily silt near the tar pond at the head of the creek. It was expected that a gradient of PAH levels existed along this creek.

Sediment Chemistry

Samples for PAH analysis were extracted by a base digestion method based upon that of Cretney *et al*. (2). Samples were analysed on a Finnigan 9500/3200 gas chromatograph-mass spectrometer (GC/MS), with a Finnigan 6100 data system using the following conditions:

Column:	30 m x 0.25 mm BP-1 bonded phase fused silica column (S.G.E.)
Carrier Gas:	helium
Injector Flow Rate:	60 mL min^{-1}
Injector Pressure:	17 p.s.i.g.

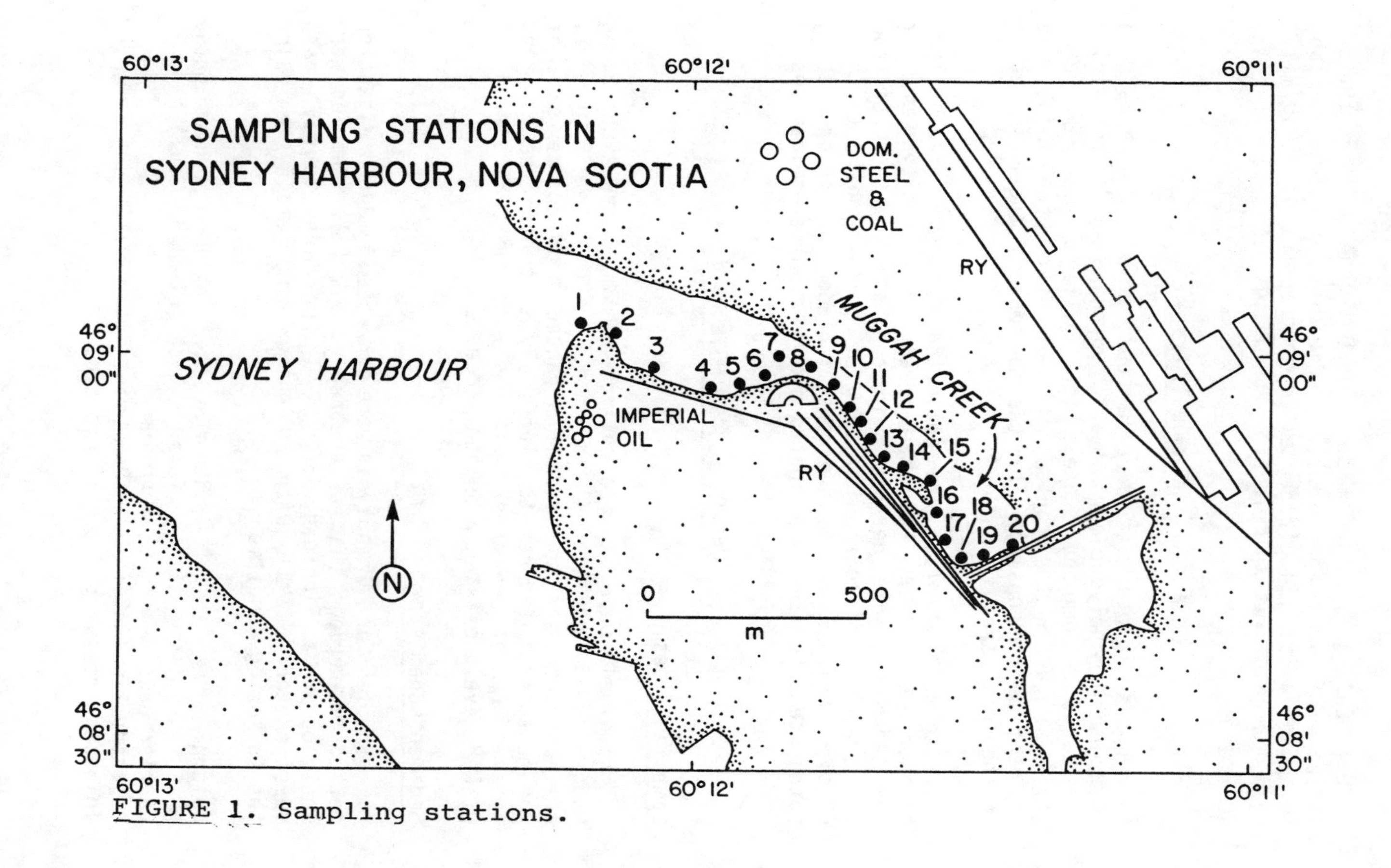

FIGURE 1. Sampling stations.

Column Flow	40 cm s^{-1}
Split Ratio:	40:1 (approximately)
Injector Temperature:	260°C
Injection Sequence:	splitless injection at room temperature, splitting resumed at 1 minute, 100°C at 2 minutes and 10° min^{-1} at 6 minutes to 280°C and hold for 10 minutes. 0.5 uL injections.
Mass Spectrometer:	electron impact source
Source Emission:	0.50 mA
Electron Energy:	40 eV
Operating Pressure:	1 x 10^{-5} torr
Multiplier Voltage:	2000 V (gain > 10^6)
Data System:	data acquired in the "selected ion monitor" mode
Acquisition Rate:	one scan/sec; four ions per scan in five clusters of four ions per run.

The extraction of chlorinated hydrocarbons and phthalate esters was based upon the procedure of Inland Waters Directorate (3). The analysis of chlorinated hydrocarbons was performed on a Hewlett Packard 5830A gas chromotagraph using 30 m x 0.25 mm DB5 Durabond Fused Silica capillary column (J. & W. Scientific), a ^{63}Ni electron capture detector and the following conditions:

Carrier Gas:	Hydrogen: 60 mL/min, approximate 40:1 split ratio
Make-up Gas:	Argon with 5% methane; 30 mL/min
Injector Temperature:	250°C
Detector Temperature:	300°C
Oven Temperature:	Programmed 100°C to 180°C at 20°/min with a 1 min initial hold at 100°C then 10°/min to 270°C, and held for 5 min.
Syringe:	Hamilton 10 uL syring
Injection Mode:	Grob splitless injection, stream splitting resumed after three minutes
Injection Pressure:	12 psi (split), 37 psi (splitless)

A composite pesticide standard and a PCB standard (Aroclor 1254) are run daily before samples to verify instrument performance, establish current retention times and determine current response factors.

The phthalate esters were analysed in the third Florisil fraction (F3) from the PCB-organochlorine extract by GC-MS (Finnegan 9500/3200/6100) using selected ion monitoring.

SEDIMENT MUTAGENICITY AND TOXICITY

The extraction of chlorophenols was based on the procedure of Renburg (4). The analysis was carried out on a Hewlett Packard 5830A gas chromatagraph under the same conditions as the chlorinated hydrocarbon analysis except for the following:

Oven Temperature:	Programmed 80° to 140°C at 10°/min with a 1 min initial hold at 80°C and a hold at 140°C for 3 min.
Injector Pressure:	16 psi (split), 37 psi (splitless)
Splitless Time:	1 min.

Ames Test

Sediment Extraction. Extractions were performed on sediments from five sites along Muggah Creek: Site #1, 11, 15, 18, 20. 125 mL methanol was added to 125 gm, wet weight, of sediment in a 250 mL teflon jar. This mixture was shaken on a paint shaker for 15 min and vacuum filtered through a Whatman GF/A glass microfibre filter. This methanol extraction was repeated twice, after which the sediment was extracted in a similar fashion with 9:1 dichloromethane:methanol three times. The extracts were combined and filtered. The dichloromethane extract from site 20 would not filter and had to be centrifuged at 8000 g for 5 min in order to remove the particulates.

Sample Fractionation. Extracts were fractionated into neutral, acid and base fractions. A volume of water equal to 40% of the total extract volume was added to the extract and pH adjusted to 7.0 with 1 NHCl or 20% NaOH. After shaking 100 times the organic layer was drawn off (neutral fraction). A volume of dichloromethane equal to one half the neutral fraction volume was then added and the pH adjusted to 2.0. This was shaken and the organic layer drawn off (acid fraction). After dichloromethane was again added the pH was adjusted to 11.0. This was shaken and the organic layer was drawn off (base fraction).

The extracts were dried with anhydrous Na_2SO_4 and reduced in volume by rotary evaporation to between 5 and 20 mL. The remainder of the solvent was removed by passing nitrogen over the extract. The extract was made up in dimethyl sulfoxide (DMSO) to a volume of 10.0 mL. A subsample of 1.0 mL (4.3%) of the Site 20 neutral extract was dissolved in 9.0 mL DMSO.

Mutagenicity Testing. The Ames tests were performed as recommended in Maron and Ames (5). The neutral, acid and base extracts from each site were tested on the four standard Ames strains: TA97a, TA98, TA100 and TA102. All tests were done with and without activation by S9 liver microsomes (obtained from Litton Bionetics - Aroclor 1254 induced male Sprague-Dawley rats). Positive and negative controls were run with each test. Two hundred uL of extract were used per plate.

Toxicity Tests

Culture of Blepherisma. Stock cultures of ciliates were maintained in 50 mL conical flasks containing enriched seawater growth media (see Table 1). Bioassays were carried out with strains derived from single organisms. Six Blepherisma were placed in each of 15 disposable petri dishes (52 mm dia.) along with 4.0 mL of test solution and a 5x5 mm square of eel grass. The dishes were placed in a constant temperature water bath and maintained at 20°C. Fifteen controls were run simultaneously and growth was monitored over six days.

TABLE 1

CULTURE MEDIA FOR BLEPHARISMA

KNO_3	250 mg
$NaH_2PO_4H_2O$	34.5 mg
Na_2SiO_3 $9H_2O$	84.0 mg
Thiamine HCl	0.5 mg
Biotin	1 ug
B_{12}	2 ug
Buffer:	
TRIS•HCL (41.3 mM)	200 mL
Seawater (33 °/oo) to 1 liter	

Sediment Extraction. Seawater extracts of the sediment were prepared by first adding 20 mL of sediment to 80 mL of seawater in a 250 mL flask and stirring for 15 minutes. This mixture was passed through a coarse filter and then a #54 Whatman filter and the filtrate collected.

RESULTS

Sediment Chemistry

The results of the PAH determinations are given in Table 2. There is apparently a steep gradient of PAH levels along Muggah Creek, ranging from 5.4 ug g^{-1} dry weight at the mouth of Muggah Creek to 911 ug g^{-1} at the head of the creek. At Site 20, the levels of napthalene, phenanthrene, fluoranthene and pyrene were all over 100 ug g^{-1}.

TABLE 2

POLYCYCLIC AROMATIC HYDROCARBONS (ug.g^{-1}, dry weight)

Compound	Site 1	Site 11	Site 20
naphthalene	0.06	9.8	106
2 methyl naphthalene	0.1	1.9	16
1 methyl naphthalene	<0.1	1.5	14
fluorene	0.05	2.2	43
phenanthrene	0.65	13	174
anthracene	0.31	5.0	58
fluoranthene	1.0	36	152
pyrene	0.83	21	110
benz(a)anthracene	0.54	16	56
chrysene	0.63	11	38
benzo(e)pyrene	0.46	9.5	56
benzo(a)pyrene	0.59	12	68
perylene	0.23	32	20
Total	5.4	142	911

The results of the chlorinated hydrocarbon determination also indicate a steep gradient along Muggah Creek. PCB values are: Site 1, 20 ng g^{-1} dry weight; Site 11, 1730 ng g^{-1}; Site 20, 52,000 ng g^{-1}. The pesticides, dieldrin, heptachlor epoxide and trans-chlordane were also detected although not above the level of 13.0 ug g^{-1}.

The main species of phthalate ester identified in the sediment samples was bis(2-ethylhexyl) phthalate. The values for the total monitored phthalate ester are: Site 1, 80 ng g^{-1}; Site 11, 7100 ng g^{-1}; Site 20, 6080 ng g^{-1}.

None of the five chlorophenol species monitored was detected in Site 1 sediment, whereas all five were detected in Site 11 sediment. The highest value detected was 98 ng g^{-1} for 2,3,4,5-tetrachlorophenol in Site 11 sediment.

Ames Test

The quantity of material in the sediment extracts for the Ames test were estimated for the site with the least (Site 1) and the most (Site 20) extractible material. Ten mL subsamples of the neutral, acid and base extracts were air dried overnight in a fume hood and the residue weighed. The amount of extractable material from 125 gm of wet sediment ranged widely, from approximately 0.53 gm at Site 1 to 25.62 gm at Site 20. The neutral fraction contained far more material than the acid or base fractions yet it was these latter extracts which were mutagenically active. For the mutagenicity tests, the amount of extracted material applied per plate in the neutral, acid and base fractions were: 9.9 mg, 0.28 mg and <0.26 mg respectively for Site 1. The corresponding values for Site 20 were 17.8 mg, 74 mg and 2.4 mg. Note that the neutral extract for Site 20 was reduced to 4.3% of the initial extract quantity.

Of the four strains of *S. typhimurium* bacteria used in the mutagenicity tests, only the two strains sensitive to frameshift mutation, TA97a and TA98, indicated significant mutagenicity in the sediment extracts (see Table 3 and Figure 2). All of the tests using the strains sensitive to base pair mutation (TA100 and TA102) were negative. Tests were considered significantly mutagenic if they had a doubling of reversion frequency over controls. Some of the mutagens required activation by liver microsomes, others were deactivated by the addition of liver microsomes. Sediments from all five sites were mutagenic and in general, the further up Muggah Creek the greater the mutagenicity of the activated base extracts. Site 20 does not fit this trend, however, it was extracted differently from the other sites and it is likely that the extraction efficiency was relatively poor.

In the nonactivated mutagenicity tests only the TA98 strain had a significantly increased mutation frequency and this occurred only for the acid fraction of Site 15 and 18 sediment extracts. The mutagenicity of these extracts was abolished by the addition of liver microsomes (deactivated). None of the nonactivated neutral fractions were significantly mutagenic, although the Site 15 and 18 extracts almost caused a doubling of reversion frequency in the tests with TA98. All

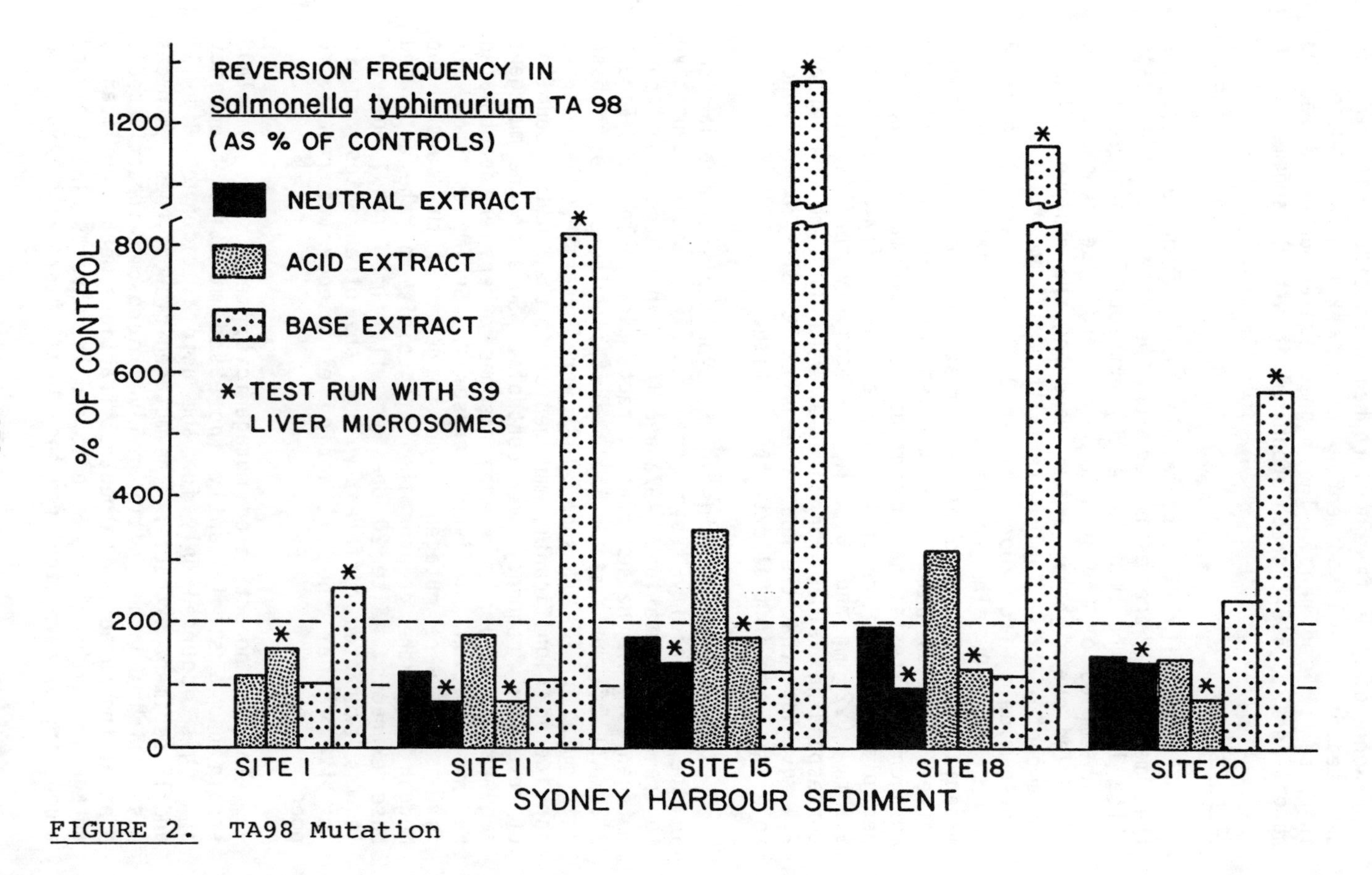

FIGURE 2. TA98 Mutation

of the tests with Site 1 extract were toxic to the test bacteria. None of the nonactivated base fractions were significantly mutagenic although the Site 20 reversion frequency was almost doubled in the test with TA98.

TABLE 3

AMES TEST OF SEDIMENT EXTRACTS (AS % OF CONTROLS, AVERAGE OF 3 PLATES

Extract Fraction	Site	S. typhimurium Strain TA97a	TA97a+S9	TA98	TA98+S9
Neutral	1	T	T	T	T
	11	T	T	129	74
	15	162	146	178	134
	18	112	125	184	94
	20**	140	150	147	144
Acid	1	121	117	129	162
	11	T	T	186	70
	15	150	134	347*	177
	18	117	115	310*	128
	20**	119	104	145	70
Base	1	96	157	105	260*
	11	103	242*	103	826*
	15	126	327*	125	1266*
	18	113	254*	114	1160*
	20**	144	255*	191	573*

T: extract was toxic to bacteria
*: significantly mutagenic (>2x control)
**: extraction method differed somewhat for Site 20

In contrast to the unactivated tests, the activated tests were positive in the base fractions and not the acid fractions. The activated base extracts were significantly mutagenic in all five Sydney Harbour sites tested with strain TA98. With strain TA97a, all base fractions except those from Site 1 were mutagenic. These mutagens all required metabolic activation. No mutagens were detected in the activated tests of the acid fraction or neutral fraction. The neutral extract from Site 1 was again toxic to all strains of the test bacteria.

SEDIMENT MUTAGENICITY AND TOXICITY

Toxicity Test

The results of the Blepharisma bioassay are given in Table 4. As in previous experiments, the survivorship drops during the first day and then increases. The sediments from Sites 7, 11, 15 and 20 caused a marked reduction in growth compared to controls. Sites 1 and 18 were not toxic. A bioassay of Site 18 was repeated to check that the sediment is indeed non-toxic. The second bioassay indicated growth enhancement over the controls.

DISCUSSION

The sediment along Muggah Creek was found to be contaminated by: PAH, PCB, pesticides, phthalate esters and chlorophenols. The PAH and PCB were found in the highest levels and these compounds appeared to have a steep gradient along Muggah Creek. The mutagenicity testing also indicated a range of mutagen concentrations along Muggah Creek, with the lowest at Site 1 and higher levels further up the creek.

The mutagenicity of Muggah Creek Sediments may appear to relate to PAH and PCB concentrations, however, the mutagenicity was detected in the acid and base fractions and not the neutral fraction. The neutral fraction contained the great majority of the extractable material and the PAH, PCB and aliphatics would recide in this fraction. Mutagenic PAH such as benzo(a)pyrene were detected in the sediment and further fractionation of the neutral extract may be required to unmask this mutagenicity. The high mutagenicity of the base fractions occurred at a maximum of 2.4 ug per plate. A useful extension to this work would be the identification of the highly mutagenic compound(s).

Of the three sites where chemistry, mutagenicity and toxicity data are available, there appears to be an association between the toxicity of the sediment elutriate and the presence of mutagens and chemical contaminants in the sediment. Site 1 elutriate is not toxic to Blepharisma and this contains the lowest level of chemical contamination and mutagenicity. Site 11 and 20 elutriates are both toxic and both have high levels of chemical contamination and mutagenicity. The mutagenicity data from Sites 15 and 18 complicate this picture. While both Site 15 and 18 are highly mutagenic, only Site 15 is toxic. This suggests that the mutagen are not the toxic components or that they are not available to the Blepharisma at neutral pH. The toxicity test of 18 was repeated and indicated a growth enhancement.

TABLE 4

GROWTH OF BLEPHARISMA EXPOSED TO SEAWATER CONTROL (SC) AND SEDIMENT ELUTRIATE (SE) FROM SYDNEY HARBOUR, NOVA SCOTIA

OF BLEPHARISMA SP/TEST CONTAINER ($\bar{x}$ ± S.E., n=14 OR 15)

Site		Day 0	1	3	4	6
1	SC	6	1.93±0.27		3.07±0.43	5.07±0.80
	SE	6	2.33±0.32		3.93±0.43	6.80±0.85
	SE/SC		121%		128%	119%
7	SC	6	1.93±0.27		3.07±0.43	5.07±0.80
	SE	6	0.67±0.21		0.13±0.09	0.60±0.21
	SE/SC		35%		4%	12%
11	SC	6	2.86±0.46	4.71±0.44		6.93±0.89
	SE	6	3.12±0.33	1.76±0.76		0.60±0.13
	SE/SC		109%	37%		9%
15	SC	6	2.86±0.46	4.71±0.44		6.93±0.89
	SE	6	2.00±0.39	0.80±0.20		0.33±0.13
	SE/SC		70%	17%		5%
18	SC	6	1.87±0.35		4.13±0.47	8.80±0.80
	SE	6	1.60±0.34		4.07±0.53	6.13±0.99
	SE/SC		86%		98%	70%
18	SC	6	1.27±0.36		2.79±0.96	3.64±0.97
	SE	6	1.40±0.25		4.00±0.75	6.57±1.05
	SE/SC		110%		143%	180%
20	SC	6	1.27±0.36		2.79±0.96	3.64±0.97
	SD	6	2.14±0.42		1.07±0.30	0.79±0.28
	SE/SC		168%		38%	22%

ACKNOWLEDGEMENTS

This work was funded through the Canadian Department of Supply and Services contract number OSB83-00507. Support was provided by: Department of Environment; Department of Fisheries and Oceans; Government of British Columbia.

REFERENCES

1. Hildebrand, L.P. (1982): Environmental Quality in Sydney and Northeast Industrial Cape Breton, Nova Scotia. EPS-5-AR-82-3, Environmental Protection Service, Environment Canada, Atlantic Region, 89 pp.
2. Cretney, W.J., C.S. Wong, P.A. Christensen, B.W. McIntyre and B.R. Fowler (1980): Quantification of polycylic aromatic hydrocarbons in marine environmental samples. In: Hydrocarbons and Halogenated Hydrocarbons in the Aquatic Environment, edited by B.K. Afgan and D. McKay, Environmental Science Research Series. Plenum Press, New York.
3. Inland Waters Directorate (1979): Analytical Methods Manual. Environment Canada - Inland Waters Directorate - Water Quality Branch.
4. Maron, D.M. and B.N. Ames (1983): Revised methods for the Samonella mutagenicity test, Mutation Res., 113:173-215.
5. Renberg, L. (1974): Ion exchange technique for the determination of chlorinated phenols and phenoxy acids in organic tissue, soil and water, Anal. Chem., 46:459-461.

IDENTIFICATION OF A PROMUTAGENIC COMPOUND FORMED BY THE ACTION OF NEAR-ULTRAVIOLET LIGHT ON A PRIMARY AROMATIC AMINE

R.T. OKINAKA[1], T.W. WHALEY[2], U. HOLLSTEIN[3], J.W. NICKOLS[1], AND G.F. STRNISTE[1]. Genetics[1] and Toxicology[2] Groups, Life Sciences Division, Los Alamos National Laboratory, Los Alamos, NM 87545 and Department of Chemistry[3], University of New Mexico, Albuquerque, NM 87131.

INTRODUCTION

The events leading to deleterious health effects by environmental pollutants include a residency time between the release of suspect carcinogenic chemicals and their absorption by living organisms. For a number of years we have been concerned with photochemical changes that may occur during these residency periods and thereby influence the genotoxic behavior of these pollutants [1,2,3]. We have demonstrated that natural sunlight and artificial sources of near ultraviolet light (UVA) can potentiate the conversion of certain nitrogen-containing polycyclic aromatic hydrocarbons, the aromatic amines, into potent bacterial mutagens [4,5,6]. This report reviews the critical photochemical transformations that have been identified and that account for the increased direct-acting mutagenicity (mutagens detected in the absence of exogenous metabolic enzymes) of solutions of 2-aminofluorene (2-AF) after exposure to UVA. In addition, we present newly acquired data on the characterization of a promutagenic photoproduct (*i.e.,* a mutagen requiring exogenous metabolic enzymes for its activation) that is also found in irradiated solutions of 2-AF.

MATERIALS AND METHODS

Chemicals

2-Aminofluorene [CAS registry number 153-78-6], 2-aminofluoren-9-one [CAS registry number 3096-57-9] and 2-nitrofluorene [CAS registry number 607-57-8] were obtained from Aldrich Chemical Company, Milwaukee, WI. 2-Nitrosofluorene [CAS registry number 2508-20-5] was synthesized from 2-aminofluorene by *m*-chloroperoxybenzoic acid oxidation and purified by silica gel chromatography (F.A. Beland, National Center for Toxicological Research, Jefferson, AK, personal communication and reference [5]).

UVA Irradiation

Two parallel 15-Watt blacklights (GE F15T8 BLB) positioned above a benchtop served as the artificial source of UVA radiation (320-400 nm wavelength). The incident fluence through a glass Petri dish cover was adjusted to average 30 $kJ/m^2/hr$ as measured by an Eppley thermopile (Eppley Laboratory, Inc., Newport, RI; also see reference [2]). For each dose point a four ml sample of 2-AF (5 mM in DMSO) was placed in a 60 mm glass Petri dish and exposed to the UVA light source with glass cover in place as described previously [5,6,7].

HPLC Analysis

Analytical scale high performance liquid chromatography (HPLC) was performed as previously described [5,7] with a Beckman Model 334 Gradient Liquid Chromatograph System (Beckman Instruments, Inc., Palo Alto, CA) fitted with a 10 μm Radial PAK C_{18} cartridge and an RCSS Guard-PAK C_{18} precolumn insert contained in a Radial Compression Separation System Z-module (Waters and Associates, Milford, MA). A 20 μl aliquot of the irradiated 2-AF solution was applied to the column and eluted with a programmed linear gradient of triethylammonium bicarbonate buffer (1 mM, pH 8.3):acetonitrile (95:5 to 0:100) for 30 minutes at a flowrate of 2 ml/min. The absorptivity of the eluate was monitored continuously at 254 nm and the resultant HPLC profile was compared to those obtained for reference standards.

Ames/Salmonella Bioassay

Mutagenic activity of the irradiated samples was determined in the histidine reversion test utilizing *Salmonella typhimurium* strains TA98 and TA1538 as described by Ames *et al.* [8] and Maron and Ames [9]. Experiments designed to measure direct-acting mutagenicity were performed in the absence of exogenous metabolic enzymes. Promutagenic components were detected by adding to each plate 0.5 ml of an S9 mixture containing 15 μl rat liver homogenate (Aroclor-induced, Litton Bionetics, Inc., Kensington, MD) and appropriate cofactors as described by Maron and Ames [9].

RESULTS AND DISCUSSION

The primary aromatic amines (PAA) are a class of compounds that are used in significant quantities in the production of many dyes and rubber products. Although epidemiological studies have convincingly demonstrated that aromatic amines are involved in the induction of bladder tumors for workers in specialized workplaces (see Scott [10] for an early review), PAA have never been considered to be ubiquitous environmental pollutants. However, recent reports demonstrating the presence of PAA as mutagenic components in coal- and oil shale-derived synfuels [11,12,13] suggest that potential environmental occurrances of PAA could be greater than thought previously.

Two years ago at this symposium [4] we reported our initial observations concerning the phototransformation of selected chemicals. Several model PAH and PAA were dissolved in DMSO and exposed to either natural sunlight or an artificial source of UVA. These irradiated solutions were tested for direct-acting mutagenic activity in the Ames/*Salmonella* histidine reversion test. Results of this study revealed that one aromatic amine, 2-aminofluorene (2-AF), was particularly sensitive to exposure to UVA. Furthermore, there was a UVA dose-dependent increase in the mutagenicity of these irradiated solutions. More recent experiments have shown that 2-AF is photochemically oxidized into a variety of products including three direct-acting mutagens; 2-nitrofluorene, 2-nitrosofluorene, and 2-nitrofluoren-9-one [5,7,14]. These photochemical changes are illustrated in Figure 1, which shows the HPLC analysis of a 2-AF solution exposed to a relatively low dose (150 kJ/m^2) of UVA.

It was concluded from these previous studies that oxidation of the exocyclic nitrogen is critical in converting 2-AF and other primary aromatic amines into direct-acting mutagens [4,5,6,7]. There is an analogy between these photochemical results and numerous reports which indicate that the *metabolic* conversion of 2-AF to the ultimate mutagenic form proceeds via N-oxidation to the hydroxylamine [15,16,17]. The nitroso- and nitro-containing photoproducts are mutagenic in bacteria due to their reduction (presumably to the hydroxylamine [18,19]) by nitroreductase enzymes. However, the role of these enzymes in converting nitroaromatic compounds to reactive mutagenic/carcinogenic species in mammalian cells is not established [19]. As a consequence we began studies to determine whether metabolic enzymes (rat liver S9 homogenates) could affect the genotoxic properties of these mutagens or other photoproducts of 2-AF [20].

To investigate the presence of "promutagenic" activity, a 250 μl sample of an irradiated 2-AF solution (150 kJ/m^2)

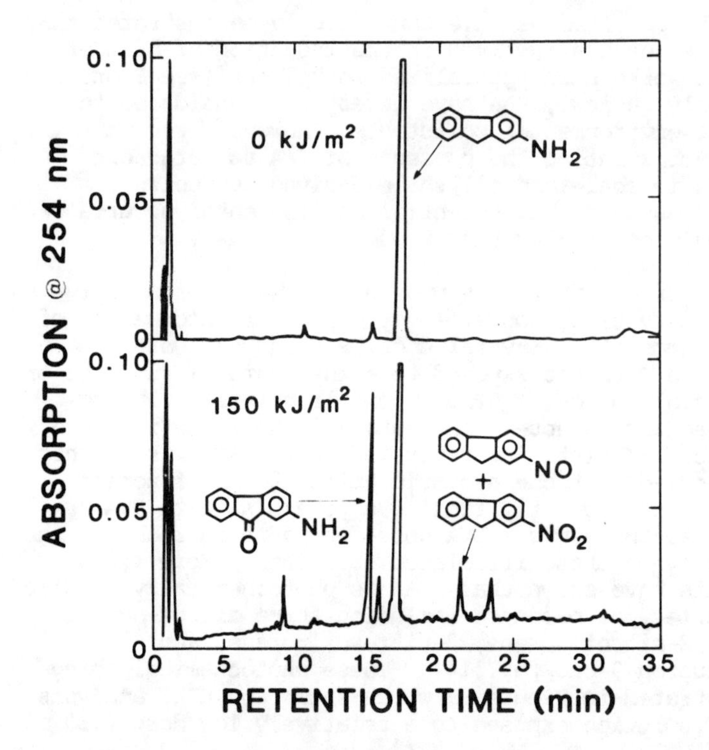

FIGURE 1. Analysis by HPLC for unirradiated (0 kJ/m^2) and 150 kJ/m^2 UVA irradiated 2-AF solutions. Aliquots of 20 μl were chromatographed as described in Materials and Methods and the eluates were monitored at 254 nm. Reference standards were run separately and their elution positions are as noted.

was fractionated by HPLC (defined as a semi-preparative experiment [7]) into ten samples (see Figure 1 for the analytical profile). The collected sample volumes varied from 4-7 ml to accomodate the selection of specific absorption peaks. Each of these samples was then bioassayed in the Ames test in the presence or absence of S9 proteins (see Table 1). As expected, most of the recoverable direct-acting mutagenic activity resides in the fractions collected between 18-23.5 min. This corresponds to the region of the chromatogram where 2-nitrofluorene and 2-nitrosofluorene co-elute. When S9 proteins are included in the bioassay, three distinct regions of the chromatogram contain most (96%) of the recovered mutagenic activity (recovery of mutagenicity in this experiment was 60%). The largest region of activity is found in the area where there

TABLE 1

MUTAGENICITY OF HPLC-FRACTIONATED, UVA-IRRADIATED (150 kJ/m^2) 2-AF IN THE PRESENCE OR ABSENCE OF S9 MIXTURES

Sample	Fraction interval (min)	Total number of his^+ revertants per fraction[a]	
		-S9	+S9
1	0-3	240.0	360.0
2	3-6	240.0	300.0
3	6-9	300.0	420.0
4	9-12	0.0	120.0
5	12-14	150.0	160.0
6	14-16	0.0	6000.0
7	16-18	330.0	60400.0
8	18-21	1260.0	720.0
9	21-23.5	4250.0	2100.0
10	23.5-27	700.0	800.0

[a] The number of his^+ revertants per fraction was obtained by multiplying the mutagenicity of a particular fraction (calculated from linear regression analysis of a dose-response curve) by the total volume of the fraction.

is a significant quantity of the parental 2-AF present (16-18 min) that exists after the short exposure of 150 kJ/m^2 of UVA. The nitrofluorene/nitrosofluorene region (21-23.5 min) is also active although the activity is diminished when compared to the direct assay. The third mutagenic region (14-16 min) is in an area where a 2-aminofluoren-9-one (2-AF-9-one) standard elutes, and represents approximately 9% of the total mutagenicity recovered in this HPLC experiment. It is important to reiterate that the mutagenic activity in this third region, in addition to that attributed to the parent 2-AF, is dependent upon exogenous metabolic enzymes for its mutagenic expression. These results demonstrate that UVA exposure, even at relatively low doses, induces at least one promutagenic species (presumptively identified as 2-AF-9-one). The remaining "promutagenic activity" can be accounted for by the presence of the direct-acting mutagens 2-nitro- and 2-nitrosofluorene.

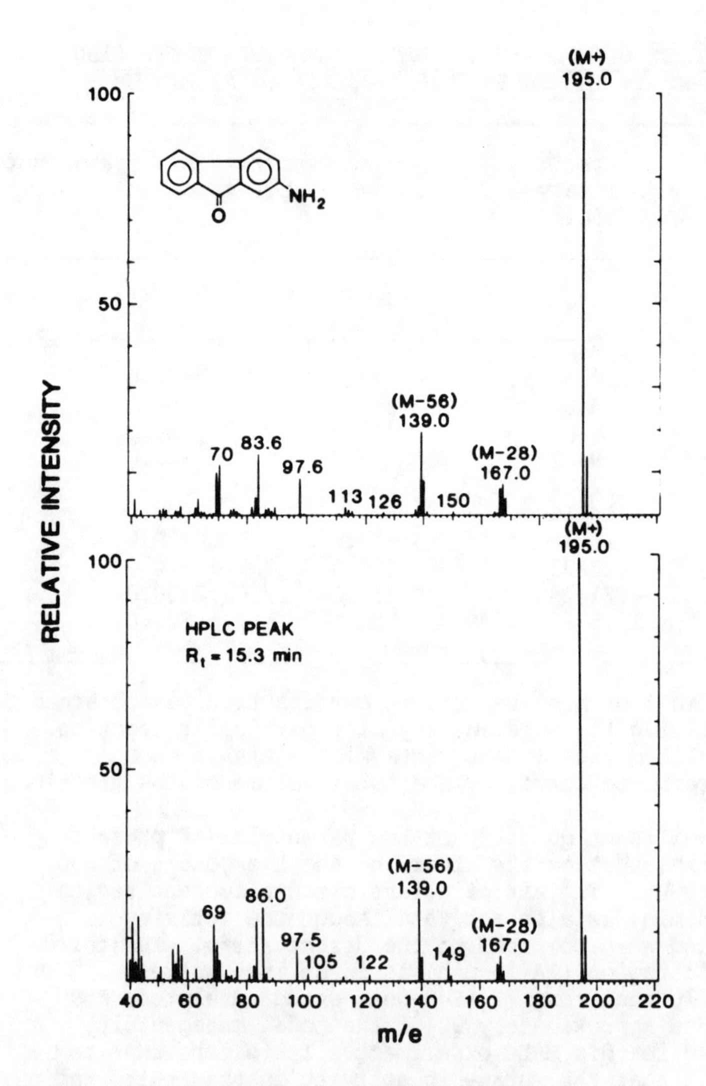

FIGURE 2. Electron impact mass spectra of 2-aminofluoren-9-one reference standard (upper panel) and pooled HPLC fractionated material eluting between 14.9-15.7 min for a 150 kJ/m^2 UVA-irradiated 2-AF sample. The eluates from 7 separate semi-preparative HPLC fractionation experiments were combined, dried by rotary evaporation and analyzed on a Finnegan mass spectrometer (Model 4600, Finnegan Corp., San Jose, CA) operated at 70 eV using a direct insertion probe.

It was plausible that photochemical oxidation of the 9 position of the fluorene ring occurred because this position can be chemically oxidized [21]. Previous results had suggested that 2-AF-9-one was an early oxidation product of UVA-irradiated 2-AF [5,7,14]. Tentative identification of 2-AF-9-one as the promutagenic species in the above experiments comes from the observation that the components in the region at approximately 15.3 min behaved identically to that observed for 2-AF-9-one in both HPLC and thin-layer chromatographic experiments utilizing different solvent systems [5,7]. In irradiated solutions of 2-AF, 2-AF-9-one also appeared to be the most likely precursor to the direct-acting mutagen, 2-nitrofluoren-9-one, a recently identified mutagenic photoproduct [14]. More definitive proof of structure, however, comes from mass spectral comparisons of a commercially prepared standard of 2-AF-9-one and a sample of presumptive 2-AF-9-one obtained by pooling 14.9-15.7 min fractions from 7 semi-preparative HPLC experiments (Figure 2). The spectra are nearly identical with intense molecular ion peaks ($M^+ = 195$) and fragments at m/e = 167 (M - 28) and m/e = 139 (M - 56). These primary fragmentation patterns are characteristic for an aromatic amine that also contains a carbonyl group; i.e., the loss of a CNH_2 group (m/e = 28) and/or the loss of a CO group (m/e = 28).

Table 2 contains data comparing the mutagenicity of 2-AF and a commercial sample of 2-AF-9-one assayed either in the presence or absence of S9 proteins. The rationale for this experiment was to determine the influence that oxidation of the 9 position of the fluorene molecule has on the mutagenic activity derived from the amino group at the 2 position. In the absence of S9, 2-AF appears to be considerably more mutagenic than 2-AF-9-one (10-fold higher). However, the genotoxicological importance of this difference would seem to be minimal since the relative mutagenicities are quite low. In the presence of S9 proteins, however, 2-AF-9-one is significantly more mutagenic than 2-AF (> 3-fold).

These results indicate that UVA light can potentially increase the promutagenic activity of 2-AF by oxidizing the 9 position of the fluorene ring. Strniste et al. [20] have recently demonstrated that the promutagenic activity of UVA-irradiated 2-AF solutions decreases as a function of increasing exposure. These findings suggest that the increase in 2-AF-9-one related promutagenic activity of the irradiated solutions is negated by the rapid conversion of 2-AF and other photoproducts to non-mutagenic species.

TABLE 2

COMPARISON OF THE MUTAGENIC ACTIVITIES OF 2-AMINOFLUORENE AND 2-AMINOFLUOREN-9-ONE IN THE PRESENCE OR ABSENCE OF S9 MIXTURE.

Compound[a]	S9	Salmonella typhimurium tester strain[b] his+ revertants per nmol[c]	
		TA98	TA1538
2-Aminofluorene	−	0.22 ± 0.01	0.29 ± 0.02
	+	85.4 ± 0.2	94.3 ± 0.4
2-Aminofluoren-9-one	−	0.02 ± 0.004	0.01 ± 0.005
	+	315.0 ± 18.0	302.0 ± 11.5

[a] Dilutions used in the mutagenicity test were made into DMSO from 5 mM stock solutions of each compound prepared freshly in spectroscopic grade DMSO.

[b] Tester strain spontaneous reversion frequencies to his^+ were: TA98; 22 ± 4 and TA1538; 16 ± 3. Solvent controls were not cytotoxic or mutagenic on either tester strain over the dose ranges used in the experiments.

[c] Calculated by regression analysis of dose-response curves. Numbers represent averages of results obtained from 2-10 independent experiments.

Three explanations can account for the increased promutagenic potential of 2-AF-9-one over 2-AF: (1) the metabolic conversion of the amino group to the penultimate mutagenic form, i.e., the hydroxyamino group, during the bioassay may be more favorable for 2-AF-9-one than for 2-AF; (2) the reaction of the ultimate mutagenic species (presumably a nitrenium ion) with the target DNA may proceed more efficiently when the 9-position of the fluorene ring is oxidized; and (3) the rate of removal of adducts from the altered DNA may be affected by the presence of the oxygen at the 9 position of the fluorene nucleus. We tend to discount the second explanation since an electron-withdrawing carbonyl group would be predicted to destabilize a nitrenium ion at the 2 position of the fluorene ring. This prediction is supported by our observation (unpublished results) that 2-(acetoxyacetylamino)fluoren-9-one is unreactive toward 2'-deoxyguanosine under conditions where 2-(acetoxyacetylamino)fluorene reacts to afford the expected C-8 adduct. Experiments currently in progress are attempting to evaluate the other explanations.

In summary, our data indicate that photooxidation of the nitrogen group is critical in the conversion of 2-AF and other primary aromatic amines [6,22] to direct-acting mutagens. UVA-irradiation also results in the oxidation of 2-AF at its carbon-9 position. The resulting 2-AF-9-one can be further photooxidized to the direct-acting mutagen, 2-nitrofluoren-9-one [14]. In addition, 2-AF-9-one is a more potent promutagenic species than the parental 2-AF when metabolically activated by rat hepatic enzymes. These results demonstrate that photochemical transformation of 2-AF induced by wavelengths of light in the near ultraviolet range can yield unique photoproducts with potent mutagenic or promutagenic properties.

ACKNOWLEDGEMENTS

This work was performed under the auspices of the U.S. Department of Energy under contract No. W-7405-ENG-36. The authors thank Ms. M. Fink and Ms. G. Coffin for the processing of this manuscript.

REFERENCES

1. Strniste, G.F. and Brake, R.J. (1981): Cytotoxicity in human skin fibroblasts induced by photoactivated polycyclic aromatic hydrocarbons. In: Polynuclear Aromatic Hydrocarbons: Chemical Analysis and Biological Fate, edited by M. Cooke and A.J. Dennis, pp. 109-118, Battelle Press, Columbus, OH.
2. Strniste, G.F. and Chen, D.J. (1981): Cytotoxic and mutagenic properties of shale oil byproducts. I. Activation of retort process waters with near ultraviolet light. Environ. Mutagenesis, 3:221-231.
3. Okinaka, R.T., Bingham, J.M., MacInnes, M.A. and Strniste, G.F. (1984): Light activation of a complex mixture: Effects of UV excision repair on the modulation

of genotoxic and molecular events in Chinese hamster cells. Photochem. Photobiol., 39:353-358.

4. Okinaka, R.T., Nickols, J.W., Whaley, T.W. and Strniste, G.F. (1983): Phototransformation of polycyclic aromatic hydrocarbons into stable, mutagenic components. In: Polynuclear Aromatic Hydrocarbons: Mechanisms, Methods and Metabolism, edited by M. Cooke and A.J. Dennis, pp. 961-971, Battelle Press, Columbus, OH.
5. Okinaka, R.T. Nickols, J.W., Whaley, T.W. and Strniste, G.F. (1984): Phototransformation of 2-aminofluorene into N-oxidized mutagens. Carcinogenesis, 5:1741-1743.
6. Okinaka, R.T., Nickols, J.W., Whaley, T.W. and Strniste, G.F. (1985): Photochemical transformation of primary aromatic amines into direct-acting mutagens. In: Polynuclear Aromatic Hydrocarbons: Chemistry, Characterization and Carcinogenesis, Vol. 9, edited by M. Cooke and A.J. Dennis, in press, Battelle Press, Columbus, OH.
7. Strniste, G.F., Nickols, J.W. and Okinaka, R.T. (1985): Photochemical oxidation of 2-aminofluorene: Correlation between the induction of direct-acting mutagenicity and the formation of nitro and nitroso aromatics. Mutat. Res., 151:15-24.
8. Ames, B.N. McCann, J. and Yamasaki, E. (1975): Methods for detecting carcinogens and mutagens with the Salmonella/mammalian-microsome mutagenicity test. Mutat. Res., 31:347-364.
9. Maron, D.M. and Ames, B.N. (1983): Revised methods for the Salmonella mutagenicity test. Mutat. Res., 113:173-215.
10. Scott, T.S. (1962): Carcinogenic and chronic toxic hazards of aromatic amines. In: Elsevier Monographs on Toxic Agents, edited by E. Browning, pp. 1-208, Elsevier Press, Amsterdam.
11. Guerin, M.R., Ho, C.-H., Rao, T.K., Clark, B.R. and Epler, J.L. (1980): Polycyclic aromatic primary amines as determinant chemical mutagens in petroleum substitutes. Environ. Res., 23:42-53.
12. Later, D.W., Andros, T.G. and Lee, M.L. (1983): Isolation and identification of amino polycyclic aromatic hydrocarbons from coal-derived products. Anal. Chem., 55:2126-2132.
13. Wilson, B.W., Pelroy, R. and Cresto, J.T. (1980): Identification of primary aromatic amines in mutagenically active subfractions from coal liquefaction materials. Mutat. Res., 79:193-202.
14. Strniste, G.F., Nickols, J.W., Okinaka, R.T. and Whaley, T.W. (1985): 2-Nitrofluoren-9-one: A unique mutagen formed in the photooxidation of 2-aminofluorene. Carcinogenesis, in press.

15. Kriek, E. and Westra, J.G. (1979): Metabolic activation of aromatic amines and amides and interactions with nucleic acids. In: Chemical Carcinogens and DNA, Vol. II, edited by P.L. Grover, pp. 1-28, CRC Press, Boca Raton, FL.
16. Miller, E.C. (1978): Some current perspectives on chemical carcinogenesis in humans and experimental animals. Cancer Res., 38:1479-1496.
17. Miller, J.A. (1970): Carcinogenesis by chemicals: An overview - G.H.A. Clowes Memorial Lecture. Cancer Res., 30:559-576.
18. Rosenkranz, E.J., McCoy, E.C., Mermelstein, R. and Rosenkranz, H.S. (1982): Evidence for the existence of distinct nitroreductases in Salmonella typhimurium: role in mutagenesis. Carcinogenesis, 3:121-123.
19. Rosenkranz, H.S. and Mermelstein, R. (1983): Mutagencity and genotoxicity of nitroarenes: All nitro-containing chemicals were not created equal. Mutat. Res., 114:217-267.
20. Strniste, G.F., Nickols, J.W., Okinaka, R.T. and Whaley, T.W. (1985): Mutagenicity of photochemically transformed polycyclic aromatic amines. 24th Hanford Life Sciences Symposium, Battelle Pacific Northwest Laboratories, Richland, WA, in press.
21. Rieveschi, G. and Ray, F.E. (1955): Fluorenone-2-carboxylic acid. In: Organic Synthesis, Collected Vol. 3, edited by E.C. Horning, pp. 420-422, John Wiley and Sons, NY.
22. Okinaka, R.T., Nickols, J.W., Whaley, T.W. and Strniste, G.F. (1985): 1-Nitropyrene: a mutagenic product induced by the action of near ultraviolet light on 1-aminopyrene. Mutat. Res. Lett., in press.

ANALYSIS OF POLYCYCLIC AROMATIC COMPOUNDS IN SELECTED BITUMEN AND BITUMEN FUMES.

CONNY E. ÖSTMAN and ANDERS L. COLMSJÖ
University of Stockholm
Department of Analytical Chemistry
S-106 91, Stockholm, Sweden

ABSTRACT

A method is presented for the study of bitumen fumes with respect to the content of polycyclic aromatic compounds. Analysis of selected bitumens for road paving is presented, including a two dimensional HPLC separation route, utilizing aminopropylsilane and polymeric octadecylsilane coated silica. The efficiencies of the two dimensional separation are shown and further use of this method for petroleum analysis is outlined.

INTRODUCTION

The term bitumen represents a group of products derived from crude oil, which are manufactured by a series of succesive distillations, blowing and solvent precipitation. Blending of the distillates can then be made in order to obtain the required properties of the bitumen products, e.g. penetration grades. Due to their waterproofing and strongly adhesive properties, bitumens are used in a large number of applications within road and house construction.

Bitumen has to be heated when used, for example, in road paving. Due to this heating procedure, a hazardous area of bitumen fumes is thus produced in the vicinity of the bitumen. Therefore, during different handling procedures, exposure to bitumen fumes can occur. Few investigations have been made [1-10], especially regarding the contents of polycyclic aromatic compounds both in the fumes and in the bitumen itself.

This investigation has been initiated to develop methods for analysis of bitumen and bitumen fumes primarily with respect to the contents of polynuclear aromatic compounds (PAC).

MATERIALS AND METHODS

Generation and sampling of the bitumen fumes were performed in a apparatus containing a "cold finger" trapping device constructed at the department [11].

Liquid chromatography: A Shimadzu LC-4 with dual detection was employed. UV at 289 nm being used for the fractionation together with fluorescense compound group selective detection. Aminopropylsilane coated silica (Liqchrosorb NH_2, 10um) was used for the straight phase separation with hexane as mobile phase. Polymeric octadecylsilane coated silica (synthesized at the department according to Sander & Wise [12], 5um) was utilized for the reversed phase separation with methanol as mobile phase. Both straight and reversed phase runs were flow programmed.

Gas chromatography: A Varian 3700 with a 10 m SE-54 fused silica capillary column, temperature programmed from 70°C to 290°C with 7°C/min, was used for the GC-separation. Splitless injection during 2 minutes.

Mass spectrometry: A Hewlett Packard 5700, equipped and programmed as above, connected to a Jeol JMS-D300 mass spectrometer controlled by a Finnigan INCOS computer comprised the mass spectrometric system.

Shpol'skii fluorescence was recorded by an apparatus constructed at the department.

GENERATING AND SAMPLING THE BITUMEN FUMES

Bitumen fumes: A deposition of 1 g ± 10% of bitumen was made at the bottom of a sample chamber tube. A constantly cooled condensation area was positioned above the bitumen surface, and the temperature of the chamber was then maintained at 210°C for 1 hour. The condensate was extracted with acetone, acetone:cyclohexane (1:1) and cyclohexane. The PAC fraction of the condensate was DMF-extracted prior to analysis.

Bitumen: Approx. 0.05g of bitumen was dissolved in 2 ml of cyclohexane and sonicated for 15 minutes. The sample was then DMF-extracted.

RESULTS

Investigation of seven road bitumen fume condensates showed the presence of a complex mixture of PAC with up to five rings (fig 1). In most of the bitumens, five ring PAC were detected in the "hump" of material coeluting on the GC, by using selective ion monitoring. Quantitation of the benzopyrenes with GC-FID could be made in only one of the bitumens. No PAC structures of more than five rings could be detected in any of the seven bitumen fume condensates, which all exhibited similar patterns of PAC components. In the following, the analysis of one of the bitumens is presented.

About 180 PAC could be detected in the range from fluorene up to perylene. The PAC consisted of the unsubstituted parent compounds (fluorene, phenanthrene, anthracene, fluoranthene, pyrene, benz(a)anthracene, chrysene, benzo(j&k)fluoranthene, benzo(e)pyrene, benzo(a)pyrene, perylene) and methylsubstituted derivatives of these. The methyl derivatives ranged from monomethyl to pentamethyl. A series of sulfur containing PAC was also observed (dibenzothiophene, 4,5-epithiaphenanthrene, the three benzonaphthothiophenes and also an S-PAC of molecular weight 258, suspected to be 1,12-epithiatriphenylene) also with wide a spectra of methyl substituted derivatives.

Analysis of the bitumen itself showed the same pattern of components as in the fume condensate, figure 2. The "hump" in the chromatogram was much larger and displaced toward higher molecular weights. In table 1, the concentrations of 23 selected PAC are given. It can be observed that PAC are concentrated in the condensate. This was to be expected, since bitumen consists of asphaltenes with an average molecular weight of 2000-2500, resins with a molecular distribution of 800 to

TABLE 1

CONCENTRATIONS OF 23 SELECTED PAC IN BITUMEN AND ITS FUME CONDENSATE BASED ON TWO AND THREE MEASUREMENTS, RESPECTIVELY

Compound	Bitumen	Condensate
Dibenzothiophene	5.8	1000
Phenanthrene	12	2000
Anthracene	n.d.	150
3-methylphenanthrene	16	2400
2-methylphenanthrene	13	1700
9-methylphenanthrene	24	3500
1-methylphenanthrene	8.2	1000
Dimethyldibenzothiophene:1	14	2200
Dimethyldibenzothiophene:2	15	2000
Dimethylphenanthrene:1	37	4400
Dimethylphenanthrene:2	18	2100
Dimethylphenanthrene:3	17	2000
Fluoranthene	10	1000
Pyrene	8.8	540
Benzo(a)fluorene	9.9	540
4-methylpyrene	4.3	300
2-methylpyrene	17	820
1-methylpyrene	15	850
Benzo(a)anthracene	5.9	120
Chrysene	38	890
Benzo(e)pyrene	n.d.	n.d.
Benzo(a)pyrene	n.d.	n.d.
Perylene	n.d.	n.d

Concentrations above are given in [ug/g bitumen] and [ug/g fume condensate] respectively. Standard deviations are about 20 percent for the bitumen and 15 percent for the fume condensate analysis.

n.d. = Not quantitable with GC.

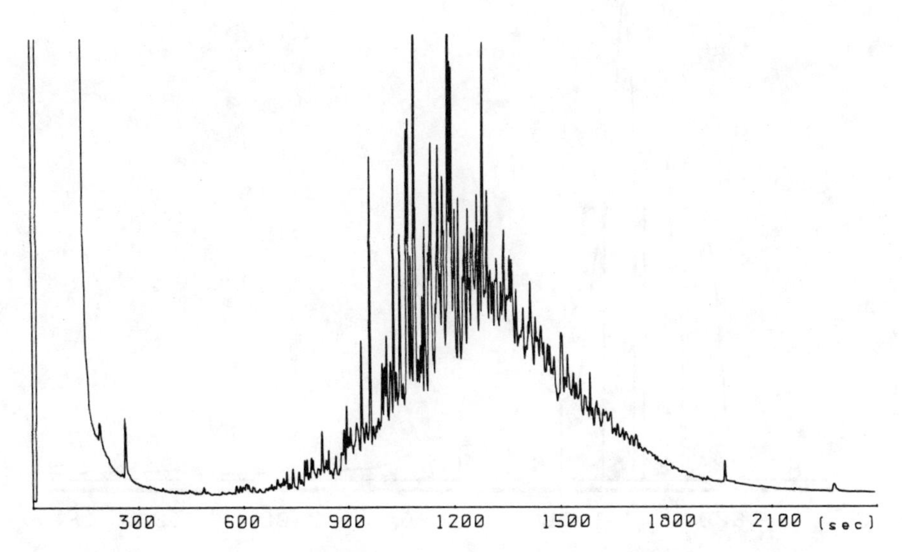

FIGURE 1. Gas chromatogram of the DMF-extracted bitumen fume condensate.

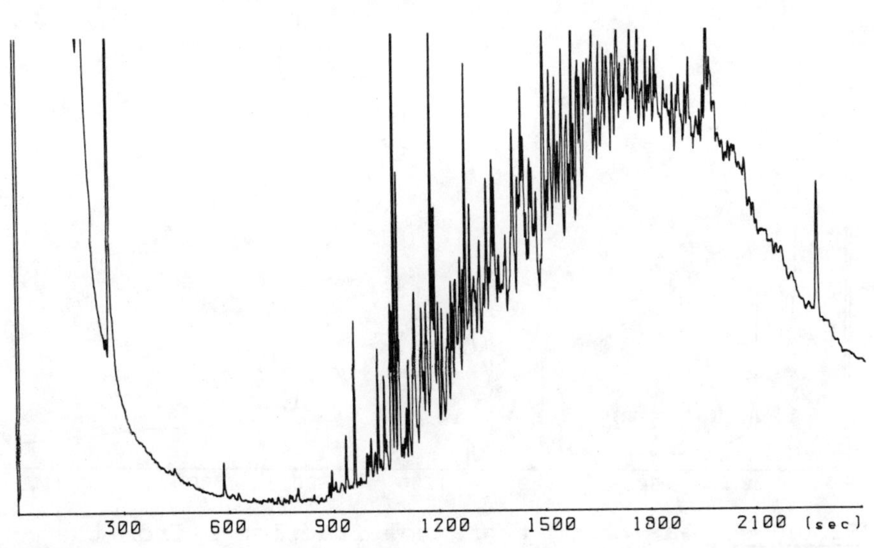

FIGURE 2. Gas chromatogram of the DMF-extract of the bitumen.

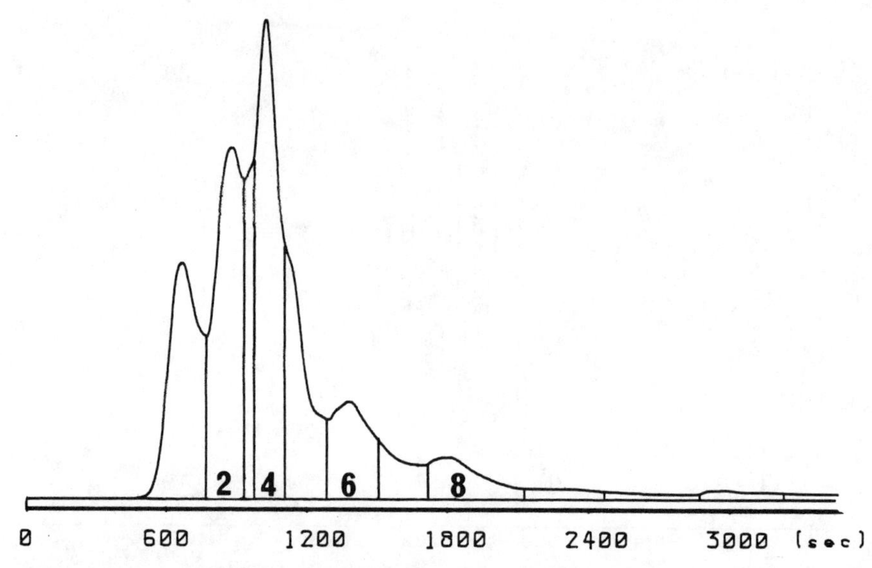

FIGURE 3. HPLC chromatogram showing the fractionation of the bitumen fume condensate on straight phase. Wavelength: 289 nm

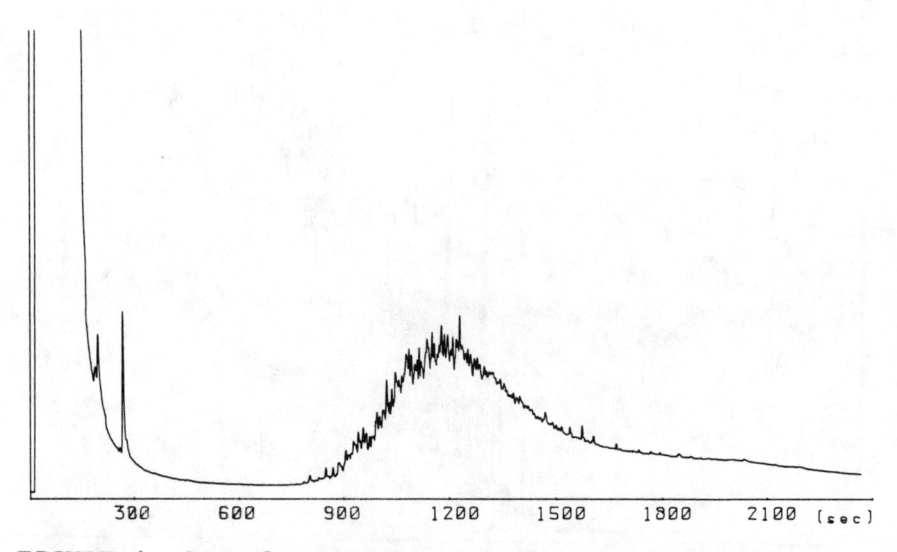

FIGURE 4. Gas chromatogram of fraction 1 from the straight phase fractionation. The "hump" contains monoaromatic material.

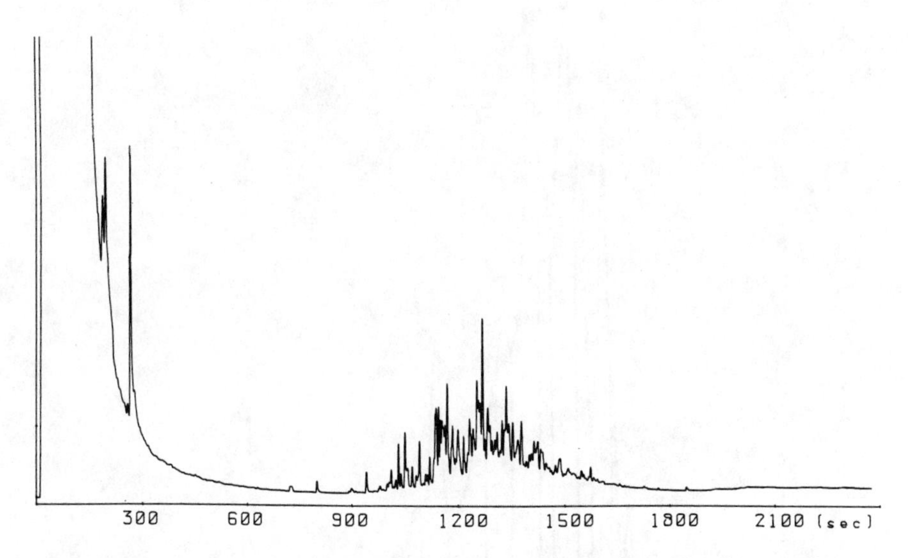

FIGURE 5. Gas chromatogram of fraction 2 from the straight phase fractionation. The peaks emanates from bicyclic material.

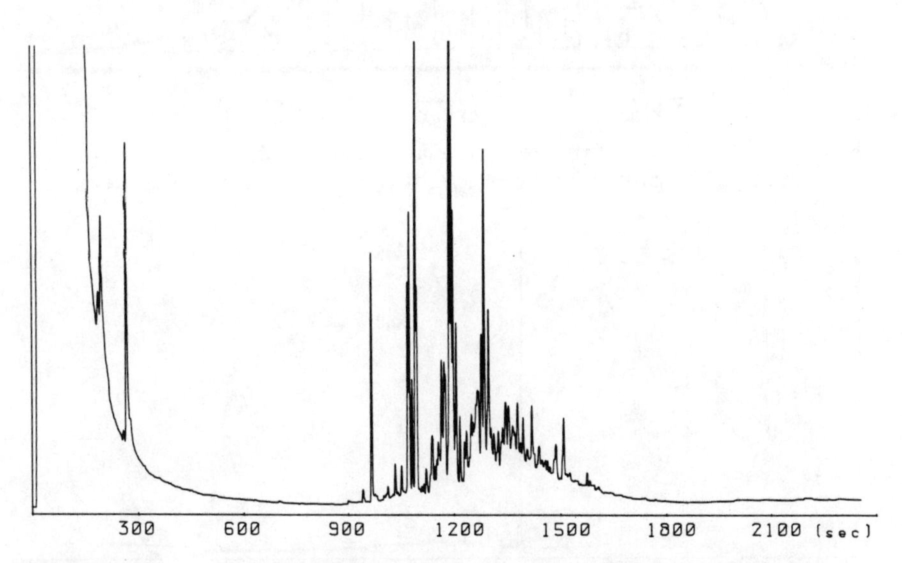

FIGURE 6. Gas chromatogram of fraction 4 from the straight phase fractionation. The peaks emanates from phenanthrene and methylated phenanthrene derivatives.

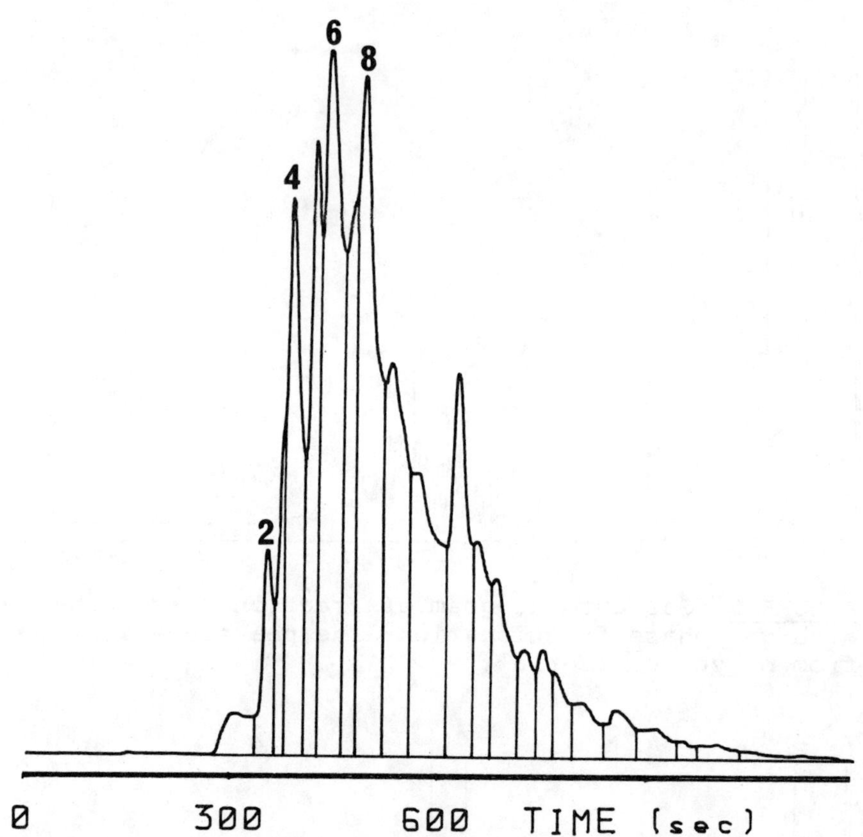

FIGURE 7. HPLC chromatogram showing fractionation of straight phase fraction 4 on reversed phase.

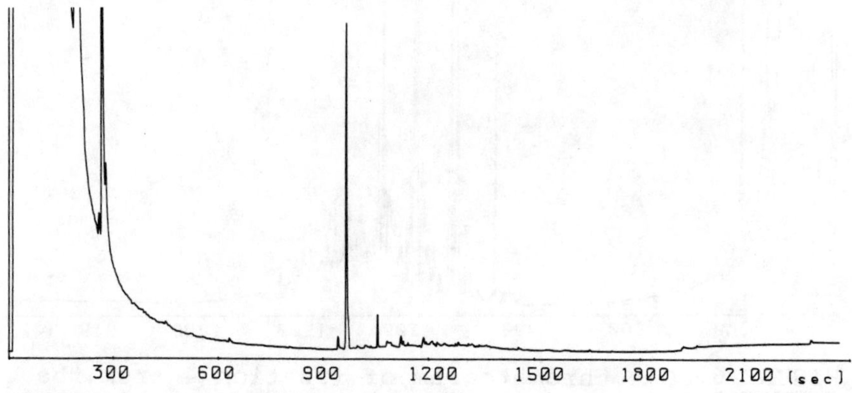

FIGURE 8. Gas chromatogram of fraction 2 from the reversed phase fractionation of straight phase fraction 4. The peak corresponds to phenathrene.

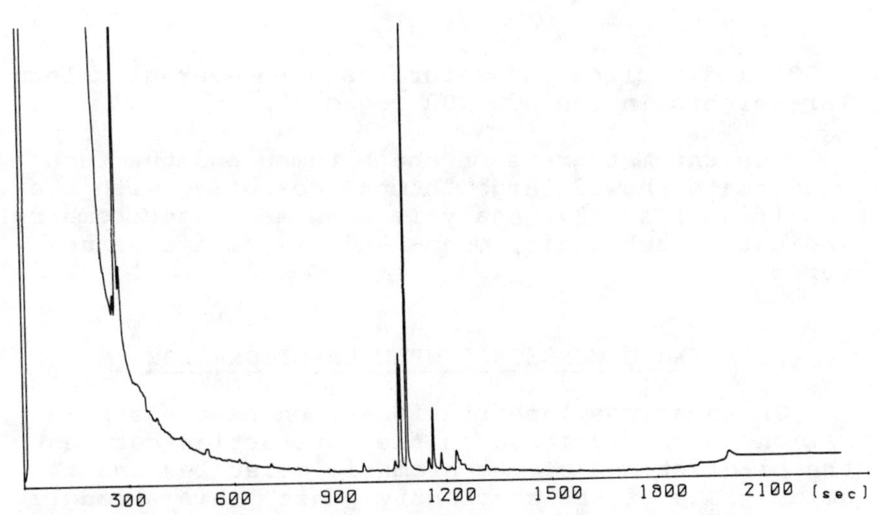

FIGURE 9. Gas chromatogram of fraction 4 from the reversed phase fractionation of straight phase fraction 4.

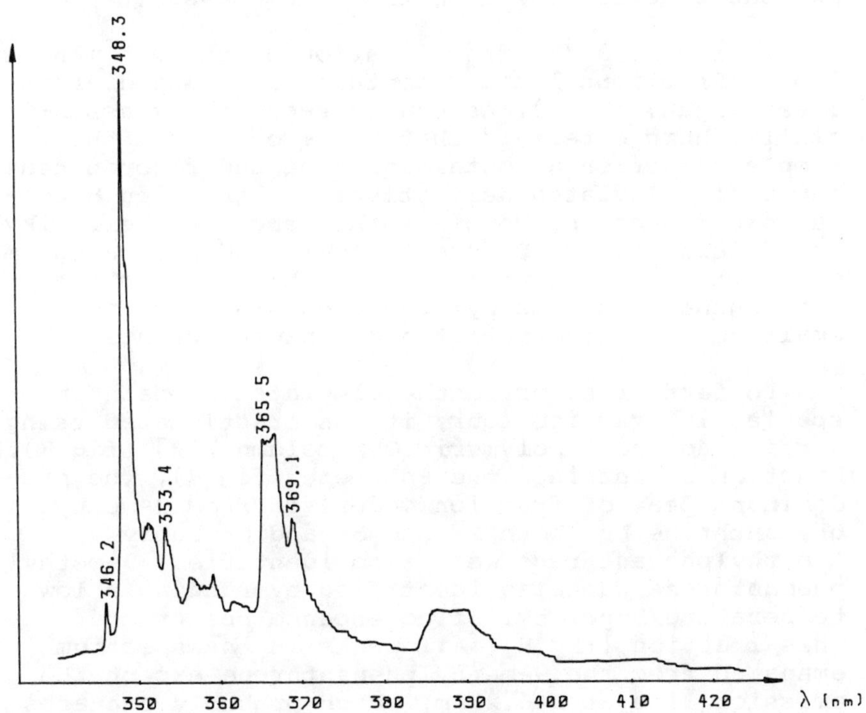

FIGURE 10. Shpol'skii low temperature fluorescence spectrum of the fraction above, showing a typical fingerprint of 9-methylphenantrene.

2000, and cyclics and saturates with average molecular weights in the 500-900 region [1].

The chromatograms of the bitumen and the fume condensate showed large "humps" coeluting with the PAC (figs 1 and 2). Analysis with mass spectrometry indicated naphthenic, mono- and diaromatic structures.

TWO DIMENSIONAL HPLC FRACTIONATION

On an aminosilane stationary phase the separation is mainly based on the interaction between the pi-electrons of the aromatic molecules and the amino group of the stationary phase. This strongly reduces the importance of alkyl subtituents on the PAC [13]. Use of this type of stationary phase makes it possible to isolate and further separate and characterize the aromatic ring classes.

In figure 3 the fractionation of the bitumen is shown. Fractions 1 and 2 contain mono- and diaromatics (figs 4 and 5). As can be seen, it is essencially "hump material" that is removed from the sample. Fraction 6 contains pyrene and fluoranthene and their alkylated derivatives and fraction 8 contains chrysene and benz(a)anthracene and their alkylated derivatives. Figure 6 shows a GC chromatogram of fraction 4. The peaks in the chromatogram derive from phenantrene, methylated phenanthrenes and small amounts of methylated dibenzothiophenes.

To further separate the alkylated phenathrene species in fraction four, it was fractionated using a reversed phase polymeric ODS column [12] (fig 7). Fraction 2 contained phenanthrene (fig 8). The predominant peak of fraction 4 derived from 9-methylphenanthrene but both 2- and 3- and probably 4-methylphenanthrene were also identified. 9-methylphenanthrene was also identified by running a low temperature Shpol'skii fluorescence spectrum of this fraction (fig 8). All peaks in the spectrum emanated from the 9-methylphenanthrene except the emission line at 346.2 nm, which probably emanates from the 0-0' transition emission line of 4-methylphenanthrene (based on the knowledge of retention times and spectra of 2-,3-and 9-methylphenanthrene.

DISCUSSION

Analysis of the bitumens and bitumen fumes has shown the presence of complex mixtures of PAC from two to five ring structures, and up to 180 compounds have been detected. PAC found to be present in the bitumen have also been found in the fume condensate in increased concentrations. Higher molecular weight PAC have lower relative concentrations in the fumes compared with the bitumen, which is an effect of their lower volatility. Earlier investigations have shown unsubstituted PAC with molecular weights in the range 276-300 to be present in the bitumen fumes [6,10]. This is probably due to the fact that bitumen from different sources have been used.

Separations on HPLC using aminosilane and polymeric octadecylsilane have shown to be efficient techniques for clean up and multicomponent analysis of bitumen. The previously used method for anlysis of petroleum based products [14] is to be combined with the HPLC separation techniques to give an efficient method for overall PAC analysis, PAC ring class analysis and analysis for identification purposes of narrow PAC fractions of bitumen and petroleum derived products in general.

ACKNOWLEDGEMENT

This investigation was kindly supported by a grant from Nynäs Petroleum. Thanks are due to Barbro Majgren and Per Redelius for helpful discussions and selecting the samples and to Beryl Holm for reviewing the manuscript.

REFERENCES

1. Concave, No. 6/84 (1984): Review of Bitumen Fume Exposures and Guidance on Measurement.
2. Concave, No. 7/82 (1982): Health Aspects of Bitumens.

3. Chipperfield E.H. (1984): IARC Review on Bitumen Carcinogenity. Bitumen Production, Properties and Uses in Relation to Occupational Exposures. Institute of Petroleum, IP 84-006.
4. NIOSH Technical Report, May 1983 (1983): Petroleum Refinery Workers Exposure to PAH at Fluid Catalytic Cracker, Coker and Asphalt Processing Units.
5. Asphalt Institute, Research Report No. 75-1 (1975): Asphalt Hot-Mix Emission Study.
6. Asphalt Institute, Research Report No. 79-2, Puzinauskas V.P. (1979): Emission From Asphalt Roofing Kettles.
7. Wallcave L. _et al_. (1971): Skin tumouogenesis in mice by petroleum asphalts and coal tar pitches of known PAH content. _Toxicology and Appl. Pharmacology_ 18, 41.
8. McKay J.F. and Latham D.R. (1973): Polyaromatic Hydrocarbons in High-Boiling Petroleum Distillates. _Anal. Chem_. 45, 1050.
9. Eldrige J.E., Shanmugam K., Bobalek E.G. and Simard G.L. (1983): PAH Emissions from Paving Asphalt in Laboratory Simulation. _Polynuclear Aromatic Hydrocarbons: Formation, Metabolism and Measurement_, Eds M. Cooke and A. Dennis, Battelle Press, 471.
10. Malaiyandi M., Benedek A., Holko A.P. and Bancsi J.J. (1982): Measurement of Potentially Hazardous Polynuclear Aromatic Hydrocarbons from Occupational Exposure During Roofing and Paving Operations. _Polynuclear Aromatic Hydrocarbonss: Physical and Biological Chemistry_, Eds M. Cooke, A. Dennis and G Fisher, Battelle Press, 471.
11. Colmsjö A. and Stenberg U. (1979): Vaccum Sublimation of Polynuclear Aromatic Hydrocarbons Separated by Thin-Layer Chromatography for Detection with Shpol'skii Low-Temperature Fluorescence. _J Chrom_ 169, 205.
12. Sander L.C. and Wise S.A. (1984): Synthesis and Characterization of Polymeric C_{18} Stationary Phases for Liquid Chromatography. _Anal Chem_ 56, 504.

13. Wise S.A., Chesler H.S., Hertz L.R. Hilpert L.R. and May W.E. (1977): Chemically-Bonded Aminosilane Stationary Phase for the High-Performance Liquid Chromatographic Separation of Polynuclear Aromatic Compounds. Anal Chem 49, 2306.
14. Östman C., Colmsjö A. and Zebühr Y. (1984): Analysis of Polycuclic Aromatic Compounds in Oil. Characterization of Heavy Crude Oils and Petroleum Residues. Edt Tissot B, Editions Technip, 253.

CYTOTOXICITY, MUTAGENICITY, AND TRANSFORMATION OF DIPLOID HUMAN FIBROBLASTS BY 1-NITROPYRENE AND 1-NITROSOPYRENE.

JOE DALE PATTON, VERONICA M. MAHER, J. JUSTIN McCORMICK
Carcinogenesis Laboratory, Michigan State University, East Lansing, Michigan 48824.

INTRODUCTION

Nitro derivatives of polycyclic aromatic hydrocarbons are produced primarily as the result of incomplete combustion processes. Among the nitroarenes, nitropyrenes are of particular interest because in microbial assays, they have been identified as primary mutagenic components of diesel emission particulate (1). Since they do not react directly with DNA (2), their biologic effects are presumably mediated through the cellular conversion of the parent compound into a species which reacts readily with DNA to form adducts. Heflich et al. (3) showed that the pathway for activation of 1-nitropyrene (1-NP) proceeds as illustrated in Figure 1. The critical first step, a rate-limiting step, is enzymatic reduction to 1-nitrosopyrene (1-NOP). This is followed by a subsequent reduction to the corresponding hydroxylamine, a species capable of undergoing acid-catalyzed decomposition to yield a nitrenium ion which reacts readily with DNA to form a covalent adduct at position 8 of guanine. Heflich et al. (3,4) showed in Salmonella and CHO cells that the adduct formed by 1-NOP is identical to that of 1-NP.

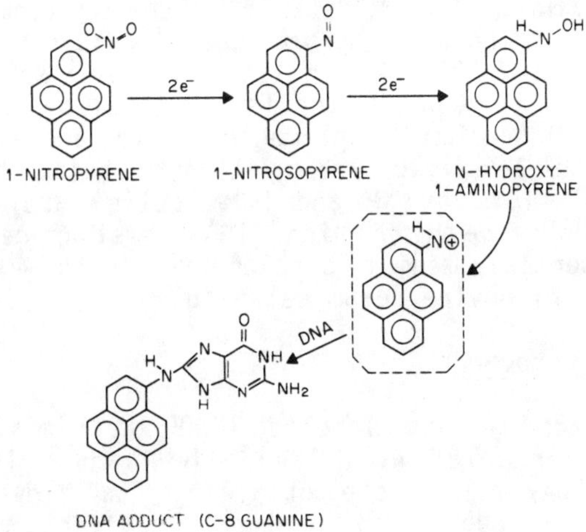

FIGURE 1. Proposed pathway for metabolic activation of 1-NP.

We compared the cytotoxic effect of 1-NP and 1-NOP in normal diploid human fibroblasts (NF), excision repair-deficient xeroderma pigmentosum (XP) fibroblasts, and fibroblasts from patients with hereditary cutaneous malignant melanoma (HCMM) (5). These HCMM cells are of interest since they are abnormally sensitive to the cytotoxic and mutagenic effects of a model nitro compound, 4-nitroquinoline-1-oxide (4NQO), but not to its reduced hydroxylamine derivative, HAQO (6), suggesting that these cells are more proficient than normal in activating (reducing) the parent nitro compound. We also investigated the mutagenicity of 1-NP and 1-NOP in normal and XP cells and the ability of these two carcinogens to cause transformation of human cells in culture. Their mutagenic effect was compared, not only on the basis of equal cytotoxicity, but also as a function of the number of residues covalently bound to the cellular DNA of NF and XP cells immediately after exposure. 1-NP and 1-NOP proved to be equally mutagenic when compared on the basis of mean lethal event and number of residues initially bound to DNA.

MATERIALS AND METHODS

Cells and Media

Normal fibroblasts were initiated in our laboratory from foreskin material. XP12BE cells (complementation group A) were obtained from the American Type Culture Collection and HCMM-3044T cells from Dr. Mark H. Greene of the National Cancer Institute. Cells were cultured in Ham's F10 medium supplemented with additional $NaHCO_3$ to bring the level to 2.2 g/l, with fetal bovine serum (10% for NF, 15% for XP and HCMM cells) and antibiotics. For selection of 6-thioguanine (TG)-resistant cells, Eagle's minimum essential medium supplemented with 40 μM TG and 10 or 15% fetal bovine serum was used.

Exposure to Compounds

1-NOP and tritium labeled 1-NOP (217 mCi/mmole) were supplied by Dr. Fred A. Beland. 1-NP and tritium labeled 1-NP (23.7 Ci/mmole) were obtained from Midwest Research Institute, through the Health Effects Institute. Exponentially growing cells were trypsinized, seeded at $10^4/cm^2$ and allowed 8-10 h to attach and elongate. The cells were washed twice with phosphate buffered saline to remove residual serum and the medium was replaced with serum-free Ham's F10. The compounds to be tested were dissolved in anhydrous dimethylsulfoxide at an appropriate concentration and delivered by micropipette. The exposure period was 1 h for 1-NOP and 10 h for 1-NP which requires metabolic activation.

Cytotoxicity Assay

Cytotoxicity was determined from the survival of colony-forming ability as described (7). Immediately following treatment, cells were plated at cloning densities. They were refed after 7 d and the colonies stained after 14 d. Cloning efficiencies ranged from 38 to 80% for NF; 20 to 54% for XP cells.

Mutagenicity Assay

The general procedures for assaying the induction of TG resistance in human cells have been described (7). For each determination, sufficient numbers of target cells were treated to ensure $>10^6$ surviving cells at the beginning of the expression period. Immediately after treatment, the cells were refed with complete medium and allowed to replicate in the original flasks. They were kept in exponential growth during the 7-10 d expression period. At selection, the cells were pooled and 1 - 2 x 10^6 cells were plated at 400 cells per cm^2, and a portion plated at cloning density to determine the cloning efficiency. This value was used to correct the observed frequency of TG resistant cells.

Transformation Protocol

Cells were treated with 1-NOP or 1-NP as above and allowed to replicate in the original flasks for 5 days. At that time they were trypsinized, pooled, and assayed for anchorage independence (AI) (ability to form colonies in soft agar). A portion of the population was replated at lower density to continue exponential growth and 1.5 to 2.0 x 10^6 cells were assayed for anchorage independence on subsequent days.

Anchorage Independence Assay

For each treated population and the untreated control, 1.5 to 2.0 x 10^6 cells were suspended in Ham's F10 medium containing 6% fetal calf serum and 0.33% agar and plated into 60-mm diameter dishes containing 5 ml of freshly-solidified bottom agar (made of the same medium but with 2% Noble agar.) Each dish contained 5 x 10^4 or 10^5 cells in the top agar layer. The next day and every 7 to 10 d later, 1 to 2 ml of agar-free culture medium was added to the top agar to make up for evaporation. The dishes were incubated at 37°C in 99% humidity and 3% CO_2, 97% air for 4 to 5 weeks. The frequency of AI cells was calculated from the number of colonies with a diameter equal to or greater than 60μm, as determined with an inverted microscope equipped with an ocular micrometer.

Analysis of Covalently Bound DNA Adducts

Cells (~ 20 x 10^6 per dose) were plated at a density of 10^4/cm², treated with tritiated compound, harvested, lysed, and their cellular DNA extracted and purified on CsCl gradients as described (4). The number of covalently bound residues was determined from the UV-absorption profile and the specific radioactivity. For HPLC analysis, the DNA was enzymatically digested and co-chromatographed on HPLC as described (8) with an authentic standard of N-(deoxyguanosin-8-yl)-1-aminopyrene supplied by Dr. Beland.

RESULTS

Cytotoxicity and Mutagenicity of 1-Nitrosopyrene

Because 1-nitrosopyrene, 1-NOP, is more readily reduced to the ultimate reactive form than is the parent compound, 1-NP, we began our comparative studies with 1-NOP. Figure 2 shows that XP12BE cells were much more sensitive than normal cells to the cytotoxic and mutagenic effect of 1-NOP. At doses which caused no significant response in the normal cells, the XP cells showed a linear, dose dependent increase in mutant frequency. The normal cells showed a shoulder on their survival curve and a threshold on their mutation curve. At doses giving equal cell killing, *e.g.*, 37% survival, the frequency of mutants induced in the XP population by 1-NOP was 2.6 times higher than in the normal.

Cytotoxicity and Mutagenicity of 1-Nitropyrene

Howard *et al*. (9) showed that 1-NP induced the transformation of human cells to anchorage independence, suggesting that human fibroblasts contain sufficient levels

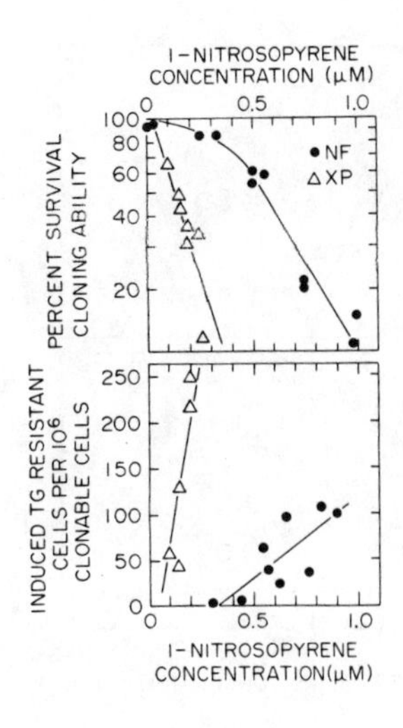

FIGURE 2. Cytotoxic and mutagenic effect of 1-NOP as a function of concentration. (From ref. 5, with permission.)

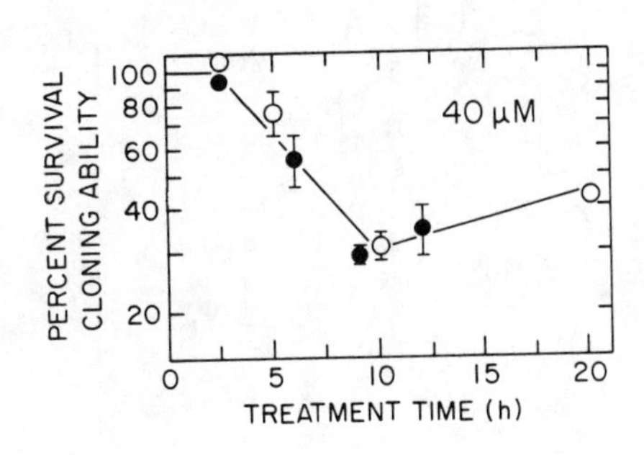

FIGURE 3. Time-dependence of the cytotoxicity of 1-NP in normal fibroblasts. (From ref. 5, with permission.)

of endogenous nitro reductases to activate the parent compound under the appropriate culture conditions. Because these investigators employed hypoxic conditions, we tested the cytotoxic effect of 1-NP using a range of concentrations (20 to 80μM) under aerobic and hypoxic conditions, but found no difference between the results obtained. Therefore, in subsequent experiments, we used aerobic conditions. The time-dependence of 1-NP-induced cytotoxicity in normal fibroblasts was investigated using a single concentration of 1-NP (Figure 3). Maximal cell killing was observed using a treatment time of 10 h. Longer exposure did not increase the cell killing, suggesting that all available compounds had been metabolized by that time or that cell replication by the surviving population was occurring which would mask additional cell killing. Therefore, a 10 h treatment time was chosen for investigating the dose-dependent behavior of this compound.

Figure 4 compares the survival of normal, XP, and HCMM fibroblasts exposed for 10 h to various concentrations

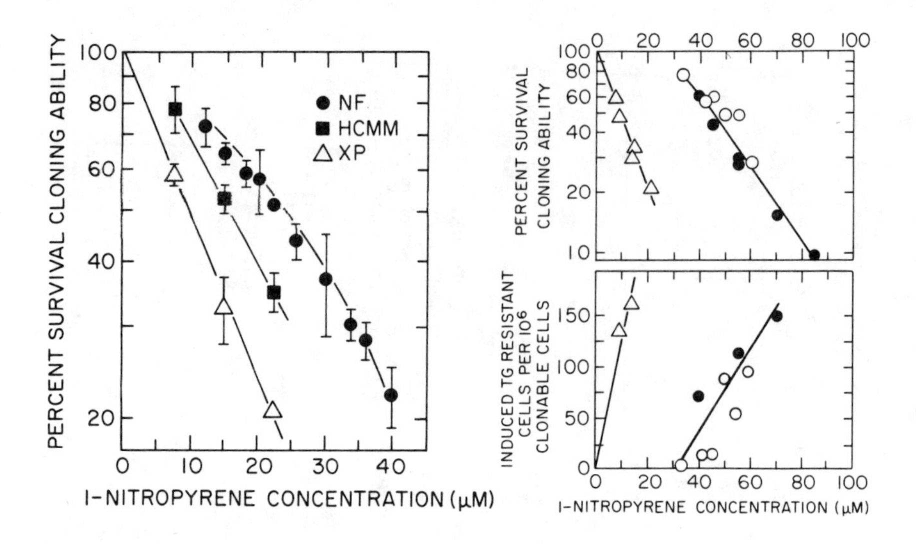

FIGURE 4. Cytotoxicity of 1-NP as a function of concentration. (From ref. 5, with permission.)

FIGURE 5. Cytotoxic and mutagenic effect of 1-NP as a function of concentration. (From ref. 5, with permission.)

of 1-NP. The XP12BE cells were ~ 2.5 times as sensitive as normal cells; the HCMM cells had a slope ~ 1.5-fold steeper than that of normal cells. The HCMM cells showed no such abnormal sensitivity to 1-NOP (data not shown). Their sensitivity to the parent compound, 1-NP, but not to 1-NOP, is predicted by the work of Howell et al. (6) who showed them to be abnormally sensitive to the cytotoxic effects of 4NQO, but not to its hydroxylamine derivative. Figure 5 shows the mutagenic effect of 1-NP, along with the corresponding survival data from the mutagenesis experiments. As was the case for 1-NOP, the XP cells were much more sensitive than the normal cells and there was a shoulder on the survival curve of the normal cells and a threshold on their mutation curve.

Comparing the Mutagenicity of 1-NP and 1-NOP

Comparing the mutagenicity of 1-NOP in Figure 2 with 1-NP in Figure 5 on the basis of equal cell killing shows

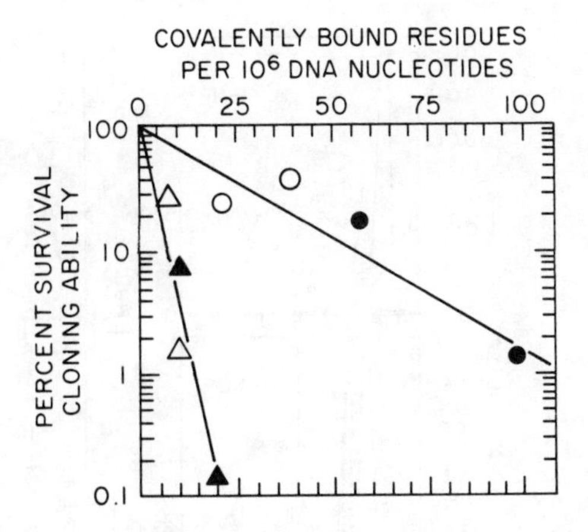

FIGURE 6. Cytotoxicity of 1-NP open symbols and 1-NOP closed symbols as a function of the initial number of DNA adducts in NF (○ , ●) and XP (△ , ▲) cells. (From ref. 5, with permission.)

that the two carcinogens are comparable. Much higher concentrations of 1-NP than 1-NOP are needed to achieve a particular amount of cytotoxicity, but the frequency of mutants induced by such adjusted concentrations of either agent are comparable. For example, in NF cells, the dose of 1-NP that lowered the survival to 37% induced ~80 TG resistant cells per 10^6; a 37% survival dose of 1-NOP induced ~60 mutants per 10^6 cells. In XP cells, with either compound, this value was ~160 mutants per 10^6 clonable cells assayed for resistance to TG.

Another way to compare agents for mutagenicity is on the basis of the initial number of adducts covalently bound to DNA at the end of the exposure. This was done using radio-labeled compound (Figure 6). Much higher concentrations of 1-NP than 1-NOP were needed to get an equal amount of each compound bound, but when the number of adducts was equal, the killing effect was also equal. For example, ~ 28 residues of 1-NP or 1-NOP bound per 10^6 nucleotides yielded 37% survival in the normal cells; 4 residues of either agent per 10^6 nucleotides were sufficient to reduce the survival of the excision repair-deficient XP12BE cells to 37%. Analyzed in this way, the data in Figures 2, 5 and 6 indicate that the two carcinogens are equal in ability to cause mutations, but 1-NOP is significantly

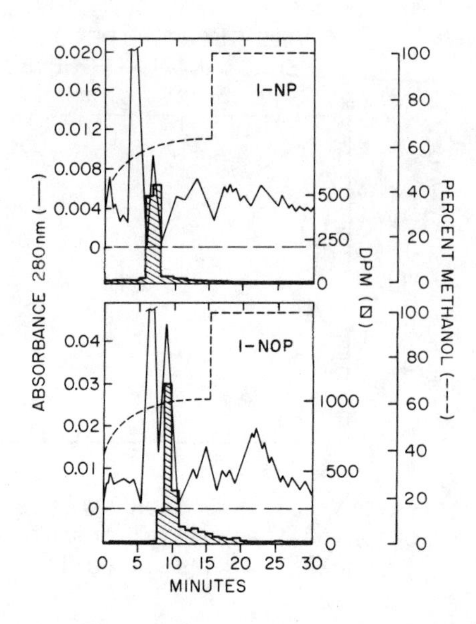

FIGURE 7. Reverse-phase HPLC profiles of DNA adducts formed in NF human cells treated with tritiated 1-NP or 1-NOP. Synthetic N-(deoxyguanosin-8-yl)-1-aminopyrene was added to the extraction as an O.D. standard.

more easily activated to a form that can bind to DNA. This result is expected if the compounds form identical DNA adducts in these human cells. HPLC analysis of DNA from 1-NP and 1-NOP-treated human cells showed that this was, indeed, the case (Figure 7). The main peak co-chromatographed with N-(deoxyguanosin-8-yl)-1-aminopyrene.

Transformation of Human Cells with 1-NP and 1-NOP

Studies measuring the ability of these two compounds to cause the transformation of human cells from anchorage dependence to anchorage independence are still in progress. Results to date indicate that both compounds cause a dose dependent increase in the frequency of AI cells, able to form large-sized colonies in soft agar. The frequencies observed were 3 to 5 times higher than the frequencies of mutants (TG resistant cells) induced in the cells by similar concentrations. This relationship between the frequency of transformants and of TG resistant human fibroblasts is consistent with what we reported previously with UV radiation as the carcinogen (10).

DISCUSSION

Although 1-nitropyrene induces mutations in Salmonella at high frequency (1,11) until now it has not been found to induce mutations in mammalian cells (4,12-14). Figures 4-6 show that, under appropriate conditions, 1-nitropyrene can be activated by human cells into a form capable of reacting with DNA to form stable covalent adducts and that it is mutagenic. The fact that 1-NP and 1-NOP exhibited similar cytotoxicity and mutagenicity in the human fibroblasts when compared on the basis of the initial number of DNA adducts formed is not unexpected since we showed that the same adduct is formed. Heflich *et al*. (4) showed that in CHO cells, both compounds form the same adducts, but the frequency of 1-NP adducts was very low (highest level, 1.9 residues per 10^6 nucleotides). As shown in Figure 6, such a low frequency of adducts would not be expected to cause a significant amount of cell killing or induction of mutations. This would explain why these authors failed to find a biologic effect of 1-NP in CHO cells.

1-NP is structurally related to compounds such as benzo[a]pyrene (BP) and 2-acetylaminofluorene (AAF). Therefore, we compared the biological activity of 1-NP to that of reactive derivatives of these carcinogens for the number of DNA adducts required to cause cell killing or mutations. The reactive forms used were the anti 7,8-diol-9,10-epoxide of BP (BPDE) and *N*-acetoxy-AAF (N-AcO-AAF). To reduce the level of survival of normal human fibroblasts to 37% with N-AcO-AAF requires ~50 initial adducts per 10^6 DNA nucleotides (15). With BPDE only 8 adducts per 10^6 nucleotides are required (16). 1-NP requires ~28 adducts per 10^6 nucleotides (Figure 6). This intermediate value is closer to that of N-AcO-AAF than to BPDE. Similarly, at the dose which results in 37% survival, N-AcO-AAF induces mutants at a frequency of ~60 x 10^{-6} (17), whereas BPDE gives a frequency of 120 x 10^{-6} (17,18). 1-NP induces mutants at an intermediate frequency of 80 x 10^{-6}, which is closer to that of N-AcO-AAF than of BPDE. This result is consistent with the fact that the adduct induced by N-AcO-AAF is formed on the C8 position of guanine like that of 1-NP and 1-NOP (19).

ACKNOWLEDGEMENTS

We thank Dr. F. A. Beland for supplying us with 1-NOP and the chromatography standard for HPLC. The excellent technical assistance of R. C. Corner, M. Howell, and P. A. Ryan is gratefully acknowledged. Research described in this article was conducted under contract to the Health Effects Institute (HEI), an organization jointly funded by the U.S. Environmental Protection Agency (EPA) (Assistance Agreement X812059) and automotive manufacturers. It is currently under review by the Institute. The contents of this article do not necessarily reflect the views of the HEI, nor do they necessarily reflect the policies of EPA, or automotive manufacturers.

REFERENCES

1. Rosenkranz, H.D. (1982): Direct-acting mutagens in diesel exhausts: magnitude of the problem. Mutation Res., 101: 1-10.
2. Howard, P.C., and Beland, F.A. (1982): Xanthine oxidase catalyzed binding of 1-nitropyrene to DNA. Biochem. Biophys. Res. Commun., 104: 727-732.
3. Heflich, R.H., Howard, P.C., and Beland, F.A. (1985): 1-Nitrosopyrene: An intermediate in the metabolic activation of 1-nitropyrene to a mutagen in Salmonella typhimurium TA1538. Mutation Res., 149: 25-32.
4. Heflich, R.H., Beland, F.A., Fullerton, N.F., and Howard, P.C. (1984): Analysis of the metabolites, DNA adducts and mutations produced by 1-nitropyrene in Chinese hamster ovary cells. Environ. Mut., 6: 456-457.
5. Patton, J.E., Maher, V.M., and McCormick, J.J. (1986): Cytotoxic and mutagenic effects of 1-nitropyrene and 1-nitrosopyrene in diploid human skin fibroblasts. Carcinogenesis, 7: in press.
6. Howell, J.N., Greene, M.H., Corner, R.C., Maher, V.M., and McCormick, J.J. (1984): Fibroblasts from patients with hereditary cutaneous malignant melanoma are abnormally sensitive to the mutagenic effect of simulated sunlight and 4-nitroquinoline-1-oxide. Proc. Nat. Acad. Sci. USA, 81: 1179-1183.
7. McCormick, J.J., and Maher, V.M. (1981): Measurement of colony-forming ability and mutagenesis in diploid human cells. In: DNA Repair, A Laboratory Manual of Research Procedures, edited by E.C. Friedberg, and P.C. Hanawalt, pp. 501-521, Marcel Dekker, NY.

8. Howard, P.C., Heflich, R.H., Evans, F.E., and Beland, F.A. (1983): Formation of DNA adducts in vitro and in Salmonella typhimurium upon metabolic reduction of the environmental mutagen 1-nitropyrene. Cancer Res., 43: 2052-2058.
9. Howard, P.C., Gerrard, J.A., Milo, G.E., Fu, P.P., Beland, F.A., and Kadlubar, F.F. (1983): Transformation of normal human skin fibroblasts by 1-nitropyrene and 6-nitrobenzo[a]pyrene, Carcinogenesis, 4: 353-355.
10. Maher, V.M., Rowan, L.A., Silinskas, K.C., Kateley, S.A., and McCormick, J.J. (1982): Frequency of UV-induced neoplastic transformation of diploid human fibroblasts is higher in xeroderma pigmentosum cells than in normal cells. Proc. Natl. Acad. Sci. U.S.A. 79: 2913-2617.
11. Pederson, T.C., and Siak, J.S. (1981): The role of nitroaromatic compounds in the direct-acting mutagenicity of diesel particulate extracts. J. Appl. Toxicol., 1: 54-60.
12. Nakayasu, M., Sakamoto, H., Wakabayashi, K., Terada, M., Sugimura, T., and Rosenkranz, H.C. (1982): Potent mutagenic activity of nitropyrenes on Chinese hamster lung cells with diphtheria toxin resistance as a selective marker. Carcinogenesis, 37: 917-922.
13. Takayama, S., Machiko, T., Katoh, Y., Masaaki, T., and Sugimura, T. (1983): Mutagenicity of nitropyrenes in Chinese hamster V79 cells. Gann, 74: 338-341.
14. Ball, J.C., Zacmanididis, P., and Salmeen, I. (1985): The reduction of 1-nitropyrene to 1-aminopyrene does not correlate with the mutagenicity of 1-nitropyrene in V79 Chinese hamster cells. In: Polynuclear Aromatic Hydrocarbons, Methods and Mechanisms, edited by M.W. Cook, and A.J. Dennis, pp. 113-120, Battelle Press, Columbus, Ohio.
15. Heflich, R.H., Hazard, R.M., Lommel, L., Scribner, J.D., Maher, V.M., and McCormick, J.J. (1980): A comparison of the DNA binding, cytotoxicity, and repair synthesis induced in human fibroblasts by reactive derivatives of aromatic amide carcinogens. Chem. Biol. Interact., 29: 43-56.
16. Yang, L.L., Maher, V.M., and McCormick, J. J. (1980): Error-free excision of the cytotoxic, mutagenic N^2-deoxyguanosine DNA adduct formed in human fibroblasts by (±)-7β,8α-dihydroxy-9α,10α-epoxy-7,8,9,10-tetrahydrobenzo[a]pyrene. Proc. Nat. Acad. Sci. USA, 77: 5933-5937.

17. Aust, A.E., Drinkwater, N.R., Debien, K., Maher, V.M., and McCormick, J.J. (1984): Comparison of the frequency of diphtheria toxin and thioguanine resistance induced by a series of carcinogens to analyze their mutational specificities in diploid human fibroblasts. Mutation Res., 125: 95-104.
18. Yang, L.L., Maher, V.M., and McCormick, J.J. (1982): Relationship between excision repair and the cytotoxic and mutagenic effect of 'anti' 7,8-diol-9,10-epoxide of benzo[a]pyrene in human cells. Mutation Res., 94: 435-447.
19. Poirier, M.C., Williams, G.M., and Yuspa, S.H. (1980): Effect of culture conditions, cell type and species of origin on the distribution of acetylated and deacetylated deoxyguanosine C-8 adducts of N-acetoxy-2-acetylaminofluorene. Mol. Pharmacol. 18: 234-240.

DIFFERENCES IN THE INFLUENCE OF π PHYSICAL BINDING INTERACTIONS WITH DNA ON THE REACTIVITY OF BAY VERSUS K-REGION HYDROCARBON EPOXIDES

A. S. PRAKASH, R. G. HARVEY[1] and P. R. LEBRETON
Department of Chemistry, University of Illinois at Chicago, Chicago, Illinois 60680 and (1) Ben May Laboratory for Cancer Research, University of Chicago, Chicago, Illinois 60637.

INTRODUCTION

Much study has indicated that the reactivity of epoxide containing metabolites of carcinogenic aromatic hydrocarbons is strongly influenced by π electronic structure (1). This influence forms the basis for the bay region theory of hydrocarbon carcinogenesis. The theory is founded on the observation that π structures which stabilize intermediate benzylic carbocations enhance epoxide reactivity. The bay region theory relies on the assumption that epoxide reactions with DNA, which is thought to be the most important biological target, take place via an S_{N1} mechanism or at least a transition state with significant carbocation character. In general there is good correlation between the predictions of epoxide reactivity based on π electron quantum mechanical calculations and hydrocarbon mutagenicity, hydrocarbon carcinogenicity and hydrocarbon epoxide rates of hydrolysis (1). However recent studies indicate that, in addition to reactive interactions between hydrocarbon metabolites and DNA, reversible binding interactions may also be important to mechanisms of hydrocarbon carcinogenesis (2-9).

Recent kinetic investigations (10,11) of the reactions of the ultimate carcinogen (±)trans-7,8-dihydroxy-anti-9,10-epoxy-7,8,9,10-tetrahydrobenzo[a]pyrene[+] (BPDE) in the presence of DNA, have pointed out that intercalated complexes form prior to the onset of reactions leading to the formation of BPDE-DNA adducts.

[+]Abbreviations: BPDE, (±) trans-7,8-dihydroxy-anti-9,10-epoxy-7,8,9,10-tetrahydrobenzo[a]pyrene; BP, benzo[a]pyrene; 7,8-di(OH)H_2BP, trans-7,8-dihydroxy-7,8-dihydrobenzo[a]pyrene; tetrol, 7,8,9,10-tetrahydroxy-7,8,9,10-tetrahydrobenzo-[a]pyrene; DMBA, 7,12-dimethylbenz[a]anthracene; DMBO, 7,12-dimethylbenz[a]anthracene-5,6-oxide; 5,6-di(OH)H_2DMBA, trans-5,6-dihydroxy-5,6-dihydro-7,12-dimethylbenz[a]anthracene

Kinetic studies of BPDE hydrolysis to a tetrol indicate that DNA, synthetic polynucleotides and mononucleotides catalyze the hydrolysis reaction (10-14). In native calf thymus DNA the hydrolysis is accelerated as much as 80 times (12). The hydrolysis, when studied in the pH range 6.0 to 6.6, is similarly catalyzed by 5'-GMP monoanions. In $H_2PO_4^-$ in the same pH range no appreciable increase in the rate of BPDE hydrolysis occurs (14). This result and studies of DNA catalyzed BPDE hydrolysis (10-12) indicate that physical complexes involving π interactions between BPDE and nucleic acid bases play an important role in enhancing BPDE hydrolysis. The close parallel between reactions leading to BPDE adduct formation and to nucleotide catalyzed hydrolysis suggest that the two pathways involve similar intermediates (10), presumably hydrocarbon-nucleotide base π complexes.

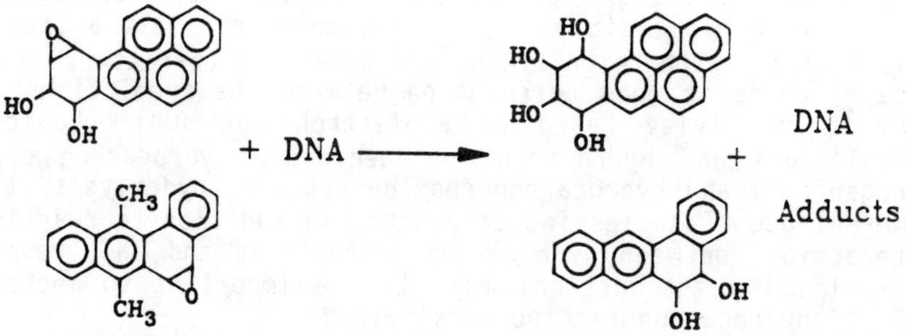

HO OH CH3 CH3 + DNA → HO HO HO OH + DNA Adducts OH OH

FIGURE 1. Structures of BPDE and of 7,12-dimethylbenz[a]anthracene-5,6-oxide (DMBO) and major pathways for reactions involving DNA.

Previous investigations from this laboratory indicate that differences in the π structures of carcinogenic hydrocarbon metabolites cause significant differences in the ability of these molecules to intercalate into DNA (2-9). This is demonstrated by previous studies with 7,12-dimethylbenz[a]anthracene (DMBA) metabolite analogues. For example, in 15% methanol the bay region metabolite analogue 1,2,3,4-tetrahydro-DMBA has an intercalation binding constant with calf thymus DNA which is 4.4 times greater than that for the K region metabolite analogue 5,6-dihydro-DMBA (9). Previous results also suggest that benzo[a]pyrene (BP) bay region metabolites intercalate better than DMBA K region metabolites. This is demonstrated by the observation that in 15%

methanol the bay region analogue 7,8,9,10-tetrahydro-BP binds to calf thymus DNA 4.2 times better than the K region analogue 5,6-dihydro-DMBA (2,4). These earlier studies point out that π structure not only influences epoxide reactivity but also influences reversible binding interactions with nucleic acids.

The goal of the present investigation has been to examine the effects of DNA on the reactivities of epoxide containing hydrocarbon metabolites which have a varying noncovalent affinity for DNA. Specifically this study examines whether DNA influences the reactivity of an epoxide with strong reversible binding characteristics differently than it influences the reactivity of an epoxide which is a weak physical binding partner. In the present studies BPDE was chosen as an example of an epoxide with strong reversible binding characteristics while 7,12-dimethylbenz[a]anthracene-5,6-oxide (DMBO) was chosen as an example of an epoxide with low noncovalent affinity for DNA. The structures of these epoxides and the major reaction pathways which they follow are shown in Fig. 1.

MATERIAL AND METHODS

Nonreactive analogues of the reactive epoxides in Fig. 1 were used to qualitatively compare the physical binding of the bay region epoxide BPDE and the K-region epoxide DMBO. The advantage of using nonreactive analogues to obtain information about the physical binding properties of epoxides is that the diols do not participate in reactions which complicate and undermine the reliability of physical binding studies. The analogues employed were trans-7,8-dihydroxy-7,8-dihydrobenzo[a]pyrene (7,8-di(OH)H_2BP) and trans-5,6-dihydroxy-5,6-dihydro-DMBA (5,6-di(OH)H_2DMBA). Structures of the diols used as epoxide analogues are shown in Fig. 2.

Binding studies and kinetic studies were performed with a Perkin Elmer 650-10 Fluorescence Spectrometer. All binding and kinetic studies were performed at 20 $\pm$ 1°C. Samples of BPDE, DMBE, 7,8-di(OH)H_2BP and 5,6-di(OH)H_2DMBA were prepared using methods described elsewhere (15,16). Calf thymus DNA was purchased from Sigma Chemical Company.

The binding studies of 7,8-di(OH)H_2BP were carried out in a solvent system consisting of double distilled water and methanol (15% by volume). Binding studies with 5,6-

di(OH)H_2DMBA and the kinetic studies of BPDE and DMBO were carried out in a solvent system without methanol. All solutions were maintained at a pH of 7.1 with 10^{-3} M sodium cacodylate. DNA concentrations are reported in terms of PO_4^- molarity calculated from an average base pair molecular weight of 617.8. Corrections in the calculation of DNA concentrations have been made for the amounts of H_2O and Na^+ reported by the supplier for each DNA batch. Denatured DNA was prepared by heating native DNA to 95°C for 5 minutes.

For 7,8-di(OH)H_2BP and 5,6-di(OH)H_2DMBA Stern-Volmer plots used to obtain binding constants were measured at excitation wavelengths of 372 nm and 308 nm respectively. The emission wavelengths were 402 nm and 372 nm. The Stern-Volmer plots were constructed by measuring the ratio of the hydrocarbon emission intensity (I_0) observed without DNA to the intensity (I) observed with DNA. This ratio is plotted versus the DNA concentration. Studies of the effects of Mg^{+2} on the quenching of the diols were carried out at the same emission and excitation wavelengths used to measure the Stern-Volmer plots.

In the kinetic studies the reaction leading to adduct formation and the hydrolysis reaction were not separately observed. Only the total reaction was monitored. The epoxides have negligible fluorescence quantum yields; the covalent adducts have an intermediate quantum yield; the unbound hydrolysis products have the highest fluorescence quantum yield. The reactions were followed by measuring, at 30 sec intervals, the increase in the fluorescence emission intensity which accompanies product formation. The excitation and emission wavelengths used in the kinetic studies were the same as those employed in the diol binding studies.

The kinetic experiments were initiated by injecting 20μℓ of a tetrahydrofuran solution containing one of the epoxides into 1 ml of solvent. The initial epoxide concentrations in the aqueous solutions were 10^{-5} to 10^{-6}M. The concentration of DNA, when used in the kinetic experiments, was 2.0 x 10^{-4} M. The wavelengths used to monitor the epoxide reactions were the same as those used to study the binding of the corresponding diols.

RESULTS

Stern-Volmer Plots.

Stern-Volmer plots for 7,8-di(OH)H_2BP and 5,6-di(OH)H_2DMBA are given in Fig. 2 along with association constants. Previous fluorescence lifetime and dialysis experiments with these molecules indicate that the slopes of the Stern-Volmer plots are equal to association constants for hydrocarbon intercalation into DNA (2,4).

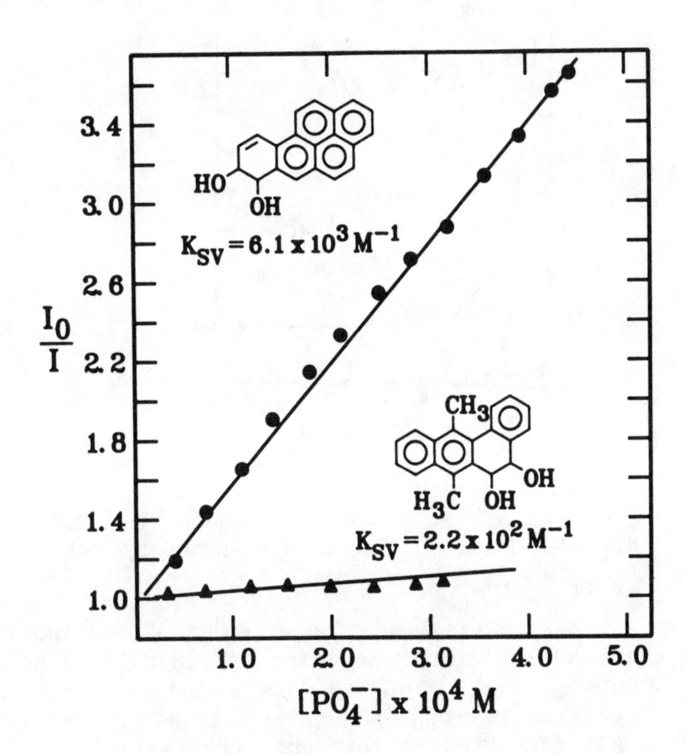

FIGURE 2. Stern-Volmer plots and association constants derived from the fluorescence quenching of 7,8-di(OH)H_2BP and 5,6-di(OH)H_2DMBA by native DNA.

The results of Fig. 2 demonstrate that 7,8-di(OH)H_2BP in 15% methanol has a binding constant which is 28 times greater than that of 5,6-di(OH)H_2DMBA in aqueous solution. The results of Fig. 2 provide evidence that BPDE binds reversibly to DNA bases much better than DMBO does.

Mg^{+2} Effects on Diol Binding.

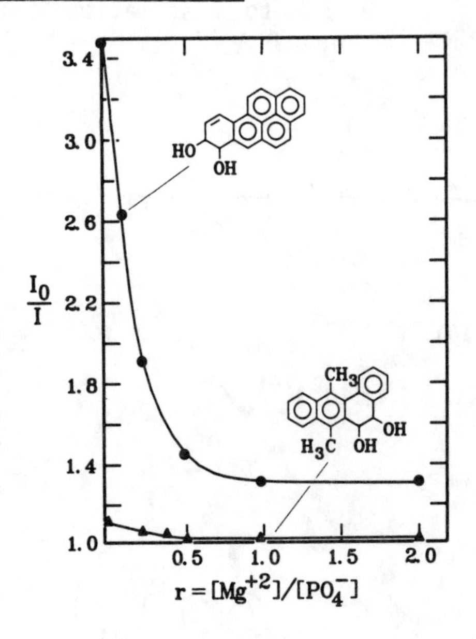

FIGURE 3. Effects of Mg^{+2} on the quenching of 7,8-di(OH)H_2BP and 5,6-di(OH)H_2DMBA by native DNA ($[PO_4^-]$ = 5.0 x 10^{-4}M).

Figure 3 shows how the addition of Mg^{+2}, a DNA stabilizer, affects the fluorescence quenching of 7,8-di(OH)H_2BP and 5,6-di(OH)H_2DMBA by DNA. The results demonstrate that for both diols quenching is reduced when Mg^{+2} is added. For 5,6-di(OH)H_2DMBA all quenching is lost when the ratio (r) of the Mg^{+2} concentration to the DNA concentration is greater than 0.3. For 7,8-di(OH)H_2BP the results are different. In this case residual quenching occurs even for r values above 2.0. At an Mg^{+2} concentration of 10^{-3} M the association constant for 7,8-di(OH)H_2BP is approximately 8.0 x 10^2 M^{-1}.

DNA Effects on the Reactivity of BPDE and DMBO.

Figure 4 contains results of kinetic studies of DNA effects on BPDE and DMBO reactivity. The results for BPDE in Fig. 4 indicate that without DNA, hydrolysis occurs with an apparent first order rate constant of 1.1×10^{-3} sec^{-1}. With native DNA, the fluorescence intensity rises much faster than without DNA. This occurs because of DNA catalyzed hydrolysis and DNA adduct formation. The fluorescence intensity after the reaction is completed is quenched compared to that measured without DNA. Quenching is primarily due to intercalation of the tetrol. The tetrol which accounts for approximately 90% of the reation product (10) has negligible fluorescence quantum yield when intercalated into DNA (4).

The effect of the weak DNA stabilizer Na^+ on the BPDE reactions is negligible.

When the BPDE reactions occur in the presence of denatured DNA or in the presence of native DNA stabilized with Mg^{+2}, the fluorescence emission again rises rapidly; however, the final products are not strongly quenched. The loss of quenching is due primarily to the decreased tetrol binding which occurs when the DNA is denatured or when Mg^{+2} is added. Denaturing the DNA, like adding Mg^{+2}, reduces physical binding. Previous studies (4) indicate that the association constant for 7,8-di(OH)H_2BP binding to denatured DNA is more than 10 times lower than binding to native DNA. Hydrocarbon association constants for binding to denatured DNA are expected to lie between those for binding to native DNA and those for binding to mononucleotides. In 5% methanol the association constant for pyrene binding to 5'-GMP is 45 M^{-1} (17).

The results of studies of BPDE reactions carried out in Mg^{+2} stabilized DNA and in denatured DNA demonstrate that the physical binding required to effectively enhance BPDE reactivity is low.

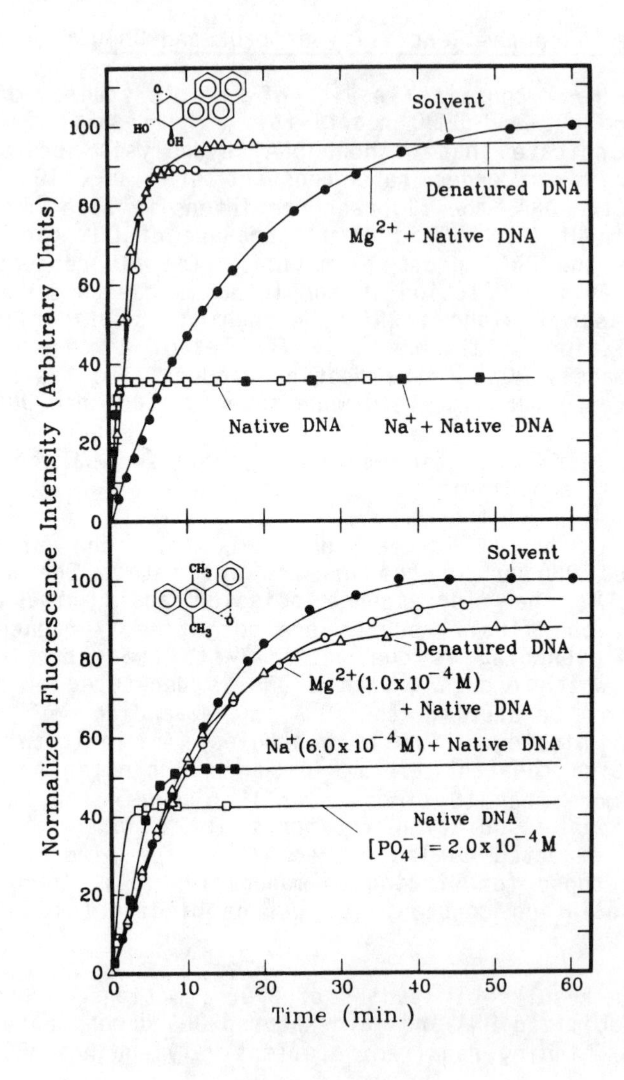

FIGURE 4. Effects of DNA complex formation on the reactivity of the bay region epoxide BPDE and the K region epoxide DMBO. In experiments with Mg^{+} and Na^{+} the same cation concentrations were used for studies of BPDE as for studies of DMBO.

The results for DMBO are in some ways different than those for BPDE. Without DNA, DMBO hydrolysis proceeds with an apparent first order rate constant of 1.5×10^{-3} sec^{-1}. With native DNA in the absence of Mg^{+2}, DMBO reactions are accelerated and, like BPDE, the final reaction products are quenched. However for DMBO, unlike BPDE, the product quenching is not due primarily to intercalation of the hydrolysis product, 5,6-di(OH)H_2DMBA, but to adduct formation. Figure 2 demonstrates that the hydrolysis product, 5,6-di(OH)H_2DMBA, is only slightly quenched by native DNA. The observation that DMBO more readily forms adducts with naked DNA than does BPDE is consistent with results from earlier studies (18).

For DMBO with native DNA, Na^+ influences reactivity only slightly.

With denatured DNA and with Mg^{+2} stabilized native DNA the most interesting differences occur between the results for BPDE and DMBO. For DMBO reactions carried out in the presence of denatured DNA or in the presence of Mg^{+2} stabilized DNA, the increase in fluorescence emission intensity with time occurs in a manner similar to that observed in the hydrolysis reaction without DNA. For DMBO, unlike BPDE, denatured DNA and Mg^{+2} stabilized native DNA do not significantly enhance reactivity.

The conclusion suggested by the BPDE results, namely that weak physical binding is sufficient to enhance reactivity, is supported by the results for DMBO with native DNA without Mg^{+2}. In this case, the physical binding of DMBO which participates only weakly in π interactions with the DNA bases is sufficient to enhance reactivity. However the results with DMBO also suggest that there is a threshold below which π interactions with DNA bases do not greatly enhance epoxide reactivity. The interactions of the weak intercalating agent DMBO with denatured DNA or with Mg^{+2} stabilized DNA lie below this threshold.

DISCUSSION

The present results point out that, while DMBO forms adducts more readily with naked native DNA than does BPDE, the DNA influenced reactivity of DMBO is much more sensitive to DNA structure and environment than the reactivity of BPDE. In future studies with Mg^{+2} stabilized DNA and with denatured DNA it will be important to separately monitor the pathway to adduct formation. In these experiments it will be interesting to compare the ability of BPDE and DMBO to form adducts under conditions in which DNA structure and environment are altered. The results of the present experiments suggest that DNA structure and environment may play an important role in determining the carcinogenic activity of hydrocarbon epoxides. In an *in vivo* environment DNA is exposed to millimolar concentrations of DNA stabilizers such as polyamines and Mg^{+2}. Under these conditions DMBO has much lower reactivity than BPDE. The relative insensitivity of the DNA enhanced reactivity of BPDE to DNA structure and environment may be an important feature which makes this epoxide a potent ultimate carcinogen.

ACKNOWLEDGEMENTS

Support for this work by the National Institutes of Health, the American Cancer Society (Grants #IN-159 and #BC-132) and the Research Board of the University of Illinois is gratefully acknowledged.

REFERENCES

1. Lehr, R. E., Kumar, S., Levin, W., Wood, A. W., Chang, R. L., Conney, A. H., Yagi, H., Sayer, J. M., and Jerina, D. M. (1985): The bay region theory of polycyclic aromatic hydrocarbon carcinogenesis. In: Polycyclic Hydrocarbons and Carcinogenesis, ACS Symposium Series 283, edited by R. G. Harvey, American Chemical Society, Washington, pp. 63-84 and references cited therein.
2. Zegar, I. S., Prakash, A. S., Harvey, R. G., and LeBreton, P. R. (1985): Stereoelectronic aspects of the intercalative binding properties of 7,12-dimethylbenz-[a]anthracene metabolites with DNA. J. Am. Chem. Soc., in press.
3. Paulius, D. E., Prakash, A. S., Harvey, R. G., Abramovich, M., and LeBreton, P. R. (1986): 1-Alkyl substitution effects on the DNA intercalation of benzo[a]-pyrene metabolites. In: Polynuclear Aromatic Hydrocarbons: Chemistry and Carcinogenesis, edited by M. Cooke and A. G. Dennis, Battelle Press, Columbus, OH, in press.
4. Abramovich, M., Prakash, A. S., Harvey, R. G., Zegar, I. S., and LeBreton, P. R. (1985): A comparison of the intercalative binding of nonreactive benzo[a]pyrene metabolites and metabolite model compounds to DNA. Chemico Biol. Inter., in press.
5. LeBreton, P. R. (1985): The intercalation of benzo[a]pyrene and 7,12-dimethylbenz[a]anthracene metabolites and metabolite model compounds into DNA. In: Polyclic Hydrocarbons and Carcinogenesis, ACS Symposium Series 283, edited by R. G. Harvey, American Chemical Society, Washington, pp. 209-238.
6. Abramovich, M., Zegar, I. S., Prakash, A. S., Harvey, R. G., and LeBreton, P. R. (1985): A comparison of the DNA intercalative binding of bay versus K region metabolites of benzo[a]pyrene. In: Molecular Basis of Cancer, Part A: Macromolecular Structure, Carcinogens, and Oncogenes, edited by R. Rein, Alan R. Liss, New York, pp. 217-225.
7. Zegar, I. S., Prakash, A. S., and LeBreton, P. R. (1984): Intercalative DNA binding of model compounds derived from metabolites of 7,12-dimethylbenz[a]anthracene. J. Biomol. Struct. and Dynamics, 2: 531-542.
8. Prakash, A. S., Zegar, I. S., Shahbaz, M., and LeBreton, P. R. (1983) Stereoelectronic aspects of the intercalative binding of benz[a]anthracene metabolite models to DNA. Int. J. Quantum Chem: Quantum Biol. Symp., 10: 349-356.

9. Shahbaz, M., Harvey, R. G., Prakash, A. S., Boal, T. R., Zegar, I. S., and LeBreton, P. R. (1983): Fluorescence and photoelectron studies of the intercalative binding of ben[a]anthracene metabolite models to DNA. Biochem. Biophys. Res. Commun., 112: 1-7.

10. Geacintov, N. E., Hibshoosh, H., Ibanez, V., Benjamin, M. J., and Harvey, R. G. (1984): Mechanisms of reactions of benzo[a]pyrene-7,8-diol-9,10-epoxide with DNA in aqueous solutions. Biophys. Chem., 20: 121-133.

11. MacLeod, M. C. and Selkirk, J. C. (1982): Physical interactions of isomeric benzo[a]pyrene diol epoxides with DNA. Carcinogenesis, 3: 287-292.

12. Geacintov, N. E., Yoshida, H., Ibanez, V. and Harvey, R. G. (1982): Noncovalent binding of 7β,8α-dihydroxy-9α,10α-epoxytetrahydrobenzo[a]pyrene to deoxyribonucleic acid and its catalytic effect on the hydrolysis of the diol epoxide to tetraol. Biochemistry, 21: 1864-1869.

13. MacLeod, M. C. and Zachary, K. L. (1985): Involvement of the exocyclic amino group of deoxyguanosine in DNA-catalyzed carcinogen detoxification. Carcinogenesis, 6: 147-149.

14. Gupta, S. C., Pohl, T. M., Friedman, S. L., Whalen, D. L., Yagi, H., and Jerina, D. M. (1982): Guanosine-5'-monophosphate catalyzed hydrolysis of diastereomeric benzo[a]pyrene-7,8-diol-9,10-epoxides. J. Am. Chem. Soc. 104: 3101-3104.

15. Harvey, R. G. and Fu, P. P. (1978): Synthesis and reaction of diol epoxides and related metabolites of carcinogenic hydrocarbons. In: Polycyclic Hydrocarbons and Cancer, Vol 1, edited by H. Gelboin and P. O. P. Tso, Academic Press, New York, pp. 133-165.

16. Harvey, R. G., Goh, S. H., and Cortez, C. (1975): K-region oxides and related oxidized metabolites of carcinogenic aromatic hydrocarbons. J. Am. Chem. Soc., 97: 3468-3479.

17. Lianos, P. and Georghiou, S. (1979): Complex formation between pyrene and the nucleotides GMP, CMP, TMP and AMP. Photochem. and Photobio., 29: 13-21.

18. Blobstein, S. H., Weinstein, I. B., Dansette, P., Yagi, H., and Jerina, D. M. (1976): Binding of K and non-K-region arene oxides and phenols of polycyclic hydrocarbons to polyguanylic acid. Cancer Res., 36: 1293-1298.

MULTIPLE MECHANISMS OF ACTIVATION OF BENZO[A]PYRENE TO DNA-BINDING METABOLITES IN EARLY PASSAGE WISTAR RAT EMBRYO CELL CULTURES

DONNA PRUESS-SCHWARTZ AND WILLIAM M. BAIRD
Department of Medicinal Chemistry and Pharmacognosy, School of Pharmacy and Pharmacal Sciences, Purdue University, West Lafayette, IN 47907.

INTRODUCTION

The metabolism and binding to DNA of the environmental carcinogenic polycyclic aromatic hydrocarbon (PAH) benzo[a]-pyrene (BaP) has been investigated in a number of animal and human cell systems (1,2,3). BaP can undergo metabolic activation in cells to form an ultimate carcinogenic and mutagenic metabolite, BaP-7,8-dihydrodiol-9,10-oxide (BaPDE) (1,2,3) which exists as a pair of diasteromers: anti-BaPDE (benzylic hydroxyl and epoxide on opposite faces of the plane of the molecule) and syn-BaPDE (benzylic hydroxyl and epoxide on the same face of the plane of the molecule). Each diastereomer consists of two enantiomers: (+) and (-). The major DNA adduct detected in cells exposed to BaP results from attachment of the 10-position of (+)-anti-BaPDE to the exocyclic amino group of deoxyguanosine (dGuo) (1,2,3). Previous studies (4,5) have demonstrated that a number of BaP-DNA adducts are formed in cells in culture and that rat embryo cell cultures contain one of the most complex mixtures of adducts. These studies have shown that (+)-anti-BaPDE-dGuo is not the only major adduct detected in rat cell cultures that have been exposed to [^{3}H]BaP for various lengths of time (4,5,6,7,8). Phillips et al. (8) reported the presence of seven BaP-DNA adducts in primary cultures of Wistar rat mammary cell cultures: none of these adducts were formed from ($\pm$)-anti-BaPDE. We have previously shown (4,5) that the DNA from BaP-treated Wistar rat embryo cell cultures contained several BaP-DNA adducts. The (+)-anti-BaPDE-dGuo adduct was almost undetectable after 5 h of exposure, but increased with time; however, (+)-anti-BaPDE-dGuo was never the BaP-DNA adduct present in the largest amount even after 4 days of exposure to BaP. To investigate the identity of the adducts formed in Wistar rat embryo cells exposed to [^{3}H]BaP, we have subjected the BaP-deoxyribonucleoside adduct peaks to acid hydrolysis under conditions which give BaP-purine and BaP-tetraol products. These hydrolysis products were then chromatographed with known standards to provide further information about the BaP metabolite(s) involved in BaP-DNA adduct formation.

MATERIALS AND METHODS

Treatment of Cells with [^{3}H]BaP, [^{3}H]BaP-7,8-diol, and [^{3}H]-3-OH BaP and Isolation of BAP-DNA Adducts

Primary embryo cell cultures were prepared from 16th day Wistar rat embryos (Harlan Sprague Dawley, Inc., Indianapolis, IN) as described previously (9). Third passage cultures were treated with [G-^{3}H]BaP (Amersham, Arlington Heights, IL) at a final concentration of 0.5 μg BaP/ml medium and at the specific activities stated in Results. For experiments with BaP-7,8-diol, tertiary cultures were treated with [G-^{3}H]-(±)-trans-7,8-dihydroxy-BaP (Radiochemical Repository, Division of Cancer Etiology, National Cancer Institute:specific activity 364 mCi/mmol) at a final concentration of 0.5 g/ml medium and for experiments with 3-OH BaP, tertiary cultures were treated with [G-^{3}H]-3-OH BaP (Radiochemical Repository, Division of Etiology, National Cancer Institute:specific activity 186 mCi/mmol) at a final concentration of 0.5 μg/ml medium. After the lengths of time stated in Results, the cells were harvested with trypsin-Versene and pelleted by centrifugation (4). The nuclei were isolated by homogenization of the cell pellets in a hypotonic buffer containing Triton-X (4) and stored at -80°C. The DNA was isolated from the nuclear pellet by treatment with RNase A and Proteinase K (Sigma Chemical Corp., St. Louis, MO) followed by chloroform:isoamyl alcohol extraction and ethanol precipitation as described previously (5). The DNA was enzymatically degraded to deoxyribonucleosides (10) and then applied to Sep-Pak C-18 cartridges (Waters Associates, Milford, MA). Unreacted deoxyribonucleosides were eluted in water and the BaP-deoxyribonucleoside adducts were eluted in methanol.

Immobilized Boronate Chromatography

The BaP-modified deoxyribonucleosides that contained cis-vicinal hydroxyl groups were separated from the other BaP-deoxyribonucleoside adducts by chromatography on a column of [N-[N'-m-(dihydroxyboryl)phenyl]succinamyl]amino ethyl cellulose as described previously (4). The BaP-deoxyribonucleoside adducts not containing cis-vicinal hydroxyl groups were eluted with 1 M morpholine, pH 9, buffer; then those with cis-vicinal hydroxyl groups were eluted in the morpholine buffer containing 10% sorbitol. The adducts in each buffer fraction were pooled and concentrated on Sep-Pak C-18 cartridges, and the individual BaP-deoxyribonucleosides were analyzed by reverse-phase HPLC.

HPLC Analysis of BaP-DNA Adducts

The BaP-deoxyribonucleoside adducts were chromatographed by HPLC on a 25 cm x 4.6 mm Ultrasphere octyl reverse-phase column (Beckman Instruments, Inc., St. Louis, MO). The column was eluted with methanol:water (46:54) for 34 min at a flow rate of 1.0 ml/min, then for 10 min with a linear gradient of methanol:water (46:54 to 55:45), followed by 24 min with methanol:water (55:45). Fifteen 1.0-min fractions followed by one-hundred forty-five 0.3 min fractions were collected in scintillation vials. Radioactivity was determined by liquid scintillation counting.

Individual BaP-deoxyribonucleoside adducts for hydrolysis studies were prepared by immobilized boronate chromatography and reverse-phase HPLC without the addition of the [^{14}C]-(±)-anti-BaPDE-dGuo marker. A 10-20 μl aliquot of each HPLC fraction was analyzed for radioactivity by liquid scintillation counting and the fractions containing each peak of radioactivity were then pooled.

Hydrolysis of the Glycosidic Bond of BaP-deoxyribonucleoside Adducts

Cleavage of the hydrocarbon-deoxyribonucleoside bond was performed using a modification of the procedure described by Osborne et al. (11). A portion of the pooled HPLC fractions containing a BaP-deoxyribonucleoside adduct was evaporated under a N_2 stream to 5 μl. An aliquot of [^{14}C]-(+)-anti-BaPDE-dGuo was added and the sample was reevaporated to 5 μl. 50 μl 0.1 N HCl was added and the sample was incubated for 24 h at 37°C. The sample was then neutralized to pH 7 with 0.5 N NaOH and rechromatographed by reverse-phase HPLC column as described above.

Hydrolysis of BaP-deoxyribonucleoside Adducts to Tetraols

BaP-deoxryibonucleosides were hydrolyzed to tetraols using a modification of the procedure described by Shugart et al. (12). A portion of the pooled HPLC column eluant fractions containing a BaP-deoxyribonucleoside adduct was evaporated under a N_2 stream to 5 μl in a mini-reaction vial. 50 μl of 0.1 N HCl was added and the sample was incubated for

6 h at 80°C. The sample was then neutralized to pH 7 with 0.5 N NaOH and rechromatographed by reverse-phase HPLC with a mixture of (±)-syn-BaPDE-tetraols and (±)-anti-BaPDE-tetraols obtained from the Chemical Repository, Division of Cancer Etiology, National Cancer Institute.

RESULTS

BaP-DNA Adducts in Wistar Rat Embryo Cell Cultures

Wistar rat embryo cell cultures were exposed to [G-^{3}H]BaP (0.5 μg/ml medium; specific activity 5 Ci/mmol) for 72 h. Analysis of the BaP-deoxyribonucleoside adducts by immobilized boronate chromatography demonstrated that 37% of the radioactivity eluted in the morpholine buffer and 63% in the morpholine:sorbitol buffer. The morpholine buffer fractions contained three adduct peaks that were resolved by reverse-phase HPLC (Figure 1A; M1, M2, and M3). Peak M2 eluted in the same position relative to the [^{14}C]-(+)-anti-BaPDE-dGuo marker as a syn-BaPDE-dGuo marker. Peak M1 eluted prior to the [^{14}C]-(+)-anti-BaPDE-dGuo and peak M3 eluted later. Reverse-phase HPLC of the morpholine:sorbitol buffer fractions demonstrated the presence of three adducts (Figure 1B; MS1, MS2, and MS3). Peak MS2 coeluted with the [^{14}C]-(+)-anti-BaPDE-dGuo marker. Peak MS1 eluted prior to the [^{14}C]-(+)-anti-BaPDE-dGuo marker and peak MS3 eluted later.

Hydrolysis of (-)-Anti-BaPDE-dGuo and (+)-Anti-BaPDE-dGuo

Analysis of a mixture of [^{3}H]-(-)-anti-BaPDE-deoxyribonucleosides and [^{14}C]-(+)-anti-BaPDE-dGuo by reverse-phase HPLC resulted in the elution of [^{3}H]-(-)-anti-BaPDE-dGuo 20 fractions prior to [^{14}C]-(+)-anti-BaPDE-dGuo (Figure 2A). A mixture of [^{3}H]-(-)-anti-BaPDE-deoxyribonucleosides and [^{14}C]-(+)-anti-BaPDE-dGuo was treated with 0.1 N HCl for 24 h at 37°C to cleave the glycosidic bonds and form enantiomers. Reverse-phase HPLC analysis of the acid hydrolysis products demonstrated that the tritium and carbon-14 coeluted and the major hydrolysis product formed eluted later than either starting material (Figure 2B). A small proportion of the hydrolysis products eluted in the same relative position as an anti-BaPDE-tetraol. Thus, acid hydrolysis of the diastereomeric BaP-deoxyribonucleoside adducts formed by reaction of the (-)- and (+)-enantiomers of anti-BaPDE with DNA results in the formation of enantiomers that coelute upon reverse-phase HPLC analysis.

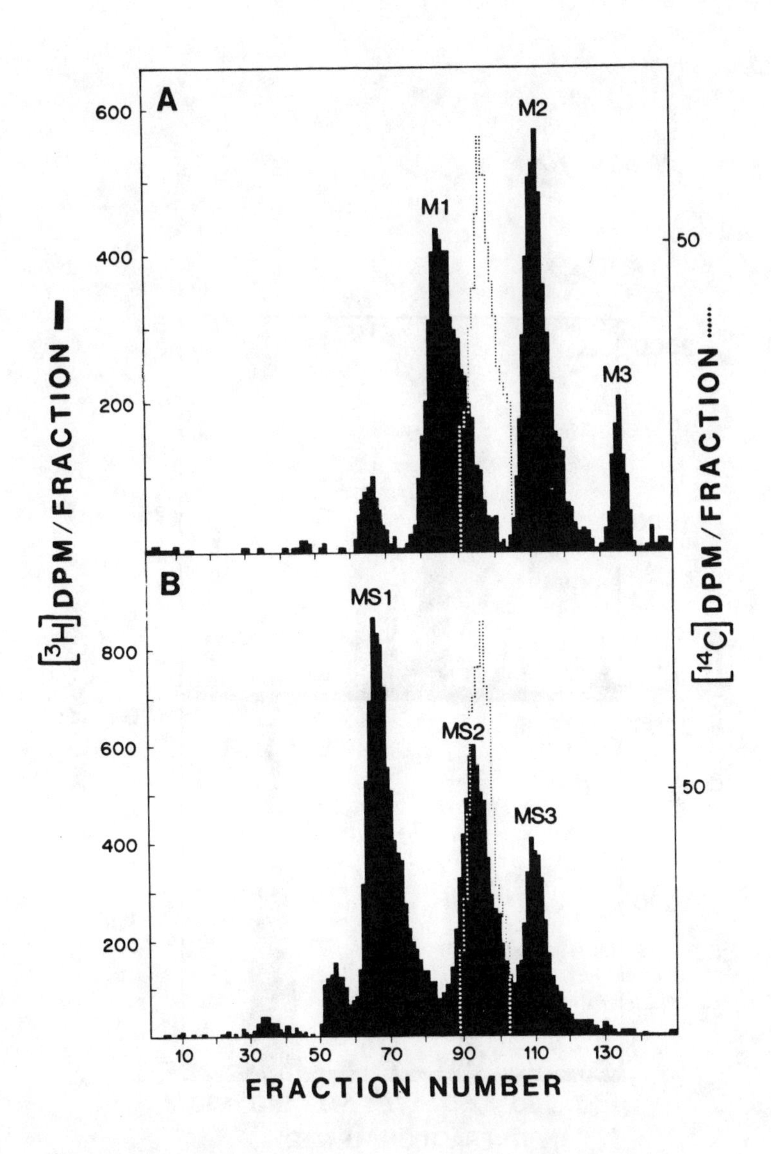

FIGURE 1. Reverse-phase HPLC elution profile of BaP-DNA adducts formed in Wistar rat embryo cell cultures after exposure to [^{3}H]BaP for 72 h. A, adducts present in the 1 M morpholine fractions and B, adducts present in the 1M morpholine, 10% sorbitol fractions. The dotted line represents the elution position of [^{14}C]-(+)-anti-BaPDE:dGuo.

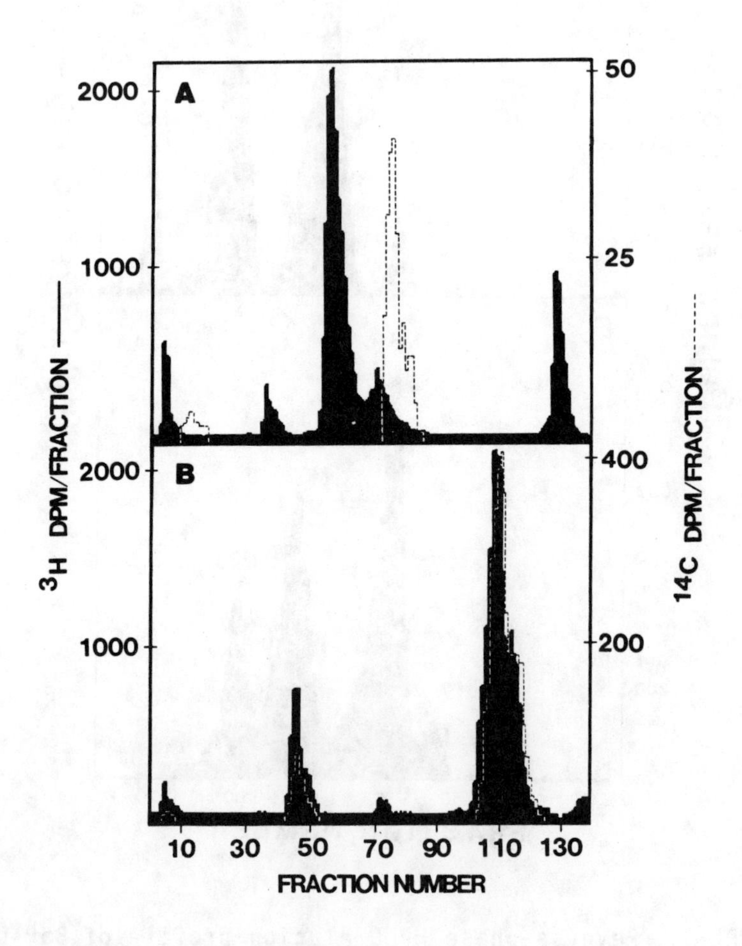

FIGURE 2. Reverse-phase HPLC elution profile of [^{3}H]-(-)-anti-BaPDE-deoxyribonucleosides (----) and [^{14}C]-(+)-anti-BaPDE-dGuo (----) (A). The mixture was exposed to 0.1 N HCl for 24 h at 37°C and rechromatographed (B).

Hydrolysis of MS1

To determine if MS1 was formed by reaction of (-)-anti-BaPDE with deoxyguanosine, a mixture of [^{3}H]MS1, isolated from Wistar rat embryo cells that had been exposed to [^{3}H]BaP (0.5 μg/ml medium; specific activity 20.2 Ci/mmol) for 24 h and [^{14}C]-(+)-anti-BaPDE-dGuo was treated with 0.1 N HCl for 24 h at 37°C. Upon neutralization and reverse-phase HPLC analysis the tritium and carbon-14 hydrolysis products did not coelute (Figure 3A). Two tritium peaks were seen: one eluted with a maximum at fraction 57 corresponding to unhydrolyzed BaP-deoxyribonucleoside adduct MS1. A later-eluting tritium peak was also present with a maximum at fraction 92; this presumably resulted from the loss of deoxyribose from MS1. This hydrolysis product did not coelute with [^{14}C]-(+)-anti-BaPDE-guanine which eluted at fraction 105 (Figure 3A). To verify that MS1 was not formed by reaction of (-)-anti-BaPDE with DNA, MS1 was hydrolyzed to tetraols by treatment with 0.1 N HCl for 6 h at 80°C. Analysis of the hydrolysis products by reverse-phase HPLC demonstrated that the major hydrolysis product formed from MS1 does not coelute with either the BaPDE-tetraols formed from (±)-anti-BaPDE or (±)-syn-BaPDE (Figure 4A). Thus, MS1 is not formed by binding to DNA of the (-)-enantiomer of anti-BaPDE nor any other enantiomeric form of BaPDE.

DNA Adducts Formed in Wistar Rat Embryo Cell Cultures Exposed to [^{3}H]BaP-7,8-Diol or [^{3}H]-3-OH BaP

Wistar rat embryo cell cultures were exposed to [^{3}H]BaP-7,8-diol (0.5 μg/ml medium; specific activity 364 mCi/mmol) for 24 h or to [^{3}H]-3-OH BaP (0.5 μg/ml medium; specific activity 186 mCi/mmol) for 24 h. Reverse-phase HPLC analysis of the DNA adducts present in the BaP-7,8-diol treated cultures demonstrated that MS2 was the only major adduct detected (Figure 5A). Reverse-phase HPLC analysis of the 3-OH BaP-deoxyribonucleoside adducts demonstrated the presence of only a small amount of one adduct (Figure 5B), which eluted later than MS1 and earlier than MS2. Thus, MS1 is not formed by activation of either BaP-7,8-diol or 3-OH BaP to a DNA-binding metabolite in rat embryo cells.

Hydrolysis of MS2

To demonstrate that MS2 was (+)-anti-BaPDE-dGuo, a mixture of [^{3}H]MS2, isolated from Wistar rat embryo cells that had been exposed to [^{3}H]BaP (0.5 μg/ml medium; specific activity 20.2 Ci/mmol) for 24 h and [^{14}C]-(+)-anti-BaPDE-dGuo

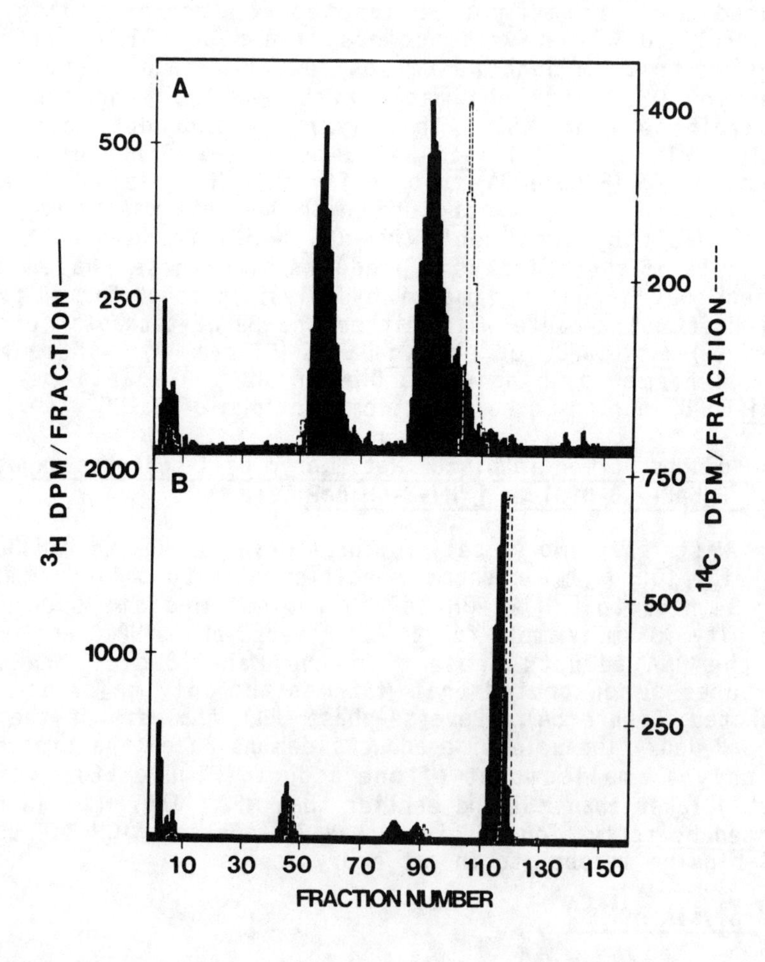

FIGURE 3. Reverse-phase HPLC analysis of the 37°C acid hydrolysis products obtained from BaP-DNA adducts MS1 (A) and MS2 (B).

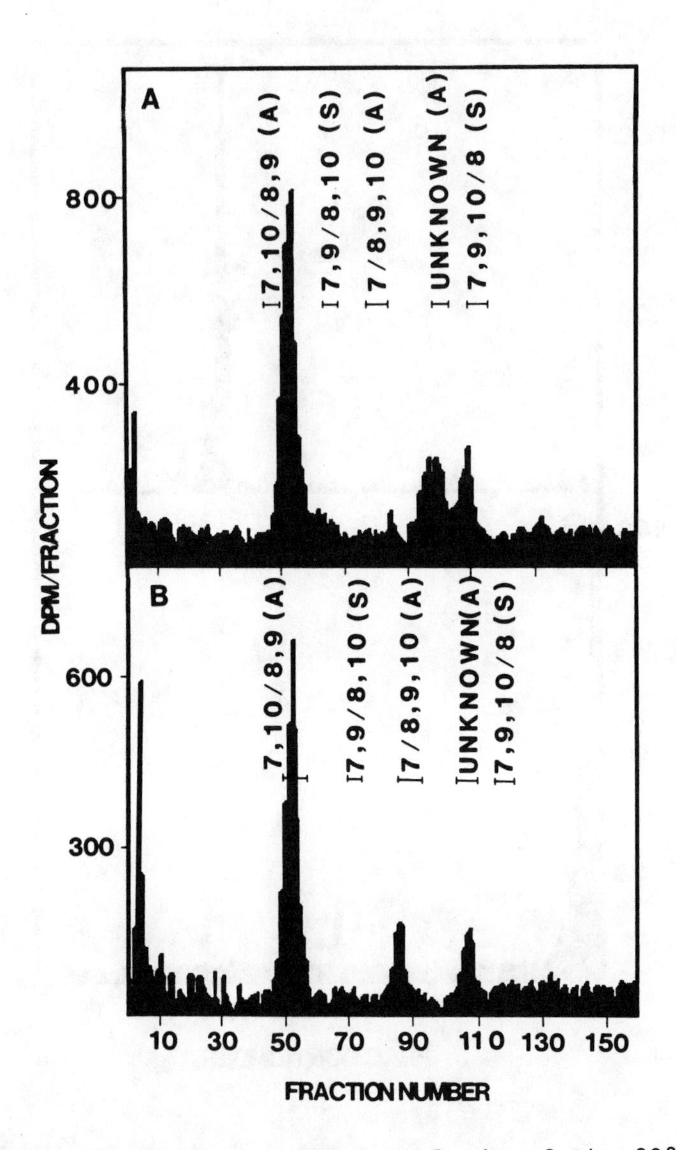

FIGURE 4. Reverse-phase HPLC analysis of the 80°C acid hydrolysis products obtained from BaP-DNA adducts MS1 (A) and MS2 (B). The brackets indicate the elution position of the UV absorbing tetraols obtained from hydrolysis of (±)-*syn*-BaPDE (indicated by S) and (±)-*anti*-BaPDE (indicated by A).

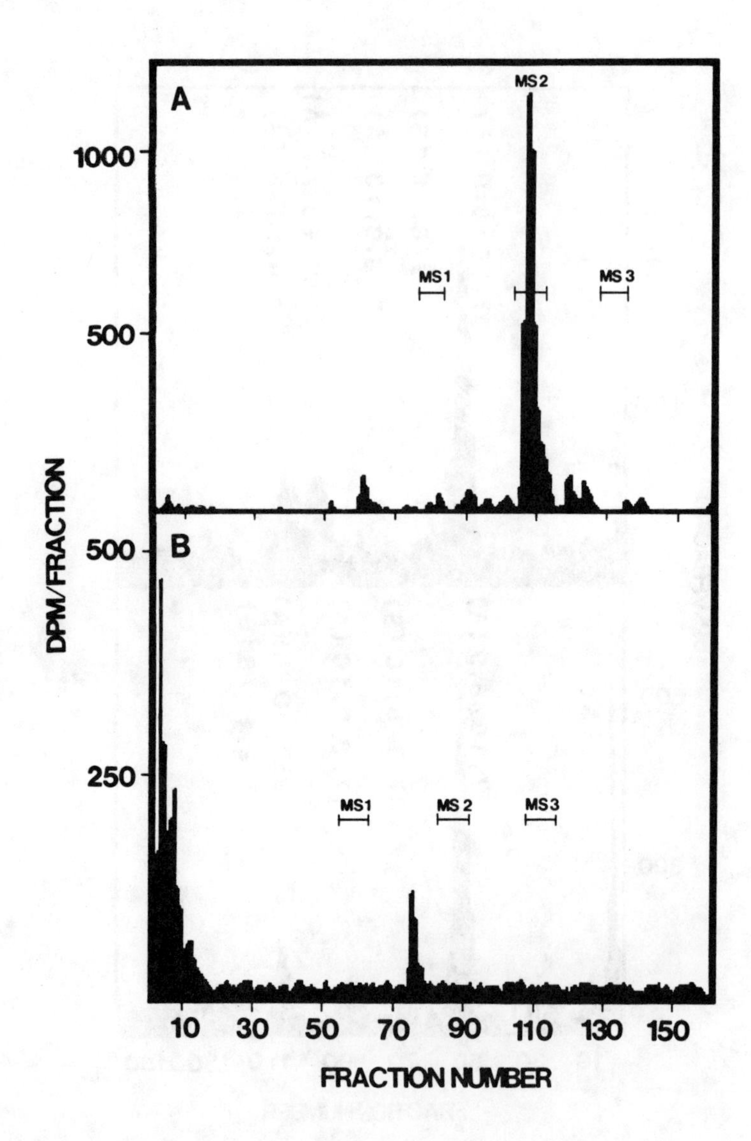

FIGURE 5. Reverse-phase HPLC analysis of the DNA adducts formed in Wistar rat embryo cell cultures after exposure to (A) [^{3}H]BaP-7,8-diol or (B) [^{3}H]-3-OH BaP for 24 h. The brackets indicate the relative elution positions of the BaP-DNA adducts MS1, MS2, and MS3.

was exposed to 0.1 N HCl for 24 h at 37°C. Upon neutralization, reverse-phase HPLC analysis of the hydrolysis products demonstrated that both the carbon-14 and tritium anti-BaPDE-guanine products (fractions 108-122) and the carbon-14 and tritium anti-BaPDE-tetraol products (fractions 42-49) coeluted (Figure 3B). The finding that MS2 was formed by reaction of (+)-anti-BaPDE with deoxyguanosine, was also demonstrated by hydrolysis of MS2 to tetraols by exposure to 0.1 N HCl for 6 h at 80°C. Reverse-phase HPLC analysis of the hydrolysis products with known (±)-syn-and (±)-anti-BaPDE-tetraols demonstrated that MS2 formed anti-BaPDE-tetraols upon acid hydrolysis (Figure 4B). Thus, MS2 is formed from the (+) enantiomer of anti-BaPDE.

DISCUSSION

Six major BaP-deoxyribonucleoside adducts were detected by immobilized boronate chromatography and reverse-phase HPLC in Wistar rat embryo cell cultures that were exposed to [^{3}H]BaP. Three contained cis-vicinal hydroxyl groups (Figure 1; MS1, MS2 and MS3) and three did not (Figure 1; M1, M2, and M3). These adduct peaks were further characterized by the use of acid hydrolysis. The structure of the adducts formed from the (+)- and (-)-enantiomers of anti-BaPDE with deoxyguanosine are shown in Figure 6. The relative stereochemistry of these adducts is opposite at every position except C-1 of the deoxyribose moiety, thus they are diastereomers. Diastereomers can be separated by standard reverse-phase HPLC techniques: (+)-anti-BaPDE dGuo and (-)-anti-BaPDE-dGuo are resolved on a reverse-phase column by 20 fractions (Figure 2A). Upon acid hydrolysis at 37°C glycosidic bond cleavage occurs converting these diastereomers into enantiomers as both purine hydrolysis products have opposite relative stereochemistry at every position (Figure 6). Enantiomers are not normally separable by standard reverse-phase HPLC techniques: (+)-anti-BaPDE-guanine and (-)-anti-BaPDE-guanine coelute on a reverse-phase column (Figure 2B). Thus, comparison of purine hydrolysis products can determine if two adducts are formed from enantiomeric diol-epoxides. Acid hydrolysis using more rigorous conditions results in cleavage of the hydrocarbon-deoxyribonucleoside bond forming hydrocarbon tetraols (Figure 7). Analysis of the tetraol hydrolysis products provides information on the origin of the hydrocarbon metabolite responsible for forming the hydrocarbon-deoxyribonucleoside adduct.

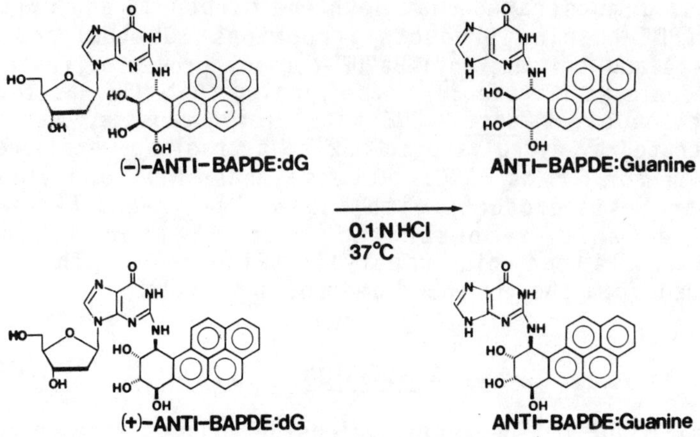

FIGURE 6. Schematic representation of 37°C acid hydrolysis of (-)-anti-BaPDE-dGuo and (+)-anti-BaPDE-dGuo.

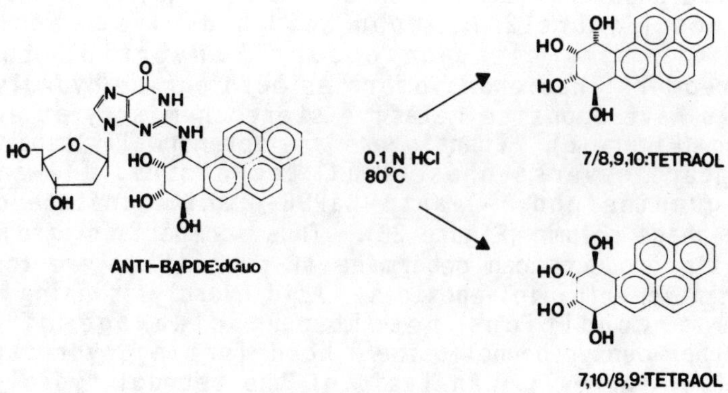

FIGURE 7. Schematic representation of 80°C acid hydrolysis of anti-BaPDE-dGuo.

The DNA adduct that has been identified by cochromatography as (-)-anti-BaPDE-dGuo (6,7) was not detected in Wistar rat embryo cell cultures. MS1 was shown not to be formed from (-)-anti-BaPDE by acid hydrolysis techniques. The purine hydrolysis products derived from MS1 did not coelute with those formed from [^{14}C]-(+)-anti-BaPDE-dGuo (Figure 3A), demonstrating that the purine hydrolysis product of MS1 was not enantiomeric with (+)-anti-BaPDE-guanine. The tetraol hydrolysis products of MS1 did not coelute with tetraols formed from (±)-anti-BaPDE or (±)-syn-BaPDE (Figure 4A). Thus, adduct MS1 does not contain a BaPDE hydrocarbon moiety.

MS1 was not detected in the DNA of Wistar rat embryo cell cultures exposed to [^{3}H]BaP-7,8-diol (Figure 5A), the precursor to BaPDE in its metabolic activation. The only adduct present in the HPLC profile was MS2, the adduct that coelutes with (+)-anti-BaPDE-dGuo. This is consistent with the above data that adduct MS1 does not contain an anti-BaPDE moiety. These results suggest that BaP can undergo metabolic activation to a DNA binding metabolite that is not formed from BaP-7,8-diol. The finding that MS1 does bind to the immobilized boronate column and therefore contains cis-hydroxyls suggests that this binding metabolite is most likely a diol-epoxide in which the BaP moiety is first metabolized to a substituted intermediate prior to 7,8-diol and diol-epoxide formation. Phillips et al. (8) have also suggested that a substituted BaPDE metabolite is involved in BaP-DNA adduct formation in Wistar rat mammary cell cultures. Hulbert and Grover (13) have predicted that phenolic substitutions distal to a diol-epoxide moiety on a PAH may enhance the reactivity of the epoxide. When Wistar rat embryo cell cultures were exposed to [^{3}H]-3-OH BaP, the phenol present in the largest quality in metabolism samples from these cells (9), only one adduct was detected and it eluted later than MS1 and earlier than MS2 (Figure 5B). Thus, MS1 is not identical to (-)-anti-BaPDE-dGuo and is not formed by activation of either BaP-7,8-diol or 3-OH BaP to a DNA binding metabolite in Wistar rat embryo cells.

The structure of the DNA adduct that coelutes with (+)-anti-BaPDE-dGuo, MS2, has been verified by acid hydrolysis techniques. The purine hydrolysis product derived from MS2 coeluted with the purine hydrolysis product of [^{14}C]-(+)-anti-BaPDE-dGuo (Figure 3B), demonstrating that both MS2 and [^{14}C]-(+)-anti-BaPDE-dGuo contained enantiomeric hydrocarbon moieties. The tetraol hydrolysis products of MS2 coeluted with tetraols formed from (±)-anti-BaPDE, but not with tetraols formed from (±)-syn-BaPDE (Figure 4B). Thus, MS2 is formed from the (+) enantiomer of anti-BaPDE.

The use of acid hydrolysis techniques has allowed confirmation of the identity of the metabolites responsible for the formation of the BaP-deoxyribonucleoside adducts present in rat embryo cells. One striking finding is that an adduct frequently identified as the (-)-anti-BaPDE-dGuo adduct based on chromatographic properties actually results from an entirely different metabolite of BaP than BaPDE. Due to the limited amount of material available from cells, the identity of this metabolite has not yet been established. However, future identification of adducts eluting with the (-)-anti-BaPDE-dGuo marker as being an (-)-anti-BaPDE adduct should include acid hydrolysis data to verify this identification. Analysis of radioisotope-labeled hydrocarbon-DNA adducts by chromatography with known standards at the PAH-deoxyribonucleoside, PAH-purine and PAH-tetraol levels provides both better identification of the metabolites responsible for adduct formation and the opportunity to resolve adducts that are inseparable by reverse-phase HPLC as PAH-deoxyribonucleosides.

ACKNOWLEDGMENTS

The authors thank Constance G. Moynihan, Cynthia P. Salmon, and Mark Ferin for excellent technical assistance and Marilyn Hines for typing the manuscript. This investigation was supported by grants CA40228 and CA28825 from the National Cancer Institute, DHHS.

REFERENCES

1. Pelkonen, O., and Nebert, D.W. (1982): Metabolism of polycyclic aromatic hydrocarbons: etiologic role in carcinogenesis, Pharmacol. Rev., 34:189-222.
2. Cooper, C.S., Grover, P.L., and Sims, P. (1983): The metabolism and activation of benzo(a)pyrene, Prog. Drug Metab., 7:295-396.
3. Dipple, A., Moschel, R.C., and Bigger, C.A.H. (1984): Polynuclear Aromatic Carcinogens. In: Chemical Carcinogens, edited by C.E. Searle, pp. 41-163, ACS Monograph 182, Vol 1., American Chemical Society, Washington, D.C.
4. Pruess-Schwartz, D., Sebti, S.M., Gilham, P.T., and Baird, W.M. (1984): Analysis of benzo(a)pyrene: DNA

adducts formed in cells in culture by immobilized boronate chromatography, Cancer Res., 44:4104-4110.

5. Sebti, S.M., Pruess-Schwartz, D., and Baird, W.M. (1985): Species- and length of exposure-dependent differences in the benzo(a)pyrene: DNA adducts formed in embryo cell cultures from mice, rats, and hamsters, Cancer Res., 45:1594-1600.
6. Autrup, H., Wefald, F.C., Jeffrey, A.M., Tate, H., Schwartz, R.D., Trump, B.F., and Harris, C.C. (1980): Metabolism of benzo(a)pyrene by cultured tracheobronchial tissues from mice, rats, hamsters, bovines, and humans. Int. J. Cancer, 25:293-300.
7. Daniel, F.B., Schuf, H.A.J., Sandwisch, D.W., Schenck, K.M., Hoffmann, C.O., Patrick, J.R., and Stoner, G.D. (1983): Interspecies comparisons of benzo(a)pyrene metabolism and DNA-adduct formation in cultured human and animal bladder and tracheobronchial tissues, Cancer Res., 43:4723-4729.
8. Phillips, D.H., Hewer, A., and Grover, P.L. (1985): Aberrant activation of benzo(a)pyrene in cultured rat mammary cells in vitro and following direct application to rat mammary glands in vivo, Cancer Res., 45:4167-4174.
9. Baird, W.M., O'Brien, T.G., and Diamond, L. (1981): Comparison of the metabolism of benzo(a)pyrene and its activation to biologically active metabolites by low-passage hamster and rat embryo cells, Carcinogenesis (Lond.), 2:81- 88.
10. Baird, W.M., and Brookes, P. (1973): Isolation of the hydrocarbon-deoxyribonucleoside products from the DNA of mouse embryo cells treated in culture with 7-methylbenz-(a)anthracene-^{3}H, Cancer Res., 33:2378-2385.
11. Osborne, M.R., Beland, F.A., Harvey, R.G., and Brookes, P. (1976): The reaction of (±)-7α,8β-dihydroxy-9β,10β-epoxy-7,8,9,10-tetrahydrobenzo(a)pyrene with DNA, Int. J. Cancer, 18:362-368.
12. Shugart, L., Rahn, R.O., and Holland, J.M. (1983): Quantifying benzo(a)pyrene binding to DNA by fluorescent analysis. In: Polynuclear Aromatic Hydrocarbons: Formation, Metabolism and Measurement, edited by M. Cooke and A.J. Dennis, pp. 1087-1097, Battelle Press, Columbus, OH.
13. Hulbert, P.B., and Grover, P.L. (1983): Chemical rearrangement of phenol-epoxide metabolites of polycyclic aromatic hydrocarbons to quinone-methides, Biochem. Biophys. Res. Commun., 17:129-134.

STUDIES OF INHIBITION OF CHROMOSOMAL BREAKAGE IN MOUSE INDUCED BY POLYNUCLEAR AROMATIC HYDROCARBONS AND OTHER GENOTOXIC AGENTS

A.S.RAJ* and MORRIS KATZ[1]
Department of Biology and (1) Department of Chemistry, York University, 4700 Keele Street, North York (Toronto), Ontario, M3J 1P3, Canada.

INTRODUCTION

It has been estimated that 60% or more of the incidences of cancer in man and animals is due to the action of environmental chemicals or pollutants. Polynuclear aromatic hydrocarbons (PAH) are a particularly important group of carcinogenic pollutants in the environment in addition to many other potential carcinogenic substances. These PAH compounds occur in fuel, combustion products, airborne particulate matter, motor vehicle exhaust, cigarette smoking, high boiling petroleum distillates, sediments, fumes from coke ovens, the products of incineration refuse and processed foods.

PAH are relatively inert *per se* until they are metabolized to mammalian monooxygenase enzymes to biologically active products to become carcinogens and mutagens. These active compounds or ultimate carcinogens are diol epoxides in which the epoxide moiety forms part of the bay region of the hydrocarbons (1,2).

Extensive reviews of the activity of many chemical carcinogens have been published during the past few years (3,4,5,6,7,8,9). It is well established that the microsomal mixed-function oxidase system of monooxygenase enzymes is involved in the activation and detoxification of cytotoxic, carcinogenic or mutagenic chemicals (10,11,12,13). According to Okey and Nebert (14), an Ah gene complex regulates the induction of cytochrome P_1-450 and at least two dozen other associated enzyme activities that metabolize chemicals.

The exposure to PAH and other genotoxic agents may be mitigated by certain anticarcinogenic substances that are available, some of them particularly in foods. The effectiveness of various classes of chemicals as inhibitors or modifiers of carcinogenesis has been reviewed by Wattenberg (15), Wiebel (10) and Slaga (9).

This paper presents a study to determine the effects of a number of potential inhibitors in preventing the chromosomal breakage in mouse bone marrow tissue induced by 7,12-dimethylbenz(a)anthracene (DMBA), benzo(a)pyrene (BaP), cyclophosphamide (CP) and mitomycin C (MMC). The compounds studied as inhibitors were metyrapone, thioacetamide, β-carotene, glutathione, sodium selenite and fumaric acid.

MATERIALS AND METHODS

Mice

Female mice of strain B6C3F1 (Charles River, Canada, Inc.) of between 8-10 weeks old were used in all our experiments for the *in vivo* bone marrow micronucleus assay. These B6C3F1 mice were classified as homozygous responsive in respect to their ability to induce the aryl hydrocarbon hydroxylase (AHH) enzyme system. The mice were provided with Purina Laboratory Rodent Chow and water *ad libitum*.

Chemicals

The sources of the chemicals used in this study were as follows: DMBA from Eastman Kodak; BaP, CP, metyrapone, thioacetamide, β-carotene, glutathione, sodium selenite and fumaric acid from Aldrich; and MMC from Boehringer Mannheim. These were used at the doses of 30mg/Kg (DMBA), 150 mg/Kg (BaP), 45 mg/Kg (CP), 1 mg/Kg (MMC), 100 mg/Kg (metyrapone), 50 mg/Kg (thioacetamide), 100 mg/Kg (β-carotene in food), 100 mg/Kg (glutathione), 2mg/Kg (sodium selenite), and 1% (fumaric acid in food). DMBA and BaP were dissolved in dimethylsulfoxide (DMSO) whereas CP and MMC were dissolved in physiological saline. Glutathione and sodium selenite were dissolved in sterile distilled water. All the chemicals dissolved in solvents were injected to mice intraperitoneally (i.p.). β-carotene and fumaric acid were mixed in powdered food and the mice were fed on these diets (p.o.).

Experimental Protocol

In separate experiments, either metyrapone or sodium selenite were injected twice to mice at 48 and 24 hr prior to exposure to a single injection of the mutagenic agent. Mice were injected only once with the inhibitor glutathione 24 hr prior to exposure to a single injection of the mutagenic agent. Thioacetamide was injected to mice one hour before mutagenic treatment in one case and one hour after mutagenic treatment in another case. In other instances where the mice were given the inhibitors orally, p.o., (e.g., β-carotene and fumaric acid), the mice were fed on the powdered food containing the above inhibitors for one week before treating with a mutagen. Bone marrow samples were collected at various time intervals and the slides were prepared, stained and scored for polychromatic erythrocytes (PCE) and micronucleated polychromatic erythrocytes (MNPCE) according to methods developed in our laboratory by Salamone et al. (16) and Heddle and Salamone (17). Bone marrow samples were collected from 5 mice at each sampling in each treatment.

RESULTS

The results from the evaluation of chemical inhibitors of chromosomal breakage induced by PAH and other clastogenic agents are summarized in Table 1.

Metyrapone

Metyrapone or metopirone is 2-methyl-1,2-di-3-pyridyl-1-propanone. Metyrapone has been successfully used to treat the hypercortisolism that

TABLE 1

EVALUATION OF CHEMICAL INHIBITORS OF CHROMOSOMAL BREAKAGE INDUCED BY PAH AND OTHER CLASTOGENIC AGENTS

Mutagen or Inhibitors	Dose (mg/Kg)	No. PCE Scored	Time of Peak Occurrence of MNPCE (h)	MNPCE/PCE Scored ± S.E.	Inhibitor	Inhi-bition
					Metyrapone	
DMBA	30	500	48	7.0±1.7	2.3±0.6	67%
BaP	150	500	48	8.0±1.7	3.6±0.6	55%
Metyrapone	100	500	48	1.0±0.7	-	
					Thioacetamide	
DMBA	30	1000	48	16.5±3.4	8.2±1.7	50%
BaP	150	1000	48	12.0±1.5	15.2±2.3	NSE
Thioacetamide	50	1000	48	1.8±0.7	-	
					β-carotene	
DMBA	30	1000	48	18.3±4.4	12.0±2.7	32%
BaP	150	1000	48	12.7±2.3	5.0±0.4	60%
CP	45	1000	48	15.8±1.8	7.0±2.0	56%
MMC	1	1000	48	35.6±2.0	19.8±3.3	48%
β-carotene	100	1000	48	3.8±1.4	-	-

(continued)

TABLE 1. (continued)

Mutagen or Inhibitors	Dose (mg/Kg)	No. PCE Scored	Time of Peak Occurrence of MNPCE (h)	MNPCE/PCE Scored ± S.E.	Inhibitor	Inhibition
					Glutathione	
DMBA	30	1000	48	18.3±1.8	10.8±1.6	41%
BaP	150	1000	48	9.6±2.0	8.8±1.2	-
CP	45	1000	24	21.3±0.6	10.5±3.8	51%
MMC	1	1000	24	33.2±1.5	18.1±3.9	45%
Glutathione	100	1000	24	1.9±0.5	-	-
	100	1000	48	2.1±0.3	-	-
					Sodium selenite	
DMBA	30	1000	48	18.2±1.5	8.4±0.7	54%
BaP	150	1000	48	6.6±1.8	4.6±0.8	30%
Sodium Selenite	2	1000	48	1.0±0.6	-	-
					Fumaric acid	
MMC	1	1000	24	23.4±2.8	15.2±1.9	35%
Fumaric Acid	1%	1000	24	0.6±0.3	-	-

NSE = No significant effect.

results from adrenal neoplasms, and as an adjunctive therapy in Cushing's disease (18). It has also been shown to be a cytochrome P-450 inhibitor (19,20). However, in certain cases, metyrapone enhanced the microsomal activities induced in rat liver by phenobarbital (21). The above studies encouraged us to study the possible inhibitory effect of metyrapone against the formation of chromosomal breaks caused by different mutagens.

Mice were injected with metyrapone 48 and 24 hr before injecting either with DMBA or BaP and the bone marrow samples were obtained at various intervals. Metyrapone, by itself, produced only about 1 MNPCE/500 PCE at all sampling points. DMBA- treated mice showed, on averge, a peak number of 7.0 MNPCE/500 PCE at 48 hr and metyrapone pretreated mice that received DMBA showed about an average of 2.3 MNPCE/500 PCE at the same sampling time indicating 67% inhibition against DMBA. BaP-treated mice showed about 8 MNPCE/500 PCE at 48 hr and pretreatment with metyrapone followed by treatment with BaP showed about 3.6 MNPCE/500 PCE, which indicates about 55% inhibition against BaP.

Thioacetamide

Thioacetamide (TAA), or ethanethioamide, was originally used as a medium for the control of orange decay (22). Later, several reports indicated that TAA was hepatotoxic and carcinogenic (23,24,25,26) and produces marked nuclear changes. Having noted these various effects caused by TAA, it was considered desirable to investigate whether TAA would act as an inhibitor or promoter of chromosomal breaks induced by the promutagens DMBA and BaP in mice.

Experimental mice were injected with TAA i.p. one hour after injecting either with DMBA or BaP in one experiment. In another experiment, a group of mice were injected i.p. with TAA one hour before injecting either with DMBA or BaP. Bone marrow samples were collected at 24, 48 and 72 hr after the last injection. In experiments with DMBA, maximum occurrence of MNPCE occurred at 48 hr sampling period. The number of MNPCE were 24.8/1000 PCE in one experiment, and 16.5/1000 PCE in another experiment. At the same sampling time where the mice received TAA in addition to the mutagen, the occurrence of 10.0 MNPCE/1000 PCE or 8.2 MNPCE/1000 PCE indicates 60% and 50% inhibition, respectively. In the case of BaP, the maximum occurrence of MNPCE also occurred at 48 hr sampling period in which the number of MNPCE were 17.6/1000 PCE in one instance and 12.0/1000 PCE in another instance. However, in both these experiments, treatment with TAA in addition to a mutagen did not show any inhibitory effect. The mice that received only TAA showed MNPCE ranging from 2.4-3.6/1000 PCE. The TAA treatment, whether it was pre- or post-mutagenic treatment, did not make any difference.

Even though TAA was reported as carcinogenic and hepatotoxic, it did not show any appreciable mutagenic effect in the *in vivo* bone marrow micronucleus assay. The ineffectiveness of TAA by itself in the present experiment also indicates that TAA could be tissue specific in its action. TAA poisoning could be the reason for the inhibitory effect against DMBA-induced chromosomal breaks. TAA poisoning generally results in decreased activity in microsomal mixed-function oxygenases at or beyond the level of the NADPH-cytochrome C reductase (27).

β-carotene

β-carotene, an important constitutent of the provitamins A, is widely distributed in dark green leafy vegetables, carrots and certain red and yellow fruits. Epidemiological studies have indicated that the incidence of cancer may be slightly lowered among the individuals whose diet included the intake of β-carotene and other carotenoids (28,29,30,31). Since β-carotene is an antioxidant that functions as a radical trapping agent (32) and lowers the incidence of cancer risk in humans and also tumorous growth in rodents, we were interested to find whether β-carotene would act as an anticlastogenic agent against mutagens such as DMBA, BaP, CP and MMC.

β-carotene was mixed thoroughly in powdered food and the mice were kept on this diet for one week prior to the treatment with either DMBA, BAP, CP or MMC. Bone marrow samples were collected at various intervals. From our results, which are presented in Table 1, can be seen the inhibitory action of β-carotene against mutagen-induced clastogenicity. Maximum inhibition (60%) was observed against BaP-induced clastogenicity and the least (32%) against DMBA-induced clastogenicity. A detailed paper has been published elsewhere regarding the inhibitory action of β-carotene against BaP- and MMC-induced clastogenicity (33). 56% and 48% inhibition was observed against CP and MMC-induced clastogenicity, respectively.

Glutathione

Glutathione is a widely-distributed sulphur-containing tripeptide that consists of glutamic acid, cystein and glycine in that order. The functional group in the molecule is the third group and it is customary to represent reduced glutathione by the abbreviation GSH. Several reports (34,35,36) indicated how glutathione reduced tumors and liver carcinogenesis in rodents.

Experiments were conducted with mice pretreated with glutathione for 24 h prior to treating with either DMBA, BaP, CP or MMC. Bone marrow samples were collected at various intervals. At the peak induction period of clastogenicity, about 51% inhibition against CP, 45% against MMC, and 41% against DMBA was observed. No appreciable inhibitory effect was observed against BaP-induced clastogenicity. The inhibitory action of glutathione can be to due to scavenging the reactive forms of mutagens as suggested by Wattenberg (37).

Sodium selenite

Even though, in the earlier days, selenium and its compounds were linked to the occurrence of certain types of cancer, no conclusive evidence was drawn from those studies. However, there are several reports indicating selenium as a potential chemotherapeutic agent (38) as well as an antitumorgenic agent (39,40). Based on these studies, experiments have been conducted in our laboratory to check whether selenium would act as an inhibitor against clastogenicity induced by DMBA and BaP. Experimental mice received sodium selenite by i.p. 24 hr before mutagen treatment either with DMBA or BaP. Bone marrow samples were collected at various intervals. At 48 hr sampling time where there was a peak occurrence of MNPCE in mutagen-treated mice, 54% and 30% of inhibition was observed in the mice that were pretreated with sodium selenite against DMBA and BaP, respectively. Sodium selenite _per se_ is not an antioxidant (41), but it could be a precursor of a compound or complex capable of carrying out antioxidant functions.

Fumaric Acid

Fumaric acid occurs in many plants such as _capsella bursa-pastoris_ which can be used as green salad, and in many other plants including edible mushrooms. Kuroda and Takagi (42) and Kuroda _et al._ (43) have indicated that the extract of _capsella-bursa pastoris_ has various kinds of pharmacologic qualities such as diuretic oxytocic, antiinflamatory, antiulcerative and anticarcinogenic properties. Fumaric acid, isolated from the above herb, was found to be responsible for various inhibitory effects in rodents (44,45), especially against the tumors induced by nitrogen-containing carcinogens. We wanted to find whether similar inhibitory effects of fumaric acid would be observed against another nitrogen-containing mutagen, MMC. At the same time, we have also tested FA aginst the DMBA-induced chromosoaml breaks. In this way, one might know whether the inhibitory reaction of FA, if any, is due to the inhibition of the monooxygenase enzyme aryl hyrocarbon hydroxylase. Since no effect was observed against DMBA, the details are not presented in detail.

The mice were fed with powdered food containing fumaric acid for one week prior to the treatment with MMC. Bone marrow samples were collected at various intervals. A maximum number of 23.4 MNPCE on average per 1000 PCE were observed in the 24 hr sample from MMC-treated mice. However, at the same sampling time, a reduction in the number of MNPCE was observed (15.2 MNPCE/1000 PCE) in the presence of fumaric acid indicating a 35% inhibitory effect.

DISCUSSION

A tabulated summary of the results of *in vivo* experiments with various potential inhibitors against mutagens is presented in Table 2. From the table it is gathered that metyrapone, β-carotene, and sodium selenite were effective as inhibitors against DMBA and BaP-induced chromosomal breaks. In addition, β-carotene showed inhibitory effects against CP as well as MMC. Thioacetamide, a suspected promoter, turned out to be an inhibitor against DMBA-induced clastogenicity. Glutathione acted as an inhibitor against DMBA-, CP- and MMC-induced chromosomal breaks. Fumaric acid showed inhibitory effects against only a nitrogen-containing mutagen, MMC, but not against DMBA. The results reflect the complexity of enzymatic and cytoplasmic reactions involved in the processes of inhibitions and detoxification of mutagenic agents.

It has been well established that the microsomal mixed-function oxidase (MFO) system of monooxygenase enzymes is involved in the activation and detoxification of cytotoxic, carcinogenic or mutagenic chemicals. Gelboin (46) has suggested that the monooxygenase enzymes (cytochromes P-450) act only as the initial enzyme interface between environmental xenobiotics. Further metabolic processing is carried out by a number of metabolically related enzymes, such as glutathione transferase, epoxide hydrase, sulfotransferase and uridine diphospho-glucuronyl transferase.

Wattenberg (37) has proposed a classification of inhibitors comprising three categories: (a) compounds that prevent the formation of carcinogens from precursor substances; (b) compounds that act as "blocking agents" to prevent carcinogenic substances from reaching or reacting with critical target sites in the tissues; and (c) a third category of inhibitors that are "suppressive agents" and can act subsequent to an exposure to a carcinogenic agent. However, there is insufficient evidence as yet to indicate whether this

TABLE 2

SUMMARY OF THE EFFECTS OF POTENTIAL INHIBITORS AGAINST MUTAGENS

Inhibitor	Mutagens			
	DMBA	BaP	CP	MMC
Metyrapone	67% I at 48 hr	55% at 48 hr	NT	NT
Thioacetamide	50%-60% I at 48 hr	No effect	NT	NT
β-Carotene	32% I at 48 hr	60% I at 48hr	56% I at 48hr	48% I at 48hr
Glutathione	41% I at 48 hr	No effect	51% I at 24 hr	45% I at 24 hr
Sodium Selenite	54% I at 48 hr	30% I at 48 hr	NT	NT
Fumaric Acid	No effect	NT	NT	35% I at 24 hr

I = Inhibition
NT = Not tested.

classification, or the model by Okey and Nebert (14), is applicable to inhibitors that reduce or prevent chromosomal breakage by genotoxic agents.

Our results indicate that there is no single universal inhibitory mechanism that can be applicable to all cases. Some inhibitors act only against promutagens (DMBA and BaP) but not against others; whereas FA acted as an inhibitor against only direct-acting agent, MMC, but not against promutagen DMBA. β-carotene acted as an inhibitor against all four mutagens tested. Thioacetamide acted as an inhibitor against promutagen DMBA but not another promutgen, BAP, indicating different mechanisms involved in their mutagenic action.

DMBA, as well as BaP, requires metabolic activation to express mutagenic activity. If FA acts as an inhibitor of the monooxygenase enzyme, arylhydrocarbon hydroxylase (AHH), it should have reduced the mutagenicity of DMBA. However, in our present studies, no inhibitory action was noticed. This indicates that FA probably did not interfere with the second monooxygenase-catalysed reaction and allowed the formation of diol epoxides. The inhibitory effect of FA against MMC-induced chromosomal breakage is probably due to this transforming action of MMC into a derivative that is less toxic to the animals.

Some of these inhibitors (β-carotene, glutathione and FA) tested are present in vegetables and fruits. Mostly they are antixoidants and act as trapping agents. Similar observations were reported from our laboratory earlier dealing with caffeic acid (47), α-naphthoflavone (48), corn oil and its minor constituents (49) and many other inhibitors (50). The effectiveness of inhibitors in preventing genotoxic action may depend upon the extent to which they can prevent the formation of active forms of mutagens from precursors that would bind covalently to DNA, RNA or proteins.

ACKNOWLEDGEMENTS

This research was supported, in part, by a grant from the Ontario Ministry of the Environment, Provincial Lottery Fund Project No. 81-055-33. We gratefully acknowlege the technical assistance of Mr. Anthony Wilson and the secretarial assistance of Ms. Bette Kosmolak.

REFERENCES

1. Conney, A.H., Levin, W., Wood, A.W., Yagi, H., Lehr, R.E. and Jerina, D.M. (1978): Biological activity of polycyclic hydrocarbon metabolites and the bay region theory. Adv. Pharmacol. Ther., 9:41-52.
2. Jerina, D.M., Sayer, J.M., Thakker, D.R., Yagi, H., Levin, W., Wood, A.W., and Conney, A.H. (1980): Carcinogenesis: Fundamental Mechanisms and Environmental Effects, edited by B. Pullman, P.O.P. Ts'O and H.V. Gelboin, pp.1-12, D. Reidel, Dordrecht, Holland.
3. Heidelberger, C. (1975): Chemical Carcinogenesis, Ann. Review Biochem., 44:79-121.
4. Searle, C.E. (1976): Chemical Carcinogens, ACS Monograph 173, American Chemical Society, Washington, D.C.
5. Miller, E.C. and Miller, J.A. (1974): In: Molecular Biology of Cancer, edited by H. Busch, pp.377-402, Academic Press, New York.
6. National Academy of Sciences (1972): Particulate Polycyclic Organic Matter. Committee on Biological Effects of Atmospheric Pollutants, Division of Medical Sciences, National Research Council, pp.138-141, Washington, D.C.
7. Gelboin, H.V. and Ts'O, P.O.P., editors (1978): Polycyclic Hydrocarbons and Cancer, Academic Press, New York.
8. International Agency for Cancer Research (1972-1978): Evaluation of the Carcinogenic Risk of Chemicals to Man. A series of 7 monographs, Lyon, France.
9. Slaga, T.J., editor (1980): Carcinogenesis, Vol.5, Modifiers of Chemical Carcinogenesis, Raven Press, New York.
10. Wiebel, F.J. (1980): Activation and inactivation of carcinogens by microsomal monooxygenases: Modifications by benzoflavones and polycyclic aromatic hydrocarbons. In: Carciogenesis, Vol.5, Modifiers of Chemical Car-

cinogenesis, edited by T.J. Slaga, pp.57-84, Raven Press, New York.

11. Gelboin, H.B. (1967): Carcinogens, enzyme induction, and gene action. Adv. Cancer Res., 10:1-81..
12. Gelboin, H.V., Kinoshita, N. and Wiebel, F.J. (1972): Microsomal hydroxylases: Induction and role in polycyclic hydrocarbon carcinogenesis and toxicity. Fed. Proc. 31:1298-1309.
13. Miller, J.A. (1970): Carcinogenesis by chemicals: An overview, Cancer Res., 30:559-576.
14. Okey, A.B. and Nebert, D.W. (1981): The Ah Gene Complex: A model for studying specific interactions of carcinogens and toxic chemicals. Workshop on the Combined Effects of Xenobiotics, pp.37-66, National Research Council Canada, Publication No. 18978, Ottawa, June 22-23.
15. Wattenberg, L.W. (1979): Inhibitors of carcinogenesis, In: Carcinogens: Identification and Mechanisms of Action, edited by A.C. Griffin and C.R. Shaw, pp.299-316, Raven Press, New York.
16. Salamone, M.F., Heddle, J.A., Stuart, E. and Katz, M. (1980): Towards an improved micronucleus test. Studies on 3 model agents, Mitomycin C, cyclophosphamide and dimethylbenzanthracene. Mutat. Res., 74:347-356.
17. Heddle, J.A. and Salamone, M.F. (1981): The micronucleus assay I: In vivo. In: Short-Term Tests for Chemical Carcinogens, edited by H.F. Stich and R.H.C. San, pp.243-249, Springer-Verlag, New York.
18. Orth, D.N. (1978): Metyrapone is useful only as adjunctive therapy in Cushing's disease. Ann. Intern. Med. 89:128-130.
19. Hildebrandt, A.G. (1971): The binding of metyrapone to cytochrome P-450 and its inhibitory action on liver microsomal mixed function oxidase reactions. Biochem. J. 125:6p-8p.
20. Izzo, N.J., Singer, H.A., Saye, J.A. and Peach, M.J. (1983): Cytochrome P-450 inhibitors block endothelial-dependent aortic relaxation responses. Fed. Am. Soc. Exp. Biol. 3:81 (abstract).
21. Waxman, D.J. and Walsh, C. (1983): Cyto-

chrome P-450 isozyme I from phenobarbital-induced rat-liver: Purifications, characterizations, and interactions with metyrapone and cytochrome b_5, Biochemistry, 22:4846-4855.
22. Childs, J.F.L. and Siegler, E.A. (1945): Scientific apparatus and laboratory methods: Compounds for control of orange decays. Science, 102:68.
23. Ambrose, A.M., De Eds, F. and Rather, L.F. (1949): Toxicity of thioacetamide in rats. J. Inc. Hyg. Toxicol. 31:158-161.
24. Fitzhugh, O.G. and Nelson, A.A. (1948): Liver tumors in rats fed thiourea or thioacetamide. Science, 108:626-628.
25. Gupta, D.N. (1955): Production of cancer of the bile ducts with thioacetamide. Nature, 175:257.
26. Kleinfeld, R.G. (1957): Early changes in rat liver and kidney cells induced by thioacetamide. Cancer Res. 17:954-962.
27. Barker, E.A. and Smuckler, E.A. (1972): Altered microsome function during acute thioacetamide poisoning. Mol. Pharmacol., 8:318-326.
28. Peto, R., Doll, R., Buckley, J.D. and Sporn, M.B. (1981): Can dietary beta-carotene materially reduce human cancer rates? Nature, 290:201-208.
29. Shekelle, R.B., Liu, S., Raynor Jr., W.J., Lepper, M., Maliza, C., Rossof, A.H., Paul, O., Shryock, A.M. and Stamler, J. (1981): Dietary vitamin A and risk of cancer in the western electric study. The Lancet, pp.1185-1190.
30. Mettlin, C., Graham, S. and Swanson, M. (1979): Vitamin A and lung cancer. J. Natl. Cancer Inst. 62:1435-1438.
31. Stich, H.F., Stich, W., Rosin, M.P. and Vallejera, M.O. (1984): Use of the micronucleus test to monitor the effect of vitamin A, beta-carotene and canthaxanthin on the buccal mucosa of betel nut/tobacco chewers. Int. J. Cancer, 34:745-750.
32. Burton, G.W. and Ingold, K.U. (1984): β-carotene: An unusual type of lipid antioxidant. Science, 224:569-573.
33. Raj, A.S. and Katz, M. (1985): β-carotene as

an inhibitor of benzo(a)pyrene and mitomycin C induced chromosomal breaks in the bone marrow of mice. Can. J. Gen. Cytol. 27(5):598-602.

34. Mitchell, J.R., Jollow, D.J., Potter, W.Z., Gillette, J.R. and Brodie, B.R. (1973): Acetaminophen induced hepatic necrosis. IV. Protective role of glutathione. J. Pharm. Exp. Therap. 187:211-217.
35. Novi, A.M. (1980): Regression of aflatoxin B -induced hepatocellular carcinomas by reduced glutathione. Science 212:541-542.
36. Hinson, J.A., Pohl, L.R., Monks, T.J. and Gillette, J.R. (1981): Acetaminophen-induced hepatotoxicity. Life Sciences, 29:107-116.
37. Wattenberg, L.W. (1983): Inhibition of neoplasia by minor dietary constituents. Cancer Res. 43:2448S-2453S.
38. Klayman, D.L. (1973). In: Organic Selenium Compounds: Their Chemistry and Biology, edited by D.L. Klayman and W.H.H. Gunther, pp.727, Wiley, New York.
39. Shamberger, R.J. (1970): Relationship of selenium to cancer, I. Inhibitory effect of selenium on carcinogenesis. J. Nat. Cancer Inst. 44:931-936.
40. Jacobs, M.M., Jansson, B. and Griffin, A.C. (1977): Inhibitory effects of selenium on 1,2-dimethylhydrazine and methylazoxymethanol. Cancer Lett. 2:133.
41. Hoekstra, W.G. (1975): Biochemical function of selenium and its relation to vitamin E. Fed. Proc. 34:2083.
42. Kuroda, K. and Takagi, K. (1969): Studies on Capsella bursa-pastoris, II. Diuretic anti-inflammatory and anti-ulcer action of ethanol extract of the herb. Arch. Int. Pharmacodyn. Ther. 178:392-399.
43. Kuroda, K., Akao, M., Kanisawa, M. and Miyaki, K. (1974): Inhibitory effect of Capsella bursa-pastoris on hepatocarcinogenesis induced by 3 -methyl-4-(dimethylamino) azobenzene in rats. Gan. 65:317-321.
44. Kuroda, K., Akao, M., Kanisawa, M. and Miyaki, K. (1976): Inhibitory effect of Capsella bursa-pastoris extract on growth of Ehrlich solid tumor in mice. Cancer Res. 36:1900-1903.

45. Kuroda, K. and Akao, M. (1977): Inhibitory effect of fumaric acid and dicarboxylic acid on gastric ulceration in rats. Arch. Int. Pharmacodyn. Ther. 226:324-330.
46. Gelboin, H.V. (1983): Carcinogens, drugs and cytochromes P-450, New Engl. J. Med. 309:105-107.
47. Raj, A.S., Heddle, J.A., Newmark, H.L. and Katz, M. (1983): Caffeic acid as an inhibitor of DMBA-induced chromosomal breakage in mice assessed by bone marrow micronucleus test. Mutat. Res. 124:247-253.
48. Raj, A.S. and Katz, M. (1983): Inhibitory effect of 7,8-benzoflavone on DMBA- and BaP-induced bone marrow micronuclei in mouse. Mutat. Res. 110:337-342.
49. Raj, A.S. and Katz, M. (1983): Corn oil and its minor constituents as inhibitors of DMBA-induced chromosomal breaks in vivo. Mutat. Res. 136:247-253.
50. Katz, M. and Raj, A.S. (1983): Inhibitors of chromosomal breakage in mouse induced by polynuclear aromatic hydrocarbons and other genotoxic agents. In: Polynuclear Aromatic Hydrocarbons: Mechanisms, Methods and Metabolism, edited by W.M. Cooke and A.J. Dennis, pp.697-711, Battelle Press, Columbus, Ohio.

DETERMINATION OF NITROFLUORANTHENES AND NITROPYRENES IN AMBIENT AIR AND THEIR CONTRIBUTION TO DIRECT MUTAGENICITY

THOMAS RAMDAHL*, JANET A. SWEETMAN, BARBARA ZIELINSKA, WILLIAM P. HARGER, ARTHUR M. WINER, ROGER ATKINSON
Statewide Air Pollution Research Center, University of California, Riverside, California 92521, USA.

INTRODUCTION

Nitrated polycyclic aromatic hydrocarbons (nitro-PAH) have been recognized as one of the most important classes of PAH derivatives, since they are direct acting mutagens (1,2) and/or carcinogens (3,4). Among combustion products such as gasoline engine exhaust (5,6), aluminum smelting effluent (7), coal fly ash (8), wood smoke (6) and cigarette smoke condensates (9), exhaust from diesel engines is the most well characterized source of nitro-PAH (6,10,11). Approximately 100 nitro-compounds have been identified in diesel exhaust alone, with the most abundant nitro-PAH in these samples being 1-nitropyrene (1-NP).

Another possible pathway to production of nitro-PAH in the environment is atmospheric transformations involving the reactions of adsorbed or gas-phase PAH with oxides of nitrogen and/or nitric acid and other atmospherically important reactive species, including possibly the OH radical (12,13). Evidence for atmospheric formation of nitro-PAH has been presented very recently by the identification of 2-nitrofluoranthene (2-NF) (14) and 2-nitropyrene (2-NP) (14,15) in ambient air. These nitro-PAH isomers have not been observed in direct emissions, nor in laboratory experiments investigating the interactions of adsorbed PAH with oxides of nitrogen. Thus it appears that these two nitro-PAH must be formed in the atmosphere during transport (14,15).

The contributions of 1-nitropyrene and 1,6- and 1,8-dinitropyrene to the direct-acting mutagenicity of ambient air samples have been calculated to range from <1% in a rural area up to 26% in an industrialized region (6,16-18). We have observed that 2-nitrofluoranthene may be up to ten times more abundant in ambient air than 1-nitropyrene (14). This suggests that the contribution of 2-nitrofluoranthene to the overall mutagenicity may be important relative to that from the nitro-PAH mentioned above, and that the formation of nitro-PAH during atmospheric transport may be important with respect to their contributions to the observed direct acting mutagenicity.

In this work we have studied the diurnal variations of PAH and nitro-PAH concentrations, and mutagenicity during a photochemical air pollution episode. The contribution of the individual nitro-PAH species to the total mutagenicity is also calculated.

MATERIALS AND METHODS

Sampling

Samples were collected from 0600 on Saturday, September 14, through 0600 on Sunday, September 15, 1985 in Claremont, a site ~25 miles northeast of downtown Los Angeles. Particulate collections were made at 6-hr intervals with the SAPRC-designed and constructed ultra-high volume "mega-sampler" (19). This sampler has an inlet with a 50% cutoff point of 20 μm, and has a total flow rate of approximately 640 scfm through four 16-in. x 20-in. filters. The Teflon-coated glass fiber filters (Pallflex T60A20) used were prewashed by sequential Soxhlet extraction with dichloromethane (DCM) and methanol. The particle loadings after sampling were determined by differential weighing of the filters.

Extraction and Chemical Analyses

After particulate collection, two 16-in. x 20-in. filters from each time period were Soxhlet extracted for 18 hr with DCM. Pyrene-d_{10} at ~3 μg, perylene-d_{12} at ~1 μg, and 2-nitrofluoranthene-d_9, 1-nitropyrene-d_9, 1,3-, 1,6- and 1,8-dinitropyrene-d_8 at ~0.3 μg each, were added as internal standards to these DCM extracts. The extracts were precleaned by open-column silica (5% H_2O) chromatography to remove aliphatic hydrocarbons (eluted with hexane) and polar material (left on the column). The fractions of interest (eluted with DCM) were further fractionated by normal phase high performance liquid chromatography (HPLC) using a semi-prep (25 cm x 10 mm) Ultrasphere Si column (Altex). The HPLC system consisted of a Spectra-Physics Model 8100 chromatograph, Model 4100 computing integrator, Model 8400 variable wavelength UV/vis detector and an ISCO fraction collector. The solvent program comprised a starting composition of 5% DCM in hexane for 10 min isocratic, then a linear gradient to 100% DCM over 15 min. The solvent flow was 3 ml min^{-1}. As established by chromatography of standard mixtures, the PAH and nitro-PAH of interest were eluted in the 6.5-15.5 min (PAH), 21.5-24.5 min (mononitro-PAH) and 24.5-27.7 min (dinitro-PAH) fractions, respectively.

The HPLC fractions were analyzed by multiple ion detection gas chromatography-mass spectrometry (MID GC-MS) using a Finnigan Model 3200 MS operating in the electron impact (EI) mode, and interfaced to a Teknivent Data System. The GC was equipped with an on-column injector (J&W Scientific, Inc.) and a DB-5 fused silica capillary column (60 m x 0.32 mm, 0.25 μm film thickness) directly eluting into the ion source. Helium was used as a carrier gas. Injection was made with the column at $100^{o}C$, followed by a temperature program to $350^{o}C$ at $8^{o}C\ min^{-1}$. Quantification was performed by comparing the ratios of the areas of the molecular ion peaks to those of the deuterated internal standards.

Mutagenicity Assay

One 16-in. x 20-in. filter from each time period was extracted for 18 hr with DCM for mutagen assay. All samples were transferred to dimethyl sulfoxide and tested using the Ames *Salmonella* mutagenicity test (20) using the TA98, TA98NR and TA98/1,8-DNP_6 tester strains according to our standard protocol (21). Eight doses were tested in triplicate and the mean of the three responses (net revertants = total revertants - spontaneous background revertants) was used to determine the dose-response curve. Standard compounds tested were 2-nitrofluoranthene, 1-nitropyrene and 2-nitropyrene. Positive controls were 2-nitrofluorene, 1,8-dinitropyrene and quercetin. The activities of these compounds are given in Table 1.

RESULTS

Figure 1 shows the MID traces of the HPLC fraction containing the mononitro-PAH for the particulate organic matter (POM) collected during the 1800-2400 hr time period. The molecular ions and the characteristic $[M-NO]^+$ fragment ions for the deuterated internal standards and the non-deuterated species are shown. As can be seen from the figure, the sample contained 2-NF, 1-NP and 2-NP. The identification of 2-NF, eluting just ahead of 3-NF, is based on its characteristic low 217 $[M-NO]^+$ fragment ion abundance in addition to the retention time (14,22).

Figure 2 shows the MID traces of the molecular ion of nitrofluoranthene and nitropyrene, m/z 247, for all four time periods analyzed. The concentrations of nitro-PAH for these time periods are given in Table 2 in addition to PAH concentrations, average temperature and particulate data. In all but the 1200-1800 hr time periods 2-nitrofluoranthene is the

TABLE 1

MUTAGENICITY OF STANDARD COMPOUNDS (rev μg^{-1})

Compound	TA98	TA98NR	TA98/1,8-DNP_6
2-Nitrofluoranthene	3,900	930	580
1-Nitropyrene	2,300	310	1,300
2-Nitropyrene	16,000	1,600	2,300
2-Nitrofluorene	450	60	70
1,8-Dinitropyrene	900,000	990,000	21,000
Quercetin	12	12	6

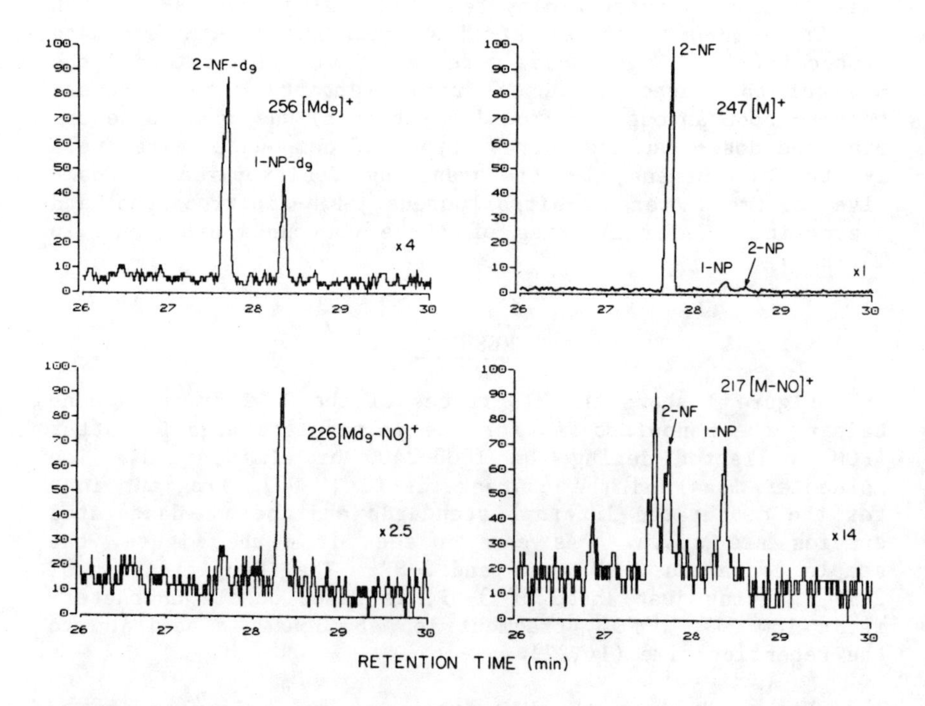

FIGURE 1. MID traces of the nitro-PAH fraction from ambient particles collected 1800-2400 hr, September 14, 1985 at Claremont, CA. Injection was made at 50°C, then 8°C min^{-1} to 350°C.

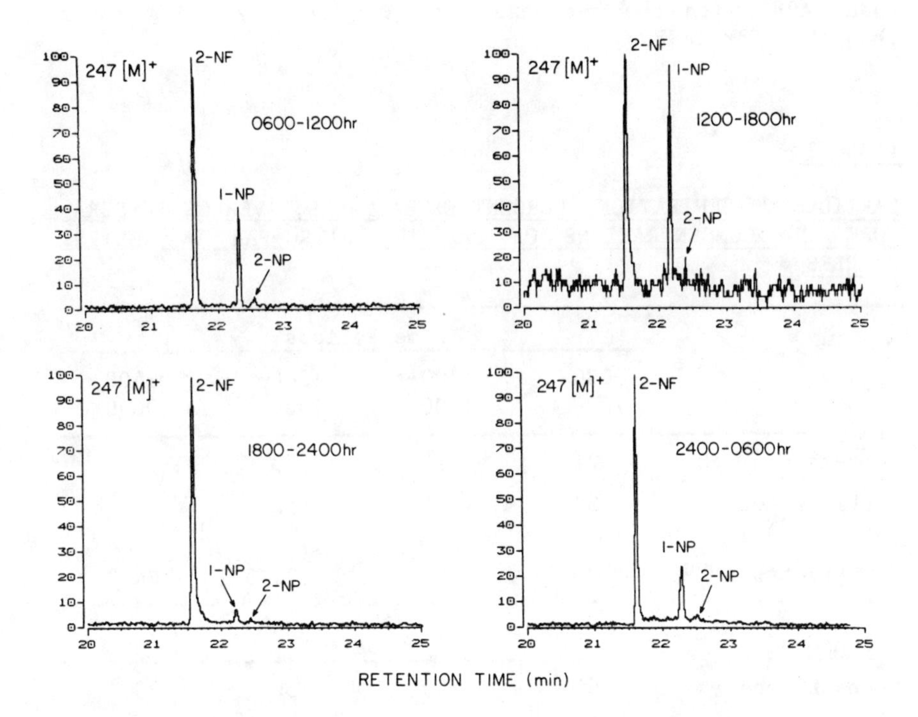

FIGURE 2. MID traces of the molecular ion (m/z 247) for nitrofluoranthene and nitropyrene isomers in the nitro-PAH fraction from ambient particles collected September 14-15, 1985, at Claremont, CA, during the periods indicated.

most abundant nitro-PAH present, and the concentration of this compound varies from slightly less than 1-nitropyrene in the 1200-1800 hr period to almost ten times more than 1-nitropyrene in the 1800-2400 hr period. No dinitroarenes were detected in any of the samples, at least in part due to low recovery of the deuterated dinitropyrene standards.

The mutagenicity of the extracts of the samples using the various Salmonella tester strains are given in Table 3. Also given in Table 3 are the calculated contributions of the nitro-PAH to the mutagenicity of the samples, using the activities of the standard compounds given in Table 1. The contribution to the mutagenicity from nitro-PAH in these samples varies from 1.5 to 6% of the total mutagenicity.

Both nitroreductase deficient strains showed a lower response than TA98, with the response for TA98/1,8-DNP_6 being lower than that for TA98NR.

TABLE 2

PARTICLE CONCENTRATION, PERCENT EXTRACTABLE, AVERAGE TEMPERATURE AND CONCENTRATIONS OF PAH AND NITRO-PAH IN AMBIENT SAMPLES

	Time Periods			
	0600-1200	1200-1800	1800-2400	2400-0600
Temperature, ^{o}C	21	32	23	17
Particle Conc. ($\mu g\ m^{-3}$)	89	71	52	47
% Extractable	20.9	17.1	22.5	24.3
Compound	$pg\ m^{-3}$			
Phenanthrene	60	40	50	20
Anthracene	5	2	5	2
Fluoranthene	110	70	130	50
Pyrene	140	80	140	60
Benzo[ghi]-fluoranthene	90	50	100	40
Benz[a]anthracene	60	30	60	20
Chrysene/ Triphenylene	170	130	190	80
Benzo[b,j&k]-fluoranthenes	580	260	680	250
Benzo[e]pyrene	280	130	360	130
Benzo[a]pyrene	90	40	140	30
Perylene	20	6	30	6
Indeno[1,2,3-cd]-pyrene	370	140	450	170

(continued)

TABLE 2 (continued)

	Time Periods			
	0600-1200	1200-1800	1800-2400	2400-0600
Benzo[ghi]-perylene	1,100	360	1,600	500
Total PAH	3,100(35)[a]	1,300(18)[a]	3,900(77)[a]	1,300(29)[a]
2-Nitro-fluoranthene	80(0.9)[a]	30(0.4)[a]	500(10.0)[a]	100(2.1)[a]
1-Nitropyrene	60(0.7)[a]	40(0.5)[a]	60(1.1)[a]	40(0.9)[a]
2-Nitropyrene	5(0.05)[a]	2(0.02)[a]	10(0.2)[a]	3(0.06)[a]

[a] $\mu g\ g^{-1}$ particulate

TABLE 3

MUTAGENICITY OF AIR SAMPLES

	TA98			TA98NR	TA98/1,8-DNP$_6$
Time Period	rev μg^{-1} extract	rev m^{-3}	% from nitro-PAH[a]	rev μg^{-1} extract	rev μg^{-1} extract
0600-1200	1.9	35	1.5	0.9	0.2
1200-1800	1.0	15	1.5	0.4	0.08
1800-2400	3.4	40	6.0	1.8	0.5
2400-0600	1.6	20	3.0	0.8	0.2

[a] Per cent contribution of measured 2-NF, 1-NP and 2-NP concentration to the mutagenicity determined in TA98

DISCUSSION

Chemical Data

The occurrence and identity of many nitro-PAH in diesel exhaust emissions are well documented in the literature. However, there are only a few reports of nitro-PAH in ambient air (6,14-18,23,24), some of the reported work being designed to detect specific mono- and dinitropyrenes, but not other isomers, i.e., nitrofluoranthenes (6,16-18). In several of these studies (14,15,23,24) nitrofluoranthene and nitropyrene isomers have been detected in addition to 1-NP. Thus Nielsen et al. (15) reported the presence of 2-NP, 1-NP and 3-nitrofluoranthene (3-NF) in an ambient air sample. However, due to a small difference in gas chromatographic retention time between the 3-NF standard and the nitrofluoranthene isomer in this air sample, a re-analysis is being conducted to check if the reported 3-NF is actually 2-NF (26). Pitts et al. (14) also detected 2-NP and 1-NP, together with 2-NF, in a Riverside, CA, ambient air sample. 2-NF was the most abundant nitro-PAH in all of these samples (14) and only traces of 3-NF were detected in one of four samples analyzed. Ramdahl et al. (23) reported 3-NF as the most abundant nitro-PAH in the National Bureau of Standards - Standard Reference Material (NBS SRM) 1648 (urban air particles from St. Louis, MO), but this compound was mis-identified. This isomer has recently been shown, in fact, to be 2-NF (25). 2-NF is also the most abundant nitro-PAH in the NBS SRM 1649 (urban air particles from Washington, DC) (25). Morita et al. (24) have also reported a nitrofluoranthene/pyrene isomer which is more abundant than 1-NP in air samples from Japan, without identifying the isomer. All of these data demonstrate that 2-NF is often the most abundant nitro-PAH in ambient air, regardless of the sampling location.

That 2-NF is the most abundant isomer in ambient air POM has also been observed in the present study, as shown in Figure 2 and Table 2. Since neither 2-NF nor 2-NP are directly emitted from combustion sources and are not formed as artifacts on filters during sampling, their presence in ambient air provides evidence for atmospheric formation of these compounds (14,15). Possible reaction pathways leading to these isomers have been discussed in detail elsewhere, and may involve gas phase reactions, possibly involving radical species (14,15,25,27). These gas-phase reaction products then presumably condense on atmospheric particles, to be detected by analysis of filter samples.

The ratios of 2-NF, 1-NP and 2-NP to benzo[e]pyrene (BeP) [a fairly non-reactive (28) PAH of approximately the same volatility] are presented in Figure 3. The ratios of 1-NP and 2-NP to BeP are reasonably constant over the four time periods, as is the ratio between benzo[ghi]fluoranthene

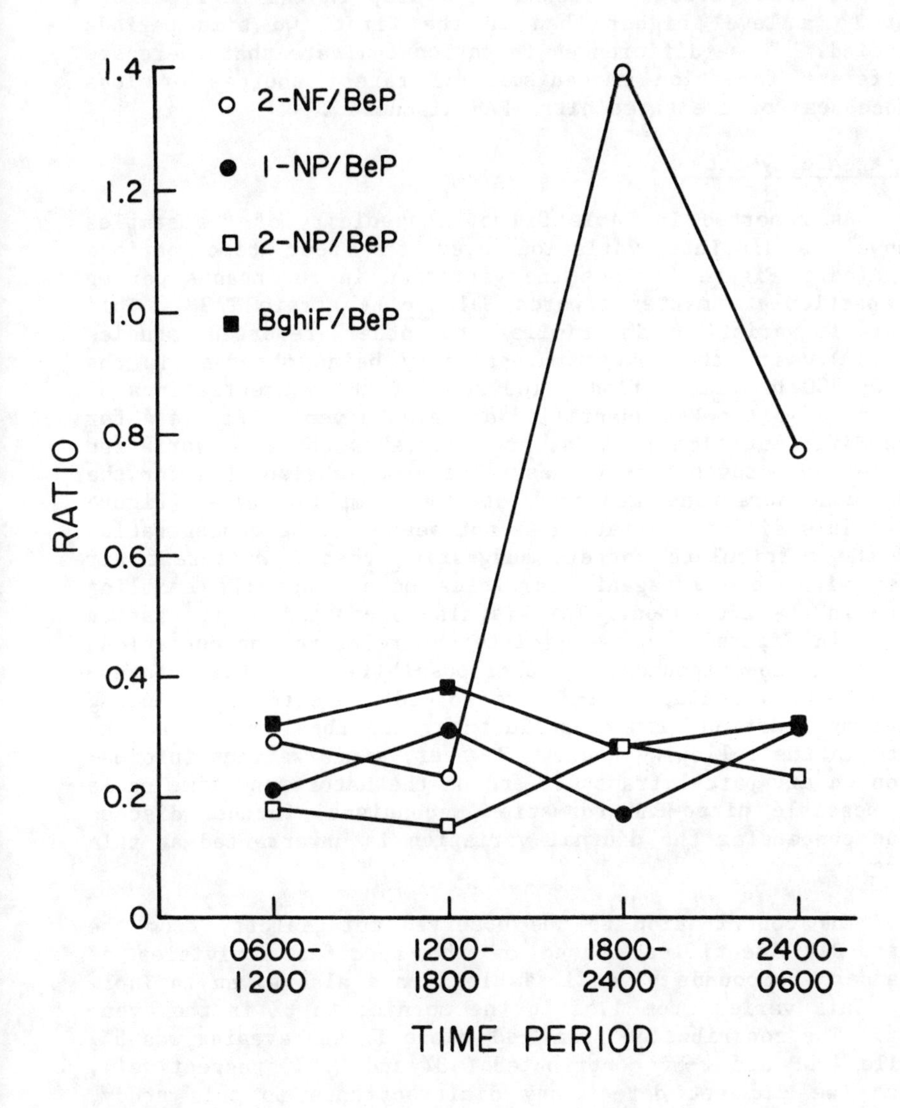

FIGURE 3. Ratios of individual nitro-PAH and benzo[ghi]-fluoranthene to benzo[e]pyrene for the different collection periods.

(BghiF) and BeP. This suggests that 1-NP (and possibly 2-NP) is associated with the particles as a constant fraction relative to the PAH over the 24-hr period studied, with no major net loss or formation. On the contrary, the 2-NF/BeP ratio varies widely, with a dramatic increase in the 1800-2400 hr time period, followed by a drop in the next period, but to a level higher than in the first two time periods studied. These differences in ratios indicate that there are different formation mechanisms and rates, sources or loss processes for the three nitro-PAH discussed.

Mutagenicity Data

As reported in Table 3 the mutagenicity of the samples showed a distinct variation over the four time periods studied. Figure 4 shows the variation in revertants per mg of particulate matter towards *Salmonella* strain TA98. This diurnal variation is similar to other reported studies (29,30), with the lowest mutagenicity being observed in the 1200-1800 hr time period. The sums of the concentrations of the three nitro-PAH quantified are also given in Figure 4 for the different time periods, and they show the same variation as do the mutagenicity values. The same is also true for the NO_2 concentrations measured at the sampling site (Figure 4). This diurnal variation is not seen for the concentration of the particulate matter, suggesting that a different air mass with more mutagenic particles moved into the sampling site in the afternoon. This is also supported by the ratios given in Figure 3, showing different relative concentrations of the nitro-compounds. Another possibility is that mutagenic compounds including 2-NF are formed close to the sampling site by reaction with co-pollutants, and then sampled at the site in the following hours. However, since we lack information on air parcel transport and on the nature and time scale of possible nitro-PAH formation mechanisms, further discussion concerning the diurnal variation is unwarranted at this time.

The contribution to the observed mutagenicity from the nitro-PAH identified, based on the specific activities of standard compounds given in Table 1, are also given in Table 3. This varies from 1.5% in the morning to 6% in the evening. The contribution from 2-NF alone in the evening was 5%, while 1-NP and 2-NP contributed 0.3% and 0.5%, respectively. Since we did not detect any dinitropyrenes in this study, their contribution to the mutagenicity could not be determined for these samples. Siak *et al.* (17) have quantified 1,6- and 1,8-dinitropyrenes in ambient air samples, and the contributions of these two highly mutagenic compounds (2) to

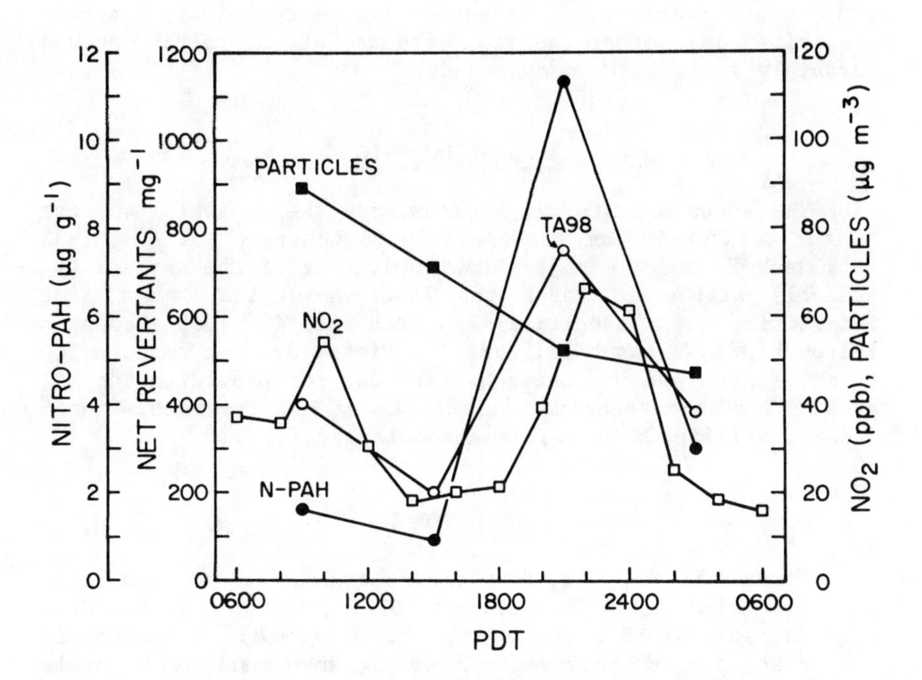

FIGURE 4. Diurnal variations of the collected particulate matter (μg m^{-3}), nitrogen dioxide concentration (ppb), nitro-PAH concentration (sum of 2-NF, 1-NP and 2-NP, μg g^{-1} particulate matter) and direct mutagenicity (TA98, revertants mg^{-1} particulate matter) for the September 14-15, 1985, sampling period.

the total mutagenicity in these samples were two to three times the contributions from 1-NP. This suggests that the mutagenicity caused by 2-NF in the ambient samples we have studied could still be higher than any other single nitro-PAH isomer.

As shown in Table 3, the nitroreductase deficient strains TA98NR and TA98/1,8-DNP_6 exhibited lower mutagenic response than TA98, the latter strain lowest of the two. This is consistent with the presence of nitroarenes in these samples, although the reduction in response cannot be used to estimate the amount of additional nitro-compounds present (31,32).

Since 2-NF and 2-NP have comparable mutagenicity to 1-NP and are often present in greater or equal amounts in ambient POM, future research on the metabolism, mutagenicity, carcinogenicity and other health effects of nitro-PAH should certainly include these compounds.

ACKNOWLEDGMENTS

The authors gratefully acknowledge the financial support of the California Air Resources Board Contract No. A4-081-32 (Dr. Jack K. Suder, Project Monitor). One of the authors (T. Ramdahl) wishes to thank the Royal Norwegian Council for Scientific and Industrial Research for a 1985 Research Fellowship. We thank Dr. James N. Pitts, Jr. for encouraging this research and Dr. Heinz W. Biermann for providing the NO_2 data. The able technical assistance of Mr. Travis M. Dinoff and Mr. Phillip C. Pelzel is acknowledged.

REFERENCES

1. Pitts, J. N., Jr., Van Cauwenberghe, K. A., Grosjean, D., Schmid, J. P., Fitz, D. R., Belser, W. L., Jr., Knudson, G. B., and Hynds, P. M. (1978): Atmospheric reactions of polycyclic aromatic hydrocarbons: Facile formation of mutagenic nitro derivatives. Science, 202:515-519.
2. Rosenkranz, H. S. and Mermelstein, R. (1983): Mutagenicity and genotoxicity of nitroarenes: All nitro-containing chemicals were not created equal. Mutat. Res., 114:217-267.
3. Hirose, M., Lee, M.-S., Wang, C. Y., and King, C. M. (1984): Induction of rat mammary gland tumors by 1-nitropyrene, a recently recognized environmental mutagen. Cancer Res., 44:1158-1162.
4. Ohgaki, H., Negishi, C., Wakabayashi, K., Kusama, K., Sato, S., and Sagimura, T. (1984): Induction of sarcomas in rats by subcutaneous injection of dinitropyrenes. Carcinogenesis, 5:583-585.
5. Tejada, S. B., Zweidinger, R. B., and Sigsby, J. E., Jr. (1982): Analysis of nitroaromatics in diesel and gasoline car emissions. SAE paper 820775, Society of Automotive Engineers, Pittsburgh, PA.
6. Gibson, T. L. (1982): Nitroderivatives of polynuclear aromatic hydrocarbons in airborne and source particulate matter. Atmos. Environ., 16:2037-2040.
7. Oehme, M., Manö, S., and Stray, H. (1982): Determina-

tion of nitrated polycyclic hydrocarbons in aerosols using capillary gas chromatography combined with different electron capture detection methods. J. High Resol. Chromatogr. Chromatogr. Commun., 5:417-423.

8. Harris, W. R., Chess, E. K., Okamoto, D., Remsen, J. F., and Later, D. W. (1984): Contribution of nitropyrene to the mutagenic activity of coal fly ash. Environ. Mutagen., 6:131-144.
9. McCoy, E. C. and Rosenkranz, H. S. (1982): Cigarette smoking may yield nitroarenes. Cancer Lett., 15:9-13.
10. Schuetzle, D., Lee, F. S.-C., Prater, T. J., and Tejada, S. B. (1981): The identification of polynuclear aromatic hydrocarbon (PAH) derivatives in mutagenic fractions of diesel particulate extracts. Int. J. Environ. Anal. Chem., 9:93-144.
11. Paputa-Peck, M. C., Marano, R. S., Schuetzle, D., Riley, T. L., Hampton, C. V., Prater, T. J., Skewes, L. M., Jensen, T. E., Ruehle, P. H., Bosch, L. C., and Duncan, W. P. (1983): Determination of nitrated polynuclear aromatic hydrocarbons in particulate extracts by capillary column gas chromatography with nitrogen selective detection. Anal. Chem., 55:1946-1954.
12. Nielsen, T., Ramdahl, T., and Bjørseth, A. (1983): The fate of airborne polycyclic organic matter. Environ. Health Perspect., 47:103-114.
13. Pitts, J. N., Jr. (1983): Formation and fate of gaseous and particulate mutagens and carcinogens in real and simulated atmospheres. Environ. Health Perspect., 47:115-140.
14. Pitts, J. N., Jr., Sweetman, J. A., Zielinska, B., Winer, A. M., and Atkinson, R. (1985): Determination of 2-nitrofluoranthene and 2-nitropyrene in ambient particulate organic matter: Evidence for atmospheric reactions. Atmos. Environ. 19: 1601-1608.
15. Nielsen, T., Seitz, B., and Ramdahl, T. (1984): Occurrence of nitro-PAH in the atmosphere in a rural area. Atmos. Environ., 18:2159-2165.
16. Gibson, T. L., and Tironi, G. (1985): Polynuclear aromatic compounds and the bacterial mutagenicity of airborne particulate matter. In: Polynuclear Aromatic Hydrocarbons. Mechanisms, Methods and Metabolism, edited by M. Cooke and A. J. Dennis, pp. 463-474, Battelle Press, Columbia, Ohio.
17. Siak, J., Chau, T. L., Gibson, T. L., and Wolff, G. T. (1985): Contribution to bacterial mutagenicity from nitro-PAH compounds in ambient aerosols. Atmos. Environ., 19:369-376.

18. Tokiwa, H., Kitamori, S., Nakagawa, R., Horikawa, K., and Matamala, L. (1983): Demonstration of powerful mutagenic dinitropyrene in airborne particulate matter. Mutat. Res. 121:107-116.
19. Fitz, D. R., Doyle, G. J., and Pitts, J. N., Jr. (1983): An ultrahigh volume sampler for the multiple filter collection of respirable particulate matter. J. Air Pollut. Control Assoc. 33:877-879.
20. Ames, B. N., McCann, J., and Yamasaki, E. (1975): Methods for detecting carcinogens and mutagens with the Salmonella/ mammalian-microsome mutagenicity test. Mutat. Res. 31:347-364.
21. Belser, W. L., Jr., Shaffer, S. D., Bliss, R. D., Hynds, P. M., Yamamoto, L., Pitts, J. N., Jr., and Winer, J. A. (1981): A standardized procedure for quantification of the Ames Salmonella/mammalian-microsome mutagenicity test. Environ. Mutagen. 3:123-139.
22. Ramdahl, T., Sweetman, J. A., Zielinska, B., Atkinson, R., Winer, A. M., and Pitts, J. N., Jr. (1985): Analysis of mononitro-isomers of fluoranthene and pyrene by high resolution capillary gas chromatography/mass spectrometry. J. High Resol. Chromatogr. Chromatogr. Commun. (in press).
23. Ramdahl, T., Becher, G., and Bjørseth, A. (1982): Nitrated polycyclic aromatic hydrocarbons in urban air particles. Environ. Sci. Technol., 16:861-865.
24. Morita, K., Fukamachi, K., and Tokiwa, H. (1983): Studies on the aromatic nitrocompounds in air. III. Determination of 1-nitropyrene in airborne particulate matter and automobile emissions. Eisei Kagaku., 29:199-205.
25. Sweetman, J. A., Zielinska, B., Atkinson, R., Ramdahl, T., Winer, A. M., and Pitts, J. N., Jr. (1985): A possible formation pathway for the 2-nitrofluoranthene observed in ambient particulate organic matter. Atmos. Environ. (in press).
26. Nielsen, T. and Ramdahl, T. (1985): (in preparation).
27. Sweetman, J. A., Zielinska, B., Atkinson, R., Winer, A. M., and Pitts, J. N., Jr. (1985): Nitration products from the reaction of fluoranthene and pyrene with N_2O_5 and other nitrogenous species in the gaseous, adsorbed and solution phases: implications for atmospheric transformations of PAH. Presented at the 10th Int. Symp. on PAH, Battelle, Columbus, Ohio.
28. Nielsen, T. (1984): Reactivity of polycyclic aromatic hydrocarbons towards nitrating species. Environ. Sci. Technol., 18:157-163.

29. Pitts, J. N., Jr., Sweetman, J. A., Harger, W., Fitz, D. R., Paur, H.-R., and Winer, A. M. (1985): Diurnal mutagenicity of airborne particulate organic matter adjacent to a heavily travelled West Los Angeles Freeway. J. Air Pollut. Control Assoc., 35:638-643.
30. Pitts, J. N., Jr., Harger, W. P., Lokensgard, D. M., Fitz, D. R., Scorziell, G. M., and Mejia, V. (1982): Diurnal variation in the mutagenicity of airborne particulate organic matter in California's south coast air basin. Mutat. Res. 104:35-41.
31. Rosenkranz, H. S., McCoy, E. C., Mermelstein, R., and Speck, W. T. (1981): A cautionary note on the use of nitroreductase-deficient strains of Salmonella typhimurium for the detection of nitroarenes as mutagens in complex mixtures including diesel exhaust. Mutat. Res., 91:103-105.
32. Alfheim, I., Bjørseth, A., and Möller, M. (1984): Characterization of microbial mutagens in complex samples - Methodology and application. CRC Crit. Rev. Environ. Control, 14:91-150.

DETOXICATION MECHANISMS OF BENZO(A)PYRENE AS STUDIED IN THE CHO/HGPRT ASSAY.

LESLIE RECIO* AND ABRAHAM W. HSIE
University of Kentucky Graduate Center for Toxicology, Lexington, Kentucky 40536 and Health and Safety Research Division, Oak Ridge National Laboratory, Oak Ridge, Tennessee 37831

INTRODUCTION

The biotransformation of benzo(a)pyrene (BaP) in mammals is a sequential and integrated enzymatic process which results in predominantly water-soluble excretory products (1). Since the discovery that the mixed-function oxidase (MFO) system can transform BaP to biologically reactive metabolites, numerous studies have been carried out to elucidate the mechanisms and biological consequences of MFO bioactivation (2). Many short-term mutational assays utilize a rat liver homogenate preparation supplemented with cofactors necessary for bioactivation of BaP by the MFO system (S9 mix); however, the effects of enzymatic detoxication pathways on BaP-induced cytotoxic and mutagenic effects have been rarely studied. Since the biological activity of BaP *in vivo* is likely to be affected by the relative balance between bioactivation and detoxication pathways, the use of S9 mix to activate promutagens to their active form(s) without the inclusion of cofactors necessary for key enzymes involved in conjugation and detoxication can result in a distortion in the relative quantity of toxic products formed.

Conjugation and detoxication of MFO-mediated BaP metabolites with glucuronic acid and glutathione (GSH) are major pathways of BaP elimination and ultimately excretion *in vivo* (3). UDP-glucuronyltransferases (UDPGTs) and GSH-S-transferases (GSHTs) exist in multiple forms that can catalyze the conjugation of a wide variety of substrates with glucuronic acid and GSH respectively, and are inducible by numerous substances (4). Uridine diphosphate α-D-glucuronic acid (UDPGA), the endogenous donor of glucuronic acid, is the cofactor necessary for the synthesis of glucuronide conjugates *in vitro* by UDPGTs located in the microsomal fraction of tissue homogenate preparations (5). GSH is the cofactor necessary for the synthesis of GSH conjugates *in vivo* and *in vitro* by GSHTs predominantly located in the cystolic fraction of tissue homogenate preparations (6). Glucuronide and GSH conjugates of BaP are usually detoxified water-soluble excretory products.

Biotransformation of BaP by the MFO system results in numerous phenols, dihydrodiols, and epoxides most of which are substrates for conjugation and elimination with glucuronic acid or GSH (1). Biochemical analysis by high performance liquid chromatography has shown that the addition of BaP to rat liver microsomal preparations containing MFO system cofactors and UDPGA results in a selective reduction in the amount of organic-solvent soluble phenols and quinones formed without affecting the amount of dihydrodiols formed (7). GSH and GSHTs have been shown to inhibit the DNA binding of biologically reactive BaP and BaP 7,8-diol metabolites (8). Due to the important roles that glucuronide and GSH conjugation can play in the detoxication and elimination of BaP *in vivo*, we have studied the effects of UDPGA and GSH on BaP-induced cytotoxicity and mutagenicity in cultured mammalian cells. The S9-mix used in our mammalian cell gene mutational assay, Chinese hamster ovary cells/hypoxanthine-guanine phosphoribosyl transferase (CHO/HGPRT) (9), was supplemented with UDPGA or GSH and in certain experiments GSH and purified GSHTs, to study the biological effects of glucuronide and GSH detoxication mechanisms on BaP-induced cytotoxicity and mutagenicity.

MATERIALS AND METHODS

Chemicals

BaP (99 + % purity) was obtained from Aldrich Chemical Co. (Milwaukee, WI). BaP 7,8-diol was purchased from the NCI Chemical Carcinogen Reference Standard Respository, Division of Cancer Cause and Prevention, NCI, NIH (Bethesda, MD). Other biochemicals and purified GSHTs were bought from Sigma Chemical Co. (St. Louis, MO). GSHTs had an activity of 0.06 units/mg protein for the substrate 1,2-epoxy-3-(p-nitrophenoxy)-propane according to the accompanying technical data (Sigma, lot # 104F-7830). Stock solutions were prepared in spectrophotometric grade dimethylsulfoxide (DMSO) purchased from Schwarz/Mann (Spring Valley, NY).

Cell Culture

CHO cells (strain K_1-BH_4) were cultured in a monolayer in Ham's F12 medium (K. C. Biological Inc., Lenexa, KS) as previously described (10). Cell number was determined by use of a Model B Coulter Counter (Coulter Electronics Inc., Hialeah, FL).

Preparation of Rat Liver Homogenate S9 and S9 Mix

The polychlorinated biphenyl (Aroclor-1254)-induced rat (male Sprague Dawley) liver homogenate (9,000g soluble fraction) was used as the exogeneous metabolic source (S9) and prepared according to a standard method (11). The S9 reaction mixture (S9 mix) consists of Na_2HPO_4 (25 mM; pH 8.0), Tris-HCl (62.5 mM; pH 7.0), KCl (30 mM), $MgCl_2$ (10 mM), $CaCl_2$ (10 mM), glucose 6-phosphate (5 mM), NADP (4 mM), and S9 protein (1 mg/ml) (12).

Treatment of Cell Cultures with Chemicals

We followed a standard protocol for treatment of cells with chemical mutagens and for determination of cytotoxicity and mutagenicity developed in our laboratory (13). Briefly, exponentially growing cells, 1.0-1.5 x 10^6/25 cm^2 flask, containing 4 ml of serum-free Ham's F12 medium and 1 ml of S9 mix with the appropriate enzyme cofactor (UDPGA or GSH) or GSH and purified GSHTs, were treated with either BaP or BaP-7,8-diol dissolved in DMSO. UDPGA was dissolved in Saline G. GSH was dissolved in 1M HEPES (pH 7.5) buffer. 100 μl aliquots of either UDPGA or GSH containing 1 and 10 mg respectively were added to each flask. One mg of GSHTs dissolved in Saline G were added in 100 μl aliquots. The cells were treated for 5 hrs at 37°C in 5% CO_2 in air in a 100% humidified incubator. The treatment medium in each flask was then removed and the cells were rinsed 3X with Saline G. Five ml of Ham's F12 containing 5% fetal calf serum (F12FCM5) was then added to each flask and the flasks were incubated for an additional 16 hr.

Determination of Cytotoxicity and Mutagenicity

For the estimation of cytotoxicity, depending on the expected survival, 200-2,000 cells were plated in triplicate in 60-mm dishes in 5 ml of F12FCM5. After 7 days of incubation the colonies were fixed, stained, and counted. Cytotoxicity is expressed as cloning efficiency relative to the untreated controls (9).

For mutagenicity determinations, approximately 10^6 cells from each flask were plated in 100-mm dishes in 10 ml of

F12FCM5. To fully allow phenotypic expression (13), cells were subcultured at 2-day intervals and on the eighth day of subculture mutant cells were selected by plating 10^6 cells in five 100-mm dishes in Ham's F12 medium without hypoxanthine containing 6-thioguanine (TG) (10 μM) and 2.5% fetal calf serum. Aliquots of 200 cells were plated in triplicate in 60-mm dishes in the same medium without TG to determine cloning efficiency of the cells in the absence of selection. After 7 days of incubation the colonies were fixed, stained, and counted. Mutation frequency is calculated by dividing the total number of mutant colonies by the total number of cells selected (1×10^6) corrected for cloning efficiency and is expressed as mutants/10^6 clonable cells (9).

RESULTS

The CHO/HGPRT assay has been used to quantify mutations at the hgprt locus and cytotoxicity induced by a wide variety of chemical and physical agents. S9 mix containing cofactors of an NADPH generating system necessary for MFO activity is used to determine the cytotoxic and mutagenic effects of chemicals requiring bioactivation (9). All experiments presented here were performed with S9 mix with and without the addition of the appropriate enzyme cofactors UDPGA or GSH, or GSH with purified GSHTs.

After establishing optimal conditions for UDPGA concentration and the buffer system based on detoxication of BaP-induced cytotoxicity, the cytotoxicity (Fig. 1A) and mutagenicity (Fig. 1B) of BaP was assayed with S9 mix in the presence and absence of UDPGA. In the absence of UDPGA, BaP is cytotoxic exhibiting a dose-dependent cellular lethality, as we reported earlier (12). The detoxication of BaP by UDPGA is shown by a broad shoulder in the cytotoxicity curve and a reduction of lethality; at 8 and 20 μM of BaP in the absence of UDPGA the relative survival is approximately 50% and 8%, respectively, and with UDPGA the corresponding survival is 90% and 65%, respectively. However, UDPGA had no effect on BaP-induced mutagenicity in the concentration range (1-20 μM) used.

BaP 7,8-diol is the proximate metabolite to the ultimate mutagenic metabolite of BaP, BaP 7,8-diol-9,10-epoxide (8). Biochemical studies have shown that BaP 7,8-diol is a poor substrate for UDPGTs (7). To assess the relative balance between MFO bioactivation of BaP 7,8-diol to cytotoxic and

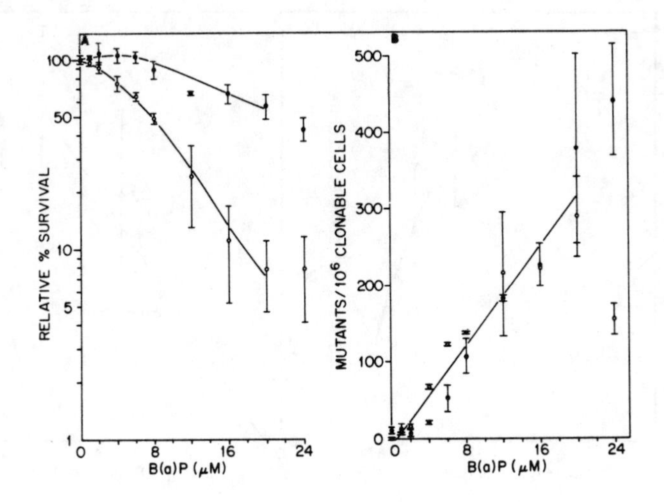

FIGURE 1. The effects of UDPGA on BaP-induced cytotoxicity (A) and mutagenicity (B). O, without UDPGA; ●, with UDPGA. The bars indicate one standard error from the mean (12).

mutagenic metabolites and detoxication by UDPGTs, the cytotoxicity and mutagenicity of BaP 7,8-diol was assayed with S9 mix in the presence and absence of UDPGA (Fig. 2). BaP 7,8-diol is highly cytotoxic and mutagenic with S9 mix in the presence and absence of UDPGA. The dose-dependent cellular lethality and mutagenicity of BaP 7,8-diol was assayed over a concentration range one fourth the molarity of the parent compound BaP. The survival curves for BaP 7,8-diol with and without UDPGA overlapped and are represented as a single line over the concentration of 0-6 μM; UDPGA had no effect on BaP 7,8-diol-induced mutagenicity and cytotoxicity (12).

Biochemical studies have shown that BaP 7,8-diol bioactivated by the MFO system can be conjugated with GSH catalyzed by GSHTs and results in a reduction in the amount

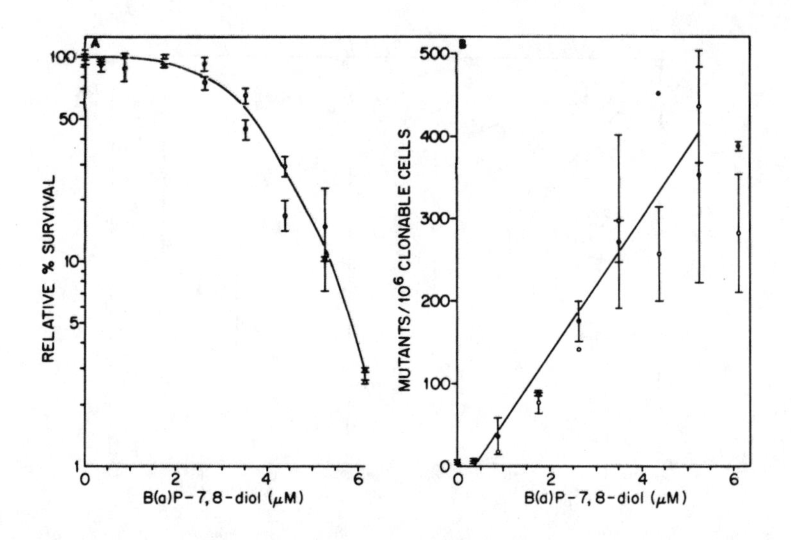

FIGURE 2. The effects of UDPGA on BaP 7,8-diol-induced cytotoxicity (A) and mutagenicity (B). O,without UDPGA; ●,with UDPGA. The bars indicate one standard error from the mean (12).

of covalently bound BaP 7,8-diol-9,10-epoxide to DNA (8). After establishment of an appropriate concentration of GSH in the treatment medium of 6.5 mM, based on an inhibition of BaP 7,8-diol-induced cytotoxicity, BaP 7,8-diol was assayed for cytotoxicity and mutagenicity with S9 mix in the absence and presence of GSH (Table 1). In the absence of GSH, BaP 7,8-diol-induced a dose-dependent reduction of cell survival and an increase in mutant frequency. GSH inhibited the cytotoxicity of BaP 7,8-diol, however, with the exception of the lowest concentration assayed (3.5 μM) GSH did not inhibit the mutagenicity of BaP 7,8-diol. At 3.5 μM of BaP 7,8-diol in the presence of GSH, there was an apparent four-fold reduction in the mutagenicity observed with respect to the mutagenicity of BaP 7,8-diol in the absence of GSH.

TABLE 1

EFFECT OF GLUTATHIONE ON B(A)P 7,8-DIOL-INDUCED CYTOTOXICITY AND MUTAGENICITY

BaP 7,8-diol	Relative % Survival	
μM	-GSH	+GSH
0.0	100	100
3.5	84 ± 4	89 ± 2
7.0	43 ± 7	103 ± 16
10.5	19 ± 6	52 ± 1
14.0	6	28
	Mutants/10^6 Clonable Cells	
0.0	5 ± 5	1 ± 1
3.5	99 ± 30	18 ± 1
7.0	231 ± 89	193 ± 68
10.5	393 ± 76	359 ± 132
14.0	605	770

Absolute cloning efficiencies in the absence and presence of GSH are 59 and 62%, respectively.

Since there was an indication of an inhibition of mutagenicity at a concentration of 3.5 μM BaP 7,8-diol not apparent at higher concentrations, we felt that it was necessary to characterize the shapes of the dose-response curves in the absence and presence of GSH at low concentrations of BaP 7,8-diol. Therefore, BaP 7,8-diol was assayed for cytotoxicity and mutagenicity with S9 mix in the absence and presence of GSH (6.5 mM) using a number of non-cytotoxic concentrations of BaP 7,8-diol which yielded a significant number of mutants in the absence of GSH (Figure 3). In the absence of GSH there is an apparent linear increase in the number of mutants over the concentration

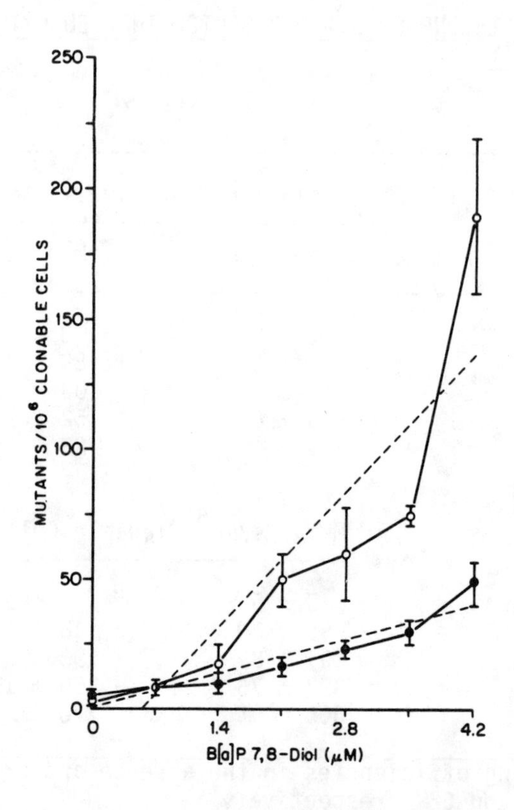

FIGURE 3. The effects of GSH on BaP 7,8-diol-induced mutagenicity. O, without GSH; ●, with GSH. The bars indicating one standard error from the mean.

range assayed (0.7-4.2 μM). However, in the presence of GSH there is a significant reduction of mutants in the concentration range of 2.1-4.2 μM. These data confirm our initial observation of an inhibition of mutagenicity at a low concentration (3.5 μM) of BaP 7,8-diol not observed at higher concentrations we previously assayed (Table 1).

In the next experiments, we added purified GSHTs to the treatment medium to further enhance the activity of GSHTs present in S9 mix. BaP 7,8-diol was assayed with S9 mix and 6.5 mM GSH in the treatment medium in the absence and presence of purified GSHTs (Table 2). In the presence of GSH alone BaP 7,8-diol induced a dose-dependent reduction of cell survival and an increase in mutant frequency. GSHTs inhibited both the cytotoxic and mutagenic effects of BaP 7,8-diol over the entire concentration range assayed. These data indicate that both the availability of GSH and the enzymatic activity of GSHTs play an important role in the detoxication of BaP 7,8-diol-induced cytotoxicity and mutagenicity.

TABLE 2

EFFECT OF GLUTATHIONE-S-TRANSFERASES ON BaP 7,8-DIOL-INDUCED CYTOTOXICITY AND MUTAGENICITY

BaP 7,8-diol μM	Relative % Survival	
	-GSH-trans	+GSH-trans
0.0	100[a]	100
2.1	100	116
4.2	75	81
7.0	40	95
10.5	14	80
14.0	6	58
	Mutants/10^6 Clonable Cells	
0.0	3	2
2.1	27	8
4.2	81	23
7.0	276	70
10.5	728	78
14.0	279	141

The absolute cloning efficiency was 76%. The data shown were reproduced in an independent experiment.

DISCUSSION

The mechanisms and biological effects of MFO system bioactivation of BaP and other polycyclic aromatic hydrocarbons (PAH) has been extensively investigated by numerous researchers in diverse biological systems. However, the biological effects and mechanisms of enzymatic detoxication of BaP have been rarely studied. These detoxication enzymes result in water-soluble products and are the primary mechanisms by which the biological effects of BaP can be eliminated and is excreted from the animal body. A continued emphasis on bioactivation of BaP and other PAH to biologically reactive metabolites can greatly influence our perception of these molecules particularly when assessing risk from exposure to these ubiquitous environmental pollutants. This is apparent in the design and interpretation of data from many genotoxicity bioassays in which bioactivation by MFO enzymes takes precedent over the available detoxication enzyme systems also present in the liver. We believe that these detoxication enzyme systems can affect the biological activity of BaP as profoundly as MFO system bioactivation especially since these processes result in the major metabolites of biotransformation of BaP *in vivo*.

We have investigated the biological effects of glucuronide and glutathione conjugation on BaP-induced cytotoxicity and mutagenicity. Our studies demonstrate that glucuronide conjugation of BaP results in a reduction of cytotoxicity without affecting mutagenicity (12). This is likely due to an elimination of cytotoxic phenols and quinones without affecting the formation and bioactivation of BaP 7,8-diol to the ultimate mutagenic form of BaP. GSH conjugation of BaP catalyzed by GSHTs inhibits the cytotoxicity and mutagenicity of BaP 7,8-diol in a concentration-dependent manner. These are consistent with and supported by biochemical studies on the biotransformation of BaP and DNA binding by bioactivated BaP (7,8). We conclude that these two enzyme systems have overlapping and complimentary roles in the detoxication of BaP. The activity of these enzyme systems *in vivo* can profoundly affect the biological activity of BaP and other PAH.

ACKNOWLEDGEMENTS

This work was supported by EPA and ORNL under contract DE-AC05-84OR21400 with Martin Marietta Energy Systems, Inc. Leslie Recio is an ORAU Graduate Fellow, and Abraham W. Hsie is supported in part by EPA as an EPA Distinguished Visiting

Scientist. The assistance of Faye Young in the preparation of this manuscript is appreciated.

REFERENCES

1. Jefcoate, C. R. (1983): Integration of xenobiotic metabolism in carcinogen activation and detoxication. In: Biological Basis of Detoxication, edited by J. Caldwell and W. B. Jakoby, pp. 31-76, Academic Press, New York.
2. Gelboin, H. V. (1980): Benzo(a)pyrene metabolism, activation, and carcinogenesis: Role and regulation of mixed-function oxidases and related enzymes. Physiol. Rev. 60: 1107-1166.
3. Boroujerdi, M., Kung, H., Wilson, A. G. E., and Anderson, M. W. (1981): Metabolism and DNA binding of benzo(a)pyrene in vivo in the rat. Cancer Res. 41: 951-957.
4. Jakoby, W. B. (1980): Enzymatic Basis of Detoxication, Vols. I and II. Academic Press, New York.
5. Mehendale, H. M. and Dorough, H. W. (1971); Glucuronidation mechanisms in the rat and their significance in the metabolism of insecticides. Pest. Biochem. Physiol. 1: 307-318.
6. Jakoby, W. B. and Habig, W. H. (1980): Glutathione Transferases. In:Enzymatic Basis of Detoxication, Vol. II, edited by W. B. Jakoby, pp. 63-94. Academic Press, New York.
7. Nemoto, N. and Takayama, S. (1977): Modification of benzo(a)pyrene metabolism with microsomes by addition of uridine-5'-diphosphoglucuronic acid. Cancer Res. 37: 4125-4129.
8. Hesse, S. and Jernstrom, B. (1984): Role of glutathione-S-transferases: Detoxification of reactive metabolites of benzo(a)pyrene-7,8-dihydrodiol by conjugation with glutathione. In: Biochemical Basis of Chemical Carcinogenesis, edited by H. Greim, R. Jung, M. Kramer, H. Marquardt, and F. Oesch, pp. 5-12. Raven Press, New York.
9. Hsie, A. W., Casciano, D. A., Couch, D. B., Krahn, D. G., O'Neill, J. P. and Whitfield, B. L. (1981): The use of Chinese hamster ovary cells to quantify specific locus mutation and to determine mutagenicity of chemicals: A report of the GENE-TOX program. Mutat. Res. 86: 193-214.
10. Hsie, A. W., Brimer, P. A., Mitchell, T. J. and Gosslee, D. G. (1975). The dose-response relationship for ethyl methane sulfonate-induced mutations at the hypoxanthine-

guanine phosphoribosyltransferase locus in Chinese hamster ovary cells. Somat. Cell Genetics 1: 247-261.
11. Ames, B. N., McCann, J. and Yamasaki, E. (1975). Methods for detecting carcinogens and mutagens with the Salmonella/mammalian-microsome mutagenicity test. Mutat. Res. 31: 347-364.
12. Recio, L. and Hsie, A. W. (1984). Glucuronide conjugation reduces the cytotoxicity but not the mutagenicity of benzo(a)pyrene in the CHO/HGPRT assay. Terato., Carcino. and Mutagen. 4: 391-402.
13. O'Neill, J. P., Brimer, P. A., Machanoff, R., Hirsch, G. P. and Hsie, A. W. (1977). A quantiative assay of mutation induction at the hypoxanthine-guanine phosphoribosyl transferase locus in Chinese hamster ovary cells: Development and definition of the system. Mutat. Res. 45: 91-101.

STRUCTURAL REQUIREMENTS FAVORING MUTAGENIC ACTIVITY AMONG METHYLATED PYRENES IN S. TYPHIMURIUM

JOSEPH E. RICE*, NORA G. GEDDIE, MARIANNE C. DeFLORIA, AND EDMOND J. LaVOIE
Naylor Dana Institute for Disease Prevention, American Health Foundation, Valhalla, NY 10595.

INTRODUCTION

Pyrene (Figure 1) and methylated pyrenes are detected along with other polycyclic aromatic hydrocarbons (PAH)

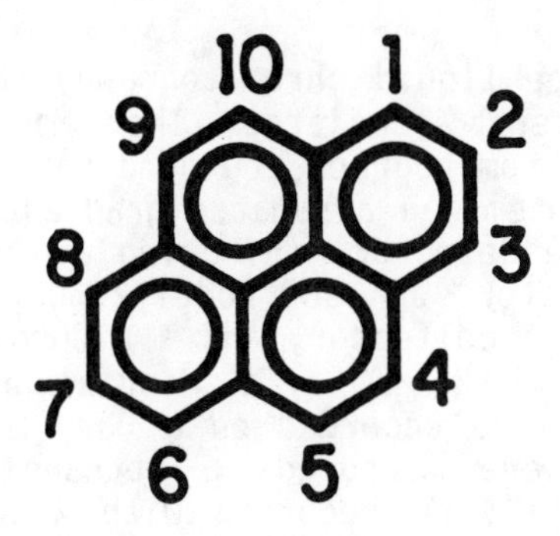

FIGURE 1. Structure and numbering of pyrene.

throughout the environment. Methylpyrenes have been detected in cigarette smoke condensate, diesel engine soot, emissions from cereal straw burning, drinking water, and in energy-related materials (1-9). Although the biological activity of pyrene has been thoroughly reviewed in a recent IARC Monograph (10) only a limited number of studies have been reported concerning the biological activity of methylated pyrenes. A number of methylated pyrenes, including 1-, 2-, and 4-methylpyrene, 1,6- and 4,5-dimethylpyrene, 1,3,6-trimethylpyrene, and 1,3,6,8-tetramethylpyrene, were tested and found to be inactive as complete carcinogens on mouse skin (11). 1-Methylpyrene and 2-methylpyrene were also reported to be inactive as tumor initiators on mouse skin (12,13).

Studies on the metabolism of pyrene in rats and rabbits have shown that the K-region dihydrodiol is a major metabolite (14,15). This dihydrodiol presumably forms upon enzymatic hydrolysis of a K-region epoxide. Recent studies from our laboratory suggest that for methylated anthracenes, a *peri*-methyl substituent adjacent to a simple arene oxide greatly enhances its mutagenic and tumorigenic potency (16,17). Based upon these studies, we have investigated the mutagenic activity of a series of mono- and dimethylpyrenes.

MATERIALS AND METHODS

Apparatus

High performance liquid chromatography (HPLC) was performed using a Waters Associates, Inc. Model ALC/GPC-204 high speed liquid chromatograph equipped with a Model 6000A solvent delivery system, an automated gradient controller, a Model 440 UV/visible detector monitoring at 254 nm, a Model U6K septumless injector, an ISCO Model 2150 peak separator, an ISCO FOXY fraction collector, and a LiChrosorb 10 μm RP-18 column, 4.6 cm x 25 cm (MCB Manufacturing Chemists, Inc.). UV spectra were recorded on a Cary Model 118 spectrophotometer and were measured in methanol or methanol-water. Mass spectra were recorded with a Hewlett-Packard Model 5982A instrument. 90 MHz NMR spectra were recorded on a JEOL FX-90Q instrument in deuterochloroform with internal tetramethylsilane as a reference.

Chemicals

Pyrene and 2-methylpyrene were obtained from Aldrich Chemical Co., Milwaukee, Wisc. 1-Methylpyrene was purchased from Cambridge Chemical Co., Milwaukee, Wisc. Pyrene was treated with ethyl diazoacetate in the presence of copper (18). The resulting 4,5-dihydrocyclopropa[e]pyrene yielded ethyl 4-pyrenylacetate upon treatment with iodotrimethylsilane. Basic hydrolysis to the carboxylic acid followed by copper catalyzed decarboxylation in refluxing quinoline yielded 4-methylpyrene. 1,6- and 1,8-Dimethylpyrene were prepared from 1,6- and 1,8-dibromopyrene (19) by treatment with n-butyllithium followed by dry ice. The dicarboxylic acids were esterified with ethanol and reduced to the bis-(hydroxymethyl) derivatives with lithium aluminum hydride. Acetylation of the alcohols and hydrogenolysis over palladium-on-charcoal yielded, respectively, 1,6- and 1,8-dimethylpyrene. Treatment of 1-methyl-1,2-dihydrophenanthren-4(3H)-one with diethyl carbonate and sodium hydride yielded a β-ketoester which was alkylated with methyl iodide in the presence of sodium ethoxide. Heating this compound with aqueous sodium hydroxide formed the 1,3-dimethylketone. Treatment with the ethyl acetate enolate, followed by hydrogenolysis, and alkaline hydrolysis yielded a tetrahydrophenanthryl acetic acid which was cyclized with methanesulfonic acid to 1,3-dimethyl-1,2,3,3a-tetrahydropyren-5(4H)-one. 1,3-Dimethylpyrene was prepared from this ketone by Clemmensen reduction followed by aromatization over palladium-on-charcoal. 2,7-Dimethylpyrene was prepared from 2,7-dimethyl-1,2,6,7-tetrahydropyren-3,8-dione by sodium borohydride reduction followed by acid-catalyzed dehydration. The expected dihydropyrene product aromatized during silica gel chromatography. The required tetrahydropyrene-dione intermediate was prepared from 1,5-dimethylnaphthalene by N-bromosuccinimide bromination of the two methyl groups followed by displacement by the enolate of ethyl propionate. The ester groups were hydrolyzed and the resultant acids cyclized upon treatment with methanesulfonic acid. 4,9-Dimethylpyrene was prepared from 4,9-bis(carbomethoxy)pyrene (20) by lithium aluminum hydride reduction to the bis(hydroxymethyl) derivative and hydrogenolysis of the diacetate. 1-Hydroxymethylpyrene was prepared from 1-pyrenecarboxaldehyde by treatment with sodium borohydride. Osmylation of 1-methylpyrene yielded a mixture of the <u>cis</u>-4,5- and 9,10-dihydrodiols. These were separated by preparative TLC (two developments) using a solvent system of chloroform:methanol:ethyl acetate (19:1:1). The faster moving diol was identified as the <u>cis</u>-9,10-dihydrodiol of 1-methylpyrene on the basis of its NMR spectrum. The NMR spectrum of the slower moving diol was consistent with it being the <u>cis</u>-4,5-dihydrodiol of 1-methylpyrene. Each diol was oxidized to the corresponding quinone with manganese dioxide and reduced to the <u>trans</u>-dihydrodiol upon treatment with potassium borohydride. The <u>cis</u>-dihydrodiols were also converted to the corresponding epoxides by treatment with trimethylorthoacetate, followed by chlorotrimethylsilane, and sodium methoxide (21). All hydrocarbons were recrystallized from either methanol or ethanol to a purity > 99% as determined by capillary gas chromatography.

<u>Metabolism Studies In Vitro</u>

Metabolism studies were performed by dissolving 200 μg of the pyrene derivative in 50 μl of dimethylsulfoxide and adding this solution to 2 ml of S-9 mix in a 25 ml Erlenmeyer flask. The mixture was shaken for 20 min at 37°C in a Dubnoff metabolic shaking incubator. Incubations were terminated by the addition of 2 ml of ice-cold acetone. The mixture was extracted five times with 10 ml portions of ethyl acetate which were combined, dried over sodium sulfate, and evaporated in vacuo below 40°C.

The S-9 mix which was used for these metabolism studies and for the mutagenicity assays was prepared from the livers of male Fischer-344 rats. Rats weighing between 250 and 300 g were injected five days prior to sacrifice with 500 mg/kg body weight Aroclor 1254 in corn oil. Each milliliter of S-9 mix contained 50 µmol potassium phosphate buffer (pH 7.4), 8.0 µmol $MgCl_2$, 33 µmol KCl, 5.0 µmol glucose-6-phosphate, 4.0 µmol $NADP^+$, and 0.5 ml of the 9000 x g supernatant of the rat liver homogenate.

Mutagenicity Assays

The mutagenic activity of the methylpyrenes was evaluated in *S. typhimurium* TA98 with and without metabolic activation using the procedure of Ames, *et al*. (22). Briefly, the test compounds were dissolved in 50 µl of dimethylsulfoxide and were added to 0.1 ml of an overnight nutrient broth culture of the bacterial tester strain. After the addition of 2 ml of molten top agar at 45°C, the contents were mixed and poured onto minimal glucose agar plates. Those assays performed with metabolic activation contained 200 µl of S-9 mix per plate. Compounds were considered to be mutagenic when the average number of histidine revertants obtained at any dose level assayed was at least twice that of the solvent control.

RESULTS AND DISCUSSION

The results of the mutagenicity testing of pyrene and 1-, 2-, and 4-methylpyrene in *S. typhimurium* TA98 are presented in Figure 2. None of the compounds tested were mutagenic in the absence of S-9 mix. When tested with metabolic activation only 1-methylpyrene gave a positive response.

A series of dimethylpyrenes was synthesized and evaluated for mutagenic activity in *S. typhimurium* TA98. The results of this assay are shown in Figure 3. While none of these compounds exhibited any direct-acting mutagenic activ-

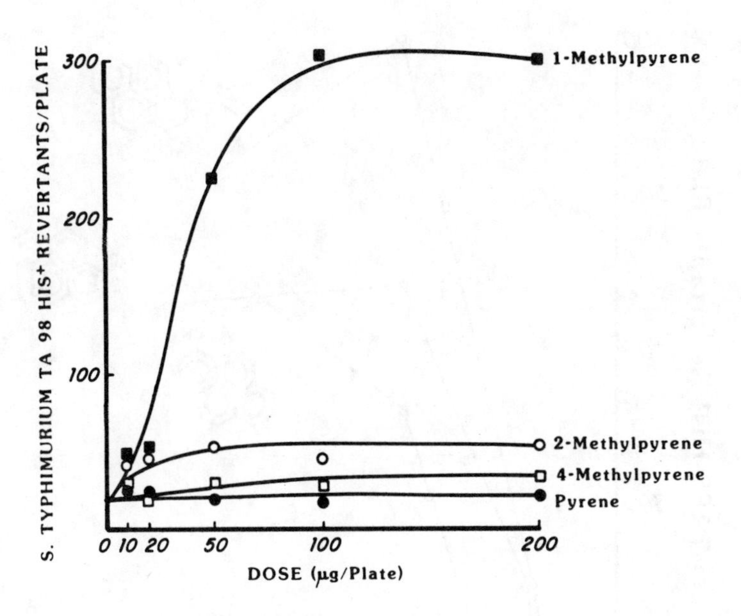

FIGURE 2. Mutagenic activity of the isomeric monomethylpyrenes in *S. typhimurium* TA98 with S-9.

ity, 1,3- and 1,6-dimethylpyrene were potent mutagens in the presence of S-9 mix. Both of these compounds attained a maximum mutagenic response at a dose of 50 μg/plate. At higher doses, 1,6-dimethylpyrene became cytotoxic towards *S. typhimurium*. While 1,8-dimethylpyrene exhibits a mutagenic response at the higher doses tested, it is much less active than 1,3- and 1,6-dimethylpyrene at low doses. Substitution of a methyl group on each K-region, as in 4,9-dimethylpyrene, resulted in a slight mutagenic response. 2,7-Dimethylpyrene was found to have no significant mutagenic activity under these bioassay conditions. From these studies it can be seen that the presence of a methyl group at a *peri*-position adjacent to each K-region results in greatly enhanced mutagenic activity among methylated pyrenes.

A similar structure-activity relationship has been found to exist among methylated anthracenes (16,17). It was shown that the presence of a methyl group at both the 9- and 10-positions of anthracene is a requirement for the expression of mutagenic activity. The presence of a methyl group at each of these positions ensures that metabolism to an

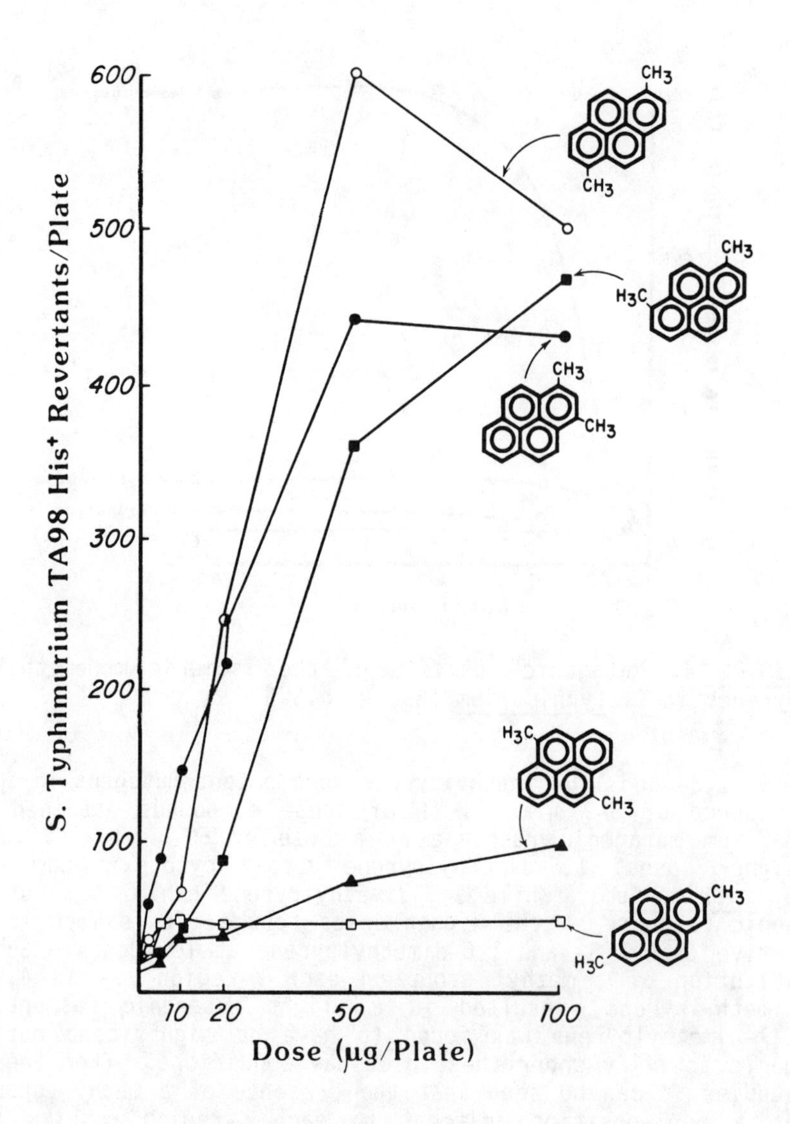

FIGURE 3. Mutagenic activity of dimethylpyrenes in S. typhimurium TA98 with metabolic activation.

arene oxide will occur at a bond adjacent to a peri-methyl group. Evidence has been presented which shows that such epoxides exhibit enhanced biological activity as compared to epoxides which are not adjacent to a peri-methyl substituent. Metabolism studies on 9-methylanthracene have shown that there is preferential formation of the 3,4-dihydrodiol as compared to the more sterically hindered 1,2-dihydrodiol (16,17). These results indicate that in a symmetrical molecule a *peri*-methyl group can inhibit metabolism at those positions to which it is adjacent.

The possibility that a similar metabolic preference could account in part for the decreased mutagenic activity of 1-methylpyrene as compared to either 1,3- or 1,6-dimethylpyrene was explored. An HPLC profile of the ethyl acetate extractable metabolites of 1-methylpyrene as formed *in vitro* with S-9 mix identical to that used for the mutagenicity assays is shown in Figure 4. Metabolite A was shown to be a dihydrodiol on the basis of its mass spectrum. In order to characterize this metabolite, both the 4,5- and 9,10-dihydrodiols of 1-methylpyrene were synthesized. One of the synthetic diols was found to coelute with Peak A. The other diol eluted earlier on reverse-phase HPLC. The vicinal hydroxyl groups of the 9,10-diol are expected to adopt a diaxial conformation due to the steric effect of the *peri*-methyl substituent. Diaxial dihydrodiols have been shown to be more polar than diequatorial diols and therefore elute earlier on reverse-phase HPLC (23). On this basis, metabolite A was identified as the *trans*-4,5-dihydrodiol of 1-methylpyrene. The UV spectrum of Peak A, however, revealed the presence of a second metabolite that coelutes with the 4,5-diol. This metabolite has a UV spectrum similar to 1-hydroxypyrene (Figure 5). This metabolite was tentatively identified as either the 3-, 6-, or 8-phenol of 1-hydroxymethylpyrene on the basis of its mass spectrum. Metabolism of 1-hydroxymethylpyrene under the same conditions employed for 1-methylpyrene resulted in the formation of this same metabolite. Peak B was identified as a quinone of 1-methylpyrene on the basis of its mass spectrum. The UV spectrum of this peak closely resembled a spectrum of the 1,8-quinone of pyrene. Metabolite B was therefore tentatively identified as the 3,6-quinone of 1-methylpyrene. Metabolite C was identified as 1-hydroxymethylpyrene by comparison of its UV and mass spectrum and HPLC retention time with a synthetic standard. Peak D has a UV spectrum which closely resembles that of 1-hydroxypyrene. Based upon this data and its mass spectrum, metabolite D was tentatively identified as either 3-, 6-, or 8-hydroxy-1-methylpyrene. As in the case of 9-methylanthracene, the primary site of dihydrodiol formation for 1-methylpyrene was found to be at a site distal to the methyl substituent. The decreased mutagenic activity of 1-methylpyrene relative to 1,3- and 1,6-di-

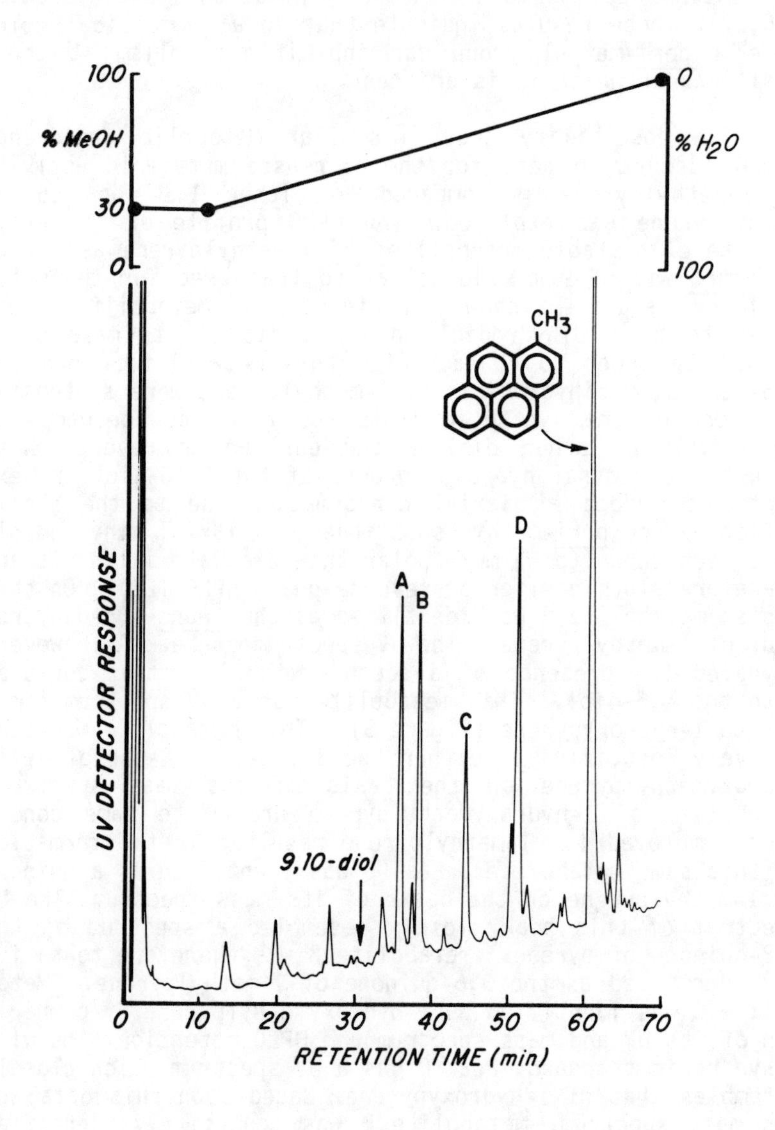

FIGURE 4. Reverse-phase HPLC profile of the ethyl acetate-extractable metabolites of 1-methylpyrene as formed *in vitro* in liver homogenate from Aroclor pretreated rats.

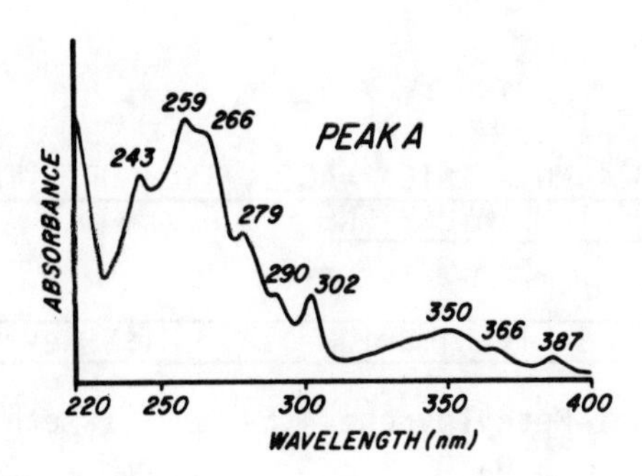

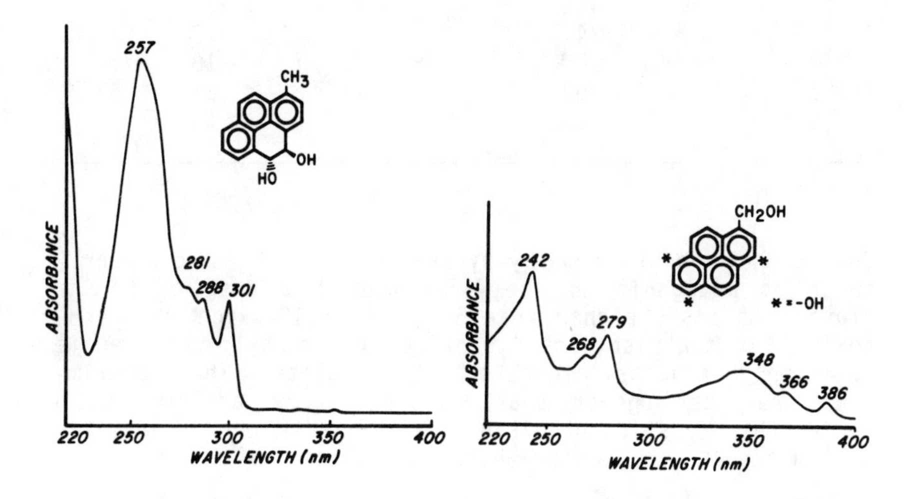

FIGURE 5. The UV spectrum of Peak A compared to spectra of the synthetic *trans*-4,5-dihydrodiol of 1-methylpyrene and a major metabolite of 1-hydroxymethylpyrene which coelutes with Peak A.

methylpyrene could be related to the fact that the less hindered arene oxide is formed metabolically in S-9 mix. For 1,3- and 1,6-dimethylpyrene, K-region oxide formation must occur adjacent to a *peri*-methyl substituent. In order to substantiate this hypothesis, the 4,5- and 9,10-oxides of 1-methylpyrene were synthesized and evaluated for direct-acting mutagenic activity in *S. typhimurium* TA98 (Table 1).

TABLE 1

DIRECT-ACTING MUTAGENIC ACTIVITY OF THE 4,5- AND 9,10-OXIDES OF 1-METHYLPYRENE

	S. Typhimurium TA98 His$^+$ Revertants/Plate	
Dose (μg)	1-Methylpyrene-4,5-oxide	1-Methylpyrene-9,10-oxide
10	415	337
5	241	473
2	109	218
1	50	67
0	19	19

The 9,10-oxide of 1-methylpyrene was found to be nearly twice as mutagenic as the 4,5-oxide at a dose of 5 μg/plate. At doses higher than 5 μg the 9,10-oxide was cytotoxic. In contrast, the 4,5-oxide of 1-methylpyrene became toxic only at doses in excess of 20 μg/plate. These preliminary results suggest that an arene oxide adjacent to a *peri*-methyl substituent exhibits enhanced mutagenic activity as compared to an epoxide which is not similarly situated.

ACKNOWLEDGEMENT

These studies were supported in part by Department of Energy Contract No. DE-AC-2-8ER60080. The authors would like to thank Professor Albert Padwa of Emory University, Atlanta, GA. for a generous gift of 4,9-bis(carbomethoxy)-pyrene.

REFERENCES

1. Kubota, H., Griest, W.H., Guerin, M.R. (1975): Determination of carcinogens in tobacco smoke and coal-derived samples-trace polynuclear aromatic hydrocarbons. In: Trace Substances in Environmental Health - IX, edited by D.D. Hemphill, pp. 281-289, University of Missouri, Columbia, MO.
2. Lee, M.L., Novotny, M., Bartle, K.D. (1976): Gas Chromatography/mass spectrometric and nuclear magnetic resonance spectrometric studies of carcinogenic polynuclear aromatic hydrocarbons in tobacco and marijuana smoke condensates. Anal. Chem. 48: 405-416.
3. Severson, R.F., Schlotzhauer, W.D., Chortyk, O.T., Arrendale, R.F., Snook, M.E. (1979): Precursors of polynuclear aromatic hydrocarbons in tobacco smoke. In: Polynuclear Aromatic Hydrocarbons, edited by P.W. Jones and P. Leber, pp. 277-298, Ann Arbor Science Publishers, Ann Arbor, MI.
4. Yu, M.L., Hites, R.A. (1981): Identification of organic compounds on disel engine soot. Anal. Chem. 53: 951-954.
5. Ramdahl, T., Moller, M. (1983): Chemical and biological characterization of emissions from a cereal straw burning furnace. Chemosphere 12: 23-34.
6. Olufsen, B. (1980): Polynuclear aromatic hydrocarbons in Norwegian drinking water resources. In: Polynuclear Aromatic Hydrocarbons: Chemistry and Biological Effects. edited by A. Bjorseth and A.J. Dennis, pp. 333-343, Battelle Press, Columbus, OH.
7. Grimmer, G., Böhnke, H. (1975): Profile analysis of polycyclic aromatic hydrocarbons and metal content in sediment layers of a lake. Cancer Letters 1: 75-84.
8. Sporstol, S., Gjos, N., Lichtenthaler, R.G., Gustavsen, K.O., Urdal, K., Oreld, F., Skei, J. (1983): Source identfication of aromatic hydrocarbons in sediments using GC/MS. Environ. Sci. Technol. 17: 282-286.
9. Griest, W.H., Tomkins, B., Epler, J.L., Rao, T.K. (1979): Characterization of multialkylated polycyclic aromatic hydrocarbons in energy-related materials. In: Polynuclear Aromatic Hydrocarbons, edited by P.W. Jones and P. Leber, pp. 395-409, Ann Arbor Science Publishers, Ann Arbor, MI.

10. IARC Monographs on the Evaluation of the Carcinogenic Risk of Chemicals to Humans. Vol. 32 (1983), pp. 431-445, IARC, Lyon, France.
11. Dannenberg, H. (1959): Beitrag zur krebserzeugenden Wirkung aromatischer Kohlenwasserstoffe und verwandter Verbindungen. Z. Krebsforsch. 62: 102-117.
12. VanDuuren, B.L., Sivak, A., Segal, A., Orris, L., Langseth, L. (1966): The tumor-promoting agents of tobacco leaf and tobacco smoke condensate. J. Natl. Cancer Inst., 37: 519-526.
13. Barry, G., Cook, J.W., Haslewood, G.A.D., Hewett, C.L., Hieger, I., Kennaway, E.L. (1935): Proc. Roy. Soc. London B 117: 318-351.
14. Boyland, E., Sims, P. (1964): Metabolism of polycyclic hydrocarbons. 23. The metabolism of pyrene in rats and rabbits. Biochem. J. 90: 391-398
15. Sims, P. (1970): Qualitative and quantitative studies on the metabolism of a series of aromatic hydrocarbons by rat-liver preparations. Biochem. Pharm. 19: 795-818.
16. LaVoie, E.J., Coleman, D.T., Rice, J.E., Geddie, N.G., Hoffmann, D. (1985): Tumor-initiating activity, mutagenicity, and metabolism of methylated anthracenes. Carcinogenesis. In press.
17. LaVoie, E.J., Coleman, D.T., Tonne, R.L., Hoffmann, D. (1983): Mutagenicity, tumor-initiating activity, and metabolism of methylated anthracenes. In: Polynuclear Aromatic Hydrocarbons: Formation, Metabolism, and Measurement, edited by M. Cooke and A.J. Dennis, pp. 785-798, Battelle Press, Columbus, OH.
18. Badger, G.M., Cook, J.W., and Gibb, A.R.M. (1951): The reaction of ethyl diazoacetate with anthracene, 1,2-benzanthracene, and pyrene. J. Chem. Soc., 3456-3459.
19. Lock, G. (1937): Über Abkömmbinge des Pyrenes. Berichte 70B: 926-930.
20. Padwa, A., Doubleday, C. and Mazzu, A. (1977): Photocyclization reactions of substituted 2,2'-divinylbiphenyl derivatives. J. Org. Chem. 42: 3271-3279.
21. Dansete, P. and Jerina, D.J. (1974): A facile synthesis of arene oxides at the K-regions of polycyclic hydrocarbons. J. Am. Chem. Soc., 96: 1224-1225.
22. Ames, B.N., McCann, J. and Yamasaki, E. (1975): Methods for detecting carcinogens and mutagens with the Salmonella/mammalian-microsome test. Mutat. Res. 31:347-364.

23. Zacharias, D.E., Glusker, J.P., Fu, P.P., and Harvey, R.G. (1979): Molecular structures of the dihydrodiols and diol epoxides of carcinogenic polycyclic aromatic hydrocarbons. X-ray crystallographic and NMR analysis. J. Am. Chem. Soc. 101: 4043-4051.

STUDIES ON THE MUTAGENICITY AND TUMOR-INITIATING ACTIVITY OF PAH WITH METHYLENE-BRIDGED BAY REGIONS

JOSEPH E. RICE*[1], MARIANNE C. DEFLORIA[1], ALBERTO A. LEON[2], AND EDMOND J. LAVOIE[1]
Naylor Dana Institute for Disease Prevention, American Health Foundation, Valhalla, NY 10595 (1). University of New Mexico, School of Medicine, Alburquerque, NM 87131 (2).

INTRODUCTION

Polycyclic aromatic hydrocarbons (PAH) which possess a methyl group in the bay region often exhibit enhanced tumor-initiating activity as compared to their parent hydrocarbons (1-8). Although methylated PAH are prevalent as environmental pollutants, bay region methyl substituted PAH are often the least dominant isomers detected. Studies in our laboratories have shown that bay region methyl substituted PAH cyclize to form methylene-bridged bay region PAH (9) at temperatures at which PAH are pyrosynthesized from cellulose (600-1000°C). Recent studies by Coombs (10,11) have demonstrated that the methylene-bridged bay region compound 15,16-dihydro-1,11-methanocyclopenta[a]phenanthren-17-one retains much of the potent tumorigenic activity of its bay region methyl analog.

Several studies have recently shown that polycyclic aromatic ketones (PAK) can be detected in high levels from a variety of sources such as wood combustion, cereal straw incineration, and diesel particulate matter (12,13). One such compound, 4H-cyclopenta[def]phenanthren-4-one (CPK) was detected at levels almost seven times that of benzo[a]pyrene from wood combustion (13). A possible mechanism for the formation of PAK in the environment would involve oxidation at the methylene-bridge of bridged bay region PAH. Fluorene has been shown previously to oxidize readily to fluorenone (14). Oxidation of methylene-bridged bay region PAH to PAK could occur while adsorbed on particulates in the atmosphere or on high volume air sampling filters.

In this study the mutagenic activity in S. typhimurium of a series of methylene-bridged and keto-bridged bay region derivatives of phenanthrene and chrysene were evaluated. Their relative tumor-initiating activities on CD-1 mouse skin were examined.

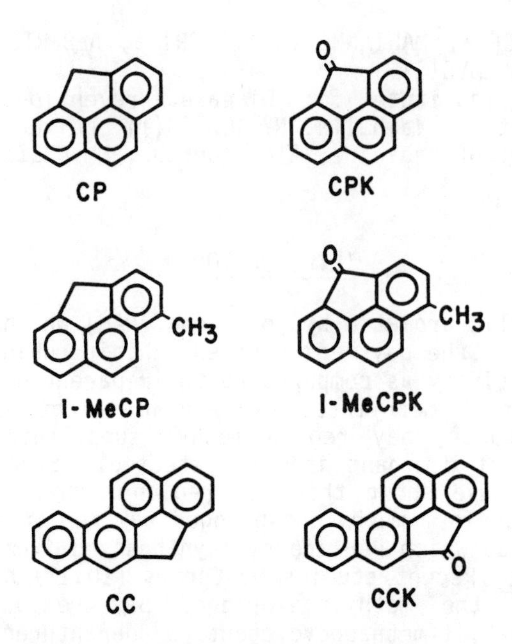

FIGURE 1. Structures of the methylene-bridged and keto-bridged bay region PAH under investigation.

MATERIALS AND METHODS

Chemicals

The structures of the six methylene-bridged bay region PAH under investigation are shown in Figure 1. 4H-Cyclopenta[def]phenanthrene (CP) was purchased from Aldrich Chemical Co., Milwaukee, Wisc. The keto derivative of this compound, 4H-cyclopenta[def]phenanthren-4-one (CPK) was prepared from the hydrocarbon by treatment with Triton B (benzyltrimethylammonium hydroxide) in the presence of oxygen (15). 1-Methyl-4H-cyclopenta[def]phenanthrene (1-MeCP) was prepared from 3,4-dihydrocyclopenta[def]phenanthren-1(2H)-one (16) by treatment with methylmagnesium bromide followed by acid catalyzed dehydration and aromatization by heating over 10% palladium-on-charcoal catalyst. The keto derivative of this compound, 1-MeCPK, was prepared as described above for CPK. 4H-Cyclopenta[def]chrysene (CC), and 4H-cyclopenta[def]chrysen-4-one (CCK) were prepared as described previously (17). The synthesis of 1H-cyclopenta[pqr]benz[a]anthracene was attempted starting with 12-methylbenz[a]anthracene. Bromination of 12-methylbenz[a]-anthracene with N-bromosuccinimide yielded the 12-bromomethyl derivative (18). Attempted Friedel-Crafts ring closure using either aluminum chloride or zinc chloride failed to give the desired cyclic compound. Hydrolysis of 12-bromomethylbenz[a]anthracene to the 12-hydroxymethyl derivative was readily accomplished upon treatment with water. This compound also failed to cyclize upon treatment with polyphosphoric acid, sulfuric acid, or methanesulfonic acid. Efforts are currently underway to synthesize this methylenebridged benz[a]anthracene derivative by an alternate route. The synthesis of 10H-cyclopenta[mno]benzo[a]-pyrene is currently in progress starting from 12-methylbenz-[a]anthracene (18). Each compound used for bioassay had a purity $>$ 99% as determined by capillary gas chromatography.

Mutagenicity Assays

The mutagenic activity of each of the compounds shown in Figure 1 was evaluated in *S. typhimurium* TA98 and TA100 with and without metabolic activation using the procedure of Ames *et al*. (19). In summary, the test compounds were dissolved in 50 µl of dimethyl sulfoxide and were added to 0.1 ml of an overnight nutrient broth culture of the bacterial tester strain. After the addition of 2 ml of molten top agar at 45°C, the contents were mixed and poured onto minimal glucose agar plates. For those assays which were performed with metabolic activation, each plate contained 200 l of S-9 mix. The S-9 mix used in these assays was prepared from the livers of male Fischer-344 rats weighing between 250 and 300 g. Each rat was injected with 500 mg/kg body weight of Aroclor 1254 in corn oil five days prior to sacrifice. Each milliliter of S-9 mix contained: 50 µmol of potassium phosphate buffer (pH 7.4), 8µmol of $MgCl_2$, 33 µmol of KCl, 5 µmol of glucose 6-phosphate, 4 µmol of $NADP^+$, and 0.5 ml of the 9000 X g supernatant from the rat liver homogenate.

Compounds were considered to be mutagenic when the average number of histidine revertant colonies (His^+ rev.) obtained at any dose level assayed were at least twice that of the solvent control.

Tumor-Initiation Bioassay

Compounds that were evaluated as tumor initiators were applied to the skin of Crl:CD/1 (ICR)BR (Charles River Breeding Laboratories, Portage, MI) outbred albino female mice. Animals were housed in solid-bottomed polycarbonate cages (5 to a cage) and fed Purina Lab Chow *ad libitum*. Animals were kept under standard conditions (22 ± 2°C; 50 ± 10% relative humidity; light-dark cycle, 12-12 hrs).

A solution of the test compound (100 g) in acetone (100 l) was applied *via* biopipette to the shaved backs of groups of 25 mice during the early part of the second telogen phase (50-55 days of age) of the hair cycle. Solutions were applied every other day for a total of 10 subdoses (1.0 mg total initiating dose). Benzo[a]pyrene (BaP) was employed as a positive control at a total initiating dose of 0.3 mg. Promotion with 12-O-tetradecanoylphorbol-13-acetate (TPA; 0.0025% in 100 l acetone) commenced 10 days after the final dose of initiator. TPA was applied three times weekly for 20 weeks. Mice were shaved as necessary and tumors were counted weekly. Significance was evaluated using the χ^2 test.

RESULTS AND DISCUSSION

The results of the mutagenicity testing of methylene-bridged bay region derivatives of phenanthrene in *S. typhimurium* TA98 and TA100 are presented in Figure 2. Among these derivatives only 1-MeCP exhibits significant mutagenic activity in both tester strains. 4H-Cyclopenta[def]phenanthren-4-one displays a weak mutagenic response only in TA100. These data can be compared with previous studies on the mutagenic activity of methylated phenanthrenes. The structural requirements for mutagenic activity among methylated phenanthrenes have been shown to be the presence of a methyl group at or adjacent to the K-region (the 9- and 10-positions), and an unsubstituted angular ring adjacent to a free *peri*-position (20). The structure of 1-methyl-4H-cyclopenta[def]phenanthrene fits these requirements except that it has no unsubstituted angular ring. This compound was found to be mutagenic in both tester strains employed. 4H-Cyclopenta[def]phenanthrene, however, has no methyl group on or adjacent to the K-regon and is not active as a mutagen in either tester strain. Although the molecular basis for the weak mutagenic activity of CPK in *S. typhimurium* TA100

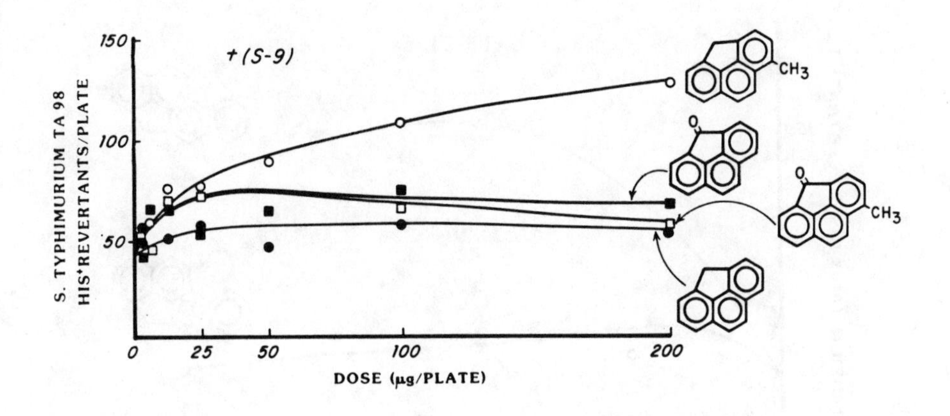

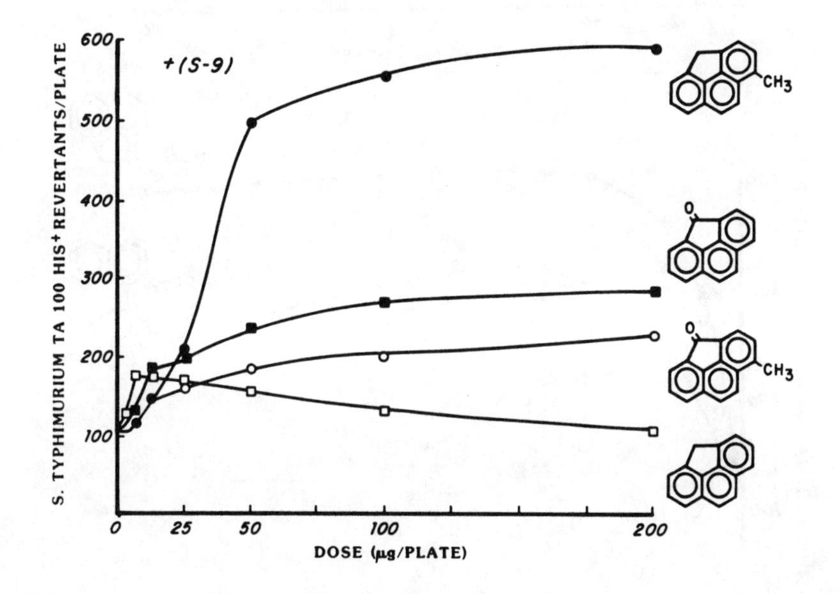

FIGURE 2. Results of the mutagenicity assays in *S. typhimurium* TA98 and TA100 of methylene-bridged and keto-bridged derivatives of phenanthrene. These assays were conducted using 200 µl of S-9 mix/plate.

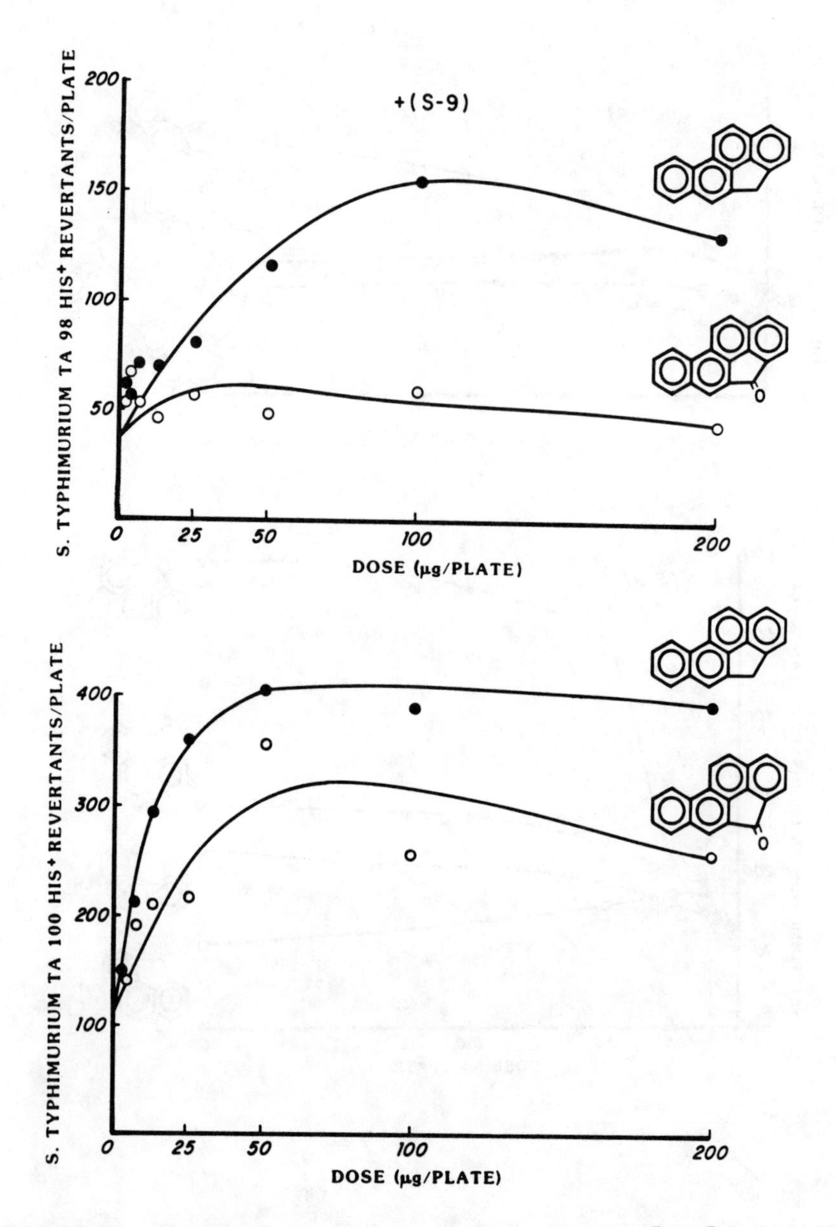

FIGURE 3. Mutagenic response of 4H-cyclopenta[def]chrysene and 4H-cyclopenta[def]chrysen-4-one in S. typhimurium TA98 and TA100 in the presence of S-9 mix.

is not known for certain, the inactivity of 1-MeCPK is more easily understood. It has been shown that the major mutagenic metabolite of 1-methylphenanthrene is either the 3,4- or 5,6-dihydrodiol (20). Substitution of the 4- and 5-positions of 1-methylphenanthrene with a carbonyl group as in 1-MeCPK would be expected to inhibit metabolism to dihydrodiols at both the 3,4- and 5,6-positions. The results of the mutagenicity assays of 4H-cyclopenta[def]chrysene and 4H-cyclopenta[def]chrysen-4-one in S. typhimurium TA98 and TA100 are presented in Figure 3. 4H-Cyclopenta[def]chrysene exhibited a strong mutagenic response in both tester strains. In contrast to this, the keto-derivative CCK was mutagenic only in S. typhimurium TA100.

Four of the compounds tested for mutagenic activity were also evaluated for tumor-initiating activity on CD-1 mouse skin (Figure 4). 4H-Cyclopenta[def]phenanthrene was tested previously in our laboratories and was found to be inactive as a tumor-initiator on mouse skin (unpublished results). Among the compounds tested for tumorigenic activity, 4H-cyclopenta[def]chrysene was the most potent, resulting in a 100% incidence of tumor-bearing mice with 5.63 papillomas/mouse. This compound has been shown previously to be moderately active as a carcinogen when administered to C_3H mice by subcutaneous injection (4). It was, however, less active in that assay than 5-methylchrysene. The keto-derivative, CCK, was shown to have weak tumor-initiating activity ($P<0.05$) on CD-1 mouse skin. At a dose of 1.0 mg CCK induced tumors in 36% of the animals with 0.52 tumors/animal. The presence of an electron-withdrawing substituent at the 4-position of chrysene would be expected to inhibit formation of a diol epoxide in that ring. The weak tumor-initiating activity of CCK might, therefore, be due to the presence of a second bay region in the molecule at which metabolic activation to a diol-epoxide might occur. Among the phenanthrene derivatives assayed, only 1-MeCP exhibited a significant ($P<0.01$) tumorigenic response resulting in a 44% incidence of tumor-bearing mice with 0.56 tumors/mouse. 1-MeCP is the methylene-bridged bay region derivative of 1,4-dimethylphenanthrene, which is also active as a tumor-initiator on mouse skin (21). The structure of 1-MeCP correlates well with the requirements for tumorigenic activity among methyl phenanthrenes (21) with two exceptions. The angular ring of 1-MeCP is not unsubstituted and there is a methylene-bridge, not a methyl group, in the bay region. While the metabolic activation pathway for 1-MeCP has not been investigated, it is postulated that it is analogous to

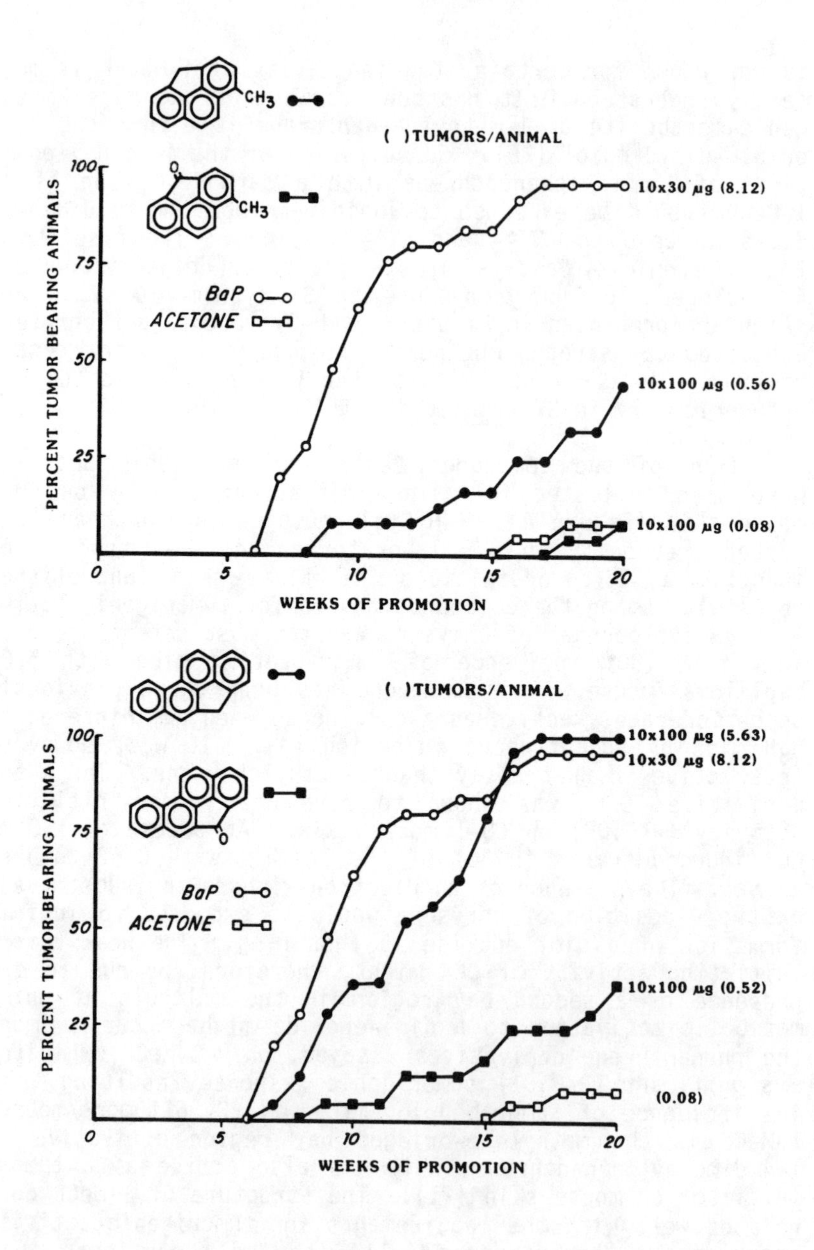

FIGURE 4. Tumor-initiating activity of 1-methyl 4H-cyclopenta[def]phenanthrene, 4H-cyclopenta[def]chrysene, and their keto-derivatives on CD-1 mice.

that of 1,4-dimethylphenanthrene which is metabolized to the 7,8-dihydrodiol. A recent study by Coombs et al. (11,22) has demonstrated that 15,16-dihydro-1,11-methanocyclopenta[a]-phenanthren-17-one is metabolized to a dihydrodiol which could serve as a precursor for a bay region diol-epoxide. A structurally similar diolepoxide has been shown to be an ultimate carcinogenic metabolite of 15,16-dihydro-11-methyl-cyclopenta[a]phenanthren-17-one (23). Metabolic activation at a substituted carbon has also been shown for 8-methyl-benz[a]anthracene and 7-methylbenzo[a]pyrene which are metabolized to the 8,9- and 7,8-dihydrodiols, respectively (24,25).

Methylene-bridged and keto-bridged bay region PAH are prevalent in the environment and generally occur at higher levels than their bay-region methyl substituted counterparts. Further testing is therefore necessary to more accurately assess the relative tumorigenic potency of these compounds as compared with their bay region methylated analogs and their parent hydrocarbons

ACKNOWLEDGEMENTS

This work was supported, in part, by the Department of Energy, Contract No. DE-AC-2-8ER60080.

REFERENCES

1. Coombs, M.M. and Croft, C.J. (1969): Carcinogenic cyclopenta[a]phenanthrenes. In: Progress in Experimental Tumor Research, Vol. 11, edited by F. Homburger, pp. 69-85, S. Karger, Basel, Switzerland.
2. Coombs, M.M., Bhatt, T.S., and Croft, C.J. (1973): Correlation between carcinogenicity and chemical structure in cyclopenta[a]phenanthrenes. Cancer Res., 33: 832-837.
3. DiGiovanni, J., Diamond, L., Harvey, R.G. and Slaga, T.J. (1983): Enhancement of the skin tumor-initiating activity of polycyclic aromatic hydrocarbons by methyl-substitution at the non-benzo 'bay region' positions. Carcinogenesis, 4: 403-407.
4. Dunlop, C.E. and Warren, S. (1943): The carcinogenic activity of some new derivatives of aromatic hydrocarbons. I. Compounds related to chrysene. Cancer Res., 3: 606-607.

5. Heidelberger, C., Baumann, M.E., Griesbach, L., Ghobar, A., and Vaughan, T.J. (1962): The carcinogenic activities of various derivatives of dibenzanthracene. Cancer Res., 22: 78-83.
6. Hoffmann, D., Bondinell, W.E., and Wynder, E.L. (1974): Carcinogenicity of methylchrysenes. Science, 183: 215-216.
7. Pataki, J. and Huggins, C. (1969): Molecular site of substituents of benz[a]anthracene related to carcinogenicity. Cancer Res., 29: 506-509.
8. Slaga, T.J., Iyer, R.P., Lyga, W., Secrist, A., Daub, G.H., and Harvey, R.G. (1980): Comparison of the skin tumor-initiating activities of dihydrodiols, diol-epoxides, and methylated derivatives of various polycyclic aromatic hydrocarbons. In: Polynuclear Aromatic Hydrocarbons: Chemistry and Biological Effects, edited by A. Bjorseth and A.J. Dennis, pp. 753-769, Battelle Press, Columbus, OH.
9. Adams, J.D., LaVoie, E.J., and Hoffmann, D. (1982): Analysis of methylated polynuclear aromatic hydrocarbons by capillary gas chromatography. Influence of temperature on the pyrosynthesis of anthracene, phenanthrene, and their methylated derivatives. J. Chromatogr. Sci., 20: 274-277
10. Bhatt, T.S., Coombs, M.M., Kissonerghis, A.M., Clayton, A.F.D., and McPartlin, M. (1979): A carcinogen with a bridged bay region: synthesis, x-ray structure, and biological activity of 15,16-dihydro-1,11-methanocyclopenta[a]phenanthren-17-one. J. Chem Soc. Chem. Commun., 433-434.
11. Coombs, M.M., Hadfield, S.T., and Bhatt, T.S. (1983): 15,16-Dihydro-1,11-methanocyclopenta[a]phenanthren-17-one: a carcinogen with a bridged bay region. In: Polynuclear Aromatic Hydrocarbons: Formation, Metabolism, and Measurement, edited by M. Cooke and A.J. Dennis, pp. 351-363, Battelle Press, Columbus, OH.
12. Ramdahl, T. (1983): Polycyclic aromatic ketones in environmental samples. Environ. Sci. Technol., 17: 666-670.
13. Ramdahl, T. (1985): Polycyclic aromatic ketones in source emissions and ambient air. In: Polynuclear Aromatic Hydrocarbons: Mechanisms, Methods, and Metabolism, edited by M. Cooke and A.J. Dennis, pp. 1075-1087, Battelle Press, Columbus, OH.
14. Cabaniss, G.E. and Linton, R.W. (1981): The in situ determination of the chemistry of PAH on particle surfaces. In: Polynuclear Aromatic Hydrocarbons: Chemical

Analysis and Biological Fate, edited by M. Cooke and A.J. Dennis, pp. 277-286, Battelle Press, Columbus, OH.

15. Sprinzak, Y. (1958): Reactions of active methylene compounds in pyridine solution. The ionic autoxidation of fluorene and its derivatives. J. Am. Chem. Soc., 80: 5449-5455.
16. Bachmann, W.E. and Sheenan, J.C. (1941): The synthesis of 1-methyl-, 1-ethyl-, and 3-ethyl-4,5-methylenephenanthrene. J. Am. Chem. Soc., 63: 2598-2600.
17. Lee-Ruff, E., Kruk, H., and Katz, M. (1984): A short synthesis of 4,5-methanochrysene and 6-oxo-7-oxabenzo[a]pyrene, two benzo[a]pyrene analogs. J. Org. Chem., 49: 553-555.
18. Leon, A.A., Daub, G.H., and VanderJagt, D.L. (1985): Synthesis of 4-, 5-, 11-, and 12-(chloromethyl)benzo[a]pyrene. J. Org. Chem., 50: 553-556.
19. Ames, B.N., McCann, J., and Yamasaki, E. (1975): Methods for detecting carcinogens and mutagens with the Salmonella/mammalian microsome mutagenicity test. Mutat. Res. 31: 347-364.
20. LaVoie, E.J., Tulley-Freiler, L., Bedenko, V., and Hoffmann, D. (1981): Mutagenicity, tumor-initiating activity, and metabolism of methylphenanthrenes. Cancer Res., 41: 3441-3447.
21. LaVoie, E.J., Bedenko, V., Tulley-Freiler, L., and Hoffmann, D. (1982): Tumor-initiating activity and metabolism of polymethylated phenanthrenes. Cancer Res., 42: 4045-4049.
22. Hadfield, S.T., Bhatt, T.S., and Coombs, M.M. (1984): The biological activity and activation of 15,16-dihydro-1,11-methanocyclopenta[a]phenanthren-17-one, a carcinogen with an obstructed bay region. Carcinogenesis, 5: 1485-1491.
23. Wiebers, J.L., Abbott, P.J., Coombs, M.M., and Livingston, D.C. (1981): Mass spectral characterization of the major DNA carcinogen adduct formed from the metabolically activated carcinogen 15,16-dihydro-11-methylcyclopenta[a]phenanthren-17-one. Carcinogenesis, 2: 637-643.
24. Yang, S.K. Chou, M.W., and Fu, P.P. (1981): Microsomal oxidation of methyl-substituted and unsubstituted aromatic hydrocarbons of monomethylbenz[a]anthracenes. In: Polynuclear Aromatic Hydrocarbons, edited by M. Cooke and A.J. Dennis, pp. 253-264, Battelle Press, Columbus, OH.

25. Kinoshita, T., Konieczay, M., Santella, R., and Jeffrey, A.M. (1982): Metabolism and covalent binding to DNA of 7-methylbenzo[a]pyrene. Cancer Res., 42: 4032-4038.

THE GLUTATHIONE CONJUGATION OF BENZO(A)PYRENE DIOL-EPOXIDE BY HUMAN GLUTATHIONE TRANSFERASES

IAIN G.C. ROBERTSON, CLAES GUTHENBERG[1], BENGT MANNERVIK[1], BENGT JERNSTRÖM*
Department of Toxicology, Karolinska Institute, S-104 01 Stockholm; (1) Department of Biochemistry, Arrhenius Laboratory, University of Stockholm, S-106 91 Stockholm, Sweden.

INTRODUCTION

It is well known that polycyclic aromatic hydrocarbons (PAH) such as benzo(a)pyrene are highly suspected as etiological factors in human chemical carcinogenesis. These compounds are believed to be mutagenic and carcinogenic only after metabolic activation to reactive intermediates which subsequently bind to DNA (1). The ultimate carcinogenic metabolite of benzo(a)pyrene is thought to be 7β,8α-dihydroxy-9α,10α-7,8,9,10-tetrahydrobenzo(a)pyrene (anti-BPDE), in particular, the (+)-enantiomer (1-3). This metabolite may be detoxified by enzymatic conjugation with glutathione (GSH). Both the DNA binding of anti-BPDE in rat hepatocytes and the cytotoxicity of anti-BPDE in C3H/101/2 cells are reduced by conjugation with GSH (4,5). Further, the cytosol fraction catalyzes the GSH conjugation of synthetic (±)-anti-BPDE (6-8) and individual isoenzymes of GSH transferase from rat liver and lung differ widely in their rate of conjugation of this compound (9,10).

Three distinct groups of human GSH transferase have been identified so far and are referred to as basic (α-ε), near-neutral (μ), and acidic (π) transferases according to their isoelectric points (11).

Significant individual differences in the isoenzyme patterns have been established in human subjects. Thus, the number of demonstrable forms of hepatic basic GSH transferase varies, and transferase π is not usually detectable in liver samples (12). The most prominent difference, however, is the complete absence of transferase μ in the liver in approximately 40 % of the (Caucasian) population examined (12, 13). This finding is of particular significance because this type is the most active in the conjugation of styrene-7,8-oxide and benzo(a)pyrene-4,5-dihydro-4,5-oxide (BP-4,5-oxide) with GSH (13). We have therefore determined the efficiency of these three types of human GSH transferase in the conjugation of anti-BPDE with GSH.

MATERIALS AND METHODS

Chemicals

Tritiated (+)- and (-)-anti-BPDE and BP-4,5-oxide and unlabelled racemic anti-BPDE and BP-4,5-oxide were obtained from Midwest Research Institute, (Kansas City, MO, USA); and GSH from Sigma Chemical Co. (St. Louis, MO, USA).

Purification of GSH transferases

GSH transferases α-ε and μ were isolated from human liver and transferase π from human placenta as described previously (14,15), and identified by determination of their substrate specificities, apparent sub-unit Mr and reactivity with specific antibodies. Protein concentrations were determined by the method of Lowry *et al*. (16).

Incubation of the GSH transferases with GSH and (±)-anti-BPDE

The purified GSH transferases were dialysed against 2.5 mM KCl, 0.5 mM EDTA, and 50 mM Tris/HCl, pH 7.5, and stored at -20°C. The activities with 1-chloro-2,4-dinitrobenzene were determined immediately before each experiment. Incubations were made for 30 or 60 sec in the same buffer (37°C) in a total volume of 0.25 ml, essentially as previously described (9).

The BPDE-GSH conjugates were quantified by fluorimetry after separation by high performance liquid chromatography (HPLC) on a μBondapak-NH_2 ion exchange column (Waters Associates, Milford, MA, USA), as previously described (4,7). A Zorbax-C_8 analytical column (Du Pont, Wilmington, DE, USA) was used for separation of the GSH conjugates of the (+) and (-)-enantiomers of (±)-anti-BPDE. Elution was carried out at 2 ml per min with methanol/25 mM ammonium acetate (37:63, v/v, pH 3.5 with acetic acid) under isocratic conditions. The peak fractions were identified by comparison with authentic standards [(+)-anti-BPDE-GSH, retention time 12.6 min, and (-)-anti-BPDE, retention time 13.7 min; as demonstrated in Fig. 1].

The GSH conjugation of 1-chloro-2,4-dinitrobenzene and BP-4,5-oxide were measured at 30°C as described (17,18). Kinetic data were analyzed by use of non-linear regression analysis (19).

RESULTS AND DISCUSSION

The highest activity towards (±)-anti-BPDE was obtained with transferases μ and π, whereas transferase α-ε had weak activity only. Calculated kinetic parameters are given in Table 1. Linear Lineweaver-Burk plots were obtained (plots not shown). In order of Vmax values the transferases can be ranked π>μ>>α-ε. However, the Km value obtained for transferase μ was approximately half that for transferase π. Calculation of the catalytic efficiency (kcat/Km) thus resulted in a value for transferase μ 1.4 times that for transferase π and approximately 50 times that for transferase α-ε. Lower Km values (13-30 μM) and a narrower range of values for catalytic efficiency (1.5-10 $mM^{-1}\cdot s^{-1}$) were obtained in an earlier study with purified rat liver GSH transferases (at 1 mM GSH) (9). However, we have recently isolated an enzyme, trans-

TABLE 1

KINETIC PARAMETERS WITH ANTI-BPDE AS SUBSTRATE

Transferase	Substrate	Vmax ($nmol\cdot min^{-1}\cdot mg^{-1}$)	Km (μM)	kcat/Km[a] ($mM^{-1}\cdot s^{-1}$)	Enantioselectivity[b] (%)
α-ε	(±)-anti-BPDE	38	88	0.4	59
μ	(±)-anti-BPDE	570	27	18	60
μ	(+)-anti-BPDE	690	50	11	
μ	(-)-anti-BPDE	500	45	9	
π	(±)-anti-BPDE	825	54	13	90
π	(+)-anti-BPDE	2900	85	28	
π	(-)-anti-BPDE	N.D.[c]			

The values for Km and Vmax are the computed estimate from non-linear regression analysis of the combined results of two separate experiments at 5 mM GSH and 5 to 120 μM anti-BPDE.

a Catalytic efficiency derived from $Vmax\cdot Km^{-1}\cdot[enzyme]^{-1}$.

b The percentage of GSH conjugate formed from the (+)-enantiomer.

c N.D. = no activity detected.

ferase 7-7, from rat lung and kidney with the highest activity so far observed with (±)-anti-BPDE (Vmax 5000 $nmol \cdot min^{-1} \cdot mg^{-1}$) (10).

No transferase-dependent increase in formation of tetrols was observed either in the presence or absence of GSH with any of the three transferases (results not shown). However, a high enzyme concentration was necessary with transferase α-ε to give measurable activity. The concentration of active sites thus approximated the lowest substrate concentrations which would significantly lower the free substrate concentration resulting in an overestimation of the Km value. The concentrations of transferases μ and π used were ≦8 % of the lowest (±)-anti-BPDE concentration used.

The anti-BPDE substrate used in the above experiments was racemic. The conjugates formed with (±)-anti-BPDE and GSH were resolved on a Zorbax-C_8 column, as shown in Fig. 1 for transferases μ and π. The percentages of (+)-anti-BPDE-conjugate are also given in Table 1. In the absence of enzyme, approximately equal amounts of conjugates were formed from the (+) and (-)-enantiomers of anti-BPDE, as expected for the racemic sub-

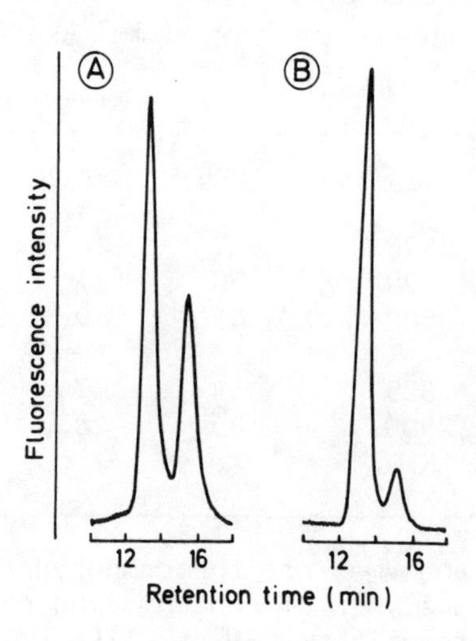

FIGURE 1. HPLC separation of the GSH conjugates of the (+)- and (-)-enantiomers of anti-BPDE. A, transferase μ; B, transferase π.

strate. The percentage of the (+)-anti-BPDE-conjugate was somewhat higher than 50 % for both transferase α-ε and μ, but approximately 90 % for the transferase π, after subtraction of the non-enzymatic values. These percentages were constant over the entire range of anti-BPDE concentrations tested (results not shown). In experiments with the separate enantiomers (see below) the (-)-enantiomer did not give measurable activity with transferase π. The degree of enantioselectivity thus appears to be greater than the value of 90 % estimated from the separation of conjugates formed with the racemic substrate. Also, a selectivity of ≧97 % for the (+)-enantiomer has been found for transferase 7-7 and all other rat transferases with relatively high activity so far tested (10).

Kinetic parameters have been estimated with the separate (+)- or (-)-enantiomers of anti-BPDE as substrates for transferase μ and the (+)-enantiomer as substrate for transferase π (Table 1). For transferase μ, the relative Vmax values obtained with the separate enantiomers closely reflect the percentages of the respective enantiomer conjugates obtained with (±)-anti-BPDE and higher Km values were obtained with the separate enantiomers (Table 1). These results are in close agreement with that expected for an enzyme working on a racemic mixture where both enantiomers are substrates. Thus it can be shown under the assumption of Michaelis-Menten kinetics, that if the Km values for the (+)-enantiomer equals that for the (-)-enantiomer, then the Km for the racemic substrate will be half that obtainable in experiments with the separate enantiomers. Under the same conditions the Vmax value for the racemic substrate will equal the average of the Vmax values for the separate enantiomers.

In contrast to the results with transferase μ, only the (+)-enantiomer was a substrate for transferase π. The results are consistent with the (-)-enantiomer acting as a competitive inhibitor of transferase π. It can be shown that under such conditions the true Vmax and Km values for the (+)-enantiomer would be lowered by the same factor (1 + Km/Ki), where Ki is the inhibition constant for the inhibitory (-)-enantiomer, when the racemic substrate is used. Thus, the ratio Vmax/Km, is constant when the values determined for transferase π are compared (Table 1). (The Km value in Table 1 has to be multiplied by 0.5 to obtain the proportion of the (+)-enantiomer in the mixture.) Using the above factor the Ki value for the (-)-enantiomer can be calculated as approximately 30 μM for transferase π. Inhibition by this enantiomer would thus not be of significance at the concentrations expected in vivo.

Apparent Km and Vmax values have also been estimated when

the concentration of GSH was varied at a single non-saturating concentration of (±)-anti-BPDE. Linear Lineweaver-Burk plots were obtained (plots not shown). The results were as follows: for transferase α-ε Vmax = 20 $nmol\cdot min^{-1}\cdot mg^{-1}$, Km = 0.4 mM; for transferase μ Vmax = 450 $nmol\cdot min^{-1}\cdot mg^{-1}$, Km = 0.7 mM; and for transferase π Vmax = 470 $nmol\cdot min^{-1}\cdot mg^{-1}$, Km = 0.1 mM. These results contrast with those with the rat hepatic and pulmonary isoenzymes (9,10), where non-linear Lineweaver-Burk plots were obtained. The apparent Km values obtained with transferases α-ε and μ were similar to the higher Km values obtained with the rat isoenzymes whereas that for transferase π was 4 to 7-fold lower and similar to the lower Km values obtained with the rat isoenzymes. Transferase π and the rat isoenzymes would thus appear to be more efficient at low GSH concentrations. However, the low Km values for GSH, in comparison to intracellular GSH concentrations, indicates that variations in GSH concentration would have little effect on the efficiency of these transferases under normal physiological conditions.

Finally we have also compared the activity of transferase α-ε, μ and π with anti-BPDE as substrate with those obtained with styrene 7,8-oxide and BP-4,5-oxide (Table 2, ref. 13). Transferase α-ε had low activity and transferase μ high activity with all substrates. It is particularly noteworthy that the activity of transferase π is more than 20-fold higher with (+)-anti-BPDE than with the other two epoxides. This distinct substrate specificity and enantioselectivity of transferase π suggests that the structure of the active site has evolved to

TABLE 2

ACTIVITIES OF THE HUMAN GSH TRANSFERASES AT SATURATING CONCENTRATIONS OF EPOXIDE SUBSTRATES

Substrate	Transferase[a]		
	α-ε	μ	π
(+)-anti-BPDE	~50[b]	690	2900
(-)-anti-BPDE	~50[b]	500	N.D.[c]
BP-4,5-oxide	50	660	90
Styrene-7,8-oxide	20	2600	140

a Values are given as $nmol\cdot min^{-1}\cdot mg^{-1}$.
b Estimated from experiment using (±)-anti-BPDE.
c N.D. = no activity detected.

participate in the biotransformation of compounds sharing the structural features of bay-region diol-epoxides.

Transferase π is the dominant or exclusive human enzyme form present in placenta, lung, brain, erythrocytes, and the lens of the eye. It is absent from or present in small amounts in the adult adrenal gland and liver but is a major form in all fetal organs tested including adrenal and liver tissues (11). In adult individuals, on the other hand, transferase μ has been demonstrated in the liver (but only at a frequency of 60 % in the Caucasian population) and in the adrenal gland, but is absent from a variety of other organs as well as from fetal tissues (11). Therefore, in fetal and a number of adult extrahepatic tissues including lung, the dominant transferase in the GSH conjugation of (+)-anti-BPDE would be transferase π. In the adult liver, generally considered the major site of xenobiotic metabolism, the predominant form appears to be transferase μ, in those individuals having this form. In individuals lacking transferase μ, transferase α-ε, in spite of its low activity, may be important. It is noteworthy that the rat GSH transferase 7-7, with the highest activity so far observed in the GSH conjugation of (+)-anti-BPDE, is present in rat lung and kidney but not in normal liver, although in hyperplastic liver nodules (20,21). Transferase π and 7-7 appear to be closely related by several criteria (22). They also have strongly similar amino acid compositions and antibodies against transferase π give a positive precipitin reaction with transferase 7-7 (22).

The preferential conjugation of the (+)-enantiomer of anti-BPDE by transferase π is of particular interest in that generally, (+)-anti-BPDE and other bay-region diol-epoxides with R,S,S,R absolute configuration have been shown to be the most carcinogenic intermediates of PAH (1-3,23,24). While the (-)-enantiomer is also a comparatively good substrate for transferase μ, transferase π does not catalyze the conjugation of (-)-anti-BPDE. In contrast, the latter enantiomer appears to be a competitive inhibitor of transferase π in the conjugation of the (+)-enantiomer. In the rat liver, at least, the BPDE formed is predominantly the (+)-anti isomer (25,26). To our knowledge, no information is available on the relative amounts of (+) and (-)-anti-BPDE that are formed in human preparations.

None of the human GSH transferases has been assayed for the GSH conjugation of other bay-region diol-epoxides. However, transferase 4-4, purified from rat liver, has been shown to catalyze the conjugation of the respective epoxides of benzanthracene and chrysene (Robertson and Jernström, submitted).

Similar profiles to those for transferase π with (±)-anti-BPDE (Fig. 1) are obtained indicating preferential conjugation of the (+)-enantiomer.

No information is as yet available on the importance *in vivo* of the GSH catalyzed conjugation of BPDE in human cells or tissues. However, enzyme catalyzed conjugation of (±)-*anti*-BPDE resulted in reduced DNA binding in rat hepatocytes (4) and reduced cytotoxicity in C3H/10T1/2 cells (5). We suggest, by analogy, that also in human tissues, the GSH transferases are important in protection against mutagenic and carcinogenic bay-region diol-epoxides.

ACKNOWLEDGEMENTS

We gratefully thank Ms. M. Staaff for typing this manuscript. This work was supported by the Swedish Cancer Society and the Swedish Tobacco Company.

REFERENCES

1. Cooper, C.S., Grover, P.L. and Sims, P. (1983): The metabolism and activation of benzo(a)pyrene, *Prog. Drug Metab.*, 7:295-396.
2. Buening, M.K., Wislocki, P.G., Levin, W., Yagi, H., Thakker, D.R., Akagi, H., Koreeda, M., Jerina, D.M. and Conney, A.H. (1978): Tumorigenicity of the optical enantiomers of the diastereomeric benzo(a)pyrene-7,8-diol-9,10-epoxides in newborn mice: exceptional activity of (+)-7β,8α-dihydroxy-9α,10α-epoxy-7,8,9,10-tetrahydrobenzo(a)pyrene, *Proc. Natl. Acad. Sci. USA*, 75:5358-5361.
3. Slaga, T.J., Bracken, W.J., Gleason, G., Levin, W., Yagi, H., Jerina, D.M. and Conney, A.H. (1979): Marked differences in the skin tumor-initiating activities of the optical enantiomers of the diastereomeric benzo(a)pyrene-7,8-diol-9,10-epoxides, *Cancer Res.*, 39:67-71.
4. Jernström, B., Babson, J.R., Moldéus, P., Holmgren, A. and Reed, D.J. (1982): Glutathione conjugation and DNA-binding of (±)-7β,8α-dihydroxy-9α,10α-epoxy-7,8,9,10-tetrahydrobenzo(a)pyrene in isolated rat hepatocytes, *Carcinogenesis*, 3:861-866.
5. Ho, D. and Fahl, W.E. (1984): Modification of glutathione levels in C3H/10T1/2 cells and its relationship to benzo-(a)pyrene *anti*-7,8-dihydrodiol 9,10-epoxide-induced cytotoxicity, *J. Biol. Chem.*, 259:11231-11235.
6. Cooper, C.S., Hewer, A., Ribeiro, O., Grover, P.L. and Sims, P. (1980): The enzyme-catalyzed conversion of *anti*-

benzo(a)pyrene-7,8-diol-9,10-oxide into a glutathione conjugate, Carcinogenesis, 1:1075-1080.

7. Jernström, B., Dock, L. and Martinez, M. (1984): Metabolic activation of benzo(a)pyrene-7,8-dihydrodiol and benzo(a)-pyrene-7,8-dihydrodiol-9,10-epoxide to protein-binding products and the inhibitory effect of glutathione and cysteine, Carcinogenesis, 5:199-204.
8. Jernström, B., Martinez, M., Svensson, S-Å. and Dock, L. (1984): Metabolism of benzo(a)pyrene-7,8-dihydrodiol and benzo(a)pyrene-7,8-dihydrodiol-9,10-epoxide to protein-binding products and glutathione conjugates in isolated rat hepatocytes, Carcinogenesis, 5:1079-1085.
9. Jernström, B., Martinez, M., Meyer, D.J. and Ketterer, B. (1985): Glutathione conjugation of the carcinogenic and mutagenic electrophile (±)-7β,8α-dihydroxy-9α,10α-oxy-7,8,9,10-tetrahydrobenzo(a)pyrene catalyzed by purified rat liver glutathione transferases, Carcinogenesis, 6:85-89.
10. Robertson, I.G.C., Jensson, H., Mannervik, B. and Jernström, B. (1986): Glutathione transferases in rat lung: the presence of a form, transferase 7-7, highly efficient in the conjugation of glutathione with the carcinogenic (+)-7β,8α-dihydroxy-9α,10α-oxy-7,8,9,10-tetrahydrobenzo-(a)pyrene, Carcinogenesis, in press.
11. Mannervik, B. (1985): The isoenzymes of glutathione transferase, Adv. Enzymol., 57:357-417.
12. Warholm, M., Guthenberg, C., Mannervik, B., von Bahr, C. and Glaumann, H. (1980): Identification of a new glutathione S-transferase in human liver, Acta Chem. Scand. (B), 34:607- 601.
13. Warholm, M., Guthenberg, C. and Mannervik, B. (1983): Molecular and catalytic properties of glutathione transferase μ from human liver: an enzyme efficiently conjugating epoxides, Biochemistry, 22:3610-3617.
14. Warholm, M., Guthenberg, C., Mannervik, B. and von Bahr, C. (1981): Purification of a new glutathione S-transferase (transferase μ) from human liver having high activity with benzo(a)pyrene-4,5-oxide, Biochem. Biophys. Res. Commun., 98:512-519.
15. Mannervik, B. and Guthenberg, C. (1981): Glutathione transferase (human placenta), Meth. Enzymol., 77:231-235.
16. Lowry, O.H., Rosebrough, N.J., Farr, A.L. and Randall, R.J. (1951): Protein measurement with Folin phenol reagent, J. Biol. Chem., 193:265-275.
17. Habig, W.H. and Jakoby, W.B. (1981): Assays for differentiation of glutathione S-transferases, Meth. Enzymol., 77:398-405.
18. Muhktar, H. and Bend, J.R. (1977): Serum glutathione S-transferases: perinatal development, sex difference, and effect of carbon tetrachloride administration on enzyme

activity in the rat, Life Sciences, 21:1277-1296.
19. Mannervik, B. (1982): Regression analysis, experimental error, and statistical criteria in the design and analysis of experiments for discrimination between rival kinetic models, Meth. Enzymol., 87:370-390.
20. Kitahara, A., Satoh, K., Nishimura, K., Ishikawa, T., Ruike, K., Sato, K., Tsuda, H. and Ito, N. (1984): Changes in molecular forms of rat hepatic glutathione S-transferase during chemical hepatocarcinogensis, Cancer Res., 44:2698-2703.
21. Jensson, H., Eriksson, L.C. and Mannervik, B. (1985): Selective expression of glutathione transferase isoenzymes in chemically induced preneoplastic rat hepatocyte nodules, FEBS Lett., 187:115-120.
22. Mannervik, B., Ålin, P., Guthenberg, C., Jensson, H., Tahir, M.K., Warholm, M. and Jörnvall, H. (1985): Identification of three classes of cytosolic glutathione transferase common to several mammalian species. Correlation between structural data and enzymatic properties, Proc. Natl. Acad. Sci. USA, in press.
23. Chang, R.L., Levin, W., Wood, A.W., Yagi, H., Tada, M., Vyas, K.P., Jerina, D.M. and Conney, A.H. (1983): Tumorigenicity of enantiomers of chrysene 1,2-dihydrodiol of the diastereomeric bay-region chrysene 1,2-diol-3,4-epoxides on mouse skin and in newborn mice, Cancer Res., 43:192-196.
24. Levin, W., Chang, R.L., Wood, A.W., Yagi, H., Thakker, D.R., Jerina, D.M. and Conney, A.H. (1984): High stereoselectivity among the optical isomers of the diastereomeric bay-region diol-epoxides of benz(a)anthracene in the expression of tumorigenic activity in murine tumor models, Cancer Res., 44: 929-933.
25. Yang, S.K., McCourt, D.W., Roller, P.P. and Gelboin, H.V. (1970): Enzymatic conversion of benzo(a)pyrene leading predominantly to the diol-epoxide r-7,t-8-dihydroxy-t-9,10-oxy-7,8,9,10-tetrahydrobenzo(a)pyrene through a single enantiomer of r-7,t-8-dihydroxy-7,8-dihydrobenzo(a)pyrene, Proc. Natl. Acad. Sci. USA, 73:2591-2598.
26. Thakker, D.R., Yagi, H., Akagi, H., Koreeda, M., Lu, A.Y.H., Levin, W., Wood, A.W., Conney, A.H. and Jerina, D.M. (1977): Metabolism of benzo(a)pyrene. VI. Stereoselective metabolism of benzo(a)pyrene and benzo(a)pyrene 7,8-dihydrodiol to diol-epoxides, Chem.-Biol. Interactions, 16:281-300.

ESTIMATION OF MUTAGENIC AND DERMAL CARCINOGENIC ACTIVITIES OF PETROLEUM FRACTIONS BASED ON POLYNUCLEAR AROMATIC HYDROCARBON CONTENT

T. A. ROY, S. W. JOHNSON, G. R. BLACKBURN, R. A. DEITCH,
C. A. SCHREINER, AND C. R. MACKERER
Mobil Environmental and Health Science Laboratory, Box 1029,
Princeton, New Jersey 08540.

INTRODUCTION

In vitro mutagenicity tests, such as the *Salmonella*/-Microsomal Activation Assay of Ames *et al*. (1) have been widely accepted as cost-effective, short-term screening methods for the evalutation of potential chemical carcinogenicity. Recent modifications to the standard *Salmonella* assay have permitted the testing of water-insoluble complex mixtures such as petroleum refinery streams (2). Blackburn *et al*. (3) were able to show excellent correlation between the mutagenic potency of 13 petroleum derived oils and their carcinogenic potency previously determined from dermal carcinogenicity bioassays. One of the major modifications to the standard *Salmonella* assay used by Blackburn *et al*. was the pre-assay enrichment of the polynuclear aromatic hydrocarbon (PAH) fraction of the oils into dimethyl sulfoxide (DMSO); this PAH-enriched extract can be efficiently delivered to the S-9 metabolizing enzymes and tester bacteria.

To date, a number of *Salmonella* assays and mouse skin-painting studies have shown that many individual PAH are mutagens and/or carcinogens exhibiting dose-response relationships (4-10). Results from similar studies on fossil fuel-derived mixtures indicate that it is the PAH components present that are responsible for the mixture's carcinogenic activity (11-14). This association has been supported by the observation that solvent refining, which quantitatively removes PAH from oils, eliminates carcinogenic activity (12,15-17). However, attempts at correlating the mutagenicity or carcinogenicity of these mixtures with levels of a specific, or group of specific PAH of known activity have met with limited success (18-23). It has been suggested that this lack of correlation may be due to the interaction of the PAH with the other components in the mixture or to a lack of bioavailability of the PAH from the mixture to the tester bacteria or to mixture pre-fractionation schemes which may selectively effect mutagenesis.

This study describes the isolation and characterization of the PAH[1] in 21 petroleum distillates (all with boiling points >500°F) which have been subjected to a variety of post-distillation treatments (Table 1). These materials were previously assayed in the modified Ames *Salmonella*/Microsomal Activation Assay in our laboratory. The sample preparation procedures employed followed, as closely as possible, those used by Blackburn *et al.* (24) in the modified Ames assay of the same materials. This was done in an attempt to assure that the profile and quantity of PAH determined analytically was identical to that supplied to the tester bacteria in the modified Ames assay. The results of these analyses were examined along with those obtained from the modified Ames *Salmonella*/Microsomal assays and those obtained in mouse skin-painting assays for 12 of the 21 petroleum distillate fractions.

EXPERIMENTAL

Petroleum distillate fractions

Table 1 describes the petroleum distillate fractions used in this study. All are distillation fractions produced either by standard refining techniques or laboratory-simulated refining techniques. A number of the samples were subjected to a variety of post-distillation treatments (*e.g.*, solvent-refining, hydrotreating, etc.) to produce lubricating base stocks. They contain no additives or other components not derived from the original crude oil. In each case, the material assayed was a sample of the same material tested in the modified Ames *Salmonella* microsomal mutagenesis assay and/or in a mouse skin-painting dermal carcinogenicity bioassay.

[1] The term PAH, as used in this paper, also encompasses those nitrogen-and sulfur-PAH isolated from the petroleum distillates during sample preparation.

TABLE 1

PETROLEUM DISTILLATE FRACTIONS ASSAYED IN MUTAGENICITY AND CARCINOGENICITY STUDIES AND ANALYZED FOR POLYNUCLEAR AROMATIC HYDROCARBON CONTENT

SAMPLE [a]	BOILING RANGE F	CAS NUMBER	CI [b]	MI [c]	WEIGHT PERCENT [d] TOTAL PAH	3-7 RING PAH
1. Aromatic Subfraction of a Naphthenic Distillate (Aromatics-C4)	550-770	---------	1.7	4.9	17	11
2. Heavy Naphthenic Distillate (C5)	700-1070	64741-53-3	2.0	7.1	7.8	7.6
3. Vacuum Residuum (D6)	>1070	64741-55-6	1.4	0	2.3	2.1
4. Re-refined oil	600-1000	68476-77-7	1.3	1.6	1.4	1.2
5. Light Paraffinic Distillate	600-900	64741-50-0	2.9	10	14	12
6. Hydrotreated Heavy Naphthenic Distillate	700-900	64742-52-5	1.8	5.9	8.8	7.9
7. Heavy Naphthenic Distillate Blend	700-900	64741-53-3	2.1	4.1	4.2	4.1
8. Solvent-Refined Heavy Paraffinic/Light Naphthenic Distillate Blend	600-900	64741-52-2	1.6	1.2	1.2	1.0
9. Heavy Naphthenic/ Solvent-Refined Heavy Paraffinic Blend	700-900	64741-88-4	1.6	2.1	1.5	1.5
10. Solvent-Refined - Hydrotreated Heavy Paraffinic Distillate	700-900	64742-54-7	1.1	0	0.8	0.7
11. Solvent-Refined - Hydrotreated Heavy Naphthenic Distillate	700-900	64742-52-5	NA [e]	0	1.2	1.1
12. Chemically Neutralized Hydrotreated Heavy Naphthenic Distillate	700-900	64742-52-5	1.9	4.5	4.9	4.6

Table 1 (continued)

13.Chemically Neutralized Hydrotreated Heavy Naphthenic Distillate	700-900	64742-52-5	ND[f]	5.1	5.3	4.7
14.Heavy Naphthenic Distillate 500"	600-1000	64741-53-3	ND	4.7	2.2	2.2
15.Heavy Paraffinic Distillate-Saturate Subfraction	700-1070	---------	ND	0	0.7	0.7
16.Hydrotreated Heavy Naphthenic	650-800	64742-52-5	ND	1.0	2.6	2.3
17.100" Mild Solvent-Refined Naphthenic	650-850	64741-96-4	ND	5.1	4.7	4.2
18.100" Naphthenic Distillate	650-800	64741-53-3	ND	11	8.0	7.5
19.2400" Heavy Naphthenic	>800	64741-53-3	ND	4.5	2.9	2.9
20.300" Hydrotreated Heavy Naphthenic	700-850	64742-52-5	ND	1.6	3.0	2.8
21.500" Hydrotreated Heavy Naphthenic	700-850	64742-53-3	ND	1.4	2.0	1.9

a) samples 1,3 and 15 were obtained from the American Petroleum Institute; the designations in parentheses are API code names for fractions prepared from crudes "C" (domestic, high-sulfur naphthenic) and "D" (foreign, low-sulfur, paraffinic; samples 1 and 15 are chromatographically separated subfractions selected to be representative of refinery fractions from two chemically distinct crude oils; samples 2 and 3 represent laboratory distillation cuts. b) CI = carcinogenicity index (see experimental for calculation)
c) MI = mutagenicity index = the slope of the linear portion of the mutagenic activity dose-response curve in the modified Ames assay (see experimental). d) weight percents (wt/wt) are the average of at least two samplings; 3-7 ring values exclude 1- and 2-ring aromatics as determined by GC/FID.
e) NA = not applicable; since this oil sample exhibited no significant biological response in the skin-painting study, no CI value is calculated. f) ND = not determined; skin-painting bioassays have not been performed for these samples.

Mutagenesis assays

The modified Ames *Salmonella*/microsomal mutagenesis testing procedure used to assay the materials in Table 1 is described in detail elsewhere (24). Briefly, modifications to the standard Ames assay are 1) the use of a DMSO extract of the oil sample as opposed to the neat or solvent-diluted oil sample, and 2) the use of an eight-fold increase in rat liver S-9. The extraction procedure consists of dissolving 2 ml oil in 3 ml cyclohexane and extracting the solution once with 10 ml of DMSO. These extracts are stored at 4°C until tested (<1 week). This extraction procedure is analogous to several solvent refining processes which are designed to selectively remove PAH from petroleum distillate fractions (25-26).

Each of the materials was tested across a range of doses that provided an initial linear dose response. The slopes (revertants/μl) of the ascending linear portions of the mutagenicity curves (determined by using appropriate Statistical Analysis System (SAS) procedures (SAS Institute, Inc., Cary, N.C.(27)) were taken as a measure of the relative mutagenic potencies of the samples tested. The slopes or mutagenicity indicies (MI) for the twenty-one samples assayed are shown in Table 1.

Dermal carcinogenicity assays

Dermal carcinogenicity assays were performed on samples 1-12 in Table 1 during the period from August 1979 to May 1984 at the Kettering Institute, Cincinnati, OH. Details of the testing procedure are described elsewhere (24). The following equation which considers both time to tumor and tumor incidence was used to calculate a carcinogenicity index (CI) for each sample:

$$CI = 1/\text{MEAN LATENT PERIOD} \times 100$$

Mean Latent Period is the mean time to tumor formation (weeks) in the study. The factor of 100 was used to minimize decimal indices. Blackburn *et al.* (24) have evaluated data from these dermal carcinogenicity assays for estimates of carcinogenic potency and found that the Mean Latent Period provides the best correlation with measured mutagenic potency. The CI values for samples 1-12 are shown Table 1.

Isolation of DMSO - extractable PAH (DEPAH)

A nominal 1-5 gram sample of the petroleum distillate fraction was accurately weighed in a tared 30-60 ml separatory funnel. The sample was then diluted with 6 ml of cyclohexane and fortified with anthracene-d10 (Aldrich 98+ atom %D) and perylene-d12 (Isotopes 98%) at the 100 part per million (ppm) level. The sample was immediately extracted with 2 x 10 ml DMSO. The combined DMSO extracts were diluted 2:1 with 40 ml distilled water and extracted with 3 x 30 ml pentane in a 125 ml separatory funnel. The pentane extracts were combined and the pentane evaporated under a gentle stream of dry nitrogen to approximately 5 ml in a graduated centrifuge tube. Approximately 0.5-1 ml of distilled water was added to the concentrate which was then thoroughly mixed and centrifuged. The water layer (containing any residual DMSO) was removed with a Pasteur pipet and the remaining pentane layer passed through a small column of methylene chloride-washed anhydrous sodium sulfate (Fisher, anhydrous, granular) and collected in a pre-weighed 12 ml graduated centrifuge tube. The 40 ml centrifuge tube was washed with 2 x 2 ml pentane and the rinsings passed through the sodium sulfate column. The column was rinsed with an additional 2 ml of fresh pentane. All the rinsings were collected in the 12 ml graduated centrifuge tube and concentrated to 2 ml in preparation for GC/FID and GC/MS analysis.

Following GC/FID and GC/MS analysis, the DEPAH concentrates were immersed in a 35-40°C water bath and evaporated to dryness with dry nitrogen. The residues were weighed after allowing the tubes to equilibrate to room temperature. The extracts were then reconstituted with 2 ml pentane and reanalyzed by GC/FID. The overall efficiency of PAH extraction and recovery was determined by a comparison of the integrated areas of the internal standards in the extracts with a prepared standard of the same nominal concentration. All samples were prepared in duplicate in an area free from direct flourescent lighting and stored at 4°C pending analysis.

Analysis of DMSO - extractable PAH (DEPAH)

The DEPAH were analyzed both by GC/FID and GC/MS. The GC/FID used was a Hewlett-Packard model 5880A equipped with a J&W (Rancho Cordova, CA) 30 meter x 0.32 mm i.d. DB-5 (0.25 µm film thickness) fused silica capillary column. The oven temperature was held at 100°C for 5 min and then programmed to 300°C at a rate of 3°C/min and held at the final temperature for 20 min. A split mode of injection (50:1) was used. Other general conditions of analysis were as follows: He carrier gas, 35-40 cm/sec; detector 320°C, injection port 320°C. The GC/FID chromatograms of the extracts were integrated in the standard peak integration mode and in the slice mode. A PAH standard containing naphthalene, phenanthrene, pyrene, benzo[a]pyrene, benzo[ghi]perylene and coronene was used to define the boundaries of the retention time windows for PAH containing two- through seven-rings, respectively. The GC/FID chromatogram of a high percentage PAH petroleum distillate extract is shown in figure 1 with an overlay of the PAH standard.

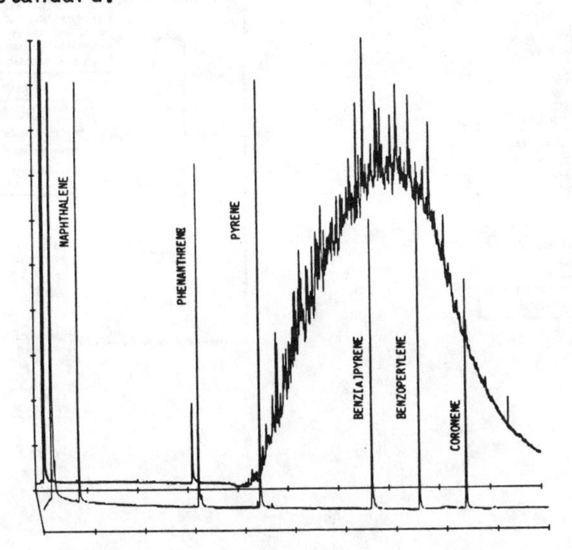

FIGURE 1. GC/FID chromatogram of a high percentage PAH petroleum distillate extract with overlay of PAH standard chromatogram.

The composition of the DEPAH in terms of the percentage of two- through seven-ring PAH, was determined using a BASIC algorithm and the retention windows defined by the PAH standard. The flowchart for the algorithm is shown in figure 2.

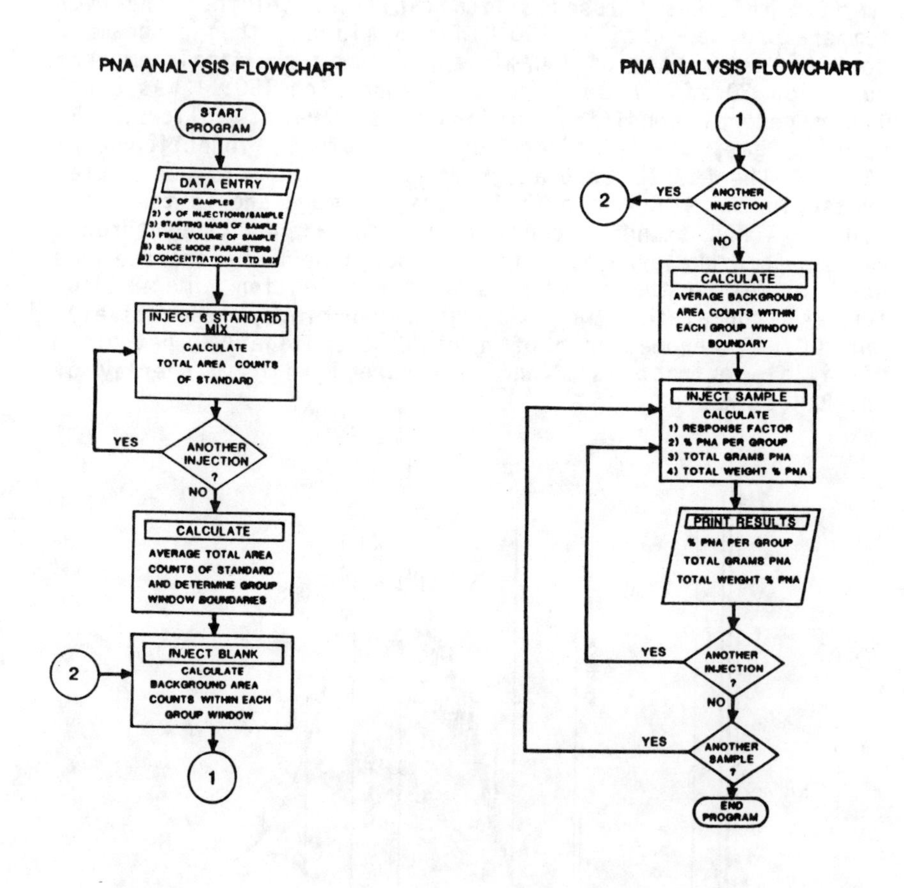

FIGURE 2. Flowchart for BASIC algorithm used to determine weight percent two-through seven-ring PAH contained in petroleum distillates.

The GC/MS used was a Hewlett-Packard 5985B GC/MS system with a J&W 30 meter x 0.32 mm i.d. DB-5 (0.25 μm film thickness) fused silica capillary column operated in the split mode (50:1). The oven temperature profile was identical to that used for GC/FID analysis. Helium was used as the carrier gas at 10 psi head pressure and the column was directly coupled to the ion source of the mass spectrometer. The mass spectrometer was operated in the electron impact mode scanning from 50 to 350 amu. Other general conditions of analysis were as follows: scan speed 3 sec/ 300 amu; GC injection port 300°C, MS source 200°C. The mass spectra of the individual peaks in the mass chromatogram of the DEPAH were compared to GC/MS computer mass spectra library using an automated search program, ALIBII.

RESULTS

PAH profile of petroleum distillate fractions

The 3-7 ring PAH content (weight percent) of the DEPAH of the 21 petroleum distillate fractions assayed is shown in Table 1. The values are derived from duplicate gravimetric weight determinations of the DEPAH followed by both GC/FID and GC/MS characterization of the 1- through 7-(fused) ring aromatic content.

Correlation between mutagenicity and carcinogenicity of oil samples

The mutagenic activity measured by the modified Ames test and expressed as a mutagenicity index (MI) and the carcinogenic activity index (CI) derived from the mouse skin-painting studies (see experimental section) for the oil samples are shown in Table 1. The rank-order correlation between the mutagenicity and carcinogenicity indices is 0.84 for the eleven oil samples tested that elicited a carcinogenic response. A direct measure of the correlation between mutagenic and carcinogenic activities for these same eleven samples shows a correlation of 0.91.

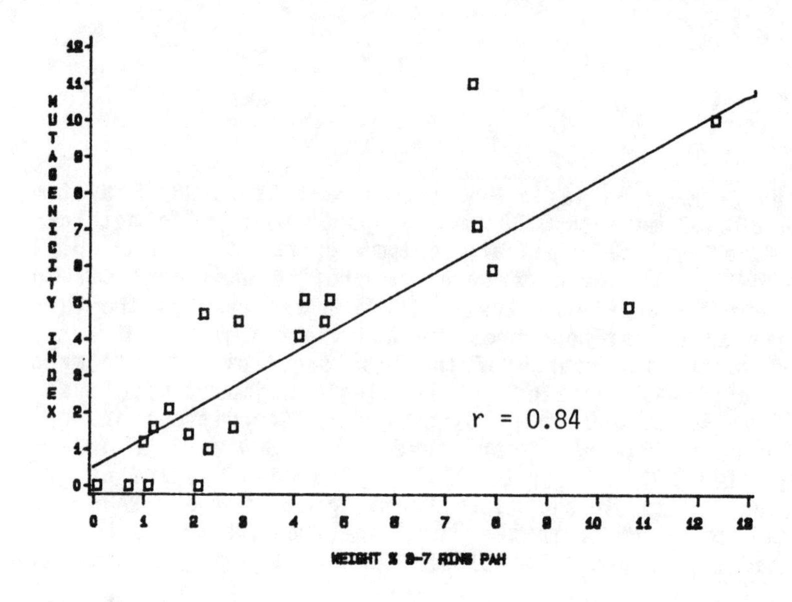

FIGURE 3. Direct Correlation of 3-7 Ring PAH Content and Mutagenicity Indices for 21 Oil Samples.

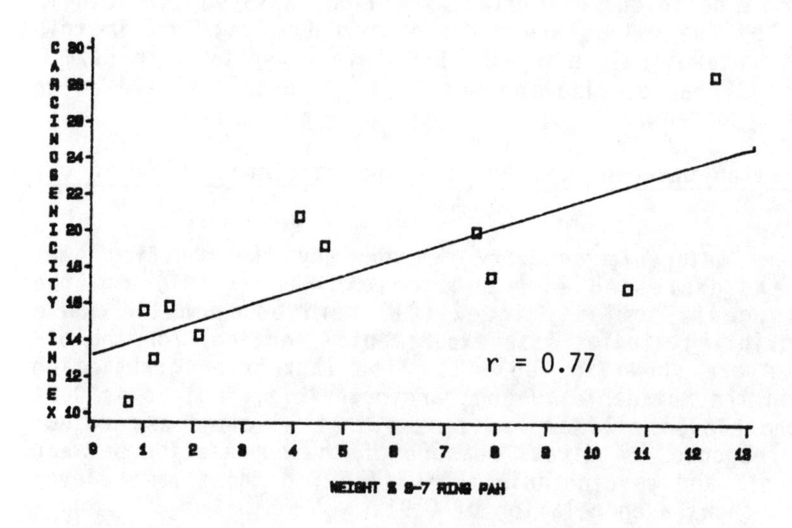

FIGURE 4. Direct Correlation of 3-7 Ring PAH Content and Carcinogenicity Indices for 11 Oil Samples.

Correlation between PAH content and mutagenicity of oil samples

The correlation coefficient for the ranked total PAH content versus the ranked mutagenicity indices for the 21 oil samples is 0.85; the rank-order correlation does not change significantly when only the weight of the 3-7 ring PAH ($r = 0.88$) in the oil samples are considered. A direct measure of the correlation between the total weight of the PAH fractions and MI shows a correlation of $r = 0.72$; this value increases to 0.84 when only the weight of 3-7 ring PAH in the fractions is considered. Figure 3 shows a plot of 3-7 ring PAH content versus MI with the least squares regression line for the 21 oil samples.

Correlation between PAH content and carcinogenic potency of oil samples

The correlation coefficient for the total PAH content versus the ranked carcinogenicity for the eleven oil samples that elicited a carcinogenic response is 0.74; the rank order correlation increases slightly to 0.78 when only the 3-7 ring PAH in the oil samples are considered. A direct measure of the correlation between the total weight of the PAH fractions and their carcinogenicity indices ($n=11$) shows a correlation of 0.62; this value increases to 0.77 when only the weight of 3-7 ring PAH in the fraction is considered. Figure 4 shows a plot of 3-7 ring PAH versus CI with the least squares regression line for the eleven oil samples.

DISCUSSION

Blackburn *et al.* (3,24) have shown that the modified Ames *Salmonella* microsomal mutagenesis assay permits reliable estimations of the mutagenic potency of mineral oils. They have further shown that the mutagenic potencies of petroleum distillates ranging from light fuel oil (>500°F) to vacuum residue (>1070°F) are highly correlated with their corresponding carcinogenic potencies as determined by mouse skin painting studies. The mutagenicity (modified Ames) of petroleum fractions with a mean boiling point less than 500°F does not correlate with their carcinogenic potencies which suggests that the carcinogenicity of materials in the kerosine boiling range or lower may be mediated by mechanisms (3,24) different from those of the higher boiling oils. Previous studies have indicated that the tumorigenic activity of petroleum distillation fractions is associated with the presence of PAH (14,18-19). Indeed, the excellent correlation (3,24) between mutagenicity and carcinogenicity obtained in this laboratory prompted the present study to evaluate the relationship between the PAH content of the oils tested and their measured mutagenic and carcinogenic potencies. PAH analysis of several kerosines and naphthas (mean boiling point <500°F) using the procedures described in this study shows poor correlation between PAH content and carcinogenic potency. Furthermore, analysis of several petroleum fractions boiling over 1070°F also shows poor correlation between PAH content and both mutagenic and carcinogenic potency. Thus, only samples boiling in the range of 500-1070°F have been included in the PAH correlations; the PAH content of samples outside of this range may not be a valid indicator of their mutagenic or carcinogenic potency.

It is evident that a good estimate of mutagenic potency ranking can be made for the group of 21 oils based on their total weight percent PAH content ($r = 0.85$). The correlation decreases when a direct comparison is made between total PAH and the measured mutagenic potency ($r = 0.72$); when only the weight percent of 3-7 ring PAH are considered, the direct comparison correlation increases to 0.84 (re: Figure 3). The correlation for the direct comparison of PAH content with carcinogenic potency also increases from 0.62 to 0.77 (re: Figure 4) when only the weight percent of 3-7 ring PAH are considered. These direct comparisons provide a more exacting and quantitative comparison of the data. Because the carcinogenicity of PAH has mainly been observed with tri-, tetra-, penta-, and hexacyclic compounds, it is believed that the increased correlation observed when considering only 3-7 fused ring compounds is a strong indication that these groups of PAH are responsible for the mutagenic and carcinogenic activity of the oil extracts. Interestingly, both the PAH versus mutagenicity and carcinogenicity correlations decrease when only the 4-7 ring PAH content is considered indicating that the 3-fused-ring PAH make a significant contribution to the observed activity.

Blackburn (24) showed that only one of 22 oils tested in both the modified Ames assay and carcinogenicity bioassay was misclassified by the modified Ames mutagenicity test. A review of the mutagenicity and carcinogenicity data included in the present study indicates that oils containing more than 1% percent of PAH with 3-7 fused rings are likely to elicit definitive positive responses in the modified Ames Salmonella/microsomal mutagenesis and mouse skin-painting carcinogenicity assays. Chi-square analysis of a PAH/skin-painting contingency table for samples 1-12 (cutoffs: 1.0% 3-7 ring PAH and n=2 tumors) shows significance at the 90% level ($0.5 < P < 0.1$). However, when the 1.0 percent PAH level is used by itself to predict the carcinogenic potential of the 12 samples, one of the samples is misclassified as non-carcinogenic (false negative); the modified Ames test correctly predicts (at an MI = 1.0 cutoff) the sample to be a carcinogen.

In conclusion, this paper presents new data which correlates the content of 3-7 ring PAH of mineral oils boiling between 500° and 1070°F with both mutagenic and carcinogenic activity. Oils containing less than 1.0% 3-7 ring PAH by weight were inactive while oils containing more than 1.0% were usually carcinogenic and mutagenic. These new findings support the concept that PAH are responsible for both mutagenic and carcinogenic activities of heavier distillate fractions and also serve to back-up the predictions made from the modified Ames test. Thus, oils which give a positive response in the modified Ames test and contain more than 1.0% 3-7 ring PAH would be very likely to elicit a carcinogenic response in a mouse bioassay. However, oils giving a negative response in the modified Ames test and containing less than 1.0% PAH would be very unlikely to be carcinogenic. Samples showing discordant results between the two tests would be classified as borderline or be retested. The use of both of these tests substantially increases our ability to reliably predict carcinogenic activity over the use of either test alone.

ACKNOWLEDGEMENTS

The authors thank Debra Greta for assistance in the preparation of this manuscript.

REFERENCES

1. Ames, B.N., McCann, J. and Yamasaki, E. (1975): Methods for detecting carcinogens and mutagens with the Salmonella microsome mutagenicity test, Mutat. Res., 31: 347-363.
2. Blackburn, G.R., Deitch, R.A., Schreiner C.A. and Mackerer, C.R. (1983): Modification of the Ames Salmonella/microsome assay for testing complex hydrocarbon mixtures., Environ. Mutagen., 5: 401.
3. Mackerer, C.R., Blackburn, G.R. Deitch, R.A., Dooley, J.F., Schreiner, C.A. and Mehlman, M.A., Correlation of In Vivo Mutagenicity with Dermal Carcinogenicity for Complex Petroleum Derived Mixtures, presented at the Skin Carcinogenic Symposium, Centre Henri Becquerel, Rouen, France, June 20, 1984.
4. Blackburn, G.R., Deitch, R.A., Schreiner, C.A., Mehlman, M.A. and Mackerer, C.R. (1984): Estimation of the dermal carcinogenic activity of petroleum fractions using a modified Ames assay, Cell Biol. Toxicol., 1: 67-80.
5. Coombs, M.M., Dixon, C. and Kissonerghis, A.M. (1976): Evaluation of the mutagenicity of compounds of known carcinogenicity, belonging to the benz[a]anthracene, chrysene, and cyclopenta[a]phenanthrene series, using Ames test, Cancer Res., 36: 4525-4529.
6. McCann, J., Choi, E., Yamasaki, E. and Ames, B.N. (1975): Detection of carcinogens as mutagens in the Salmonella microsome test: assay of 300 chemicals, Proc. Natl. Acad. Sci. U.S.A., 72: 5135-5139.
7. Andrews, A.W., Thibault, L.H. and Lijinsky, W. (1978): The relationship between carcinogenicity and mutagenicity of some polynuclear hydrocarbons, Mutat. Res., 51: 311-318.
8. Rinkus, S.J. and Legator, M.S. (1979): Chemical characterization of 465 known or suspected carcinogens and their correlation with mutagenic activity in the Salmonella typhimurium system, Cancer Res., 39: 3289-3318.
9. Purchase, I.F.H., Longstaff, E., Ashby, J., Styles, J.A., Anderson, D., LeFevre, P.A. and Westwood, F.R. (1976): Evaluation of six short-term tests for detecting organic chemical carcinogens and recommendations for their use, Nature (London), 264: 624-627.

10. Purchase, I.F.H., Longstaff, E., Ashby, J. Styles, J.A., Anderson, D. LeFevre, P.A. and Westwood, F.R. (1978): An evaluation of six short-term tests for detecting organic chemical carcinogens, Br. J. Cancer, 37: 873902.
11. Bingham, E., Horton, A.W. and Tye, R. (1965): The carcinogenic potency of certain oils, Arch. Environ. Health, 10: 449-451.
12. Bingham, E. and Horton, A.W. (1969): Environmental carcinogenesis: experimental observations related to occupational cancer. Advances in Biology of the Skin, Vol. III, pp. 183-193, Pergamon Press, Elmsford, New York.
13. Grimmer, G., Dettbarn B., G., Brune , N., Deutsch-Wenzel, R. and Misfeld, J. (1982): Quantification of the carcinogenic effect of polycyclic aromatic hydrocarbons in used engine oil by topical application on the skin of mice, Int. Arch. Occup. Environ. Health, 50: 95-100.
14. Gradiski, D., Vinot, J. Zissu, D., Limasset, J.C. and LaFontaine, M. (1983): The carcinogenic effect of a series of petroleum-derived oils on the skin of mice, Environ. Res., 32: 258-268.
15. IRAC (1984) IRAC Monographs on the Evaluation of the Carcinogenic Risk of Chemicals to Humans. Polynuclear Aromatic Compounds, Vol. 33, Mineral Oils, Lyons, pp. 87-168.
16. Halder, C.A., Warne, T.M., Little, R.Q. and Garvin, P.J. (1984): Carcinogenicity of petroleum lubricating oil distillates: effects of solvent refining, hydroprocessing, and blending, Am. J. Ind. Med., 5: 265-274.
17. Kane, M.L., Ladov, E.N., Holdsworth, C.E. and Weaver, N.K. (1984): Toxicological characteristics of refinery streams used to manufacture lubricating oils, Am. J. Ind. Med., 5: 183-200.
18. Holland, J.M., Rahn, R.O., Smith, L.H., Clark, B.R., Chang, S.S. and Stephens, T.J. (1979): Skin carcinogenicity of synthetic and natural petroleums, J. Occup. Med., 21: 614-618.
19. Cragg, S.T., Conaway, C.C. and MacGregor, J.A. (1985): Lack of concordance of the Salmonella/microsome assay with the mouse dermal carcinogenesis bioassay for complex petroleum hydrocarbon mixtures, Fund. Appl. Toxicol., 5: 382-390.
20. Carver, J.H., MacGregor, J.A. and King, R.W. (1984): Mutagenicity and chemical characterization of two petroleum distillates, J. Appl. Toxicol., 4: 163-169.

21. Macgregor, J.A., Conaway, C.C. and Cragg, S.T. (1982): Predicitivity of the Salmonella/microsome assay for carcinogenic and noncarcinogenic complex petroleum hydrocarbon mixtures. In: The Toxicology of Petroleum Hydrocarbons, edited by H.N. MacFarland, C.E. Holdsworth, J.A. MacGregor, R.W. Call and M.L. Kane, pp. 149-161, American Petroleum Institute, Washington, D.C.
22. Epler, J.P. (1980): The use of short-term tests in the isolation and identification of chemical mutagens in complex mixtures. In: Chemical Mutagenesis: Principles and Methods for Their Detection, Volume 6, edited by F.J. DeSerra and A. Hollander, pp. 239-270 Plenum Press, New York and London.
23. Epler, J.L., Clark, B.R., Ho, C., Guerin, M.R. and Rao, T.K. (1978): Short-term bioassay of complex organic mixtures: part II, mutagenicity testing. In: Applications of Complex Environmental Mixtures: Environmental Science Research, Volume 15, edited by M.D. Walters, S. Nesnow, J.L. Huisingh, S.S. Sandhu and L. Claxton, pp. 269-289, Plenum Press, New York and London.
24. Blackburn, G.R., Deitch, R.A., Schreiner, C.A. and Mackerer, C.R. (1986): Predicting tumorigenicity of petroleum distillation fractions using a modified Salmonella mutagenicity assay, Cell Biol. Toxicol., 2(1): 63-84.
25. The Petroleum Handbook. Sixth Edition 264-266 (1983). Elsevier Science Publishing Company. Inc., New York, N.Y.
26. Speight, J.G. (1980): The Chemistry and Technology of Petroleum, Chapter 5 (1980). J.G. Marcel Dekker, Inc., New York, N.Y.
27. Freund, R.J. and Littell, R.C.)(1981): Regression. SAS for Linear Models, pp. 9-45, SAS Institute, Inc., Cary .C.

CYCLOPENTA-FUSED PAH ISOMERS OF CATA-ANNELATED BENZENOID SYSTEMS

R. SANGAIAH AND AVRAM GOLD
Department of Environmental Sciences and Engineering, School of Public Health, University of North Carolina at Chapel Hill, Chapel Hill, NC 27514

INTRODUCTION

Cyclopenta-fused polycyclic aromatic hydrocarbons (PAH) are a unique class of PAH present in the environment. The stabilization energies ($\Delta E_{deloc}/\beta$) derived from the corresponding ring-opened cyclopenta epoxides, which are predictors of biological activity, are in general larger than those of other peripheral arene oxide derivatives. The predicted high level of activity for many members of this class compounds prompted us to investigate their metabolism to elucidate mechanisms of activation and to determine structure-activity relationships for active metabolites.

We have already synthesized and studied the biological activity of aceanthrylene (1,2,3), acephenanthrylene (2,3) and some of their benz-annelated derivatives (4-6). We concluded from these studies that oxidation on the cyclopenta-ring is a primary pathway for metabolic activation of most of the above compounds. In our continuing studies on cyclopenta-fused PAH we undertook synthesis of benz(d)aceanthrylene (I) and benz(k)aceanthrylene (II) because these two isomers have among the largest $\Delta E_{deloc}/\beta$ values (-1.087) theoretically possible for isomeric five-ring PAH. We have also synthesized and characterized benz(j)acephenanthrylene(III), a previously unknown non-alternant PAH. We wished to investigate the effect of cyclopenta ring fusion on the bay region activity for this compound.

MATERIALS AND METHODS

Synthesis of Benz(d)aceanthrylene (BdAA, I)

1,2,3,4-Tetrahydronaphthacene-6,11-quinone. This quinone was prepared in two steps from tetralin by published procedures (7,8).

1,2,3,4-Tetrahydronaphthacene. The above quinone (4.0 g) was heated at reflux with HI (57%, 15 mL) and glacial acetic acid (185 mL) for 6h. The reaction mixture was poured into 1% sodium bisulfite solution, and the precipitate, collected by filtration, was purified by recrystallization from benzene to give colorless needles of tetrahydronaphthacene (3.37 g, 95%); mp 236°.

5,6-Dioxo-1,2,3,4,5,6-hexahydrobenz(d)aceanthrylene. A mixture of tetrahydronaphthacene (1.3 g), oxalyl chloride (2.5 g), CS_2 (8 mL) and $AlCl_3$ (0.8 g) was stirred at 0° for 2 h. Further addition of $AlCl_3$ (0.7 g) and CS_2 (8 mL) was followed by stirring for 4 h at 0° and overnight at room temperature. The reaction mixture was poured into ice-water, heated to distill the CS_2, cooled and extracted with $CHCl_3$. Evaporation of solvent followed by purification of the crude product by column chromatography on silica gel with $CHCl_3$ eluant gave the red colored pure product (1.31 g, 82%); mp 248-250° (dec.).

5,6-Dioxo-1,2,3,4,5,6-hexahydrobenz(d)aceanthrylene. The above diketone (1.20 g) was reduced with sodium borohydride (3.80 g) in ethanol (260 mL) and water (45 mL) by stirring the reaction mixture overnight at room temperature. Work-up afforded the diol (1.21 g, 100%) which was used directly in the next step.

5-Oxo-1,2,3,4,5,6-hexahydrobenz(d)aceanthrylene (IV). A mixture of diol (1.2 g) and p-toluenesulfonic acid (120 mg) in dry benzene (150 mL) was heated under reflux for 2 h. The benzene solution was washed with brine (2 x 100 mL), dried (Na_2SO_4) and evaporated to give the crude ketone which was purified by chromatography on alumina with benzene to obtain the yellow crystalline solid IV (0.80 g, 71%); mp. 189-190° (dec.).

1,2,3,4,5,6-Hexahydrobenz(d)aceanthrylene. A mixture of ketone IV (636 mg), diethylene glycol (30 mL), hydrazine monohydrate (1 mL) and KOH (1 g) was refluxed for 6 H, cooled to room temperature, and poured into excess water. The product was extracted with CH_2Cl_2 (2 x 50 mL) and the organic extract washed with water (100 mL), and dried (Na_2SO_4). The crude product, obtained after evaporation of solvent, was purified by chromatography on silica gel with chloroform-benzene (3:1) to give hexahydro I (420 mg, 70%), mp 142-143°.

Benz(d)aceanthrylene (I). The hexahydro compound (182 mg) was dehydrogenated by refluxing with DDQ (480 mg) in dry benzene (100 mL) under nitrogen for 2.5 h. The cooled, filtered solution was passed through an alumina column and evaporated. The dark purple residue was chromatographed on alumina with benzene-hexane (1:9) eluant and BdAA collected as a violet, non-fluorescent band (103 mg, 58%). mp >300°. For UV and ^{1}H NMR, see Fig. 1.

Synthesis of Benz(k)aceanthrylene (BkAA, II)

Diethyl 2-(1-indanyl)succinate. The triester V (27 g), obtained by condensation of 1-bromoindane (9) with triethyl 1,1,2-ethanetricarboxylate, was refluxed with NaCl (8.2 g), water (4 g) and DMSO (160 mL) at 178-183° for 6 h. The resulting diester was purified by chromatography on alumina with benzene (18.40 g, 85%).

2a, 3-Dihydroacenaphthene-3,4-dicarboxylic Acid diethyl ester. To a suspension of powdered sodium (1.37 g) in dry ether (25 mL) was added a solution of absolute ethanol (3.46 mL) in dry ether (15 mL) and the mixture refluxed for 10 h. The reaction mixture was cooled to -10° to -15° and a solution of the above diester (8 g) and ethyl formate (4.7 mL) in dry ether (20 mL) added with vigorous stirring. The mixture was stirred for 4 h at -10° and 72 h at room temperature. The reaction mixture was added to ice-water (200 mL) and the aqueous layer extracted with ether to remove the unreacted diester. After acidification with dilute H_2SO_4, the aqueous layer was extracted with ether (3 x 200 mL). Evaporation of the ether gave the formyl derivative (4.814 g, 55%) which was used directly in the next step.

A mixture of 90% H_3PO_4 (17.3 mL) and 98% H_2SO_4 (3.7 mL) cooled to -10° was added to the formyl derivative (4.814 g) precooled to -10°. The reaction mixture was allowed to warm to 0° with stirring which was continued for 2 h at 0 - 10°. The reaction mixture was poured into ice-water (100 mL), partially neutralized with NaOH (40%, 52 mL) (with cooling), and the product extracted into ether (3 x 100 mL). The ether extract was washed with water, dried and evaporated to give dihydroacenaphthene dicarboxylic ester (4.313 g, 95%).

Acenaphthene-3,4-dicarboxylic acid. Diester (4.30 g) obtained in the previous step was refluxed with DDQ (3.25 g) in dry benzene (100 mL) for 2 h. After filtration, the reaction mixture was chromatographed on alumina with benzene as eluant to afford pure acenaphthene-3-4-dicarboxylic ester (3.76 g, 88%) mp 94-95°.

The diester (3.60 g) was hydrolyzed by refluxing with KOH (45%, 40 mL) and methanol (160 mL) for 2h to give the dicarboxylic acid (2.8 g, 96%), mp 244-246° (remelts and dec. 255-256°).

Acenaphthene-3,4-dicarboxylic acid anhydride. The 3,4-diacid (2.76 g) was refluxed with acetic anhydride (20 mL) for 2 h. Evaporation of the acetic anhydride under vacuum yielded the anhydride (2.56 g, 100%), mp 256-257° (dec.).

1,2-Dihydrobenz(k)aceanthrylene-7,12-quinone (VI). A mixture of the anhydride (2.54 g) and anhydrous $AlCl_3$ (3.14 g) in dry benzene (10 mL) was heated at 100° for 8 h. Work-up afforded a mixture of keto acids (2.91 g, 85 %) which was used directly in the next step.

The keto acids (2 g), $AlCl_3$ (20 g) and NaCl (4 g) were heated and the resulting melt stirred vigorously at 130-140° for 1 h. The reaction mixture, cooled to room temperature, was treated with ice-water (200 mL) and extracted with $CHCl_3$ (3 x 100 mL). Evaporation of the solvent followed by purification of the crude product by chromatography on alumina with $CHCl_3$ gave pure quinone VI (1.03 g, 55%), mp 225-226° (dec.).

1,2,7,12-Tetrahydrobenz(k)aceanthrylene. The quinone (1.0 g) in ethanol (100 mL) was stirred with NaOH (20%, 10 mL) and zinc dust (5 g) under reflux for 6 h. The reaction mixture was cooled to room temperature, filtered and, after distillation of ethanol, the aqueous filtrate was extracted with benzene. The benzene extract was washed with water, evaporated to give tetrahydro BkAA (0.77 g, 85%) which was used directly in the next step.

Benz(k)aceanthrylene (II). A solution of the tetrahydro compound (174 mg) and DDQ (308 mg) in dry benzene (100 mL) was refluxed under nitrogen for 2 h. The reaction mixture was cooled and flash chromatographed on alumina with benzene. The solution was evaporated to dryness and the purple solid obtained was rechromatographed on alumina with heptane-benzene (8:2). The violet, non-fluorescent fraction afforded benz(k)aceanthrylene (0.108 g, 63%); mp >300°; for UV and ^{1}H NMR, see Fig. 2.

Synthesis of Benz(j)acephenanthrylene (BjAP, III)

7-Oxo-4,5,7,8,9,10-hexahydroacephenanthrylene. This ketone was prepared in three steps from acenaphthene (Aldrich Chemical Co.), as described by Krishnan and Hites (10).

Ethyl (7-acephenanthryl)acetate. Reformatsky reaction on the above ketone (3.0 g) with Zn dust (7.5 g) and ethyl bromoacetate (3 mL) afforded a β-hydroxy ester which was dehydrated by refluxing with p-toluenesulfonic acid (150 mg) in dry benzene (300 mL) for 2 h to a mixture of exo- and endocyclic olefinic esters (3.35 g, 85%), which were used directly in the next step.

The olefinic esters (3.35 g) and DDQ (2.72 g) were refluxed in dry benzene (400 mL) for 2 h, cooled to room temperature, and filtered. The filtrate was chromatographed on alumina with benzene eluant to yield pure ethyl (7-acephenanthryl)acetate (2.86 g, 86%).

4-(7-Acephenanthryl)butanoic acid (VIII). The acephenanthryl ester (2.80 g) in dry ether (100 mL) was added to a stirred suspension of LAH (0.92 g) in dry ether (150 mL) to obtain the hydroxy compound (2.27 g, 95%) after work-up.

The reduction product (2.25 g) was dissolved in dry benzene (10 mL) and treated with PBr_3 (1.8 g) in dry benzene (10 mL). After stirring overnight at room temperature, the reaction mixture was poured into ice-water (100 mL). The benzene layer was separated, washed with brine, dried and evaporated to give the bromide VII (2.31 g, 82%) which was used directly in the next step.

Absolute alcohol (2 mL) was added to a suspension of sodium (0.4 g) in dry benzene (5 mL), the reaction mixture refluxed to dissolve the sodium and then cooled to room temperature. After adding ethyl malonate (4.22 mL), the reaction mixture was refluxed for 0.5 h and cooled. The bromide VII was added and the reaction mixture refluxed for 15 h to yield the malonic ester derivative (2.55 g, 88%).

The malonic ester derivative was hydrolyzed by refluxing with conc. HCl (40 mL) and glacial acetic acid (10 mL) for 16 h. The hydrolysis mixture was diluted with ice-water (100 mL) and the precipitated (acephenanthryl)butanoic acid (VIII) (1.61 g, 85%), mp 123-125°, was filtered and dried.

10-Oxo-4,5,7,8,9,10-hexahydrobenz(j)acephenanthrylene (IX). Acephenanthryl butanoic acid (1.45 g) was stirred with anhydrous HF (100 mL) for 15 h at room temperature. The product was purified by chromatography on alumina with benzene to furnish pure yellow ketone IX (1.18 g, 87%), mp 212-213° (dec.).

Benz(j)acephenanthrylene (BjAP, III. Ketone IX (688 mg) was reduced with sodium borohydride (600 mg) in THF (70 mL) and MeOH (36 mL) to give the colorless solid alcohol (675 mg, 97%), mp 189-191°.

The alcohol was refluxed with p-toluenesulfonic acid (25 mg) and dry benzene (100 mL) for 2 h. The benzene solution was washed with brine (2 x 100 mL), dried (Na_2SO_4), and evaporated to give tetrahydro BjAP (580 mg, 92%).

The tetrahydro derivative (512 mg) and DDQ (1.0 g) were refluxed in dry benzene (50 mL) for 2 h, cooled at room temperature, and filtered. The filtrate was chromatographed on alumina with benzene eluant to yield BjAP (433 mg, 86%), mp 170-171° (hexane); for UV and ^{1}H NMR, see Fig. 3.

All the intermediate compounds were characterized by their ^{1}H NMR spectra which were obtained at 250 MH_z, on a Bruker WM 250 Spectrometer. Mass spectra were obtained on a VG7070F Micromass Mass Spectrometer with an electron impact source at 70 eV, and melting points (uncorrected) were done on a Fisher-Johns melting point apparatus.

RESULTS AND DISCUSSION

The benz-annelated derivatives of aceanthrylene and acephenanthrylene I-III have been synthesized for the first-time and synthetic routes are described in Schemes 1 - 3. For the synthesis of BdAA (Scheme 1), we initially attempted Friedel-Crafts condensation of oxalyl chloride with naphthacene itself. As this reaction gave only naphthacene-5-carboxylic acid, we undertook the same condensation on 1,2,3,4-tetrahydronaphthacene because oxalyl chloride had been reported to condense at the peri positions of the anthracene nucleus to yield a cyclopenta-fused derivative. The structure of the resulting diketone was assigned from its ^{1}H NMR spectrum which has two aromatic singlets at δ 7.90 and 8.55 for H_{12} and H_{11} and a down-field doublet at δ 8.92 for the peri-hydrogen (H_7) adjacent to the five-membered ring. The mono-ketone IV was obtained by a previously described sequence of reactions (11,12). Wolff-Krishner reduction of IV gave hexahydro-BdAA which was dehydrogenated with 3 mole-equivalents of DDQ to give the desired PAH, BdAA. The mass spectrum was consistent with composition and aromatic character of BdAA, while its structure was confirmed by ^{1}H NMR (Fig. 1). The required C_2 symmetry is evident in the appearance of the cyclopenta ring olefinic protons (H_5, H_6) and both <u>meso</u> protons (H_{11}, H_{12}) as two-proton singlets at 7.63 and 8.75, respectively.

For the synthesis of BkAA (Scheme 2), we envisaged quinone (VI) as the potential intermediate. Since the products of condensation of phthalic anhydride with acenaphthene or its 5-substituted derivatives failed to give the required BkAA skeleton, we approached the synthesis through the intermediate acenaphthene-3,4-dicarboxylic acid anhydride which we prepared according to Scheme 2. Friedel-Crafts reaction of the anhydride with benzene afforded a keto-acid mixture which, on cyclization in an $AlCl_3$-NaCl melt gave the desired quinone (VI). The quinone, on reduction with Zn and NaOH followed by dehydrogenation of the product with 2 mole-equivalents of DDQ, furnished BkAA. The structure of BkAA was assigned from its mass spectrum and ^1H NMR spectrum, characterized by the presence of three aromatic singlets at δ8.94, 8.91 and 9.07 for H_6, H_7 and H_{12}, an AX quartet for the olefinic protons (H_1 and H_2) at δ 7.90 (H_1) and 7.18 H_2), and the lack of a true bay region feature reflected in absence of the typically low-field resonances below 9 ppm.

Scheme 3 provides a straight-forward approach to pure BjAP using a known ketone (10) as the starting material. A two-carbon unit was introduced on the periperhy at C_7 by a Reformatsky reaction with ethyl bromoacetate and the resulting β-hydroxyester was converted to the bromoethylacephenanthrene derivative (VII) by the four step sequence described in Scheme 4. A second two-carbon unit was added by diethyl malonate condensation with bromide VII. Cyclization of acid VIII with anhydrous HF afforded the ketone IX which was converted to BjAP by a reduction, dehydration and dehydrogenation sequence in overall 26% yield. ^1H NMR spectrum of BjAP shows a two proton singlet for the olefinic cyclopenta protons (H_4 and H_5) at δ7.25 resulting from accidental magnetic equivalence. Of four bay region protons: a singlet at δ 9.03 is assigned to H_6; a doublet at δ8.72 to H_{12} and two doublets at δ8.86 and 8.53, broadened by long-range couplings, to H_1 and H_7, which have not at present been distinguished.

The physico-chemical properties of these three cyclopenta-fused PAH are of interest, as well as their potential biological activity. Like other cyclopenta-PAH (4,13,14), the final products do not fluoresce under long-wavelength UV light (360 nm) and are highly colored. Compounds I and II are violet, while III is orange-yellow.

The mass spectrometric fragmentation patterns of I, II and III are typical of PAH (15), with the molecular ions (M^+) as base peaks and fragments corresponding to $(M-H_2)^+$, M^{2+}, and $(M-H_2)^{2+}$.

Scheme 1

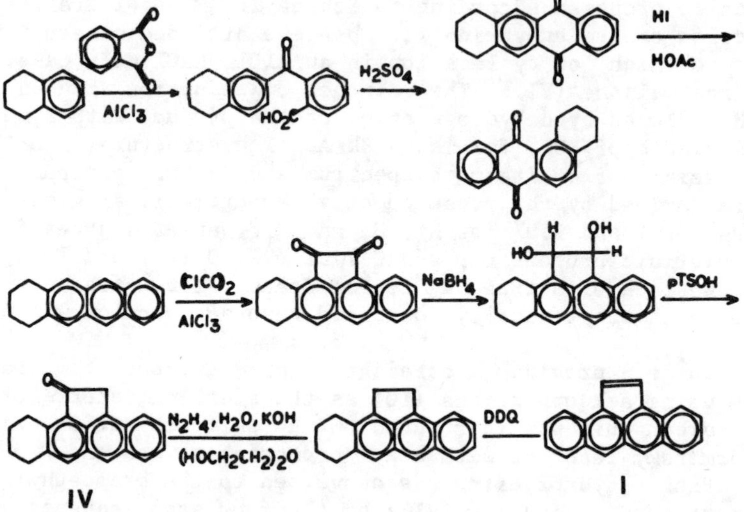

AlCl$_3$ · H_2SO_4 · HI / HOAc

(ClCO)$_2$ / AlCl$_3$ · NaBH$_4$ · pTSOH

IV · N_2H_4, H_2O, KOH / $(HOCH_2CH_2)_2O$ · DDQ · I

Scheme 2

$AlCl_3$ · $AlCl_3$ + NaCl · 1) Zn/NaOH 2) DDQ

II · VI

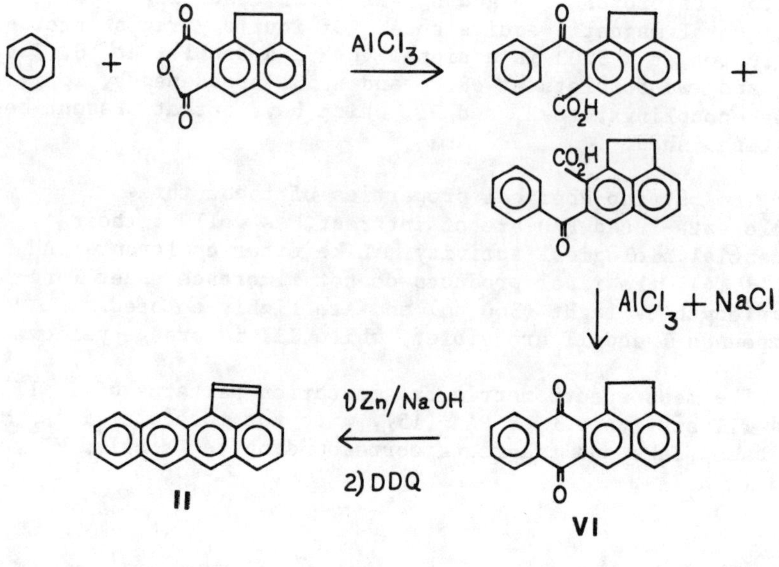

Scheme 3

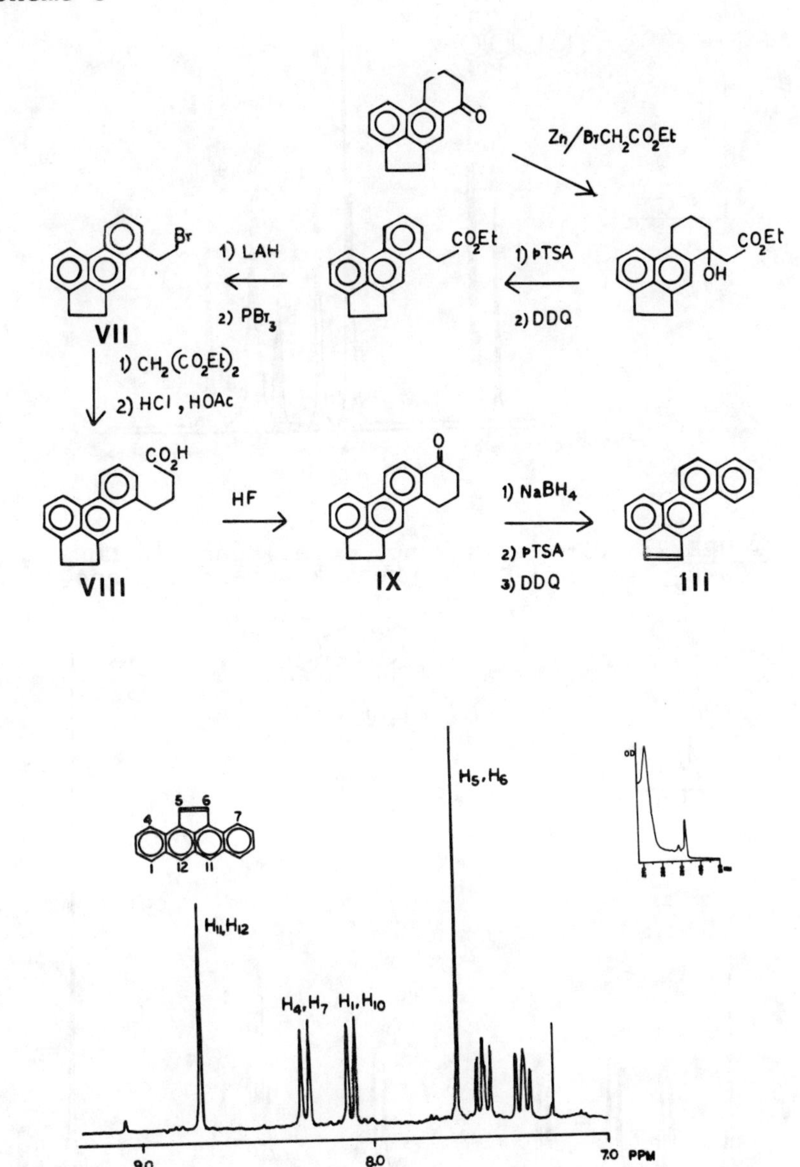

FIGURE 1. UV-VIS (hexane, inset) and ^{1}H NMR (250 MH, acetone - d_6) of benz(d)aceanthrylene.

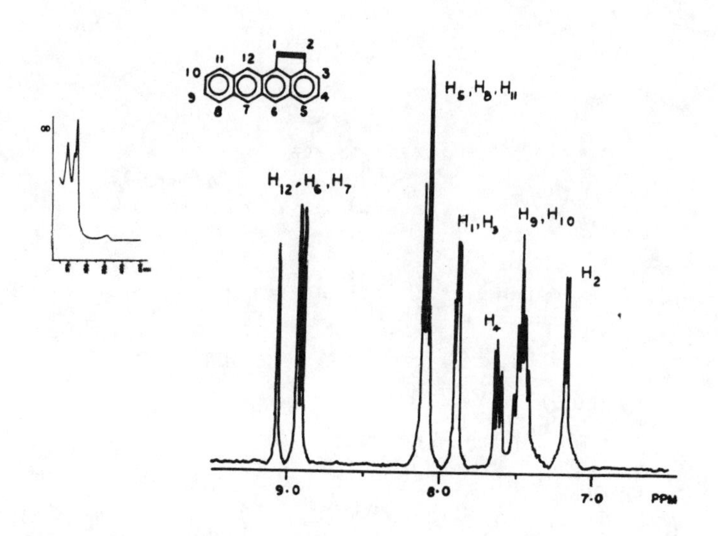

FIGURE 2. UV-vis (hexane, insel) and ^1H NMR ($250MH_z$, acetone-d_6) of benz(k) aceanthrylane.

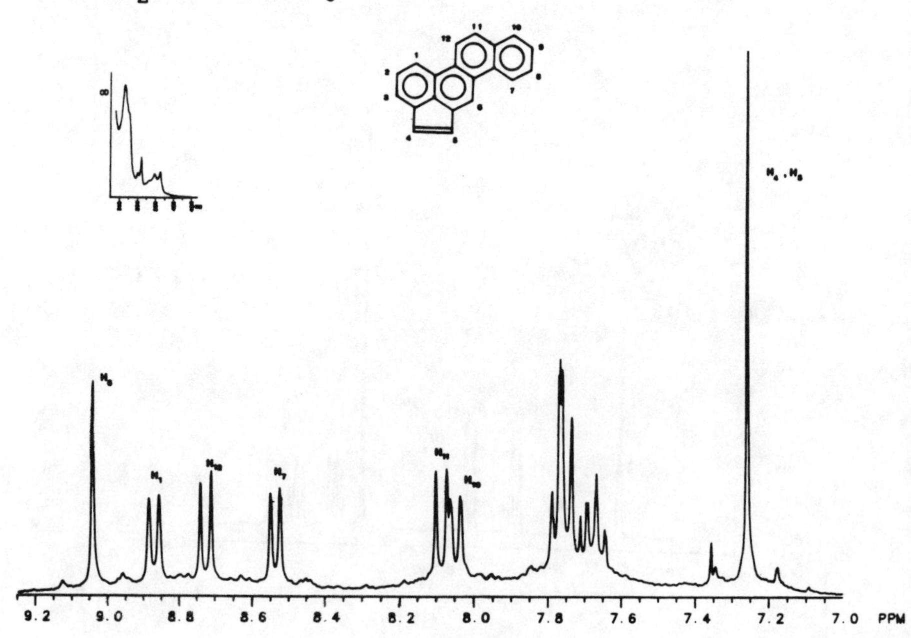

FIGURE 3. UV-vis (hexane, inset) and ^1H NMR ($200MH_z$, acetone-d_6) benz(j)acephenanthrylene.

ACKNOWLEDGEMENTS

This work was supported in part by USPHS Grant ES 03433-04A and in part by EPA Grant CR811817-01-0. This report has been reviewed by the Health Effects Research Laboratory, U.S. Environmental Protection Agency, and approved for publication. Approval does not signify that the contents necessarily reflect the views and policies of the Agency, nor does mention of trade names or commercial products constitute endorsement or recommendation for use.

REFERENCES

1. Sangaiah, R. and Gold, A. (1985). **Org. Prep. Proced. Inter., 17**: 53.
2. Kohan, M.J., Sangaiah, R., Ball, L.M. and Gold, A. (1985): **Mutat. Res. 155**: 95.
3. Sangaiah, R., Gold, A., Ball, L.M., Kohan, M.J., Bryant, B.J., Rudo, K., Claxton, L. and Nesnow, S. (1985): **Proc. Nineth Int. Symp. on PAH**, Battelle Press, Columbus, OH (in press).
4. Sangaiah, R., Gold, A. and Toney, G.E. (1983). **J. Org. Chem. 48**, 1632.
5. Nesnow, S., Leavitt, S., Easterling, R., Watts, R., Toney, S.H., Claxton, L., Sangaiah, R., Toney, G.E., Wiley, J., Fraher, P. and Gold, A. (1984). **Cancer Res. 44**: 4993.
6. Nesnow, S., Gold, A., Sangaiah, R., Triplett, L.L. and Slaga, T.J. (1984). **Cancer Lett. 22**: 263.
7. Fieser, L.F. (1931). **J. Am. Chem. Soc. 53**: 2329.
8. Stepan, V. and Vodehnal, J. (1971). **Collection Czech. Chem. Commun. 36**: 3964.
9. Alder, K., Pascher, F. and Vagt, H. (1942). **Ber. 75B**: 1501.
10. Krishnan, S. and Hites, R.A. (1981). **Anal. Chem. 53**: 342.
11. Plummer, B.F., Al-Saigh, Z.Y. and Arfan, M. (1984). **J. Org. Chem. 49**: 2069.
12. Becker, H.D., Hansen, L. and Anderson, K. (1985). **J. Org. Chem. 50**: 277.
13. Gold, A., Eisenstadt, E. and Schultz, J. (1978). **Tetrahedron Lett.**, p. 4491.
14. Eisenstadt, E. and Shpizner, B. and Gold, A. (1981). **Biochem. Biophys. Res. Commun. 100: 965.**
15. Benyon, J.H. (1968). **The Mass Spectra of Organic Molecules**. Elsevier, New York, 129 pp.

CATION RADICALS AND OXY-RADICALS FROM BENZO(A)PYRENE AND DERIVATIVES

PAUL D. SULLIVAN*, FOUAD BANNOURA, XIN-HUA CHEN
Department of Chemistry, Ohio University, Athens, Ohio 45701

INTRODUCTION

There is a continuing interest in the possibility that free radicals are involved as intermediates in the metabolism of polycyclic hydrocarbons in general, and of benzo(a)pyrene (BaP) in particular. The BaP cation radical, formed via a single electron oxidation of BaP, was first hypothesized as a reactive metabolic intermediate of BaP by Wilk *et al.* (1) and Fried and Schumn (2). Since that time much circumstantial evidence has been presented and recently reviewed by Cavalieri and Rogan (3) for the biological involvement of cation radicals in aromatic hydrocarbon carcinogenesis. Unfortunately, the reactivity of the BaP cation radical has precluded its direct observation in either *in vivo* or *in vitro* systems and it has only been directly observed in model systems. On the other hand the 6-oxy-BaP radical is a much more stable species and has been directly observed in extracts from *in vitro* metabolisms of BaP (4,5). It is believed to be an intermediate on the pathway from the enzymatically formed 6-hydroxy-BaP to the product BaP-quinones (4-7).

The analysis of the EPR spectra of the radicals from BaP and its derivatives is of interest because of the information it provides regarding the spin density distribution in the highest occupied molecular orbital. This information could be of use to explain some of the metabolic products of BaP, as well as indicate the possible electronic effects of substituents on these pathways. In this paper EPR studies on methylated BaP cation radicals and oxy radicals as well as ^{13}C labeled BaP's have been used to assign splitting constants in the BaP cation radical and 6-oxy-BaP radical. EPR and ENDOR spectra of 6-substituted BaP cation radicals have also been studied to provide information on conformational preferences as well as reactivities of these species.

MATERIALS AND METHODS

The 1-,2-,3-,4-,5-,8-,9-,11- and 12-monomethyl BaP's and the 6-hydroxy-BaP were obtained from the National Cancer Institute Chemical Repository. BaP was a commercially available sample of high purity (Eastman Kodak), 6-, 7- and 10-methyl BaP's and 6-hydroxymethyl-BaP were prepared by literature methods (8,9). The singly labeled ^{13}C BaP at each of the 12 protonated carbon atoms were kindly supplied by Dr. G. H. Daub (10-12). Cation radicals were prepared by one of two methods: (i) 0.2-0.5 mg of the hydrocarbon was dissolved in 0.2-0.5 mL of 98% H_2SO_4. After approximately 2 min about 50μL of the solution was drawn into a capillary tube and placed in the EPR cavity. (ii) A few drops of 0.8 M thallium (III) tristrifluoroacetate (TTFA) in trifluoroacetic acid (TFA) were added to a solution of the appropriate hydrocarbon in degassed TFA, samples were sealed and degassed under vacuum. Oxy-radicals were prepared by dissolving 0.1-0.5 mg of the hydrocarbon in 250 μL benzene or dichloromethane, 10 μL of TFA and 10 μL of 30% H_2O_2 were then added. The mixture was shaken for 5 min and the sample was transferred to an EPR tube where it was degassed and evacuated.

EPR spectra were recorded at room temperature (H_2SO_4) or at low temperature, -10° to -35° (TFA), in a Varian E-9 EPR spectrometer under various conditions of modulation amplitude and frequency. ENDOR experiments were carried out at the National Biomedical ESR Center, Medical College of Wisconsin, using a Varian ENDOR spectrometer, with the assistance of Dr. R. C. Sealy. Computer simulations of spectra were carried out with previously described programs (13) by using a modified procedure which downloads the simulated data points from the IBM 370/158 mainframe to an Apple II+ computer which then plots the spectrum on a chart recorder. UV-vis spectra were measured on a Hewlett-Packard 8451A diode array spectrometer. HPLC separations were carried out on an Tracor liquid chromatograph by reverse phase chromatography on a Waters Associates μ Bondapack C_{18} column using gradient elution.

RESULTS AND DISCUSSION

The BaP Cation Radical

The EPR spectrum of the BaP cation radical in H_2SO_4 was first observed over 25 years ago (14) and has since been investigated by several groups (5,15,16), however, only recently has a complete analysis been proposed (17-19). The interpretation of the BaP cation radical EPR spectrum was made difficult due to the asymmetry of the molecule and attempts to simplify the spectrum using ENDOR were unsuccessful. However, with the availability of all 12 monomethylated BaP's a method to estimate and assign the splitting constants in BaP was suggested (17). The cation radicals of the monomethylated BaP's were therefore investigated and methyl splittings were obtained for each compound (17,18). Once the methyl splittings were available they could be used as a starting point for the analysis of the unsubstituted BaP cation radical. Using high resolution spectra of the wings of the BaP spectrum in H_2SO_4 it was possible to extract 9 splitting constants which were confirmed by spectral simulation. The 3 largest splittings were obtained from a fully deuterated BaP sample (19). Initial assignments of the 12 splittings to the individual positions were made by comparison with the absolute magnitudes of the methyl splittings in the mono-methylated BaP cation radicals. Further study indicated that for various reasons (19) the proton splitting in the 1 position should be the 2nd largest proton splitting despite the fact that the 1-methyl splitting is only the 5th largest methyl splitting. The other splittings were assigned in the same relative order and are as shown in Table 1. From the proton splittings, the absolute values of the spin densities at each position may then be calculated (Table 1).

TABLE 1

EPR DATA FOR BaP AND MONOMETHYL-BaP CATION RADICALS.

Position	Methyl Splitting in Mono-Methylated BaP Cation Radical(G)	Assigned Proton Splittings of BaP Cation Radical(G)	Spin Density Calc. from Proton Splitting[a]	Calc. Proton Splittings[b]
1	3.08	4.57	0.163	4.35
2	0.67	0.54	0.019	-1.13
3	4.56	3.77	0.135	3.92
4	0.26	0.37	0.013	2.49
5	2.26	2.11	0.074	2.40
6	7.47	6.63	0.237	7.37
7	2.80	2.23	0.079	2.44
8	0.0	0.19	0.007	-0.48
9	3.57	2.95	0.105	1.63
10	1.15	1.94	0.069	0.40
11	0.92	0.82	0.029	0.40
12	3.43	2.75	0.097	3.03

a. Using McConnell's equation $|\rho| = |a|/Q$, Q = 28G
b. Calc. using a modified Hückel Molecular Orbital Method

Further justification for these assignments was obtained from measurements of ^{13}C splittings (19) (Table 2). Theoretically (20), the values of the ^{13}C splitting constants depend upon the spin densities at adjacent carbon atoms as well as on the spin density at the carbon atom itself. Using the spin densities derived from the proton splittings, assuming negative spin densities at positions 2, 8 and 11, and ignoring the spin densities at the blind positions (carbons to which no protons are attached), the ^{13}C splittings could be calculated (Table 2 column A). The agreement between these calculated splittings and the experimental values is clearly consistent with the assignments of the proton splittings. Even better agreement can be obtained if the spin densities at the blind positions are appropriately adjusted (column B in Table 2).

TABLE 2

EPR DATA FOR CARBON-13 SPLITTINGS IN THE BaP CATION RADICAL

Position	Exptl. C-13 Splittings(G)	Calc. C-13 Splittings(A)	Calc. C-13 Splittings(B)
1	6.01	6.07	5.92
2	4.55	-4.83	-4.83
3	4.44	5.07	4.58
4	0.93	-0.56	-1.11
5	1.32	2.47	1.45
6	8.30	8.43	8.43
7	3.74	2.92	3.93
8	2.91	-2.79	-2.79
9	3.01	2.87	2.73
10	1.29	1.01	1.29
11	1.96	-2.39	-1.97
12	3.60	3.86	3.71

The results are sufficiently self consistent to enable reasonable confidence to be placed in the spin densities given in Table 1 as a true reflection of the HOMO of BaP. These values can now be used to test more sophisticated theoretical calculations as well as to rationalize the chemical and biological reactions of BaP. Table 1 also lists the results of a modified Hückel molecular orbital calculation of the proton splittings. Considering the approximate nature of the calculation the overall agreement with the general features of the experimental values is quite good. Thus, the calculation reproduces the order and magnitude of the largest splittings, 6>1>3, and indicates that the 2, 8 and 11 positions have the smallest splittings. However, the calculation appears to underestimate the spin density at the 9 and 10 positions while overestimating at the 4 position. Since the spin densities represent approximately one half of the electron density in the HOMO, and since the latter have been related to the nucleophilic reactivity at the carbon atoms (21,22), the observed deviations between the calculated and experimental values could have a significant effect on interpretations or rationalizations of metabolic pathways based on calculated values. For example, the high spin density observed at the 9 position could indicate that direct oxygen insertion to form 9-hydroxy-BaP may account for part of the observed metabolic production of this phenol. Additionally the inequivalence of the 4 and 5 positions is supported by both the results on the methylated BaP's and by the ^{13}C splittings. This inequivalence, which is not reproduced by any method of calculation so far investigated, would explain the selective production of 5-hydroxy-BaP from the isomerization of BaP-4,5-oxide or from the acid dehydration of BaP-4,5-diol (23). It is also consistent with the breaking of the C_4-O bond in the enzymatic hydrolysis of BaP-4,5-oxide (25).

6-Methyl-, 6-Trifluoroacetoxy and 6-Fluoro-BaP Cation Radicals

6-Methyl-BaP. Oxidation of 6-methyl-BaP (6-CH_3-BaP) with either H_2SO_4 or TTFA/TFA leads to the production of a well resolved EPR spectrum. The spectrum from H_2SO_4 was unstable at room temperature, changing over a period of two hours from a well resolved spectrum with a width of *ca* 43G to a somewhat narrower (∿37G) less well resolved spectrum (24). The EPR spectrum in TTFA/TFA (25) was found to be relatively stable for several hours at temperatures below 0°, and had an overall width of *ca* 43G. It has been previously shown (25) that BaP and several substituted BaP's undergo reactions in TTFA/TFA which lead to the formation of a 6-trifluoroacetoxy cation radical. This reaction does not appear to occur for 6-CH_3-BaP since when the most reactive positions are blocked from reaction, as shown by 9,10-dimethylanthracene (26), one can observe a spectrum from the cation radical of the parent compound. Analysis of the EPR spectrum of 6-CH_3-BaP in TTFA/TFA was made possible by the observation of an ENDOR spectrum

(24). Splitting constants from the ENDOR spectrum were compared to the wing lines of the EPR spectrum and after many simulations led to an analysis in terms of 10 non-equivalent proton splittings and 3 equivalent methyl proton splittings. A comparison with the splittings for $BaP^{\dot{+}}$ indicates considerable similarity if the splittings of $6\text{-}CH_3\text{-}BaP^{\dot{+}}$ are assigned by analogy (Table 3). The EPR spectral changes of $6\text{-}CH_3$-BaP in H_2SO_4 were paralleled by changes in the UV absorption spectra. After 1 hr the EPR and UV spectra of $6\text{-}CH_3$-BaP in H_2SO_4 were identical to the EPR and UV spectra of freshly prepared 6-hydroxymethyl-BaP ($6\text{-}CH_2OH$-BaP) in H_2SO_4. Additionally the major product found on extraction and HPLC separation of a solution of $6\text{-}CH_3$-BaP in H_2SO_4 was found to be $6\text{-}CH_2OH$-BaP. It therefore appears that $6\text{-}CH_3$-BaP reacts in H_2SO_4 to initially produce the cation radical of the parent compound which undergoes further reaction to produce $6\text{-}CH_2OH$-BaP.

6-Trifluoroacetoxy-BaP. The EPR spectrum observed (27) when BaP is reacted with TTFA/TFA has been attributed to the 6-trifluoroacetoxy-BaP ($6\text{-}CF_3CO_2$-BaP) cation radical by analogy with the behavior of anthracene in TTFA/TFA (26). The trifluoroacetoxylation at the 6-position results in a decrease in the EPR spectral width from 28.8G in H_2SO_4 to ca 23.0G in TTFA/TFA due to the replacement of the proton at the 6 position (splitting 6.63G) with the 3 fluorines from the trifluoroacetoxy group. An ENDOR spectrum of BaP in TTFA/TFA shows a pair of lines on either side of the free fluorine frequency with a splitting of 0.23G (24). Eleven proton splittings are also observable from the ENDOR spectrum. The splittings are tabulated and compared to $BaP^{\dot{+}}$ and $6\text{-}CH_3\text{-}BaP^{\dot{+}}$ in Table 3. If the splittings are assigned by analogy with the parent compound the values indicate that the spin density distribution in the rest of the molecule is little changed by substitution at the 6-position with CH_3 or CF_3CO_2. If the assignments are correct this would imply that changes in metabolic pathways between BaP and $6\text{-}CH_3$-BaP are likely to be unaffected by electronic factors. Differences most likely arise because of the steric effects of the methyl group and the possible reaction pathways involving oxidation of the CH_3 group.

6-Fluoro-BaP. The EPR spectrum of 6-F-BaP in H_2SO_4, D_2SO_4 and TTFA/TFA consists of a broad doublet of 17.5G. On the basis of previous studies with fluorinated naphthalene cation radicals (28), it is expected that the fluorine splitting should be much larger than the splitting of the proton it is replacing. The large doublet is therefore assigned as the

TABLE 3

COMPARISON OF SPLITTINGS FOR BaP, 6-CH_3-BaP, and 6-CF_3CO_2-BaP CATION RADICALS

$BaP^{\dot{+}}$		6-CH_3-$BaP^{\dot{+}}$		6-CF_3CO_2-$BaP^{\dot{+}}$
Splitting Constant(G)	Position Assigned	EPR Splittings	ENDOR Splittings	ENDOR Splittings
6.63	6	7.38(CH_3)	∿7.5	0.23(CF_3)
4.57	1	4.45	4.54	4.68
3.77	3	3.63	3.71	3.73
2.95	9	3.06	3.07	2.93
2.75	12	2.47	2.52	2.67
2.23	7	2.21	2.25	2.34
2.11	5	1.81	1.89	1.80
1.94	10	1.70	1.68	1.45
0.82	11	0.79	0.79	1.27
0.54	2	0.44	0.38	0.70
0.37	4	0.34	0.38	0.43
0.19	8	--	--	0.23

splitting of the fluorine in the 6 position. The spectrum of 6-F-BaP in H_2SO_4 is, however, time dependent. Over a period of several hours the spectrum changes from a broad doublet with little resolution into a much narrower spectrum with considerable resolution. A comparison of the spectrum after 10 hours with a freshly prepared solution of 6-OH-BaP in H_2SO_4 showed that the two spectra were identical. UV spectral changes further confirmed the similarity. When the products of the reaction of 6-F-BaP in H_2SO_4 were separated by HPLC the formation of 1,6-, 3,6- and 6,12-BaP quinones was confirmed.

The reactions of 6-CH_3-BaP and 6-F-BaP in H_2SO_4 may be of some significance to proposed metabolic pathways for these compounds. Cavalieri (3) has proposed that cation radicals can be generated from polycyclic hydrocarbons by one electron oxidation with an Fe(V) form of cytochrome P-450. It was further proposed that the cation radicals would then react with nucleophiles, including DNA, at the positions of highest charge density. The metabolic formation of quinones (3,29) from 6-F-BaP is hypothesized to proceed via a cation radical intermediate which is then attacked by a nucleophilic oxygen atom. Adducts of 6-F-BaP to DNA in model systems have also been identified which are consistent with reaction at the 6-position of BaP. Circumstantial evidence for the intermediacy of cation radicals in the metabolism of 6-CH_3-BaP comes from the identification of adducts in a variety of systems in which the BaP-6-CH_2 group is bound to the 2 amino group of quanine (30) and the formation of 6-CH_2OH-BaP as a major product of metabolism (31). The present studies seem to have demonstrated rather directly that the formation of cation radicals from 6-CH_3-BaP and 6-F-BaP could lead to the observed reaction products. In H_2SO_4, the first produced one electron oxidation products appear to react with residual water to give 6-CH_2OH-$BaP^{+\cdot}$ and 6-OH-$BaP^{+\cdot}$, the former yielding 6-CH_2OH-BaP on workup and the latter autoxidizing to give the three BaP-quinones.

Oxy-BaP Radicals

The 6-oxy-BaP radical is the most common radical form of BaP, it is produced from BaP both enzymatically and chemically in the presence of molecular oxygen (4-7,27) and is stable in nonpolar solvents in the absence of oxygen for long periods of time. This unusual stability of an unhindered phenoxy radical may be due to the fact that a more appropriate representation of this radical should be as an O^- substituted BaP cation radical (B), rather than as a phenoxy radical (A).

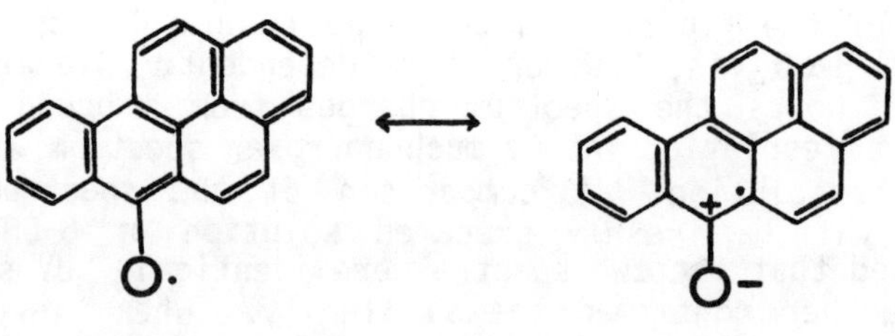

(A) (B)

Many of the published EPR spectra of this radical are poorly resolved and complete analysis is not possible. The most completely resolved spectrum has been obtained by light irradiation of a BaP solution in dichloromethane in the presence of a pure O_2 atmosphere, followed by several freeze-pump-thaw degassing cycles (27). This spectrum was analyzed in terms of eleven different splitting constants, as expected for a 6-oxy-

BaP radical, with values of 0.11, 0.26, 1.097, 1.102, 1.243, 1.615, 2.591, 2.95, 4.15, 5.15 and 5.65G. A computer simila-tion of the spectrum is in good agreement with the experi-mental spectrum. The splitting constants were not assigned to specific positions although simple molecular orbital calcu-lations indicated that the 4 largest splittings should be associated with the 1, 3, 4 and 12 positions.

With the availability of the ^{13}C labeled BaP's and the monomethylated BaP's the assignment of the proton splittings in the 6-oxy-BaP radical was reexamined. EPR spectra of 6-oxy-radicals from the ^{13}C labeled BaP's were obtained by oxidation with TFA/H_2O_2 in CH_2Cl_2 at -10 C. The spectra were compared with that of the unlabeled BaP under the same condi-tions and the increase in the total width was measured to obtain the C-13 splittings at each position. (See Fig. 1 and Table 4). The C-13 splittings were also measured using ben-zene as a solvent at room temperature and were found to be

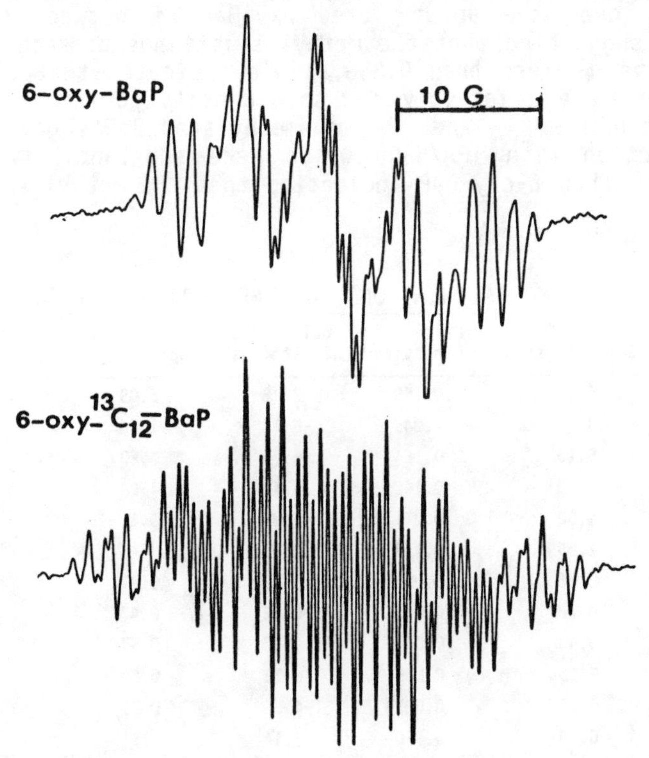

FIGURE 1. EPR spectra of 6-oxy-BaP and C-13 labeled (position 12) 6-oxy-BaP.

somewhat solvent dependent. Qualitatively the large C-13 splittings at the 1 through 5 and 11 and 12 positions indicate high spin density in these regions and the small splittings at the 7 through 10 positions indicate regions of low spin density.

Attempts were made to use the width of the monomethylated oxy-BaP's to estimate the methyl splittings, and hence the proton splittings at the different positions, using similar assumptions to those made for the monomethylated BaP cation radicals (17,18). Unfortunately the reaction with TFA/H_2O_2 to generate the oxy radicals from the methylated BaP's was not totally successful. The 2-, 5-, 7-, 8-, 9-, 10- and 11-methyl BaP's behaved as expected, giving EPR spectra which were consistent with 6-oxy radicals, and giving products on extraction and HPLC separation which were consistent with methylated 1,6-, 3,6- and 6,12-BaP quinones. The width of the oxy-BaP spectra in each case was measurable and showed increases over the unmethylated oxy-BaP of between 0.60 and 1.70G, suggesting that the methyl splittings at each of these positions is less than 0.85G. This indicates that all these positions are relatively low spin density positions. On the other hand 3-, 4- and 12- monomethylated BaP's gave spectra on reaction with TFA/H_2O_2 which were substantially greater in width than 6-oxy-BaP indicating that the methyl splittings

TABLE 4

EPR DATA AND HMO CALCULATIONS FOR THE 6-OXY-BaP RADICAL

Position	Exptl. C-13 Splittings[a] (G)	Calc. Spin Densities[b]	Calc. C-13 Splittings[c]	Calc. Proton Splittings[c]	Exptl. Proton Splittings[e]
1	6.50	0.169	7.18	5.09	5.65
2	4.95	-0.049	-6.38	-1.46	2.59
3	5.70	0.164	6.76	4.93	5.15
4	5.30	0.142	5.13	4.24	2.95
5	4.50	0.011	-2.99	0.32	0.11
6	2.95	0.117	2.97	--	--
7	0.15	0.032	0.88	0.97	1.61
8	0.30	0.015	-0.16	0.47	0.26
9	0.40	0.019	0.05	0.57	1.09
10	0.55	0.030	0.73	0.89	1.24
11	6.50	-0.023	-4.58	-0.71	1.10
12	6.60	0.150	6.14	4.51	4.15
0	4.75[d]	0.0692	4.75[d]		

a. Solvent is CH_2Cl_2 at -10°; b. Parameters used h_O = 1.70, k_{OC} = 1.22;
c. Calculated from calculated spin densities; d. Exptl. and Calc 0-17 splittings.
e. Assigned by comparison with calculated proton splittings.

at these positions are relatively large (ca 3.72, 3.55 and 4.27G respectively). However, the spectra were less stable than those from the other methylated BaP's and HPLC separation of products indicated that further reactions had occurred. 1-methyl-BaP behaved anomalously, giving a spectrum with a width considerably less than 6-oxy-BaP suggesting that secondary reactions had occurred. This behavior of the 1,3,4 and 12 monomethyl-BaP's may be due to the fact that these positions are thought to be the highest spin density positions and electron donating substituents at these positions may make the 6-oxy radicals less stable. It was not therefore possible to specify complete proton assignments on the basis of the methylated oxy-BaP radicals.

In order to proceed further a comparison was made with the results of some modified Hückel molecular orbital calculations (32). To carry out a calculation for the 6-oxy-BaP radical one must specify values for the Coulomb and resonance integral parameters for the oxygen atom. Previous studies have indicated that these parameters reflect the environment of the oxygen atom (33). Our approach was to take the known O-17 splitting constant of 6-oxy-BaP (7) and the measured C-13 splitting at the 6-position and to find a set of parameters which gave agreement for these two splittings using known equations relating O-17 and C-13 splittings to calculated spin densities (33-35). The spin densities so calculated were then used to calculate proton and C-13 splittings at all 12 measured positions (see Table 4). If the proton splittings are assigned according to these calculations i.e. 1>3>12>4>2>7>10>11>9>8>5 the agreement between the calculated and experimental values is only fair (Table 4). Unfortunately the most that can be said is that all the results are consistent with the 1, 3, 12 and 4 positions being the high spin density positions, the 11, 9, 8 and 5 positions seem to be the lowest spin density positions with the 2, 7 and 10 positions being of slightly higher values. More definite assignments cannot be justified at this point.

ACKNOWLEDGEMENTS

We are indebted to the late Prof. Guido H. Daub for his painstaking synthesis of the carbon-13 labeled BaP's and for his generous gift of these compounds. This work was supported in part, by Grant # CA-34966 awarded by the National Cancer Institute, DHEW, to P.D.S.

REFERENCES

1. Wilk, M., Bez, W., and Rochlitz, J. (1966): Neue Reaktionen der carcinogenen kohlenwasserstoffe 3,4 benzpyren, 9,10-dimethyl, 1,2-benzanthracen und 20-methylcholanthren. Tetrahedron, 22: 2599-2608.
2. Fried, J., and Schumn, D.E. (1967): One electron transfer oxidation of 7,12-dimethyl-benzo(a)anthracene, a model for the metabolic activation of carcinogenic hydrocarbons. J. Amer. Chem. Soc., 89: 5508-5509.
3. Cavalieri, E., and Rogan, E. (1985): The role of radical cations in aromatic hydrocarbon carcinogenesis. Env. Health Perspectives, in press.
4. Lesko, S., Caspary, W., Lorenzten, R., and Ts'o, P.O.P. (1975): Enzymic formation of 6-oxobenzo(a)pyrene radical in rat liver homogenates from carcinogenic benzo(a)pyrene. Biochemistry, 14: 3978-3984.
5. Nagata, C., Inomata, M., Kodama, M., and Tagashira, Y. (1968): Electron spin resonance study on the interaction between the chemical carcinogens and tissue components III. Determination of the structure of the free radical produced either by stirring 3,4-benzopyrene with albumin or incubating it with liver homogenates. Gann. 59: 289-298.
6. Lorenzten, R.J., Caspary, W.J., Lesko, S.A., and Ts'o P.O.P. (1975): The autoxidation of 6-hydroxybenzo(a)pyrene and 6-oxobenzo(a)pyrene radical, reactive metabolites of benzo(a)pyrene. Biochemistry, 14: 3970-3977.
7. Rispin, A.S., Kon, H., and Nebert, D.W. (1976): Electron spin resonance study of ^{17}O-enriched oxybenzo(a)pyrene radical. Mol. Pharmacol. 12: 476-482.
8. Dewhurst, F., and Kitchen, D.A. (1972): Synthesis and properties of 6-substituted benzo(a)pyrene derivatives. J. Chem. Soc. Perkin Trans I, 710-712.
9. Newman, M.S., and Kumar, S. (1977): New synthesis of benzo(a)pyrene, 7,10-dimethylbenzo(a)pyrene. J. Org. Chem. 42: 3284-3286.
10. Bodine, R.S., Hylarides, M., Daub, G.H., and VanderJagt, D.L. (1978): ^{13}C labeled benzo(a)pyrene and derivatives 1. Efficient pathways to labeling the 4,5,11 and 12 positions. J. Org. Chem. 43: 4025-4028.
11. Simpson, J.E., Daub, G.H., and VanderJagt, D.L. (1980): Carbon-13 labeled benzo(a)pyrenes and derivatives 2. The synthesis of benzo(a)pyrene-6-^{13}C. J. Labeled Compd. Radiopharm. 17: 895-900.
12. Unkefer, C.J., London, R.E., Whaley, T.W., and Daub, G.H. (1983): ^{13}C and ^{1}H NMR analysis of isotopically labeled benzo(a)pyrenes. J. Am. Chem. Soc., 105: 733-735.
13. Forbes, W.F., Sullivan, P.D., and Wang, H.M. (1967): An

electron spin resonance study of the radical cations of some p-dialkoxybenzenes. J. Am. Chem. Soc. 89: 2705-2711.
14. Kon, H., and Blois, M.S. (1958): Paramagnetism of hydrocarbon-conc H_2SO_4 systems. J. Chem. Phys. 28: 743-744.
15. Forbes, W.F., Robinson, J.C., and Wright, G.V. (1967): Free radicals of biological interest I. Electron spin resonance of tobacco smoke condensates. Can. J. Biochem. 45: 1087-1098.
16. Elmore, J.J., and Forman, A. (1975): EPR of free radical Intermediates in oxidation of carcinogenic polycyclic hydrocarbons. Cancer Biochem. Biophys. 1: 115-120.
17. Sullivan, P.D. (1983): EPR Studies of methylated benzo(a)-pyrene cation radicals. J. Magnetic Res. 54: 314-318.
18. Sullivan, P.D., Bannoura, F., and Roach, S. (1985): Cation radicals of methylated benzo(a)pyrenes. In: Polynuclear Aromatic Hydrocarbons: Mechanisms, Methods and Metabolisms, edited by M. Cooke and A.J. Dennis, pp 1273-1283, Battelle Press, Columbus, Ohio.
19. Sullivan, P.D., Bannoura, F., and Daub, G.H. (1985): ^{13}C and ^{1}H EPR analysis of the benzo(a)pyrene cation radical. J. Am. Chem. Soc. 107: 32-35.
20. Karplus, M., and Fraenkel, G.K. (1961): Theoretical interpretation of carbon-13 hyperfine interactions in electron spin resonance spectra. J. Chem. Phys. 35: 1312-1323.
21. Loew, G.H., Wong, J., Phillips, J., Hjelmeland, L., and Pack, G. (1978): Quantum chemical studies of the metabolism of benzo(a)pyrene. Cancer Biochem. Biophys. 2: 123-130.
22. Fu, P.P., Harvey, R.G., and Beland, F.A. (1978): Molecular orbital theoretical prediction of the isomeric products formed from reactions of arene oxides and related metabolites of polycyclic aromatic hydrocarbons. Tetrahedron 34: 857-866.
23. Yang, S.K., Roller, P.P., and Gelboin, H.V. (1977): Enzymatic mechanism of benzo(a)pyrene conversion to phenols and diols and an improved hplc separation of benzo(a)pyrene derivatives. Biochemistry 16: 3680-3687.
24. Sullivan, P.D., Ocasio, I.J., Chen, X-H., and Bannoura, F. (1985): EPR and ENDOR studies of 6-substituted benzo(a)pyrene cation radicals. J. Am. Chem. Soc., in press.
25. Sullivan, P.D., Ellis, L.E., Calle, L.M., and Ocasio, I.J. (1982): Chemical and enzymatic oxidation of alkylated benzo(a)pyrenes. Chem. Biol. Interactions 40: 177-191.
26. Sullivan, P.D., Menger, E.M., Reddoch, A.H., and Paskovich, D.H. (1978): Oxidation of anthracene by thallium (III) trifluoroacetate. Electron spin resonance and structure of the product cation radicals. J. Phys. Chem. 82: 1158-1160.

27. Menger, E.M., Spokane, R.B., and Sullivan, P.D. (1976): Free radicals derived from benzo(a)pyrene. Biochem. Biophys. Res. Communs. 71: 610-616.
28. Thomson, C., and MacCulloch, W.J. (1970): ESR spectra of the cation radicals of some highly fluorinated naphthalenes. Mol. Phys. 19: 817-832.
29. Cavalieri, E., Cremonesi, P., Warner, C., Tibbels, S. and Rogan, E. (1984): One-electron oxidation of 6-fluorobenzo(a)pyrene in quinone formation and carcinogenesis. Proc. Am. Assoc. Cancer Res. 25: 124.
30. Rogan, E.G., Hakam, A. and Cavalieri, E.L. (1983): Structure elucidation of a 6-methylbenzo(a)pyrene-DNA adduct formed by horseradish peroxidase in vitro and mouse skin in vivo. Chem-Biol. Interactions 47: 111-122.
31. Hamernik, K.L., Chiu, P-L, Chou, M.W., Fu, P.P., and Yang, S.K. (1983): Metabolic activation of 6-methylbenzo(a)pyrene. In: Polynuclear Aromatic Hydrocarbons: Formation, Metabolism and Measurement, edited by M. Cooke and A.J. Dennis, pp 583-597, Battelle Press, Columbus, Ohio.
32. McLachan, A.D. (1960): Self-consistent field theory of the electron spin distribution in π-electron radicals. Mol. Phys. 3: 233-252.
33. Sullivan, P.D., Bolton, J.R., and Geiger, W.E. (1970): Oxygen-17 and carbon-13 hyperfine interactions in the electron paramagnetic resonance spectrum of the hydroquinone cation radical. J. Chem. Soc. 92: 4176-4180.
34. Das, M.R., and Fraenkel, G.K. (1965): Electron spin resonance of semiquinones: spin density distribution and carbonyl sigma-pi parameters. J. Chem. Phys. 42: 1350-1360.
35. Broze, M., and Luz, Z. (1969): Carbonyl carbon-13 σ-π polarization constants. J. Chem. Phys. 51: 749-753.

NITRATION PRODUCTS FROM THE REACTION OF FLUORANTHENE AND PYRENE WITH N_2O_5 AND OTHER NITROGENOUS SPECIES IN THE GASEOUS, ADSORBED AND SOLUTION PHASES: IMPLICATIONS FOR ATMOSPHERIC TRANSFORMATIONS OF PAH

JANET A. SWEETMAN*, BARBARA ZIELINSKA, ROGER ATKINSON, ARTHUR M. WINER, JAMES N. PITTS, JR.
Statewide Air Pollution Research Center, University of California, Riverside, CA 92521, USA

INTRODUCTION

In addition to being present in primary emissions (1-4), it has been proposed that certain of the nitrated polycyclic aromatic hydrocarbons (NO_2-PAH) detected in extracts from ambient particles may arise from reactions of their parent PAH with gaseous copollutants in the atmosphere and/or during collection of particulate organic matter (POM) (5-7). While the formation of NO_2-PAH from the reaction of gaseous NO_2 (containing traces of HNO_3) with PAH adsorbed on combustion-generated particles has been well documented (8-10), little is known about the reaction of PAH with N_2O_5 and other reactive nitrogenous species.

Gaseous NO_3 radicals and N_2O_5 are formed in the atmosphere at night from the reactions

$$NO_2 + O_3 \rightarrow NO_3 + O_2 \qquad (1)$$

$$NO_3 + NO_2 \overset{M}{\rightleftarrows} N_2O_5 \qquad (2,-2)$$

Ambient atmospheric concentrations of the NO_3 radical during nighttime hours have been measured at a variety of locations in southern California and elsewhere by <u>in situ</u> longpath differential optical absorption spectroscopic techniques, and observed to range from <1 up to 430 ppt (11-14). The corresponding N_2O_5 concentrations can be calculated from observed NO_3 radical and NO_2 concentrations, and the equilibrium constant for reactions (2,-2). These calculated N_2O_5 concentrations are highest during early evening hours and range from ~5 ppt to ~15 ppb (14).

REACTION OF FLUORANTHENE AND PYRENE WITH N_2O_5

Recently, 2-nitrofluoranthene ($2-NO_2$-FL) and 2-nitropyrene ($2-NO_2$-PY) have been shown to be present in extracts of ambient POM (15-18). In contrast to $1-NO_2$-PY and 3- and $8-NO_2$-FL, the 2-nitro-isomers of FL and PY have not been reported to be present in diesel exhaust particulate (19,20), nor are they products of the reaction of FL and PY with NO_2/HNO_3 during transport or collection of POM (5-10). For these reasons it has been proposed that the presence of $2-NO_2$-PY (15,16) and $2-NO_2$-FL (16,17) in ambient POM extracts must be the result of atmospheric transformations of the parent PY and FL.

We report here on the reactions of N_2O_5 with FL and PY in solution and adsorbed on filters, and with FL in the gas phase, and discuss the implications for atmospheric transformation of FL and PY.

MATERIALS AND METHODS

Chemicals

FL and PY were purchased from Aldrich Chemical Co. (98% purity). N_2O_5 was prepared by collection at 196 K of the products of the reaction of NO_2 with O_3 (21).

Reaction of N_2O_5 with FL and PY in CCl_4 Solution

Complete details of the solution phase reactions and product identification are given elsewhere (22,23). Briefly, the experiments were carried out at room temperature and in the dark in a dry box under 1 atm of N_2. Freshly prepared N_2O_5 was dissolved in CCl_4 (Spectra AR, Mallinckrodt, stored over 5Å molecular sieve) and maintained at liquid N_2 temperature prior to use. Typically, $\lesssim$1 millimolar solutions of FL or PY were prepared in CCl_4 and the N_2O_5 solution was added with stirring. For N_2O_5 to PAH ratios less than or equal to one, the reaction was rapid and products were analyzed after ~5 min.

Chamber Exposures of FL and PY Adsorbed on Filters and FL in the Gas Phase

Teflon impregnated glass fiber (TIGF) filters (Pallflex T60A20) and glass fiber (GF) filters (Gelman type AE) were cleaned and coated with toluene solutions of FL or PY as described elsewhere (17,22,24). After evaporation of the toluene solvent, the filters were hung from a Teflon line in a collapsible 6400-l all-Teflon chamber, and the reactive gases introduced, with rapid mixing being ensured by use of a Teflon-coated fan rated at 900 l sec^{-1}. Three 45 min simultaneous exposures of FL and PY (each coated on both GF and TIGF filters) were made, as well as an additional 30 min exposure of FL alone. All exposures were carried out in the dark. The three exposures of FL and PY were to: 10 ppm NO_2 + 1.5 ppm HNO_3, 5 ppm N_2O_5 and 5 ppm N_2O_5 + 10 ppm NO_2.

The exposure of FL alone was to 5 ppm N_2O_5, and crystalline FL was added to the chamber along with the FL-coated filters. At the end of this exposure, ~75% of the chamber volume was sampled through a polyurethane foam (PUF) plug (25) to collect gas-phase reaction products. The PUF plug was doped with FL-d_{10} to serve as a control for reactions occurring on the PUF during sampling. The chamber was then partially refilled with pure air and the filters were removed.

Product Analyses

The exposed filters were Soxhlet extracted with CH_2Cl_2 and the NO_2-FL and NO_2-PY products quantified by HPLC as described previously (22). The nitro-isomers were identified by GC/MS analysis with multiple ion detection (MID) using a Finnigan 3200 GC/MS operated in the EI mode and fitted with a cool on-column injector and a 60 m DB-5 capillary column (0.25 μm film thickness, 0.32 mm i.d.) eluting directly into the ion source. Injections were made at 50°C followed by programming at 8°C min^{-1} to 350°C.

The PUF plug was Soxhlet extracted with CCl_4 and, prior to GC/MS analysis of the extract, semi-preparative HPLC was performed (Altex ultrasphere ODS column; isocratic 80:20/ methanol:water) to remove interfering polar products arising from the reaction of the PUF plug with the gaseous nitrogenous species in the chamber.

RESULTS

Reactions of N_2O_5 with FL and PY in CCl_4 Solution

For a PY to N_2O_5 ratio of ~1:1, 90% of the PY was converted to 1-NO_2-PY with small amounts (~5%) of di-NO_2-PYs and other products, leaving ~5% of the PY unreacted. Thus the ambient temperature reaction of N_2O_5 with PY in CCl_4 produced the same isomer (1-NO_2-PY) as reported for electrophilic substitution reaction involving the nitronium ion (26). In contrast, the ambient temperature reaction of FL with N_2O_5 in

CCl_4 produced 2-NO_2-FL, an isomer either not formed by the usual direct nitration methods (27,28), or at most formed only as a minor product (29). The 2-NO_2-FL isomer was the only mononitro-isomer formed and, for an ~1:1 ratio of FL to N_2O_5, it was formed in ~30% yield. A single dinitro-isomer, identified as 1,2-dinitrofluoranthene (23), was also formed (in ~15% yield). Approximately 15% of the fluoranthene remained unreacted and the other 40% formed polar products, including hydroxynitrofluoranthenes (23).

When 2-NO_2-FL was further reacted with N_2O_5, a single dinitro-isomer (not 1,2-di-NO_2-FL) and lesser amounts of a trinitro-isomer were formed (23). Reaction of N_2O_5 with a solution containing equimolar amounts of PY and FL produced mainly 1-NO_2-PY showing that PY is the more reactive of these PAH in CCl_4.

Chamber Exposures of FL and PY Adsorbed on Filters

Table 1 gives the mono-NO_2-PAH yields observed after 45 min for each of the three exposures in which GF and TIGF filters coated with FL or PY were simultaneously exposed. Figure 1 shows MID traces of the m/z = 247 molecular ion of NO_2-FL and NO_2-PY for extracts from the PY- and FL-coated TIGF filters exposed to 10 ppm NO_2 + 1.5 ppm HNO_3 and to 5 ppm N_2O_5. The amount of nitration of pyrene by NO_2/HNO_3 was higher on the TIGF than on the GF filters which may reflect the lower basicity of the TIGF filters (pH of water extract ~7 for TIGF versus ~10 for GF filters). The amounts of nitration by N_2O_5 and N_2O_5/NO_2 were more nearly equal for the two filter types though slightly lower for the TIGF filters. More mononitro-products were extracted from the GF filters (see Table 1), however, the ratios of dinitro- to mononitro-products were higher for the TIGF filter extracts.

Figure 1A shows that only 1-NO_2-PY was produced on the PY-coated filters exposed to NO_2 + HNO_3. In Figure 1B, the extract from a FL-coated TIGF filter has been concentrated ~20-fold over the extract shown in Figure 1A to allow the NO_2-FL isomer distribution to be seen. Therefore, PY is much more reactive towards NO_2 + HNO_3 than is FL (see also Table 1). The distribution of the NO_2-FL isomers is similar to that observed for nitration by the nitronium ion, namely: (3- > 8- >> 7- > 1-) (27). The 1-NO_2-PY peak in Figure 1B is probably due to traces of PY impurity in the FL-coating solution or, less likely, from gas-phase PY depositing onto the filter.

TABLE 1

HPLC QUANTIFICATION OF THE MONO-NO_2-PAH YIELDS (%) FROM EXPOSURES OF FLUORANTHENE AND PYRENE-COATED FILTERS

Species Added to Teflon Chamber	Mono-NO_2-FL		Mono-NO_2-PY	
	TIGF[a]	GF[b]	TIGF[a]	GF[b]
NO_2 (10 ppm)/HNO_3 (1.5 ppm)	<1	<1	17	5
N_2O_5 (5 ppm)	38	58	48	69
N_2O_5 (5 ppm)/NO_2 (10 ppm)	42	59	53	68

[a]Teflon Impregnated Glass Fiber Filter.
[b]Glass Fiber Filter.

Figure 1C shows that on PY-coated filters exposed to N_2O_5 traces of 4-NO_2-PY were observed in addition to 1-NO_2-PY. Figure 1D illustrates the nearly equal distribution of the 1-, 3-, 7-, and 8-NO_2-FL isomers produced by exposing FL-coated filters to N_2O_5, as observed in our previous such experiments (22,17). In contrast to the higher reactivity of PY relative to FL towards NO_2 + HNO_3, the amounts of nitration produced by exposure to N_2O_5 and N_2O_5 + NO_2 are more nearly equal for FL and PY coated on filters.

Upon introduction of N_2O_5 into an environmental chamber, minor amounts of NO_2 and HNO_3 are present (30). The low yield of nitration from the exposure of FL- and PY-coated filters to 10 ppm NO_2 + 1.5 ppm HNO_3 eliminates these as the nitrating species in the N_2O_5 exposure. The isomer distributions from the N_2O_5 and N_2O_5 + NO_2 exposures are similar, as are the amounts of PY and FL nitration in these exposures (see Table 1). Since the addition of 10 ppm NO_2 will drive the equilibrium between NO_3 radicals, NO_2, and N_2O_5 [reactions (2,-2)] toward N_2O_5, the similar amounts of nitration in these two exposures demonstrates that the nitrating species is not the NO_3 radical, but must rather be N_2O_5 (22).

In addition to the mononitro-isomers shown in Figure 1, the reaction of N_2O_5 and N_2O_5 + NO_2 with FL and PY adsorbed on filters also produced dinitro products. The MID traces in Figure 2A and 2B are for the m/z = 292 and m/z = 300 molecular ions of di-NO_2-PY and di-NO_2-PY-d_8, respectively.

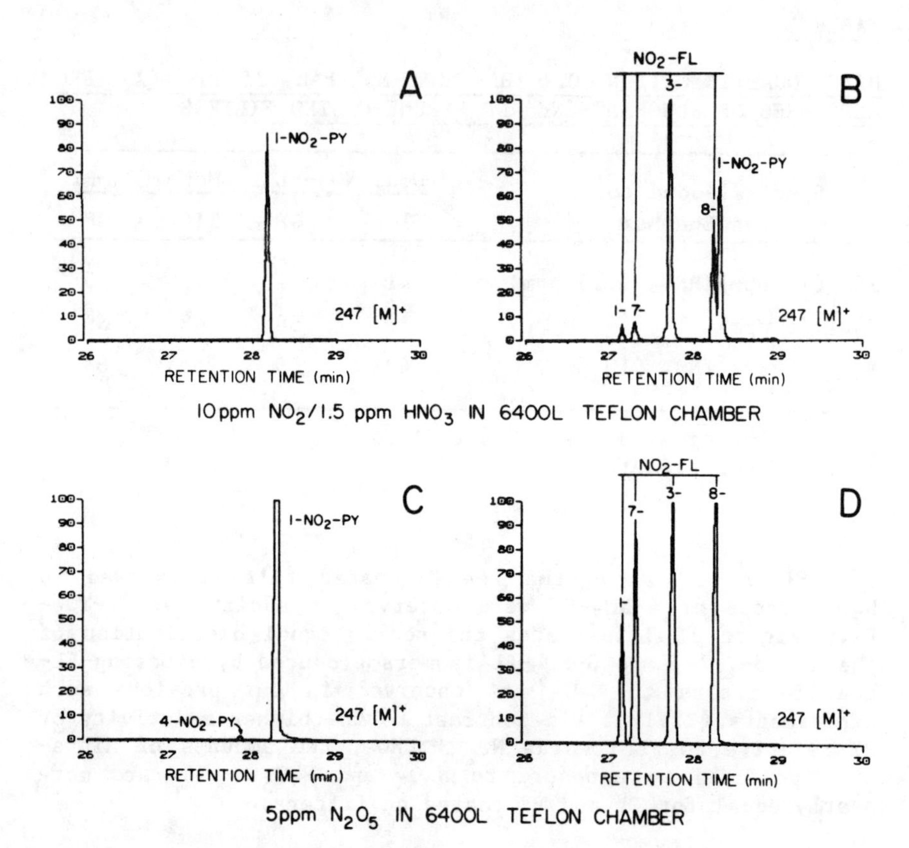

FIGURE 1. GC/MS MID traces of the m/z=247 molecular ion for NO_2-PY and NO_2-FL using a 60 m DB-5 capillary column (injection at 50°C, programmed at 8°C min^{-1} to 350°C). Shown in A and B are extracts from TIGF filters simultaneously exposed to a 10 ppm NO_2 + 1.5 ppm HNO_3 mixture. A: PY-coated filter extract. B: FL-coated filter extract [more highly concentrated (by a factor of ~20) than A]. Shown in C and D are extracts from TIGF filters simultaneously exposed to 5 ppm N_2O_5. C: PY-coated filter extract. D: FL-coated filter extract.

Shown are MID traces of the extract of a PY-coated filter exposed to N_2O_5, spiked with a solution of di-NO_2-PY-d_8 isomers [produced from the reaction of PY-d_{10} with nitric acid in glacial acetic acid (31)]. The deuterated NO_2-PAH elute just prior to their non-deuterated analogs on the DB-5 column employed, and therefore, as can be seen from Figure 2A and 2B, 1,3-, 1,6- and 1,8-di-NO_2-PY were present on the filter, as well as one additional, as yet unidentified, di-NO_2-PY isomer.

Figure 2C and 2D give the corresponding MID traces for the injection of the extract of a FL-coated filter exposed to N_2O_5 which has been spiked with a standard solution of deuterated 1,2-di-NO_2-FL. While at least eleven di-NO_2-FL isomers were produced on the FL-coated filter exposed to N_2O_5, the 1,2-di-NO_2-FL isomer which, as described above is the sole di-NO_2-FL isomer produced from FL by reaction with N_2O_5 in CCl_4 at ambient temperature, was not formed.

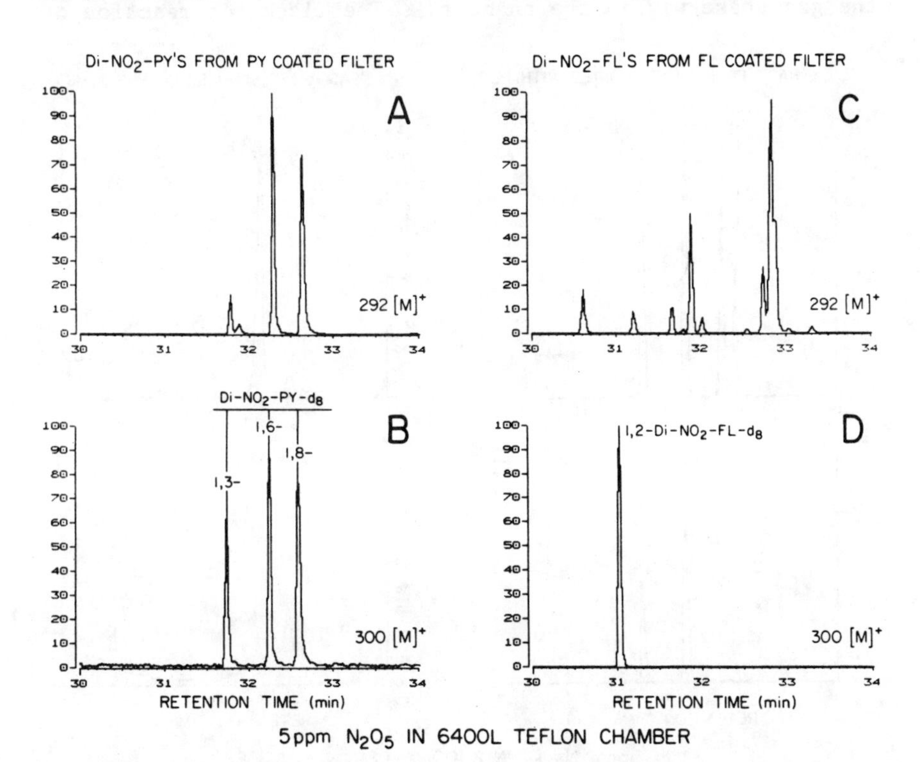

FIGURE 2. GC/MS MID traces of the m/z = 292 molecular ion for di-NO_2-PY and di-NO_2-FL and the m/z = 300 molecular ion for di-NO_2-PY-d_8 and di-NO_2-FL-d_8. A and B are from the injection of the PY-coated filter extract shown in Figure 1C spiked with deuterated 1,3-, 1,6- and 1,8-di-NO_2-PY. C and D are from the injection of the FL-coated filter extract shown in Figure 1D spiked with deuterated 1,2-di-NO_2-FL.

Chamber Exposure of FL Adsorbed on Filters and in the Gas Phase

As shown from the MID traces in Figure 3A and 3B, the 30 min N_2O_5 exposure of FL alone produced the usual distribution of the 1-, 3-, 7- and 8-NO_2-FL isomers on the filters, as discussed above. The NO_2-FL-containing fraction from the HPLC separation of the PUF plug extract produced the MID traces shown in Figure 3C and 3D. The peak eluting at 27.5 min has been identified as the 2-NO_2-FL isomer on the basis of its retention time and the low m/z = 217 $[M-NO]^+$ fragment ion abundance (15). Approximately 5 µg of FL was also collected on the PUF plug, confirming the presence of FL in the gas phase within the chamber. The lack of reaction of

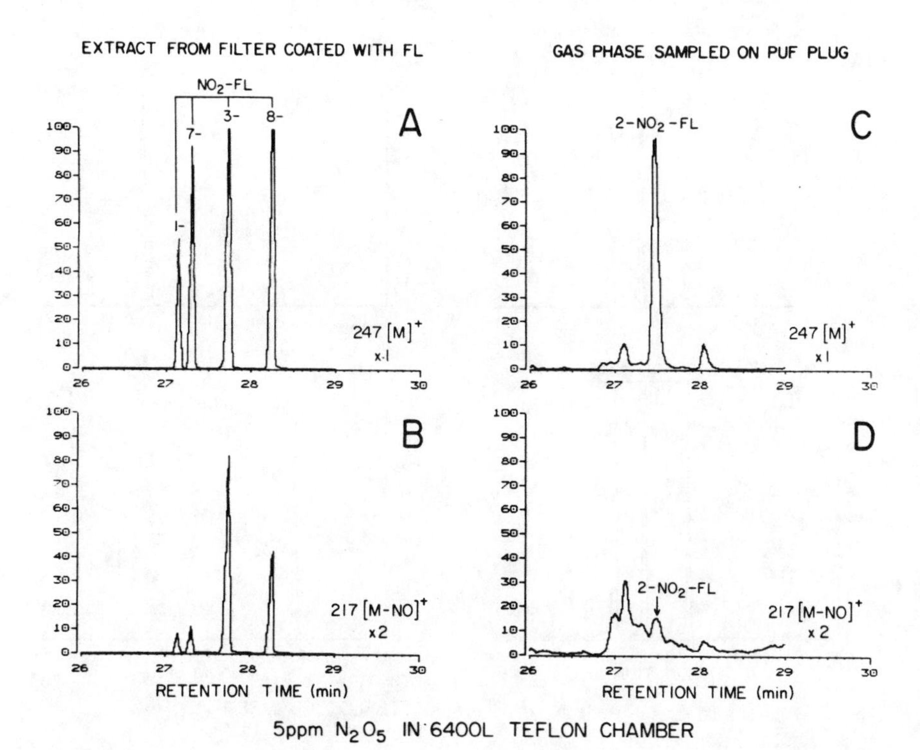

FIGURE 3. GC/MS MID traces of the m/z=247 $[M]^+$ and m/z = 217 $[M-NO]^+$ ions for: A and B, the extract from a FL-coated TIGF filter exposed for 30 min to 5 ppm N_2O_5; C and D, the PUF plug extract from sampling ~75% of the chamber gas volume after 30 min of reaction. Column conditions as given in Figure 1.

the FL-d_{10} spiked onto the PUF plug showed that the observed 2-NO_2-FL was not formed during collection. Thus, 2-NO_2-FL was formed in the gas phase from the reaction of FL with N_2O_5, confirming our earlier report (17). In this earlier work we have shown that, as with naphthalene in the gas phase (32), N_2O_5, and not the NO_3 radical, is responsible for the observed nitration products.

DISCUSSION

Implications for Atmospheric Transformations of PAH

As discussed earlier, the presence of the 2-nitro-isomers of FL and PY in ambient POM extracts is evidence for the existence of atmospheric transformation reactions of these PAH since, to date, neither of these 2-nitro-isomers has been reported to be present in direct emissions of POM.

The 2-NO_2-FL observed in ambient POM is consistently present in as high or higher concentration than 1- or 2-NO_2-PY (15-18). Since FL is a factor of two more volatile than PY (33), and both of these PAH exist partially in the gas phase at the ambient temperatures encountered in these previous studies, this suggests that gas-phase reactions of PY and FL forming the nitro-PAH may be a dominant source of these nitro-PAH in ambient POM.

The room temperature reaction of PY with N_2O_5 in CCl_4 solution, as well as the reaction of PY adsorbed on filters with N_2O_5, produces 1-nitropyrene. Therefore, since the anticipated gas-phase reaction product of PY with N_2O_5 is also 1-NO_2-PY, reaction of N_2O_5 with PY in the atmosphere is an unlikely pathway for the formation of the 2-NO_2-PY observed in ambient POM extracts.

For FL, the sole mononitro-PAH product of the room temperature reaction with N_2O_5 in CCl_4 solution [in which N_2O_5 is reported to be non-ionized (34)] is 2-NO_2-FL. This is also the only nitrofluoranthene isomer formed from the gas-phase reaction of N_2O_5 with FL. It has recently been recognized that a substantial fraction of nighttime ambient NO_x can be in the form of N_2O_5, with calculated concentrations for N_2O_5 ranging up to ~15 ppb (14). Since a significant portion of the FL emitted from combustion sources will be present in the gas phase, the nighttime reaction of gaseous FL with N_2O_5 followed by condensation onto POM is thus one possible formation pathway for the 2-nitrofluoranthene observed in ambient POM extracts.

However, further work is clearly necessary to investigate the NO_2-PAH isomers formed from the other expected reaction pathways operative under atmospheric conditions, for example, reaction with OH radicals in the presence of NO_x. Based upon the limited data set available, any gas-phase N_2O_5 reactions are probably relatively slow, with half-lives with respect to the parent PAH of days or more (32), in contrast to the gas-phase OH radical reactions where the PAH lifetimes are measured in hours (35).

ACKNOWLEDGMENTS

We thank Dr. Thomas Ramdahl for helpful discussions and Travis M. Dinoff, Sara M. Aschmann and Tricia McElroy for very able technical assistance. We also thank the U. S. Department of Energy (Project Officers, Drs. David A. Ballantine and George Stapleton) for support of this research through Contract No. DE-AM03-76SF00034, Project No. DE-AT03-79EV10048.

REFERENCES

1. Schuetzle, D., Riley, T. L., and Prater, T. J. (1982): Analysis of nitrated polycyclic aromatic hydrocarbons in diesel particulates. Anal. Chem., 54:265-271.
2. Xu, X. B., Nachtman, J. P., Jin, Z. L., Wei, E. T., Rappaport, S. M., and Burlingame, A. L. (1982): Isolation and identification of mutagenic nitro-PAH in diesel-exhaust particulates. Anal. Chim. Acta., 136:163-174.
3. Pitts, J. N., Jr., Lokensgard, D. M., Harger, W., Fisher, T. S., Mejia, V., Schuler, J. J., Scorziell, G. M., and Katzenstein, Y. (1982): Mutagens in diesel exhaust particulate: Identification and direct activities of 6-nitrobenzo[a]pyrene, 9-nitroanthracene, 1-nitropyrene and 5H-phenanthro[4,5-bcd]pyran-5-one. Mutation Res., 103:241-249.
4. Gibson, T. L. (1983): Sources of direct-acting nitroarene mutagens in airborne particulate matter. Mutat. Res., 122:115-121.
5. Pitts, J. N., Jr., Van Cauwenberghe, K. A., Grosjean, D., Schmid, J. P., Fitz, D. R., Belser, W. L., Jr., Knudson, G. B., and Hunds, P. M. (1978): Atmospheric reactions of polycyclic aromatic hydrocarbons: Facile formation of mutagenic nitroderivatives. Science, 202:515-519.

6. Pitts, J. N., Jr. (1979): Photochemical and biological implications of the atmospheric reactions of amines and benzo(a)pyrene. Phil. Trans. R. Soc. Lond., 290:551-576.
7. Brorström, E., Grennfelt, P., and Lindskog, A. (1983): The effect of nitrogen dioxide and ozone on the decomposition of particle-associated polycyclic aromatic hdyrocarbons during sampling from the atmosphere. Atmos. Environ., 17:601-605.
8. Hughes, M. M., Natusch, D. F. S., Taylor, D. R., and Zeller, M. V. (1980): Chemical transformations of particulate polycyclic organic matter. In: Polynuclear Aromatic Hydrocarbons: Chemistry and Biological Effects, edited by A. Bjørseth and A. J. Dennis, pp. 1-8, Battelle Press, Columbus.
9. Jäger, J. and Hanus, V. (1980): Reaction of solid-carrier adsorbed polycyclic aromatic hydrocarbons with gaseous low-concentrated nitrogen dioxide. J. Hyg. Epidemiol. Microbiol. Immunol., 24:1-12.
10. Butler, J. D. and Crossley, P. (1980): Reactivity of polycyclic aromatic hydrocarbons adsorbed on soot particles. Atmos. Environ., 15:91-94.
11. Platt, U., Perner, D., Winer, A. M., Harris, G. W., and Pitts, J. N., Jr. (1980): Detection of NO_3 in the polluted troposphere by differential optical absorption. Geophys. Res. Lett., 7:89-92.
12. Pitts, J. N., Jr., Biermann, H. W., Atkinson, R., and Winer, A. M. (1984): Atmospheric implications of simultaneous nighttime measurements of NO_3 radicals and HONO. Geophys. Res. Lett., 11:557-560.
13. Pitts, J. N., Jr., Winer, A. M., and Sweetman, J. A. (1984): Particulate and gas phase mutagens in ambient and simulated atmospheres. CARB A3-049-032, California Air Resources Board, Sacramento, CA.
14. Atkinson, R., Winer, A. M., and Pitts, J. N., Jr. (1985): Estimation of nighttime N_2O_5 concentrations from ambient NO_2 and NO_3 radical concentrations and the role of N_2O_5 in nighttime chemistry. Atmos. Environ. (in press).
15. Pitts, J. N., Jr., Sweetman, J. A., Zielinska, B., Winer, A. M., and Atkinson, R. (1985): Determination of 2-nitrofluoranthene and 2-nitropyrene in ambient particulate organic matter: Evidence for atmospheric reactions. Atmos. Environ., 19:1601-1608.
16. Nielsen, T., Seitz, B., and Ramdahl, T. (1984): Occurrence of nitro-PAH in the atmosphere in a rural area. Atmos. Environ., 18:2159-2165.
17. Sweetman, J. A., Zielinska, B., Atkinson, R., Ramdahl, T., Winer, A. M., and Pitts, J. N., Jr. (1985): A

possible formation pathway for the 2-nitrofluoranthene observed in ambient particulate organic matter. Atmos. Environ. (in press).

18. Ramdahl, T., Sweetman, J. A., Zielinska, B., Harger, W. P., Winer, A. M., and Atkinson, R. (1985): Determination of nitrofluoranthenes and nitropyrenes in ambient air and their contribution to direct mutagenicity. Presented at the 10th Int. Symp. on PAH, Battelle, Columbus, Ohio.
19. Paputa-Peck, M. C., Marano, R. S., Schuetzle, D., Riley, T. L., Hampton, C. V., Prater, T. J., Skewes, L. M., Jensen, T. E., Ruehle, P. H., Bosch, L. C., and Duncan, W. P. (1983): Determination of nitrated polynuclear aromatic hydrocarbons in particulate extracts by capillary column gas chromatography with nitrogen selective detection. Anal. Chem., 55:1946-1954.
20. Campbell, R. M. and Lee, M. L. (1984): Capillary column gas chromatographic determination of nitro polycyclic aromatic compounds in particulate extracts. Anal. Chem., 56:1026-1030.
21. Schott, G. and Davidson, N. (1985): Shock waves in chemical kinetics: The decomposition of N_2O_5 at high temperatures. J. Am. Chem. Soc., 80:1841-1853.
22. Pitts, J. N., Jr., Sweetman, J. A., Zielinska, B., Atkinson, R., Winer, A. M., and Harger, W. P. (1985): Formation of nitroarenes from the reaction of polycyclic aromatic hdyrocarbons with dinitrogen pentoxide. Environ. Sci. Technol. (in press).
23. Zielinska, B., Sweetman, J. A., Atkinson, R., Ramdahl, T., Winer, A. M., and J. N. Pitts, Jr. (1985): The reaction of dinitrogen pentoxide with fluoranthene (in preparation).
24. Pitts, J. N., Jr., Zielinska, B., Sweetman, J. A., Atkinson, R., and Winer, A. M. (1985): Reactions of adsorbed pyrene and perylene with gaseous N_2O_5 under simulated atmospheric conditions. Atmos. Environ., 19:911-915.
25. Thrane, K. E. and Mikalsen, A. (1981): High volume sampling of airborne PAH using glass fiber filters and polyurethane foam. Atmos. Environ., 15:909-918.
26. Dewar, M. J. S., Mole, T., and Warford, E. W. T. (1956): Electrophilic substitution - VI. The nitration of aromatic hydrocarbons, partial rate factors and their interpretation. J. Chem. Soc. 3581-3586.
27. Streitwieser, A., Jr. and Fahey, R. C. (1962): Partial rate factors for nitration of fluoranthene. J. Org. Chem., 27:2352-2355.
28. Kloetzel, M. C., King, W., and Menkes, J. H. (1956): Fluoranthene derivatives. III. 2-Nitrofluoranthene and

2-aminofluoranthene. J. Am. Chem. Soc., 78:1165-1168.

29. Svendsen, H., Rønningsen, H.-P., Sydnes, L. K., and Greibrokk, T. (1983): Separation and characterization of mononitroderivatives of phenanthrene, pyrene, chrysene, fluoranthene and triphenylene. Acta Chem, Scand. B., 37:833-844.
30. Tuazon, E. C., Sanhueza, E., Atkinson, R., Carter, W. P. L., Winer, A. M., and Pitts, J. N., Jr. (1984): Direct determination of the equilibrium constant at 298 K for the $NO_2 + NO_3 \rightleftarrows N_2O_5$ reactions. J. Phys. Chem., 88:3095-3098.
31. Vollman, H., Becker, H., Correll, M., and Streeck, H. (1937): Beiträge zur Kenntnis des pyrens und seiner derivate. Justus Liebigs Ann. Chem., 53:1-159.
32. Pitts, J. N., Jr., Atkinson, R., Sweetman, J. A., and Zielinska, B. (1985): The gas-phase reaction of naphthalene with N_2O_5 to form nitronaphthalenes. Atmos. Environ., 19:701-705.
33. Sonnefeld, W. J., Zoller, W. H., and May, W. E. (1983): Dynamic coupled column liquid chromatographic determination of ambient temperature vapor pressure of polynuclear aromatic hydrocarbons. Anal Chem., 55:275-280.
34. Schofield, K. (1980): Aromatic Nitration. Cambridge University Press, Cambridge, U.K., 376 pp.
35. Biermann, H. W., Mac Leod, H., Atkinson, R., Winer, A. M., and Pitts, J. N., Jr. (1985): Kinetics of the gas-phase reactions of the hydroxyl radical with naphthalene, phenanthrene, and anthracene. Environ. Sci. Technol., 19:244-248.

PAH CHARACTERIZATION IN HAZARDOUS WASTE FROM COKE PROCESSING PLANTS

ARTHUR F. TUCCI
ERCO, A Division of ENSECO, Inc., Cambridge, Massachusetts 02138.

INTRODUCTION

The process of high temperature carbonization of coal in the formation of coke and other related products yields a tremendous amount of waste.[1,2] The process not only produces products vital to the steel and construction industry, but also the sole source of many chemicals used in laboratories and research studies, as well as many petroleum fuel products.

It is of great importance that the waste generated by this process be characterized and deemed hazardous or not. A waste is deemed hazardous if it exhibits one of the following criteria: ignitability, reactivity, EP (Extraction Procedure) toxicity, or if the waste is specifically listed as hazardous. Two wastes from the coking process - ammonia lime still bottom sludge and decantor tar sludge - are two such listed wastes.[3,4] PAH determination can help characterize the waste and act as a guide in deciding the eventual disposition of the waste in question. The method described here provides a quick and effective means for PAH determination of such wastes.

MATERIALS AND METHODS

Chemicals

Hexane and dichloromethane used were distilled in glass grade from Caledon, Laboratories, Ltd. Alumina 85-325 mesh was purchased from MCB Manufacturing Chemicals, Inc.

Sampling

Sludge samples were obtained from filter process tar decanter tank sludge. A pile sampler was used to collect random core samples of the sludge insuring objective, representative sampling.

PAH CHARACTERIZATION

Sample Preparation

A 1 to 2 g aliquot of sludge was Soxhlet extracted with 300 ml of dichloromethane for 16 hours. The extract is back extracted at pH > 12 and again at pH < 2[5]. The acid extractions can be used to determine levels of phenols and other acid constituents. The base extraction used to recover the PAH is passed through sodium sulfate and concentrated to 10 ml. To recover the PAH, the dichloromethane extract is passed through anhydrous sodium sulfate and concentrated to 10 ml. The extract was solvent exchanged to hexane and concentrated to 1.0 ml. This hexane extract is charged to an alumina column and separated into aliphatic and aromatic fractions. The aromatic fraction is concentrated and analyzed by GC/MS.

Alumina Column Chromatography

Column chromatography using a 1 cm i d x 50 cm glass column packed with 10 g alumina oxide was activated overnight at 120°C. The hexane extract was charged to the column and eluted with 15 ml of hexane yielding the aliphatic fraction, then 100 ml dichloromethane yielding the aromatic fraction. The aromatic fraction was concentrated to 1 ml and analyzed by GC/MS.

Instrumentation

GC/MS was performed on a Finnigan 4530 GC/MS/DS with the column interfaced directly into the ion source. The mass spectrometer was operated at an electron energy of 70 ev, 0.32 mA emission current, and the ion source at 120°C. The instrument was operated in the electron impact mode with spectra obtained at 1 second intervals scanning from m/z 45 to 450.

The extract was injected onto a 30 m x .32 mm id SE-54 fused silica capillary column with helium used as a carrier gas. The GC oven was programmed from 30°C (4 min) to 270°C (10 min) at a rate of 10°C/min.

GC/MS data were acquired and processed using a Finnigan/ Incos Data System. Compound identifications were made by spectral interpretation and/or matching the mass spectra with the NBS mass spectral library.

RESULTS

Table 1 lists some selected PAH that are present in waste generated from coke processing plants. Since these may be present at low levels, removal of interference is necessary for instrument detection. Since the major interference is caused by the presence of aliphatic hydrocarbon, the alumina column chromatography proved extremely effective in reducing this aliphatic hydrocarbon interference content and thus making detection of the PAH(s), even at low parts-per-million levels, possible. Extensive method validation studies show that not only are the results reproducible, but the recovery of spiked compounds was extremely good.

TABLE 1

PAH SUSPECTED IN COKE PROCESSING WASTE

Naphthalene	Pyrene
α-Naphthylamine	Benzo(a)anthracene
β-Naphthylamine	Chrysene
Pyridine	Benzo(b)fluoranthene
Quinoline	Benzo(e)pyrene
Aniline	Benzo(a)pyrene
Acenaphthylene	Indeno(1,2,3-cd)pyrene
Acenaphthene	Dibenzo(a,h)anthracene
Fluorene	Dibenzo(a,j)acridine
Phenanthrene	Benzo(g,h,i)perylene
Anthracene	Coronene
Acridine	Carbazole
Benzoquinoline	Dibenzofuran
Fluoranthene	Dibenzothiophene

A series of replicate extractions was performed on a waste sludge. Care was taken to maintain sample homogeneity. The results of some selected compounds are presented in Table 2. In most instances, there is good reproducibility as demonstrated by the coefficient of variance (C_v).

A series of spike and recovery studies were also performed on a waste sludge. The compounds were spiked into

TABLE 2

CONCENTRATION OF SELECTED COMPOUNDS IN REPLICATE ANALYSES ON WASTE SLUDGE

	Conc (ppm)						$\overline{x}$	$\%C_v$
Benzo(a)pyrene	3	3	4	5	4	4	3.8	19%
Benzo(b)fluoranthene	3	3	4	5	4	4	3.8	20%
Benzo(g,h,i)perylene	0.8	0.6	0.8	1.1	0.8	1.1	0.9	23%
Naphthalene	410	500	310	540	510	330	430	23%
Acenaphthene	64	100	62	85	84	72	78	19%
Pyrene	23	31	27	34	31	32	30	13%
Chrysene	21	26	27	32	25	35	28	18%
Phenanthrene	196	324	214	262	301	264	260	19%

TABLE 3

% RECOVERY OF SELECTED COMPOUNDS SPIKED INTO SLUDGE SAMPLE

Dibenzofuran	101 ± 16
Dibenzothiophene	78 ± 30
Quinoline	79 ± 14
Acridine	94 ± 12
Carbazole	22 ± 16
α-Naphthylane	52 ± 12
Phenanthrene	82 ± 57
Anthracene	64 ± 10
Pyrene	75 ± 13
Benzo(a)anthracene	100 ± 18
Chrysene	70 ± 16
Benzo(b)fluoranthene	79 ± 17
Benzo(a)pyrene	87 ± 10
Dibenzo(a,h)anthracene	93 ± 13

the sample at 50 ppm level. The recoveries are listed in Table 3.

DISCUSSION

PAH characterization of coke processing waste is achieved using the method described. The preparation is relatively simple and straight forward. Variation in the waste matrix has often complicated the analysis. Various extraction procedures, back extractions, solvent partitioning, and numerous column clean-up steps have been used to prepare the sample for analysis.[5] Thus, the method is applicable to a wide range of sample matrices. The PAH are mostly contained in the aromatic fraction (the aliphatic fraction can be set aside for other analyses). GC/MS is the chosen analytical finish for several reasons. It is highly sensitive; most compounds can be detected at low ppm levels. GC/MS offers greater selectivity than either GC/FID or HPLC (two other commonly used analytical finishes). One can quantitate on a single, representative ion, thus allowing detection of compounds of interest in spite of any background masking and also allowing detection of those compounds that coelute. Using high resolution capillary columns with appropriate standards, isomers such as benzo(e)pyrene and benzo(a)pyrene can be resolved.[6,7] Lastly, with the aid of library searching and standard spectral information, identification of unknown species is made possible.

Therefore, the method described here is an effective and economical way to characterize waste for PAH content. The method shows good reproducibility and the capability for detecting compounds present even at low ppm levels.

Complete characterization requires more than just determination of PAH content. For such characterization, a much greater range of analytical studies must be undertaken on the waste in question.

ACKNOWLEDGEMENTS

I wish to thank Dr. Kin Chiu for his help and support, Dr. John Maney, and Dr. Dallas Wait for their technical assistance and the production department for preparation of this manuscript.

REFERENCES

1. United States Environmental Protection Agency: "Benzene emissions from coke by-product recovery plants - back-

ground information for proposed standards," October 1981.
2. Lebowitz, H., et al.: "Potentially hazardous emissions from the extraction and processing of coal and oil," EPA-650/2-75-038, April 1975.
3. Perch, Michael and Muder, Richard E.: "Coal carbonization and recovery of coal chemicals," Handbook of Industrial Chemistry, James A. Kent, Ed. Van Nostrand Reinhold Co., NY, 1974.
4. EPA Guidelines for Hazardous Waste Management, Part 261 of RCRA regulations.
5. Method 3530, SW846 Test Methods for Evaluating Solid Waste, Physical/Chemical Methods, Government Printing Office Publication 055-002-B1001-2, 1982.
6. Lee, M.L., Vassilards, D.L., White, C.M., Novotny, M., (1981): Analytical Chemistry, 53:768.
7. Wiley, C., Iworo, M., Castle, R.N., Lee, M.L., (1980): Analytical Chemistry, 53:400.

STUDY OF PAH METABOLISM IN SINGLE LIVING CELLS: SIMULTANEOUS DETERMINATION OF KINETIC PARAMETERS CHARACTERISTIC OF THE ACTIVATION STEP AND OF SOME PATHWAYS OF THE DETOXIFICATION STEP.

PIERRE VIALLET*, DOMINIQUE LAUTIER, BERNARD ANTHELME, SYLVIE LAHMY, JEAN-MARIE SALMON
Group de Microfluorimétrie Quantitative et Pharmacocinétique Cellulair, Laboratoire de Chimie Physique, Université de Perpignan, 66025 Perpignan Cedex, France.

INTRODUCTION

Conventional techniques are able to detect very small amounts of a great variety of metabolites in cell culture media -- or in microsomal supernatant -- but are unable to give any information about possible accumulation in cells. Nothing can be said about the rates of the successive steps, or the different pathways involved in complex metabolism.

Quantitative analysis of cell fluorescence is now possible due to microspectrofluorimetry (1,2,3). Resolution of complex cell fluorescence spectra into its components enables us not only to measure the amount of the fluorescent parent compound remaining in each cell but also to characterize the main fluorescent metabolites which accumulate in every cell and to evaluate their relative amounts. This resolution supposes that a library of the fluorescence spectra of these metabolites exists. Microspectrofluorimetric techniques cannot be effective without the support of conventional methodologies but they have been proved to be more informative on many points:

- As it has been stated above, fluorescent metabolites can be identified as long as they accumulate in cells. From a kinetic point-of-view, even the lack of a fluorescent metabolite is informative; it is proof that the compound undergoes further metabolism as soon as it is produced. Of course, nothing can be said about the non-fluorescent metabolites of the parent compound. That means special care has to be taken in the choice of the parent compound so that it can be transformed into an effective fluorescent metabolite through the pathway under study.

- With microspectrofluorimetric techniques, the time decay of the intensity of the characteristic parent compound fluorescence spectrum (benzo(*a*)pyrene (B(*a*)P) for instance) can be used to monitor the rate constant of the activation step of metabolism; *i.e.*, the step which is supposed to transform the parent compound into an epoxide, as long as no exclusion process interferes. In the same way, when fluorescent metabolites accumulate in the cells, the kinetics of the time decay of the intensity of each characteristic fluorescence spectrum gives information about the rates of both accumulation and exclusion processes of each compound.

- Moreover, all this information is obtained on living material with intact microstructures. Furthermore, they are obtained on single living cells. Due to the fast increase of computer efficiency, data on relatively large numbers of single cells will be soon accessible so that histograms can be used to check heterogeneity of cell populations.

Heterogeneity of enzymatic activity between cells belonging to the same "homogeneous" line was first demonstrated using cloning techniques (4). More recently a very elegant, powerful and time-saving method has been proposed. Using cell sorting equipment, these authors have been able to characterize and isolate from a mouse hepatoma (Hepa 1c1c7) cell line new variants in B(*a*)P metabolism (5). Unfortunately this very attractive method is limited by the short time of the cell exposure to irradiation. If a great number of cells are checked, the amount of information obtained on each cell is very poor. Observations are generally performed at only one wavelength. In this case the method is only convenient to study cells in which no fluorescent metabolites are able to accumulate. As a consequence, only the activation step of the polycyclic aromatic hydrocarbon (PAH) metabolism can be studied.

Furthermore cells have to be suspended in PBS which is far from a biological environment.

We have previously demonstrated that when RTG2 cells are incubated with B(a)P, two phenolic intermediate metabolites, 3-OH-B(a)P and 9-OH-B(a)P, were easily identified in the cells. These findings suggested that the mechanism responsible for the metabolism of B(a)P could be different in these cells from the one existing in other cells such as 3T3 cells. The difference results from difference in the dynamic equilibrium between the activation step and the detoxification step. But it could also relate to the nature of the enzymes involved. We intend to present here the results of these investigations.

MATERIALS AND METHODS

Cells

RTG2 cells (Rainbow Trout gonad cells, American Type Culture Collection n°CCL55) were cultivated as monolayers (Falcon 25 cm^2) in Minimum Essential Medium (MEM, Boehringer) with 10% decomplemented Foetal Calf Serum (FCS, Gibco) at 18°C. Confluent cells were resuspended after trypsin treatment (Trypsin 0.25%, without Ca^{2+} nor Mg^{2+}, Gibco). Sykes-Moore chambers with upper lamella removed, were incubated with 30,000 cells per chamber at 18°C in 5% CO_2 atmosphere. The medium was removed every 24 hours. All experiments were carried out 72 hours after incubation in order to eliminate the trypsin effects.

Murine 3T3 fibroblasts (A.T.C.C.) were cultivated as above at 37°C, and resuspended after dispase treatment (Dispase Grad II, Boehringer).

Solutions

In order to avoid possible effects of any solvent on the characteristics of the cell membranes, ethanolic solutions of each PAH and each inhibitor were first evaporated; resulting microcrystals were then dissolved with gentle agitation in the culture medium (10% FCS, 1% HEPES). Dissolution was obtained through interactions with proteins and lipoproteins (6). Concentrations were determined spectrophotometrically after ethyl acetate extraction. They remained in the range of 1 μM for each PAH under study. Indomethacin (Ind) was used at 1 and 10 μM and 6-amino-chrysene (6-AC) at 1 μM.

Recording and Data Analysis

The microspectrofluorometer used has been described in detail elsewhere (7,8). Spectra were recorded during the whole experiment with the best signal/noise ratio in order to get experimental data valuable for numerical treatment. A rather complex resolution method with computer assistance was necessary to quantify the participation of each component (intrinsic cellular fluorescence, parent compound, and metabolites fluorescence) to the whole recorded cell fluorescence spectrum.

All experiments were performed at 18°C on RTG2 and at 37°C on 3T3. Due to the use of a stepping motor stage, it has been possible to follow the simultaneous evolution with time of the fluorescence spectra of ten individual cells belonging to the same culture chamber (9). This improvement gives a greater statistical signification to our results and shortens the duration of the experiments.

Precautions were always taken to guard against potential artifacts resulting from the high sensitivity of the equipment and from the light sensitivity of drugs and living material.

Kinetic Studies

As has been previously stated, when cells were incubated with PAH containing medium, the cell fluorescence intensity increased due to the uptake of the PAH. With our experimental conditions, about 15 minutes were necessary for a stabilization of the cell fluorescence intensity due to the competition between the intake process and the metabolism of the drug. When this plateau was reached, the PAH containing medium was removed and the culture chambers were quickly washed three times with fresh medium. Finally, a predetermined amount of fresh medium was added and the experiment initiated.

In order to study the effects of inhibitors, cells were incubated for 30 min with B(a)P and indomethacin or 6-AC, then washed so that the concentration of inhibitor remained constant in the medium during kinetic study.

RESULTS

Our equipment enabled following simultaneously on the same cell, the variation of the fluorescence intensity of each compound which participated in the recorded cell fluorescence spectrum. Specifically, the decay with time of the PAH fluorescence intensity gives information about the activation step kinetics of metabolism, whereas the same fluorescence intensity decay due to phenolic compounds gave information about the kinetics of the pathways involved in further metabolism, *i.e.*, mainly their excretion as conjugates, but also as substrates of mixed-function oxidases (MFO). In both cases, enzymatic activities are characterized by an experimental rate constant "K".

Though this paper is mainly devoted to the metabolism of B(*a*)P by RTG2 cells, some comparisons will be made with results involving other PAH compounds; benzo(*k*)fluoranthene (B(*k*)P), cyclopentano(*c,d*)pyrene (CPAP), and other cell lines (3T3 cells).

Heterogeneity of Cellular Enzymatic Responses when PAH were Used as Substrate

The histograms presented in Figure 1 illustrate the difference between responses of RTG2 and 3T3 cells. Histograms D and E are good examples of a gaussian distribution whereas histograms A, B, and C are representative of a more complex distribution in which a number of cells show poor enzymatic activities. Histogram F shows that this situation can also be found to a lesser extent when CPAP was used as a substrate for RTG2 cells.

These complex responses suggested that the disappearance of the parent compounds could occur through two competing enzymatic systems: one involving the MFO and the other involving some lipoxygenases. This hypothesis has been tested using blocking agents and the results are presented at the end of this paper.

Heterogeneity of the Responses When Phenolic Compounds Were Used as the Substrate

Because 3-OH-B(*a*)P and 9-OH-B(*a*)P were easily identified in RTG2 cells when B(*a*)P was used as the substrate, the metabolism of these compounds has been studied, each compound used separately for incubating cells. These results are shown in Figure 2, Part A. Typically the shape of histograms

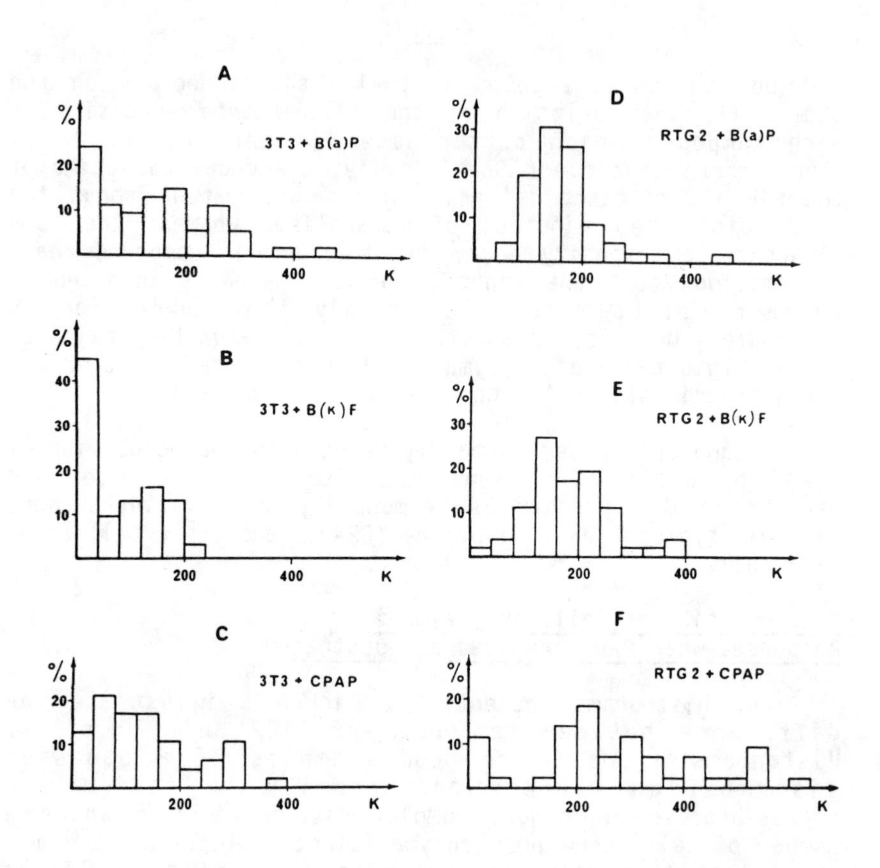

FIGURE 1. Distribution of activation rate constants (K) expressed in 10^{-5} s^{-1}: A,B,C: 3T3, respectively treated with B(a)P, B(k)F and CPAP. D,E,F: RTG2 respectively treated with B(a)P, B(k)F and CPAP.

which characterize both compounds is not very different from that obtained with the parent compound. It is interesting to notice that, even with experimental conditions where a phenolic compound could be used as the substrate, not only for glucuronyl and sulfotransferases, but also for oxygenases, the rate constants were always lower for these primary metabolites than for the parent compound. These results suggest that these compounds are less efficient as substrates for oxygenases when they have to compete with

B(a)P. This agrees with the fact that these primary metabolites are able to accumulate in RTG2 cells when B(a)P is used as the substrate.

Heterogeneity of the Responses Versus the Metabolism of Phenolic Compounds When B(a)P Was Used as the Substrate

When the RTG2 cells were incubated with B(a)P, resolution techniques enabled us to calculate the metabolism rate constants of both 3-OH-B(a)P and 9-OH-B(a)P, and to compare these results with data simultaneously obtained for B(a)P (Figure 2, Part B).

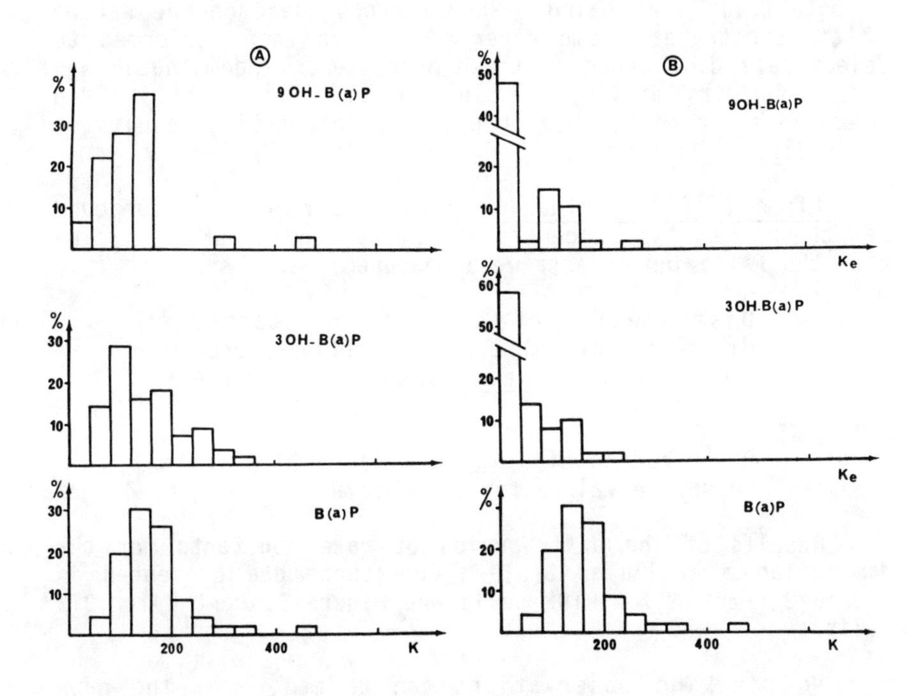

FIGURE 2. A: Distribution of activation rate constants (K) expressed in 10^{-5} s^{-1} with RTG2 treated respectively with B(a)P, 3-OH-B(a)P, and 9-OH-B(a)P.
B: Distribution of activation rate constants (K) of B(a)P and excretion rate constants (Ke) expressed in 10^{-5} s^{-1} of phenolic metabolites using RTG2 treated with B(a)P.

It is obvious that these results are lower than the corresponding values in Part A and the shapes of these histograms are different. In this case, a large number of cells seem to be very inefficient in the metabolism of 3-OH-B(a)P and/or 9-OH-B(a)P.

All these results suggest that under experimental conditions, phenolic compounds were acting mainly as substrates for conjugation. For that reason, the corresponding rate constants are referenced as Ke.

Influence of Blocking Agents

Some of our results suggest that the metabolism of B(a)P can proceed through two competitive processes, the importance in both cell lines being eventually dependent on the nature of the substrate. Some experiments have been performed to selectively block each of these processes. Indomethacin was chosen for its ability to block lipoxygenase activity and 6-AC was selected because it produced interesting results on 3T3 cells.

Effects of Indomethacin. The effect of this compound was checked at two concentrations, 1 μM and 10 μM. In each case the following effects were measured:

- Distribution of the rate constants, diminution of cellular B(a)P fluorescence intensity, *i.e.*, activation rate constants.

- Distribution of Ke values for 9-OH-B(a)P versus Ke values for 3-OH-B(a)P.

Results of the distribution of rate constants for the diminution of cellular B(a)P fluorescence are presented in Figure 3, Part A for RTG2 cells and Figure 3, Part B for 3T3 cells.

No striking observations can be made from the data obtained with the RTG2 cell line. Surprisingly indomethacin at low levels is more efficient than at high levels. Indomethacin appears more efficient in 3T3 cells with some dose/effect correlations. The reasons for these findings in the case of RTG2 may be that too low doses of indomethacin have been used in the past. Nevertheless, these doses were found to be large enough to modify the Ke values for 3-OH-B(a)P and 9-OH-B(a)P as shown in Figure 5.

Distribution plots of Ke values for 9-OH-B(a)P versus Ke values for 3-OH-B(a)P are presented in Figure 5, Parts A and B, and give a representation of the relative activity of each cell versus the phenolic compounds. It can be seen that when cells were used without blocking agents, most of the cells appeared to metabolize only one of the phenols. It should be noticed in order to clarify Figure 5, the number of cells showing practically no activity versus both of the compounds, has been under-represented. In fact, these cells represent

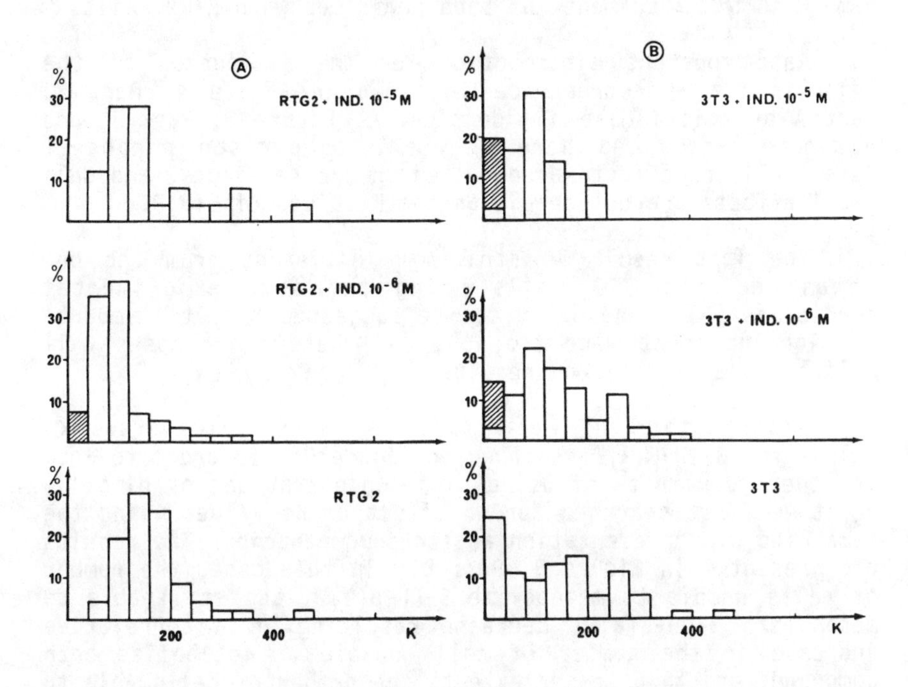

FIGURE 3. Distribution of activation rate constants (K) expressed in 10^{-5} s^{-1}: A: RTG2 respectively treated with B(a)P for 30mn, with B(a)P and indomethacin 1 μM and with B(a)P and indomethacin 10 μM (indomethacin remaining constant in the medium during kinetic studies); B: The same for 3T3.

30% of the "control population, 11.5% of indomethacin-treated cells, and 42.6% of 6-AC-treated cells. Only a relatively low number of cells have shown a measurable activity versus both compounds. This number was strongly increased when cells were incubated with indomethacin even at micromolar level. Ke values for 9-OH-B(a)P seem to be more affected by indomethacin than for 3-OH-B(a)P suggesting that further experiments have to be undertaken in order to clarify both the meaning of Ke and the effects of indomethacin.

Effects of 6-AC. We previously demonstrated the effect of exposure to micromolar doses of 6-AC on the rate constant of the B(a)P metabolism (2). Using the 3T3 cell line, more striking results have been obtained when the concentration of the blocking agent was maintained constant in the cells. The same kind of experiment has been performed with RTG2 cells.

Rate constant distributions of the diminution of the cellular B(a)P fluorescence is shown in Figure 4. Part B. Part A of this figure is identical to Figure 2, Part B, and has been reproduced here only for comparison purposes. Essentially no effect was detected on the Ke values, and only small effects are noticeable on the K values of B(a)P.

The last result is strikingly different from the one obtained with 3T3 cells using the same experimental conditions (2). Again, it can be suggested that the amounts of blocking agent were too low, and that higher doses will deliver more significant results.

Distribution of Ke values for 9-OH-B(a)P versus Ke values for 3-OH-B(a)P is shown in Figure 5. In order to look for the opportunity of using high concentrations of blocking agent we first searched for an effect on Ke values using the same kind of representation as for indomethacin. The results are presented in Figure 5, Part C. In this case, the number of cells unable to metabolize 9-OH-B(a)P, and still able to metabolize 3-OH-B(a)P, decrease resulting in a correlative increase in the number of cells unable to metabolize both compounds and to a lesser extent, the number of cells able to metabolize each of them.

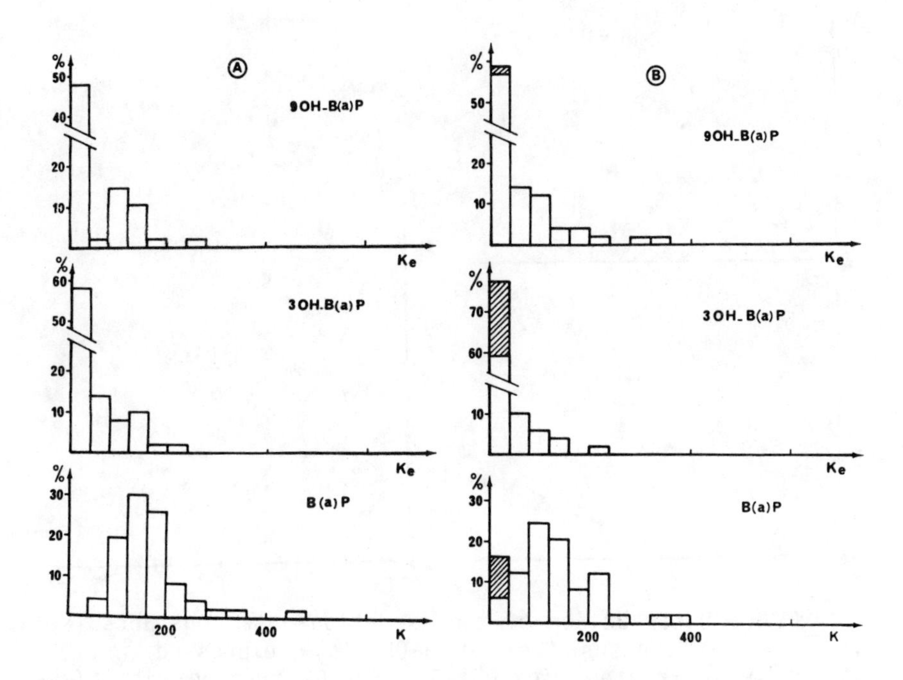

FIGURE 4. Distribution of activation rate constants (K) of B(a)P and excretion rate constants (Ke) expressed in 10^{-5} s^{-1} of phenolic metabolites: A: RTG2 treated with B(a)P for 30mn; B: RTG2 treated with B(a)P and 6-AC 1 μM (6-AC remaining constant in the medium during kinetic studies).

DISCUSSION

Microfluorometric techniques afford us the opportunity to collect much information about the enzymatic activity in single living cells. Because the data can be expressed in terms of histograms, and due to the increasing power of relatively low-cost computers, the statistical significance of these results will be more important in the future. Data interpretation, on the other hand, very often is not as conclusive as for usual biological experiments. It is the belief of the authors that this situation is not due to limits in the potential of the method but to the complexity of intact living material. The experiments above were selected in order to illustrate this effect.

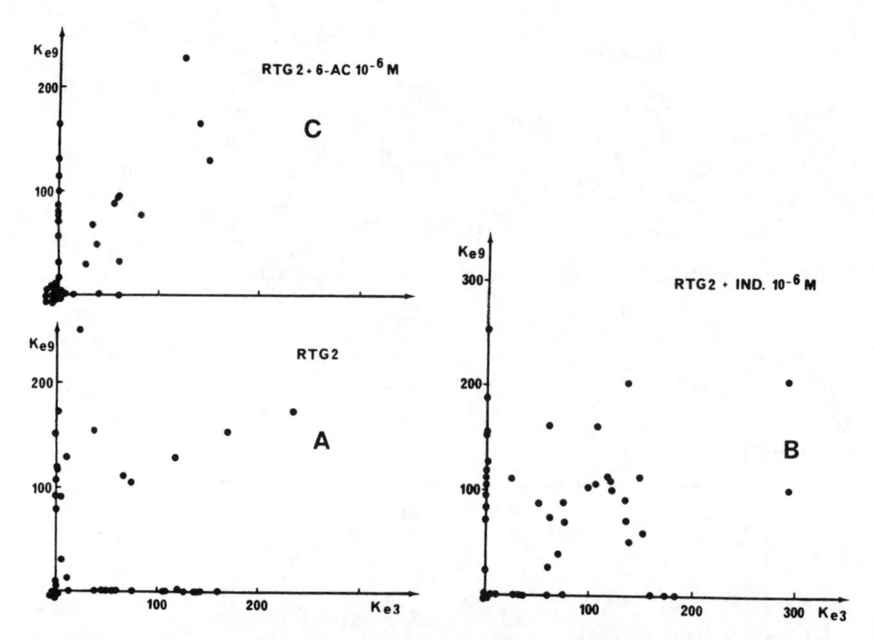

FIGURE 5. Distribution of the excretion rate constants (Ke) of 9-OH-B(a)P versus (Ke) of 3-OH-B(a)P expressed in 10^{-5} s^{-1}: A: RTG2 treated with B(a)P; B: RTG2 treated with B(a)P and indomethacin 1 µM; C: RTG2 treated with B(a)P and 6-AC 1 µM.

Three main results can be pointed out:

- Neither 3T3 cells nor RTG2 cells seem to metabolize B(a)P through a single enzymatic pathway. It is clear from previous experiments that the activity of the first cell line tested with B(a)P is very sensitive to 6-AC; but dose/effect relationships can be detected using indomethacin.

  It is also clear that the RTG2 cell line is not sensitive to 6-AC but a slight effect is detectable. Indomethacin, however, appears controversial at first sight. Nevertheless we cite quinones which are proven to result from lipoxygenase or prostaglandin synthetase activity and are believed to block monooxygenase activity. If this is true, then it is possible that the dose of indomethacin could result in an increase, or a decrease, of the rate constant in the complex enzymatic system responsible for the metabolism of B(a)P. This explanation holds only if we agree that both the enzymatic pathways - MFO and lipoxygenase - are effective in the cells under study.

- Compounds which decrease the activity of any enzyme may also be efficient against another enzyme involved in another step of the process or in another pathway, and these effects can be dose-dependent. This is clear with both cell lines: the so-called Ke constants are sensitive to indomethacin and/or 6-AC. Such results are not easy to interpret. An increase in Ke can be the result of an increase in the rate of excretion of the phenols from the cell through the conjugation processes. But because intermediate metabolites are also potential substrates for the enzymes involved in the first pathway, it can also result from some hypothetical specificity of the blocking effect of the drug. Fortunately in our case the shape of the histograms describing the distribution of either K or Ke for the phenolic compounds are so different that they can be evaluated from results such as those given in Figure 5. The shifts in Ke values are due to an increase in the speed of excretion.

Results presented in Figure 1 show that the shape of the histograms which characterize the distribution of the cells versus the values of the rate constant K are strongly dependent on the chemical nature of the PAH. This strongly suggests that a mixture of enzymes is involved in the first step of their metabolism. This agrees with the hypothesis developed in the second paragraph of this conclusion.

REFERENCES

1. Salmon, J.M., Thierry, C., Serrou, B., and Viallet, P. (1981): Slower step of the polycyclic aromatic hydrocarbons metabolism: Kinetic data from microspectrofluorimetric techniques, Biomedicine, 34:102-107.
2. Lahmy, S., Salmon, J.M., and Viallet, P. (1984): Quantitative microspectrofluorimetry study of the "blocking effect" of 6-aminochrysene on benzo(a)pyrene metabolism using single living cells, Toxicology, 29:345-356.
3. Viallet, P., Boiret, M., Lahmy, S., and Salmon, J.M. (1984): Determination of kinetic parameters of MFO activity on single living cells using microspectrofluorimetric techniques. In: Polynuclear Aromatic Hydrocarbons, Chemistry, Characterization, and Carcinogenesis, edited by M. Cooke and A. J. Dennis, pp. 1329-1342, Battelle Press, Columbus, Ohio.
4. Whitlock, J.P., Gelboin, H.V., and Coon, H.G. (1976): Variation in aryl hydrocarbon hydroxylase activity in heteroploid and predominantly diploid rat liver cells in culture, J. Cell. Biol., 70:217-225.
5. Miller, A.G. and Whitlock, J.P. (1982): Heterogeneity in the rate of benzo(a)pyrene metabolism in single living cells: Quantitative using flow cytometry, Mol. Cell. Biol., Volume 2, n°6, pp. 625-632.
6. Avigan, J. (1959): The interaction between carcinogenic hydrocarbons and serum lipoproteins, Cancer Res., 19:831-834.
7. Salmon, J.M. (1980): Réalisation d'un microspectrofluorimétre. Application à l'etude de quelques mécanismes cellulaires, Thése d'état.
8. Salmon, J.M., Vigo, J., and Viallet, P. (1981): Microspectrofluorimétrie sur cellule vivante isolée: Couplage d'un microspectrofluorimétre à un microordinateur. Innov. Tech. Biol. Méd, 2:679-686.
9. Allegre, J.M., Salmon, J.M., Commalonga, J., Savelli, M., and Viallet, P. (1985): Microspectrofluorimétrie quantitative. Automatisation d'un microspectrofluorimétre équipé simultané sur différentes cellules d'une population cellulaire, de cinétiques intracellulaires, Innov. Tech. Biol. Méd., (in press).

FIBEROPTICS IMMUNOFLUORESCENCE SPECTROSCOPY FOR CHEMICAL AND BIOLOGICAL MONITORING

T. VO-DINH*[1], G. D. GRIFFIN[1], K. R. AMBROSE[1], M. J. SEPANIAK[2], and B. J. TROMBERG[1]. (1)Advanced Monitoring Development Group, Health and Safety Research Division, Oak Ridge National Laboratory, Oak Ridge, Tennessee 37831; (2) Department of Chemistry, University of Tennessee, Knoxville, Tennessee 37916.

INTRODUCTION

There is increasing interest for human health protection against environmental pollutants such as the carcinogenic polycyclic aromatic hydrocarbons (PAH), which are produced in many occupational and residential activities as a result of incomplete combustion of organic matter (1). As the concern over long-term health effects due to low-level and chronic exposures has increased, so has the need for more effective and practical monitors. Biological monitoring, if conducted as an adjunct to chemical exposure monitoring, can provide a critical tool to determine the "absorbed dose" of chemicals, and thus to improve our abilities to effectively assess the effects of hazardous chemicals, and ultimately to provide better human health protection. Simple and practical tools for monitoring complex biological systems have been recently identified as one of the major research needs in monitoring instrumentation for occupational health research (2).

Among the various techniques for trace organic detection, immunoanalytical techniques can provide powerful tools to detect and quantify individual exposure to carcinogens. Such techniques involve determination of the presence of carcinogens, their metabolites, and carcinogen-DNA adducts in human tissue and cells, and in body fluids. Recently, immunoassays have become widely used for the analysis of a wide variety of chemical and biological species including proteins, hormones, drugs, and viruses. High specificity of the assays is assured by the inherent selectivity of immunological reactions. Sensitivity is determined by the detection limit of the label. Radioimmunoassays (RIA) use radioactive labels (1) and are currently the most widely utilized techniques for immunoassays. RIA techniques, however, suffer from several disadvantages: radioisotopes have limited shelf life (60 days for ^{125}I half-life); instruments for detecting and reagents are relatively

expensive; and special precautions are required for the shipping, handling, and waste disposal of radioactive materials. Although sensitivities achieved by RIA are adequate, alternative techniques are being developed to overcome the need for radioactive labels. Recent developments in fluorescence instrumentation, biotechnology and fiberoptics research have opened new possibilities for the development of sensitive and selective tools to measure human exposure to hazardous polynuclear aromatic compounds. This work presents the basic principle of immunofluorescence spectroscopy based on fiberoptics sensors for use in the detection of trace amounts of biological compounds in complex samples. Special focus is on a unique fluoroimmuno-sensor (FIS) which derives its analytical selectivity through the specificity of antibody-antigen reactions. Antibodies are immobilized at the terminus of a fiberoptic within the FIS for use in both in-vitro and in-vivo fluorescence assays. High sensitivity is provided by laser excitation and optical detection techniques. Data illustrating the feasibility, analytical capabilities, and applications of the FIS are discussed. Discussion of the analytical methodologies and instrumentation associated with FIS devices is presented.

BASIC PRINCIPLE AND METHODOLOGY

Molecular Recognition of PAHs: The Antigen-Antibody Reaction

The basis for the specificity of immunological techniques is the antigen-antibody (Ag-Ab) binding reaction, which is the primary mechanism by which the body's immune system detects and eliminates foreign matter. The occurrence of foreign matter in the body induces the production and release into the bloodstream of proteins known as antibodies. In the design of an immuno-sensor, antibodies can be developed to the carcinogenic PAHs themselves, to their metabolites, or to the carcinogen-DNA adducts; these antibodies are then used to determine the amount of antigens present in biological materials of interest (3,4). A substance capable of inducing a response of the immune system with the production of antibodies is called "immunogenic." To be immunogenic, a substance must meet certain requirements in molecular size and complexity. Proteins with molecular weights >5000 are generally immunogenic whereas smaller compounds such as PAH molecules are not immunogenic. The design of immunosensors for PAHs requires the development of antibodies specifically directed against these PAHs. In order to produce anti-PAH antibodies, the PAH are coupled to a carrier protein,

then used to immunize a laboratory animal in order to elicit the animal's immunoresponse. We have recently produced antibodies directed against benzo(a)pyrene (BaP) by coupling BaP to bovine serum albumine (5). Further experimental work is being conducted to produce better quality antibodies in larger quantities using the monoclonal antibody technique (6). In this technique an antibody producing spleen cell of a laboratory animal is fused with a tumore cell. The resulting hybridoma cells can produce large quantities of antibodies that are identical in binding affinity and chemical structure.

Immunofluorescence Detection

In spite of the excellent sensitivity of RIA techniques (7), recent research efforts have been increasingly directed towards the development of alternative nonradioisotopic procedures. The basic principles of radioisotopic and nonradioisotopic immunoassays are similar. In competitive binding immunoassays, an unlabelled antigen (Ag) competes with a labelled antigen (Ag*) for binding sites on an antibody (Ab) directed against the antigen. In the RIA, the label is a radioisotope whereas, in fluoroimmunoassay (FIA), the label is a fluorescent molecule. A variety of fluorescent molecules have been utilized as fluorotags in FIA, with the most common being fluorescein isothiocyanate. Several molecules used as fluorescent-labels are shown in Figure 1. In FIA, the fluorescence signal from the labelled antigen bound to the antibody (Ag*-Ab) decreases as a function of increasing concentration of unlabelled antigen. This functional relationship is used to establish calibration curves for quantitative determination of the antigens in a sample of interest. Alternatively, direct immunoassay of a naturally fluorescent antigen or hapten can be performed. This procedure is particularly straight forward for PAHs, since the increase in signal is directly proportional to antigen (PAH) concentration and the covalent attachment of a fluorescent label is unnescessary.

To develop an efficient FIA system, it is important to distinguish or isolate the emission of the bound AG*(-Ab) from the emission of free Ag*. In many situations, one must resort to a separation procedure for Ag* and Ag*(-Ab), e.g. precipitation, electrophoresis, chromatography, or physical adsorption onto a solid phase (8). In some cases, the formation of the Ab-Ag* antibody complex can be studied by fluorescence methods without the necessity for separation of free Ag* from bound antigen. Referred to as a homogeneous immunoassay this

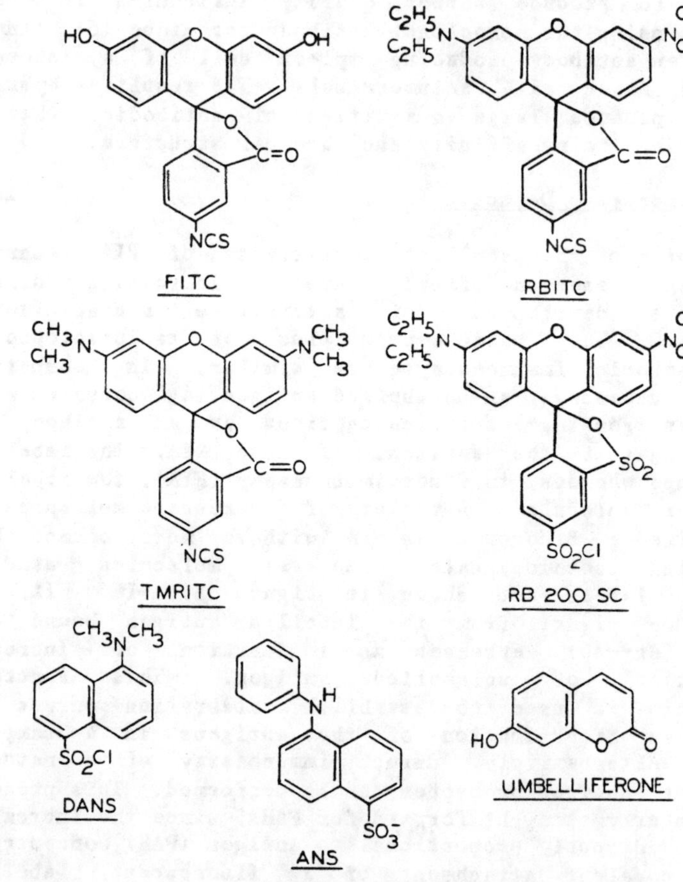

FIGURE 1. **Several Fluorescent Labels Used in Fluoroimmunoassays. (FITC = Fluorescein Isothiocyanate; RBITC = Rhodamine B Isothiocyanate; TMRITC = Tetramethylrhodamine Isothiocyanate; RB 200 SC = Lissamine Rhodamine B Sulfonylchloride: DANS - Dansyl Chloride; ANS = Anilinonaphthalenesulfonic Acid.)**

procedure relies on techniques detecting differences in fluorescence polarization, energy transfer, quenching, and enhancement between the bound and unbound antigen. In the fluorescence polarization method, a labelled Ag* is excited with polarized light and the degree of polarization of the emission is measured. As the antigen binds to the antibody its rotation slows down and the degree of polarization increases. This method is relatively simple but has several limitations: the sensitivity is restricted because of the background fluorescence, and the polarization signal has a limited dynamic range.

INSTRUMENTATION AND MEASUREMENTS

Principle of the Fluoroimmuno-Sensor (FIS)

Fiberoptics technology has been recently used in the development of a variety of chemical sensors and portable monitors (9-11). In this work an FIS device is developed using quartz or glass optical fibers with a core of diameters of 100-500 μm, which are etched to increase the surface area of the fibers at their termini. Antibodies to a specific compound to be monitored are then covalently bonded to the fiber termini. Briefly, the FIS functions for competitive binding assays in the following manner: the FIS probe (i.e., fiber termini) with antibodies are placed in concentrated labelled antigen solutions. After equilibrium has been established, the FIS probe is removed and washed to remove non-specifically bound antigen. The probe prepared in this fashion is ready for use by placing it in a solution of antigens to be quantified. Fluorescence signals are measured before and after the fiber probe is placed in the sample. The antigens are allowed to react with the antibodies and displace the labelled antigens from the antibodies due to competitive binding. The resulting decrease in fluorescence signal, due to the displacement of fluorescent labelled antigens, will be related via a calibration plot to antigen concentration. The measurement procedures are schematically illustrated in Figure 2.

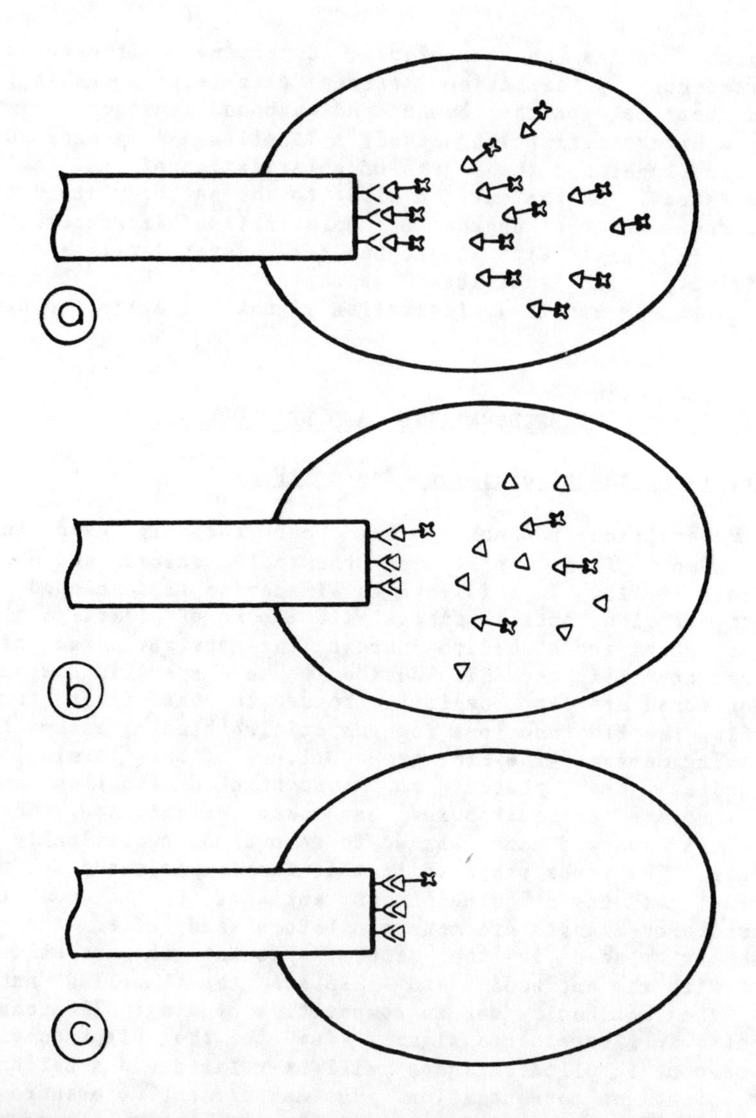

FIGURE 2. **Measurement Procedures Using the Fluoroimmuno-Sensor (FIS).**

a) Place the probe in a solution of tagged antigens; remove and measure fluorescence signal.

b) Place the probe in *in vivo* sample for predetermined period; untagged antigens compete with tagged antigens for sites.

c) Remove the probe; measure the diminished fluorescence signal.

The properties of the optical fibers are important factors determining device capabilities. Quartz tends not to be as effectively etched as glass. The material of the optical fibers also determines the usable spectral range. Fused-silica optical fibers permit measurements in the ultraviolet down to 220 nm, but are relatively expensive. Glass is less costly but is only appropriate for measurement in the visible. Plastic fibers are less expensive still, but are restricted to spectral ranges above 450 nm. Fiberoptics devices may consist of either bifurcated or single optical fibers as illustrated in Figure 3. In a device based on bifurcated design, separated fibers carry the excitation and emission radiation. In the single-fiber device, a dischroic filter is generally used to separate the excitation and emission radiation.

An essential requirement of a competitive binding assay is to be able to distinguish between bound and free antigens. In conventional FIA techniques, this is achieved by immobilizing one component, i.e., by linking the antigen to a substrate (plastic plate, beads), and subsequently, physically separating the substrate from the supernatant to measure the amount of bound fluorescent antigen. Typical configurations of the FIS probe are illustrated in Figure 4. The antibodies can be bound either to the termini of the optical fiber, or to the internal surface of a hollow cylinder, or to the wall of a microcuvette attached to the end of the fiber.

Time-Resolved Fluorescence: Strategy for Improved Sensitivity

In many FIA procedures, the sensitivity of the detection is often limited by the background fluorescence of the biological samples. A common method adopted to minimize background fluorescence consists of isolating the fluorescence of interest by means of optical filters or monochromators. An additional approach relies on distinguishing the fluorescence emission of interest on the basis of its temporal characteristics. The fluorescence associated with many body fluids such as serum proteins has lifetimes on the order of approximately 10 nsec. The lifetimes of the fluorescent labels commonly employed (see Figure 1) are also of the order from 5-100 nsec. In contrast, the fluorescent lanthanide chelates, such as those of europium and terbium, provide a unique class of fluorescent labels because of their long fluorescence decay times, which are in the 10^3 - 10^4 nsec range, thus allowing easy time resolution between the background emission and the analyte fluorescence. The relatively long fluorescence lifetimes of the lanthanides provides the basis for the development of simple time-resolved luminescence monitor. Figure 5 shows the schematic design of such a fiberoptics-based luminescence monitor (12).

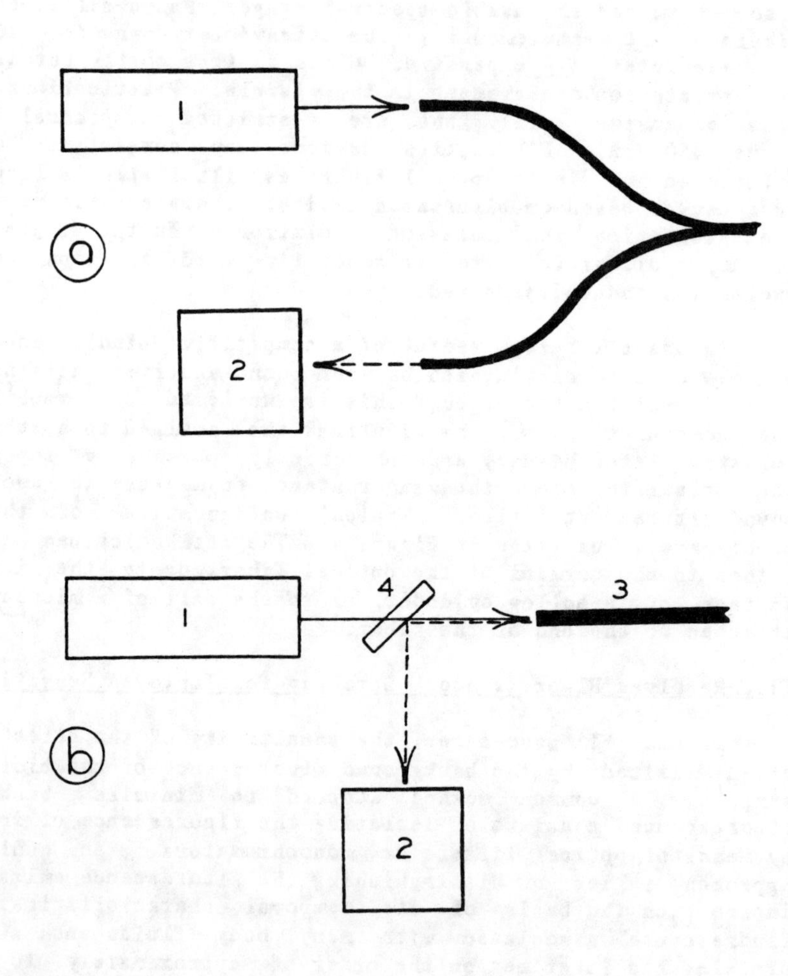

FIGURE 3. Two Optical Configurations for the FIS System.
a) System with bifurcated fiber.
b) System with single fiber. (1 = light source; 2 = detector; 3 = optical fiber, and 4 = dichroic filter.)

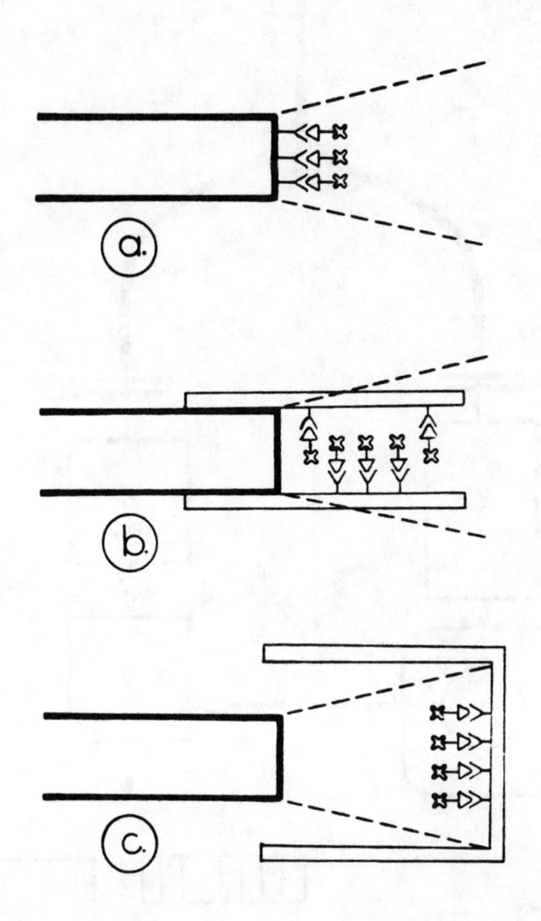

FIGURE 4. Several Configurations for the FIS Probe.
a) Bare-end optical fiber sensor.
b) Cylinder-type sensor.
c) Microwell-type sensor.

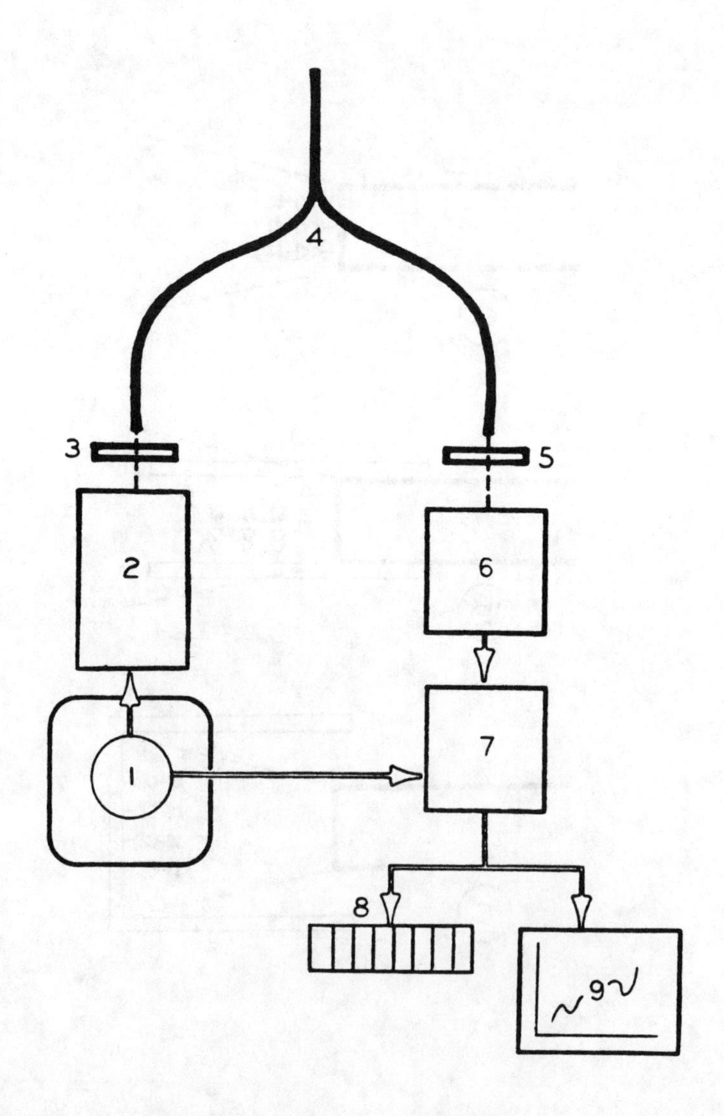

FIGURE 5. **Schematic Diagram of a Time-Resolved Fiberoptics Monitor.**
(1 = pulsing unit; 2 = light source; 3 = optical filter; 4 = bifurcated optical fiber; 5 = optical filter; 6 = photon counter; 7 = gating device; 8 = digital output; 9 = analog output.)

The Prototype Instrument

Apparatus. A schematic diagram of the apparatus is shown in Figure 6. The laser source was an argon ion laser (Spectra Physics, Mountain View, CA 94042, Model 171) operated at 488 nm. The laser radiation was reflected with a Corion (Corion Corp., Holliston, MA 01746, Model CR-500-25D) dichroic filter. This filter transmitted approximately 90% of the Stokes-shifted fluorescence emission. A Corion P10-520-S 10 nm bandpass filter and an Instruments SA, Metuchen, NJ 08840, Model H-20 monchromater, with a 6 nm bandpass were used to isolate the 520 nm emission. The fluorescence was detected with an RCA IP28B photomultiplier tube operated at 800 volts, and processed with a Pacific Precision Instruments, Concord, CA 94518, Model 126 quantum photometer. Data points were recorded using a strip-chart recorder.

A 50 mm focal-length f/2 lens was used to focus the laser radiation onto the end of a 600 μm core diameter, 25° acceptance angle, quartz optical fiber (Math Associates Inc., Port Washington, NY 11050, Model QSF-600). A single fiber was used to transmit the excitation radiation into the sample and collect the fluorescence emission. An optical fiber positioner was used to accurately situate the fiber's incident end along the X, Y, and Z axes. This allowed us to optimize transmission of the focused laser beam to the sample.

Sensors. Two sensor designs were employed in our initial studies. They are illustrated in Figure 6.

A bare fiber sensor was prepared by stripping about 10 mm of cladding from the end of the fiber. The fiber terminus was then scored and broken to provide a flat face, perpendicular to the fiber's sides. The bare quartz was etched in HF and derivatized with a 10% 3-Glycidoxypropyltrimethoxysilane (GOPS) solution at pH 3, 90°C for two hours. The derivatized fibers were dried overnight at 105°C. Activation of the GOPS derivatized fibers was accomplished by oxidation w/0.04 M HIO_4 for 30 min. At this point, the fiber's surface aldehyde groups formed from the oxidation step were allowed to react with Rabbit IgG (Cooper Biomedical, Malvern, PA 19355) for 24 hrs. at 4°C. The resulting Schiff bases were reduced with 1% $NaBH_4$ for 15 min. This produced protein covalently bound to the bare fiber. The sensor was stored at 4°C in phosphate buffered saline until used.

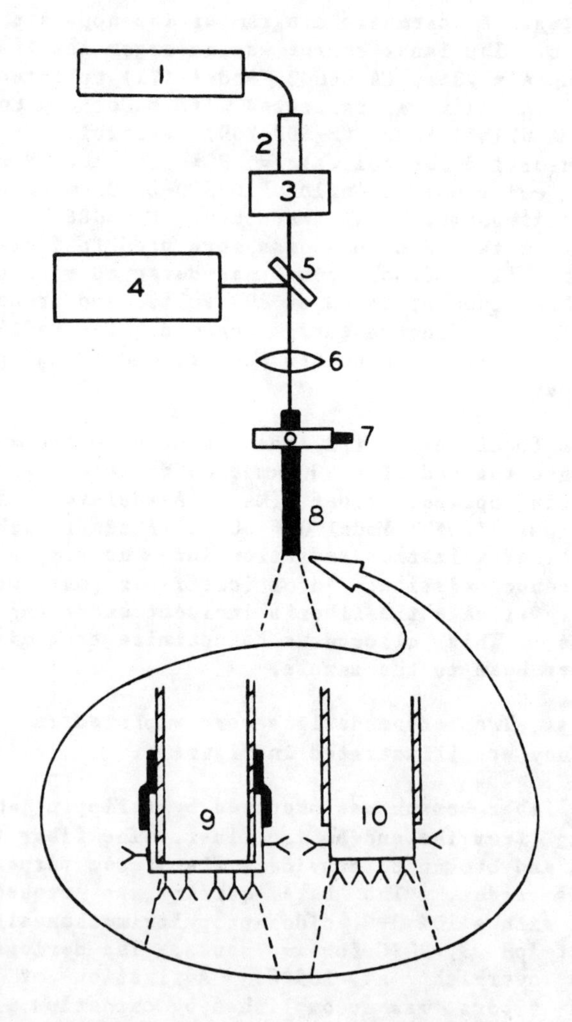

FIGURE 6. Experimental Set-Up of the Prototype FIS Instrument.
(1 = photometer; 2 = photomultiplier tube; 3 = emission filter; 4 = laser; 5 = dichroic filter; 6 = lens; 7 = optical fiber positioner; 8 = optical fiber; 9 = membrane drum sensor; 10 = bare etched fiber sensor.)

A second sensor, using a membrane drum, was developed because it was anticipated that the cellulose membranes might have a greater effective surface area and more protein binding sites than the derivatized quartz. The sensor was prepared by stretching a section of clear cellulose dialysis sheet across the flat surface of the fiber and fixing it in place with heat shrink tubing. The cellulose was activated by oxidation with 0.1 M HIO_4 for 60 min. The activated sensor was then reacted with Rabbit IgG for 24 hrs at 4°C and reduced with 1% $NaBH_4$ to form covalently bound protein.

<u>Measurements</u>

An initital demonstration of the capabilities of the bare fiber FIS is shown in Figure 7. Rabbit IgG was immobilized on the fiber surface. The sensor was then incubated with 1% Bovine Serum Albumin (BSA), in order to minimize the effects of non-specific binding, and 1.8 mg/mL anti-Rabbit IgG labelled with Fluorecein isothiocyanate (FITC), (Cooper Biomedical, Malvern, PA 19355). Measurements for each incubation time were made in Public Broadcasting System after a 30-second rinse. After approximately 85 minutes, the sensor was placed in 3M KSCN/PBS-a chaotropic reagent designed to disrupt the antigen-antibody association. At this point, the fluorescence signal fell nearly to baseline. Continuation of the incubation in 1.8 mg/mL anti IgG-FITC produced a steady rise in fluorescence signal to nearly the same level as that produced by the first incubation.

A closer examination of the protein covalent binding capacities of derivatized bare fiber vs. cellulose membrane indicated the latter to have a greater capacity. By utilizing an FIS with more immobilized protein, larger fluorescence signals and lower limits of detection in competitive binding assays would be expected.

Accordingly, a membrane FIS was evaluated in a competitive binding experiment. The results are illustrated in Figure 8. Rabbit IgG was immobilized on the FIS and the sensor was incubated in 2 mg/mL anti-Rabbit IgG-FITC. After approximately 60 minutes, a fluorescence plateau was reached. The FIS was rinsed well and placed in <u>unlabelled</u> 5 x 10^{-6}M anti-IgG, (Cooper Biomedical, Malvern, PA 19355). The unlabelled anti-IgG displaced the labelled anti-IgG-FITC and the signal gradually diminished. Addition of 3M KSCN dropped the signal to baseline. The displacement of 1 x 10^{-15} moles of anti-IgG-FITC by unlabelled anti-IgG are detectable with this sensor.

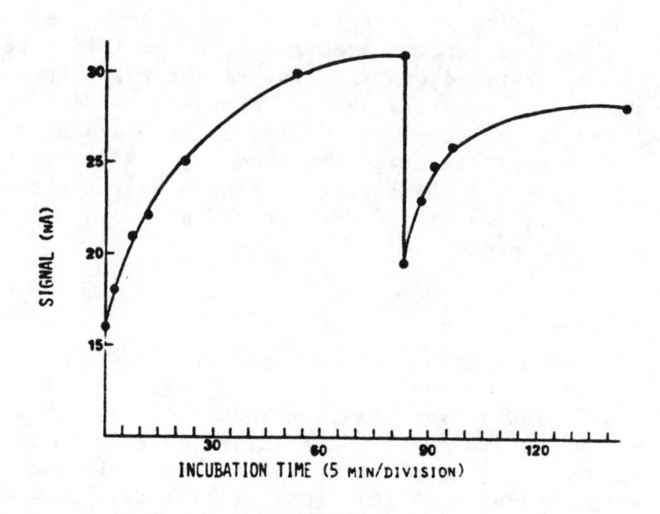

FIGURE 7. **Measurements of Anti-Rabbit IgG using the FIS. Device having the Bare-etched fiber.**

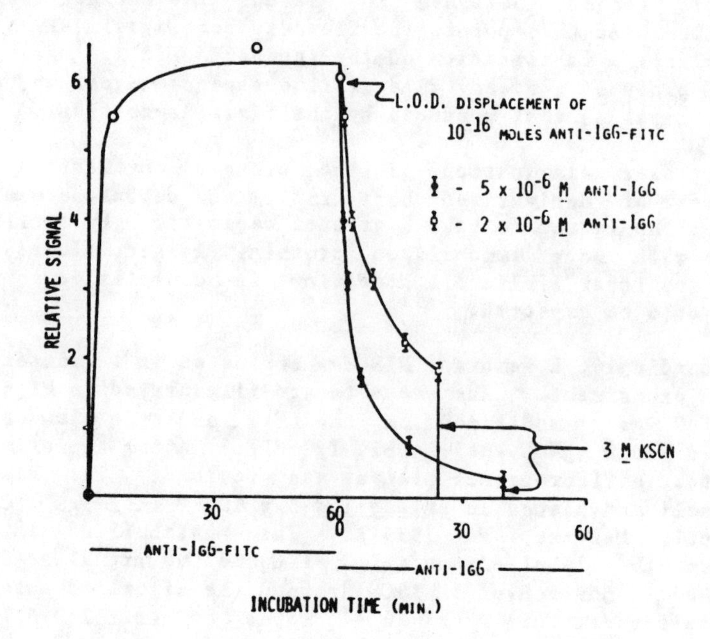

FIGURE 8. **Measurements with the Membrane-Drum FIS Sensor.**

CONCLUSION

In this work we have successfully demonstrated the feasibility of the FIS concept with IgG proteins tagged with FITC. We have also developed antibodies to BaP (5) for future incorporation onto the FIS and other antibody-based monitors.

In view of the biotechnological advances in the area of monoclonal antibodies and the production of anti-hapten sera we believe that fiberoptics-based FIS could be generally useful in a wide spectrum of biochemical and clinical analyses. Specifically, the FIS could be employed in the assessment of an individual's exposure to chemical PAH carcinogens, response to drug therapy, or in the characterization of naturally occuring biochemicals. For example, studies investigating the presence of carcinogens and carcinogen-DNA adducts in tissues and body fluids could be performed with the FIS using immunochemical techniques. In this research we plan to integrate FIA principles with laser-based fiberoptics systems in order to develop FIS which should be capable of providing reliable analytical data for such studies. Other analytical methodologies (e.g., measurements of alkylation of hemoglobin, mutagenicity assays of urine) have been used to monitor environmental carcinogens. Complementation of such methods with the immunological techniques discussed in this work will undoubtedly provide a new generation of biological monitors for human exposure to toxic PAH chemicals.

ACKNOWLEDGEMENTS

This research is jointly sponsored by the National Science Foundation (Grant No. 01R-8413145) and the Office of Health and Environmental Research, U.S. Department of Energy under contract DE-AC05-84OR21400 with Martin Marietta Energy Systems, Inc.

REFERENCES

1. Grimmer, G., Editor (1983), Environmental Carcinogens: Polycyclic Aromatic Hydrocarbons, CRC Press, Boca Raton, Florida.
2. Vo-Dinh, T. and Hawthorne, A. R. (1984): Monitoring Instrumentation, in Proceedings of the DOE-OHER Workshop on Monitoring and Dosimetry in an Occupational Health Research Program, U. S. Department of Energy, Report No. CONF-8403150: 45-62.
3. Poirier, M. C. (1981): Antibodies to Carcinogen-DNA Adducts. J. Natl. Cancer Inst., 67: 515-113.
4. Miller, R. and Rajewsky, M. F. (1981): Antibodies Specific to DNA Components Structurally Modified by Chemical Carcinogens. J. Cancer Res. Clin. Corp., 102: 99-113.
5. Griffin, G. D., Ambrose, K., Thomason, R., Murchison, C. M., McManis, M., St. Wecker, P. G. R., and Vo-Dinh, T. (1985): Production and Characteriation of Antibodies to Benzo(a)pyrene. Proceedings of the 10th International Symposium on Polycyclic Aromatic Hydrocarbons, Battelle, Columbus, Ohio, October 21-24, 1985.
6. Kennett, R. H., McKearn, T. J., and Bechtol, K. B., (1982): Monoclonal Antibodies-Hybridomas: A New Dimension in Biological Analyses. Plenum Press, New York.
7. Yalow, R. S. and Berson, S. A. (1960): Immunoassay of Endogenous Plasma Insulin in Man. J. Clin. Invest., 39: 1157.
8. Lidofsky, J. D., Imasaka, T., and Zare, R. N. (1979): Laser Fluorescence Immunoassay of Insulin. Anal. Chem., 51: 1602-1605.
9. Seitz, W. R. (1984): Chemical Sensors Based on Fiberoptics. Anal. Chem., 56: 16A-34A.
10. Tromberg, B. J., Eastham, J. F., and Sepaniak, M. (1984): Optical Fibers Fluoreprobes for Biological Measurements. Appl. Spectrosc., 38 (1): 38-42.
11. Vo-Dinh, T. and Gammage, R. B. (1981): The Use of a Fiberoptics Skin Contamination Monitor in the Workplace, in Chemical Hazards in the Workplace, Choudhary G. Ed, ACS, Washington, D.C., pp. 269-281.
12. Vo-Dinh, T. (1981): Improved Fiberoptics Portable Luminoscope. Patent Disclosure CNID No. 4112, Oak Ridge National Laboratory, Oak Ridge, Tennessee.

INHIBITION OF MUTAGENICITY OF POLYCYCLIC AROMATIC HYDROCARBONS AND AFLATOXIN B_1 BY CHINESE TEA AND AN INVESTIGATION OF POSSIBLE MECHANISMS

ZHI-YUAN WANG*[1], ZONG-CAN ZHOU[2], YIN WU[3], FEI-LAN WANG[1], GUO-HUA WANG[1], YUAN-KAI CHENG[4]
(1) Institute of Labor Protection, Ministry of Labor and Personnel, Box 4711, Beijing; (2) Beijing Medical University; (3) Beijing Normal University; (4) Institute of Industrial Health, Anshan Iron and Steel Co., The People's Republic of China.

INTRODUCTION

Although we are surrounded by various carcinogens and mutagens, fortunately, a considerable number of naturally occurring and synthetic compounds have been found to be antimutagens and anticarcinogens in our environment (1-4). Removal of environmental carcinogens is a primary goal of cancer prevention, but this is likely to be very difficult. For example, PAH which are widely present in our environment and food-stuff are hard to eliminate. Therefore chemoprevention plays an important role in cancer control. Recently we were interested in screening the least toxic and most effective antimutagens from traditional Chinese herbs (5-8). In this paper, several kinds of Chinese tea and several constituents of tea that inhibited the mutagenicity resulting from the metabolic activation of PAH and aflatoxin B_1 towards *S. typhimurium* strain TA 98 with dose-dependent response, were reported, and their possible mechanisms were discussed. Tea is not only a daily beverage for Chinese people, but also for people of many countries of the world. Without doubt it is very important to evaluate the role of tea on chemoprevention of cancer.

MATERIALS AND METHODS

Chemicals

Chemicals and their sources were as follows: D-catechin (ICN Pharmaceutical, Inc., Plainview, NY); ellagic acid, aflatoxin B_1 (AFB_1) (Aldrich Chemical Co.); ferulic acid, caffeic acid, gallic acid, rutin, hesperidin and quercetin (Beijing Chemical Co.) were repurified and tested as pure chemicals. Instant tea A and B (prepared with Chinese green tea), black tea, green tea, uron tea (a half green tea and a half black tea) were purchased from a Beijing supermarket. Catechol tannin TF, TR and TB were extracted from Chinese black tea. Others were: ^{3}H-benzo(a)pyrene (^{3}H-B(a)P) (47.0 ci/m mol, New England Nuclear) and NADPH (Sigma Chemical Co.).

Preparation of Water Extract of Tea

Water extract of tea (WET) was prepared with 2 g green tea (or black tea, or uron tea) and 20 ml boiling water extracting for 2 min and after filtration used immediately.

Bacterial Mutagenicity Assay

Histidine-dependent *Salmonella typhimurium* strain TA 98 was a gift from Dr. B. Ames (University of California, Berkeley, CA). Mutagenicity assay utilizing rat hepatic microsomes as the source of metabolic activation enzymes was based on the procedure described by Ames *et al.* (9) and our previous work (1). Inhibition of mutagenicity was assayed by preincubation for 20 min at 37°C with the PAH or AFB_1, various amounts of test substances, $2x10^8$ bacteria suspension and S_9 mixture before the addition of top agar.

Monooxygenase Assay

The radioactive assay for monooxygenase activity was described by Van Contfort *et al.* (10) and our previous work (6). Radioactivity of B(a)P metabolites were quantified by a liquid scintillation spectrometer (FJ 2100 Model, made in China).

RESULTS

Antimutagenicity of Several Kinds of Chinese Tea

Coal tar pitch is a useful, but hazardous industrial product, which contains a large amount of PAH. Recently we found coal tar pitch could significantly induce rat lung cancer (11). In order to accord with the practical human exposure, the methanol extract of coal tar pitch was selected as the major sample for mutagenicity experiments. In addition, B(a)P and AFB_1 were used as positive mutagens.

The metabolites of 320 μg of the methanol extracts of coal tar pitch, or 0.5 μg of AFB_1 per plate, activated by rat liver microsome resulted in a strong mutagenic response to strain TA 98. These metabolites were inhibited in a dose-dependent manner by instant tea A and B, and WET of green tea (Figure 1).

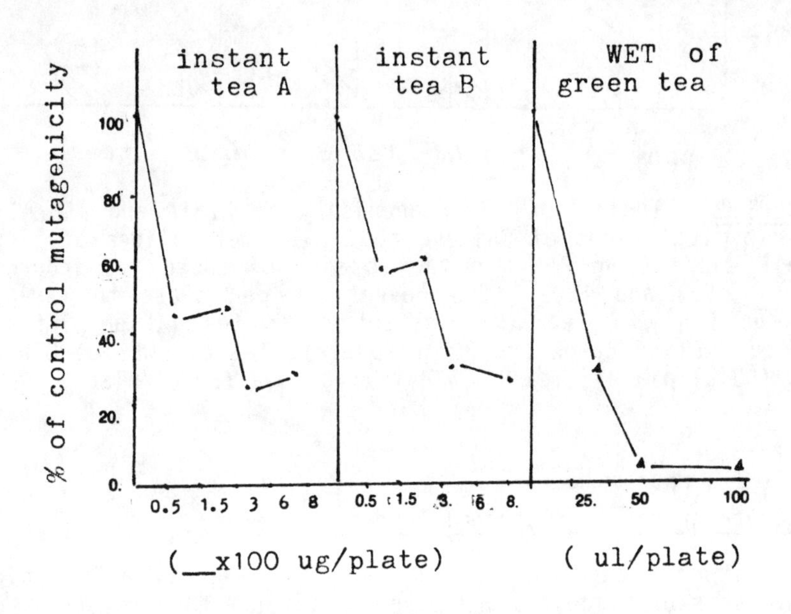

FIGURE 1. Antimutagenicity of instant tea A and B and WET green tea on methanol extracts of coal tar pitch (•) and AFB_1 (▲). The metabolites of 320 μg methanol extracts of coal tar pitch and 0.5 μg AFB_1 in the absence of added tea induced 476 ± 68 and 1802 ± 345 revertants per plate to TA 98, respectively.

Figure 2 shows the inhibition of mutagenicity of B(a)P and AFB_1 by WET of four typical kinds of Chinese tea. It seemed that there were no significant differences of antimutagenicity among these teas.

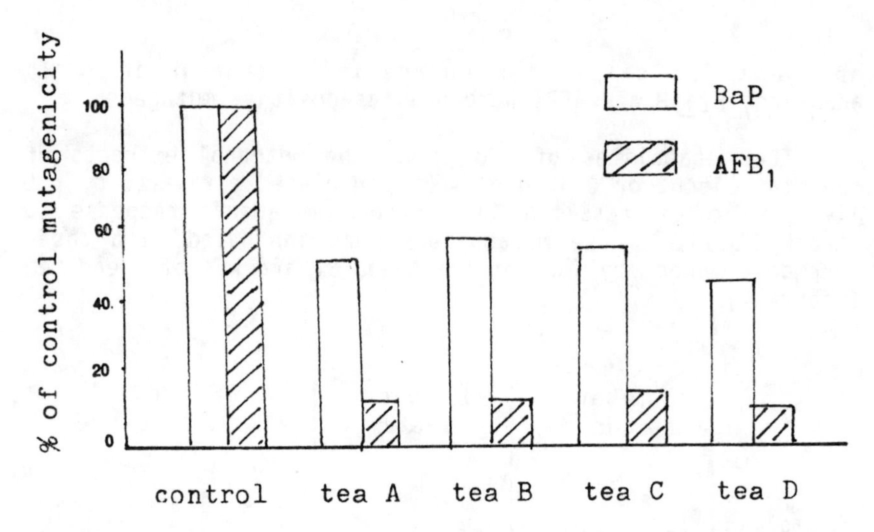

FIGURE 2. Inhibition of mutagenicity of B(a)P and AFB_1 by four typical kinds of Chinese tea. The metabolites of 1 μg B(a)P and 0.5 μg AFB_1 in the absence of added tea induced 532 ± 46.3 and 1229 ± 273 revertants per plate to TA 98, respectively. Tea A: instant green tea (10 μg/plate); Tea B: WET of green tea (50 μl/plate); Tea C: WET of black tea (50 μl/plate); Tea D: WET of uron tea (50 μl/plate).

Antimutagenicity of Catechol Tannins

Catechol tannins (TF, TR, and TB) are characteristic components of black tea and possess valuable bioactivity. We found all of these three catechol tannins inhibited the mutagenicity of methanol extracts of coal tar pitch towards TA 98 with a dose-dependence (Figure 3).

Antimutagenicity of Tea Flavonoids

Tea contains rich hydroxylated tea flavonoids. Most of them are catechins, such as catechin, epigallocatechin, epigallocatechin-3-gallate, etc. In addition, there are amounts of flavonols and anthocyanidins in tea. Figure 4 shows that three typical tea flavonoids, i.e., catechin, rutin, and hesperidin, inhibited the mutagenicity of coal tar pitch. Catechin antagonized the mutagenicity of AFB_1, also.

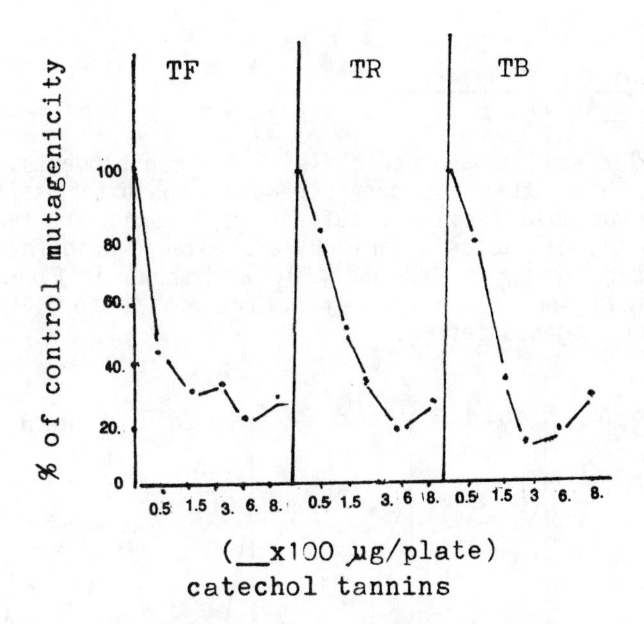

FIGURE 3. Antimutagenicity of catechol tannins, TF, TR, and TB). The metabolites of 320 µg methanol extracts of coal tar pitch in the absence of added catechol tannins induced 476 ± 68 revertants per plate to TA 98.

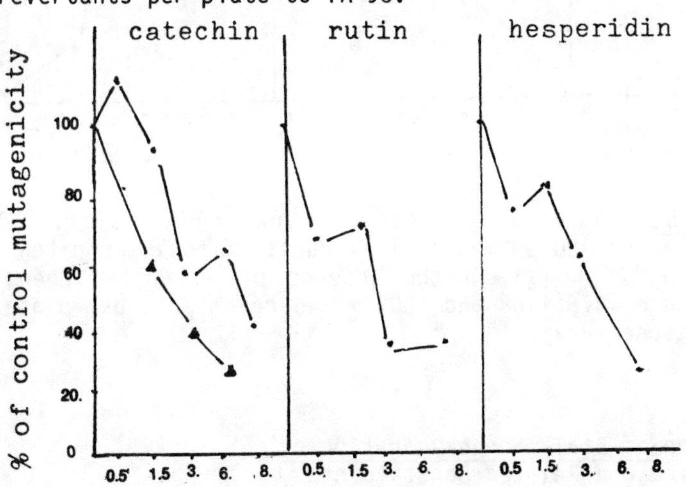

FIGURE 4. Antimutagenicity of tea flavonoids. The metabolites of 320 µg methanol extracts of coal tar pitch (•) and 0.5 µg AFB_1 (▲) in the absence of added flavonoids induced 476 ± 68 and 1802 ± 345 revertants per plate to TA 98, respectively.

Antimutagenicity of Several Tea Phenolic Acids

Phenolic acid is another class of tea constituents. Our results indicate that the dose-dependent inhibitory effects of chlorgenic acid (acid A), caffeic acid (acid B), ferulic acid (acid C), gallic acid (acid D) and ellagic acid (acid E) on the mutagenicity of PAH and AFB_1 are shown in Figure 5. Ellagic acid was the most effective inhibitor of the antimutagenic agents tested.

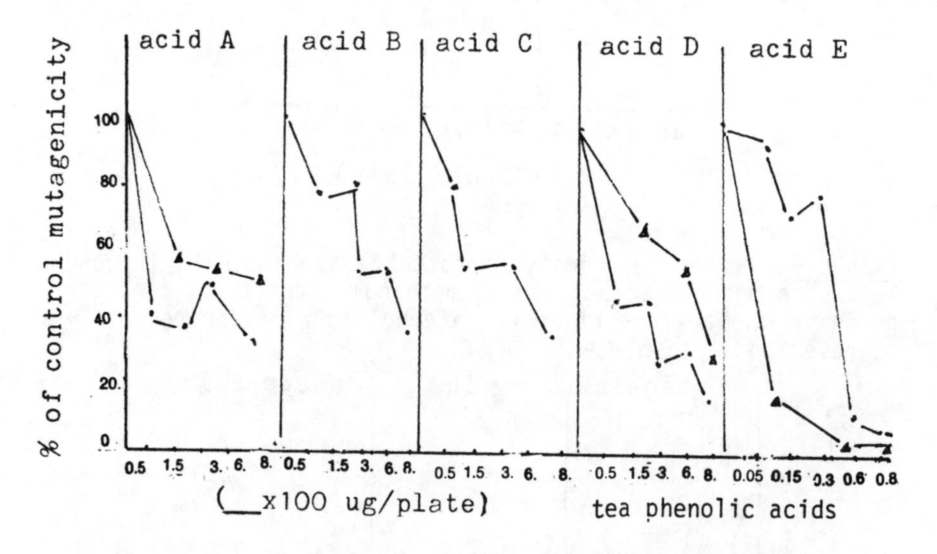

FIGURE 5. Antimutagenicity of phenolic acids. The metabolites of 320 µg methanol extracts of coal tar pitch (•) and 0.5 µg AFB_1 (▲) in the absence of added tea phenolic acids induced 476 ± 68 and 1802 ± 345 revertants per plate to TA 98, respectively.

Inhibition of B(a)P Monooxygenation by Chinese Tea and Related Constituents

Figure 6 indicates that tea and related contents inhibited B(a)P monooxygenation with a dose-dependence. It seemed that the mentioned substances caused considerable loss of microsomal monooxygenase activity via formation of enzyme

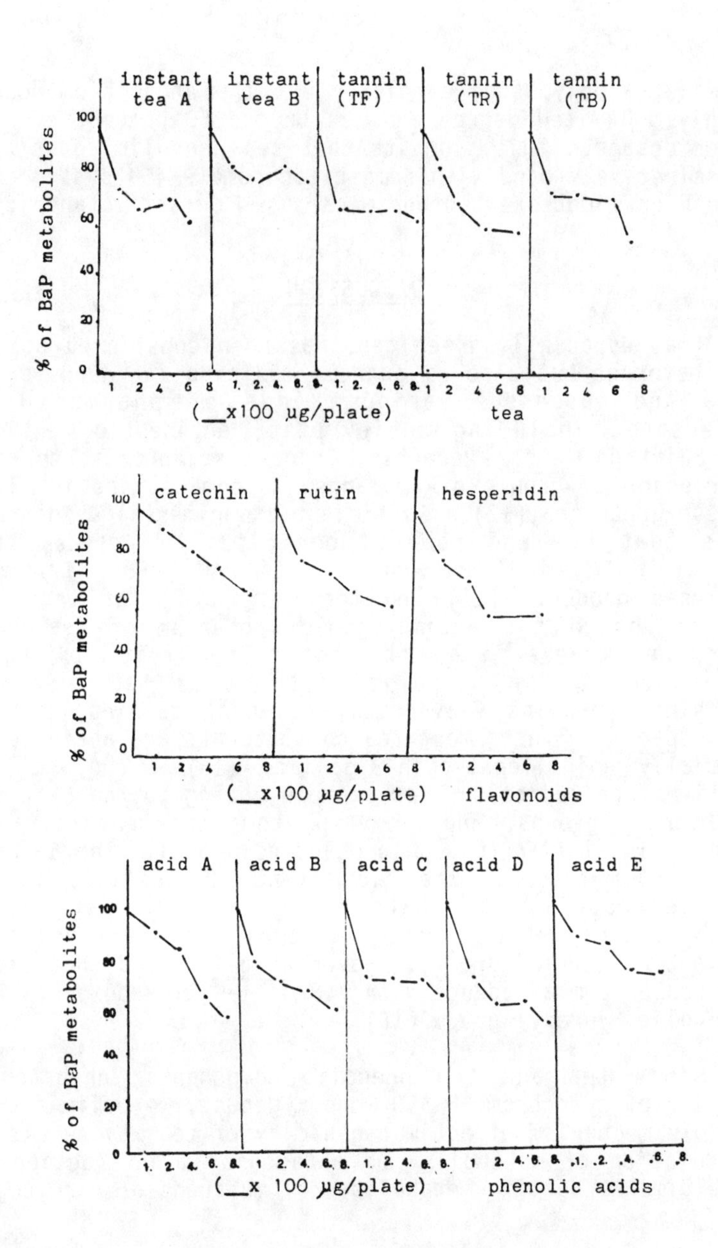

FIGURE 6. Inhibition of B(a)P monooxygenation by the addition of tea and relative effective components to 3-MC induced rat liver microsomes. Results were expressed as percentage of control (2.7 ± 13 n mol B(a)P/mg protein/min). Acid A, B, C, D, and E were chlorgenic acid, caffeic, acid, ferulic acid, gallic acid, and ellagic acid, respectively.

protein-tea polyphenol complex. But the inhibition occurred to only a limited degree and not beyond 50 percent even at a higher dosage. This suggests that tea phenolic constituents may selectively bind with some cytochrome P-450 system enzyme protein or might exert other mechanisms on antimutagenicity.

DISCUSSION

Tea, especially green tea, has been considered not only as a beverage but also as a crude medicine for 4000 years in China and possesses various kinds of pharmaceutical properties, including antipyretic, antiseptic, antidote, antiradiation, *etc.* Recently, Chinese scientists reported a suppression of growth of implanted sarcoma tumors in mice by Chinese green tea (12). Sufficient experimental evidence now shows that tea and related constituents possess strong potency to inhibit mutagenicity of PAH, AFB_1, tryptophan pyrolysis products (13) and the nitrosation of methylurea (14). Though the antimutagenic mechanism of tea is not clear, we believed that the antimutagenicity of tea is attributed to the phenolic contents of tea, including catechins, phenolic flavonoids, catechol tannins, phenolic acids, *etc*. In fact, most tea constituents are antioxidants, especially epigallocatechin-3-gallate (15). They may trap reactive free radicals. For example, owing to the 5- and 7,- two hydroxy groups, and 1-oxygen atom, the benzoic ring of catechin is similar to a phloroglucinol model. The 6- and 8- position of catechin are two strong nucleophilic centers. The electrophilic ultimate metagenic or carcinogenic metabolites may attack these nucleophilic centers and form a covalent adduct (Figure 7). Sayer *et al.* proved that ellagic acid could form an adduct with 7,8-diol-9, 10-epoxy-B(*a*)P via a phenolic hydroxyl group (16).

Since tea and tea phenolic components inhibited the activity of cytochrome P-450 monooxygenase, we believe that a possible mechanism of antimutagenicity of tea may result from a modulation of metabolism or antagonistic interaction with the ultimate reactive metabolites of mutagens or carcinogens.

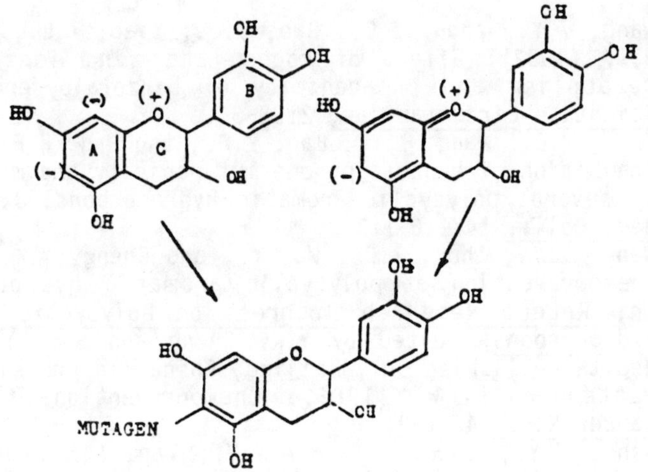

FIGURE 7. A probable model for reaction of a catechin adduct with a mutagen.

Sparnins et al., (17) reported that instant tea and tea leaves could markedly affect the levels of GSH S-transferase activity in tissues of test animals. There is not doubt that the enhancement of conjugating enzyme activity by tea will surely lead to detoxifying chemical mutagens and carcinogens. In the future we will explore if tea can prevent carcinogenicity in animal models and evaluate the overall implications of tea use to public health in the human population.

ACKNOWLEDGEMENTS

We thank professor Shen K. Yang (Uniformed Services University of the Health Sciences, US) for valuable direction and support, and thank Dr. Su J. Cheng (Institute of Cancer, Chinese Academy of Medical Sciences) for his kind help.

REFERENCES

1. Wang, Z.Y., Zhou, Z.C., Bao, Z.P., Xiao, X.L., and Liu, P.T. (1982): Effect of copper and zinc ions on the metabolism and mutagenicity of benzo(a)pyrene, Acta. Scientise Circumtantiae, 2:28-36.
2. Zhou, Z.C., Wang, Z.Y., Bao, Z.P., and Hsu, H.E. (1982): Inhibition of benzo(a)pyrene mutagenicity and metabolism by several polycyclic aromatic hydrocarbons, J. Beijing Med. Coll., 14:110-113.
3. Wang, Z.Y., Zhou, Z.C., Wu, Y., and Cheng, Y.K. (1985): Chemoprevention of polycyclic aromatic hydrocarbons. In: Recent Research Progress on Polycyclic Aromatic Hydrocarbons, edited by Y.K. Cheng and S.K. Yang, The Health Publishing Co., Beijing, China (in press).
4. Wattenberg, L.W. (1985): Chemoprevention of cancer, Cancer Res., 45:1-8.
5. Wang, Z.Y., Zhou, Z.C., Wu, Y., Yan, L.S., and Wang, F.L. (1983): Inhibition of benzo(a)pyrene metabolism in rat liver microsomes by Chinese naturally occurring quinones. In: Proceedings of the Eighth International Symposium on Polynuclear Aromatic Hydrocarbons, edited by M. Cooke and A.J. Dennis, Battelle Press, Columbus, Ohio.
6. Hou, F.Z., Bao, Z.P., Wang, Z.Y., and Liu, P.T. (1983): Inhibitory effects of phenolic and hydroxylated compounds on the formation of the metabolites of benzo-(a)pyrene. Presented at the International Workshop on Environmental Mutagenesis, Carcinogenesis and Teratogenesis, Shanghai, China (May).
7. Sun, W.J., Wang, Z.Y. (1985): Antimutagenicity of Ginsenosides (in press).
8. Wang, Z.Y., Cheng, Y.K., Wu, Y., Yan, L.S., Wang, F.L., and Wang, G.H. (1984): Inhibition of mutagenicity of methanol extracts of coal tar pitch by some naturally occurring plant flavonoids. In: Proceedings of 1984 International Chemical Congress of Pacific Basin Societies, Honolulu, Hawaii (December).
9. Ames, B.N., McCann, J., and Yamasaki, E. (1975): Methods for detecting carcinogens and mutagens with the Salmonella/mammalian microsomes mutagenicity test, Mutation Res., 31:347-364.
10. Van Contfort, J., De Graeve, J., and Gielen, J.E. (1977): Radioactive assay for aryl hydrocarbon hydroxylase, improved method and biological importance, Biochem. Biophys. Res. Commun., 79:505-512.

11. Cheng, Y.K. (1985): Squamous cell carcinoma of lung induced by intratracheal instillation of coal tar pitch particulates in Wistar rat, Ind. Hyg. Occup. Dis., II:65-68.
12. Yi, Z.C., Yi, C.S., Li, Z.P., and Wu, P.Y. (1984): Suppression of growth of implanted sarcoma tumors in mice by Chinese tea, Tumor (China), 4:128.
13. Kada, T. (1982): Mechanisms and genetic implications of environmental antimutagens. In: Environmental Mutagens and Carcinogens, edited by T. Sugimura, S. Kondo, and H. Takebe, pp. 355-359, University of Tokyo Press, Tokyo, and Alan R. Liss, Inc., NY.
14. Stich, H.F., Rosin, M.P., and Bryson, L. (1982): Inhibition of mutagenicity of a model nitrosation reaction by naturally occurring phenolics, coffee, and tea, Mutation Res., 95:119-128.
15. Matsuzaki, T. and Hara, Y. (1985): Antioxidative activity of tea leaf catechins, Nippon Nogeikaga-ku Kaishi, 59:129-134.
16. Sayer, J.M., Yagi, H., Wood, A.W., Conney, A.H., and Jerina, D.M. (1982): Extremely facile reaction between the ultimate carcinogen benzo(a)pyrene 7,8-diol-9,10-epoxide and ellagic acid, J. Am. Chem. Soc., 104:5562-5664.
17. Sparnins, V.L., Venegas, P.L., and Wattenberg, L.W. (1982): Glutathione S-transferase activity: enhancement by compounds inhibiting chemical carcinogenesis and by dietary constituents, J. Natl. Cancer Inst., 68:493-496.

BENZO[A]PYRENE METABOLISM IN VIVO FOLLOWING

INTRATRACHEAL ADMINISTRATION

ERIC H. WEYAND, DAVID R. BEVAN
Virginia Polytechnic Institute and State University,
Blacksburg, Virginia, 24061.

INTRODUCTION

Benzo[a]pyrene (B[a]P) metabolism has been studied extensively *in vitro* using microsomes (1,2), purified enzymes (3), cell cultures (4,5), and isolated perfused organs (6-8). A great deal of information about B[a]P activation to reactive metabolites such as the 7,8-diol-9,10-epoxide as well as covalent binding of this and other metabolites to cellular macromolecules such as DNA has been delineated with the use of these systems. Fewer studies have reported on B[a]P disposition and metabolic activation in intact animal systems. Most studies that have investigated B[a]P metabolism *in vivo* have been limited to partial characterization of distribution of B[a]P and its metabolites among organs (9-11).

In our investigation we have characterized in detail disposition of radioactivity in rats following intratracheal instillation of [^{3}H]-B[a]P. In addition, the types of B[a]P metabolites in several organs and in excreta were identified at various times following [^{3}H]-B[a]P administration.

MATERIALS AND METHODS

Chemicals

[G-^{3}H]-Benzo[a]pyrene (B[a]P) was obtained from Amersham Corp. (Arlington Heights, Ill.) and purified by the method of Van Cantfort *et al.* (12). Standard B[a]P metabolites were obtained from IIT Research Institute (Chicago, Ill.). β-Glucuronidase (type H-1, from *Helix pomantia*) and aryl sulfatase (type V, from limpets) were purchased from Sigma (Sigma Chemical Co., St. Louis, Mo.).

Animals and Cannula Implantation

Male Sprague-Dawley rats weighing 200-250 g were used in all experiments. Cannulas were placed in tracheas following the method of Shanker (13). In experiments in which bile was collected, the common bile duct was cannulated using polyethylene tubing (PE 10) prior to tracheal cannulation. Bile flow ranged from 0.02 to 0.03 ml/min. Saline (1.0 to 2.0 ml) was administered hourly by subcutaneous injection to prevent dehydration. Body temperatures were monitored and maintained at 37°C.

Administration of [^{3}H]-B[a]P and Tissue Analysis

[^{3}H]-B[a]P (3.85 nmole/kg) dissolved in triethylene glycol was administered to animals at the bronchial bifurcation by injection through the tracheal cannula. Animals were sacrificed at selected times following [^{3}H]-B[a]P administration. Blood was collected in heparinized tubes, and lung, liver, kidneys, stomach, heart/thymus, spleen and testes were quickly excised, weighed and placed on ice. Intestines were removed, flushed with water to obtain intestinal contents, weighed and placed on ice. Tissues were minced and homogenized in water using a polytron homogenizer. The remaining carcass was digested in 1.5 N NaOH at 50°C for 24 hours. Homogenates were solubilized with 1 ml Beckman Tissue Solubilizer (BTS-450) at 50°C for 2 hours. Solubilized samples were decolorized with tert-butyl hydroperoxide and the pH was adjusted to neutrality with an appropriate volume of glacial acetic acid. Radioactivity in each sample was determined by liquid scintillation counting.

Benzo[a]pyrene Metabolite Analysis

Metabolites of B[a]P in lung, liver, and intestinal contents were identified using procedures similar to those reported by others (14). Aliquots of ethyl acetate extract and remaining aqueous phase were counted by LSC to determine amounts of ethyl acetate extractable and non-extractable radioactivity.

Metabolites extracted into ethyl acetate were characterized further by high pressure liquid chromatography (HPLC) as described by Selkirk (15). The effluent was monitored by U.V. absorption at 254 nm and was collected in 0.5 ml fractions for liquid scintillation counting. Metabolites were identified by elution times corresponding to those determined by U.V. absorption for authentic standards.

Conjugates of B[a]P metabolites in intestinal contents and bile were characterized by their sensitivity to enzyme hydrolysis as reported by MacLeod *et al.* (16). After incubation with β-glucuronidase, arylsulfatase or in the absence of these enzymes, ethanol was added (3:1, v/v), and precipitated material removed by centrifugation. The supernatant was removed and evaporated to a small volume under nitrogen. Samples with and without treatment with β-glucuronidase and aryl sulfatase were chromatographed on silica gel with a solvent of ethyl acetate/methanol/water/-formic acid (100:25:20:1) (16). Lanes were sectioned (0.5cm/section) and the amount of radioactivity in each determined by liquid scintillation counting. Metabolites insensitive to enzyme hydrolysis are probably thioether conjugates derived from conjugation of B[a]P metabolites with glutathione. Unmetabolized B[a]P and non-conjugated metabolites migrated at the solvent front.

RESULTS

[^{3}H]-B[a]P Distribution Following Intratracheal Instillation

Elimination of radioactivity from lung was rapid following intratracheal administration of [^{3}H]-B[a]P. Within 5 min after instillation, 40 percent of the dose was cleared from lung while at 360 min more than 94 percent of the dose was cleared (Fig. 1). The amount of radioactivity in liver increased rapidly, with the maximum amount being found 10 min after administration of [^{3}H]-B[a]P. Subsequently, radioactivity in liver slowly decreased (Fig. 2). The sum of radioactivity in several other tissues increased rapidly to about 4 percent of the administered dose and remained at that level over the course of these experiments (Fig. 2). Notably, radioactivity in carcass accounted for a large fraction of the administered dose as early as 5 min after instillation (Fig. 2). Amounts of radioactivity in the GI tract accounted for 57 percent of the total dose by 360 min after dose administration (Fig. 1). The majority of radioactivity in the GI tract was in intestinal contents while amounts in stomach and intestines contributed much less to the total.

Analysis of B[a]P Metabolites

Types of B[a]P metabolites in lung, liver, and intestinal contents were characterized initially by their extractability into ethyl acetate (Table 1). As expected, relative amounts of metabolites extracted into ethyl acetate decreased in lung and liver at longer times after instillation. The majority of radioactivity in intestinal contents

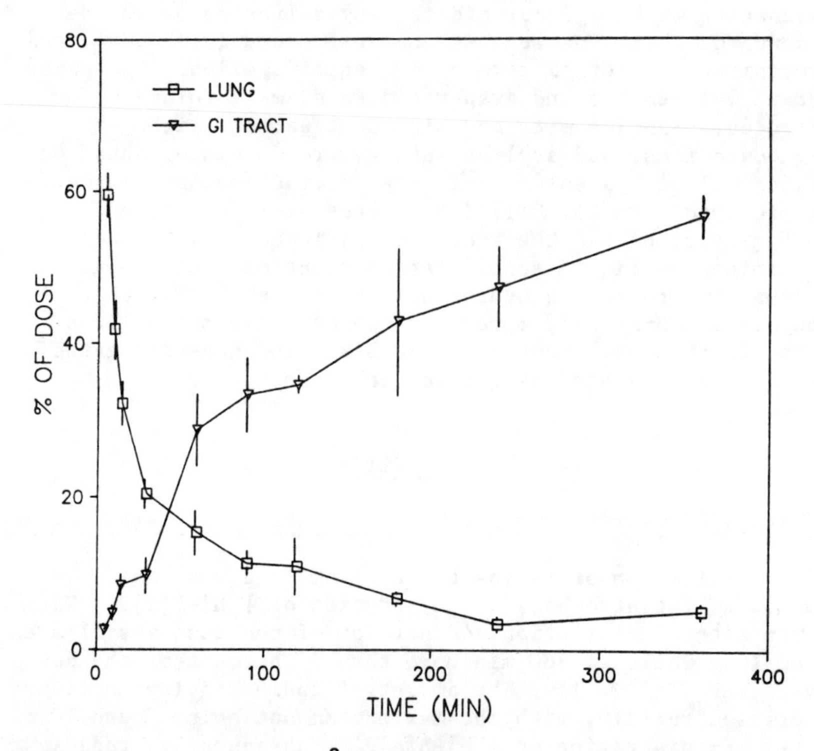

FIGURE 1. Clearance of [^{3}H]-radioactivity from lung and disposition to gastrointestinal tract (GI tract) following intratracheal administration of [^{3}H]-B[a]P. GI tract data represents the sum of [^{3}H]-radioactivity in stomach, intestines and intestinal contents. Results are mean percent of dose ± S.E. for 3 animals at each time point.

was non-extractable at 60 and 360 min after instillation of [^{3}H]-B[a]P, reflecting the formation of conjugated metabolites.

Unmetabolized B[a]P accounted for the majority of ethyl acetate extractable material in lung at 5 min after instillation (Fig. 3). Metabolites in highest concentration were B[a]P-quinones, which accounted for 12 percent of the total extractable material. Amounts of unmetabolized B[a]P and B[a]P metabolites at 360 min were considerably less than amounts detected at 5 min. However, when expressed as percent of total extractable material, B[a]P metabolites accounted for 22 percent of total radioactivity at 5 min and

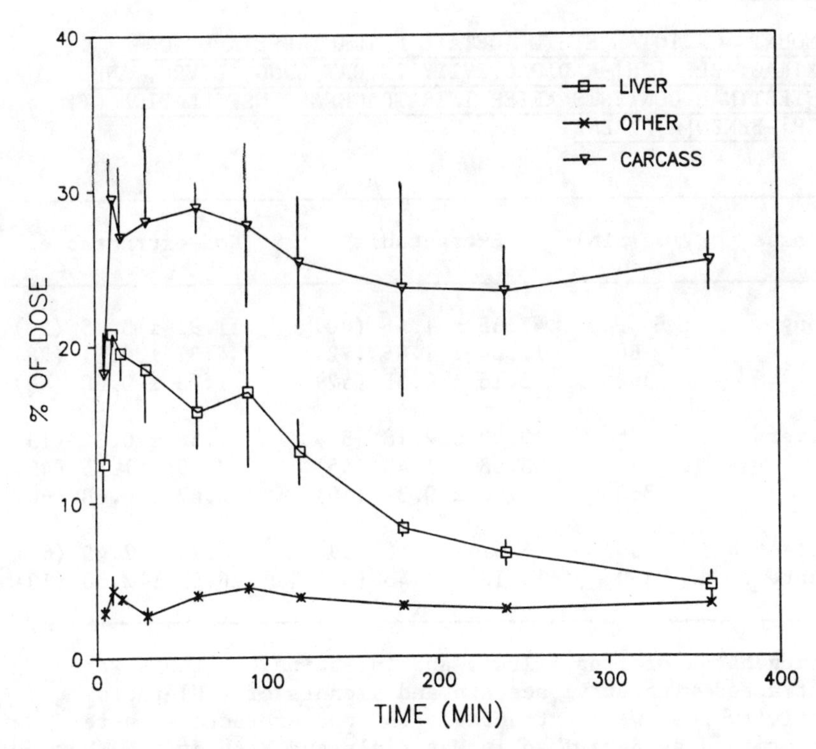

FIGURE 2. Disposition of [^{3}H]-radioactivity in hepatic and extrahepatic tissue following intratracheal administration of [^{3}H]-B[a]P. "Other" tissues represent the sum of total [^{3}H]-radioactivity in kidneys, spleen, heart/thymus, and testes while "Carcass" data represent the sum of total [^{3}H]-radioactivity in carcass, urine, and blood. Results are mean percent of dose ± S.E., for 3 animals at each time point.

43 percent at 360 min. Metabolite analysis of extractable material in liver showed a pattern of metabolites similar to that observed for lung (data not shown).

Conjugates of B[a]P metabolites in intestinal contents were identified by enzyme hydrolysis. The major conjugates of B[a]P metabolites were those resulting from conjugation with glutathione which accounted for approximately 63 percent of material in intestinal contents and 25 percent of the administered dose (Fig. 4). B[a]P glucuronides and sulfate

TABLE 1

AMOUNTS OF TOTAL ETHYL ACETATE EXTRACTABLE AND NON-EXTRACTABLE [^{3}H]-RADIOACTIVITY IN RAT LUNG, LIVER, AND INTESTINAL CONTENTS AFTER INTRATRACHEAL INSTILLATION OF [^{3}H]-BENZO[A]PYRENE

Tissue	Time (min)	Extractable	Non-extractable
Lung	5	47.65 ± 1.46 (80)	11.91 ± 3.25 (20)
	60	11.04 ± 1.85 (72)	4.37 ± 0.81 (28)
	360	2.16 ± 0.35 (57)	1.83 ± 0.68 (43)
Liver	5	10.90 ± 2.18 (87)	1.56 ± 0.32 (13)
	60	8.68 ± 1.43 (55)	7.15 ± 1.55 (45)
	360	1.80 ± 0.32 (40)	2.67 ± 0.68 (60)
Intestinal	60	5.51 ± 1.66 (39)	8.75 ± 2.45 (61)
Content	360	18.18 ± 1.43 (41)	26.48 ± 2.80 (59)

Homogenates of lung, liver, and intestinal contents were extracted with ethyl acetate and amounts of [^{3}H]-radioactivity that were extractable and non-extractable were determined as described in Materials and Methods. Numbers in parentheses are percent of total [^{3}H]-radioactivity in each tissue.

conjugates accounted for 9 and 7 percent of the total metabolites, respectively.

Excretion of B[a]P metabolites in bile was also characterized. Radioactivity in bile from animals with a biliary cannula accounted for 74 percent of the dose 360 min after [^{3}H]-B[a]P instillation. As was observed in studies of intestinal contents, thioether conjugates accounted for the majority of material which was excreted (Fig. 4). Amounts of glucuronides in bile were much greater than in intestinal contents whether expressed as percent of dose or as percent of excreted material. Amounts of sulfate conjugates and non-conjugated metabolites were similar in intestinal contents and bile.

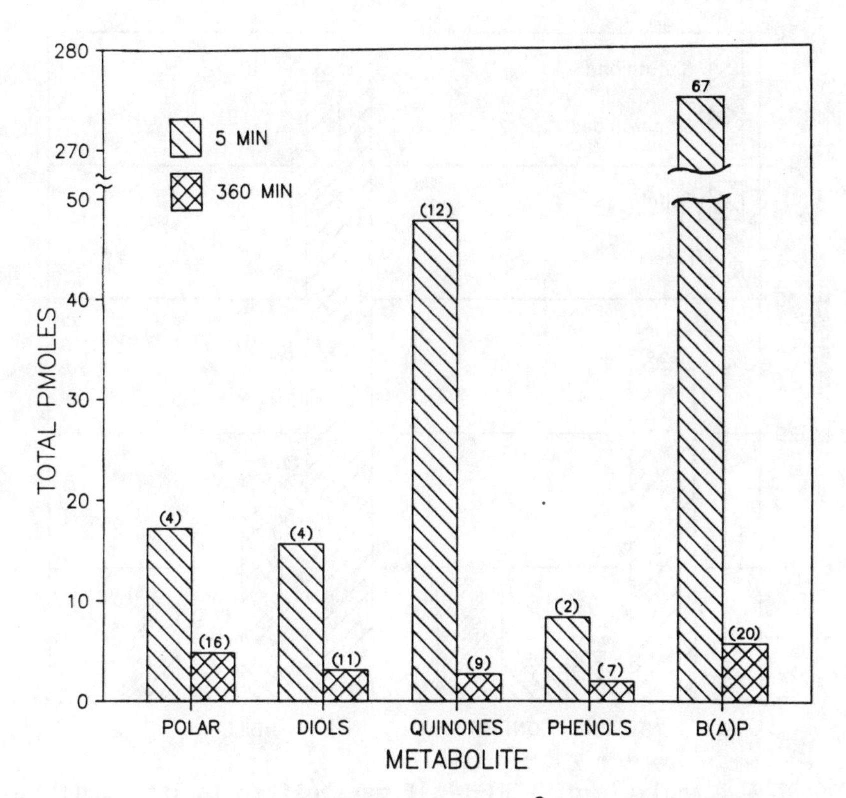

FIGURE 3. Ethyl acetate extractable [^{3}H]-B[a]P metabolites in lung at 5 and 360 min following [^{3}H]-B[a]P administration. HPLC analysis was as described in Materials and Methods. Results are expressed as total pmoles for each class of metabolites. Data are from 3 animals at each time point. Numbers in parentheses are percent of total metabolites at each time.

DISCUSSION

In this investigation we have shown that [^{3}H]-B[a]P dissolved in triethylene glycol is rapidly distributed throughout animals when administered by intratracheal instillation. Radioactivity was detected in all organs, with highest amounts being detected in lung, liver, carcass and the gastrointestinal tract.

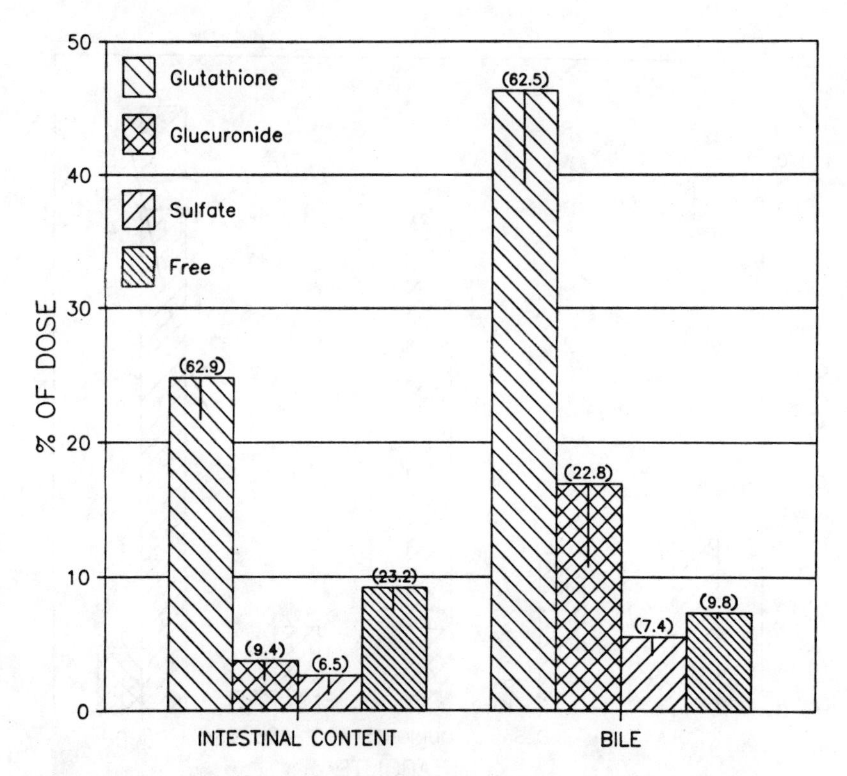

FIGURE 4. Analysis of [^{3}H]-B[a]P metabolites in intestinal contents and bile at 360 min following [^{3}H]-B[a]P administration. Material sensitive to β-glucuronidase and sulfatase hydrolysis represents B[a]P glucuronides and sulfate conjugates respectively. Material insensitive to enzyme hydrolysis is considered to be B[a]P-thioether conjugates resulting from conjugation with glutathione. Results are mean percent of dose ± S.E. for 3 animals at each time. Numbers in parentheses are amount of each metabolite expressed as percent of total material in intestinal contents or bile.

Elimination of radioactivity from lung was more rapid in our studies than in some previous reports. Kotin et al. (9) studied distribution of radioactivity in rats after intratracheal administration of [^{14}C]-B[a]P. When B[a]P was administered in triethylene glycol, highest amounts of radioactivity were in lungs (17.7%), feces (17.2%), and intestines (44.2%) at 4 hours after administration. In our experiments, only about 5 percent of total radioactivity remained in lungs after 4 hours. Another discrepancy between our results and those of Kotin et al. is evident when comparing amounts of radioactivity in carcass. We observed 25-30 percent of administered dose in carcass between 5 and 360 min after administration whereas Kotin and coworkers did not report any radioactivity associated with carcass even though their recoveries of radiolabel approached 100 percent. Reasons for these differences are not readily apparent. Mitchell (10) studied distribution of radiolabel in rats after inhalation exposure to [^{3}H]-B[a]P, and he also observed slower elimination of radiolabel from lungs than we did. In this case, the discrepancies could result from differences between inhalation and intratracheal exposure as well as differences in doses of B[a]P.

Our studies of metabolites of B[a]P in various organs produced results consistent with those of Mitchell (10). That is, unmetabolized B[a]P and B[a]P quinones were prominent shortly after exposure while material more polar than B[a]P-9,10-diol predominated at longer times after exposure.

Biliary excretion of radioactivity following intratracheal administration of [^{3}H]-B[a]P in this study was similar to excretion reported in other studies when B[a]P was administered by intravenous injection. Boroujerdi et al. (11) reported that 68 percent of the dose was excreted in bile 6 hours after an intravenous injection of 1.0 μmol/kg of [^{3}H]-B[a]P. We detected 74 percent of the dose in bile 6 hours after the intratracheal administration of 3.8 nmol/kg [^{3}H]-B[a]P. In addition, the types of conjugated metabolites of B[a]P were similar, with B[a]P metabolites conjugated to glutathione being the major type of conjugates formed. The rapid absorption of B[a]P from lungs in our system may account for the similarities in B[a]P excretion via the bile when our results are compared with those obtained when B[a]P was injected intravenously. It may be possible to alter rates of absorption of B[a]P from lungs and hence alter patterns of metabolism and rates of excretion by using different vehicles for dissolution or dispersion of B[a]P (9).

ACKNOWLEDGEMENTS

This work was supported in part by EPA grant R811484 and by Hatch Project No. 612458. We thank Karen Dove for her assistance in the preparation of this manuscript.

REFERENCES

1. Seifried, H. E., Birkett, D. J., Levin, W., Lu, A. Y. H., Conney, A. H., and Jerina, D. M. (1977) Metabolism of Benzo[a]pyrene. Arch. Biochem. Biophys., 178: 256-263.
2. Dock, L., Cha, Y., Jernstrom, B., and Moldéus, P. (1982) Differential Effects of Dietary BHA on Hepatic Enzyme Activities and Benzo[a]pyrene Metabolism in Male and Female NMRI Mice. Carcinogenesis, 3: 15-19.
3. Depierre, J. W., and Ernster, L. (1978) The Metabolism of Polycyclic Hydrocarbons and its Relationship to Cancer. Biochem. Biophys. Acta., 473; 149-186.
4. Hirakawa, T., Nemoto, H., Yamada, M., and Takayama, S. (1979) Metabolism of Benzo[a]pyrene and the Related Enzyme Activities in Hamster Embryo Cells. Chem-Biol. Interac., 25; 189-195.
5. Burke, M. D., Vadi, H., Jernström, B., and Orrenius, S. (1977) Metabolism of Benzo[a]pyrene with Isolated Hepatocytes and the Formation and Degradation of DNA-binding Derivatives. J. Biol. Chem., 252; 6424-6431.
6. Foth, H., Molliere, M., Kahl, R., Jahnchen, E., and Kahl, G. (1984) Covalent Binding of Benzo(a)pyrene in Perfused Rat Lung Following Systemic and Intratracheal Administration. Drug Metab. and Disp., 12; 760-766.
7. Ball, L. M., Plummer, J. L., Smith, B. R., and Bend, J. R. (1979) Benzo[a]pyrene Oxidation, Conjugation, and Disposition in the Isolated Perfused Rabbit Lung: Role of the Glutathione-S-Transferases. Med. Biol., 57; 298-305.
8. Kahl, G. F., Klaus, E., Legraverend, C., Nebert, D. W., and Pelkonen, O. (1979) Formation of Benzo(a)pyrene Metabolite-Nucleoside Adducts in Isolated Perfused Rat and Mouse Liver and in Mouse Lung Slices. Biochem. Pharmacol., 28: 1051-1056.
9. Kotin, P., Falk, H. L., and Busser, R. (1959) Distribution, Retention, and Elimination of C^{14}-3,4-Benzopyrene After Administration to Mice and Rats. J. Nat. Cancer Inst., 23: 541-555.
10. Mitchell, C. E. (1982) Distribution and Retention of Benzo(a)pyrene in Rats After Inhalation. Toxicol. Lett., 11: 35-42.
11. Boroujerdi, M., Kung, H. C., Wilson, A. E. G., and Marshall, M. W. (1981) Metabolism and DNA Binding of Benzo(a)pyrene in vivo in the Rat. Cancer Res., 41: 951-957.

12. Van Cantfort, J., DeGraeve, J., and Gielen, J. E. (1977) Radioactive Assay for Aryl Hydrocarbon Hydroxylase. Improved Method and Biological Importance. Biochem. Biophys. Res. Commun., 79; 505-512.
13. Shanker, L. S. (1978) Drug Absorption from the Lung. Biochem. Pharmacol., 27: 381-385.
14. Smith, B. R., Philpot, R. M., and Bend, J. R. (1978) Metabolism of Benzo(a)pyrene by the Isolated Perfused Rabbit Lung. Drug Metab. Dispos., 6: 425-431.
15. Selkirk, J. K. (1976) High-Pressure Liquid Chromatography: A New Technique for Studying Metabolism and Activation of Chemical Carcinogens, In: Advances in Modern Toxicology. Vol. 3: Environmental Cancer, Kraybill, H. F., and Mehlman, M. A. (eds), Hemisphere Publ., New York, pp. 1-25.
16. MacLeod, M. C., Moore, C. J., and Selkirk, J. K. (1979) Analysis of Water-soluble Conjugates Produced by Hamster Embryo Cells Exposed to Polynuclear Aromatic Hydrocarbons, In: Polynuclear Aromatic Hydrocarbons: Chemistry and Biological Effects, A. Bjorseth and A. J. Dennis (eds.), Battelle Press, Columbus, pp. 9-23.

AUTHOR INDEX

KEY WORD INDEX